Petroleum Biodegradation and Oil Spill Bioremediation

Karuna K. Arjoon
KKA Associates
California, Trinidad

and

James G. Speight
CD&W Inc.
Laramie, USA

CRC **CRC Press**
Taylor & Francis Group
Boca Raton London New York

CRC Press is an imprint of the
Taylor & Francis Group, an **informa** business

A SCIENCE PUBLISHERS BOOK

First edition published 2023
by CRC Press
6000 Broken Sound Parkway NW, Suite 300, Boca Raton, FL 33487-2742

and by CRC Press
4 Park Square, Milton Park, Abingdon, Oxon, OX14 4RN

Library of Congress Cataloging-in-Publication Data (applied for)

ISBN: 978-0-367-48739-3 (hbk)
ISBN: 978-1-032-41115-6 (pbk)
ISBN: 978-1-003-04263-1 (ebk)

DOI: 10.1201/9781003042631

Typeset in Times New Roman
by Radiant Productions

Preface

It is not the purpose of this book to condemn the crude oil refining industry as the bad actor in terms of environmental pollution. In fact, there are many areas of the chemical industry that are responsible for the release of pollutants into the environment. The refining industry is one such industry that has seen inadvertent spillage of unrefined crude oil and crude oil products. The continuing question, since the beginning of the environmental movement in the 1960s, relates to the relative condition of the environment.

However, it is the purpose of the book to identify effective clean-up methods for the spills of crude oil and crude oil products (as well as the spills of other chemicals) that is both economically and ecologically suitable is of paramount urgency for continued existence of humanity. Thus far, bioremediation has been gaining momentum as a valuable clean-up method of many environmental hazards over the past three-to-four decades.

In this book, other than the terms *crude oil* and *crude oil product*, the term *oil* is often substituted to describe a broad range of hydrocarbon-based chemicals which includes substances that are commonly thought of as oils, such as crude oil and refined petroleum products. Each type of oil has distinct physical and chemical properties. These properties affect the way oil will spread and break down, the hazard it may pose to aquatic and human life, and the likelihood that it will pose a threat to natural and man-made resources.

Bioremediation is the use of natural organisms to eliminate many components of crude oil and crude oil products from the environment. The concern is whether or not bioremediation technologies can accelerate this natural process enough to be considered practical, and, if so, whether they might find a niche as replacements for, or adjuncts to, other crude oil-spill response technologies.

Crude oil (also called petroleum) is a complex mixture of thousands of different chemical compounds. In addition, the composition of each accumulation of oil is unique, varying in different producing regions and even in different unconnected zones of the same formation. The composition of crude oil also varies with the amount of refining. Significantly, the many constituents of crude oil differ markedly in volatility, solubility, and susceptibility to biodegradation—some constituents are susceptible to microbial biodegraded while others non-biodegradable.

Furthermore, the biodegradation of different crude oil constituents occurs simultaneously but at very different rates. This leads to the sequential disappearance of individual components of crude oil over time and, because different species of microbes preferentially attack different compounds, to successional changes in the degrading microbial community. Thus, to evaluate the effectiveness of biodegradation, through the application of bioremediation technologies it is necessary to know the molecular effects of the process starting with the molecular composition of the contaminants.

This book introduces the reader to the science and technology of bioremediation—a key process for the removal of crude oil and crude oil based contaminants at spill sites. The contaminants of concern in the molecularly-variable crude oil and crude oil products can be degraded under appropriate conditions. But the success of the process depends on the ability to determine the necessary conditions and establish them in the contaminated environment.

Although the prime focus of the book is to determine the mechanism, extent, and efficiency of biodegradation processes it is necessary to know the composition of the original crude oil or crude oil product. The laws of science dictate what can or cannot be done with crude oil and crude oil products to insure that biodegradation (hence, bioremediation) processes are effective. The science of the composition of crude oil and crude oil products is at the core of understanding the chemistry of biodegradation and bioremediation processes. Hence, inclusion of crude oil analyses and properties along with crude oil product analyses and properties is a necessary part of this text.

It is the purpose of this book to present to the reader an introduction to the science and technology of bioremediation as applied to the spills of crude oil. Bioremediation is a key process that can be applied to the removal of contaminants that arise from the spill of crude oil or crude oil products. The contaminants of concern in the molecularly-variable constituents of crude oil and crude oil products can be degraded under appropriate conditions. However, the success of the bioremediation process depends on the conditions at the spill site and the ability to establish the necessary conditions to decontaminate the site.

To accomplish this goal, the book focuses on the various aspects of environmental science and engineering as applied to the crude oil and crude oil products. The initial section (Part I) presents an introduction to crude oil and crude oil products. The second section (Part II) introduces the reader to the nature of oil spills and the concept if bioremediation and biodegradation and presents descriptions of site evaluation, methods for bioremediation, the analytical methods that can be used to determine the extent of the cleanup, and the various cleanup methods, and degradation methods, recommendations for oil spillage prevention, control, and the future of bioremediation.

Each chapter includes a copious reference section and the book is further extended and improved by the inclusion of an extensive Glossary. In addition, a basic knowledge of the organic chemistry of the constituents of crude oil and crude oil products is necessary and, for those readers not fully familiar with this area of chemistry, an explanation of the chemistry of crude oil constituents is included for reference (Appendix A).

Karuna K. Arjoon, MPhil
California, Trinidad and Tobago

James G. Speight PhD, DSc, PhD
Laramie, Wyoming, USA

Contents

Preface   iii

Glossary   xiii

PART I: Crude Oil and Crude Oil Products

1. Crude Oil Composition and Properties   **3**
  1.1   Types of Crude Oil   7
      1.1.1 Conventional Crude Oil   9
      1.1.2 Heavy Crude Oil   11
      1.1.3 Extra Heavy Crude Oil   12
      1.1.4 Tar Sand Bitumen   12
  1.2   Crude Oil Composition   15
      1.2.1 Elemental Composition   18
      1.2.2 Chemical Composition   18
      1.2.3 Composition by Volatility   21
      1.2.4 Composition by Fractionation   22
      1.2.5 Composition by Spectroscopy   23
           1.2.5.1 Infrared Spectroscopy   23
           1.2.5.2 Nuclear Magnetic Resonance Spectroscopy   24
           1.2.5.3 Mass Spectrometry   24
           1.2.5.4 Other Techniques   24
  1.3   Crude Oil Properties   24
      1.3.1 Density and Specific Gravity   28
      1.3.2 Elemental Analysis   28
      1.3.3 Chromatographic Fractionation   29
      1.3.4 Liquefaction and Solidification   30
      1.3.5 Metals Content   31
      1.3.6 Surface Tension and Interfacial Tension   31
      1.3.7 Viscosity   32
      1.3.8 Volatility   33
  1.4   Summary   34
References   35

2. Crude Oil Products   **37**
  2.1   Refinery Products   40
  2.2   Bulk Products   42
      2.2.1 Liquefied Petroleum Gas   43
      2.2.2 Naphtha, Gasoline, and Solvents   44

2.2.3	Kerosene and Diesel Fuel	48
2.2.4	Fuel Oil	49
2.2.5	Lubricating Oil	51
2.2.6	White Oil, Insulating Oil, Insecticides	52
2.2.7	Grease	54
2.2.8	Wax	54
2.2.9	Asphalt	55
2.2.10	Coke	60
2.3	Petrochemical Products	61
2.4	Occurrence and Uses of Common Hydrocarbon Derivatives	63
2.5	Refinery Waste	67
2.5.1	Acid Sludge	71
2.5.2	Spent Acid	72
2.5.3	Spent Catalyst	73
2.5.4	Spent Caustic	74
2.5.5	Sulfonic Acids	74
2.5.6	Product Blending	75
2.5.7	Waste by Process	76
2.5.7.1	Dewatering and Desalting	78
2.5.7.2	Gas Processing	80
2.5.7.3	Distillation	80
2.5.7.4	Visbreaking and Thermal Cracking	82
2.5.7.5	Coking Processes	82
2.5.7.6	Fluid Catalytic Cracking	83
2.5.7.7	Hydrocracking and Hydrotreating	84
2.5.7.8	Catalytic Reforming	86
2.5.7.9	Alkylation	86
2.5.7.10	Isomerization and Polymerization	87
2.5.7.11	Deasphalting	88
2.5.7.12	Dewaxing	89
2.5.8	Types of Waste	89
2.5.9	Waste Toxicity	96
2.6	Entry into the Environment	96
2.6.1	Storage and Handling of Crude Oil and Crude Oil Products	97
2.6.2	Release into the Environment	98
2.6.2.1	Dispersion	98
2.6.2.2	Dissolution	99
2.6.2.3	Emulsification	99
2.6.2.4	Evaporation	99
2.6.2.5	Leaching	100
2.6.2.6	Sedimentation or Adsorption	100
2.6.2.7	Spreading	100
2.6.2.8	Wind	100
References		100
3. Test Methods for Crude Oil and Crude Oil Products		**102**
3.1	The Need for Test Methods	104
3.2	Chemical and Physical Properties of Crude Oil and Crude Oil Products	107
3.2.1	Adhesion	108
3.2.2	Biological Oxygen Demand	109
3.2.3	Boiling Point Distribution	109

3.2.4 Chemical Dispersability 110
3.2.5 Density, Specific Gravity, and API Gravity 111
3.2.6 Emulsion Formation 112
3.2.7 Evaporation 113
3.2.8 Flash Point and Fire Point 114
3.2.9 Fractionation 115
3.2.10 Leachability and Toxicity 116
3.2.11 Metals Content 116
3.2.12 Pour Point and Cloud Point 117
3.2.13 Solubility in Aqueous Media 117
3.2.14 Sulfur Content 117
3.2.15 Surface Tension and Interfacial Tension 118
3.2.16 Total Petroleum Hydrocarbons 118
3.2.17 Viscosity 119
3.2.18 Volatility 120
3.2.19 Water Content 120
3.2.20 Weathering Processes 121
3.3 Petroleum Group Analysis 122
3.3.1 Thin Layer Chromatography 123
3.3.2 Immunoassay 124
3.3.3 Gas Chromatography 124
3.3.4 High Performance Liquid Chromatography 127
3.3.5 Gas Chromatography-Mass Spectrometry 128
3.4 Other Analytical Methods 129
3.4.1 Infrared Spectroscopy 129
3.4.2 Gravimetry 130
3.5 Properties and Analysis of Crude Oil Products 131
3.5.1 Gaseous Products 131
3.5.2 Liquid Products 133
3.5.3 Solid Products 133
References 137

4. The Nature of Oil Spills **141**
4.1 Environmental Effects of Crude Oil Refining 141
4.1.1 Gases 142
4.1.2 Liquids 145
4.1.3 Solids 148
4.2 Understanding Spills of Crude Oil and Crude Oil Products 150
4.2.1 Types of Spills 154
4.2.2 Composition of a Spill 154
4.2.3 Dangers of a Spill in the Workplace 155
4.2.4 Causes of Crude Oil Spills 156
4.2.4.1 Transportation 156
4.2.4.2 Errors by Personnel 156
4.2.4.3 Equipment Breakdown 157
4.2.4.4 Natural Disasters 157
4.2.4.5 Acts of Terrorism, War, Vandalism, or Illegal Dumping 157
4.3 Entry into the Environment and Toxicity 158
4.3.1 Entry into the Environment 160
4.3.1.1 Dispersion 160
4.3.1.2 Dissolution 160

        4.3.1.3 Emulsification   161

4.3.1.3   Emulsification   161
4.3.1.4   Evaporation   161
4.3.1.5   Leaching   161
4.3.1.6   Sedimentation or Adsorption   161
4.3.1.7   Spreading   162
4.3.1.8   Wind   162
4.3.2   Toxicity   162
4.3.2.1   Lower Boiling Constituents   163
4.3.2.2   Higher Boiling Constituents   165
4.3.2.3   Wastewater   166
4.4   General Methods for Soil and Groundwater Remediation   167
4.4.1   Bioremediation   168
4.4.2   Containment   170
4.4.3   Dredging or Excavation   170
4.4.4   *In situ* Oxidation   171
4.4.5   Metals Removal   172
4.4.6   Nanoremediation   173
4.4.7   Pump and Treat   174
4.4.8   Soil Vapor Extraction   174
4.4.9   Solidification and Stabilization   175
4.4.10 Surfactant Enhanced Aquifer Remediation   176
4.4.11 Thermal Desorption   176
4.5   Remediation Management   178
References   179

5. Overview of Oil Spill Clean Up Methods   **181**
5.1   Types of Effluents   184
5.1.1   Gaseous Effluents   185
5.1.1.1   Liquefied Petroleum Gas   187
5.1.1.2   Natural Gas   188
5.1.1.3   Refinery Gas   189
5.1.1.4   Sulfur Oxides, Nitrogen Oxides, Hydrogen Sulfide, Carbon Dioxide 191
5.2   Environmental Effects   191
5.2.1   Gaseous Effluents   191
5.2.2   Liquid Effluents   192
5.2.2.1   Naphtha   193
5.2.2.2   Wastewater   201
5.2.3   Solid Effluents   202
5.2.3.1   Residua and Asphalt   204
5.2.3.2   Coke   207
5.2.3.3   Particulate Matter   209
5.3   Oil Spill Cleanup at Sea   209
5.3.1   Oil Booms   209
5.3.2   Skimmers   210
5.3.3   Sorbents   211
5.3.4   Burning *In situ*   212
5.3.5   Dispersants   214
5.3.6   Hot Water and High-Pressure Washing   214
5.3.7   Chemical Stabilization   215

5.4   Methods for Oil Spill Cleanup on Land   215
    5.4.1   Physical Methods   216
    5.4.2   Chemical Methods   216
5.5   Bioremediation   217
    5.5.1   Importance of Bioremediation   218
5.6   Issues Related to Use of Bioremediation Technologies   219
    5.6.1   Environmental   219
    5.6.2   Health   220
    5.6.3   Process Evaluation   220
        5.6.3.1   Feasibility Assessment   221
        5.6.3.2   Bioremediation Services   222
        5.6.3.3   Barriers to Commercialization   224
        5.6.3.4   Supporting Research and Development   224
        5.6.3.5   Technical Regulations   225
        5.6.3.6   Economic Assessment   225
        5.6.3.7   Potential for Future Implementation   226
References   227

PART II: Bioremediation, Biodegradation, and Site Cleanup

6. Bioremediation and Biodegradation   **231**
6.1   The Origin of Bioremediation   238
6.2   The Mechanism of Bioremediation   240
6.3   Types of Bioremediation   248
    6.3.1   Natural Bioremediation   248
    6.3.2   Traditional Bioremediation   249
    6.3.3   Enhanced Bioremediation   250
    6.3.4   Monitored Natural Attenuation   251
6.4   Factors that Affect Bioremediation   252
    6.4.1   Chemistry   253
    6.4.2   Types of Microbes   254
    6.4.3   Types of Contaminants   256
        6.4.3.1   Physical Properties   257
        6.4.3.2   Chemical Properties   258
        6.4.3.3   Site Specific Issues   259
            6.4.3.3.1   Geology of the Site   260
            6.4.3.3.2   Chemistry of the Site   260
6.5   Site Remediation   261
    6.5.1   Method Parameters   261
    6.5.2   *In situ* and *ex situ* Bioremediation   261
    6.5.3   Biostimulation and Bioaugmentation   264
    6.5.4   Miscellaneous Processes   265
        6.5.4.1   Bioslurping   265
        6.5.4.2   Biosparging   265
        6.5.4.3   Biosurfactant Treatment   266
        6.5.4.4   Bioventing   266
        6.5.4.5   Rhizosphere Bioremediation   266
    6.6   Bioremediation of Land Ecosystems   267
    6.7   Bioremediation of Water Ecosystems   273
References   276

7. Site Evaluation and the Impact of an Oil Spill   **282**
   7.1   Site Evaluation   285
          7.1.1   Land Ecosystems   286
                  7.1.1.1   Soil Evaluation   288
                  7.1.1.2   Physical Properties   289
                  7.1.1.3   Chemical Properties   290
                  7.1.1.4   Biological Properties   291
                  7.1.1.5   Temperature   292
                  7.1.1.6   Acidity and Alkalinity   293
                  7.1.1.7   Salinity   293
          7.1.2   Water Ecosystems   294
                  7.1.2.1   Biodegradation   296
                  7.1.2.2   Bioremediation   297
                  7.1.2.3   Temperature   298
                  7.1.2.4   Effect of Oxygen   299
                  7.1.2.5   Effect of Nutrients   299
                            7.1.2.5.1   Marine Environments   300
                            7.1.2.5.2   Freshwater Environments   300
                            7.1.2.5.3   Soil Environments   300
                  7.1.2.6   Effect of Spill Characteristics   301
                  7.1.2.7   Effect of Prior Exposure   301
                  7.1.2.8   Effect of Dispersants   301
                  7.1.2.9   Effect of Flowing Water   302
                  7.1.2.10 Effect of a Deep-Sea Environment   302
   7.2   Effects on Flora and Fauna   303
          7.2.1   Effect on the Biosphere   303
          7.2.2   Effect on Micro-organisms   304
          7.2.3   Effect on Plants   305
          7.2.4   Effect on Animals   306
          7.2.5   Effect on Humans   307
          7.2.6   Effect on the Economy   308
   7.3   Risk Analysis   309
          7.3.1   Oil Spill Risk Assessment   309
          7.3.2   Guidelines for Oil Spill Risk Assessment   312
          7.3.3   Oil Spill Risk Analysis Model   314
   7.4   The Characteristics of a Spill Response   315
References   316

8. Biodegradation of the Constituents of Crude Oil and Crude Oil Products   **323**
   8.1   Biodegradability   324
          8.1.1   Conditions for Biodegradation   326
          8.1.2   Effect of Nutrients   328
          8.1.3   Effect of Temperature   329
          8.1.4   Effect of Dispersants   330
          8.1.5   Effect of Weathering   331
   8.2   Biodegradation of Specific Constituents   332
          8.2.1   Alkanes   332
          8.2.2   Aromatic Hydrocarbons   333
          8.2.3   Polynuclear Aromatic Hydrocarbons   334
          8.2.4   Phenolic Compounds   338
          8.2.5   Chlorinated Compounds   339

8.3 Biodegradation of Extra Heavy Crude Oil 340
    8.3.1 Microbial Enhanced Oil Recovery 341
    8.3.2 Biotransformation 342
    8.3.3 Biodegradation and Bioconversion 344
        8.3.3.1 Biodesulfurization 344
        8.3.3.2 Biodenitrogenation 346
8.4 Biodegradation of Other Products 349
    8.4.1 Acid Sludge 349
    8.4.2 Spent Acid 350
    8.4.3 Spent Catalyst 350
    8.4.4 Spent Caustic 350
    8.4.5 Wastewater 351
8.5 Rates of Biodegradation 351
8.6 Application to Spills 353
References 356

9. **Methods Used to Determine the Progress of Bioremediation** **371**
9.1 The Spilled Material 375
9.2 Sample Collection and Preparation 377
    9.2.1 Crude Oil and Crude Oil Products 378
    9.2.2 Sample Collection and Preparation 379
        9.2.2.1 Sample Collection 380
            9.2.2.1.1 Volatile Compounds 381
            9.2.2.1.2 Condensate Releases 382
            9.2.2.1.3 Semi-Volatile and Non-Volatile Compounds 383
            9.2.2.1.4 Solids 386
        9.2.2.2 Extract Concentration 387
        9.2.2.3 Sample Cleanup 389
        9.2.2.4 Measurement 389
            9.2.2.4.1 Accuracy 390
            9.2.2.4.2 Precision 391
            9.2.2.4.3 Method Validation 392
        9.2.2.5 Quality Control and Quality Assurance 394
            9.2.2.5.1 Quality Control 394
            9.2.2.5.2 Quality Assurance 395
            9.2.2.5.3 Method Detection Limit 396
9.3 Sampling in the Field 396
    9.3.1 Sampling Strategies 398
    9.3.2 Acquiring a Representative Sample 399
9.4 Group Analyses 399
    9.4.1 Gas Chromatography 400
    9.4.2 Gas Chromatography-Mass Spectrometry 403
    9.4.3 High Performance Liquid Chromatography 403
    9.4.4 Immunoassay 404
    9.4.5 Infrared Spectroscopy 404
    9.4.6 Thin Layer Chromatography 405
9.5 Gravimetric Analysis 405
9.6 Microbiological Analysis 406
    9.6.1 Chemical Analysis of Nutrients 407
    9.6.2 Chemical Analysis of Crude Oil and Oil Constituents 408

9.7   Biomarkers   410
       9.7.1   Types   410
       9.7.2   Commonly Used Biomarkers   411
9.8   Fractionation of the Spilled Material   411
9.9   Leachability and Toxicity   411
9.10   Monitoring General Site Background Conditions   413
       9.10.1 Oxygen   413
       9.10.2 Acidity-Alkalinity   413
       9.10.3 Temperature   414
       9.10.4 Salinity   414
9.11   Assessment of the Methods   414
References   417

10. Recommendations for Oil Spill Prevention and Control   423
10.1   Refinery Products   424
       10.1.1 Bulk Products   424
       10.1.2 Petrochemicals   424
       10.1.3 Refinery Waste   424
10.2   Environmental Impact of Crude Oil and Crude Oil Products   426
       10.2.1 Air Pollution   427
       10.2.2 Water Pollution   429
       10.2.3 Soil Pollution   430
10.3   Pollution Prevention   431
       10.3.1 Options   433
               10.3.1.1 Operating Practices   433
               10.3.1.2 Process Modifications   434
               10.3.1.3 Material Substitution Options   436
10.4   Adoption of Pollution Reduction Options   437
10.5   The Future of Bioremediation   439
       10.5.1 Conventional Bioremediation   440
       10.5.2 Enhanced Bioremediation   441
       10.5.3 Bioremediation in Extreme Environments   443
10.6   Advantages and Disadvantages of Bioremediation   445
10.7   Conclusion   446
References   447

Appendix: The Chemistry of Crude Oil and Crude Oil Products   452

A1   Abstract   452
A2   Introduction   452
A3   Crude Oil and Crude Oil Products   452
     A3.1 Chemical Composition   453
     A3.2 Physical Properties   455
A4   Analytical Techniques for Oil Spill Sample Analysis   455
A5   Challenges   456
Further Reading   457

Conversion Factors   **458**

Index   **461**

About the Authors   **465**

Glossary

The following list contains a selection of definitions that are commonly used in reference to crude oil and crude oil products.

Abiotic: Not associated with living organisms; synonymous with abiological or non-biological.

Abiotic transformation: The process in which a substance in the environment is modified by non-biological mechanisms.

Absorption: The penetration of atoms, ions, or molecules into the bulk mass of a substance.

Acidophiles: Metabolically active in highly acidic environments, and often have a high heavy metal resistance.

Acyclic: A compound with straight or branched carbon-carbon linkages but without cyclic (ring) structures.

ABN separation: A method of fractionation by which petroleum is separated into acidic, basic, and neutral constituents.

Acid catalyst: A catalyst having acidic character; alumina is an example of such a catalyst.

Acid deposition: Acid rain; a form of pollution depletion in which pollutants, such as nitrogen oxides and sulfur oxides, are transferred from the atmosphere to soil or water; often referred to as atmospheric self-cleaning. The pollutants usually arise from the use of fossil fuels.

Acidity: The capacity of an acid to neutralize a base such as a hydroxyl ion (OH^-).

Acid number: A measure of the reactivity of petroleum with a caustic solution and given in terms of milligrams of potassium hydroxide that are neutralized by one gram of petroleum.

Acid rain: The precipitation phenomenon that incorporates anthropogenic acids and other acidic chemicals from the atmosphere to the land and water (see Acid deposition).

Acid sludge: The residue left after treating petroleum oil with sulfuric acid for the removal of impurities; a black, viscous substance containing the spent acid and impurities.

Acid treating: A process in which unfinished petroleum products, such as gasoline, kerosene, and lubricating-oil stocks, are contacted with sulfuric acid to improve their color, odor, and other properties.

Adhesion: The degree to which crude oil or a crude oil product will coat a surface, expressed as the mass of oil adhering per unit area. A test has been developed for a standard surface that gives a semi-quantitative measure of this property.

Adsorption: The retention of atoms, ions, or molecules on to the surface of another substance.

Aerobe: An organism that needs oxygen for respiration and hence for growth.

Aerobic: In the presence of, or requiring, oxygen; an environment or process that sustains biological life and growth, or occurs only when free (molecular) oxygen is present.

Aerobic bacteria: Any bacteria requiring free oxygen for growth and cell division.

Aerobic conditions: Conditions for growth or metabolism in which the organism is sufficiently supplied with oxygen.

Aerobic respiration: The process whereby micro-organisms use oxygen as an electron acceptor.

Air sparging: See Sparging.

Aliphatic hydrocarbon: Any organic compound of hydrogen and carbon characterized by a linear-chain or branched-chain of carbon atoms; three subgroups of such compounds are alkanes, alkenes, and alkynes; hydrocarbon derivatives that contain saturated and/or single unsaturated bonds and elute during chromatography using non-polar solvents such hexane; includes alkane derivatives and alkene derivatives, but not aromatic derivatives.

Alkalinity: The capacity of a base to neutralize the hydrogen ion (H^+).

Alkalitolerants: Organisms that are able to grow or survive at pH values above 9, but their optimum growth rate is around neutrality or less.

Alkaliphiles: Organisms that have their optimum growth rate at least 2 pH units above neutrality.

Alkane (paraffin): A group of hydrocarbon derivatives composed of only carbon and hydrogen with no double bonds or aromaticity. They are said to be "saturated" with hydrogen. They may by straight-chain (normal), branched or cyclic. The smallest alkane is methane (CH_4), the next, ethane (CH_3CH_3), then propane ($CH_3CH_2CH_3$), and so on.

Alkene (olefin): An unsaturated hydrocarbon derivative that contains only hydrogen and carbon with one or more double bonds ($>C = C<$), but having no aromaticity; alkene derivatives are not typically found in crude oils, but can occur as a result of thermal processes.

Alcohol: Compounds in which a hydroxy group (–OH) is attached to a saturated carbon atom (e.g., ethyl alcohol, C_2H_5OH).

Aldehydes: Compounds in which a carbonyl group is bonded to one hydrogen atom and to one alkyl group [RC(=O)H].

Aliphatic compounds: A broad category of hydrocarbon compounds distinguished by a straight, or branched, open chain arrangement of the constituent carbon atoms, excluding aromatic compounds; the carbon-carbon bonds may be either single or multiple bonds—alkanes, alkenes, and alkynes are aliphatic hydrocarbons.

Alkanes: The homologous group of linear (acyclic) aliphatic hydrocarbons having the general formula C_nH_{2n+2}; alkanes can be straight chains (linear), branched chains, or ring structures; often referred to as paraffins.

Alkenes: Acyclic branched or unbranched hydrocarbons having one carbon-carbon double bond ($>C = C<$) and the general formula C_nH_{2n}; often referred to as olefins.

Alkylation: The process in which isobutane reacts with olefins such as butylene to produce a gasoline range alkylate as a blend stock for gasoline production.

Alkyl group: A hydrocarbon functional group (C_nH_{2n+1}) obtained by substituting a hydrogen from fully saturated compound; e.g., methyl (-CH_3), ethyl (-CH_2CH_3), propyl (-$CH_2CH_2CH_3$), or isopropyl [$(CH_3)_2CH$-].

Alkyl radicals: Carbon-centered radicals derived formally by removal of one hydrogen atom from an alkane, for example the ethyl radical (CH_3CH_2).

Alkynes: The group of acyclic branched or unbranched hydrocarbons having a carbon-carbon triple bond ($-C \equiv C-$).

Ambient: The surrounding environment and prevailing conditions.

Amendments: Additives to the bioremediation process that are intended to improve the existing conditions in the subsurface for enhanced bioremediation, may include nutrients, terminal electron acceptors, and surfactants.

Amphiphile: A chemical compound possessing both hydrophilic and lipophilic properties; common amphiphilic substances are soaps, detergents, and lipoproteins.

Anaerobe: An organism that does not need free-form oxygen for growth. Many anaerobes are even sensitive to free oxygen.

Anaerobic: A biologically-mediated process or condition not requiring molecular or free oxygen; relating to a process that occurs with little or no oxygen present.

Anaerobic bacteria: Any bacteria that can grow and divide in the partial or complete absence of oxygen.

Anaerobic respiration: The process whereby micro-organisms use a chemical other than oxygen as an electron acceptor; common substitutes for oxygen are nitrate, sulfate, and iron.

Analyte: The component of a system to be analyzed—for example, chemical elements or ions in groundwater sample.

Aniline point: The minimum temperature for complete miscibility of equal volumes of aniline and a test sample; this test is an indication of the paraffinic nature and the ignition quality of a fuel such as diesel fuel.

Anoxic: An environment without oxygen.

API Gravity: An American Petroleum Institute measure of *density* for crude oil and crude oil products:

API Gravity = [141.5/(specific gravity at 15.6°C) – 131.5]

Fresh water has a gravity of 10 °API. The scale is commercially important for arbitrarily ranking crude oil quality: heavy oils are typically < 20° API; medium oils are 20 to 35° API; light oils are 35 to 45° API.

Aquifer: a water-bearing layer of soil, sand, gravel, rock or other geologic formation that will yield usable quantities of water to a well under normal hydraulic gradients or by pumping.

Archaea: A domain of single-celled organisms which lack cell nuclei.

Aromatic compound: An organic cyclic compounds that contain one or more benzene rings; these can be monocyclic, bicyclic, or polycyclic hydrocarbons and their substituted derivatives; in aromatic ring structures, every ring carbon atom possesses one double bond; aromatic derivatives include such compounds as the BTEX group (benzene, toluene, ethylbenzene and the three xylene isomers), polycyclic aromatic hydrocarbon derivatives (PAHs, such as naphthalene), and some heterocyclic aromatic derivatives such as the di-benzothiophene derivatives and pyridine derivatives.

Asphaltene fraction (asphaltenes): a complex mixture of heavy organic compounds precipitated from crude oil and bitumen by natural processes or in laboratory by addition of excess n-pentane, or n-heptane; after precipitation of the asphaltene fraction, the remaining oil or bitumen consists of saturates, aromatics, and resins.

Assay: qualitative or (more usually) quantitative determination of the components of a material or system.

ATSDR: Agency for Toxic Substances and Disease Registry.

ASTM distillation: Standardized laboratory batch distillation for naphtha and middle distillate at atmospheric pressure.

ASTM International: Formerly the American Society for Testing and Materials (ASTM); the official organization in the United States for designing standard tests for petroleum and other industrial products.

Atmospheric equivalent boiling point (AEBP): A mathematical method of estimating the boiling point at atmospheric pressure of non-volatile fractions of petroleum.

Atmospheric residuum: A residuum (*q.v.*) obtained by distillation of a crude oil under atmospheric pressure and which boils above 350°C (660°F).

Atmospheric tower: The distillation unit operated at atmospheric pressure.

Attenuation: The set of human-made or natural processes that either reduce or appear to reduce the amount of a chemical compound as it migrates away or is disposed from one specific point towards another point in space or time; for example, the apparent reduction in the amount of a chemical in a ground-water plume as it migrates away from its source; degradation, dilution, dispersion, sorption, or volatilization are common processes of attenuation.

Barrel: The unit of measurement of liquids in the petroleum industry; equivalent to 42 US standard gallons or 33.6 imperial gallons.

Base number: The quantity of acid, expressed in milligrams of potassium hydroxide per gram of sample that is required to titrate a sample to a specified end-point.

Basic nitrogen: Nitrogen (in petroleum) that occurs in pyridine form.

Basic sediment and water (BS&W, BSW): the material that collects in the bottom of storage tanks usually composed of oil, water, and foreign matter; also called bottoms, bottom settlings.

Bbl: see Barrel.

Benthic zone: The ecological region at the lowest level of a body of water such as an ocean or a lake, including the sediment surface and some sub-surface layers; organisms living in this zone (benthos or benthic organisms) generally live in close relationship with the substrate bottom; many such organisms are permanently attached to the bottom; because light does not penetrate very deep ocean-water, the energy source for the benthic ecosystem is often organic matter from higher up in the water column which sinks to the depths.

Bioaccumulation: The gradual accumulation of substances, such as pesticides or other chemicals, in an organism. Bioaccumulation occurs when an organism absorbs a substance at a rate faster than that at which the substance is lost or eliminated by catabolism and excretion.

Bioattenuation: A process in which biodegradation occurs naturally without human intervention; the biodegradation process is determined by the metabolic potential of microorganisms to detoxify or transform the pesticide molecule, which is dependent on both accessibility and bioavailability.

Bioaugmentation (Bio-augmentation): A process in which acclimated micro-organisms are added to soil and groundwater to increase biological activity. Spray irrigation is typically used for shallow contaminated soils, and injection wells are used for deeper contaminated soils.

Bioavailability: The degree (or rate) to which a substance is absorbed into a system or made available at the site of physiological activity, as compounds must be released into solution before the organisms can degrade the compounds.

Biocomposting (bio-composting): The purposeful biodegradation of organic matter by micro-organisms with the help of macro-organisms under controlled aerobic conditions; although microbes as primary consumers, initiate the composting, it is the secondary and tertiary consumers including insects, worms, snails and associated species that feed upon this semi-converted organic matter, keep the compost pile cleaner, enhance the composting process, and convert humus into stable and cured compost.

Biodegradability: The ability of a contaminant to biodegradation as measured in a standard test procedure.

Biodegradation: The process whereby bacteria or other micro-organisms chemically alter and break down organic molecules; the breakdown or transformation of a chemical substance or substances by micro-organisms using the substance as a carbon and/or energy source; the degradation (or breakdown) of (in this case) contaminants into more benign environmentally acceptable products such as water, carbon dioxide, and any form of carbonaceous products that are not harmful to the environment.

Biodegradation, primary: The modification of some of the physical and chemical properties of the contaminant caused by the activity of microorganisms.

Biodegradation, ultimate: Total utilization of the contaminant resulting in the conversion of the contaminant into carbon dioxide (CO_2) or methane (CH_4), water (H_2O), mineral salts and microbial cellular constituents (biomass).

Biogeochemical cycle: Any of the natural pathways by which essential elements of living matter are circulated; the pathway by which a chemical substance cycles the biotic and the abiotic compartments of Earth—the biotic compartment is the biosphere and the abiotic compartments are the atmosphere, hydrosphere and lithosphere; the term is a contraction that refers to the consideration of the biological, geological, and chemical aspects of each cycle.

Bioleaching: A process in mining and biohydrometallurgy (natural processes of interactions between microbes and minerals) that extracts valuable metals from a low-grade ore with the help of micro-organisms such as bacteria or archaea; sometimes referred to as biomining.

Biological marker (biomarker): complex organic compounds composed of carbon, hydrogen, and other elements which are found in oil, bitumen, rocks, and sediments and which have undergone little or no change in structure from their parent organic molecules in living organisms; typically, biomarkers are isoprenoids, composed of isoprene subunits. Biomarkers include pristane, phytane, triterpane derivatives, sterane derivatives, and porphyrin derivatives.

Biological oxidation: The oxidative consumption of organic matter by bacteria by which the organic matter is converted into gases.

Biomagnification: The concentration of toxins in an organism as a result of its ingesting other plants or animals in which the toxins are more widely disbursed.

Biomass: Biological organic matter.

Biopile: A technology in which excavated soils are piled and typically constructed in a treatment area that consists of a leachate collection and aeration system; commonly applied to reduce concentrations of crude oil constituents in soil through utilizing the process of biodegradation.

Biopiling: A hybrid of land farming and composting, it is essentially engineered cells that are constructed as aerated composted piles.

Bioremediation: The act of treating waste or pollutants by the use of micro-organisms (as bacteria) that can break down the undesirable substances.

Bioreactor: A reaction chamber equipped with a mixing mechanism, a system that supplies oxygen and nutrients and influent and effluent pumps; an *ex situ* bioremediation technology that offers the direct control of environmental/nutritional factors (such as oxygen, moisture, nutrients, pH and even microbial population) that influence biodegradation.

Bioreactor (slurry phase): A bioreactor in which the reaction chamber is filled with excavated hydrocarbon-polluted soil mixed with liquid waste saturated with microbes to form a slurry and then mechanically agitated to encourage aerobic biodegradation.

Bioremediation: A treatment technology that uses biological activity to reduce the concentration or toxicity of contaminants: materials are added to contaminated environments to accelerate natural biodegradation.

Bioremediation (intrinsic): See Intrinsic bioremediation.

Bioslurry reactor: A reactor that can provide rapid biodegradation of contaminants due to enhanced mass transfer rates and increased contaminant-to-microorganism contact; this type of unit is capable of aerobically biodegrading aqueous slurries created through the mixing of soils or sludge with water.

Biostimulation (Bio-stimulation): A process by which nutrients and suitable physiological conditions are provided for the growth of the indigenous microbial populations.

Biota: Living organisms.

Bioslurping: The adaptation and application of vacuum-enhanced dewatering technologies to remediate hydrocarbon-contaminated sites; utilizes elements of both, bioventing and free product recovery, to address two separate contaminant media.

Biosparging: An *in situ* remediation technology that uses indigenous micro-organisms to biodegrade organic constituents in the saturated; the injection of air into the groundwater to provide oxygen for groundwater remediation.

Bioventing: A technology that stimulates the natural *in situ* biodegradation of any aerobically degradable compounds in soil by providing oxygen to existing soil micro-organisms; the aeration of the unsaturated vadose zone to stimulate aerobic biodegradation.

Bitumen: A complex mixture of hydrocarbonaceous constituents of natural or pyrogenous origin or a combination of both; a semi-solid to solid hydrocarbonaceous material found filling pores and crevices of sandstone, limestone, or argillaceous sediments.

Bituminous: Containing bitumen or constituting the source of bitumen.

Bituminous rock: See Bituminous sand.

Bituminous sand: A formation in which the bituminous material (see Bitumen) is found as a filling in veins and fissures in fractured rocks or impregnating relatively shallow sand, sandstone, and limestone strata; a sandstone reservoir that is impregnated with a heavy, viscous black petroleum-like material that cannot be retrieved through a well by conventional production techniques.

Black acid: A mixture of the sulfonates found in acid sludge that is insoluble in naphtha, benzene, and carbon tetrachloride; very soluble in water but insoluble in 30 per cent sulfuric acid; in the dry, oil-free state, the sodium soaps are black powders.

Boiling point: The temperature at which a liquid begins to boil—that is, it is the temperature at which the vapor pressure of a liquid is equal to the atmospheric or external pressure. The boiling point distributions of crude oil and crude oil products may be in a range from 30 to in excess of 700°C (86 to 1290°F).

Breakdown product: A compound derived by chemical, biological, or physical action on a chemical compound; the breakdown is a process which may result in a more toxic or a less toxic compound and a more persistent or less persistent compound than the original compound.

Bromine number: The number of grams of bromine absorbed by 100 g of oil which indicates the percentage of double bonds in the material.

Brown acid: The oil-soluble petroleum sulfonate derivatives found in acid sludge that can be recovered by extraction with naphtha solvent; brown-acid sulfonates are somewhat similar to mahogany sulfonates but are more water-soluble. In the dry, oil-free state, the sodium soaps are light-colored powders.

Brownfield site: A polluted sites that can be (or has been) cleaned and redeveloped.

BTEX: The collective name given to benzene, toluene, ethylbenzene and the xylene isomers (*p*-, *m*-, and *o*-xylene); a group of volatile organic compounds (VOCs) found in crude oil and crude oil products, such as gasoline, and other common environmental contaminants.

C_1, C_2, C_3, C_4, C_5 fractions: A common way of representing fractions containing a preponderance of hydrocarbons having 1, 2, 3, 4, or 5 carbon atoms, respectively, and without reference to hydrocarbon type.

Carbon: Element number 6 in the periodic table of elements.

Carbon preference index (CPI): The ratio of odd to even *n*-alkanes; odd/even CPI *alkanes* are equally abundant in crude oil and crude oil products but not in biological material; a CPI near 1 is an indication of crude oil.

Carcinogen: A chemical with the capacity to cause cancer in humans; a carcinogen may be natural such as (1) aflatoxin, which is produced by a fungus and sometimes found on stored grains or manmade, (2) such as asbestos or tobacco smoke; carcinogens work by interacting with the DNA of the cell and induce genetic mutations.

Catabolism: The breakdown of complex molecules into simpler ones through the oxidation of organic substrates to *provide* biologically available energy—ATP (adenosine triphosphate) is an example of such a molecule.

Catalysis: The process where a catalyst increases the rate of a chemical reaction without modifying the overall standard Gibbs energy change in the reaction.

Catalyst: A substance that alters the rate of a chemical reaction and may be recovered essentially unaltered in form or amount at the end of the reaction.

Catalytic cracking: The process of by which higher molecular weight constituents of crude oil are thermally decomposed in the presence of a catalyst to produce lower molecular weight products.

Cation exchange: The interchange between a cation in solution and another cation in the boundary layer between the solution and surface of negatively charged material such as clay or organic matter.

Cation exchange capacity (CEC): The sum of the exchangeable bases plus total soil acidity at a specific pH, usually 7.0 or 8.0. When acidity is expressed as salt extractable acidity, the cation exchange capacity is called the effective cation exchange capacity (ECEC), because this is considered to be the CEC of the exchanger at the native pH value. It is usually expressed in centimoles of charge per kilogram of exchanger (cmol/kg) or millimoles of charge per kilogram of exchanger.

CERCLA: Comprehensive Environmental Response, Compensation, and Liability Act. This law created a tax on the chemical and crude oil and crude oil products industries and provided broad federal authority to respond directly to releases or threatened releases of hazardous substances that may endanger public health or the environment.

Check standard: An analyte with a well-characterized property of interest, e.g., concentration, density, and other properties that is used to verify method, instrument and operator performance during regular operation; *check standards* may be obtained from a certified supplier, may be a pure substance with properties obtained from the literature or may be developed in-house.

Chemical bond: The forces acting among two atoms or groups of atoms that lead to the formation of an aggregate with sufficient stability to be considered as an independent molecular species.

Chemical dispersion: In the context of oil spills, this term refers to the creation of oil-in-water *emulsions* by the use of chemical dispersants made for this purpose.

Chemical induction (coupling): When one reaction accelerates another in a chemical system there is said to be chemical induction or coupling. Coupling is caused by an intermediate or by-product of the inducing reaction that participates in a second reaction; chemical induction is often *observed* in oxidation-reduction reactions.

Chemical reaction: A process that results in the interconversion of chemical species.

Chemical waste: Any gaseous, liquid, or solid material discharged from a process and that may pose substantial hazards to human health and environment.

Chlorinated solvent: A volatile organic compound containing chlorine; common solvents are trichloroethylene, tetrachloroethylene, and carbon tetrachloride.

Chromatographic adsorption: The selective adsorption on materials such as activated carbon, alumina, or silica gel; liquid or gaseous mixtures of hydrocarbons are passed through the adsorbent in a stream of diluent, and certain components are preferentially adsorbed.

Chromatography: A method of separation based on selective adsorption; see also Chromatographic adsorption.

Chromatogram: The resultant electrical output of sample components passing through a detection system following chromatographic separation.

Cis trans isomers: The difference in the positions of atoms (or groups of atoms) relative to a reference plane in an organic molecule; in a *cis-isomer*, the atoms are on the same side of the molecule, but are on opposite sides in the *trans*-isomer; sometimes called stereoisomers; these arrangements are common in alkenes and cycloalkanes.

Clean Water Act: The Clean Water Act establishes the basic structure for regulating discharges of pollutants into the waters of the United States. It gives EPA the authority to implement pollution control programs such as setting wastewater standards for industry; also continued requirements to set water quality standards for all contaminants in surface waters and makes it unlawful for any person to discharge any pollutant from a point source into navigable waters, unless a permit was obtained under its provisions.

Cloud point: The temperature at which a haze appears in a sample which is attributed to the formation of wax crystals.

Coke: A gray to black solid carbonaceous material produced from petroleum during thermal processing; characterized by having a high carbon content (95% + by weight) and a honeycomb type of appearance and is insoluble in organic solvents.

Cometabolism (co-metabolism): The process by which compounds in crude oil and crude oil products may be enzymatically attacked by micro-organisms without furnishing carbon for cell growth and division; a variation on biodegradation in which microbes transform a contaminant even though the contaminant cannot serve as the primary energy source for the organisms. To degrade the contaminant, the microbes require the presence of other compounds (primary substrates) that can support their growth.

Complex modulus: A measure of the overall resistance of a material to flow under an applied stress, in units of force per unit area. It combines *viscosity* and elasticity elements to provide a measure of "stiffness", or resistance to flow. The *complex modulus* is more useful than *viscosity* for assessing the physical behavior of very non-Newtonian materials such as *emulsions*.

Compost bioremediation: The use of a biological system of micro-organisms in a mature, cured compost to sequester or break down contaminants in water or soil.

Composting: A technique that involves combining contaminated soil with nonhazardous organic materials such as manure or agricultural wastes; the presence of the organic materials allows the development of a rich microbial population and elevated temperature characteristic of composting.

Concentration: Composition of a mixture characterized in terms of mass, amount, volume or number concentration with respect to the volume of the mixture.

Conservative constituent or compound: One that does not degrade, is unreactive, and its *movement* is not retarded within a given environment (aquifer, stream, contaminant plume).

Constituent: An essential part or component of a system or group (that is, an ingredient of a chemical mixture); for example, benzene is one constituent of gasoline.

Contaminant: A substance that causes deviation from the normal composition of an environment; the contaminant may be added to the environment as a non-indigenous contaminant or the contaminant may be an indigenous chemical (a non-contaminant) that is released into the environment in amounts that are in excess of the indigenous level of the chemical and, therefore, becomes classed as a contaminant.

Covalent bond: A region of *relatively* high electron density between atomic nuclei that results from sharing of electrons and that *gives* rise to an attractive force and a characteristic internuclear distance; carbon-hydrogen bonds are covalent bonds.

Crude assay: A series of test methods for determining the general distillation and quality characteristics of crude oil.

Culture: The growth of cells or micro-organisms in a controlled artificial environment.

Cut point: The temperature on the whole crude true boiling point (TBP) curve that represents the limits (upper and lower) of a fraction to be produced (yield of a fraction).

Cycloalkanes (naphthene, cycloparaffin): A saturated, cyclic compound containing only carbon and hydrogen. One of the simplest *cycloalkanes* is cyclohexane (C_6H_{12}). Sterane derivatives and triterpane derivatives are branched naphthenes consisting of multiple condensed five- or six-carbon rings.

Daughter product: A compound that results directly from the degradation of another chemical.

Dead oil: Crude oil from which the most volatile components have been evaporated so that they do not interfere with gravimetric determination of the proportions of the separated fractions. See also: Live oil.

Deasphalting: The removal of the asphaltene fraction from petroleum by the addition of a low-boiling hydrocarbon liquid such as n-pentane or n-heptane; more correctly the removal asphalt (tacky, semi-solid) from petroleum (as occurs in a refinery asphalt plant) by the addition of liquid propane or liquid butane under pressure.

Degradation: The breakdown or transformation of a compound into byproducts and/or end products.

Dehydrohalogenation: Removal of hydrogen and halide ions from an alkane resulting in the formation of an alkene.

Delayed coking: A coking process in which the thermal reaction are allowed to proceed to completion to produce gaseous, liquid, and solid (coke) products.

Denitrification: Bacterial reduction of nitrate to nitrite to gaseous nitrogen or nitrous oxides under anaerobic conditions.

Density: The mass per unit volume of a substance. *Density* is temperature-dependent, generally decreasing with temperature. The density of crude oil (or a crude oil product) relative to water, its specific gravity, governs whether a particular oil will float on water. Most fresh crude oils and fuels will float on water. Bitumen and certain residual fuel oils, however, may have densities greater than water at some temperature ranges and may submerge in water. The density of a spilled oil will also increase with time as components are lost due to weathering and more polar oxygen functions are introduced into the spilled material.

Desorption: The reverse process of adsorption whereby adsorbed matter is removed from the adsorbent; also used as the reverse of absorption (*q.v.*).

Detection limit (in analysis): The minimum single result that, with a stated probability, can be distinguished from a representative blank value during the laboratory analysis of substances such as water, soil, air, rock, and biota.

Dichloroelimination: Removal of two chlorine atoms from an alkane compound and the formation of an alkene compound within a reducing environment.

Dihaloelimination: Removal of two halide atoms from an alkane compound and the formation of an alkene compound within a reducing environment.

Diols: Chemical compounds that contain two hydroxy (-OH) groups, generally assumed to be, but not necessarily, alcoholic; aliphatic diols are also called glycols.

Dispersant (chemical dispersant): A chemical that reduces the surface tension between water and a hydrophobic substance such as oil. In the case of an oil spill, dispersants facilitate the breakup and dispersal of an oil slick throughout the water column in the form of an oil-in-water emulsion; chemical dispersants can only be used in areas where biological damage will not occur and must be approved for use by government regulatory agencies.

Distillate: The products of distillation formed by condensing vapors.

Downgradient: In the direction of decreasing static hydraulic head.

Dredging: A form of excavation that is used underwater or partially underwater in shallow waterways or in ocean waterways; a method in which sediment and other materials from the bottom of the water are removed by pumping through pipelines and into a processing facility.

Ecosystem: A system that consists of all of the organisms and the physical environment with which the organisms interact. These biotic and abiotic components are linked together through nutrient cycles and energy flows and energy enters the system through photosynthesis and is incorporated into plant tissue.

Effluent: Any contaminating substance, usually (but not always) a liquid that enters the environment via a domestic industrial, agricultural, or sewage plant outlet.

Electron acceptor: The compound that receives electrons (and therefore is reduced) in the energy-producing oxidation-reduction reactions that are essential for the growth of micro-organisms and bioremediation—common electron acceptors in bioremediation are oxygen, nitrate, sulfate, and iron.

Electron donor: The compound that donates electrons (and therefore is oxidized). In bioremediation the organic contaminant often serves as an electron donor.

Electronegativity: The power of an atom to attract electrons to itself.

Elimination: A reaction where two groups such as chlorine and hydrogen are lost from adjacent carbon atoms and a double bond is formed in their place.

Emulsan: A polyanionic heteropolysaccharide bioemulsifier produced by *Acinetobacter calcoaceticus* RAG-1; used to stabilize oil-in-water emulsions.

Emulsion: A stable mixture of two immiscible liquids, consisting of a continuous phase and a dispersed phase. Oil and water can form both oil-in-water and water-in-oil emulsions. The former is termed a dispersion, while the term *emulsion* implies the latter. Water-in-oil emulsions formed from crude oil and crude oil products and brine can be grouped into four stability classes: stable, a formal emulsion that will persist indefinitely; meso-stable, which gradually degrade over time due to a lack of one or more stabilizing factors; entrained water, a mechanical mixture characterized by high viscosity of the crude oil and crude oil products component which impedes separation of the two phases; and unstable, which are mixtures that rapidly separate into immiscible layers.

Emulsion stability: Generally accompanied by a marked increase in *viscosity* and elasticity, over that of the parent oil which significantly changes behavior. Coupled with the increased volume due to the introduction of brine, emulsion formation has a large effect on the choice of countermeasures employed to combat a spill.

Emulsification: The process of *emulsion* formation, typically by mechanical mixing. In the environment, *emulsions* are most often formed as a result of wave action. Chemical agents can be used to prevent the formation of *emulsions* or to "break" the *emulsions* to their component oil and water phases.

Endergonic reaction: A chemical reaction that requires energy to proceed. A chemical reaction is endergonic when the change in free energy is positive.

End point: The actual terminal temperatures of a fraction produced commercially.

Engineered bioremediation: A type of remediation that increases the growth and degradative activity of micro-organisms by using engineered systems that supply nutrients, electron acceptors, and/or other growth-stimulating materials.

Enhanced bioremediation: A process which involves the addition of micro-organisms (e.g., fungi, bacteria, and other microbes) or nutrients (e.g., oxygen, nitrates) to the subsurface environment to accelerate the natural biodegradation process.

Environmental remediation: The removal of pollutants or contaminants from groundwater, surface water, soil, and sediment. See also: Environmental remediation, Water Remediation.

Enzyme: A biological catalyst; a macromolecule, mostly proteins or conjugated proteins produced by living organisms, that facilitate the degradation of a chemical compound (catalyst); in general, an enzyme catalyzes only one reaction type (reaction specificity) and operates on only one type of substrate (substrate specificity); any of a group of catalytic proteins that are produced by cells and that mediate or promote the chemical processes of life without themselves being altered or destroyed.

Environmental guidance: A document developed by a governmental agency that outlines a position on a topic or which give instructions on how a procedure must be carried out. It explains how to do something and provides governmental interpretations on a governmental act or policy. See also: Environmental policy, Environmental regulation.

Environmental policy: A requirement that specifies operating procedures that must be followed. See also: Environmental guidance, Environmental regulation.

Environmental regulation: A legal mechanism that spells determines how the policy directives of an environmental law are to be carried out. See also: Environmental guidance, Environmental policy.

Epoxidation: A reaction wherein an oxygen molecule is inserted in a carbon-carbon double bond and an epoxide is formed.

Epoxides: A subclass of epoxy compounds containing a saturated three-membered cyclic ether. See *Epoxy compounds*.

Epoxy compounds: Compounds in which an oxygen atom is directly attached to two adjacent or nonadjacent carbon atoms in a carbon chain or ring system; thus cyclic ethers.

Equipment blank: A sample of analyte-free media which has been used to rinse the sampling equipment. It is collected after completion of decontamination and prior to sampling. This blank is useful in documenting and controlling the preparation of the sampling and laboratory equipment.

Ex situ bioremediation: A process which involves removing the contaminated soil or water to another location before treatment.

Facultative anaerobes: Micro-organisms that use (and prefer) oxygen when it is available, but can also use alternate electron acceptors such as nitrate under anaerobic conditions when necessary.

Fenton's reagent: A solution of hydrogen peroxide (H_2O_2) with ferrous iron [typically iron (Fe^{2+}) sulfate, $FeSO_4$] as a catalyst that is used to oxidize contaminants or waste waters; the reagent can also be used to destroy organic compounds such as trichloroethylene ($CHCl = CCl_2$, TCE) and tetrachloroethylene (perchloroethylene, $CCl_2 = CCl_2$, PCE).

Fermentation: The process whereby micro-organisms use an organic compound as both electron donor and electron acceptor, converting the compound to fermentation products such as organic acids, alcohols, hydrogen, and carbon dioxide; microbial metabolism in which a particular compound is used both as an electron donor and an electron acceptor resulting in the production of oxidized and reduced daughter products.

Fertilization: The method of adding nutrients such as phosphorus and nitrogen to a contaminated environment to stimulate the growth of the microorganisms capable of biodegradation; limited supplies of these nutrients in nature usually control the growth of native microorganism populations; when more nutrients are added, the native microorganism population can grow rapidly, potentially increasing the rate of biodegradation; also called nutrient enrichment. See also: Seeding.

Field capacity or *in situ* (field water capacity): The water content, on a mass or volume basis, remaining in soil 2 or 3 days after having been wetted with water and after free drainage is negligible.

Fingerprint: A chromatographic signature of relative intensities used in oil-oil or oil-source rock correlations; mass chromatograms of sterane derivatives or terpane derivatives are examples of fingerprints that can be used for qualitative or quantitative comparison of oils.

Fingerprint analysis: A direct injection gas chromatographic analysis in which the detector output—the chromatogram—is compared to chromatograms of reference materials as an aid to product identification.

Finishing: The purification of various product streams by processes such as desulfurization or acid treatment of the crude oil fractions to remove impurities from the product or to stabilize the product.

Flash Point: The temperature at which the vapor over a liquid will ignite when exposed to an ignition source; a liquid is considered to be flammable if its *flash point* is less than 60°C. *Flash point* is an extremely important factor in relation to the safety of spill cleanup operations; gasoline and other light fuels can ignite under most ambient conditions and therefore are a serious hazard when spilled. Many freshly spilled crude oils also have low *flash points* until the lighter components have evaporated or dispersed.

Fluid catalytic cracking (FCC): A major process for the production of naphtha as a blend stock for the production of gasoline; the catalyst is (typically) a zeolite base for the cracking function.

Fraction: One of the portions of a chemical mixture separated by chemical or physical means from the remainder.

Fractional composition: The composition of petroleum as determined by fractionation (separation) methods.

Freeboard: The distance between the waterline and the main deck or weather deck of a ship or between the level of the water and the upper edge of the side of a small boat; also, the height above the recorded high-water mark of a structure (such as a dam) associated with the water.

Freezing point: The temperature at which a hydrocarbon liquid solidifies at atmospheric pressure.

Fuel oil: A general term applied to oil used for the production of power or heat; in a more restricted sense, the term is applied to any petroleum product that is used as boiler fuel or in industrial furnaces.

No. 1 Fuel oil: Very similar to kerosene (q.v.) and is used in burners where vaporization before burning is usually required and a clean flame is specified.

No. 2 Fuel oil: Also called domestic heating oil; has properties similar to diesel fuel and heavy jet fuel; used in burners where complete vaporization is not required before burning.

No. 4 Fuel oil: A light industrial heating oil and is used where preheating is not required for handling or burning; there are two grades of No. 4 fuel oil, differing in safety (flash point) and flow (viscosity) properties.

No. 5 Fuel oil: A heavy industrial fuel oil that requires preheating before burning.

No. 6 Fuel oil: A heavy fuel oil and is more commonly known as Bunker C oil when it is used to fuel ocean-going vessels; preheating is always required for burning this oil.

Functional group: An atom or a group of atoms attached to the base structure of a compound that has similar chemical properties irrespective of the compound to which it is a part; a means of defining the characteristic physical and chemical properties of families of organic compounds.

Gas chromatography (GC): A separation technique involving passage of a gaseous moving phase through a column containing a fixed liquid phase; it is used principally as a quantitative analytical technique for compounds that are volatile or can be converted to volatile forms.

Gaseous nutrient injection: A process in which nutrients are fed to contaminated groundwater and soil via wells to encourage and feed naturally occurring micro-organisms—the most common added gas is air in the presence of sufficient oxygen, micro-organisms convert many organic contaminants to carbon dioxide, water, and microbial cell mass. In the absence of oxygen, organic contaminants are metabolized to methane, limited amounts of carbon dioxide, and trace amounts of hydrogen gas. Another gas that is added is methane. It enhances degradation by co-metabolism in which as bacteria consume the methane, they produce enzymes that react with the organic contaminant and degrade it to harmless minerals.

Gas oil: A petroleum distillate with a viscosity and distillation range intermediate between those of kerosene and light lubricating oil.

GC-MS: Gas chromatography-mass spectrometry.

GC-TPH: GC detectable total crude oil hydrocarbons, that is the sum of all GC-resolved and unresolved hydrocarbons. The resolvable hydrocarbons appear as peaks and the unresolvable hydrocarbons appear as the area between the lower baseline and the curve defining the base of resolvable peaks.

Gravimetric analysis: A technique of quantitative analytical chemistry in which a desired constituent is efficiently recovered and weighed.

Grout curtain: A barrier that is used up-gradient of the contaminated area to prevent clean water from migrating through waste, or down-gradient, to limit migration of contaminants; generally used at shallow depths (30 to 40 ft maximum depth); constructed by drilling and grouting a linear sequence of holes.

Half-life: The time required to reduce the concentration of a chemical to 50 percent of its initial concentration; units are typically in hours or days.

Halide: An element from the halogen group, which include fluorine, chlorine, bromine, iodine, and astatine.

Halogen: Group 17 in the periodic table of the elements; these elements are the reactive nonmetals and are electronegative.

Hazardous waste: Any gaseous, liquid, or solid waste material that, if improperly managed or disposed of, may pose substantial hazards to human health and the environment. In many cases, the term chemical waste is often used interchangeably with the term hazardous waste. However, not all chemical wastes are hazardous and caution in the correct use of the terms must be exercised lest unqualified hysteria take control.

Heavy fuel oil: Fuel oil having a high density and viscosity; generally residual fuel oil such as No. 5 fuel oil and No. 6 fuel oil.

Heavy metal: Any metallic chemical element that has a relatively high density and is toxic or poisonous at low concentrations; examples of heavy metals include mercury (Hg), cadmium (Cd), arsenic (As), chromium (Cr), thallium (Tl), and lead (Pb).

Henry's law: The relation between the partial pressure of a compound and the equilibrium concentration in the liquid through a proportionality constant known as the Henry's Law constant.

Henry's law constant: The concentration ratio between a compound in air (or vapor) and the concentration of the compound in water under equilibrium conditions.

Heteroatom compounds: Chemical compounds that contain nitrogen and/or oxygen and/or sulfur and/or metals bound within their molecular structure(s).

Heterogeneous: Varying in structure or composition at different locations in space.

Heterotroph: An organism that cannot synthesize its own food and is dependent on complex organic substances for nutrition.

Heterotrophic bacteria: Bacteria that utilize organic carbon as a source of energy; organisms that derive carbon from organic matter for cell growth.

Hopane: A pentacyclic *hydrocarbon* of the *triterpane* group believed to be derived primarily from bacteriohopanoids in bacterial membranes.

Homogeneous: Having uniform structure or composition at all locations in space.

Hopane: A pentacyclic hydrocarbon derivative of the triterpane group believed to be derived primarily from bacteriohopanoids in bacterial membranes.

Humus: The organic component of soil, formed by the decomposition of organic material, leaves, and other plant material by soil microorganisms.

Hydration: The addition of a water molecule to a compound within an aerobic degradation pathway.

Hydrocarbon: A member of a very large and diverse group of chemical compounds composed only of carbon and hydrogen; the largest source of hydrocarbons is crude oil and crude oil products; the principal constituents of crude oils and refined crude oil products.

Hydrogen bond: A form of association between an electronegative atom and a hydrogen atom attached to a second, relatively electronegative atom; best considered as an electrostatic interaction, heightened by the small size of hydrogen, which permits close proximity of the interacting dipoles or charges.

Hydrogenation: A process whereby an enzyme in certain micro-organisms catalyzes the hydrolysis or reduction of a substrate by molecular hydrogen.

Hydrogenolysis: A reductive reaction in which a carbon-halogen bond is broken, and hydrogen replaces the halogen substituent.

Hydrolysis: A chemical transformation process in which a chemical reacts with water. In the process, a new carbon-oxygen bond is formed with oxygen derived from the water molecule, and a bond is cleaved within the chemical between carbon and some functional group.

Hydroxylation: Addition of a hydroxyl group to a chlorinated aliphatic hydrocarbon.

Ignitability: A characteristic of liquids whose vapors are likely to ignite in the presence of ignition source; also characteristic of non-liquids that may catch fire from friction or contact with water and that burn vigorously.

Immunoassay: A series of test methods that take advantage of an interaction between an antibody and a specific analyte; the test methods are semi-quantitative and usually rely on color changes of varying intensities to indicate relative concentrations.

Infiltration rate: The time required for water at a given depth to soak into the ground.

Inhibition: The decrease in rate of reaction brought about by the addition of a substance (inhibitor), by virtue of its effect on the concentration of a reactant, catalyst, or reaction intermediate.

Inoculum: A small amount of material (either liquid or solid) containing bacteria removed from a culture in order to start a new culture.

Inorganic: Pertaining to, or composed of, chemical compounds that are not organic, that is, contain no carbon-hydrogen bonds; examples include chemicals with no carbon and those with carbon in non-hydrogen-linked forms.

In situ: In its original place; unmoved; unexcavated; remaining in the subsurface.

In situ bioremediation: SA process which treats the contaminated water or soil where it was found.

Interfacial tension: The net energy per unit area at the interface of two substances, such as oil and water or oil and air. The air/liquid interfacial tension is often referred to as surface tension. The SI units for interfacial tension are milli-Newtons per meter (mN/m). The higher the interfacial tension, the less attractive the two surfaces are to each other and the more size of the interface will be minimized. Low surface tensions can drive the spreading of one fluid on another. The surface tension of an oil, together its viscosity, affects the rate at which spilled oil will spread over a water surface or into the ground.

Internal Standard (IS): A pure analyte added to a sample extract in a known amount, which is used to measure the relative responses of other analytes and surrogates that are components of the same solution. The *internal standard* must be an analyte that is not a sample component.

Intrinsic aerobic biodegradation: A means of remediating soil and groundwater contaminated with fuel hydrocarbons.

Intrinsic bioremediation: A type of bioremediation that manages the innate capabilities of naturally occurring microbes to degrade contaminants without taking any engineering steps to enhance the process; the combined effect of natural destructive and non-destructive processes to reduce the mobility, mass, and associated risk of a contaminant. Non-destructive mechanisms include sorption, dilution and volatilization. Destructive processes are aerobic and anaerobic biodegradation.

IUPAC: International Union of Pure and Applied Chemistry.

Lag phase: The growth interval (adaption phase) between microbial inoculation and the start of the exponential growth phase during which there is little or no microbial growth.

Land farming (landfarming): A soil bioremediation technique that involves mixing of the hydrocarbon-contaminated soil; an *ex situ* waste treatment process that is performed in the upper soil zone or in biotreatment cells; contaminated soils, contaminated sediments, or sludge are transported to the landfarming site, mixed into the soil surface, and periodically turned over to aerate the mixture; a method in which contaminated soil is excavated and spread over a prepared area and periodically tilled until the pollutants are degraded.

Live oil: A technical term used in the crude oil industry for the reservoir-derived mixture of gaseous and liquid crude oil constituents petroleum components that is affected after release from the reservoir due to the change in pressure and temperature. See also: Dead oil.

Loading rate: The amount of material that can be absorbed per volume of soil.

LTU: Land Treatment Unit; a physically delimited area where contaminated land is treated to remove/minimize contaminants and where parameters such as moisture, pH, salinity, temperature and nutrient content can be controlled.

Lubricant: A crude oil product that, when interposed between two surfaces, reduces the friction or wear between the surfaces.

Maltenes: That fraction of crude oil that is soluble in, for example, pentane or heptane; also the term arbitrarily assigned to the pentane-soluble portion of crude oil that is relatively high boiling (> 300°C, 760 mm); see also Petrolenes.

Measurement: A description of a property of a system by means of a set of specified rules, that maps the property on to a scale of specified values, by direct or mathematical comparison with specified references.

Metabolic by-product: A product of the reaction between an electron donor and an electron acceptor; metabolic by-products include volatile fatty acids, daughter products of chlorinated aliphatic hydrocarbons, methane, and chloride.

Metabolism: The physical and chemical processes by which foodstuffs are synthesized into complex elements, complex substances are transformed into simple ones, and energy is made available for use by an organism; thus all biochemical reactions of a cell or tissue, both synthetic and degradative, are included; the sum of all of the enzyme-catalyzed reactions in living cells that transform organic molecules into simpler compounds used in biosynthesis of cellular components or in extraction of energy used in cellular processes.

Metabolize: A product of metabolism.

Methanogens: Strictly anaerobic archaebacteria, able to use only a very limited spectrum of substrates (for example, molecular hydrogen, formate, methanol, methylamine, carbon monoxide or acetate) as electron donors for the reduction of carbon dioxide to methane.

Methanogenic: The formation of methane by certain anaerobic bacteria (methanogens) during the process of anaerobic fermentation.

Microcosm: A diminutive, representative system analogous to a larger system in composition, development, or configuration.

Microorganism (micro-organism): An organism of microscopic size that is capable of growth and reproduction through biodegradation of food sources, which can include hazardous contaminants; microscopic organisms including bacteria, yeasts, filamentous fungi, algae, and protozoa; a living organism too small to be seen with the naked eye; includes bacteria, fungi, protozoans, microscopic algae, and viruses.

Microbe: The shortened term for micro-organism.

Microbial inocula (the microbial materials used in an inoculation): Prepared in the laboratory from soil or groundwater either from the site where they are to be used or from another site where the biodegradation of the chemicals of interest is known to be occurring; in the process, microbes from the soil or groundwater are isolated and are added to media containing the chemicals to be degraded.

Microorganism (micro-organism, microbe): A microscopic organism, which may exist in its single-celled form or a colony of cells.

Microcrystalline wax: Wax extracted from certain petroleum residua and having a finer and less apparent crystalline structure than paraffin wax.

Mineralization: The biological process of complete breakdown of organic compounds, whereby organic materials are converted to inorganic products (e.g., the conversion of hydrocarbons to carbon dioxide and water); the release of inorganic chemicals from organic matter in the process of aerobic or anaerobic decay.

Mobile phase: In chromatography, the phase (gaseous or liquid) responsible for moving an introduced sample through a porous medium to separate components of interest.

Monoaromatic: Aromatic hydrocarbons containing a single benzene ring.

MTBE (methyl tertiary butyl ether): A fuel additive which has been used in the United States since 1979. Its use began as a replacement for lead in gasoline because of health hazards associated with lead. MTBE has distinctive physical properties that result in it being highly soluble, persistent in the environment, and able to migrate through the ground. Environmental regulations have required the monitoring and cleanup of MTBE at crude oil and crude oil products contaminated sites since February, 1990; the program continues to monitor studies focusing on the potential health effects of MTBE and other fuel additives.

Mycoremediation: A form of bioremediation in which fungi are used to decontaminate a polluted site.

Natural attenuation: A variety of physical, chemical, or biological processes that, under favorable conditions, act without human intervention to reduce the mass, toxicity, mobility, volume, or concentration of contaminants in soil or groundwater.

NCP: National Contingency Plan—also called the National Oil and Hazardous Substances Pollution Contingency Plan); provides a comprehensive system of accident reporting, spill containment, and cleanup, and established response headquarters (National Response Team and Regional Response Teams).

Nitrate enhancement: A process in which a solution of nitrate is sometimes added to groundwater to enhance anaerobic biodegradation.

Nucleophile: A chemical reagent that reacts by forming covalent bonds with electronegative atoms and compounds.

Nutrient enrichment: See Fertilization.

Nutrients: Major elements (for example, nitrogen and phosphorus) and trace elements (including sulfur, potassium, calcium, and magnesium) that are essential for the growth of organisms.

Octanol-water partition coefficient (K_{ow}): The equilibrium ratio of a chemical's concentration in octanol (an alcoholic compound) to its concentration in the aqueous phase of a two-phase octanol-water system, typically expressed in log units (log K_{ow}); K_{ow} provides an indication of a chemical's solubility in fats (lipophilicity), its tendency to bioconcentrate in aquatic organisms, or sorb to soil or sediment.

Oily sludge: A thick, viscous emulsion containing oil, water, sediment and residue that forms because of the incompatibility of certain native crude oils and strong inorganic acids used in well treatments. See also: Petroleum waste, Refinery sludge, Refinery waste.

Oleophilic: Oil seeking or oil loving (e.g., nutrients that stick to or dissolve in oil).

Order of reaction: A chemical rate process occurring in systems for which concentration changes (and hence the rate of reaction) are not themselves measurable, provided it is possible to measure a chemical flux.

Organic: Chemical compounds based on carbon that also contain hydrogen, with or without oxygen, nitrogen, and other elements.

Organic carbon (soil) partition coefficient (K_{oc}): The proportion of a chemical sorbed to the solid phase, at equilibrium in a two-phase, water/soil or water/sediment system expressed on an organic carbon basis. Chemicals with higher Koc values are more strongly sorbed to organic carbon and, therefore, tend to be less mobile in the environment.

Organic liquid nutrient injection: An enhanced bioremediation process in which an organic liquid, which can be naturally degraded and fermented in the subsurface to result in the generation of hydrogen. The most commonly added for enhanced anaerobic bioremediation include lactate, molasses, hydrogen release compounds (HRCs), and vegetable oils.

Osmotic potential: Expressed as a negative value (or zero), indicates the ability of the soil to dissolve salts and organic molecules; the reduction of soil water osmotic potential is caused by the presence of dissolved solutes.

OPA: Oil Pollution Act of 1990; an act which addresses oil pollution and establishes liability for the discharge and substantial threat of a discharge of oil to navigable waters and shorelines of the United States.

Oven dry: The weight of a soil after all water has been removed by heating in an oven at a specified temperature (usually in excess of 100°C, 212°F) for water; the temperature will vary of other solvents have been used.

Oxidation: The transfer of electrons away from a compound, such as an organic contaminant; the coupling of oxidation to reduction (see below) usually supplies energy that micro-organisms use for growth and reproduction. Often (but not always), oxidation results in the addition of an oxygen atom and/or the loss of a hydrogen atom.

Oxygen enhancement with hydrogen peroxide: An alternative process to pumping oxygen gas into groundwater involves injecting a dilute solution of hydrogen peroxide. Its chemical formula is H_2O_2, and it easily releases the extra oxygen atom to form water and free oxygen. This circulates through the contaminated groundwater zone to enhance the rate of aerobic biodegradation of organic contaminants by naturally occurring microbes. A solid peroxide product [e.g., oxygen releasing compound (ORC$^-$)] can also be used to increase the rate of biodegradation.

PAHs: Polycyclic aromatic hydrocarbons. Alkylated *PAHs* are *alkyl group* derivatives of the parent *PAHs*. The five target alkylated *PAHs* referred to in this report are the alkylated naphthalene, phenanthrene, dibenzothiophene, fluorene, and chrysene series.

Paraffin (alkane): One of a series of saturated aliphatic hydrocarbons, the lowest numbers of which are methane, ethane, and propane. The higher homologues are solid waxes.

Paraffin wax: The colorless, translucent, highly crystalline material obtained from the light lubricating fractions of paraffinic crude oils (wax distillates).

Partial pressure: The contribution of one component of a system to the total pressure of its vapor at a specified temperature and gross composition.

Pathogen: An organism that causes disease (e.g., some bacteria or viruses).

Permeability: The capability of the soil to allow water or air movement through it. The quality of the soil that enables water to move downward through the profile, measured as the number of inches per hour that water moves downward through the saturated soil.

Petroleum waste: Waste generated during the production and exploration of petroleum; includes drilling fluid, petroleum wastewater, petroleum effluent treatment plant sludge and bottom tank sludge. See also: Oily sludge, Refinery sludge, Refinery waste.

Phytodegradation: The process in which some plant species can metabolize VOC contaminants. The resulting metabolic products include trichloroethanol, trichloroacetic acid, and dichloracetic acid. Mineralization products are probably incorporated into insoluble products such as components of plant cell walls.

Phytoremediation: The use of living plants and associated soil microbes to reduce the concentrations or toxic effects of contaminants in the environment; an *in situ* treatment of pollutant contaminated soils, sediments, and water-terrestrial, aquatic and wetland plants and algae can be used for the phytoremediation process under specific cases and conditions of hydrocarbon contamination.

Phytovolatilization: The process in which VOCs are taken up by plants and discharged into the atmosphere during transpiration.

PINA analysis: A method of analysis for paraffins, *iso*-paraffins, naphthenes, and aromatics.

PIONA analysis: A method of analysis for paraffins, *iso*-paraffins, olefins, naphthenes, and aromatics.

Plasmid: A small, extrachromosomal DNA molecule within a cell that is physically separated from chromosomal DNA and can replicate independently; most commonly found as small circular, double-stranded DNA molecules in bacteria; however, plasmids are sometimes present in archaea and eukaryotic organisms; frequently used in the laboratory manipulation of genes.

PNA: A polynuclear aromatic compound; also referred to as PAH.

PNA analysis: A method of analysis for paraffins, naphthenes, and aromatics.

Polar compound: An organic compound with distinct regions of positive and negative charge. *Polar compounds* include alcohols, such as sterols, and some *aromatics*, such as monoaromatic-steroids. Because of their polarity, these compounds are more soluble in polar solvents, including water, compared to non-polar compounds of similar molecular structure.

Pollutant: A substance present in a particular location *(ecosystem)* when it is not indigenous to the location or is present in a greater-than-natural concentration; in the context of this book, the substance is a product of human activity which, by virtue of its name, has a detrimental effect on the environment, in part or *in toto*. Pollutants can also be subdivided into two classes: primary and secondary. Thus:

Source → Primary pollutant → Secondary pollutant

See also: Primary pollutant.

Polynuclear aromatic compound: An aromatic compound having two or more fused benzene rings, e.g., naphthalene, anthracene, and phenanthrene.

PONA analysis: A method of analysis for paraffins (P), olefins (O), naphthenes (N), and aromatics (A).

Polycyclic aromatic hydrocarbons (PAHs): Polycyclic aromatic hydrocarbons are a suite of compounds comprised of two or more condensed aromatic rings. They are found in many petroleum mixtures, and they are predominantly introduced to the environment through natural and anthropogenic combustion processes.

Porphyrins: The organometallic constituents of petroleum that contain vanadium or nickel; the degradation products of chlorophyll that became included in the protopetroleum.

Pour point: The lowest temperature at which an oil will appear to flow under ambient pressure over a period of five seconds—the pour point of a conventional crude oil is typically on the order of $-60°C$ to $30°C$; low-boiling oils with low viscosities generally have lower pour points.

Primary pollutant: A pollutant that is emitted directly from the source; in terms of atmospheric pollutants, examples are carbon oxides (i.e., carbon monoxide and carbon dioxide), sulfur dioxide, and nitrogen oxides from fuel combustion operations:

$2[C]_{crude\ oil} + O_2 \rightarrow 2CO$

$[C]_{crude\ oil} + O_2 \rightarrow CO_2$

$2[N]_{crude\ oil} + O_2 \rightarrow 2NO$

$[N]_{crude\ oil} + O_2 \rightarrow NO_2$

$[S]_{\text{crude oil}} + O_2 \rightarrow SO_2$

$2SO_2 + O_2 \rightarrow 2SO_3$

Other examples include the formation of the constituents of acid rain is an example of the formation of secondary pollutants:

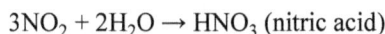

$SO_2 + H_2O \rightarrow H_2SO_3$ (sulfurous acid)

$SO_3 + H_2O \rightarrow H_2SO_4$ (sulfuric acid)

$NO + H_2O \rightarrow HNO_2$ (nitrous acid)

$3NO_2 + 2H_2O \rightarrow HNO_3$ (nitric acid)

Any pollutant, either primary or secondary can have a serious effect on the various ecological cycles and, therefore, understanding the means by which a chemical pollutant can enter these ecosystems and influence the future behavior of the ecosystem, is extremely important. See also: Pollutant.

Primary substrates: The electron donor and electron acceptor that are essential to ensure the growth of micro-organisms. These compounds can be viewed as analogous to the food and oxygen that are required for human growth and reproduction.

Propagule: Any part of a plant (e.g., bud) that facilitates dispersal of the species and from which a new plant may form.

Rate: A derived quantity in which time is a denominator quantity so that the progress of a reaction is measured with time.

Rate constant, k: See Order of reaction.

Rate-controlling step (rate-limiting step, rate-determining step): The elementary reaction having the largest control factor exerts the strongest influence on the rate; a step having a control factor much larger than any other step is said to be rate-controlling.

Recalcitrant: Unreactive, non-degradable, refractory.

Redox reaction (reduction-oxidation reaction or oxidation-reduction reaction): A chemical reaction in which oxidation and reduction occur simultaneously; in general, the oxidizing agent gains electrons in the process (and is reduced) while the reducing agent donates electrons (and is oxidized).

Reduced crude: The non-volatile product remaining after the removal, by distillation or other means, of an appreciable quantity of the more volatile components of crude oil.

Reduction: The transfer of electrons to a compound, such as oxygen, that occurs when another compound is oxidized.

Reductive dehalogenation: A variation on biodegradation in which microbially catalyzed reactions cause the replacement of a halogen atom on an organic compound with a hydrogen atom. The reactions result in the net addition of two electrons to the organic compound.

Refinery sludge: During the process of crude oil refining, large amounts of oily sludge, which contains oil, benzenes, phenols, and other odorous and toxic substances, can be produced in refineries; may be rich in hydrocarbon derivatives and have a high potential energy. See also: Oily sludge, Petroleum waste, Refinery sludge, Refinery waste.

Refinery waste: The waste products from any petroleum refinery or production process which has been dewatered; commonly called slop oil; may contain benzene and related organic compounds to make them a characteristic hazardous waste; also may contain hazardous concentrations of heavy metals, including arsenic, cadmium, chromium, lead, mercury, and selenium. See also: Oily sludge, Petroleum waste, Refinery sludge.

Reflux: The portion of the distillate returned to the fractionating column to assist in attaining better separation into desired fractions.

Reformulated gasoline (RFG): Gasoline that is designed to mitigate smog production and to improve air quality by limiting the emission levels of certain chemical compounds such as benzene and other aromatic derivatives; often contains oxygenates (*q.v.*).

Reid vapor pressure (RVP): The vapor pressure determined in a volume of air four times the liquid volume at 37.8°C (100°F).

Remediation: The process for site cleanup after a spill of crude oil or a crude oil product; the process can be achieved *ex situ* or *in situ* through various technologies—the *ex situ* method involves the removal of the contaminated soil and/or water to clean up on another surface while the *in situ* method refers to decontaminating the soil and/or water at the site of pollution.

Residues (crude oil, clay-treating filter wash): A complex residuum from the solvent washing of clay-treating filters.

Residuum (resid; *plural:* residua): The residue obtained from petroleum after nondestructive distillation has removed all the volatile materials from crude oil, e.g., an atmospheric (345°C, 650°F$^+$) residuum.

Resins: The name given to a large group of polar compounds in oil. These include hetero-substituted aromatics, acids, ketones, alcohols and monoaromatic steroids. Because of their polarity, these compounds are more soluble in polar solvents, including water, than the non-polar compounds, such as waxes and aromatics, of similar molecular weight. They are largely responsible for oil adhesion.

Respiration: The process of coupling oxidation of organic compounds with the reduction of inorganic compounds such as oxygen, nitrate (NO_3^-), iron (Fe^{3+}), manganese (Mn^{4+}), and sulfate (SO_4^{2-}).

Rhizodegradation: The process whereby plants modify the environment of the root zone soil by releasing root exudates and secondary plant metabolites. Root exudates are typically photosynthetic carbon, low molecular weight molecules, and high molecular weight organic acids. This complex mixture modifies and promotes the development of a microbial community in the rhizosphere. These secondary metabolites have a potential role in the development of naturally occurring contaminant-degrading enzymes.

Rhizosphere: The soil environment encompassing the root zone of the plant.

Risk analysis: A technique that is used to identify and assess factors that may jeopardize the success of a project or achieving a goal; performing a risk analysis includes considering the possibility of adverse events caused by either natural processes, like severe storms, earthquakes or floods, or adverse events caused by malicious or inadvertent human activities. An important part of risk analysis is identifying the potential for harm from these events, as well as the likelihood that they will occur; the process is used help organizations avoid or mitigate the risks.

Risk assessment: A way of ensuring the health and safety of workers and regulatory requirements create a need to map the environmental risk of potential spills of crude oil and to identify preparedness and response options; the first component of the four-stage contingency plan for oil spill preparedness and response, which can be used to develop and implement risk-based planning for incidents involving spills of crude oil and crude oil products.

RRF: Relative response factor.

Sampling, cluster: A process which selection of the sample units in groups. The analysis of cluster samples must take into account the intra-cluster correlation which reflects the fact that units in the same cluster are likely to be more similar than two units picked at random.

Sampling, probability: A process in which the sample in which each item has a known probability of being in the sample. As part of the sampling protocol, a simple random sample is selected so that all samples of the same size have an equal chance of being selected from the population.

Sampling, random: A process in which the sample is chosen by a method involving an unpredictable component. Random sampling can also refer to taking a number of independent observations from the same probability distribution, without involving any real population.

Sampling, stratified: A process which involves the selection of independent samples from a number of subpopulations, group or strata within the population. Great gains in efficiency are sometimes possible from judicious stratification.

Sampling, weighting (equal probability of selection method, EPSEM): A process in which every individual, or object, in the population of interest has an equal opportunity of being selected for the sample. Simple random samples are self-weighting.

Saturated hydrocarbon: A saturated carbon-hydrogen compound with all carbon bonds filled; that is, there are no double or triple bonds, as in olefins or acetylenes.

Saturates fraction: A fraction containing non-aromatic hydrocarbon derivatives; includes both normal and branched alkanes (paraffins), and cycloalkanes (naphthenes).

Saturation: The maximum amount of solute that can be dissolved or absorbed under prescribed conditions.

Saybolt Furol viscosity: The time, in seconds (Saybolt Furol Seconds, SFS), for 60 ml of fluid to flow through a capillary tube in a Saybolt Furol viscometer at specified temperatures between 70 and 210°F; the method is appropriate for high-viscosity oils such as transmission, gear, and heavy fuel oils.

Saybolt Universal viscosity: The time, in seconds (Saybolt Universal Seconds, SUS), for 60 ml of fluid to flow through a capillary tube in a Saybolt Universal viscometer at a given temperature.

Seeding: The addition of microorganisms to the existing native oil-degrading population; some species of bacteria that do not naturally exist in an area will be added to the native population; as with fertilization, the purpose of seeding is to increase the population of microorganisms that can biodegrade the spilled oil. See Fertilization.

Separation process: A physical process in which the constituents (or fractions) of crude oil are separated by different techniques.

Sheet piling: A type of type of retaining wall in which segments with indented profiles interlock to form a wall with alternating indents and outdents.

SIM (selection ion monitoring): Mass spectrometric monitoring of a specific mass/charge (m/z) ratio. The *SIM* mode offers better sensitivity than can be obtained using the full scan mode.

Site evaluation: The practice of investigating, evaluating and reporting basic soil and site conditions which apply to the onsite treatment and disposal of contaminants.

Slop oil: See Refinery waste.

Sludge: A semi-solid to solid product that results from the storage instability and/or the thermal instability of petroleum and petroleum products.

Slurry wall: A technique used to build an impermeable barrier in areas of soft earth close to open water, or with a high groundwater table; this technique is typically used to build diaphragm (water-blocking) walls surrounding contaminant spills into the environment.

Soil remediation: The strategies that are used to purify and revitalize the soil that includes topsoil, subsoil, and sediment. Soil and water remediation may be conducted separately or together, depending on the type and extent of the pollution. See also: Environmental remediation, Water Remediation.

Solubility: The amount of a substance (solute) that dissolves in a given amount of another substance (solvent). Particularly relevant to oil spill cleanup is the measure of how much and the composition of oil which will dissolve in the water column. This is important as the soluble fractions of the oil are often toxic to aquatic life, especially at high concentrations. The solubility of oil in water is very low, generally less than 1 part per million (ppm).

Soluble: Capable of being dissolved in a solvent.

Solvolysis: Generally, a reaction with a solvent, involving the rupture of one or more bonds in the reacting solute: more specifically the term is used for substitution, elimination, or fragmentation reactions in which a solvent species is the nucleophile; hydrolysis, if the solvent is water or alcoholysis if the solvent is an alcohol.

Sparging: A subsurface contaminant remediation technique that involves the injection of (for example) pressurized air into contaminated groundwater causing any hydrocarbon derivatives to change state from the dissolved state to vapor state after which the air is sent to a vacuum extraction system to remove the contaminants; also known as *in situ* air stripping and/or *in situ* volatilization.

Spent catalyst: A catalyst that has lost much of its activity due to the deposition of coke and metals.

Stable: As applied to chemical species, the term expresses a thermodynamic property, which is quantitatively measured by relative molar standard Gibbs energies; a chemical species A is more stable than its isomer B under the same standard conditions.

Steranes: A class of tetracyclic, saturated biomarkers constructed from six isoprene subunits (approximately C30 units); steranes are derived from sterols, which are important membrane and hormone components in eukaryotic organisms. Most commonly used steranes are in the range of C26 to C30 and are detected using m/z 217 mass chromatograms.

Stomata: Minute openings found in the epidermis of leaves, stems and other plant organs which allow gases such as carbon dioxide, water vapor, and oxygen to diffuse into and out of the internal tissues of the plant.

Stripping: The removal (by steam-induced vaporization or flash evaporation) of the more volatile components from a cut or fraction.

Substrate: Component in a nutrient medium, supplying micro-organisms with carbon (C-substrate), nitrogen (N-substrate) as food needed to grow.

Sulfur content: A measure of "sourness" (high sulfur content) and "sweetness" (low sulfur content) of crude oil and crude oil products as much as the regulations or the market will accept.

Surface-active agent: A compound that reduces the surface tension of liquids, or reduces interfacial tension between two liquids or a liquid and a solid; also known as surfactant, wetting agent, or detergent.

Sustainable enhancement: An intervention action that continues until such time that the enhancement is no longer required to reduce contaminant concentrations or fluxes.

Steranes: A class of tetracyclic, saturated biomarkers constructed from six isoprene subunits ($\sim C_{30}$). *Steranes* are derived from sterols, which are important membrane and hormone components in eukaryotic organisms. Most commonly used *steranes* are in the range of C_{26} to C_{30} and are detected using m/z 217 mass chromatograms.

Surrogate analyte: A pure analyte that is extremely unlikely to be found in any sample, which is added to a sample aliquot in a known amount and is measured with the same procedures used to measure other components. The purpose of a *surrogate analyte* is to monitor the method performance with each sample.

Synthetic crude oil: A product from the upgrading of extra heavy crude oil and tar sand bitumen; also referred to as syncrude.

Terminal electron acceptor (TEA): A compound or molecule is reduced by accepting an electron during metabolism (oxidation) of a carbon source; under aerobic conditions molecular oxygen is the terminal electron acceptor; under anaerobic conditions a variety of terminal electron acceptors may be used. In order of decreasing redox potential, these terminal electron acceptors include nitrate, manganic manganese, ferric iron, sulfate, and carbon dioxide; micro-organisms preferentially utilize electron acceptors that provide the maximum free energy during respiration; of the common terminal electron acceptors listed above, oxygen has the highest redox potential and provides the most free energy during electron transfer.

Terpanes: A class of branched, cyclic alkane biomarkers including *hopanes* and tricyclic compounds. They are commonly monitored using m/z 191 mass chromatograms.

Thin layer chromatography (TLC): A chromatographic technique employing a porous medium of glass coated with a stationary phase. An extract is spotted near the bottom of the medium and placed in a chamber with solvent (mobile phase). The solvent moves up the medium and separates the components of the extract, based on affinities for the medium and solvent.

Time immemorial: A term that is used to refer to a point of time in the past that was so long ago that humans have no knowledge or memory of it.

Total n-alkanes: The sum of all resolved *n-alkanes* (from C_8 to C_{40} plus pristane and phytane).

Total 5 alkylated PAH homologs: The sum of the 5 target PAHs (naphthalene, phenanthrene, dibenzothiophene, fluorene, chrysene) and their alkylated (C_1 to C_4) homologues, as determined by GCMS. These 5 target alkylated PAH homologous series are oil-characteristic aromatic compounds.

Total aromatics: The sum of all resolved and unresolved aromatic hydrocarbons including the total of BTEX and other alkyl benzene compounds, total 5 target alkylated PAH homologues, and other EPA priority PAHs.

Total n-alkanes: The sum of all resolved n-alkane derivatives (from C8 to C40 plus pristane and phytane).

Total saturates: The sum of all resolved and unresolved aliphatic hydrocarbons including the total n-alkanes, branched alkanes, and cyclic saturates.

Total alkylated PAH homologs: The sum of the 5 target PAHs (naphthalene, phenanthrene, dibenzothiophene, fluorene, chrysene) and their alkylated (C1 to C4) homologues, as determined by GCMS. These 5 target alkylated PAH homologous series are oil-characteristic aromatic compounds.

Total aromatics: The sum of all resolved and unresolved aromatic hydrocarbons including the total of BTEX and other alkyl benzene compounds, total 5 target alkylated PAH homologues, and other EPA priority PAHs.

Total saturates: The sum of all resolved and unresolved aliphatic hydrocarbons including the total n-alkane derivatives, branched alkane derivatives, and cyclic saturate derivatives.

TPH (total petroleum hydrocarbons): The total measurable amount of crude oil and crude oil products-based hydrocarbons present in a medium as determined by gravimetric or chromatographic means.

Triterpanes: A class of cyclic saturated *biomarkers* constructed from six isoprene subunits. Cyclic *terpane* compounds containing two, four, and six isoprene subunits are called monoterpane (C_{10}), diterpane (C_{20}) and *triterpane* (C_{30}), respectively.

UCM: Unresolved complex mixture of hydrocarbons on, for example, a gas chromatographic tracing; the UCM appear as the *envelope* or *hump area* between the solvent baseline and the curve defining the base of resolvable peaks.

Underground storage tank: A storage tank that is partially or completely buried in the earth.

Unsaturated zone: The zone between land surface and the capillary fringe within which the moisture content is less than saturation and pressure is less than atmospheric; soil pore spaces also typically contain air or other gases; the capillary fringe is not included in the unsaturated zone (See *Vadose zone*).

Upgradient: In the direction of increasing potentiometric (piezometric) head. See also *Downgradient*.

US EPA: United States Environmental Protection Agency.

USGS United States Geological Survey.

Vadose zone: The zone between land surface and the water table within which the moisture content is less than saturation (except in the capillary fringe) and pressure is less than atmospheric; soil pore spaces also typically contain air or other gases; the capillary fringe is included in the vadose zone.

Vapor pressure: A measure of how oil partitions between the liquid and gas phases, or the partial pressure of a vapor above a liquid oil at a fixed temperature; the force per unit area exerted by a vapor in an equilibrium state with its pure solid, liquid, or solution at a given temperature.

Visbreaking: A mild thermal cracking process (insofar as the thermal reactions are not allowed to proceed to completion) that is used to break the high viscosity and pour points of vacuum residua to the level which can be used in further downstream processes.

Viscosity: The resistance of a fluid to shear, movement or flow. The viscosity of an oil is a function of its composition. In general, the greater the fraction of *saturates* and *aromatics* and the lower the amount of *asphaltenes* and *resins*, the lower the viscosity. As oil weathers, the evaporation of the lighter components leads to increased viscosity. Viscosity also increases with decreased temperature, and decreases with increased temperature.

The viscosity of an ideal, non-interacting fluid does not change with shear rate. Such fluids are called Newtonian. Most crude oils and oil products are Newtonian. The viscosity of non-Newtonian materials may vary with shear rate, as well as duration of shear. Oils with high *wax* content are often non-Newtonian, and stable water-in-oil *emulsions* are always non-Newtonian. A material that exhibits a decrease in viscosity with shear stress is termed pseudoplastic, while those that exhibit a decrease in viscosity with time of applied shear force are referred to as thixotropic. Both effects are caused by internal interactions of the molecules and larger structures in the fluid which change with the movement of the material under applied stress. Generally, non-Newtonian oils are pseudoplastic, while *emulsions* may be either thixotropic or pseudoplastic.

In terms of oil spill cleanup, viscous oils do not spread rapidly, do not penetrate soils as rapidly, and affect the ability of pumps and skimmers to handle the oil.

VGC (viscosity-gravity constant): An index of the chemical composition of crude oil defined by the general relation between specific gravity, sg, at 60°F and Saybolt Universal viscosity, SUV, at 100°F.

VOC (VOCs): Volatile organic compound(s); volatile organic compounds are regulated because they are precursors to ozone; carbon-containing gases and vapors from incomplete gasoline combustion and from the evaporation of solvents.

Volatile: Readily dissipating by evaporation.

Volatile compounds: A relative term that may mean (1) any compound that will purge, (2) any compound that will elute before the solvent peak (usually those < C6), or (3) any compound that will not evaporate during a solvent removal step.

Volatile organic compounds (VOC): Organic compounds with high *vapor pressures* at normal temperatures. *VOCs* include light *saturates* and *aromatics*, such as pentane, hexane, *BTEX* and other lighter substituted benzene compounds, which can make up to a few percent of the total mass of some crude oils.

Water remediation is the process of removing contaminants from water, both groundwater and surface water. See also: Environmental remediation, Soil Remediation.

Water solubility: The maximum amount of a chemical that can be dissolved in a given amount of pure water at standard conditions of temperature and pressure; typical units are milligrams per liter (mg/L), gallons per liter (g/L), or pounds per gallon (lbs/gall).

Wax: A predominately straight-chain *saturates* with melting points above 20°C (generally, the *n*-alkanes C_{18} and higher molecular weight); wax of petroleum origin consists primarily of normal paraffins; wax of plant origin consists of esters of unsaturated fatty acids.

Weathered crude oil: Crude oil which, due to natural causes during storage and handling, has lost an appreciable quantity of its more volatile components; also indicates uptake of oxygen.

Weathering: Processes related to the physical and chemical actions of air, water and organisms after oil spill; the major weathering processes include evaporation, dissolution, dispersion, photochemical oxidation, water-in-oil *emulsification*, microbial degradation, adsorption onto suspended particulate materials, interaction with mineral fines, sinking, sedimentation, and formation of tar balls.

Wilting point: The largest water content of a soil at which indicator plants, growing in that soil, wilt and fail to recover when placed in a humid chamber.

Windrow: A long line of material heaped up by the wind or by a machine.

Windrow composting: The production of compost by piling organic matter or biodegradable waste, such as animal manure and crop residues, in long rows (windrows).

Zeolite: A microporous, aluminosilicate mineral commonly used as a commercial adsorbent and catalyst; by origin, a zeolite can be a natural or a synthetic materials; an aluminosilicate minerals with rigid anionic frameworks containing well-defined channels and cavities which contain metal cations, which are exchangeable, or they may also host neutral guest molecules that can also be removed and replaced; the majority of natural zeolites have a general formula $Al_2O_3xSiO_2yH_2O$; the mineral structure is based on alumina (AlO_3) and silica (SiO_2) tetrahedra, which can share 1, 2, or 3 oxygen atoms, so there is a wide variety of possible structures as the network is extended in three dimensions and this unique structural feature is a basis for their well-known microporous structure.

PART I
Crude Oil and Crude Oil Products

CHAPTER 1
Crude Oil Composition and Properties

Crude oil is a naturally occurring hydrocarbonaceous product insofar as it is composed of hydrocarbon constituents as well as other organic constituents that contain (in addition to carbon and hydrogen, hence the use of the term carbonaceous) nitrogen, oxygen, sulfur, and metals (such as nickel, vanadium, and iron). Crude oil is a type of fossil fuel that can be refined to produce usable products such as naphtha (the precursor to gasoline), kerosene (the precursor to diesel fuel), various types of fuel oil, lubricating oil, and asphalt as well as various types of petrochemical products (Parkash, 2003; Gary et al., 2007; Speight, 2014a; Hsu and Robinson, 2017; Speight, 2020a).

For this book, crude oil (also referred to as petroleum) is a native (naturally occurring) which, with further processing, will be transformed into one or more constituent fractions by distillation. In the modern refinery, the total feedstock is no longer (or very rarely) a single crude oil and is typically a blend of two or more crude oils (including heavy crude oil and even extra heavy crude oil as well as tar sand bitumen) (Parkash, 2003; Gary et al., 2007; Speight, 2014a; Hsu and Robinson, 2017; Speight, 2020a). However, the constituents of crude oil generally cause processing problems and knowledge of the behavior of these elements is essential for process improvements, process flexibility, and environmental compliance.

In terms of crude oil blends, the risks may be high because some of the more viscous feedstocks usually contain contaminants such as destabilized asphaltene constituents and metal-containing constituents. These destabilized constituents can cause considerable variations in the composition and quality of the crude oils. The more consistent the supply of crude oil to a specific refinery, the more that the refinery can tailor its operation and the composition of the saleable products (Chapter 2) to the particular supply of crude oil by reducing the potential for incompatibility of the crude oil blends and the crude oil products (Mushrush and Speight, 1995; Del Carmen García and Urbina, 2003; Bai et al., 2010; Speight, 2014a, 2015; Ben Mahmoud and Aboujadeed, 2017; Speight, 2017; Kumar et al., 2018). An essential aspect of the use of any crude oil as a refinery feedstock or the crude oil products relates to the composition of the feedstocks (Speight, 2014, 2017, 2021b).

Crude oil is typically a free-flowing to viscous liquid that was formed during the Carboniferous period (that existed approximately 300 to 400 million years ago), which includes the Pennsylvanian and Mississippian epochs) which can be shown on the geologic time scale, a system of chronological dating that classifies geological strata on the time scale of the Earth (Table 1.1). During the Carboniferous period, crude oil was formed from the decay of the remains of dead carbonaceous organisms that existed millions of years ago in a typical marine environment.

Table 1.1: Geologic Eons, Eras, and Periods.

Eon*	Era	Period	Epoch	Mya**
Phanerozoic	Cenozoic	Quaternary	Holocene	0.01
			Pleistocene	2.6
		Tertiary	Pliocene	5.3
			Miocene	23.0
			Oligocene	33.9
			Eocene	55.8
			Paleocene	65.5
	Mesozoic	Cretaceous		145.5
		Jurassic		199.6
		Triassic		251.0
	Paleozoic	Permian		299.0
		Carboniferous***		359.2
		Devonian		416.0
		Silurian		443.7
		Ordovician		488.3
		Cambrian		542.0
Proterozoic				2500
Archean				4000
Hadean				4500

*Names may vary depending upon the source.
**Approximate millions of years ago.
***Includes Pennsylvanian (318 Mya) and Mississippian (359.2 Mya)

During their lifetime, the organisms absorbed energy from the sun and stored it as carbon molecules within their bodies and when, the organisms died, their remains sank to the bottom of the oceans or riverbeds and were buried in layers of sand, mud, and rock. During the following millennia, the organic material was buried under more sediment and organic materials. The high pressure, as well as a relative increase in the temperature of the deposit with depth and the lack of oxygen, transforming the organic matter into hydrocarbon derivatives as well as smaller amounts of organic compounds containing nitrogen, oxygen, and sulfur. The liquids—now commonly referred to as proto-crude oil or proto-petroleum—collected in the pores of the sediments after which the liquids moved through the pores in the surrounding rock from an area of high pressure to low pressure leading to the collection of the crude oil collected in underground reservoirs (Parkash, 2003; Gary et al., 2007; Moreno-Letelier et al., 2012; Speight, 2014a; Hsu and Robinson, 2017).

The reservoir rocks deposited during the Tertiary, Cretaceous, Mississippian, Devonian, and Ordovician periods are particularly productive in terms of the production of crude oil. However, the variation in local geologic effects results in the production of the different types of crude oil around the world and the variations in quality and location result in the way in which the crude oil is refined as well as the behavior and effects of the crude oil (and the ensuing products formed by chemical transformation of the simple oil constituents) in the environment. During this time of chemical change, the low-boiling (gaseous components) that are commonly referred to as natural were also formed (Speight, 2014a, 2019).

As a side note for comparison, the precursors that led to the majority of the coal resources of the world were deposited during the Carboniferous period (359 to 299 million years ago) (Speight, 2013a, 2021a).

Because of the variation in the amounts of chemical types and bulk fractions (Table 1.2 and Table 1.3), it should not be surprising that crude oil exhibits a wide range of physical properties, and several relationships can be made between various physical properties and the type of crude oil (Table 1.4, Table 1.5). On the other hand, crude oil containing 9.5% w/w heteroatoms (i.e., nitrogen,

Table 1.2: Boiling fractions from crude oil (Speight, 2014a).

Product	Lower carbon limit	Upper carbon limit	Lower boiling point*	Upper boiling point*	Lower boiling point**	Upper boiling point**
Refinery gas	C1	C4	−16	−1	−259	31
Liquefied crude oil gas	C3	C4	−42	−1	−44	31
Naphtha	C5	C17	36	302	97	575
Gasoline	C4	C12	−1	216	31	421
Kerosene/diesel fuel	C8	C18	126	258	302	575
Aviation turbine fuel	C8	C16	126	287	302	548
Fuel oil	C12	> C20	216	421	> 343	> 649
Lubricating oil	> C20		> 343		> 649	
Wax	C17	> C20	302	> 343	575	> 649
Asphalt	> C20		> 343		> 649	
Coke	> C50*		> 1000*		> 1832*	

*°C
**°F

Table 1.3: Compound Types in Crude oil and Crude oil Fractions.

Class	Compound Types
Saturated hydrocarbons	n-Paraffins
	iso-Paraffins and other branched paraffins
	Cycloparaffins (naphthenes)
	Condensed cycloparaffins (including steranes, hopanes)
Unsaturated hydrocarbons	Olefin derivatives are not indigenous to crude oil but are present in products of thermal reactions
Aromatic hydrocarbons	Benzene systems
	Condensed aromatic systems
	Condensed aromatic-cycloalkyl systems
	Alkyl side chains on ring systems
Saturated heteroatomic systems	Alkyl sulfides
	Cycloalkyl sulfides
	Alkyl side chains on ring systems
Aromatic heteroatomic systems	Furans (single-ring and multi-ring systems)
	Thiophenes (single-ring and multi-ring systems)
	Pyrroles (single-ring and multi-ring systems)
	Pyridines (single-ring and multi-ring systems)
	Mixed heteroatomic systems
	Amphoteric (acid-base) systems
	Alkyl side chains on ring systems

oxygen, sulfur, and metals) insofar as the hydrocarbon constituents insofar as the constituents contain *at least one or more* nitrogen, oxygen, and/or sulfur and within the molecular structures. Coupled with the changes brought about to the feedstock constituents by refinery operations, it is not surprising that crude oil characterization is a monumental task (Speight, 2014a, 2015, 2016).

Chemically, the term crude oil covers a wide assortment of materials consisting of mixtures of hydrocarbon derivatives and other compounds containing variable amounts of sulfur, nitrogen, and oxygen, which may vary widely in volatility, specific gravity, and viscosity (Speight, 2014, 2015). Metal-containing constituents, notably those compounds that contain vanadium and nickel, usually occur in the more viscous crude oils in amounts up to several thousand parts per million and can have severe consequences for processing the feedstocks (Parkash, 2003; Gary et al., 2007; Speight, 2014a, 2015; Hsu and Robinson, 2017; Speight, 2017). Because crude oil is a mixture of widely varying constituents and proportions, its physical properties also vary widely and the color varies from colorless to black.

Table 1.4: Simplified Differentiation between Conventional Crude Oil, Heavy Crude Oil, Extra Heavy Crude Oil, and Tar Sand Bitumen.*

Conventional Crude Oil
   Mobile in the reservoir; API gravity: > 25°
   High-permeability reservoir
   Primary recovery and secondary recovery

Heavy Crude Oil
   More viscous than conventional crude oil; API gravity: 10–20°
   Mobile in the reservoir
   High-permeability reservoir
   Secondary recovery and tertiary recovery (enhanced oil recovery – EOR, e.g., steam stimulation)

Extra Heavy Crude Oil
   Similar properties to the properties of tar sand bitumen; API gravity: < 10°
   Mobile in the reservoir
   High-permeability reservoir
   Secondary recovery and tertiary recovery (enhanced oil recovery – EOR, e.g., steam stimulation)

Tar Sand Bitumen
   Immobile in the deposit; API gravity: < 10°
   High-permeability reservoir
   Mining (often preceded by explosive fracturing)
   Steam assisted gravity draining (SAGD)
   Solvent methods (VAPEX)

---

* This list is not intended for use as a means of classification.

Table 1.5: Examples of the Variation in the Distribution of the Various Fractions from Light (low Density) Crude Oil and From Heavy Crude Oil.

Component	Light oil	Heavy oil
	% v/v	% v/v
Gases	5–8	< 2
Naphtha	25–30	< 10
Middles distillate	25–40	15–20
Residua	< 10	20–40

1.1 Types of Crude Oil

Typically, crude oils (sometimes referred to as the crude oil family), especially the so-called conventional crude oils (also referred to on occasion as light crude oils), are composed of hydrocarbon derivatives and varying amounts of other organic constituents. The members of the crude oil family are refined to produce usable products such as gasoline, diesel, and various other petrochemicals (Speight, 2014a, 2017, 2019b). The family member. These are non-renewable resources that cannot be replaced naturally.

The crude oil industry often characterizes crude oils according to their geographical source, e.g., Alaska North Slope Crude. However, classification of crude oil types by geographical source is not a useful classification scheme for response personnel. This classification offers little information about general toxicity, physical state, and changes that occur with time and weathering and these characteristics are primary considerations in oil spill response (Table 1.6). Hence, a basic knowledge of the organic chemistry of the constituents of crude oil and crude oil products is necessary and is included in this text (Appendix A).

Historically, crude oil fractions (in various forms) have been known and used for millennia (Speight, 2014, 2019) and currently crude oil is produced in many countries of the world. It is typically labeled by the region where the crude oil is found and all crude oils they have specific

Table 1.6: Classification Scheme for Crude Oil as Used by the US EPA.

Class A: Light, Volatile Oils • highly fluid • often clear • spread rapidly on solid or water surfaces • have a strong odor • high evaporation rate • flammable These oils penetrate porous surfaces such as dirt and sand, and may be persistent in such a matrix. They do not tend to adhere to surfaces. Flushing with water generally removes them. Class A oils may be highly toxic to humans, fish, and other organisms. Most refined products and many of the highest quality light crudes can be included in this class.
Class B: Non-Sticky Oils These oils have a waxy or oily feel. Class B oils are less toxic and adhere more firmly to surfaces than Class A oils, although they can be removed from surfaces by vigorous flushing. As temperatures rise, their tendency to penetrate porous substrates increases and they can be persistent. Evaporation of volatiles may lead to a Class C or D residue. Medium to heavy paraffin-based oils fall into this class.
Class C: Heavy, Sticky Oils • viscous • sticky or tarry • brown or black Flushing with water will not readily remove this material from surfaces, but the oil does not readily penetrate porous surfaces. The density of Class C oils may be near that of water and they often sink. Weathering or evaporation of volatiles may produce solid or tarry Class D oil. Toxicity is low, but wildlife can be smothered or drowned when contaminated. This class includes residual fuel oils and medium to heavy crudes.
Class D: Non-fluid Oils • relatively non-toxic • do not penetrate porous substrates • are usually black or dark brown in color When heated, Class D oils may melt and coat surfaces making cleanup very difficult. Residual oils, heavy crude oils, some high paraffin oils, and some weathered oils fall into this class. These classifications are dynamic for spilled oils. Weather conditions and water temperature greatly influence the behavior of oil and refined petroleum products in the environment. For example, as volatiles evaporate from a Class B oil, it may become a Class C oil. If a significant temperature drop occurs (e.g., at night), a Class C oil may solidify and resemble a Class D oil. Upon warming, the Class D oil may revert back to a Class C oil.

Source: https://www.epa.gov/emergency-response/types-crude-oil

chemical makeup, but not all crude oils are alike. Therefore, companies use a series of tests (commonly referred to as a crude oil assay) to obtain the complete chemical breakdown of the bulk fractions and bulk properties of crude oil.

Typically, crude oil exits in the liquid phase in a several of colors that range from yellow-green to dark brown or black. Crude oil is composed principally of hydrocarbon derivatives and is typically associated with natural gas.

The physical properties of crude oils are the quantitatively measurable characteristics of crude oils which vary according to (1) the composition of the crude oil, (2) the relative abundance of the groups of hydrocarbon derivatives, (3) reservoir temperature, and (4) reservoir pressure. Each types of crude oil has its own unique characteristics (Parkash, 2003; Gary et al., 2007; Speight, 2014a, 2015; Hsu and Robinson, 2017; Speight, 2017).

The density of crude oil is often resented in the form of the API gravity. The higher the number, the less dense the crude oil or the crude oil product; any liquid with an API gravity higher than 10 is less dense than water and, when spilled, likely to float on the surface of the water until weathering (the introduction of oxygen functional groups, such as ketone, $> C = O$, and alcohol functions, -C-OH) takes over. Crude oil with an API gravity less than 10 is generally considered to be the range into which bitumen is categorized, and heavy oil typically falls into the range 10 to 20 degrees API while oil with an API gravity greater than 20 is "light crude oil" or "conventional crude oil". However, these ranges are arbitrary, and it is more correct to classify these materials by the method of recovery (United States Congress, 1976; Speight, 2014a). In addition, the term "sweet" is used to describe a crude oil with a low sulfur content which typically confers upon the crude oil a mild sweet taste and a pleasant odor. Early prospectors in the 19th Century would taste and smell small quantities of crude oil to determine its quality and value. If the sulfur content of the crude is oil greater than 0.5%, it is categorized as sour crude—some of the sulfur may not be organically-bound sulfur but exists as hydrogen sulfide (H_2S). Moreover, the complexity in refining crude oil is directly related to whether the oil is sweet or sour crude. Sour crude is usually processed into fuel oil rather than gasoline or diesel to reduce processing costs (Speight, 2014a, 2017).

Common standard methods used (among other methods, depending upon the company) as part of the assay are the determination of the viscosity of the crude oil and the volatility of the crude oil constituents. The term "viscosity" relates to the resistance of the crude oil to flow. The term "volatility" refers to the boiling point (or boiling range) of some of the constituents of the oil. In addition, crude oils are also classified based on the sulfur content and terms such as (1) sweet, which indicates a no-to-low sulfur content, (2) sour, which indicates a high sulfur content, (3) light, which indicates a low viscosity with volatile constituents, and (4) heavy, which indicates a high viscosity and a low content of volatile constituents (Parkash, 2003; Gary et al., 2007; Speight, 2014a, 2015; Hsu and Robinson, 2017; Speight, 2017). Therefore, crude oils vary in the impact on the environment as might be surmised from the varying character of the members of the crude oil family.

Put simply, crude oil is considered as a mixture comprising primarily of hydrocarbon derivatives (aliphatic, alicyclic and aromatic hydrocarbon derivatives) with dissolved gases and trace amounts of heteroatoms derivatives (i.e., compounds containing nitrogen, oxygen, and sulfur as well as suspended water, and inorganic sedimentary material. Overall, crude oil is a mixture of the following chemical types (1) hydrocarbon types, (2) nitrogen compounds, (3) oxygen compounds, (4) sulfur compounds, and (5) metallic constituents. On the other hand, crude oil products (i.e., the products manufactured from crude oil) are less well defined in terms of heteroatom compounds and are better defined in terms of the hydrocarbon derivatives that are present in the products.

In the present context, the composition of crude oil is defined in terms of (1) the elemental composition, (2) the chemical composition, and (3) the fractional composition. All three parameters are interrelated although the closeness of the elemental composition makes it difficult to relate precisely to the chemical composition and the fractional composition. The chemical composition and the fractional composition are somewhat easier to relate because of the quantities of heteroatoms (nitrogen, oxygen, sulfur, and metals) that occur in the higher boiling fractions and which contribute to the separation methods (Speight, 2014a, 2015).

The majority of the crude oil reserves that occur in the world occur in a small number of reservoirs (or fields) that are commonly referred to as "giants" that contain millions (if not billions) of oil. In fact, approximately three hundred of the largest oil fields contain almost seventy five percent of the available crude oil. Although most of the world's nations produce at least minor amounts of oil, the primary concentrations are in Saudi Arabia, Russia, the United States (chiefly Texas, California, Louisiana, Alaska, Oklahoma, and Kansas), Iran, China, Norway, Mexico, Venezuela, Iraq, Great Britain, the United Arab Emirates, Nigeria, and Kuwait.

1.1.1 Conventional Crude Oil

Conventional crude oil (often referred to simple as crude oil or sometimes as light crude oil depending on the density and degree of volatility of the constituents) is a mixture of gaseous, liquid, and solid hydrocarbon compounds that occur in sedimentary rock deposits throughout the world and also contains small quantities of nitrogen-, oxygen-, and sulfur-containing compounds as well as trace amounts of metallic constituents (Speight, 2012a, 2014a). However, the molecular boundaries of crude oil cover a wide range of boiling points and carbon numbers of hydrocarbon compounds and other compounds containing nitrogen, oxygen, and sulfur, as well as metallic (porphyrin) constituents which dictate the options to be used in a refinery (Long and Speight, 1998; Parkash, 2003; Gary et al., 2007; Speight, 2014a; Hsu and Robinson, 2017; Speight, 2017). However, the actual boundaries of such a *crude oil map* can only be arbitrarily defined in terms of boiling point and carbon number. In fact, crude oil is so diverse that materials from different sources exhibit different boundary limits, and for this reason alone it is not surprising that crude oil has been difficult to *map* in a precise manner (Long and Speight, 1998). This variation in composition is extremely important when the crude oil is spilled into the environment and plans for environmental cleanup have to be designed.

At the present time, several countries are recognized as producers of crude oil and have available reserves which have been defined on the basis of reservoir character and physical properties (Campbell, 1997). Furthermore, the crude oils vary considerably in composition and properties. In fact, with the discovery of more reservoirs the current crude oils recovered from reservoirs are somewhat more viscous than the crude boil of the pre-WW II time frame insofar as they have higher proportions of non-volatile (asphaltic) constituents. In fact, many of the crude oils currently in use would have been classified as heavy feedstocks during the first half of the 20th Century. Changes in feedstock character, such as the tendency to contain higher molecular weight constituents, require adjustments to refinery operations to handle these heavier crude oils to reduce the amount of coke formed during processing and to balance the overall product slate. Also, additional refinery options are necessary to produce the desired products (Shalaby, 2005; Speight, 2014a, 2017).

The gaseous constituents of crude oil are commonly referred to (on a collective basis) as natural gas which occurs in crude oil reservoirs and is predominantly methane, but does contain other combustible hydrocarbon compounds as well as non-hydrocarbon compounds (Mokhatab et al., 2006; Speight, 2014a, 2017, 2019, 2021b).

By way of definition and in addition to composition and thermal content (Btu/scf, Btu/ft^3), natural gas can also be characterized on the basis of the mode of the natural gas found in reservoirs. For example, non-associated natural gas often exits in reservoirs in which there is no or (at best) minimal amounts of crude oil. This type of gas is typically has a high content of methane and lesser amounts of higher-boiling hydrocarbon derivative (ethane, propane, and butane) than the associated natural gas which occurs in a crude oil reservoir as free gas or as gas dissolved in the oil. Also, the gas that occurs as a solution with the crude oil is referred to as *dissolved gas*, whereas the gas that exists in contact with the crude oil (*gas cap*) is *associated gas*. Associated gas typically has a lower content of methane than the non-associated gas but has a higher content of the higher molecular weight constituents (such as ethane, propane and butane).

However, it cannot be ignored that methane is a potent greenhouse gas (GHG) which, upon escaping into the atmosphere, acts as a blanket that insulates the Earth which absorbs energy and slows the rate at which heat leaves the Earth, which is one of the contributors to the effect commonly known as global warming or global climate change (Speight, 2020b).

Also, one other type of the so-called conventional crude oil that is worthy of mention here are the high acid crude oils. The high acid crude oils are crude oils that contain considerable proportions of naphthenic acids which, as commonly used in the crude oil industry, refers collectively to all of the organic acids present in the crude oil (Shalaby, 2005; Speight, 2014a, 2021b). In many instances, the high acid crude oils are actually the higher boiling more viscous crude oils (Speight, 2014a, 2014b, 2021b). The total acid matrix is therefore complex and it is unlikely that a simple titration, such as the traditional methods for measurement of the total acid number, can give meaningful results to use in predictions of problems. An alternative way of defining the relative organic acid fraction of crude oils is therefore a real need in the oil industry, both upstream and downstream.

By the original definition, a naphthenic acid is a monobasic carboxyl group attached to a saturated cycloaliphatic structure. For example:

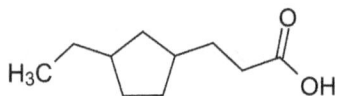

However, it has been a convention accepted in the oil industry that all organic acids in crude oil are called naphthenic acids (Shalaby, 2005; Speight, 2014a, 2021b). Naphthenic acids in crude oils are now known to be mixtures of low to high molecular weight acids and the naphthenic acid fraction also contains other acidic species. Naphthenic acids can be either (or both) water-soluble to oil-soluble depending on their molecular weight, process temperatures, salinity of waters, and fluid pressures. In the water phase, naphthenic acids can cause stable reverse emulsions (oil droplets in a continuous water phase) (Shalaby, 2005; Speight, 2014a, 2021b). In the oil phase with residual water, these acids have the potential to react with a host of minerals, which are capable of neutralizing the acids. The main reaction product found in practice is the calcium naphthenate soap (the calcium salt of naphthenic acids) (Shalaby, 2005; Speight, 2014a, 2021b). The total acid matrix is therefore complex and it is unlikely that a simple titration, such as the traditional methods for measurement of the total acid number, can give meaningful results to use in predictions of problems. An alternative way of defining the relative organic acid fraction of crude oils is therefore a real need in the oil industry, both upstream and downstream.

In addition, high acid crude oils not only cause corrosion in the refinery but can form strong bonds with the constituents of minerals when spilled into the environment thereby requiring more drastic cleanup methods for environmental remediation (Kane and Cayard, 2002; Ghoshal and Sainik, 2013; Speight, 2014b)—and occurs particularly in the atmospheric distillation unit (the first point of entry of the high acid crude oil) and also in the vacuum distillation units. In addition, overhead corrosion is caused by the mineral salts, magnesium, calcium and sodium chloride which are hydrolyzed to produce volatile hydrochloric acid, causing a highly corrosive condition in the overhead exchangers. Therefore, these salts present a significant contamination in opportunity crude oils (Speight, 2014, 2021b). Other contaminants in opportunity crude oils which are shown to accelerate the hydrolysis reactions are inorganic clays and organic acids.

Corrosion by naphthenic acids typically has a localized pattern, particularly at areas of high velocity and, in some cases, where condensation of concentrated acid vapors can occur in crude distillation units. The attack also is described as lacking corrosion products. Damage is in the form of unexpected high corrosion rates on alloys that would normally be expected to resist sulfidic corrosion (particularly steels with more than 9% Cr). In some cases, even very highly alloyed materials, i.e., 12% Cr, type 316 stainless steel (SS) and type 317 SS, and in severe cases even 6% Mo stainless steel has been found to exhibit sensitivity to corrosion under these conditions.

The corrosion reaction processes involve the formation of iron naphthenates:

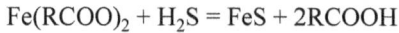

$$Fe + 2RCOOH = Fe(RCOO)_2 + H_2 \quad (4)$$

$$Fe(RCOO)_2 + H_2S = FeS + 2RCOOH$$

The iron naphthenates are soluble in oil and the surface is relatively film free. In the presence of hydrogen sulfide, a sulfide film is formed which can offer some protection depending on the acid concentration. If the sulfur-containing compounds are reduced to is hydrogen sulfide, the formation of a potentially protective layer of iron sulfide occurs on the unit walls and corrosion is reduced (Kane and Cayard, 2002; Yépez, 2005). When the reduction product is water, coming from the reduction of sulfoxides, the naphthenic acid corrosion is enhanced (Yépez, 2005).

Thermal decarboxylation can occur during the distillation process (during which the temperature of the crude oil in the distillation column can be as high as 400°C (750°F):

$$R-CO_2H \rightarrow R-H + CO_2$$

However not all acidic species in crude oil are derivatives of carboxylic acids (-COOH) and some of the acidic species are resistant to high temperatures. For example, acidic species appear in the vacuum residue after having been subjected to the inlet temperatures of an atmospheric distillation tower and a vacuum distillation tower (Speight and Francisco, 1990). In addition, for the acid species that are volatile, naphthenic acids are most active at their boiling point and the most severe corrosion generally occurs on condensation from the vapor phase back to the liquid phase.

In addition to taking preventative measure for the refinery to process these feedstocks without serious deleterious effects on the equipment, refiners will need to develop programs for detailed and immediate feedstock evaluation so that they can understand the qualities of a crude oil very quickly and it can be valued appropriately and management of the crude processing can be planned meticulously (Speight, 2021b).

1.1.2 Heavy Crude Oil

Heavy crude oil (commonly referred to as heavy oil) is a type of crude oil that has a higher density and a higher viscosity than the lower density lower viscosity conventional crude oil and, therefore, it more difficult to recovery from the reservoir. Typically, the recovery of heavy oil for the reservoir may require application of thermal or chemical processes (Parkash, 2003; Gary et al., 2007; Speight, 2014a, 2015, 2016; Hsu and Robinson, 2017; Speight, 2017). The arbitrary term "heavy oil" is often used to indicate crude oil that has a lower API gravity less than 20° as well as (but not always) a sulfur content on the order of 2% w/w or greater. Furthermore, in contrast to conventional crude oils, heavy oils are darker in color and may even be black. However, a more relevant definition of heavy oil is based on the method of recovery of the oil from the reservoir (Speight, 2016). Typically, heavy oil requires a thermal stimulation method of recovery from the reservoir and might be defined as such.

Thus, when crude oil occurs in a reservoir that allows the crude material to be recovered by pumping operations as a free-flowing dark to light colored liquid, it is often referred to as conventional crude oil. Heavy crude oils are the other types of crude oil that are different from conventional crude oil insofar as they are much more difficult to recover from the subsurface reservoir. The definition of heavy oil has been arbitrarily (and conveniently) based on the API gravity or viscosity, and the definition is quite arbitrary although there have been attempts to rationalize the definition based upon viscosity, API gravity, and density.

For example, heavy oils were considered to be those crude oils that had gravity somewhat less than 20° API with the heavy oils falling into the API gravity range 10 to 15°. For example, Cold Lake heavy crude oil has an API gravity equal to 12° and tar sand bitumen usually have an API gravity in the range 5 to 10° (Athabasca bitumen = 8° API). Residua would vary depending upon the temperature at which distillation was terminated but usually vacuum residua are in the range 2 to 8° API (Speight, 2000; Parkash, 2003; Gary et al., 2007; Speight, 2014a; Hsu and Robinson, 2017; Speight, 2017). However, the classification of heavy crude oil based on the method of recovery is a more meaningful form of classification (Table 1.4).

In addition to conventional crude oil and heavy crude oil, there are even more viscous material that offers some relief to the potential shortfalls in supply (Meyer and De Witt, 1990). These resources are often referred to as (1) extra heavy crude oil and (2) tar sand bitumen. These two other resources are not strictly members of the crude oil family (as defined by the method of recovery; Speight, 2014a, 2016) but are worthy of mention here and they are (1) extra heavy oil and (2) tar sand bitumen (Table 1.4).

1.1.3 Extra Heavy Crude Oil

The term extra heavy crude oil is a recent addition to the crude oil nomenclature but there is very little scientific justification for this terminology. This type of oil is, however, is a highly-viscous oil that cannot easily flow from production wells under normal reservoir conditions (Speight, 2014a, 2016, 2021b). Typically, extra heavy crude oil has on API gravity less than 10°. However, identification of any oil by one physical property can be suspect in terms of accuracy and general applicability.

The term extra heavy crude oil is, in effect, a non-descript term that is related to viscosity but has little scientific meaning. This type of viscous feedstock is usually compared (righty so) to tar sand bitumen, which is generally incapable of free flow under reservoir conditions. The general difference is that extra heavy oil, which may have properties similar to tar sand bitumen in the laboratory but, unlike tar sand bitumen in the deposit, has some degree of mobility in the reservoir or deposit because of the relatively high temperature of the deposit (Table 1.4) (United States Congress, 1976; Delbianco and Montanari, 2009; Speight, 2014a).

Whether or not such a material exists in the near-solid or solid state in the reservoir can be determined from the pour point and the reservoir temperature. The term is, in fact, related to the viscosity of a material that is similar to tar sand bitumen but unlike the bitumen, which is immobile in the deposit, the extra heavy oil has some mobility because of the relatively high temperature of the deposit. In other words, the temperature of the deposit exceeds the pour point of the material thereby conferring on the material a degree of mobility that the bitumen does not have (Delbianco and Montanari, 2008; Speight, 2014a, 2017, 2021b).

More specifically, an extra heavy oil can flow at a high reservoir temperature (i.e., a temperature above the pour point of the oil) and can be produced economically, without additional viscosity-reduction techniques, through variants of conventional processes such as long horizontal wells, or multilaterals. This is the case, for instance, in the Orinoco basin (Venezuela) or in offshore reservoirs of the coast of Brazil but, once outside of the influence of the high reservoir temperature, these oils are too viscous at surface to be transported through conventional pipelines and require heated pipelines for transportation. Alternatively, the oil must be partially upgraded or fully upgraded or diluted with a light hydrocarbon (such as aromatic naphtha) to create a mix that is suitable for transportation (Speight, 2014a).

1.1.4 Tar Sand Bitumen

In addition to conventional crude oil and heavy crude oil, there is an even more viscous material—this is the bitumen that occurs in the so-called tar sand deposits (Speight, 2014a, 2016, 2017).

Tar sand bitumen is the *bitumen* found in *tar sand* (*oil sand*) deposits. However, many of these reserves are only available with some difficulty and optional refinery scenarios will be necessary for conversion of these materials to liquid products (Speight, 2000, 2014a, 2021b) because of the substantial differences in character between conventional crude oil and tar sand bitumen (Table 1.4). *Tar sands*, also variously called *oil sands* or *bituminous sands*, are a loose-to-consolidated sandstone or a porous carbonate rock, impregnated with bitumen, a heavy asphaltic crude oil with an extremely high viscosity under reservoir conditions. The sand is a mixture of sand, clay, water, and bitumen. Bitumen is a black viscous oil that can form naturally in a variety of ways, usually when lighter (less dense) oil is degraded by bacteria. Bitumen has long been used in waterproofing materials for buildings, and is most familiar today as the binding agent in road asphalt. However, most of the bitumen produced from tar sands is refined and mixed with low density oils to produce synthetic crude oil that can be further refined and used in much the same way as typical crude oil.

The term *tar sand bitumen* (also, on occasion referred to as *extra heavy oil* and *native asphalt*, although the latter term is incorrect) includes a wide variety of near-black o black materials of near solid to solid character that exist in nature either with no mineral impurity or with mineral matter contents that exceed 50% by weight. Bitumen is frequently found filling pores and crevices of sandstone, limestone, or argillaceous sediments, in which case the organic and associated mineral matrix is known as *rock asphalt*.

Bitumen is also a naturally occurring material that is found in deposits that are incorrectly referred to as *tar sand* since tar is a product of the thermal processing of coal (Speight, 2013a). The permeability of a tar sand deposit low and passage of fluids through the deposit can only be achieved by prior application of fracturing techniques. Alternatively, bitumen recovery can be achieved by conversion of the bitumen to a product *in situ* (*in situ* upgrading) followed by product recovery from the deposit (Speight, 2013b, 2014a, 2016). Tar sand bitumen is a high-boiling material with little, if any, material boiling below 350°C (660°F) and the diversity of the other properties, such as the fractional composition (Table 1.7).

Table 1.7: Comparison of the Properties of Conventional Crude Oil with Tar Sand (Athabasca) Bitumen.*

Property	Athabasca bitumen	Conventional crude oil
Specific gravity	1.03	0.85–0.90
Viscosity, cp		
38°C/100°F	750,000	< 200
100°C/212°F	11,300	
Pour point, °F	> 50	ca. –20
Elemental analysis (% w/w)		
Carbon	83.0	86.0
Hydrogen	10.6	13.5
Nitrogen	0.5	0.2
Oxygen	0.9	< 0.5
Sulfur	4.9	< 2.0
Ash	0.8	0.0
Nickel (ppm)	250	< 10.0
Vanadium (ppm)	100	< 10.0
Fractional composition (% w/w)		
Asphaltenes (pentane)	17.0	< 10.0
Resins	34.0	< 20.0
Aromatics	34.0	> 30.0
Saturates	15.0	> 30.0
Carbon residue (% w/w)		
Conradson	14.0	< 10.0

* Extra heavy oil (e.g., Zuata extra heavy oil) has a similar analysis to tar sand bitumen (Table 1.5) but some mobility in the deposit because of the relatively high temperature of the deposit.

More correctly, the term tar sand is a misnomer. Tar is the term that is applied to the high boiling product that is produced by the destructive distillation of coal and other natural-occurring carbonaceous materials (Speight, 2013a). Also, the bitumen that occurs in tar sand deposits is only available with some difficulty—often requiring a mining technique for recovery and separation the bitumen from the associated sand and optional refinery scenarios will be necessary for conversion of the bitumen to saleable products (Phizackerley and Scott, 1967; Demaison, 1977; Meyer and Dietzman, 1981; Speight, 1990, 1997, 2014a, 2016, 2017).

Tar sand, also variously called oil sand or bituminous sand is any one of the several rock types (typically loose-to-consolidated sandstone or a porous carbonate rock) that contain an extremely viscous material (referred to as bitumen, which is a high boiling asphaltic material with an extremely high viscosity that is immobile under reservoir conditions and which cannot be recovered in its natural state by conventional oil well production methods including currently used enhanced recovery techniques (United States Congress, 1976). Thus, in keeping with the above definition, conventional crude oil can be recovered from the reservoirs using primary or secondary recovery methods heavy oil can be recovered from a reservoir using enhance oil recovery methods (Table 1.4).

There have been many attempts to define tar sand deposits and the bitumen contained therein. In order to define conventional crude oil, heavy oil, and bitumen, the use of a single physical parameter such as viscosity is not sufficient. Other properties such as API gravity, elemental analysis, composition, and, most of all, the properties of the bulk deposit must also be included in any definition of these materials. Only then will it be possible to classify crude oil and its derivatives.

In fact, the most appropriate definition of *tar sands* is found in the writings of the United States government (United States Congress, 1976), viz.:

Tar sands are the several rock types that contain an extremely viscous hydrocarbon which is not recoverable in its natural state by conventional oil well production methods including currently used enhanced recovery techniques. The hydrocarbon-bearing rocks are variously known as bitumen-rocks oil, impregnated rocks, oil sands, and rock asphalt.

This definition speaks to the character of the bitumen through the method of recovery. Thus, the bitumen found in tar sand deposits is an extremely viscous material that is *immobile under reservoir conditions* and cannot be recovered through a well by the application of secondary or enhanced recovery techniques. Mining methods match the requirements of this definition (since mining is not one of the specified recovery methods) and the bitumen can be recovered by alteration of its natural state such as thermal conversion to a product that is then recovered. In this sense, changing the natural state (the chemical composition) as occurs during several thermal processes (such as some *in situ* combustion processes) also matches the requirements of the definition.

By inference and by omission, conventional crude oil and heavy oil are also included in this definition. Crude oil is the material that can be recovered by conventional oil well production methods whereas heavy oil is the material that can be recovered by enhanced recovery methods. Tar sand currently recovered by a mining process followed by separation of the bitumen by the hot water process. The bitumen is then used to produce hydrocarbon derivatives by a conversion process.

Also, the term *bitumen* (also, on occasion, incorrectly referred to as *native asphalt, rock asphalt*, and more recently *extra heavy oil*) includes a wide variety of dark near-black or black materials of semisolid, viscous-to-brittle character that can exist in nature with no mineral impurity or with mineral matter (sandstone, limestone, or argillaceous sediments) contents that exceed 50% by weight (Abraham, 1945; Pfeiffer, 1950; Hoiberg, 1964). Another term—bituminous sand—is more technically correct.

Whilst on the subject of tar sand bitumen and extra heavy crude oil, it is worthy of note that synthetic crude is the output from a bitumen/extra heavy oil upgrading facility. The term may also be used in connection with the oil produced from oil shale by the thermal decomposition of the kerogen contained in the shale (2012b) or the thermal decompositon of the organic material usually referred to as coal (2013a) or from organic biomass (2011). The properties of the synthetic crude oil depend on the character of the source material and the processes used to produce the oil.

In the current context, synthetic crude oil is an intermediate product produced when an extra-heavy crude oil or tar sand bitumen is upgraded into a transportable form that can be shipped to a crude oil refinery where it is further upgraded into finished products. Synthetic crude may also be mixed, as a diluent, with heavy oil to create a product known as *synbit* which is more viscous than synthetic crude, but can also be a less expensive alternative for transporting the viscous (solid or near-solid) bitumen to a conventional refinery.

1.2 Crude Oil Composition

An important step in assessing the effects of crude oil and crude oil products that have been released into the environment is to evaluate the nature of the particular mixture and eventually select an optimum remediation technology for that mixture. As a general rule, crude oil is complex mixtures composed of the same compounds, but the quantities of the individual compounds differ in crude oils from different locations. On the other hand, crude oil products which, while also being complex mixtures, have been refined to a state where the mixture may be less complex than the original crude oil but are collections of well-defined made-to-specification mixtures which are manufactured according to the demands of the market. This general statement implies that the quantities of some compounds can be zero in a given mixture of compounds that comprise a crude oil from a specific location (Parkash, 2003; Gary et al., 2007; Speight, 2014a; Hsu and Robinson, 2017; Speight, 2017).

In the very general sense of the definition, crude oil is mixture of comparatively volatile liquid hydrocarbon derivatives (compounds composed mainly of carbon and hydrogen), but most crude oils also contain (to a lesser extent) organic derivatives of nitrogen, oxygen, and sulfur. In total, these elements form a large variety of complex molecular structures, some of which cannot be readily identified. There are however other terms which will be presented below and these are conventional crude oil and heavy crude oil as well as extra heavy crude oil and tar sand bitumen (Parkash, 2003; Gary et al., 2007; Speight, 2014a; Hsu and Robinson, 2017; Speight, 2017).

Conventional crude oil is a mixture of comparatively volatile liquid hydrocarbons (compounds composed mainly of hydrogen and carbon) and also contains constituents that contain nitrogen, oxygen, and sulfur. In short, these elements contribute to a large variety of complex molecular structures, some of which cannot be readily or conclusively identified.

The major constituents of crude oil are the determinants of the properties of the oil and the type. There are general types of crude oil (1) light crude oil, which is low-viscosity, low density crude oil crude oil that contains a significant amount of distillate, and (2) heavy crude oil, which is high viscosity, high density crude oil that has a lesser amounted distillable constituents that light crude oil and contains a relatively high proportion of high molecular weight constituents.

The main refining process used for crude oil involves fractional distillation which is the separation of the constituents of crude oil based on the relative volatility. Distillation is the first step in the refining process (after the crude oil has been cleaned and any remnants of brine removed) (Speight, 2005, 2014a, 2017).

Crude oil characterization has long been an area of concern in refining operations (Van Nes and Van Westen, 1951; Long and Speight, 1998; Speight, 2014a, 2015). However, the need to identify the chemical nature of crude has gained importance because of the complexity of refinery feedstocks. At one time, a refinery might accept one (or two) crude oil as feedstocks for the production of products. Currently, some refinery accept four or more crude oils (as a blend) in order to maintain the refinery flow.

The hydrocarbon derivatives in crude oils include (1) paraffin derivatives, which range from pentane, C_5H_{12}, to pentadecane, $C_{15}H_{32}$, (2) alkyl paraffin derivatives, which are typically referred to as branched-chin alkane derivatives, (3) naphthene derivatives, which contain a cyclic alkane system in the molecule, (4) alkylbenzene derivatives, which are alkylated derivatives of benzene systems, such as toluene, $C_6H_5CH_3$, $C_6H_5C_2H_5$, polymethyl benzene derivatives ($C_6H_{6-n}(CH_3)_n$, polyalky benzene derivatives and the like, as well as and (5) polynuclear aromatic derivatives, which are systems contained condensed aromatic ting systems. Crude oils also contain a variety of other chemical constituents comprising of sulfur, oxygen, carbon dioxide, nitrogen and trace metals (Speight, 1986, 2014a, 2015).

Crude oils are characterized by physical and chemical properties which play an important role in the understanding of the oils which form important aspects of environmental behavior. For environmental analysis, reconstruction of temperature history of the oil, and correlation of crude oils of similar geologic ages. Most of the physical properties of crude oils such as density (from which the API can be derived by a formula (Speight, 2014a, 2015) and viscosity. These properties typically depend on (1) the pressure in the reservoir, (2) the temperature in the reservoir, (3) the chemical composition of the crude oil, and (4) the amount of dissolved gaseous constituents.

The typical conventional crude oil is composed of linear alkane derivatives, branched alkane derivatives, cycloalkane derivatives, and aromatic hydrocarbon derivatives (such as toluene, ethylbenzene, xylene isomers, and their derivatives).

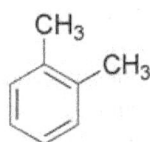
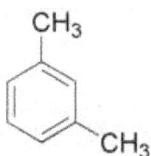

1,2-dimethylbenzene   1,3-dimethylbenzene   1,4-dimethylbenzene
o-xylene (ortho-xylene) m-xylene (meta-xylene)   para-xylene (p-xylene)

The amounts (or relative proportion) of the various hydrocarbon derivatives vary considerably from region to region. In fact, even with a single reservoir that can be a significant variation in crude oil consumption depending upon the point at which the oil is taken from the reservoir (Parkash, 2003; Gary et al., 2007; Speight, 2014a; Hsu and Robinson, 2017; Speight, 2017).

However, crude oil is a complex mixture of compounds, extending from C6 to C60, and is predominantly a mixture of straight and branched chain aliphatic hydrocarbon derivatives, ranging from gaseous methane to viscous non-volatile constituents. The most important crude oil fractions, obtained by cracking or distillation, are various hydrocarbon gases (methane, CH_4, ethane, C_2H_6, propane, C_3H_8, and butane, C_4H_{10}, as well as the corresponding (two-to-four carbon olefin derivatives), naphtha, kerosene, fuel oil, gas oil, lubricating oil, paraffin wax, and the cracked (thermally decomposed) residuum (often incorrectly referred to as "asphalt".

Crude oil, in the unrefined or crude form, like many industrial feedstock has little or no direct use and its value as an industrial commodity is only realized after the production of salable products by a series of processing steps as performed in a refinery (Figure 1.1). Each processing step is, in fact, a separate process and thus a refinery is a series of integrated processes that generate the desired products according to the market demand (Meyers, 1997; Parkash, 2003; Gary et al., 2007; Speight, 2014a; Hsu and Robinson, 2017; Speight, 2017).

A wide variety of liquid products are produced from crude oil that varies from the high-volatile naphtha to the low-volatile lubricating oil. The liquid products are often characterized by a variety of techniques that include measurement of physical properties and fractionation into group types (Speight and Arjoon, 2012).

The current trend throughout the refining industry is to produce more fuel products from each barrel of crude oil and to process those products in different ways to meet the product specifications for use in various engines (automobile, diesel, aircraft, and marine). Overall, the demand for liquid

Distillation	Atmospheric	Gas				
			Gas processing	Natural gas		
				LPG		
				Petrochemicals		
				Fuel gas		
		Naphtha		Gasoline		
				Solvents		
		Kerosene		Diesel fuel		
				Solvents		
		Gas oil	Catalytic cracking	Naphtha	Catalytic reforming	Gasoline
				Kerosene		
				Fuel oils		
			Hydrocracking	Naphtha	Gasoline	
				Kerosene	Diesel	
	Vacuum	Gas oil			Lubricating oil	
						Grease
					Waxes	
					Residuum	
			Catalytic cracking	Naphtha	Catalytic reforming	Gasoline
						Solvents
				Kerosene	Diesel	
					Solvents	
			Hydrocracking	Naphtha	Catalytic reforming	Gasoline
						Solvents
		Residuum				Asphalt
			Delayed coking			Coke

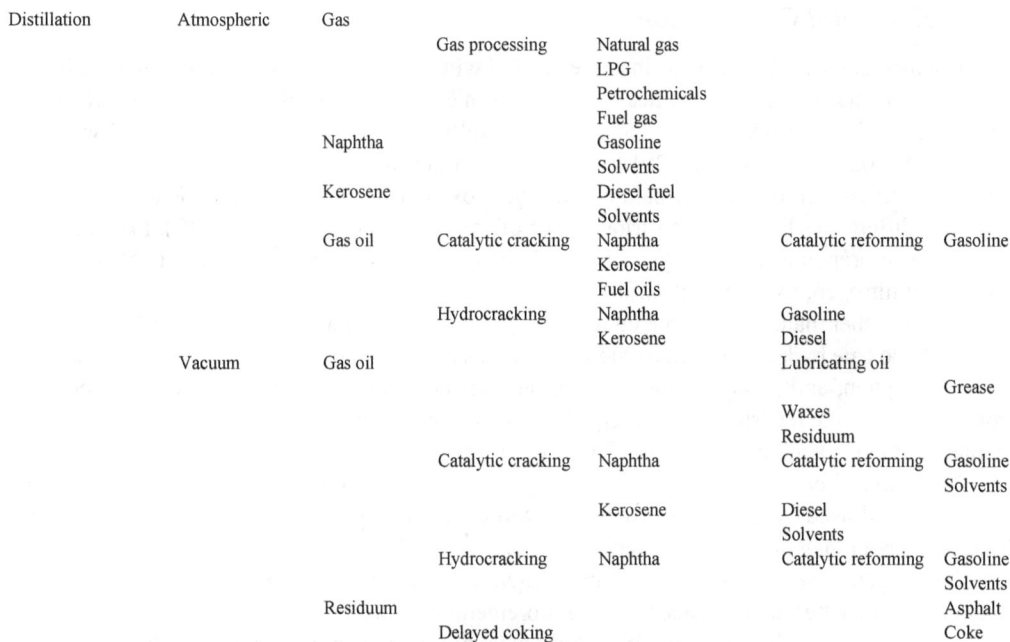

Figure 1.1: Simplified Schematic of a Crude Oil Refinery.

fuels has rapidly expanded as well as an expansion in the demand for fuel oil as well as feedstocks for the petrochemical industry.

Therefore, refineries need to be constantly upgraded (in terms of process options, i.e., modernized) to adapt and remain viable and responsive to ever changing patterns of crude supply and product market demands. As a result, refineries have been introducing increasingly complex (and expensive) processes to gain more and more low-boiling products from the high-boiling and residual portions of crude oil (Speight and Arjoon, 2012).

Typical refinery products include (1) natural gas and liquefied crude oil gas (LPG), (2) solvent naphtha, (3) kerosene, (4) diesel fuel, (5) jet fuel, (6) lubricating oil, (7) various fuel oils, (8) wax, (9) residua, and (10) asphalt. A single refinery does not necessarily produce all of these products but any refinery may also produce a variety of waste chemicals that must be disposed in an environmentally acceptable manner. Example as such products are (1) spent caustic, (2) spent acids, and (3) spent catalysts.

In very general terms, crude oil is a mixture of: (1) hydrocarbon types, (2) nitrogen compounds, (3) oxygen compounds, (4) sulfur compounds, and (5) metallic constituents. For example, the occurrence of amphoteric species (i.e., compounds having a mixed acid/base nature) is not always addressed nor is the phenomenon of molecular size or the occurrence of specific functional types that can play a major role in the distribution and effects of crude oil and crude oil products in the environment.

In the present context, crude oil composition is defined in terms of (1) the elemental composition, (2) the chemical composition, and (3) the fractional composition. All three parameters are interrelated although the closeness of the elemental composition makes it difficult to relate precisely to the chemical composition and the fractional composition. The chemical composition and the fractional composition are somewhat easier to relate because of the quantities of heteroatoms (nitrogen, oxygen, sulfur, and metals) that occur in the higher boiling fractions and which contribute to the separation methods (Parkash, 2003; Gary et al., 2007; Speight, 2014a, 2015; Hsu and Robinson, 2017; Speight, 2017).

1.2.1 Elemental Composition

On an elemental basis, the elements in crude oils fall within certain ranges regardless of the origin of the oil and the majority of the chemical components in crude oil are made up of five main elements: carbon 82 to 87% w/w, hydrogen 11 to 15% w/w/, sulfur 0 to 8% w/w, nitrogen 0 to 1% w/w, and oxygen 0.0 to 0.5% w/w (Speight, 2014a, 2017). The carbon content is relatively constant, while the hydrogen and content of the heteroatoms (nitrogen, oxygen, sulfur, and metals) is responsible for the major differences between the various types of crude oil (Speight, 2014a, 2017). In some crude oil, there is a preponderance of hydrocarbon derivatives which, as a result, consists of only trace amounts of nitrogen, oxygen, and sulfur.

On the other hand, crude oil containing a total of 9.5% w/w heteroatoms may contain essentially no true hydrocarbon constituents insofar as the constituents contain *at least one or more* nitrogen, oxygen, and/or sulfur atoms within the molecular structures. Coupled with the changes brought about to the feedstock constituents by refinery operations, it is not surprising that crude oil characterization is a monumental task (Speight, 2014a, 2015).

The analysis for oxygen is subject to interpretation because crude oil readily absorbs oxygen from the air followed by reaction to produce oxygen-containing species. There has been considerable discussion about including oxygen in crude oil analysis (Speight, 2014a, 2015). However, no matter how the oxygen arises or the origin of the oxygen, it is present in spilled crude oil and crude oil products and must be taken into account. The aforementioned presence (inclusion) of oxygen arising from interaction of spilled crude oil or spilled crude oil products to form emulsions (above) cannot be ignored and must be taken into account whenever there is a spill.

The elements are combined to form a complex mixture of organic compounds that range in molecular weight from 16 (methane; CH_4) to several hundred, perhaps even to several thousand when the constituents of the high-boiling residua are considered (Speight, 2014a, 2017). A wide range of metals is also found in trace amounts in crude oil. All metals through the atomic number 42 (molybdenum) have been found, with the exception of rubidium and niobium; a few heavier elements also have been detected.

1.2.2 Chemical Composition

Crude oil contains organic compounds: hydrocarbons (C, H, heteroatom compounds (N, O, and S)), metals in organic structures (Ni, V, Fe) and also inorganic metal salts (Na^+, Ca^{++}, Cl^-).

Crude oil	*Organic compounds*	*Inorganic compounds*
	Hydrocarbons	Metal salts (Na^+, Ca^{2+}, Cl^-)
	Heteroatom compounds (N, O, S)	
	Organo-metallic compounds (V, Ni, Fe)	

Like the elemental composition, the chemical composition of crude oil and crude oil product also varies over wide ranges molecular types with a varying range of molecular weights Appendix A).

The hydrocarbon content may be as high as 97% by weight in a conventional (lighter, low density) paraffinic crude oil or about 50% by weight in heavy crude oil and less than 30% by weight in tar sand bitumen. However, within the hydrocarbon constituents, there is also considerable variation of chemical type and molecular weight. Crude oil hydrocarbon derivatives may be paraffinic, alicyclic, or aromatic and occur in varying concentrations within the different fraction of a single crude oil. Thus, the constituents of crude oil that occur in varying amounts depend on the source and character of the oil (Parkash, 2003; Gary et al., 2007; Speight, 2014a, 2015; Hsu and Robinson, 2017).

Remembering that hydrocarbon derivatives, are (by definition) compounds containing carbon and hydrogen *only*, the hydrocarbon content may be as high as 97% by weight in a conventional (lighter, low density) paraffinic crude oil or about 50% by weight in heavy crude oil and less than

30% by weight in tar sand bitumen. However, within the hydrocarbon constituents, there is also considerable variation of chemical type and molecular weight. Crude oil hydrocarbon derivatives may be paraffinic, alicyclic, or aromatic and occur in varying concentrations within the different fraction of a single crude oil. Thus, the constituents of crude oil that occur in varying amounts depend on the source and character of the oil and are (Table 2.3) (Speight, 2001):

1. *Alkane derivatives* (also called *normal paraffin derivatives* or *n-paraffin derivatives*).

These constituents are characterized by unbranched (linear) or branched (non-linear) chains of carbon atoms with attached hydrogen atoms; alkane derivatives contain no carbon-carbon double bonds (hence the designation *saturated*), are generally insoluble in cold water—examples of alkane derivatives are pentane (C_5H_{12}) and heptane (C_7H_{16}).

2. *Cycloalkane derivatives* or *cycloparaffin derivatives* (also called *naphthene derivatives*).

These constituents are characterized by the presence of simple closed rings of carbon atoms (such as the cyclopentane ring or the cyclohexane ring); naphthenes can also contain alky moieties on the ring and are generally stable and relatively insoluble in water—examples are cyclohexane and methyl cyclohexane.

3. *Alkene derivatives* (also called *olefin derivatives*).

These constituents are characterized by the presence of a carbon-carbon double bond (> C = C <) and can be unbranched (linear) or branched (non-linear) chains of carbon atoms; typically alkene derivatives are not generally found in crude oil (having reacted over the millennia during which crude oil was formed in the ground) but are common in thermally-produced products, such as naphtha (a precursor to gasoline)—common gaseous alkene derivatives include ethylene ($CH_2 = CH_2$) and propene (also called propylene, $CH_3CH = CH_2$).

4. *Single-ring Aromatic derivatives*

Aromatic constituents are characterized by the presence of an aromatic ring with six carbon atoms and are considered to be the most acutely toxic component of crude oil constituents because of their association with chronic and carcinogenic effects; the rings can carry alkyl of naphthene substituents further distinguishes the aromatic constituents—low molecular weight aromatic constituents may have a noticeable solubility in water, increasing the potential for exposure to aquatic resources.

5. *Multi-ring Aromatic derivatives*

Aromatic constituents with two or more condensed rings are referred to as polynuclear aromatic hydrocarbon derivatives (PNAs, sometime PAHs); the most abundant aromatic hydrocarbon families in crude oil and crude oil products have two and three fused rings with one to four carbon atom alkyl group substitutions—condensed aromatic constituents with more than two condensed rings (three-to-five) are also present in the higher boiling fractions of crude oil but higher condensed systems are unlikely (Speight, 2007).

Generally, a typical conventional crude oil contains approximately 1% polynuclear aromatic hydrocarbon derivatives and typically, these derivatives are based on the naphthalene and phenanthrene types of condensed system rather than the anthracene system (Speight, 2014a, 2015).

naphthalene   anthracene

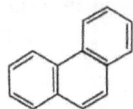

phenantherene

However, in some cases, the concentration of the total polynuclear aromatic hydrocarbon derivatives may include derivatives such as benzo(a)pyrene, range from 12 ppm to < 100 ppm.

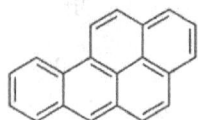

Benzo(a)pyrene

There is also the potential for this compound to be produced after the spill by oxidation of lower molecular weight alkyl derivatives of phenanthrene.

Fresh crude oil will contain a fraction of volatile hydrocarbon derivatives, some of which may pose a threat to human health such as benzene, toluene, xylenes, and other aromatics. The percent by volume (% v/v) of benzene in gasoline may range up to 3% (30,000 ppm), while the benzene content of crude oil is usually on the order of 0.2% (2000 ppm) or less. As a result of the lower percentage of volatile aromatic constituents in a product such as gasoline, vapor emissions from crude oil-contaminated soils are expected to be (but not always, depending upon the crude oil) much less than potential emissions from gasoline-contaminated soils.

In general, a typical crude oil contains approximately 1% polynuclear aromatic hydrocarbon derivatives having a variety of molecular configurations (Speight, 1986). Concentrations of total carcinogenic polynuclear aromatic hydrocarbon derivatives [such as benzo(a)pyrene] range from 12 ppm to < 100 ppm. Fresh crude oil will contain a fraction of volatile hydrocarbon derivatives, some of which may pose a threat to human health such as benzene, toluene, xylenes, and other aromatics. The percent by volume (% v/v) of benzene in gasoline may range up to 3% (30,000 ppm), while the benzene content of crude oil is usually on the order of 0.2% (2000 ppm) or less. As a result of the lower percentage of volatile aromatic constituents in a product such as gasoline, vapor emissions from crude oil-contaminated soils are expected to be (but not always, depending upon the crude oil) much less than potential emissions from gasoline-contaminated soils.

In addition to hydrocarbon derivatives, crude oil also contains compounds that contain nitrogen, oxygen, and sulfur (in the minority) as well as trace amounts of metals such as vanadium, nickel, iron, and copper. Porphyrins, the major organometallic compounds present in crude oil, are complex large cyclic carbon structures derived from chlorophyll and characterized by the ability to contain a central metal atom (trace metals are commonly found within these compounds).

By definition, the porphyrins are a group of heterocyclic macrocyclic organic compounds, composed of four modified pyrrole subunits interconnected at the α carbon atoms via methine bridges (=CH−).

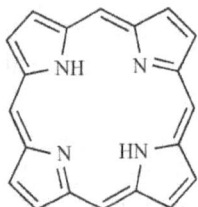

Porphine: the basic structure of the porphyrin ring system

Alkyl groups are attached to the ring systems and metals are also included in the interior of the molecule through bonding of the metal with the nitrogen atoms.

Finally, if the only constituents of crude oil were the hydrocarbon derivatives, the complexity is further illustrated by the number of potential isomers, i.e., molecules having the same atomic formula, that can exist for a given number of paraffinic carbon atoms and that increases rapidly as molecular weight increases:

Carbon atoms in the hydrocarbon	Number of potential isomers
4	2
8	18
12	355
18	60,523

The likelihood of all of these isomers occurring is minimal when consideration is given to (1) the precursors to crude oil and (2) the chemistry of the maturation processes (Speight, 2014a). Nevertheless, some isomers of the various hydrocarbon derivatives do occur hence the need for an accurate determination of the composition of the crude oil or the crude oil product by a method such as gas chromatography or mass spectrometry when the actual isomers that occur can be identified (Speight, 2014a, 2015) and appropriate action can be taken during the cleanup operations.

1.2.3 Composition by Volatility

Distillation is the separation of the constituents of a mixture based on relative volatility is a common method for the fractionation of crude oil that is used in the laboratory as well as in refineries. The technique of distillation has been practiced for many centuries, and the stills that have been employed have taken many forms (Parkash, 2003; Gary et al., 2007; Speight, 2014a, 2015; Hsu and Robinson, 2017).

In the atmospheric distillation process (Parkash, 2003; Gary et al., 2007; Speight, 2014a, 2015; Hsu and Robinson, 2017), heated crude oil is separated in a distillation column (*distillation tower, fractionating tower, atmospheric pipe still*) into streams that are then purified, transformed, adapted, and treated in a number of subsequent refining processes, into products for the refinery's market. The lower boiling products separate out higher up the column, whereas the heavier, less volatile, products settle out toward the bottom of the distillation column. The fractions produced in this manner are known as *straight-run fractions* ranging from (atmospheric tower) gas, naphtha, kerosene, gas oil, to atmospheric residuum.

The atmospheric residuum is then fed to the vacuum distillation unit at the pressure of 10 mmHg where light vacuum gas oil, heavy vacuum gas oil, and vacuum residue are the products. The fractions obtained by vacuum distillation of the reduced crude (atmospheric residuum) from an atmospheric distillation unit depend on whether or not the unit is designed to produce lubricating or vacuum gas oils. In the former case, the fractions include (1) heavy gas oil, which is an overhead product and is used as catalytic cracking stock or, after suitable treatment, a light lubricating oil, (2) lubricating oil (usually three fractions—light, intermediate, and heavy), which is obtained as a side-stream product, and (3) asphalt (or residuum), which is the non-volatile (bottom) product and may be used directly as, or to produce, asphalt, and which may also be blended with gas oils to produce a heavy fuel oil. Operating conditions for vacuum distillation are usually 350 to 410°C (660 to 770°F) 50 to 100 mm Hg (atmospheric pressure = 760 mm Hg). Although the temperature range is above the cracking temperature (350°C, 660°F) of many petroleum constituents, the residence time is adjusted to maintain cracking at a minimum.

Fractions from the atmospheric and vacuum towers are often used as feedstocks to these second stage refinery processes that break down the fractions, or cause sufficient chemical changes in the nature of a particular hydrocarbon compound to produce specific products.

However, fractionation of petroleum by volatility, informative as it might be, does not give any indication of the physical nature of petroleum. This is more often achieved by sub-division of the petroleum into bulk fractions that are separated by a variety of solvent and adsorption methods.

1.2.4 Composition by Fractionation

In the simplest sense, crude oil and (in some cases, crude oil products such as asphalt) can be considered to be composites of four major fractions (saturates, aromatics, resins and asphaltene constituents) (SARA analysis) (Figure 1.2) in varying amounts depending upon the type of crude oil and the type of crude oil product (Speight, 2014a, 2015). In fact, the lower boiling product may only be compositions of one or two fractions (saturates and aromatics).

The asphaltene constituents are substances in crude oil that are insoluble in solvents of low molecular weight such as pentane or heptane. These compounds are composed of very polynuclear aromatic and heterocyclic molecules and are solids at normal temperatures. Consequently, crude oils that have a high content of the asphaltene constituents are very viscous, have a high pour point, and are generally non-volatile in nature. During distillation, the porphyrin constituents, asphaltene constituents, and resin constituents are concentrated in the non-volatile residuum. During the weathering process, this fraction is the last to degrade, and its persistence over years has been noted (Speight, 2014a).

There are also two other operational definitions that should be noted at this point and these are the terms *carbenes* and *carboids* (Figure 1.2). Both such fractions are, by definition, insoluble in toluene (or benzene) but the carbenes are soluble in carbon disulfide whereas the carboids are insoluble in carbon disulfide (Speight, 2014a, 2015). Typically, only traces of these materials occur in crude oil and none occur in the products, unless the product is a high-boiling product of thermal treatment (such as the visbreaking process). On the other hand, oxidized crude oil and oxidized high-boiling product that have been susceptible to oxidation though a spill may contain such fractions (as defined by solubility or insolubility in solvents). But again, it must be remembered that the fraction separated by the various techniques are based on solubility or adsorption properties and not on specific chemical types (ASTM D2006, ASTM D2007, ASTM D3279, ASTM D4124) that were recognized to have selective solvency for hydrocarbon derivatives, and simple relatively low molecular weight heteroatom derivatives.

The variety of fractions isolated by these methods and the potential for the differences in composition of the fractions makes it even more essential that the method is described accurately and that it reproducible not only in any one laboratory but also between various laboratories (Speight, 2014a, 2015).

There are precautions that must be taken when attempting to separate heavy crude oil into constituent fractions. The disadvantages in using ill-defined adsorbents are that adsorbent

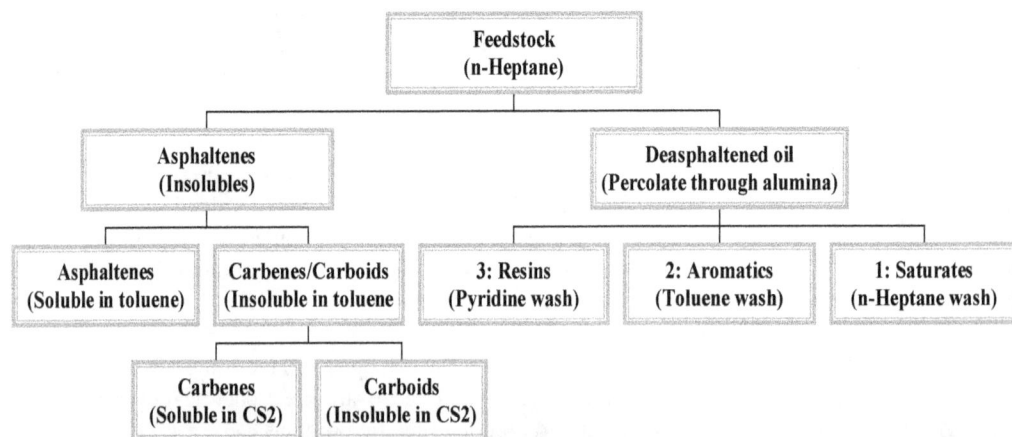

Figure 1.2: A SARA-Type Analysis (Showing Two Additional Fractions – Carbenes and Carboids

Note: Carbenes and carboids are generally produced during thermal processes and are defined by the (1) insolubility in toluene followed by (2) solubility or insolubility in a solvent such as carbon disulfide.

performance differs with the same feed and, in certain instances, may even cause chemical and physical modification of the feed constituents. The use of a chemical reactant like sulfuric acid should only be advocated with caution since feeds react differently and may even cause irreversible chemical changes and/or emulsion formation. These advantages may be of little consequence when it is not, for various reasons, the intention to recover the various product fractions *in toto* or in the original state, but in terms of the compositional evaluation of different feedstocks the disadvantages are very real.

The terminology used for the identification of the various methods might differ. However, in general terms, group-type analysis of crude oil is often identified by the acronyms for the names: PONA (paraffin derivatives, olefin derivatives, naphthene derivatives, and aromatic derivatives), PIONA (paraffin derivatives, *iso*-paraffin derivatives, olefin derivatives, naphthene derivatives, and aromatic derivatives), PNA (paraffin derivatives, naphthene derivatives, and aromatic derivatives), PINA (paraffin derivatives, *iso*-paraffin derivatives, naphthene derivatives, and aromatic derivatives), or SARA (saturate derivatives, aromatic derivatives, resin constituents, and asphaltene constituents). However, it must be recognized that the fractions produced by the use of different adsorbents will differ in content and will also be different from fractions produced by solvent separation techniques (Speight, 2014a, 2015).

1.2.5 Composition by Spectroscopy

Spectroscopic studies play an important role in the evaluation of crude oil and of crude oil products and many of the methods are now used as standard methods of analysis of crude oil and its products before and after a spill. Application of these methods to crude oil and its and products is a natural consequence for the environmental scientist and engineer.

Analytical spectroscopy-based methods can be used to investigate the reactivity, structures and associations of crude oil constituents in the environment. Because analytical environmental chemistry deals with very complex naturally occurring mixtures, new spectroscopic approaches must be utilized or further developed to fully understand global environmental processes involved in contamination by crude oil constituents and their subsequent remediation. Moreover, continually evolving environmental legislation and the introduction of stringent environmental quality standards relating to environmental contaminants continues to provide significant analytical challenges to traditional methods of analysis. Advancements in spectroscopic techniques and associated chromatographic instrumentation have become indispensable in the environmental analysis laboratory.

1.2.5.1 Infrared Spectroscopy

Conventional infrared spectroscopy yields information about the functional features of various crude oil constituents. For example, infrared spectroscopy will aid in the identification of N-H and O-H functions, the nature of polymethylene chains, the C-H out-of-place bending frequencies, and the nature of any polynuclear aromatic systems.

Infrared spectroscopy is used for the determination of benzene in motor and/or aviation gasoline (ASTM D4053) whilst ultraviolet spectroscopy is employed for the evaluation of mineral oils (ASTM D2269) and for determining the naphthalene content of aviation turbine fuels (ASTM D1840). With the recent progress of *Fourier transform infrared (FTIR) spectroscopy*, which allows the analysis of a relevant amount of compositional and structural information concerning environmental samples quantitative estimates of the various functional groups can also be made (Mecozzi et al., 2009). This is particularly important for application to the higher molecular weight solid constituents of crude oil (i.e., the asphaltene fraction).

1.2.5.2 Nuclear Magnetic Resonance Spectroscopy

Because nuclear magnetic resonance is one of the premiere techniques for studying intermolecular interactions, it has also been used extensively to investigate the environmental chemistry of humic substances through their interactions with organic contaminants and toxic metals (Nanny et al., 1997; Cardoza et al., 2004).

Nuclear magnetic resonance has frequently been employed for general studies and for the structural studies of crude oil constituents (Bouquet and Bailleul, 1982). In fact, *proton magnetic resonance* (PMR) studies (along with infrared spectroscopic studies) were, perhaps, the first studies of the modern era that allowed structural inferences to be made about the polynuclear aromatic systems that occur in the high molecular weight constituents of crude oil.

1.2.5.3 Mass Spectrometry

Mass spectrometry can play a key role in the identification of the constituents of feedstocks and products. The principal advantages of mass spectrometric methods are (1) high reproducibility of quantitative analyses; (2) the potential for obtaining detailed data on the individual components and/or carbon number homologues in complex mixtures; and (3) a minimal sample size is required for analysis.

The methods include the use of *mass spectrometry* to determine the (1) hydrocarbon types in middle distillates (ASTM D2425); (2) hydrocarbon types of gas oil saturate fractions (ASTM D2786); and (3) hydrocarbon types in low-olefin gasoline (ASTM D2789); (4) aromatic types of gas oil aromatic fractions (ASTM D3239).

Gas chromatography-mass spectrometry has become the technique of choice for volatile and semi-volatile, non-polar compounds including polycyclic aromatic hydrocarbon derivatives (PAH), polychlorinated biphenyls (PCB), pesticides, for example, organo-chlorine insecticides, dioxins, and other volatile priority pollutants.

Chemical derivatization of samples prior to analysis can be used to overcome some of these problems. Common chemical derivatization methods include: (1) silylation: used to volatilize compounds, (2) alkylation: used as the first step to further derivatization or as a method of protection of certain active hydrogen atoms, and (3) acylation, which is used to add fluorinated groups.

1.2.5.4 Other Techniques

The introduction of atmospheric pressure ionization techniques greatly expanded the applicability of liquid chromatography-mass spectrometry. In atmospheric pressure ionization liquid chromatography-mass spectrometry, the analyte molecules are ionized at atmospheric pressure.

Atmospheric pressure ionization techniques are: (1) electrospray ionization, (2) atmospheric pressure chemical ionization, and (3) atmospheric pressure photoionization.

Atmospheric pressure chemical ionization is used for molecules less than 1500 microns and is less well-suited than electrospray for analysis of large biomolecules and thermally unstable compounds. Atmospheric pressure chemical ionization is used with normal-phase chromatography more often than electrospray because the analytes are usually non-polar. Atmospheric pressure photoionization is applicable to many of the compounds that are suitable for analysis by atmospheric pressure chemical ionization. Atmospheric pressure photoionization is particularly suitable for the analysis of highly nonpolar compounds.

1.3 Crude Oil Properties

The impact of a crude oil spill or the spill of a crude oil products depends on the quantity of material released and the physical and chemical properties of the spilled material. However, because crude oil and crude oil products are extremely complex mixtures of many hundreds (even thousands) of gaseous, liquid, and solid constituents, full elucidation of their compositions at the molecular level is, to say the least, very difficult if not impossible with presently available analytical techniques.

Typically, in the refinery the most common properties used to describe crude oil are density, viscosity, and boiling point ranges. The density of crude oil is usually expressed as API (American Petroleum Institute) gravity, which is inversely related to specific gravity (also known as the specific density and the relative density). Viscosity, a measure of the internal resistance of a fluid (such as crude oil and liquids products) to flow at a given temperature and pressure, depends on the chemical composition of the crude oil or the crude oil product, including the amount of dissolved gas. The boiling point of crude oil and crude oil products is used to produce distillation fractions (cuts) of defined boiling ranges, each with a mixture of different chemical compound types.

Nevertheless best efforts to identify the constituents of the spill and the effects of these constituents on the flora and fauna of the ecosystem into which the material is spilled. For example, the spilled material can undergo chemical and physical transformation by (1) physical processes, of which evaporation, dissolution processes are examples, and/or (2) chemical process, of which photo-oxidation, microbial oxidation, selected incorporation into biomass or other forms of metabolism processes are examples. In most cases, analytical investigations are often targeted toward the chemical composition of the pre-spilled and post-spilled material rather than the physical properties. However, an investigation of the changes to the physical properties of the pre-spilled and post-spilled material must not be ignored.

Thus the focus of any post-spill investigation must be on the chemical and properties of the material prior to the spill and the chemical and physical properties of the material after the spill because of the chemical and physical changes that can occur to the spilled material.

The chemical composition of crude oil and crude oil products is complex and will change over time following release into the environment. Hence, the properties can also be expected to change once a crude oil or a crude oil product has been released into the environment. This brings about the need for an understanding of the properties of crude oil and the various products and many countries have devised a variety of standard test methods for destemming the properties of crude oils (Table 1.8).

Using the alkane derivatives as the example, there can be (1) straight-chain alkane derivatives which are chains of methyl (CH_2) groups with methyl (CH_3) groups at each end of the chain (i.e., n-alkane derivatives) or (2) chains that contain one or more alkyl side chains which are known as branched-chain alkane derivatives and/or isoprenoid alkane derivatives, or (3) alkane derivatives that contain one ring or several rings which are known as cyclic alkane derivatives and/or polycyclic alkane derivatives. There may also be several of the polycyclic saturated alkane derivatives such as the tetracyclic sterane derivatives and the pentacyclic triterpane derivatives which are also classified as biomarkers. Their presence and relative abundance as well as the stability of these molecular systems under weathering (oxidation) allow their use as fingerprints to gain information on the origin of crude oils and to distinguish crude oils from different sources (Peters et al., 2005b; Stout and Wang, 2018).

Biomarkers (also termed biological markers, molecular fossils, fossil molecules, or geochemical fossils) are organic compounds in natural waters, sediments, soils, fossils, crude oils, or coal that can be unambiguously linked to specific precursor molecules biosynthesized by living organisms (Peters et al., 2005b; Speight, 2014a, 2015; Stout and Wang, 2018; Rullkötter and Farrington, 2021).

However, biomarkers aside, the most common properties used to describe crude oil are (1) density, (2) viscosity, and (3) volatility, in terms of the boiling point ranges of the various volatile bulk fractions. The density of crude oils is usually expressed as API (American Petroleum Institute) gravity, which is inversely related to specific density. Viscosity, a measure of a fluid's internal resistance to flow at a given temperature and pressure, depends on the chemical composition of the crude oil, including the amount of dissolved gas it contains. The upstream oil industry (i.e., refineries) uses the boiling point properties of crude oils to produce distillation fractions (cuts) of defined boiling ranges, each with a mixture of different chemical compound types. These cuts, after further refinement, are the oil fractions known as (1) condensate, which includes the hydrocarbon gases and low-boiling hydrocarbon derivatives, (2) naphtha, which is sometime incorrectly referred

Table 1.8: Various Standards Organization.

AENOR	Asociacion Espanola de Normalizacion y Certificacion
AFNOR	Association Francaise de Normalisation
ANSI	American National Standards Institute
BSI	British Standards Institution
CEN	European Committee for Standardization
CSA	CSA Group
DS	Dansk Standardiseringsrad
FSA	Fluid Sealing Association
GOST	Gosudarstvennye Standarty Stae Standard Gost
ISO	International Organization for Standardization
JSA	Japanese Standards Association
KSA	Korean Standards Association
NEN	Netherlands Normalisatie-Instituut
SAI	Standards Australia International Ltd....
SIS	Standardiseringkommissio
SN	Standards Norway
SNV	Schweizerische Normen-Vereinigung
SNZ	Standards New Zealand
UNI	Ente Nazionale Italiano di Unificazione

to as gasoline, (3) kerosene, or middle distillate, (4) fuel oil, (5) lubricating oil, and (6) residuum (Speight, 2014a, 2015, 2017).

Thus, crude oil is typically described in terms of the physical properties (such as density and pour point) and chemical composition. The overall properties of a crude oil is dependent on the chemical composition of the oil and the chemical structure of the constituents. There are also impurities (i.e., non-hydrocarbon derivatives) present in crude oil which include nitrogen, oxygen, and sulfur derivatives.

The elements that make up crude oil and crude oil products are combined to form a complex mixture of organic compounds that range in molecular weight from 16 (methane; CH_4) to several hundred, perhaps even to several thousand when the constituents of the high-boiling residua are considered (Speight, 2014a). A wide range of metals is also found in trace amounts in crude oil. All metals through the atomic number 42 (molybdenum) have been found, with the exception of rubidium and niobium; a few heavier elements also have been detected. Nickel and vanadium constituents are present (to some extent) in the majority of crude oils, usually at concentrations far higher than any other metal constituents.

The analysis for oxygen is subject to interpretation because crude oil readily absorbs oxygen from the air followed by reaction to produce oxygen-containing species. There has been considerable discussion about including oxygen in crude oil analysis (Speight, 2014a, 2015). However, no matter how the oxygen arises or the origin of the oxygen, it is present in spilled crude oil and crude oil products and must be taken into account.

The amounts of the various bulk fractions (i.e., the proportions of the various classes of the hydrocarbon derivatives), the carbon number distribution of the constituents, and the concentration of hetero-elements (i.e., nitrogen derivatives, oxygen derivatives, sulfur derivatives, and metal-containing derivatives) in a crude oil determine the yields and qualities of the refined products that a refinery can produce from the crude oil.

Crude oil is typically are described in terms of the physical properties (such as density and pour point) and chemical composition (such as percent composition of various crude oil hydrocarbon derivatives, asphaltene constituents, and sulfur). Although very complex in makeup, crude can be broken down into four basic classes of crude oil hydrocarbon derivatives. Each class is distinguished on the basis of molecular composition:

Crude oils are made up of hydrocarbon compounds ranging from pentane (C_5H_{12}) to pentadecane ($C_{15}H_{32}$). There is also a variation of different groups such as the normal (straight-chain) paraffin derivatives, iso-paraffin derivatives (branched chain paraffin derivatives), alkyl paraffin derivatives, naphthene derivatives (or cycloparaffin derivatives), alkylbenzene derivatives, and polynuclear aromatic derivatives. The normal paraffin derivatives are the saturated low molecular weight hydrocarbon derivatives, including any gaseous derivatives whereas the naphthene derivatives (or cycloparaffin derivatives) are typically high molecular weight derivatives. All crude oils contain some appreciable amount of the naphthene compounds (10% w/w by composition) (Chinenyeze and Ekene, 2017).

The standard (or most common) group separation involves division of the crude oil constituents namely (1) saturated hydrocarbon derivatives, (2) aromatic hydrocarbon derivatives, (3) resin constituents, and (4) asphaltene constituents (Speight, 2014a, 2015). The saturated hydrocarbon fraction is the most abundant in light (low density, low viscosity) crude oils. The higher density and higher viscosity crude oils have greater concentrations of other components, including resin constituents and asphaltene constituents, which contain more polar compounds, often including heteroatom derivatives that contain nitrogen and/or oxygen and/or sulfur, as well as carbon and hydrogen which can influence the cleanup of any associated spillage of the oil.

Based on the nature of crude oil mixture, there are several ways to express the composition of a crude oil mixture, especially on the basis of the compound types in the oil (Speight, 2014a, 2015). Thus, the composition of crude oil may be expressed as (1) PONA, which represents the amounts of paraffin derivatives, olefin derivatives, naphthene derivatives, and aromatic derivatives, (2) PNA, which represents the amounts of paraffin derivatives, naphthene derivatives, and aromatic derivatives, (3) PIONA, which represents the amounts of paraffin derivatives, iso-paraffin derivatives, olefin derivatives, naphthene derivatives, and aromatic derivatives, (4) SARA, which represents the amounts of saturated derivatives aromatic derivatives resin constituents, and asphaltene constituents, and (5) the elemental analysis, which represents the amounts of carbon, hydrogen, nitrogen, oxygen, sulfur, and metals in the crude oil. Since most unrefined crude oil fractions are free of olefin derivatives, the hydrocarbon types can be expressed in terms of only the PINA analysis and if paraffin derivatives and iso-paraffin derivatives are combined a fraction of aromatic derivatives is simply expressed in terms of the polynuclear aromatic hydrocarbon constituents.

Generally, the properties of crude oil constituents (excluding the volatile gases and the non-volatile residuum) which are listed alphabetically below vary over the boiling range 0 to > 565°C (32 to 1050°F). Variations in density also occur over the range 0.6 to 1.3 and the pour point can vary from < 0°C to > 100°C. Although these properties may seem to be of lower consequence in the grand scheme of environmental cleanup, they are important insofar as these properties influence (1) evaporation rate, (2) ability of the crude oil constituents or crude oil product to float on water, and (3) fluidity or mobility of the crude oil or crude oil product at various temperatures.

The analysis of spilled oil in terms of its origin and transformation by physical (evaporation, dissolution) or (bio)chemical (photo-oxidation, microbial oxidation, selected incorporation into biomass or other forms of metabolism) processes is in most cases targeted toward its chemical composition rather than to the physical properties of the spilled material. The common strategy applied as a first step, after evaporation of the most volatile components (an operation that is frequently referred to as topping), is to separate the complex mixture of oil components into compound classes by polarity using liquid chromatography with various adsorbents on thin-layer plates, in gravity columns, or by medium-pressure or high-performance liquid chromatography.

The compound classes usually obtained are saturated hydrocarbons (alkanes), aromatic hydrocarbons (including some heteroaromatic species), resins, and asphaltenes (SARA). The saturates, aromatics, and resins constituents comprise a single solubility fraction, collectively called maltenes (or petrolenes) which is soluble in alkane solvents (most commonly n-pentane, C_5H_{12}, or n-heptane, C_7H_{16}), while the asphaltene fraction is isolated by virtue of the insolubility of the constituents of this fraction in n-pentane or n-heptane. Further subfractions can be obtained from each of the four bulk fractions (i.e., the saturates fraction, the aromatics fraction, the resins fraction, and the asphaltene fraction) by employing a number of additional techniques (Peters et al., 2005a, 2005b; Speight, 2014a, 2015).

1.3.1 Density and Specific Gravity

Density is the mass of a unit volume of material at a specified temperature and has the dimensions of grams per cubic centimeter (a close approximation to grams per milliliter). *Specific gravity* is the ratio of the mass of a volume of the substance to the mass of the same volume of water and is dependent on two temperatures, those at which the masses of the sample and the water are measured. When the water temperature is 4°C (39°F), the specific gravity is equal to the density in the centimeter-gram-second (cgs) system, since the volume of 1 g of water at that temperature is, by definition, 1 ml. The standard temperatures for a specific gravity in the crude oil industry in North America are 60/60°F (15.6/15.6°C).

The *density* and *specific gravity* of properties of crude oil (ASTM D287, D1298, D941, D1217, and D1555) have found wide use in the industry for preliminary assessment of the character of the crude oil. Density also is a determinant as to whether or not the crude oil or crude oil product will float on water and therefore remain susceptible to aerial oxidation with subsequent emulsion formation.

Although density and specific gravity are used extensively, the API (American Petroleum Institute) gravity is the preferred property by which crude oil is often referenced

Degrees API = 141.5/sp gr @ 60/60°F – 131.5

The specific gravity of crude oil usually ranges from approximately 0.8 (45.3° API) for the low density less viscous crude oil and heavy crude oil to over 1.0 (less than 10° API) for tar sand bitumen. This is of lesser importance in determining the behavior of crude oil or crude oil products after a spill.

Density or specific gravity or API gravity may be measured by means of a hydrometer (ASTM D287 and D1298) or by means of a pycnometer (ASTM D941 and D1217). The variation of density with temperature, effectively the coefficient of expansion, is a property of great technical importance, since most crude oil products are sold by volume and specific gravity is usually determined at the prevailing temperature (21°C, 70°F) rather than at the standard temperature (60°F, 15.6°C).

1.3.2 Elemental Analysis

Elemental analysis (also called ultimate analysis) leads to the calculation of the atomic ratios of the various elements to carbon (i.e., H/C, N/C, O/C, and S/C) are frequently used for indications of the overall character of the feedstock.

For example, *carbon content* can be determined by the method designated for coal and coke (ASTM D3178) or by the method designated for municipal solid waste (ASTM E-777). There are also methods designated for:

1. Hydrogen content (ASTM D1018, ASTM D3178, ASTM D3343, ASTM D3701, ASTM E-777);
2. Nitrogen content (ASTM D3179, ASTM D3228, ASTM D3431, ASTM E148, ASTM E258, ASTM E778);

3. Oxygen content (ASTM E-385), and

4. Sulfur content (ASTM D124, ASTM D1266, ASTM D1552, ASTM D1757, ASTM D2662, ASTM D3177, ASTM D4045 ASTM D4294).

The available data show that the proportions of the elements in crude oil vary only slightly over narrow limits. Thus:

Carbon	83.0 – 87.0% w/w
Hydrogen	10.0 – 14.0% w/w
Nitrogen	0.1 – 2.0% w/w
Oxygen	0.1 – 1.5% w/w
Sulfur	0.1 – 6.0% w/w
Metals (NI, V)	< 200 ppm

Of the ultimate analytical data, more has been made of the sulfur content than any other property. For example, the sulfur content (ASTM D124, D1552, and D4294) and the API gravity represent the two properties that have, in the past, had the greatest influence on determining the value of crude oil as a feedstock.

The sulfur content varies from about 0.1% w/w to approximately 3% w/w for the more conventional crude oils to as much as 5 to 6% w/w for heavy oil and bitumen. Residua (i.e., the non-distillable portion of crude oil, may have a sulfur content on the order of the sulfur content of extra heavy crude oil or tar sand bitumen.

The role of sulfur in producing obnoxious emission during refining and product use has been a major determinant in attempts to control sulfur emissions. There are similar cautions about nitrogen, although the amount of nitrogen in crude oil is significantly lower than the amount of sulfur. Considered alone as a crude oil constituent, these compounds are dangerous but they can react with the environment to produce secondary poisonous chemicals (specifically the respective oxides) thereby contributing to the formation of acid rain (Speight, 1996; Speight and Lee, 2000; Speight, 2005, 2014a).

1.3.3 Chromatographic Fractionation

Fractionation also informs the environmental scientist or engineer where the most environmentally recalcitrant constituents of the crude oil are located. The data also present indications of which constituents can be removed from coil by steam stripping and which constituents (or how much of the spilled material) will remain in the contaminated soil.

There are several ASTM procedures for feedstock/product evaluation. These are:

1. Determination of aromatic content of olefin-free gasoline by silica gel adsorption (ASTM D936).

2. Separation of aromatic and non-aromatic fractions from high-boiling oils (ASTM D2549).

3. Determination of hydrocarbon groups in rubber extender oils by clay-gel adsorption (ASTM D2007).

4. Determination of hydrocarbon types in liquid crude oil products by a fluorescent indicator adsorption test (ASTM D1319).

Gel permeation chromatography is an attractive technique for the determination of the number average molecular weight (Mn) distribution of crude oil fractions, especially the heavier constituents, and crude oil products.

Ion exchange chromatography is also widely used in the characterization of crude oil constituents and products. For example, cation exchange chromatography can be used primarily to isolate the nitrogen constituents in crude oil (Snyder and Buell 1965) thereby giving an indication of how the feedstock might behave during refining and also an indication of any potential deleterious effects on catalysts.

Liquid chromatography (also called *adsorption chromatography*) has been invaluable for the characterization of the group composition of crude oils and hydrocarbon products since the beginning of this century. The type and relative amount of certain hydrocarbon classes in the matrix can have a profound effect on the quality and performance of the hydrocarbon product. The *fluorescent indicator adsorption* (FIA) method (ASTM D1319) has been used to measure the paraffinic, olefinic, and aromatic content of gasoline, jet fuel, and liquid products in general.

High-performance liquid chromatography (*HPLC*) has found great utility in separating different hydrocarbon group types and identifying specific constituent types. The general advantages of high performance liquid chromatography method are: (1) each sample may be analyzed *as received* even though the boiling range may vary over a considerable range, (2) the total time per analysis is usually on the order of minutes; and, perhaps most important, and (3) the method can be adapted for any recoverable crude oil sample or product.

In recent years, *supercritical fluid chromatography* has found use in the characterization and identification of crude oil constituents and products. Currently, supercritical fluid chromatography is leaving the stages if infancy. The indications are that it will find wide applicability to the problems of characterization and identification of the higher molecular weight species in crude oil thereby adding an extra dimension to our understanding of refining chemistry. It will still retain the option as a means of product characterization although the use may be somewhat limited because of the ready availability of other characterization techniques.

1.3.4 Liquefaction and Solidification

Liquefaction is the phase-transition process of converting a gas or solid into a liquid through condensation, melting, or heating. Types of physical changes include boiling, clouding, dissolution, freezing, freeze-drying, frost, liquefaction, melting, smoke and vaporization. In some cases, the term liquefaction is also used in commercial and industrial settings to refer to the mechanical dissolution of a solid by mixing, grinding, or blending the solid with a liquid.

In the current context, the liquefaction process is a physical change process which affects the physical properties of a chemical but does not alter the chemical structure of the chemical. Also, in the current context of changes to the environment, soil liquefaction occurs when a cohesionless saturated or partially saturated soil substantially loses strength and stiffness in response to an applied stress such as during the spill of a chemical in which the soil, which ordinarily is a solid, behaves like a liquid. Also, liquefaction occurs when waterlogged, loose soil (or sand) turns into quicksand temporarily.

On the other hand, solidification (sometimes referred to as freezing) is also a phase-transition process that is caused by the withdrawal of heat from a substance to change that substance from a liquid to a solid (or a gas to a solid). The temperature must be below the substance's freezing point for the change to occur. Turning water into ice using a freezer is an example of this physical change. More technically, solidification is a process in which atoms are converted into an ordered solid state from a liquid disordered state. The conversion rate for the process of solidification can be achieved by following the kinetic laws and the movement of atoms for the conversion of liquid can be observed by these laws.

Freezing is a phase transition where a liquid turns into a solid when its temperature is lowered below its freezing point. In accordance with the internationally established definition, freezing means the solidification phase change of a liquid or the liquid content of a substance, usually due to cooling.

Crude oil (with the exception of some heavy crude oils and tar sand bitumen) and the majority of crude oil products are liquids at ambient temperature, and problems that may arise from solidification during normal use are not common. Nevertheless, the *melting point* is a test (ASTM D87 and D127) often serves to determine the state of the crude oil or the product under various weather conditions or under applied conditions, such as steam stripping of the material from the soil.

1.3.5 Metals Content

Crude oil (prior to refining) may have a considerable amount of heavy metals such as cadmium, nickel, zinc, manganese, vanadium, copper, chromium, lead, arsenic and mercury as part of the impurities present. The presence of these metals, despite the low concentrations, could still lead to serious health hazard considering their cumulative effects in floral and faunal (including humans) species. Also, unlike many other constituents of crude oil, metals are non-biodegradable and inside living tissue they tend to undergo biomagnification, which is the concentration of toxins in an organism as a result of its ingesting other plants or animals in which the toxins are more widely disbursed. Bioaccumulation of these metals in the body may result to health problems.

By way of clarification, bioaccumulation is the gradual accumulation of substances, such as pesticides or other chemicals, in an organism. Bioaccumulation occurs when an organism absorbs a substance at a rate faster than that at which the substance is lost or eliminated by catabolism and excretion.

Typically, the metals that occur in crude oil often accompany the presence of heteroatoms in the oil.

The heteroatoms (nitrogen, oxygen, sulfur, and metals) are found in the majority of conventional crude oils and the concentrations have to be reduced to convert the crude oil saleable products. The reason is that if nitrogen and sulfur are present in the final fuel during combustion, nitrogen oxides (NO_x) and sulfur oxides (SO_x) form, respectively. In addition, metals affect upgrading processes adversely by poisoning the catalysts (or in non-catalytic processes) initiating coke formation as well as causing deposits in combustion equipment. In addition, the heteroatom-containing constituents also play a major role in environmental issues and can cause the crude oil to adhere to the soil ensuring long-term contamination.

A variety of tests (ASTM D1026, ASTM D1262, ASTM D1318, ASTM D1368, ASTM D1548, ASTM D1549, ASTM D2547, ASTM D2599, ASTM D2788, ASTM D3340, ASTM D3341, and ASTM D3605) have been designated for the determination of metals in crude oil and crude oil products. This task this task can be accomplished by combustion of the sample so that only inorganic ash remains after which the ash can then be digested with an acid and the solution examined for metal species by atomic absorption (AA) spectroscopy or by inductively coupled argon plasma (ICP) spectrometry.

1.3.6 Surface Tension and Interfacial Tension

Surface tension is the tension of the surface film of a liquid caused by the attraction of the particles in the surface layer by the bulk of the liquid, which tends to minimize surface area. It is the tendency of liquid surfaces at rest to shrink into the minimum surface area possible. Surface tension is what allows objects with a higher density than water such as razor blades and insects to float on a water surface without becoming even partly submerged.

Thus, the surface tension is a measure of the force acting at a boundary between two phases. If the boundary is between a liquid and a solid or between a liquid and a gas (air) the attractive forces are referred to as surface tension, but the attractive forces between two immiscible liquids are referred to as *interfacial tension*.

Temperature and molecular weight have a significant effect on surface tension (Speight, 2015). For example, in the normal hydrocarbon series, a rise in temperature leads to a decrease in the surface tension, but an increase in molecular weight increases the surface tension. A similar trend (i.e., an increase in surface tension with an increase in molecular weight) also occurs in the acyclic series and, to a lesser extent, in the alkylbenzene series.

On the other hand, although crude oil products show little variation in surface tension, within a narrow range the *interfacial tension* of crude oil, especially of crude oil products, against aqueous solutions provides valuable information (ASTM D971) that can be sued to predict the potential for emulsion formation.

The interfacial tension of crude oil is subject to the same constraints as surface tension, that is, differences in properties such as composition and molecular weight. When oil-water systems are involved, the pH of the aqueous phase influences the tension at the interface; the change is small for highly refined oils, but increasing pH causes a rapid decrease for poorly refined, contaminated, or slightly oxidized oils.

1.3.7 Viscosity

Viscosity is the force in dynes required to move a plane of 1 cm² area at a distance of 1 cm from another plane of 1 cm² area through a distance of 1 cm in 1 s. In the centimeter-gram-second (cgs) system the unit of viscosity is the poise (P) or centipoise (1 cP = 0.01 P). Two other terms in common use are *kinematic viscosity* and *fluidity*. The former term, i.e. kinematic viscosity, is the viscosity of the crude oil crude oil product in centipoises divided by the specific gravity, and the unit is the stoke (cm²/s), although centistokes (0.01 cSt) is in more common usage. Fluidity is simply the reciprocal of viscosity.

The viscosity (ASTM D445, D88, D2161, D341, and D2270) of crude oil oils varies markedly over a very wide range less than 10 cP at room temperature to thousands of centipoises at the same temperature.

The Saybolt universal viscosity (SUS) (ASTM D88) is the time in seconds required for the flow of 60 ml of crude oil from a container, at constant temperature, through a calibrated orifice. The Saybolt furol viscosity (SFS) (ASTM D88) is determined in a similar manner except that a larger orifice is employed.

As a result of the various methods for viscosity determination, interconversion of the several scales is available, especially converting Saybolt viscosity to kinematic viscosity (ASTM D2161).

Kinematic viscosity = a x Saybolt s + b/Saybolt s

where *a* and *b* are constants.

The Saybolt universal viscosity equivalent to a given kinematic viscosity varies slightly with the temperature at which the determination is made because the temperature of the calibrated receiving flask used in the Saybolt method is not the same as that of the oil. Conversion factors are used to convert kinematic viscosity from 2 to 70 cSt at 38°C (100°F) and 99°C (210°F) to equivalent Saybolt universal viscosity in seconds. For a kinematic viscosity determined at any other temperature the equivalent Saybolt universal value is calculated by use of the Saybolt equivalent at 38°C (100°F) and a multiplier that varies with the temperature as indicate by the two following equations:

Saybolt s at 100°F (38°C) = cSt x 4.635

Saybolt s at 210°F (99°C) = cSt x 4.667

Viscosity decreases as the temperature increases and the rate of change appears to depend primarily on the nature or composition of the crude oil, but other factors, such as volatility, may also have an effect. However, the effect of temperature on viscosity is generally represented by the equation:

log log (n + c) = A + B log T

In this equation n is absolute viscosity, T is the temperature, and *A* and *B* are constants. However, the constants *A* and *B* vary widely with different crude oils and crude oil product, but c remains fixed at 0.6 for all oils having a viscosity over 1.5 cSt; it increases only slightly at lower viscosity (0.75 at 0.5 cSt). The viscosity-temperature characteristics of any oil, so plotted, thus create a straight line, and the parameters *A* and *B* are equivalent to the intercept and slope of the line. To express the viscosity and viscosity-temperature characteristics of an oil, the slope and the viscosity at one temperature must be known; the usual practice is to select 38°C (100°F) and 99°C (210°F) as the observation temperatures.

Suitable conversion tables are available (ASTM D341), and each table or chart is constructed in such a way that for any given crude oil or crude oil product the viscosity-temperature points result in a straight line over the applicable temperature range. Thus, only two viscosity measurements need be made at temperatures sufficiently different to determine a line on the appropriate *x*–*y* chart from which the approximate viscosity at any other temperature can be read.

Since the viscosity-temperature coefficient of high-boiling fractions of crude oil is an important expression of its suitability for, say, selection of a refinery process of use of the fraction as a product, hence, a viscosity index (ASTM D2270) was derived. Thus:

Viscosity index = L – U/ L – H(100)

in this equation, L and H are the viscosities of the zero and 100 index reference oils, both having the same viscosity at 99°C (210°F), and U is that of the unknown, all at 38°C (100°F).

The viscosity of crude oil fractions increases on the application of pressure, and this increase may be very large. The pressure coefficient of viscosity correlates with the temperature coefficient, even when oils of widely different types are compared. At higher pressures the viscosity decreases with increasing temperature, as at atmospheric pressure; in fact, viscosity changes of small magnitude are usually proportional to density changes, whether these are caused by pressure or by temperature.

Because of the importance of viscosity in determining the transport properties of crude oil, and this is particularly important in the migration of crude oil and crude oil products through soil, recent work has focused on the development of an empirical equation for predicting the dynamic viscosity of low molecular weight and high molecular weight hydrocarbon vapors at atmospheric pressure. The equation uses molar mass and specific temperature as the input parameters and offers a means of estimation of the viscosity of a wide range of crude oil fractions. Other work has focused on the prediction of the viscosity of blends of lubricating oils as a means of accurately predicting the viscosity of the blend from the viscosity of the base oil components (Al-Besharah et al., 1989).

1.3.8 Volatility

The volatility of a liquid or liquefied gas is a measure of the tendency of the liquid to vaporize, thereby changing from the liquid sate to the vapor state (aka the gaseous state). Because one of the three essentials for combustion in a flame is that the fuel be in the gaseous state, volatility is a primary (and necessary) characteristic of crude oil products that are generally referred to as liquid fuels.

The *flash point* of crude oil or a crude oil product is the temperature to which the product must be heated under specified conditions to give of sufficient vapor to form a mixture with air that can be ignited momentarily by a specified flame (ASTM D56, D92, and D93). On the other hand, the *fire point* of a crude oil or crude oil product is the temperature to which the product must be heated under the prescribed conditions of the method to burn continuously when the mixture of vapor and air is ignited by a specified flame (ASTM D92) (Speight, 2104, 2015).

From the viewpoint of safety, information about the flash point is of most significance at or slightly above the maximum temperatures (30 to 60°C, 86 to 140°F) that may be encountered in storage, transportation, and use of liquid crude oil products, in either closed or open containers. For products with flash point below 40°C (104°F) special precautions are necessary for safe handling. Flash points above 60°C (140°F) gradually lose their safety significance until they become indirect measures of some other quality.

A further aspect of volatility that receives considerable attention is the vapor pressure of crude oil and its constituent fractions. The temperature at which the vapor pressure of a liquid either a pure compound of a mixture of many compounds_ equals 1 atmosphere pressure (14.7 psi, absolute) is designated as the boiling point of the liquid.

1.4 Summary

In the context of this book, the severity of the spills of crude oil (and crude oil products) depends not only on the composition of the spilled material as well as the chemical properties and physical properties of the spilled material but also on the quantity of material released into the environment. In addition, and to complicate the issue, because crude oil is an extremely complex mixture of many thousands or more of gaseous, liquid, and solid constituents, full elucidation of their compositions at the molecular level is impossible with presently available analytical techniques.

Another issue that is often ignored relates not only to the analysis of the spilled material but also the transformation by of the material by physical effects (such as evaporation and dissolution) or chemical effects (such as photo-oxidation, microbial oxidation, or other forms of chemical changes) and remediation processes are, as a result, in most cases targeted toward the chemical composition of the material rather than the physical properties of the material.

The common strategy, after evaporation of the most volatile components—of ten referred to as topping—is to separate the complex mixture of the constituents of the spill into bulk compound classes by polarity often referred to as the saturates fraction using liquid chromatography with various adsorbents on thin-layer plates, in gravity columns, or by medium-pressure or high-performance liquid chromatography. The compound classes usually obtained are saturated hydrocarbon derivatives (alkane derivatives, often referred to as the saturates fraction), aromatic hydrocarbon derivatives (including some heteroaromatic species which include nitrogen and/or oxygen, and/or sulfur, and/or metals, often referred to as the aromatics fraction), the resins fraction (often a highly polar fraction that is soluble in n-pentane or n-heptane), and the asphaltene fraction, which is insoluble in n-pentane or n-heptane. The separation method is referred to as the SARA analysis— saturates, aromatics, resins, asphaltenes). Saturates, aromatics, and resin fractions comprise a single solubility fraction, collectively called maltenes or petrolenes, although the terms do have subtle differences in meaning (Speight, 2014, 2015). The maltenes (or petrolenes) are soluble in alkane solvents (most commonly *n*-pentane or *n*-heptane), while the asphaltenes are isolated by alkane solvent precipitation (insolubility). Further sub-fractions can be obtained from each of the SARA bulk fractions by employing a number of more sophisticated analytical methods.

As an additional note, several of the polycyclic saturated alkane derivatives such as the tetracyclic sterane derivatives and the pentacyclic triterpane derivatives that are classified as biomarkers (also termed biological markers, molecular fossils, fossil molecules, or geochemical fossils). The presence and relative abundance of the biomarkers as well as their remarkable stability under weathering allow them to be used as fingerprints to gain information on the origin of a crude oil and to distinguish crude oils from different sources.

Biomarkers are organic compounds in natural waters, sediments, soils, crude oils that can be unambiguously linked to specific precursor molecules biosynthesized by living organisms. The main reason for this specificity is that the bonds to the four neighboring atoms (carbon or hydrogen) are sterically oriented (tetrahedral). Thus, the rings are not planar as in aromatic hydrocarbons, but rather have a (sometimes slightly skewed) three-dimensional chair or boat configuration. Biosynthesis leads to specific steric orientation of several bonds in the biomarkers (chiral centers, optical activity). The orientation of some of these chiral centers is altered during the geothermal transformation of organic matter in petroleum source rocks (stereoisomerization) into thermodynamically more stable species, providing clues to the geothermal history (maturation) of the organic matter.

References

Abraham, H. 1945. Asphalts and Allied Substances. Van Nostrand Scientific Publishers, New York.

Al-Besharah, J.M., Mumford, C.J., Akashah, S.A., and Salman, O. 1989. Prediction of the viscosity of lubricating oil blends. Fuel., 68: 609–811.

Bai, L., Jiang, Y., Huang, D., and Liu, X. 2010. A novel scheduling strategy for crude oil blending. Chinese Journal of Chemical Engineering, 18(5): 777–78.

Ben Mahmoud, M.A.M., and Aboujadeed, A. 2017. Compatibility assessment of crude oil blends using different methods. Chemical Engineering Transactions, 57: 1705–1710.

Bouquet, M., and Bailleul, A. 1982. In Petroananlysis '81. Crump, G.B. (Ed.). Advances in Analytical Chemistry in the Petroleum Industry 1975–1982. John Wiley & Sons, Chichester, England.

Campbell, C.J. 1997. The Coming Oil Crisis. Multi-Science Publishing Company & Petroconsultants, Brentwood, Essex, United Kingdom.

Cardoza, L., Korir, A., Otto, W., Wurrey, C., and Larive, C. 2004. Applications of NMR spectroscopy in environmental science. Progress in Nuclear Magnetic Resonance Spectroscopy, 45(3-4): 209–238.

Chinenyeze, M.A.J., and Ekene, U.R. 2017. Physical and chemical properties of crude oils and their geologic significances. International Journal of Science and Research, 6(6): 1514–1521.

Delbianco, A., and Montanari, R. 2009. Encyclopedia of Hydrocarbons, Volume III/New Developments: Energy, Transport, Sustainability. Eni S.p.A., Rome, Italy.

Del Carmen García, M., and Urbina, A. 2003. Effect of crude oil composition and blending on flowing properties. Petroleum Science and Technology, 21(5-6): 863–878.

Demaison, G.J. 1977. The oil sands of Canada-venezuela. Redford, D.A., and Winestock, A.G. (Eds.). Canadian Institute of Mining and Metallurgy, Special Volume No. 17, p. 9.

Gary, J.G., Handwerk, G.E., and Kaiser, M.J. 2007. Petroleum Refining: Technology and Economics, 5th Edition. CRC Press, Taylor & Francis Group, Boca Raton, Florida.

Ghoshal, S., and Sainik, V. 2013. Monitor and minimize corrosion in high-TAN crude processing. Hydrocarbon Processing, 92(3): 35–38.

Hoiberg, A.J. 1964. Bituminous Materials: Asphalts, Tars, and Pitches. John Wiley & Sons Inc., New York.

Hsu, C.S., and Robinson, P.R. (Eds.). 2017. Handbook of Petroleum Technology. Springer International Publishing AG, Cham, Switzerland.

Kane, R.D., and Cayard, M.S. 2002. A Comprehensive Study on Naphthenic Acid Corrosion. Corrosion 2002. NACE International, Houston, Texas.

Kumar, R., Voolapalli, R.V., and Upadhyayula, S. 2018. Prediction of crude oil blends compatibility and blend optimization for increasing heavy oil processing. Fuel Processing Technology, 177: 309–327.

Long, R.B., and Speight, J.G. 1998. The composition of petroleum. In: Speight, J.G. (Ed.). Petroleum Chemistry and Refining. Taylor & Francis, Washington, DC. Chapter 2.

Mecozzi, M., Moscato, F., Pietroletti, M., Quarto, F., Oteri, F. and Cicero, A.M. 2009. Applications of FTIR spectroscopy in environmental studies supported by two dimensional correlation analysis. Global NEST Journal, 11(4): 593–600.

Meyer, R.F., and Dietzman, W.D. 1981. World geography of heavy crude oils. In: Meyer, R.F., and Steele, C.T. (Eds.). The Future of Heavy Crude and Tar Sands. McGraw-Hill, New York, p. 16.

Meyers, R.A. 1997. Handbook of Crude oil Refining Processes 2nd Edition. McGraw-Hill, New York.

Mokhatab, S., Poe, W.A., and Speight, J.G. 2006. Handbook of Natural Gas Transmission and Processing Elsevier, Amsterdam, Netherlands.

Moreno-Letelier, A., Olmedo-Alvarez, G., Eguiarte, L.E., and Souza, V. 2012. Divergence and phylogeny of firmicutes from the Cuatro Ciénegas Basin, Mexico: a window to an ancient ocean. Astrobiology, 12(7): 674–84.

Mushrush, G.W., and Speight, J.G. 1995. Petroleum Products: Instability and Incompatibility. Taylor & Francis, Philadelphia, Pennsylvania.

Nanny, M.A., Minear, R.A., and Leenheer, J.A. (Eds.). 1997. Nuclear Magnetic Resonance Spectroscopy in Environmental Chemistry. Oxford University Press, Oxford, United Kingdom.

Onwurah, I.N.E., Ogugua, V.N., Onyike, N.B., Ochonogor, A.E., and Otitoju, O.F. 2007. Crude oils spills in the environment, effects and some innovative clean-up biotechnologies. International Journal of Environmental Research, 1(4): 307–320.

Parkash, S. 2003. Refining Processes Handbook. Gulf Professional Publishing, Elsevier, Amsterdam, Netherlands.

Peters, K.E., Walters, C.C., and Moldowan, J.M. 2005a. The Biomarker Guide: Volume 1: Biomarkers and Isotopes in the Environment and Human History. Columbia University Press, Columbia University, New York.

Peters, K.E., Walters, C.C., and Moldowan, J.M. 2005b. The Biomarker Guide: Volume 2: Biomarkers and Isotopes in Petroleum Exploration and Earth History. Columbia University Press, Columbia University, New York.

Pfeiffer, J.H. 1950. The Properties of Asphaltic Bitumen. Elsevier, Amsterdam, Netherlands.

Phizackerley, P.H., and Scott, L.O. 1967. Major tar sand deposits of the world. Proceedings. 7th World Petroleum Congress, 3: 551–571.

Prescott, M.L., Harley, J.P., and Klan, A.D. 1996. Industrial microbiology and biotechnology. pp. 923–927. *In*: Microbiology 3rd Edition. Wim C Brown Publishers, Chicago.

Rullkötter, J., and Farrington, J.W. 2021. What was released? Assessing the physical properties and chemical composition of petroleum and products of burned oil. Oceanography, 34(1): 44–57. https://tos.org/oceanography/assets/docs/34-1_rullkotter.pdf.

Shalaby, H.M. 2005. Refining of Kuwait's Heavy Crude Oil: Materials Challenges. Proceedings. Workshop on Corrosion and Protection of Metals. Arab School for Science and Technology. December 3–7, Kuwait.

Short, J.W., and Heintz, R.A. 1997. Identification of exxon valdez oil in sediments and tissue from prince william sound and the north western gulf of william based on a PAH weathering model. Environ. Sci. Technol., 31: 2375–2384.

Speight, J.G. 1986. Polynuclear aromatic systems in petroleum. Preprints. Am. Chem. Soc., Div. Petrol. Chem., 31(4): 818.

Speight, J.G. 1990. Tar sands. pp 317–429. *In*: Fuel Science and Technology Handbook. Marcel Dekker Inc., New York. Chapter 12–16.

Speight, J.G. 1997. Tar sand. *In*: Kirk-Othmer Encyclopedia of Chemical Technology 4th Edition. 23: 717.

Speight, J.G., and Lee, S. 2000. Environmental Technology Handbook. 2nd Edition. Taylor & Francis, New York.

Speight, J.G. 2005. Environmental Analysis and Technology for the Refining Industry. John Wiley & Sons Inc., Hoboken, New Jersey.

Speight, J.G. (Ed.). 2011. The Biofuels Handbook. Royal Society of Chemistry, London, United Kingdom.

Speight, J.G. 2012a. Crude Oil Assay Database. Knovel, Elsevier, New York. Online version available at: http://www.knovel.com/web/portal/browse/display?_EXT_KNOVEL_DISPLAY_bookid=5485&VerticalID=0.

Speight, J.G. 2012b. Shale Oil Production Processes. Gulf Professional Publishing Company, Elsevier, Oxford, United Kingdom.

Speight, J.G. and Arjoon, K.K. 2012. Bioremediation of Crude oil and Crude oil Products. Scrivener Publishing, Beverly, Massachusetts.

Speight, J.G. 2013a. The Chemistry and Technology of Coal 3rd Edition. CRC-Taylor & Francis Group, Boca Raton, Florida.

Speight, J.G. 2013b. Heavy Oil Production Processes. Gulf Professional Publishing, Elsevier, Oxford, United Kingdom.

Speight, J.G. 2014a. The Chemistry and Technology of Petroleum 5th Edition. CRC Press, Taylor & Francis Group, Boca Raton, Florida.

Speight, J.G. 2014b. High Acid Crudes. Gulf Professional Publishing, Elsevier, Oxford, United Kingdom.

Speight, J.G. 2015. Handbook of Petroleum Product Analysis 2nd Edition. John Wiley & Sons Inc., Hoboken, New Jersey.

Speight, J.G. 2016. Introduction to Enhanced Recovery Methods for Heavy Oil and Tar Sand Bitumen 2nd Edition. Gulf Professional Publishing Company, Elsevier, Oxford, United Kingdom.

Speight, J.G. 2017. Handbook of Crude oil Refining. CRC Press, Taylor & Francis Group, Boca Raton, Florida.

Speight, J.G. 2019. Handbook of Petrochemical Processes. CRC Press, Taylor & Francis Group, Boca Raton, Florida.

Speight, J.G. 2020a. Refinery of the Future 2nd Edition. Gulf Professional Publishing, Elsevier, Cambridge, Massachusetts.

Speight, J.G. 2020b. Global Climate Change Demystified. Scrivener Publishing, Beverly, Massachusetts.

Speight, J.G. 2021a. Coal-Fired Power Generation Handbook 2nd Edition. Scrivener Publishing, Beverly, Massachusetts.

Speight, 2021b. Refinery Feedstocks. CRC Press, Taylor & Francis Group, Boca Raton, Florida.

Stout, S.A., and Wang, Z. (Eds.). 2018. Oil Spill Environmental Forensics Case Studies. Butterworth-Heinemann, Elsevier, Oxford, United Kingdom.

United States Congress. 1976. Public Law FEA-76-4. United States Library of Congress, Washington, DC.

Van Nes, K., and van Westen, H.A. 1951. Aspects of the Constitution of Mineral Oils. Elsevier, Amsterdam, Netherlands.

Wang, Z., Yang, C., Yang, Z., Brown, C.E., Hollebone, B.P., and Stout, S.A. 2016, Standard Handbook Oil Spill Environmental Forensics (Second Edition). Volume 4—Crude Oil Biomarker Fingerprinting for Oil Spill Characterization and Source Identification. Academic Press, New York. Page 131–254. https://doi.org/10.1016/B978-0-12-803832-1.00004-0.

CHAPTER 2

Crude Oil Products

The products from crude oil refining are, in contrast to petrochemicals, those bulk fractions that are derived from crude oil and have commercial value as a bulk product (Parkash, 2003; Gary et al., 2007; Speight, 2014; Hsu and Robinson, 2017; Speight, 2017).

For example, the various bulk products include (1) gaseous products, which include methane, ethane, propane, and butanes, (2) naphtha and kerosene, which the precursors to liquid fuels, (3) low boiling gas oils, which are precursors to fuel oils, (4) higher boiling gas oil, which is the precursor to lubricants and waxes and grease, and (5) resid, which is the non-volatile portion of the crude oil and is the precursor to asphalt and coke. Petrochemicals are also crude oil products but they are typically individual chemicals such as that are used as the basic building blocks of the chemical industry.

Crude oil is an extremely complex mixture of hydrocarbon compounds, usually with minor amounts of nitrogen-containing, oxygen-containing, and sulfur-containing compounds as well as trace amounts of metal-containing compounds. In addition, the properties of crude oil vary widely (Chapter 1). Thus, crude oil is not used in its raw state. A variety of processing steps is required to convert crude oil from its raw state to products that have well defined properties.

More specifically, crude oil, in the unrefined or crude form, like many industrial feedstocks has little or no direct use and its value as an industrial commodity is only realized after the production of salable products by a series of processing steps as performed in a refinery (Figure 2.1). Each processing step is, in fact, a separate process and thus a refinery is a series of integrated processes that generate the desired products according to the market demand (Speight, 2014, 2017). In general, crude oil, once refined, yields three basic groupings of distillation products that are produced when it is broken down into fractions: naphtha, middle distillates (kerosene and light gas oil), heavy gas oil/vacuum gas oil, and the residuum (Table 2.1). It is from these unrefined fractions that saleable crude oil products are produced by a series of refinery processes (Speight, 2014, 2017, 2019).

Moreover, the complexity of crude oil products has made the industry unique among the chemical industries. Indeed, current analytical techniques that are accepted as standard methods for, as an example, the aromatics content of fuels ASTM D-1319, ASTM D-2425, ASTM D-2549, ASTM D-2786, ASTM D-2789, as well as proton and carbon nuclear magnetic resonance methods, yield different information. Each method will yield the "% aromatics" in the sample but the data must be evaluated within the context of the method.

Unlike the refinery processes, products are more difficult to place on an individual evolutionary scale. Processes changed and evolved to accommodate the demand for, say, higher-octane fuels, longer-lasting asphalt, or lower sulfur coke. Another consideration that must be acknowledged is the change in character and composition of the original crude oil feedstock. In the early days of the crude oil industry several products were obtained by distillation and could be used without any further treatment. In the modern refinery, the different character and composition of the crude

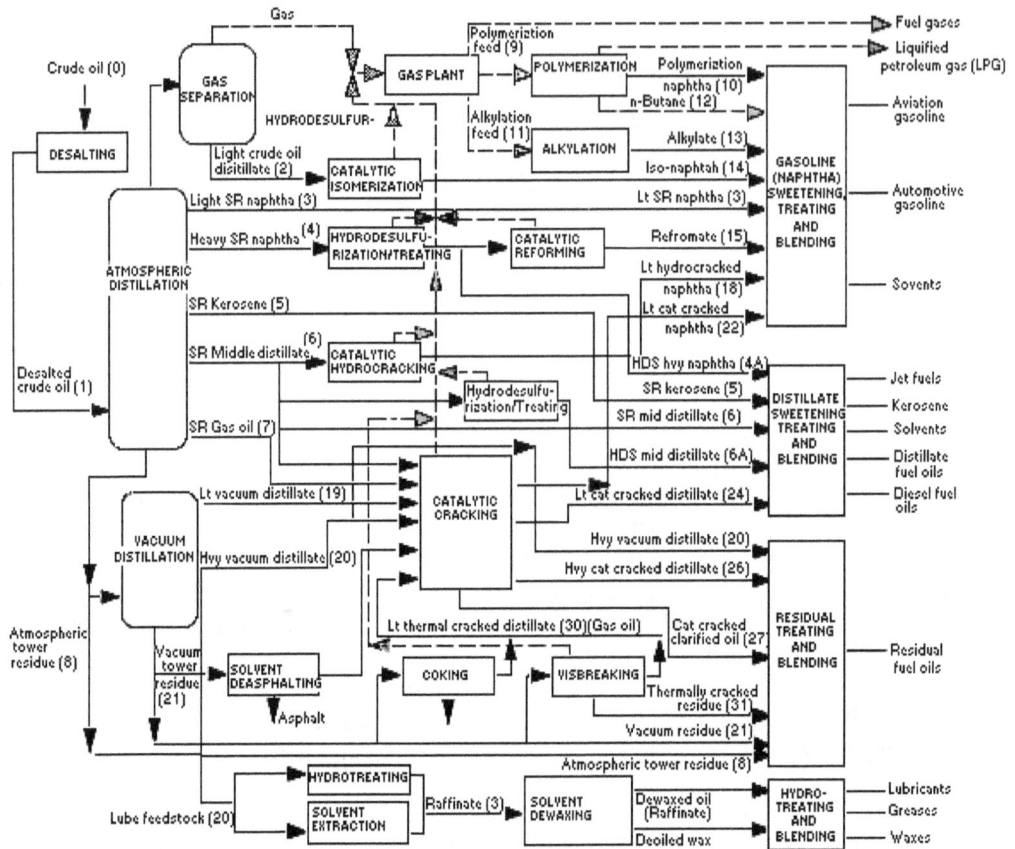

Figure 2.1: Detailed Layout of a Crude oil Refinery.
(Source: http://www.osha.gov/dts/osta/otm/otm_iv/otm_iv_2.html)

oil dictates that any liquids obtained by distillation must go through one or more of the several available product improvement processes (Parkash, 2003; Gary et al., 2007; Speight, 2014; Hsu and Robinson, 2017; Speight, 2017). Such changes in feedstock character and composition have caused the refining industry to evolve in a direction such that changes in the crude oil can be accommodated.

Refinery processes for crude oil are generally divided into two categories: (1) separation processes, of which atmospheric distillation and vacuum distillation are the prime examples, and (2) conversion processes, of which visbreaking delayed coking, fluid coking, catalytic cracking, and hydrocracking are prime examples (Table 2.2). Distillates and other fractions are usually treated further by hydrotreating to meet product specification before sale.

The current trend throughout the refining industry is to produce more fuel products from each barrel of crude oil and to process those products in different ways to meet the product specifications for use in various engines (automobile, diesel, aircraft, and marine). Furthermore, as crude oil products are shipped worldwide, they need to comply with stringent environment-related regulations prevalent in specific countries. Japan and Singapore have already implemented strict legislation/rules and many countries are likely to follow suit as they confront environmental issues such as smog. These changing rules also cause a negative impact on the market for heavy products such as fuel oil.

However, as the need for the lower boiling products developed, crude oil yielding the desired quantities of the lower boiling products became less available and refineries had to introduce conversion processes to produce greater quantities of lighter products from the higher boiling fractions. The means by which a refinery operates in terms of producing the relevant products,

Table 2.1: Crude Oil is a Complex Mixture of Different Generic Boiling Fractions.

Fraction	Boiling Range*	
	°C	°F
Refinery gas	−161–0	−259–32
Light naphtha	−1–150	30–300
Medium naphtha	−1–180	30–355
Heavy naphtha	150–205	300–400
Kerosene	205–260	400–500**
Fuel oil	216–421	343–649
Light gas oil	260–315	400–600
Lubricating oil	> 343	> 649
Wax	302–343	575–649
Heavy gas oil	315–425	600–800
Vacuum gas oil	425–510	800–950
Residuum	> 510	> 950***

* For convenience, boiling ranges are converted to the nearest 5°.
** Kerosene destined for conversion to diesel fuel; crude kerosene is often quoted as having a wider boiling range (150 to 350°C, 302 to 660°F) before separation into various fuel and/or solvent products.
*** Includes the base material for asphalt manufacture.

Table 2.2: General Examples of Separation and Conversion Processes.

(1) Separation Processes					
Process	**Action**	**Method**	**Purpose**	**Feedstock(s)**	**Product(s)**
Atmospheric distillation	Separation	Thermal	Separate fractions; no cracking	Desalted crude oil	Gas, gas oil, distillate, resid
Vacuum distillation	Separation	Thermal	Separate fractions; no cracking	Atmospheric resid	Gas oil, lube stock, resid
(2) Conversion Processes					
Process	**Action**	**Method**	**Product**	**Feedstock(s)**	**Product(s)**
Visbreaking	Decompose	Thermal	Reduce viscosity	Atmospheric resid	Distillate, tar
Coking	Polymerize	Thermal	Convert vacuum resid	Gas oil	Gasoline, petrochemical feedstock
Catalytic cracking	Alteration	Catalytic	Upgraded naphtha	Gas oil	Gasoline, petrochemical feedstock
Hydrocracking	Hydrogenate	Catalytic	Lower boiling hydrocarbons	Gas oil, resid	Lighter, higher-quality products

depends not only on the nature of the crude oil feedstock but also on its configuration (i.e., the number of types of the processes that are employed to produce the desired product slate) and the refinery configuration is, therefore, influenced by the specific demands of a market. Therefore, refineries need to be constantly upgraded (in terms of process options, i.e., modernized) to adapt and remain viable and responsive to ever changing patterns of crude supply and product market demands. As a result, refineries have been introducing increasingly complex (and expensive) processes to gain more and more low-boiling products from the high-boiling and residual portions of crude oil.

By definition in the text, crude oil products (in contrast to petrochemicals) are those bulk fractions that are derived from crude oil and have commercial value as a bulk product

(Parkash, 2003; Gary et al., 2007; Speight, 2014; Hsu and Robinson, 2017; Speight, 2017, 2019). Crude oil products are, generally speaking, hydrocarbon compounds that have various combinations of hydrogen and carbon and can take many molecular forms. Many of the combinations exist naturally in the original raw materials but other combinations are created by an ever-growing number of commercial processes for altering one combination to another.

The specifications for crude oil products (which define whether or not a product is suitable for sale and to assure that the product can perform the task for which it is intended) are based on properties such as density and boiling range (to mention only two such properties) and are used. In addition, each product has its own unique set of chemical and physical properties. As a consequence, each product can, and does, have a different effect on the environment (Speight, 2005; Speight and Arjoon, 2012).

Many of the petrochemicals are substitutes for earlier products from non-crude oil sources. Each new use often imposes additional specifications on the new product and product specifications evolve to stay abreast of advances in both product application and manufacturing methods. Like crude oil products, the petrochemicals cover such a vast range of chemical and physical properties that they also have a wide range of effects on the environment.

Therefore, it is the purpose of this chapter to describe major refinery operations and the products therefrom and focuses on the composition and the properties and the uses of these products. This presents to the reader the essence of crude oil refining, the types of feedstocks employed, and the refinery products.

The chapter also presents an indication of the types of the chemicals that can be released to the environment whenever a spill occurs. This offers the environmental analyst the ability to design the necessary test methods to examine the released chemical(s). The chapter offers to the environmental scientist and engineer the ability to start forming opinions and predictions about (1) the nature of the released chemicals, (2) the potential effects of the chemicals on the environment, and (3) the possible methods of cleanup.

2.1 Refinery Products

Like many industrial feedstocks, crude oil in the unrefined form, has little or no direct use and its value as an industrial commodity is only realized after a series of processing steps (as applied in the refinery) to the production of salable products (Figure 2.1). Each processing step is a separate process and, in fact, a refinery is a series of integrated processes that generate the desired products (Table 2.2) (Parkash, 2003; Gary et al., 2007; Speight, 2014; Hsu and Robinson, 2017; Speight, 2017).

The gases that are a part of crude oil and the process gases that are produced from crude oil refining, upgrading or natural gas processing facilities, are a category of saturated and unsaturated lower boiling hydrocarbon derivatives, predominantly the C_1 to C_4 hydrocarbon gases. Some gases may also contain inorganic compounds, such as hydrogen, nitrogen, hydrogen sulfide, carbon monoxide and carbon dioxide. As such, crude oil and refinery gases (unless produced as a salable product that must meet specifications prior to sale) are considered to be of unknown or variable composition. The site-restricted crude oil and refinery gases (i.e., those not produced for sale) often serve as fuels consumed on-site, as intermediates for purification and recovery of various gaseous products, or as feedstocks for isomerization and alkylation processes within a facility.

The processes used for refining crude oil are capable of producing gaseous pollutants which are often characterized by chemical species identification, e.g., *inorganic gases* such as sulfur dioxide (SO_2), nitrogen oxides (NO_x), and carbon monoxide (CO) or *organic gases* such as chloroform ($CHCl_3$) and formaldehyde (HCHO). The rate of release of the gases along with the type of gaseous emission greatly predetermines the applicable control technology.

The three main greenhouse gases that are products of refining are carbon dioxide, nitrous oxide, and methane. Carbon dioxide is the main contributor to climate change. Methane is generally not as abundant as carbon dioxide but is produced during refining and, if emitted into the atmosphere is a powerful greenhouse gas and more effective at trapping heat. However, gaseous emissions associated with crude oil refining are more extensive that carbon dioxide and methane and typically include process gases, petrochemical gases, volatile organic compounds (VOCs), carbon monoxide (CO), sulfur oxides (SO_x), nitrogen oxides (NO_x), particulates, ammonia (NH_3), and hydrogen sulfide (H_2S). These effluents may be discharged as air emissions and must be treated. However, gaseous emissions are more difficult to capture than wastewater or solid waste and, thus, are the largest source of untreated wastes released to the environment.

The numerous process heaters used in refineries to heat process streams or to generate steam (boilers) for heating or steam stripping, can be potential sources of sulfur oxides (SO_2, and SO_3), nitrogen oxides (NO and NO_2), carbon monoxide (CO), particulates, and hydrocarbons emissions. When operating properly and when burning cleaner fuels such as refinery fuel gas, fuel oil or natural gas, these emissions are relatively low. If, however, combustion is not complete, or heaters are fired with refinery fuel pitch or residuals, emissions can be significant.

In addition to the corrosion of equipment of acid gases, the escape into the atmosphere of sulfur-containing gases can eventually lead to the formation of the constituents of acid rain, i.e., the oxides of sulfur (SO_2 and SO_3). Similarly, the nitrogen-containing gases can also lead to nitrous and nitric acids (through the formation of the oxides NO_x, where x = 1 or 2) which are the other major contributors to acid rain. The release of carbon dioxide and hydrocarbons as constituents of refinery effluents can also influence the behavior and integrity of the ozone layer.

The majority of gas streams exiting each refinery process collected and sent to the gas treatment and sulfur recovery units to recover the refinery fuel gas and sulfur. Emissions from the sulfur recovery unit typically contain some hydrogen sulfide (H_2S), sulfur oxides, and nitrogen oxides. Other emissions sources from refinery processes arise from periodic regeneration of catalysts. These processes generate streams that may contain relatively high levels of carbon monoxide, particulates and volatile organic compounds (VOCs). Before being discharged to the atmosphere, such off-gas streams may be treated first through a carbon monoxide boiler to burn carbon monoxide and any volatile organic compounds, and then through an electrostatic precipitator or cyclone separator to remove particulates.

Refinery processes for crude oil to produce bulk products (aside from the gases described in the above paragraphs) are generally divided into two categories: (1) separation processes of which atmospheric distillation and vacuum distillation are the prime examples, and (2) conversion processes of which visbreaking, the coking processes, i.e., delayed coking and fluid coking catalytic cracking, and hydrocracking are prime examples (Table 2.2). In addition, distillates and other fractions are treated further by hydrotreating (product improvement processes) to meet product specifications before sale.

As a first step crude oil yields four types of products when separated by distillation into fractions that are: (1) naphtha, (2) middle distillates such as kerosene and light gas oil—the latter may be referred to as atmospheric gas oil—and (3), heavy gas oil, which may also be referred to as vacuum gas oil, and (4) the residuum (Table 2.1). It is from these unrefined fractions that saleable crude oil products are produced by a series of refinery processes (Parkash, 2003; Gary et al., 2007; Speight, 2014; Hsu and Robinson, 2017; Speight, 2017).

By way of definition and clarification, in this text crude oil products (in contrast to petrochemicals) are those bulk fractions (Table 2.1) that are derived from crude oil and have commercial value as a bulk product (Mushrush and Speight, 1995; Parkash, 2003; Gary et al., 2007; Speight, 2014; Hsu and Robinson, 2017; Speight, 2017). Crude oil products are, generally speaking, hydrocarbon compounds that have various combinations of hydrogen and carbon and can take many molecular forms. Many of the combinations exist naturally in the original crude oil but other combinations are created by an ever-growing number of commercial processes for producing a wide variety of products.

The specifications for crude oil products (which define whether or not a product is suitable for sale and to assure that the product can perform the task for which it is intended) are based on properties such as density and boiling range (to mention only two such properties) and are used. In addition, each product has its own unique set of chemical and physical properties. As a consequence, each product can, and does, have a different effect on the environment (Speight, 2005).

Many of the petrochemicals are substitutes for earlier products from non-crude oil sources. Each new use often imposes additional specifications on the new product and product specifications evolve to stay abreast of advances in both product application and manufacturing methods. Like crude oil products, the petrochemicals cover such a vast range of chemical and physical properties that they also have a wide range of effects on the environment.

The constituents of crude oil and crude oil products, like many organic chemical entities, have the potential to be biodegradable into the raw materials of nature and disappear into the environment. However, sustainable biodegradation of crude oil and crude oil products is not as straightforward as it may seem. Crude oil will biodegrade in its natural state, but once it is converted into saleable products, unsustainable pollution problems can arise—instead of returning to the natural cycle, these products pollute and litter the land, water, and air.

Consideration of the above factors leads to many issues which can, however, be alleviated somewhat by knowing the character of crude oil and the history of the processes by which crude oil is refined into saleable products. There are also those chemicals that are not crude oil products but as used in the production of products and as often designated as refinery chemicals or refinery waste.

Because a critical aspect of assessing the toxic effects of the release of crude oil and crude oil products is the measurement of the compounds in the environment, the approach is not only to understand the composition and properties of crude oil itself (Chapter 2) but also the processes by which crude oil is refined to products as well as and properties of the various fractions and products derived from crude oil (Speight, 2014, 2017).

Therefore, it is the purpose of this chapter to describe major refinery operations and the products therefrom and focuses on their composition and properties and uses. This presents to the reader the essence of crude oil processes, the types of feedstocks employed, and the product produced.

By so doing, the reader is also presented with a forewarning of the types of the chemicals that can be released to the environment whenever an accident occurs. Being forewarned offers the environmental analyst the ability to design the necessary test methods to examine the released chemical(s). The chapter offers, environmental scientist and engineer the ability to start forming opinions and predictions about the nature of the released chemical(s), the potential effects of the chemical(s) on the environment, and the possible methods of cleanup.

2.2 Bulk Products

Crude oil, in the unrefined or crude form, like many industrial feedstocks has little or no direct use and its value as an industrial commodity is only realized after the production of salable products by a series of processing steps as performed in a refinery (Figure 2.1). Each processing step is, in fact, a separate process and thus a refinery is a series of integrated processes that generate the desired products according to the market demand (Meyers, 1997; Speight, 2014, 2017). In general, crude oil, once refined, yields three basic groupings of distillation products that are produced when it is broken down into fractions: naphtha, middle distillates (kerosene and light gas oil), heavy gas oil/vacuum gas oil, and the residuum (Table 2.1). It is from these unrefined fractions that saleable crude oil products are produced by a series of refinery processes (Speight, 2014, 2017).

Naphtha is a precursor (or a blend stock) for gasoline and a variety of solvents as well as for a feedstock for the petrochemical industry. The term *middle distillates* refers to products from the middle boiling range of crude oil and include kerosene, diesel fuel, distillate fuel oil, and light gas oil; waxy distillate and lower boiling lubricating oils are sometimes include in the middle distillates (Parkash, 2003; Gary et al., 2007; Speight, 2014; Hsu and Robinson, 2017; Speight, 2017).

The remainder of the crude oil includes the higher boiling lubricating oil, gas oil, and residuum, which is the non-volatile fraction of the crude oil. In addition, the non-volatile fraction of the crude oil (i.e., residuum) can also produce heavy lubricating oils and waxes but is more often sued for asphalt production (Parkash, 2003; Gary et al., 2007; Speight, 2014, 2015; Hsu and Robinson, 2017; Speight, 2017).

The current trend throughout the refining industry is to produce more fuel products from each barrel of crude oil and to process those products in different ways to meet the product specifications for use in various engines (automobile, diesel, aircraft, and marine). Overall, the demand for liquid fuels has rapidly expanded and demand has also increased for fuel oils for domestic central heating and for power generation, as well as for the petrochemical industries (Parkash, 2003; Gary et al., 2007; Speight, 2014; Hsu and Robinson, 2017; Speight, 2017, 2019).

Crude oil products (in contrast to petrochemicals) are generally those bulk fractions (Table 2.1) that are derived from crude oil and have commercial value as the bulk product. Typically, these products consists of (predominantly) hydrocarbon derivatives that have various combinations of carbon and hydrogen in various molecular forms. Also the specifications for crude oil products (which define whether or not a product is suitable for sale and to assure that the product can perform the task for which it is intended) are based on properties such as density and boiling range (to mention only two such properties) (Chapter 1) and are used. In addition, each product has its own unique set of chemical and physical properties. As a consequence, each product can, and does, have a different effect on the environment (Speight, 2005; Speight and Arjoon, 2012).

2.2.1 Liquefied Petroleum Gas

Hydrocarbon derivatives with up to four carbon atoms in the hydrogen-carbon combination (i.e., methane, CH_4, ethane, C_2H_6, propane, C_3H_8, and butane, C_4H_{10}) have boiling points that are lower than room temperature. Hence these products are gases at ambient temperature and pressure— while collectively are often referred to as natural gas—are often referred to as fuel gases.

Liquefied petroleum gas (LPG, LP Gas, liquid propane gas) is a flammable mixture of gaseous hydrocarbon derivatives that is composed of propane (C_3H_8) and/or butane (C_4H_{10}) and is stored under pressure in order to keep these hydrocarbon derivatives liquefied at normal atmospheric temperatures (Mokhatab et al., 2006; Speight, 2014, 2017, 2018). Before liquefied petroleum gas is burned, it passes through a pressure relief valve that causes a reduction in pressure and the liquid vaporizes.

Varieties of liquefied petroleum gas include mixtures that are primarily propane (C_3H_8), primarily butane (C_4H_{10}) and, most commonly, mixes including both propane and butane, depending on the season—in winter the mixture will contain more propane (because of the increased volatility necessary for ignition) while in summer the mixture will contain more of the relatively less volatile butane. Propylene (C_3H_6, $CH_3CH = CH_2$) and butylenes (C_4H_8), $CH_3CH_2CH = CH_2$ and/or $CH_3CH = CHCH_3$) are usually also present in small concentration.

Natural gas which is predominantly methane (but often contains smaller amounts of ethane, propane and butane) is the lowest boiling and least complex of all hydrocarbon derivatives. Natural gas from an underground reservoir, when brought to the surface, can contain other higher boiling hydrocarbon derivatives and is often referred to as wet gas. Wet gas is usually processed to remove the entrained hydrocarbon derivatives that are higher boiling than methane and, when isolated, the higher boiling hydrocarbon derivatives can be liquefies under the correct conditions of temperature and pressure are referred to as natural gas liquids or gas condensate (Parkash, 2003; Gary et al., 2007; Speight, 2014; Hsu and Robinson, 2017; Speight, 2017). As with liquefied petroleum gas, natural gas for sales and use by the consumer also has an odorant, such as ethane thiol (ethyl mercaptan C_2H_5SH) or amyl thiol (amyl mercaptan, $C_5H_{11}SH$) added to facilitate detection of gas leaks (Mokhatab et al., 2006; Speight, 2018).

Pipestill gas (also called still gas) generally consists of low-boiling hydrocarbon constituents and is the lowest boiling fraction produced by a distillation unit in the refinery (Parkash, 2003; Gary et al., 2007; Speight, 2014; Hsu and Robinson, 2017; Speight, 2017). If the distillation unit is handling higher boiling fractions, the still gas might also contain propane ($CH_3CH_2CH_3$), butane ($CH_3CH_2CH_2CH_3$) in the form of the respective isomers (Mokhatab et al., 2006; Speight, 2014, 2018). Fuel gas is produced in considerable quantities during the different refining processes and is used as fuel for the refinery itself and as an important feedstock for the petrochemical industry (Mokhatab et al., 2006; Speight, 2014, 2018, 2019).

2.2.2 Naphtha, Gasoline, and Solvents

Naphtha is the general term that is applied to refined, partly refined, or unrefined low-boiling crude oil products which is produced by any one of several methods including (1) fractionation of distillates or even crude oil, (2) solvent extraction, (3) hydrogenation of distillates, (4) polymerization of unsaturated compounds, such as (olefin derivatives), and (5) alkylation processes. Naphtha may also be a combination of product streams from more than one of these processes (Speight, 2014, 2019).

Naphtha (often referred to as *naft* in the older literature) is actually a generic term applied to refined, partly refined, or an unrefined crude oil fraction. In the strictest sense of the term, not less than 10% of the material should distill below 175°C (345°F) and not less than 95% of the material should distill below 240°C (465°F) under standardized distillation conditions (ASTM D86).

Generally naphtha (but this can be refinery dependent), naphtha is an unrefined crude oil that distills below 240°C (465°F) and is (after the gases constituents) the most volatile fraction of the crude oil. In fact, in some specifications, not less than 10 per cent of material should distill below approximately 75°C (167°F) (Pandey et al., 2004). It is typically used as a precursor to gasoline or to a variety of solvents. Naphtha resembles gasoline in terms of boiling range and carbon number, being a precursor to gasoline.

The term aliphatic naphtha refers to naphtha containing less than 0.1% benzene and with carbon numbers from C_3 through C_{16}. Aromatic naphtha also spans the range of carbon numbers from C_6 through C_{16} but contains significant quantities of aromatic hydrocarbon derivatives such as benzene (> 0.1% v/v), toluene ($C_6H_5CH_3$), the xylene isomers ($H_3CC_6H_4CH_3$), and (on occasion) ethyl benzene ($C_6H_5CH_2CH_3$).

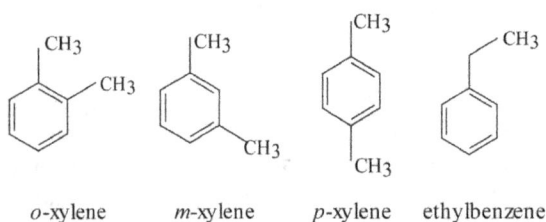

o-xylene   m-xylene   p-xylene   ethylbenzene

The main uses of naphtha fall into the general areas of (1) a blend stock for or a precursor to gasoline and other liquid fuels, (2) solvents/diluents for paints, (3) dry-cleaning solvents, (4) solvents for cutback asphalt, (5) solvents in rubber industry, and (6) solvents for industrial extraction processes.

As a side note that is of importance to any environmental cleanup efforts, the term gasoline is an archaic term from the early days of the refining industry and in the modern refinery there is no single process that produces gasoline. The final gasoline product as a transport fuel is a carefully blended mixture having a predetermined octane value (Table 2.3). Thus, gasoline is a complex mixture of hydrocarbon derivatives that typically boil below 200°C (390°F). The hydrocarbon constituents in this boiling range are those that have four to twelve carbon atoms in the molecular structure.

Gasoline varies widely in composition and even those with the same octane number may be compositionally very different (Table 2.3). The variation in the content of aromatic derivatives as

Table 2.3: Component Streams for Gasoline.

Stream		Producing Process	Boiling Range	
			°C	°F
Paraffin derivatives				
Butane		Distillation	0	32
	Conversion			
Iso-pentane		Distillation	27	81
	Conversion			
	Isomerization			
Alkylate		Alkylation	40–150	105–300
Isomerate		Isomerization	40–70	105–160
Naphtha		Distillation	30–100	85–212
Hydrocrackate		Hydrocracking	40–200	105–390
Olefin derivatives				
Catalytic naphtha		Catalytic cracking	40–200	105–390
Cracked naphtha		Steam cracking	40–200	105–390
Polymer		Polymerization	60–200	140–390
Aromatic derivatives				
Catalytic reformate		Catalytic reforming	40–200	105–390

well as the variation in the content of normal paraffin derivatives, branched paraffins derivative, cyclopentane derivatives, and cyclohexane derivatives are dependent upon the composition of the crude oil (Parkash, 2003; Gary et al., 2007; Speight, 2014; Hsu and Robinson, 2017; Speight, 2017).

Thus, automotive gasoline is a blend of low-boiling hydrocarbon compounds suitable for use in spark-ignited internal combustion engines and having an octane rating of at least 60 (Parkash, 2003; Gary et al., 2007; Speight, 2014; Hsu and Robinson, 2017; Speight, 2017). Additives that have been used in gasoline include alkyl tertiary butyl ether derivaitves such as methyl t-butyl ether (also known as MTBE) and ethanol (ethyl alcohol, C_2H_5OH). Automotive gasoline contains one hundred and fifty or more different chemical compounds and the relative concentrations of the compounds vary considerably depending on the source of crude oil, refinery process, and product specifications. Typical hydrocarbon constituents are: alkane derivatives (4 to 8% v/v), alkene derivatives (2 to 5% v/v), iso-alkane derivatives (25 to 40% v/v), cycloalkane derivatives (3 to 7% v/v), cycloalkene derivatives (1 to 4% v/v), and aromatic derivatives (20 to 50% v/v).

Reformulated gasoline (RFG) is gasoline that is blended such that hydrocarbon and air toxic emissions are significantly reduced. It is not considered to be an alternative fuel, gasohol, or part of the oxygenated fuel program.

The requirements of the reformulated gasoline program, as outlined in the Clean Air Act Amendments, are as follows: (1) an aromatic hydrocarbon content (benzene and its derivatives) no more than 25% v/v, (2) a maximum benzene content of 1% v/v, (3) an oxygen content of 2% w/w, (4) no any heavy metals, including lead and manganese, (5) detergent additives to prevent the accumulation of deposits in engines or vehicle fuel supply systems, and (6) a reduced Reid vapor pressure (a measure of how quickly fuel evaporates) during the summer.

The quality and performance of automobile gasoline is determined by means of the octane number (octane rating) which is a standard measure of the performance of automobile gasoline or aviation fuel. Generally, gasoline-type fuels with a higher octane rating are used in high-compression engines (spark ignition engines) that generally have higher performance. In contrast, fuels with low octane numbers (but high cetane numbers) are ideal for compression ignition engines (i.e., diesel engines).

Aviation fuel comes in two types: (1) aviation gasoline and (2) jet fuel. (1) Aviation gasoline, now usually found in use in light aircraft and older civil aircraft, typically has a narrower boiling range than conventional (automobile) gasoline, i.e., 38 to 170°C (100 to 340°F), compared to –1 to 200°C (30 to 390°F) for automobile gasoline. Since aircraft operate at altitudes where the prevailing pressure is less than the pressure at the surface of the earth (pressure at 17,500 ft is 7.5 psi (0.5 atmosphere) compared to 14.8 psi (1.0 atmosphere) at the surface of the earth), the vapor pressure of aviation gasoline must be limited to reduce boiling in the tanks, fuel lines, and carburetors.

Aviation gasoline consists primarily of straight and branched alkane derivatives and cycloalkane derivatives. Aromatic hydrocarbon derivatives are typically limited to 20 to 25% v/v of the total mixture because they produce smoke when burned. A maximum of 5% v/v alkene derivatives is allowed in JP-4. The approximate distribution by chemical class is: 32% v/v straight alkane derivatives, 31% v/v branched alkane derivatives, 16% v/v cycloalkane derivatives, and 21% v/v aromatic hydrocarbon derivatives (Parkash, 2003; Gary et al., 2007; Speight, 2014; Hsu and Robinson, 2017; Speight, 2017).

Jet fuel is often referred to as *aviation turbine fuel* and the and, in the specifications, ratings relative to octane number are replaced with properties concerned with the ability of the fuel to burn cleanly. Jet fuel is a light crude oil distillates that are available in several forms suitable for use in various types of jet engines. The exact composition of jet fuel is established by the United States Air Force using specifications that yield maximum performance by the aircraft. The major jet fuels used by the military are JP-4, JP-5, JP-6, JP-7, and JP-8.

Briefly, JP-4 is a fuel that has been developed for broad availability in times of need. JP-6 is a higher cut than JP-4 and is characterized by fewer impurities. JP-5 is specially blended kerosene, and JP-7 is high flash point kerosene used in advanced supersonic aircraft. JP-8 is a kerosene fraction that is modeled on Jet A-1 fuel (used in civilian aircraft). For this profile, JP-4 will be used as the prototype jet fuel due to its broad availability and extensive use.

Jet fuel comprises both gasoline and kerosene type jet fuels meeting specifications for use in aviation turbine power units and is often referred to as *gasoline-type jet fuel* and *kerosene-type jet fuel*.

Gasoline type jet fuel includes all low-boiling hydrocarbon derivatives for use in aviation turbine power units that distill between 100 to 250°C (212 to 480°F). It is obtained by blending kerosene and gasoline or naphtha in such a way that the aromatic content does not exceed 25 per cent in volume. Additives can be included to improve fuel stability and combustibility.

Solvents from crude oil (also called naphtha) are valuable because of their good dissolving power. For example, Stoddard solvent is a crude oil distillate widely used as a dry cleaning solvent and as a general cleaner and degreaser consists of linear and branched alkane derivatives (30 to 50% v/v), cycloalkane derivatives (30 to 40% v/v), and aromatic hydrocarbon derivatives (10 to 20% v/v).

The term *petroleum solvent describes* the liquid hydrocarbon fractions obtained from crude oil and used in industrial processes and formulations. These fractions are also referred to *naphtha* or as *industrial naphtha*. By definition the solvents obtained from the petrochemical industry such as alcohols, ethers, and the like are not included in this chapter. A refinery is capable of producing hydrocarbon derivatives of a high degree of purity and at the present time solvents are available covering a wide range of solvent properties including both volatile and high boiling qualities. Other crude oil products boiling within the naphtha boiling range include (1) industrial Spirit and white spirit.

Industrial spirit comprises liquids distilling between 30 and 200°C (–1 to 390°F), with a temperature difference between 5 per cent volume and 90 per cent volume distillation points, including losses, of not more than 60°C (140°F). There are several (up to eight) grades of industrial spirit, depending on the position of the cut in the distillation range defined above. On the other hand, white Spirit is an industrial spirit with a flash point above 30°C (99°F) and has a distillation range from 135 to 200°C (275 to 390°F).

Stoddard solvent is a crude oil distillate widely used as a dry cleaning solvent and as a general cleaner and degreaser. It may also be used as paint thinner, as a solvent in some types of photocopier toners, in some types of printing inks, and in some adhesives. Stoddard solvent is considered to be a form of mineral spirits, white spirits, and naphtha but not all forms of mineral spirits, white spirits, and naphtha are considered to be Stoddard solvent. Stoddard solvent consists of linear alkane derivatives (30 to 50% v/v), branched alkane derivatives (20 to 40% v/v), cycloalkane derivatives (30 to 40% v/v), and aromatic hydrocarbon derivatives (10 to 20% v/v). The typical hydrocarbon chain ranges from C_7 through C_{12} in length.

As the 20th Century evolved, there was increasing use of additives to gasoline leading to product such as *gasohol*, which is a mixture of 90 percent unleaded gasoline and 10 percent ethanol (ethyl alcohol). Gasohol burns well in gasoline engines and is a desirable alternative fuel for certain applications because of the availability of ethanol, which can be produced from grains, potatoes, and certain other plant matter. Methanol and a number of other alcohols and ethers are considered high-octane enhancers of gasoline. They can be produced from various hydrocarbon sources other than crude oil and may also offer environmental advantages insofar as the use of oxygenates would presumably suppress the release of vehicle pollutants into the air.

During the manufacture and distribution of gasoline, it comes into contact with water and particulate matter and can become *contaminated* with such materials. Water is allowed to settle from the fuel in storage tanks and the water is regularly withdrawn and disposed of properly. Particulate matter is removed by filters installed in the distribution system (ASTM D4814).

Adulteration differs from contamination insofar as unacceptable materials deliberately are added to gasoline for a variety of reasons not to be discussed here. Such activities not may not only lower the octane number but will also adversely affect volatility, which in turn also affects performance. In some countries, dyes and markers are used to detect adulteration (for example, ASTM D86 distillation testing and/or ASTM D2699/ASTM D 2700 octane number testing may be required to detect adulteration).

Additives are gasoline-soluble chemicals that are mixed with gasoline to enhance certain performance characteristics or to provide characteristics not inherent in the gasoline. Additives are generally derived from crude oil-based materials and their function and chemistry are highly specialized. They produce the desired effect at the parts-per-million (ppm) concentration range.

Oxidation inhibitors (antioxidants) are aromatic amines and hindered phenols that prevent gasoline components (particularly olefin derivatives) from reacting with oxygen in the air to form *peroxides* or *gums*. Peroxides can degrade antiknock quality, cause fuel pump wear, and attack plastic or elastomeric fuel system parts, soluble gums can lead to engine deposits, and insoluble gums can plug fuel filters. Inhibiting oxidation is particularly important for fuels used in modern fuel-injected vehicles, as their fuel recirculation design may subject the fuel to more temperature and oxygen-exposure stress.

Corrosion inhibitors are carboxylic acids and carboxylates that prevent free water in the gasoline from rusting or corroding pipelines and storage tanks. Corrosion inhibitors are less important once the gasoline is in the vehicle. The metal parts in the fuel systems of today's vehicles are made of corrosion-resistant alloys or of steel coated with corrosion-resistant coatings. More plastic parts are replacing metals in the fuel systems and, in addition, service station systems and operations are designed to prevent free water from being delivered to a vehicle's fuel tank.

Demulsifiers are polyglycol derivatives improve the water-separating characteristics of gasoline by preventing the formation of stable emulsions.

Antiknock compounds are compounds (such as tetraethyl lead) that increase the antiknock quality of gasoline. Gasoline containing tetraethyl lead was first marketed in 1920s and the average concentration of lead in gasoline gradually was increased until it reached a maximum of about 2.5 grams per gallon (*g/gal.*) in the late 1960s. After that, a series of events resulted in the use of less lead and EPA regulations required the phased reduction of the lead content of gasoline beginning in 1979. The EPA completely banned the addition of lead additives to on-road gasoline in 1996 and the amount of incidental lead may not exceed 0.05 g/gal.

Anti-icing additives are surfactants, alcohols, and glycols that prevent ice formation in the carburetor and fuel system. The need for this additive is being reduced as older-model vehicles with carburetors are replaced by vehicles with fuel injection systems.

Dyes are oil-soluble solids and liquids used to visually distinguish batches, grades, or applications of gasoline products. For example, gasoline for general aviation, which is manufactured to different and more exacting requirements, is dyed blue to distinguish it from motor gasoline for safety reasons.

Markers are a means of distinguishing specific batches of gasoline without providing an obvious visual clue. A refiner may add a marker to its gasoline so it can be identified as it moves through the distribution system.

Drag reducers are high-molecular-weight polymers that improve the fluid flow characteristics of low-viscosity crude oil products. Drag reducers lower pumping costs by reducing friction between the flowing gasoline and the walls of the pipe.

Oxygenates are carbon-, hydrogen-, and oxygen-containing combustible liquids that are added to gasoline to improve performance. The addition of oxygenates gasoline is not new since ethanol (ethyl alcohol or grain alcohol) has been added to gasoline for decades. Thus, *oxygenated gasoline* is a mixture of conventional hydrocarbon-based gasoline and one or more oxygenates. The current oxygenates belong to one of two classes of organic molecules: alcohols and ethers. The most widely used oxygenates in the United States are ethanol, methyl tertiary-butyl ether (MTBE) and tertiary-amyl methyl ether (TAME). Ethyl tertiary-butyl ether (ETBE) is another ether that could be used. Oxygenates may be used in areas of the United States where they are not required as long as concentration limits (as defined by environmental regulations) are observed.

Of all the oxygenates, methyl-t-butyl ether (MTBE) is attractive for a variety of technical reasons. It has a low vapor pressure, can be blended with other fuels without phase separation, and has the desirable octane characteristics. If oxygenates achieve recognition as vehicle fuels, the biggest contributor will probably be methanol, the production of which is mostly from synthesis gas derived from methane or from other sources (Chapter 27) (Speight, 2008).

The higher alcohols also offer some potential as motor fuels. These alcohols can be produced at temperatures below 300°C (570°F) using copper oxide-zinc oxide-alumina catalysts promoted with potassium. *Iso*-butyl alcohol is of particular interest because of its high octane rating, which makes it desirable as a gasoline-blending agent. This alcohol can be reacted with methanol in the presence of a catalyst to produce methyl-t-butyl ether. Although it is currently cheaper to make *iso*-butyl alcohol from *iso*-butylene, it can be synthesized from syngas with alkali-promoted zinc oxide catalysts at temperatures above 400°C (750°F).

2.2.3 Kerosene and Diesel Fuel

Kerosene originated as a straight-run (distilled) crude oil fraction that boiled between approximately 150 to 350°C (300 to 660°F). In the early days of crude oil refining some crude oils contained kerosene fractions of very high quality and required little additional processing. However, other crude oils, such as those oils with a high proportion of asphaltic constituents, must be purified by refining to remove aromatic compounds and sulfur compounds before a satisfactory kerosene fraction can be obtained. Chemically, the kerosene fraction is a mixture of hydrocarbon derivatives but the composition of this fraction depends on the source, but usually consists hydrocarbon derivatives each containing 10 to 16 carbon atoms per molecule as fond in a variety of alkyl benzene derivatives of which *n*-dodecane (n-$C_{12}H_{26}$) is an example, and naphthalene and its derivatives.

Diesel fuel also forms part of the kerosene boiling range (or middle distillate group of products). Diesel fuels come in two broad groups, for high speed engines in cars and trucks requiring a high quality product, and lower quality heavier diesel fuel for slower engines, such as in marine engines or for stationary power plants. An important property of diesel is cetane number (analogous to the gasoline octane number) and the cetane number determines the ease of ignition under compression.

The quality of diesel fuel is measured using the cetane number that a measure of the tendency of a diesel fuel to knock in a diesel engine and the scale, from which the cetane number is derived, is based upon the ignition characteristics of two hydrocarbon derivatives: (1) *n*-hexadecane, commonly referred to as cetane, and (2) 2,3,4,5,6,7,8-heptamethylnonane.

Kerosene type jet fuel is a medium distillate product that is used for aviation turbine power units. It has the same distillation characteristics and flash point as kerosene (between 150 and 300°C, 300 and 570°F, but not generally above 250°C, 480°F). In addition, it has particular specifications (such as freezing point) which are established by the International Air Transport Association (IATA).

The essential properties of kerosene are flash point, fire point, distillation range, burning, sulfur content, color, and cloud point. In the case of the flash point (ASTM D56), the minimum flash temperature is generally placed above the prevailing ambient temperature; the fire point (ASTM D92) determines the fire hazard associated with its handling and use.

The boiling range (ASTM D86) is of less importance for kerosene than for gasoline, but it can be taken as an indication of the viscosity of the product, for which there is no requirement for kerosene. The ability of kerosene to burn steadily and cleanly over an extended period (ASTM D187) is an important property and gives some indication of the purity or composition of the product.

The significance of the total sulfur content of a fuel oil varies greatly with the type of oil and the use to which it is put. Sulfur content is of great importance when the oil to be burned produces sulfur oxides that contaminate the surroundings. The color of kerosene is of little significance, but a product darker than usual may have resulted from contamination or aging, and in fact a color darker than specified (ASTM D156) may be considered by some users as unsatisfactory. Finally, the cloud point of kerosene (ASTM D2500) gives an indication of the temperature at which the wick may become coated with wax particles, thus lowering the burning qualities of the oil.

2.2.4 Fuel Oil

Fuel oil is often referenced in several ways but generally may be divided into two main types which are (1) distillate fuel oil and (2) residual fuel oil. These names are still employed but, of late, the terms distillate fuel oil and residual fuel oil have lost some of the significance because fuel oils are now manufactured for specific purposes and may be distillates, residuals, or mixtures of the two. The terms domestic fuel oils, diesel fuel oils, and heavy fuel oils are more indicative of the uses of fuel oils. More often than not, fuel oil is prepared by using a visbreaker unit to perform mild thermal cracking (insofar as the reactions are not allowed to proceed to completion) on a residuum on or a high boiling distillate so that the product meets specifications (Parkash, 2003; Gary et al., 2007; Speight, 2014; Hsu and Robinson, 2017; Speight, 2017).

Distillate fuel oil is vaporized and condensed during the distillation process and has a typical boiling range but does not contain high-boiling oils or asphaltic constituent. Fuel oil that contains an amount of the residuum from the distillation or thermal cracking of crude oil is typically referred to as a residual fuel oil. *Domestic fuel oil* is used primarily in the home and includes kerosene, stove oil, and furnace fuel oil. *Diesel fuel oil* is also a distillate fuel oil but residual oil has been successfully used to power marine diesel engines, and mixtures of distillates and residuals have been used on locomotive diesels. *Furnace fuel oil* is similar to diesel fuel but the proportion of cracked gas oil in diesel fuel is usually less because the high aromatic content of the cracked gas oil reduces the cetane number of the diesel fuel. Included among heavy fuel oils are various industrial oils; when used to fuel ships, heavy fuel oil is called bunker oil (often referred to as Bunker C oil).

Domestic fuel oil is fuel oil that is used primarily in the home and includes kerosene, stove oil, and furnace fuel oil (Parkash, 2003; Gary et al., 2007; Speight, 2014; Hsu and Robinson, 2017; Speight, 2017). Furnace fuel oil is similar to diesel fuel but the proportion of cracked gas oil in diesel fuel is usually less since the high aromatic content of the cracked gas oil reduces the cetane number of the diesel fuel (Parkash, 2003; Gary et al., 2007; Speight, 2014; Hsu and Robinson, 2017; Speight, 2017). Stove oil is a straight-run (distilled) fraction from crude oil whereas other fuel oils are usually blends of two or more fractions. The distillate fractions (also referred to as straight-run fractions) that are available for blending into fuel oils include high-boiling naphtha, low-boiling gas oil and high-boiling gas oil and residua.

Heavy fuel oil (high-boiling fuel oil) includes a variety of oils ranging from distillates to residual oils that must be heated to a temperature on the order of 260°C (500°F) or higher before they can be used. In general, heavy fuel oils consist of residual material that has been blended with distillate material to match the specifications as dictated by the needs. Included among heavy fuel oils are various industrial oils; when used to fuel ships, heavy fuel oil is called bunker oil.

Fuel oil that is used for heating is graded from No. 1 Fuel Oil to No. 6 Fuel Oil, and cover light distillate oils, medium distillate, heavy distillate, a blend of distillate and residue, and residue oil. For instance No. 2 and No. 3 Fuel oils refer to medium-to-light distillate grades used in domestic central heating (Table 2.4).

No. 1 fuel oil is a distillate from crude oil (such as straight-run kerosene) and consist primarily of hydrocarbon derivatives in the range C9 to C16. No 1 fuel oil is one of the most widely used types of fuel oil and is used in atomizing burners that spray fuel into a combustion chamber where the droplets of the fuel oil burn while in suspension. No 1 fuel oil is also used as (1) a carrier for pesticides, (2) as a weed killer, (3) as a mold release agent in the ceramic and pottery industry, (4) in the cleaning industry, and (5) in asphalt coatings, enamels, paints, thinners, and varnishes

No. 2 fuel oil is a crude oil distillate (which can be cracked distillate or a blend of both non-cracked and cracked distillate) that may be also referred to as domestic fuel oil or industrial fuel oil and, in many cases is used primarily for home heating and to produce diesel fuel. This type

Table 2.4: Properties of Various Types of Fuel Oil.

No. 1 Fuel Oil Similar to kerosene range oil (i.e., liquid fuel used in stoves for cooking). A distillate product intended for vaporizing in pot-type burners and other burners where a clean flame is required.
No. 2 Fuel Oil Often called domestic heating oil with properties similar to diesel and higher-boiling jet fuel. A distillate product for general purpose heating.
No. 4 Fuel oil A light (low density) industrial heating oil; preheating is not required for handling or burning. There are two grades of No. 4 fuel oil that differ primarily in safety (flash) and flow (viscosity) properties.
No. 5 Fuel oil A high density (viscous) industrial oil that often requires preheating for burning and, in cold climates, for handling.
No. 6 Fuel Oil Commonly referred to as *Bunker C oil* when it is used to fuel ocean-going vessels. A viscous oil usually containing residuum which requires preheating is for both handling and burning

of fuel oil is characterized by hydrocarbon derivatives in the C11 to C20 range, whereas diesel fuels predominantly contain a mixture of C10 to C19 hydrocarbon derivatives. The composition consists of approximately 64% aliphatic hydrocarbon derivatives (straight chain alkane derivatives and cycloalkane derivatives), 1–2% unsaturated hydrocarbon derivatives (alkene derivatives), and 35% aromatic hydrocarbon derivatives (including alkylbenzenes and 2-ring and 3-ring aromatic derivatives). No. 2 fuel oil contains less than 5% polycyclic aromatic hydrocarbon derivatives.

No. 6 fuel oil (also called Bunker C or residual fuel oil) is the residual from crude oil after the light oil, gasoline, naphtha, No. 1 fuel oil, and No. 2 fuel oil have been distilled. No. 6 fuel oil can be blended directly to heavy fuel oil or made into asphalt. The use of No. 6 fuel oil is typically limited to use in commercial and industrial burners where sufficient heat is available to fluidize the oil for pumping prior to the combustion chamber.

In summary, all fuel oils consist of complex mixtures of aliphatic and aromatic hydrocarbon derivatives. The aliphatic alkane derivatives (paraffins) and cycloalkane derivatives (naphthenes) are hydrogen saturated and compose approximately 80 to 90% v/v of the fuel oils. Aromatic derivatives (such as benzene derivatives) and olefin derivatives (such as styrene, $C_6H_5CH = CH_2$ and indene) compose 10 to 20% v/v and 1% v/v, respectively, of the fuel oils.

Styrene

Indene

No. 1 fuel oil (straight-run kerosene) is a low-boiling distillate which consists primarily of hydrocarbon derivatives in the C9 to C16 range. On the other hand, No. 2 fuel oil is a higher boiling, usually blended, distillate with hydrocarbon derivatives in the C11 to C20 range.

All of the above fuel oils contain less than 5% v/v polycyclic aromatic hydrocarbon derivatives. Fuel oil No. 4 (marine diesel fuel) is less volatile than diesel fuel oil No. 2 and may contain up to 15% v/v residual process streams, in addition to more than 5% v/v polycyclic aromatic hydrocarbon derivatives. Residual fuel oils are generally more complex in composition and impurities than distillate fuel oils; therefore, a specific composition is difficult to determine and is, at best, speculative. The sulfur content in residual fuel oils has been reported to vary from 0.18% to 4.36% w/w. Of the limited data available, No. 6 fuel oil includes aromatic derivatives (25% v/v), paraffin derivatives (15% v/v), naphthene derivatives (45% v/v), and non-hydrocarbon compounds (15% v/v). Polynuclear aromatic hydrocarbon derivatives and their alkyl derivatives and metals are important hazardous and persistent components of No. 6 fuel oil.

2.2.5 Lubricating Oil

Lubricating oil is distinguished from other fractions of crude oil by a high boiling range (> 400°C, > 750°F) as well as a high viscosity. Lubricating oil may be divided into various categories according to the types of service the oil is intended to perform such as: (1) oils that are used in intermittent service, such as motor and aviation oils and (2) oils that are designed for continuous service such as turbine oils. Hydrocarbon types ranging from C_{15} to C_{50} are found in the various types of oils, with the heavier distillates having higher percentages of the higher molecular weight constituents.

The crude oil-based lubricating oils contains wide variety of hydrocarbon derivative as well nitrogen-containing derivatives and sulfur-containing derivative. The hydrocarbon derivatives are mixtures of straight and branched chain hydrocarbon derivatives (alkane derivatives), cycloalkane derivatives, and aromatic hydrocarbon derivatives (Parkash, 2003; Gary et al., 2007; Speight, 2014; Hsu and Robinson, 2017; Speight, 2017). The components of motor oil and crankcase oil include:

(1) polynuclear aromatic hydrocarbon derivatives, (2) alkyl polynuclear aromatic hydrocarbon derivatives, and (3) metals, with the used oils typically having higher concentrations of these constituents than the new/unused oils.

In the United States, specifications for these products are defined by the Society of Automotive Engineers (SAE), which issues viscosity ratings with numbers that range from 5 to 50.

Lubricating oils are altered during service because of the breakdown of additives, contamination with the products of combustion, and the addition of metals from the wear and tear of the engine. Therefore, the precise composition of used lubricating oil is difficult to generalize in terms of the chemical constituent. Generally, the major constituents of used lubricating oil consist of aliphatic and aromatic hydrocarbon derivatives (such as naphthalene derivatives, benz(a)anthracene, benzo(a) pyrene, and fluoranthene).

Before they are used, lubricating oils consist of a base lubricating oil (a complex mixture of hydrocarbon derivatives, 80 to 90% v/v) and performance-enhancing additives (10 to 20% v/v). Generally, the major components of used lubricating oil consist of aliphatic and aromatic hydrocarbon derivatives (such as phenol, naphthalene, benz(a)anthracene, benzo(a)pyrene, and fluoranthene).

Used lubricating oil (waste crankcase oil) is the used lubricating oils removed from the crankcase of internal combustion engines (Speight and Exall, 2014).

Mineral oils are often used as lubricating oils but also have medicinal and food uses. A major type of hydraulic fluid is the mineral oil class of hydraulic fluids. The mineral-based oils are produced from heavy-end crude oil distillates. Hydrocarbon numbers ranging from C_{15} to C_{50} occur in the various types of mineral oils, with the heavier distillates having higher percentages of the higher carbon number compounds.

Crankcase oil (*motor oil*) may be either mineral-based or synthetic. The mineral-based oils are more widely used than the synthetic oils and may be used in automotive engines, railroad and truck diesel engines, marine equipment, jet and other aircraft engines, and most small 2- and 4-stroke engines. The mineral-based oils contain hundreds to thousands of hydrocarbon compounds, including a substantial fraction of nitrogen- and sulfur-containing compounds. The hydrocarbon derivatives are mainly mixtures of straight and branched chain hydrocarbon derivatives (alkane derivatives), cycloalkane derivatives, and aromatic hydrocarbon derivatives. Polynuclear aromatic hydrocarbon derivatives (and the alkyl derivatives) and metal-containing constituents are components of motor oils and crankcase oils, with the used oils typically having higher concentrations than the new unused oils. Typical carbon number chain lengths range from C_{15} to C_{50}.

2.2.6 White Oil, Insulating Oil, Insecticides

There is also a category of crude oil products known as *white oil* that generally falls into two classes: (1) *technical white oil* that is employed for cosmetics, textile lubrication, insecticide vehicles, and paper impregnation and (2) *pharmaceutical white oil* that may is employed medicinally (e.g., as a laxative) or for the lubrication of food-handling machinery.

The term white oil (white distillate) is applied to all the refinery streams with a distillation range between approximately 80 and 360°C (175 to 680°F) at atmospheric pressure and with properties similar to the corresponding straight-run distillate from atmospheric crude distillation (Parkash, 2003; Gary et al., 2007; Speight, 2014; Hsu and Robinson, 2017; Speight, 2017). There is also a category of crude oil products known as *white oil* that generally falls into two classes: (1) technical white oil that is employed for cosmetics, textile lubrication, insecticide vehicles, and paper impregnation and (2) pharmaceutical white oil that may is employed medicinally (e.g., as a laxative) or for the lubrication of food-handling machinery.

The colorless character of these oils is important in some cases, as it may indicate the chemically inert nature of the hydrocarbon constituents. Textile lubricants should be colorless to prevent the staining of light-colored threads and fabrics. Insecticide oils should be free of reactive (easily oxidized) constituents so as not to injure plant tissues when applied as sprays. Laxative oils should be free of odor, taste, and also hydrocarbon derivatives, which may react during storage and produce unwanted by-products. These properties are attained by the removal of nitrogen-containing, oxygen-containing, and sulfur-containing compounds, as well as reactive hydrocarbon derivatives by, say, sulfuric acid.

Insulating oil (transformer oil, dielectric oil) falls into two general classes: (1) oil used in transformers, circuit breakers, and oil-filled cables and (2) oil employed for impregnating the paper covering of wrapped cables. The main function of insulating oil (transformer oil) is insulating and cooling of the transformer and, accordingly, the oil should have the following properties (1) high dielectric strength, (2) low viscosity, (3) no inorganic acids, alkali, and corrosive sulfur present, (4) resistant to emulsification, (5) does not form sludge under typical operating conditions, (6) low pour point, and (7) high flash point.

Oxidation is the most common cause of oil deterioration—careful and routine vacuum dehydration to remove air and water is essential to maintaining good oil. Moisture is the main source of contamination and tends to lower the dielectric strength of the oil and promote acid formation when combined with air and sulfur. Finally, excessive heat breaks down the oil and will increase the rate of oxidation—this usually occurs when the transformer is overloaded.

Insecticides are not always derived from crude oil (Ware and Whitacre, 2004). However, insecticides are usually be applied in water-emulsion form and which have marked killing power for certain species of insects. For many applications for which their own effectiveness is too slight, the oils serve as carriers for active poisons, as in the household and livestock sprays.

Insecticides are agents of chemical or biological origin that control insects. Control may result from killing the insect or otherwise preventing it from engaging in behaviors deemed destructive. At the beginning of World War II (1939), insecticide selection was limited to several arsenic derivatives, crude oils, nicotine If only we could teach mosquitos to smoke, pyrethrum (an extract from the African Daisy, a particular type of Chrysanthemum, *Chrysanthemum cineraria folium*), rotenone (which occurs naturally in the seeds and stems of several plants, such as the jicama vine plant), sulfur, hydrogen cyanide gas, and cryolite (sodium hexafluoroaluminate, Na_3AlF_6). It was World War II that opened the modern era of chemical control with the introduction of a new concept of insect control—synthetic organic insecticides, the first of which was DDT.

Form the chemical aspects, crude oils, as such, usually applied in water-emulsion form, are still recognized as having a marked killing power for certain species of insects. For many applications for which their own effectiveness is too slight, the oils serve as carriers for active poisons, as in the household and livestock sprays.

The most extensive use of crude oil itself as a killing agent is in fruit tree sprays. The spraying of swamp waters with an oil film as a method of mosquito control has also been practiced. The fruit tree spray oils are known to be elective in the control of scale insects, leaf rollers, red spiders, tree hoppers, mites, moth eggs, and aphids. Molecular weight and structure appear to be the factors determining the insecticidal power of these oils. Olefin derivatives and aromatic derivatives are both highly toxic to insects, but they also have a detrimental effect on the plant; thus spray oils generally receive some degree of refining, especially those of the summer oil type that come into contact with foliage.

Paraffins and naphthenes are the major components of the refined spray oils, and the former appear to be the more toxic. With both naphthenic and paraffinic hydrocarbon derivatives the insecticidal effect increases with molecular weight but becomes constant at about 350 for each; the maximum toxicity has also been attributed to that fraction boiling between 240 and 300°C (465 to 570°F) at 40 mm Hg pressure.

The physical properties of crude oil-based oils, such as their solvent power for waxy coatings on leaf surfaces and insect bodies, make them suitable as carriers for more active fungicides and insecticides. The additive substance may vary from fatty acids and soaps, the latter intended chiefly to affect favorably the spreading properties of the oil, to physiologically-active compounds, such as pyrethrum, nicotine, rotenone, DDT, thiocyanates, methoxychlor, chlordane, lindane, and others. Solubility of the chlorine-containing insecticides is often aided by a an accessory solvent rich in methylnaphthalene. The hydrocarbon-base solvent used in household insecticides is generally a high-flash (66°C, 150°F) 195 to 250°C (380 to 480°F) boiling naphtha that has been heavily treated with concentrated sulfur acid. Household and livestock sprays are also made up for application from aerosol containers, in which liquefied gases (generally dichlorodifluoromethane and trichloromonofluoromethane) are used as the propelling agents.

2.2.7 Grease

Grease is lubricating oil to which a thickening agent has been added for the purpose of holding the oil to surfaces that must be lubricated. The development of the chemistry of grease formulations is closely linked to an understanding of the physics at the interfaces between the machinery and the grease. With this insight, it is possible to formulate greases that are capable of operating in increasingly demanding and wide-ranging conditions.

The term grease is applied to many different chemicals and, unlike many crude oil products has no specific formula. However, in the general context, grease is lubricating oil to which a thickening agent has been added for the purpose of ensuring that the oil adheres to surfaces that require lubrication. The most widely used thickening agents are soaps and grease manufacture is essentially the mixing of soaps with lubricating oils.

Soap is made by chemically combining a metal hydroxide with a fat or fatty acid:

$$RCO_2H + NaOH \rightarrow RCO^-_2Na^+ + H_2O$$

Fatty acid　　　　　　soap

The most common metal hydroxides used for this purpose are calcium hydroxide, lye, lithium hydroxide, and barium hydroxide. Fats are chemical combinations of fatty acids and glycerin. If a metal hydroxide is reacted with a fat, a soap containing glycerin is formed. Frequently a fat is separated into its fatty acid and glycerin components, and only the fatty acid portion is used to make soap. Commonly used fats for grease-making soaps are cottonseed oil, tallow, and lard. Among the fatty acids used are stearic acid (from tallow), oleic acid (from cottonseed oil), and animal fatty acids (from lard).

There are three basic groups of mineral oils: (1) aromatic, (2) naphthenic, and (3) paraffinic. Historically, the aromatic and naphthenic mineral oils have represented the principle volumes used in grease formulation due to the readily availability of these oils as well to their solubility characteristics. However, concerns about the carcinogenic aspects of base oil constituents containing aromatic and polyaromatic ring structures have led to their replacement by paraffinic oils as the mineral fluids of choice.

2.2.8 Wax

Paraffin wax is a solid crystalline mixture of straight-chain (normal) hydrocarbon derivatives that range from twenty to more than thirty carbon atoms per molecular. In addition, there are two general types of wax which are (1) paraffin wax in crude oil distillates and (2) microcrystalline wax in crude oil residua. There is also a third category of wax that is commonly referred to as petrolatum which is microcrystalline in nature and semi-solid at room temperature. Wax constituents are solid at ordinary temperatures (25°C, 77°F) whereas petrolatum (also commonly referred to as petroleum jelly) does contain both solid and liquid hydrocarbon derivatives. The melting point of wax is not always directly related to its boiling point, because wax contains hydrocarbon derivatives of different chemical structure.

Paraffin wax is a solid crystalline mixture of straight-chain (normal) hydrocarbon derivatives ranging from twenty to thirty (or more) carbon atoms per molecule, and even higher, which are solid at ordinary temperatures (25°C, 77°F) whereas *petrolatum* (*crude oil jelly*) does contain both solid and liquid hydrocarbon derivatives. The melting point of wax is not always directly related to its boiling point, because wax contains hydrocarbon derivatives of different chemical structure.

There are three main grades of paraffin wax: (1) fully refined, (2) semi-refined, and (3) scale which differ mainly by the degree to which entrapped oil has been removed during refining and by color. Fully refined paraffins are hard, brittle, white, odorless materials with less than 0.5% w/w oil, melting points from 46 to 68°C (115 to 155°F). on the other hand, semi-refined paraffin waxes contain more oil—0.5% to 1% w/w—and the additional oil detracts from gloss and the waxes are softer, light-colored and have a slight odor and taste. Scale waxes are softer materials with 1 to 3% w/w oil content. Color varies from white to yellow.

More generally, there are three main methods are used in modern refinery technology: (1) *solvent dewaxing* in which the feedstock is mixed with one or more solvents then the mixture is cooled down to allow the formation of wax crystals, and the solid phase is separated from the liquid phase by filtration, (2) *urea dewaxing* in which urea forms an adduct with straight chain paraffins that separated by filtration from the dewaxed oil, and (3) *catalytic dewaxing* in which straight-chain paraffin hydrocarbon derivatives are selectively cracked on zeolite-type catalysts and the lower-boiling reaction products are separated from the dewaxed lubricating oil by fractionation (Parkash, 2003; Gary et al., 2007; Speight, 2014; Hsu and Robinson, 2017; Speight, 2017).

Individual test methods are determined by molecular size and structure, chemical composition, and oil content. Paraffin wax consists mostly of straight chain hydrocarbon derivatives with 80 to 90% w/w normal paraffin content and the balance consists of branched paraffin derivatives (iso-paraffin derivatives) and cycloparaffin derivatives.

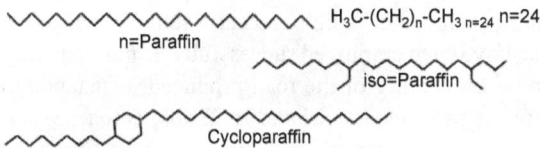

n=Paraffin

$H_3C-(CH_2)_n-CH_3$ $_{n=24}$ n=24

iso=Paraffin

Cycloparaffin

Microcrystalline waxes are produced from a combination of heavy lube distillates and residual oils. They differ from paraffin waxes in that they have poorly defined crystalline structure, darker color, and generally higher viscosity and melting points. Microcrystalline waxes (sometimes also called micro wax) tend to vary much more widely than paraffin waxes with regard to physical characteristics. Microcrystalline wax can range from being semisolid (soft) and tacky to being hard and brittle, depending on the compositional balance within the wax.

Typical test methods that can be used to determine the properties of wax include (1) melting point, (2) congealing point, (3) oil content, and (4) viscosity.

2.2.9 Asphalt

Asphalt (sometimes referred to as bitumen in some parts of the world) is produced from the distillation residuum. Asphalt has complex chemical and physical compositions that usually vary with the source of the crude oil and it is produced to certain standards of hardness or softness in controlled vacuum distillation processes (Parkash, 2003; Gary et al., 2007; Speight, 2014, 2015; Robinson, 2017; Speight, 2017).

There are wide variations in refinery operations and in the types of crude oils so different asphalts will be produced that have different environmental effects (EPA, 1996). If lubricating oils are not required, the reduced crude may be distilled in a flash drum that is similar to a distillation tower but has few, if any, trays. Asphalt descends to the base of the flasher as the volatile components pass out of the top. Asphalt is also produced by propane the deasphalting process (which typically uses liquid propane as the deasphalting liquid) and can be made softer by blending the hard extremely high viscosity semi-solid to solid asphalt with the extract obtained in the solvent treatment of lubricating oil.

Road oil is liquid asphalt materials intended for easy application to earth roads and provides a strong base or a hard surface and will maintain a satisfactory passage for light traffic. Cutback asphalt is a mixture in which hard asphalt has been diluted with a lower boiling oil to permit application as a liquid without drastic heating. Asphalt emulsions are usually the oil-in-water type emulsion which breaks on application to a stone surface or an earth surface after which the asphalt oil clings to the stone and the water disappears.

The chemical composition of asphalt is variable because it is dependent on the chemical complexity of the original crude oil and the manufacturing process. In fact, because of the precursors (residua) from which asphalt is manufactured attempts to identify individual constituents of asphalt have only led to wide speculation.

Crude oil consists of aliphatic compounds, cyclic alkane derivatives, aromatic hydrocarbon derivatives, polycyclic aromatic derivatives, and metals (nickel and vanadium with smaller amounts of iron and copper) (Parkash, 2003; Gary et al., 2007; Speight, 2014; Hsu and Robinson, 2017; Speight, 2017). The proportions of these chemicals can vary greatly because of significant differences in crude oil from oil field to oil field or even at different locations in the same oil field (Parkash, 2003; Gary et al., 2007; Speight, 2014; Hsu and Robinson, 2017; Speight, 2017). Although no two asphalts are chemically identical and chemical analysis cannot be used to define the exact chemical structure or chemical composition of asphalt, elemental analyses indicate that most asphalts contain 79 to 88% w/w carbon, 7 to 13% w/w hydrogen, traces to 8% w/w sulfur, 2 to 8% w/w oxygen, and traces to 3% w/w nitrogen.

Road Asphalt

Road oils are, as the name implies, liquid asphaltic materials intended for easy application to earth roads; they do not provide a strong base or a hard surface but maintain a satisfactory passage for light traffic. Both straight-run and cracked residua have been employed successfully. Binding quality and adhesive character are important in governing the quality of the road produced; resistance to removal by emulsification has some influence on its permanence. Liquid road oils, cutbacks, and emulsions are of recent date, but the use of asphaltic solids for paving goes back to a European practice of about 1835. The asphaltic constituents employed may have softening points up to, say, 110°C (230°F).

Cutback Asphalt

Asphalt may next be blended or *cut* (*diluted*) with a volatile solvent resulting in a product that is soft and workable at a lower temperature than pure un-cut asphalt. When the *cutback* asphalt is used for paving or construction, the volatile element evaporates when exposed to air or heat, leaving the hard asphalt. The relative speed of evaporation or volatility of the solvent determines whether cutback asphalt is classified as slow, medium, or rapid-curing. In the process for preparing cutback asphalt, heated asphalt is mixed with a gas oil type solvent from the distillation process or to produce slow-curing asphalt, the asphalt is mixed with kerosene for medium-curing, and with gasoline or naphtha for the rapid-curing asphalt.

Crude oil solvents used for dissolving binder are sometimes called distillate, diluent, or cutter stock. If the solvent used in making the cutback asphalt is highly volatile it will quickly escape by evaporation. Solvents of lower volatility evaporate more slowly. On the basis of relative speed of evaporation, cutback asphalts are divided into three types: (1) rapid-curing asphalt—a mixture of asphalt and a volatile solvent or light distillate, generally in the gasoline or naphtha boiling range, (2) medium-curing asphalt—a mixture of asphalt and a solvent of intermediate volatility or medium distillate, generally in the kerosene boiling range, and (3) slow-curing asphalt—a mixture of asphalt and an oily diluent of low volatility; slow-curing asphalt is often called *road asphalt* or *road oil* and this terminology originated in the early decades of the 20th Century when asphaltic residual oil was used to give roads a low-cost, all-weather surface.

The degree of fluidity obtained in each case depends on the grade of asphalt cement, volatility of the solvent, and proportion of solvent to binder. The degree of fluidity results in several grades of cutback asphalt. Some cutback asphalts are fluid at ordinary atmospheric temperatures and others are somewhat more viscous and may require heating to melt them enough for construction operations. Cutback asphalts can be used with cold aggregates, with a minimum of heat. Rapid curing and medium-curing types of cutback asphalts are used in a variety of highway construction. Among the more important uses are road mixing operations, stockpiling mixes, and spray applications.

Asphalt Emulsion

Asphalt may also be *emulsified* to produce a liquid product that can be pumped and transported by pipeline, mixed with aggregate, or sprayed through nozzles. To emulsify, the asphalt cement is ground into globules 5 to 10 microns and smaller (one micron is equal to one millionth of a meter), which is then mixed with water. An emulsifying agent is added, which reduces the tendency of the asphalt and water to separate. The emulsifying agent may be colloidal clay, soluble or insoluble silicate minerals, soap, or sulfonated vegetable oil.

These emulsions are normally of the oil-in-water type which reverse or break on application to a stone or earth surface, so that the oil clings to the stone and the water disappears. In addition to their usefulness in road and soil stabilization, they are useful for paper impregnation and waterproofing. The emulsions are chiefly (1) the soap or alkaline type and (2) the neutral or clay type. The former break readily on contact, but the latter are more stable and probably lose water mainly by evaporation. Good emulsions must be stable during storage or freezing, suitably fluid, and amenable to control for speed of breaking.

In the emulsification process, hot binder is mechanically (in a colloid mill) separated into minute globules and dispersed in water treated with a small quantity of emulsifying agent. The water is called the continuous phase and the globules of binder are called the discontinuous phase. The binder globules are extremely small, mostly in the colloidal size range, and by proper selection of an emulsifying agent and other manufacturing controls, emulsified asphalts are produced in several types and grades. By choice By choice of emulsifying agent, the emulsified asphalt may be: (1) anionic—the binder globules are electro-negatively charged or (2) cationic—the binder globules are electro-positively charged.

Because particles having a like electrostatic charge repel each other, asphalt globules are kept apart until the emulsion is deposited on the surface of the soil or aggregate particles. At this time, the asphalt globules coalesce through neutralization of the electrostatic charges or water evaporation. Coalescence of asphalt globules occurs in rapid and medium setting grades, resulting in a phase separation between asphalt and water. When this coalescence occurs, it is usually referred to as the break or set.

Emulsified asphalts can be used with cold as well as heated aggregates, and with aggregates that are dry, damp, or wet.

Cold Mix Asphalt

Cold mix asphalt (*cold mix asphalt concrete, cold placed mixture*) is generally a mix made with emulsified or cutback asphalt. Aggregate material may be anything from a dense-graded crushed aggregate to a granular soil having a relatively high percentage of dust. At the time of mixing, the aggregate may either be damp, air-dried, or artificially heated and dried. Cold mix asphalt may be used for surface, base, or sub-base courses if the pavement is properly designed. Cold mix surface courses are suitable for light and medium traffic; but such surface courses normally require a seal coat or hot asphalt overlay as surface protection. When used in the base or sub-base, they may be suitable for all types of traffic.

Asphalt may also be pulverized to produce powdered asphalt. In the process, the asphalt is crushed and passed through a series of fine mesh sieves to ensure uniform size of the granules. Powered asphalt can be mixed with road oil and aggregate for pavement construction. Heat and pressure in the road slowly amalgamates the powder with the aggregate and binding oil, and the substance hardens to a consistency similar to regular asphalt cement.

Briefly, aggregates (mineral aggregates) are hard, inert materials such as sand, gravel, crushed stone, slag, or rock dust. Properly selected and graded aggregates are mixed with the cementing medium asphalt to form pavements. Aggregates are the principal load-supporting components of an asphalt pavement and constitute 90 to 95% w/w or 75 to 85% v/v of the mixture.

Selection of an aggregate material for use in an asphalt pavement depends on the availability, cost, and quality of the material, as well as the type of construction for which it is intended. To determine if an aggregate material is suitable for use in asphalt construction, it should be evaluated in terms of the following properties: (1) size and grading, (2) cleanliness, (3) toughness, (4) soundness, (5) particle shape, (6) surface texture, (7) adsorption, and (8) stripping.

In terms of *size and grading*, the maximum size of an aggregate is the smallest sieve through which 100 percent of the material will pass. The end use of the asphalt-aggregate mix determines not only the maximum aggregate size, but also the desired gradation (distribution of sizes smaller than the maximum). Asphalt must also be *clean* since foreign or deleterious substances can render asphalt-aggregate mixtures unsuitable for paving. In addition, *toughness* or *hardness* is the ability of the aggregate to resist crushing or disintegration during mixing, placing, and compacting; or under traffic loading. *Soundness*, although similar to toughness, soundness is the ability of the aggregate to resist deterioration caused by natural elements such as the weather.

The *shapes of aggregate particles* influences the strength of the asphalt-aggregate mixture as well as workability and density achieved during compaction—when compacted, irregular particles such as crushed stone tend to lock together and resist displacement. Workability and pavement strength are influenced by *surface texture* of the aggregate—a rough, sandpapery texture results in a higher strength than a smooth texture. Although smooth-faced aggregates are easy to coat with an asphalt film, they are generally not as good as rough surfaces—it is more difficult for the asphalt adsorb on to a smooth surface.

The porosity of an aggregate permits the aggregate to *adsorb* asphalt and form a bond between the particle and the asphalt. A degree of porosity is desired, but aggregates that are highly absorbent are generally not used. *Stripping* occurs when the asphalt film separates from the aggregate because of the action of water. Aggregates coated with too much dust also can cause poor bonding which results in stripping. The problems of aggregates readily susceptible to stripping paving mixes can be mitigated by use of an anti-stripping agent.

For paving purposes, asphalt and aggregate (the mineral matrix) are combined in a mixing facility where they are heated, proportioned, and mixed to produce the desired paving mixture. Hot-mix facilities may be permanently located (*stationary facilities*) or it may be portable and moved from job to job. Hot-mix facilities may be classified as either a *batch facility* or a *drum-mix facility* and both can be either stationary or portable units.

Batch-type hot-mixing facilities use different size fractions of hot aggregate which are drawn in proportional amounts from storage bins to make up one batch for mixing. The combination of aggregates is dumped into a mixing chamber (pug mill) in which the asphalt, which has also been weighed, is then thoroughly mixed with the aggregate. After mixing, the material is then emptied from the pug mill into trucks, storage silos, or surge bins. The drum-mixing process heats and blends the aggregate with asphalt all at the same time in the drum mixer.

When mixing is complete, the hot-mix is then transported to the paving site and spread in a partially compacted layer to a uniform, even surface with a paving machine. While still hot, the paving mixture is further compacted by heavy rolling machines to produce a smooth pavement surface.

The quality of an asphalt product is affected by the inherent properties of the crude oil crude oil from which it was produced. Different oil fields and areas produce crude oils with very different characteristics and the refining method also affects the quality of the asphalt cement. For engineering and construction purposes, there are three important factors to consider: (1) consistency, also called the viscosity or the degree of fluidity of asphalt at a particular temperature, purity, and safety, (2) purity, and (3) ductility.

The *consistency* or viscosity of asphalt varies with temperature, and asphalt is graded based on ranges of consistency at a standard temperature. Careless temperature and mixing control can cause more hardening damage to asphalt cement than many years of service on a roadway. A standardized viscosity or penetration test is commonly specified to measure paving asphalt consistency— air-blown asphalts typically use a softening point test.

The *purity* of asphalt can be easily tested since (by definition) it is composed almost entirely of material which is soluble in carbon disulfide. Refined asphalts are usually more than 99.5% soluble in carbon disulfide and any impurities that remain are inert. Because of the hazardous flammable nature of carbon disulfide, trichloroethylene (TCE), which is also an excellent solvent for asphalt, is also used in the test for purity by solubility.

Asphalt must be free of water or moisture as it leaves the refinery. However, transports loading the asphalt may have moisture present in their tanks. This can cause the asphalt to foam when it is heated above 100°C (212°F), which is a safety hazard. Specifications typically require that asphalt does not foam at temperatures up to 175°C (347°F). Asphalt, if heated to a sufficiently high temperature, will release fumes which will flash in the presence of a spark or open flame (the *flash point*) and should be well above temperatures normally used in paving operations. Because of the possibility of asphalt foaming and to ensure an adequate margin of safety, the flashpoint of the asphalt is measured and controlled.

Ductility is another important engineering property of asphalt, which is a measure of the ability of the asphalt to be pulled, drawn, or deformed. For asphalt, the presence or absence of ductility is usually more important than the actual degree of ductility because asphalt with a high degree of ductility is also more temperature sensitive. Ductility is measured by an "extension" test, whereby a standard asphalt cement briquette molded under standard conditions and dimensions is pulled at a standard temperature (normally 25°C, 77°F) until it breaks under tension. The elongation at which the asphalt cement sample breaks is a measure of the ductility of the sample.

Finally, in keeping with environmental protection issues (Chapter 28, Chapter 29, Chapter 30) and the stringent codes limiting water flows and particulate and smoke emissions from oil refineries, asphalt processing plants are also subject to these laws. The products (or by-products) formed during asphalt production, if unchecked, create odoriferous fumes and pollutants which will stain and darken the air. Pollutants emitted from asphalt production are controlled by enclosures which capture the exhaust and then recirculate it through the heating process. This not only eliminates the pollution but also increases the heating efficiency of the process.

Higher costs of asphalt cement, stone, and sand have forced the industry to increase efficiencies and recycle old asphalt pavements. In asphalt pavement recycling, materials reclaimed from old pavements are reprocessed along with new materials. The three major categories of asphalt recycling are (1) hot-mix recycling, where reclaimed materials are combined with new materials in a central plant to produce hot-mix paving mixtures, (2) cold-mix recycling, where reclaimed materials are combined with new materials either onsite or at a central plant to produce cold-mix base materials, and (3) surface recycling, a process in which the old asphalt surface pavement is heated in place, scraped down (*scarified*), remixed, re-laid, and rolled. Organic asphalt recycling agents may also be added to help restore the aged asphalt to desired specifications.

Because of solvent evaporation and volatility, use of cutback asphalts, especially rapid-cure cutback asphalt which use gasoline or naphtha, is becoming more restricted or prohibited while emulsified asphalts (in which only the water evaporates) are becoming more popular because of various environmental regulations.

Asphalt Aging

The destruction of an asphalt pavement is due, in large part, to the aging of the asphalt binder, which made a contribution to bring down the durability and service life of asphalt pavements. The factors affecting asphalt aging included (1) characteristics of the asphalt and its content in the mix, (2) the nature of aggregate and particle size distribution, (3) the void content of the mix, and (4) production related factors such as temperature and time. The most important aging related modes of failure were traffic and thermally induced cracking, and raveling.

The main aging mechanism was an irreversible one, characterized by chemical changes of the binder, which in turn had an impact on the rheological properties. The processes contributing to this type of aging include oxidation, loss of volatile components and exudation (migration of oily components from the asphalt into the aggregate). Asphalt aging occurred during the mixing and construction process as well as during long-term service in the road. However, the circumstances at different aging stages vary considerably.

In order to evaluate the aging properties of asphalt, a number of laboratory test methods ate available such as (1) the thin-film oven test (TFOT), (2) the rolling thin-film oven test (RTFOT), (3) the pressure aging vessel (PAV) test method, and (4) the ultraviolet light (UV) aging test method. In these tests, asphalt aging is accelerated by increasing temperature, decreasing asphalt film thickness, increasing oxygen pressure, or applying various combinations of these factors.

2.2.10 Coke

Crude oil coke (also referred to as petroleum coke) is the residue left by the destructive distillation (thermal cracking or coking) of crude oil residua. The coke formed in catalytic cracking operations is usually non-recoverable because of adherence to the catalyst as it is often employed as fuel for the process.

Crude oil coke is characterized by its chemical composition and physical characteristics. The chemical composition of crude oil coke is dependent upon the composition of the feedstocks that are used in the coking process, which in turn are dependent upon the composition of the crude oil from which they are derived. The metals and sulfur composition of calcined coke is directly dependent upon the composition of the green coke from which it was produced.

The composition of crude oil coke varies with the source of the crude oil, but in general, large amounts of high-molecular-weight complex hydrocarbon derivatives (rich in carbon but correspondingly poor in hydrogen) make up a high proportion. The solubility of crude oil *coke* in carbon disulfide has been reported to be as high as 50 to 80%, but this is in fact a misnomer, since the coke is the insoluble, honeycomb material that is the end product of thermal processes.

Three physical structures of coke can be produced by delayed coking: (1) shot coke, (2) sponge coke, or (3) needle coke.

Shot coke is an abnormal type of coke resembling small balls. Due to mechanisms not well understood the coke from some coker feedstocks forms into small, tight, non-attached clusters that look like pellets, marbles or ball bearings. It usually is a very hard coke, i.e., low Hardgrove grindability index (Speight, 2013). Such coke is less desirable to the end users because of difficulties in handling and grinding. It is believed that feedstocks high in asphaltene constituents and low API favor shot coke formation. Blending aromatic materials with the feedstock and/or increasing the recycle ratio reduces the yield of shot coke. Fluidization in the coke drums may cause formation of shot coke.

Occasionally, the smaller *shot coke* may agglomerate into ostrich egg sized pieces. While shot coke may look like it is entirely made up of shot, most shot coke is not 100% shot. Interestingly, even sponge coke may have some measurement of embedded shot coke. The test for such a measurement is not precise but a low shot coke percentage is sometimes specified for anode grades of crude oil coke. In the case of fuel coke production, there is a general belief that shot structure of coke means that the coker is being run at the most economic conditions.

Sponge coke is the common type of coke produced by delayed coking units. It is in a form that resembles a sponge and has been called honeycombed. Sponge coke, mostly used for anode-grade is dull and black, having porous, amorphous structure.

Needle coke (*acicular coke*) is a special quality coke produced from aromatic feed stocks is silver-gray, having crystalline broken needle structure, and is believed to be chemically produced through cross linking of condensed aromatic hydrocarbon derivatives during coking reactions. It has a crystalline structure with more unidirectional pores and is used in the production of electrodes for the steel and aluminum industries and is particularly valuable because the electrodes must be replaced regularly.

Crude oil is insoluble on organic solvents and has a honeycomb-type appearance and can be categorized generally as either (1) green coke or (2) calcined coke. The initial product of the coking process, green coke, is used as a solid fuel. Further processing of green coke at higher temperatures and higher pressures results in the production of calcined coke, which is used in: 1) the manufacture of electrodes, (2) smelting applications, (3) graphite electrode production, or (4) for the carbonization of steel.

Typical parameters measured to define the chemical composition of crude oil coke are: (1) ash % w/w, (2) sulfur % w/w, (3) extractable material % w/w, and (4) nickel ppm and vanadium ppm. Because of the lower temperature used in its production, green or fuel-grade coke contains higher levels of residual hydrocarbon than other grades of coke. The calcining process removes essentially the entire extractable material.

The use of coke as a fuel must proceed with some caution with the acceptance by refiners, of the heavier crude oils as refinery feedstocks. The higher contents of sulfur and nitrogen in these oils mean a product coke containing substantial amounts of sulfur and nitrogen. Both of these elements will produce unacceptable pollutants—sulfur oxides and nitrogen oxides during combustion. These elements must also be regarded with caution in any coke that is scheduled for electrode manufacture and removal procedures for these elements are continually being developed.

Because crude oil coke is a solid, fairly inert material there has not been much concern about its environmental effects. The typical battery of tests used to measure a chemical's impact on the environment, such as breakdown by sunlight, stability in water, breakdown in the soil and volatility, cannot be measured for crude oil coke. In addition, there are suggestions that any residual oil associated with crude oil coke either does not come off, or is present at amounts too low to cause harmful effects.

Although the potential for coke to environmental damage is often thought to be very low, green coke, with higher oil content and the potential for leaching of metals from the coke during periods of snow melt, rain or acid rain, might be expected to have a lower hazard than calcined coke.

2.3 Petrochemical Products

A major group of products from crude oil (petrochemicals) are the basis of a major industry. They are, in the strictest sense different to crude oil products insofar as the petrochemicals are the basic building blocks of the chemical industry (Table 2.5).

A *petrochemical* is any chemical (as distinct from fuels and crude oil products) manufactured from crude oil (and natural gas) and used for a variety of commercial purposes (Parkash, 2003; Gary et al., 2007; Speight, 2014; Hsu and Robinson, 2017; Speight, 2017). The definition, however, has been broadened to include the whole range of aliphatic, aromatic, and naphthenic organic chemicals, as well as carbon black and inorganic materials such as sulfur and ammonia. Crude oil and natural gas are made up of hydrocarbon molecules, which comprise one or more carbon atoms, to which hydrogen atoms are attached. Currently, oil and gas are the main sources of the raw materials because they are the least expensive, most readily available, and can be processed most easily into the primary petrochemicals (Table 3–9). Primary petrochemicals include: olefin derivatives (ethylene, propene, and butadiene), aromatic derivatives (benzene, toluene, and the isomers of xylene), and methanol.

The petrochemical industry is a large and complex source category that is very difficult to define because its operations are intertwined functionally or physically with (in addition to the crude oil refining industry) other industries. Petrochemical feedstocks can be classified into three general

Table 2.5: Illustration of the Production of Petrochemical Starting Materials from Natural Gas and Crude oil.

Feedstock	Process	Product
Natural gas	refining	methane
		ethane
		propane
		butane
Crude oil	Distillation	Light ends
		methane
		ethane
		propane
		butane
	Catalytic cracking	ethylene
		propylene
		butylenes
		higher olefin derivatives
	Catalytic reforming	Benzene
		Toluene
		Xylenes
	Coking	Ethylene
		Propylene
		Butylene isomers
		higher olefin derivatives

groups: (1) olefin derivatives, (2) aromatic derivatives, and (3) methanol. A fourth group includes inorganic compounds and synthesis gas (mixtures of carbon monoxide and hydrogen). Thence, the petrochemical industry is involved in the production of several chemicals that fit into one or more of the following three categories: (1) basic raw materials, (2) chemical intermediates, and (3) end products.

Petrochemical intermediates are generally produced by chemical conversion of primary petrochemicals to form more complicated derivative products. Petrochemical derivative products can be made in a variety of ways: directly from primary petrochemicals; through intermediate products that still contain only carbon and hydrogen; and, through intermediates that incorporate chlorine, nitrogen, or oxygen in the finished derivative. In some cases, they are finished products; in others, more steps are needed to arrive at the desired composition.

The petrochemical industry also includes the treatment of hydrocarbon streams from the crude oil refining industry and natural gas liquids from the oil and gas production industry. Some of the raw materials used in the petrochemical industry include crude oil, natural gas, ethane, hydrocarbon derivatives, naphtha, heavy fractions, kerosene, and gas-oil. Natural gas and crude oil are the main feedstocks for the petrochemical industry. That is why about 65 percent of petrochemical facilities are located at or near refineries.

In the strictest sense, petrochemicals are also crude oil products but they are typically individual chemicals such as that are used as the basic building blocks of the chemical industry and include chemicals such as ethylene ($CH_2 = CH_2$), propylene ($CH_3CH = CH_2$), the butylene isomers (butene-1, $CH_3CH_2CH = CH_2$, butene-2 $CH_3CH = CHCH_3$) butadiene ($CH_2 = CHCH = CH_2$), and a variety of aromatic hydrocarbon derivatives, which include such as benzene (C_6H_6), toluene (C_6H5CH_3), ethylene benzene ($C_6H_5CH_2CH_3$), and the xylene isomers (dimethyl benzene derivatives, 1,2-dimethyl benzene (ortho-xylene, o-xylene), 1.3-dimethyl benzene (meta-xylene, m-xylene), and

14-dimethyl benzene (para-xylene, p-xylene), $H_3CC_6H_4CH_3$ which are used to produce polymers, plastics, and pharmaceuticals as examples of the potential products than can be produced from these monomers.

The structural formulas of the constituents of the BTEX fraction are as follows and all of the BTEX derivatives are liquid art room temperature and exhibit the following properties:

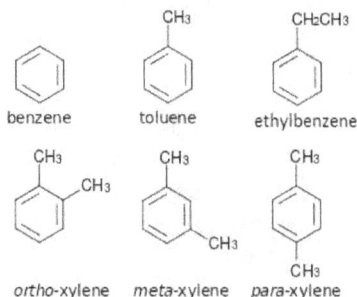

	Benzene	**Toluene**	**Ethylbenzene**	***p*-Xylene**	***m*-Xylene**	***o*-Xylene**
Molecular formula	C_6H_6	C_7H_8	C_8H_{10}	C_8H_{10}	C_8H_{10}	C_8H_{10}
Molecular mass, $g \cdot mol^{-1}$	78.12	92.15	106.17	106.17	106.17	106.17
Boiling point, °C	80.1	110.6	136.2	138.4	139.1	144.4
Melting point, °C	5.5	−95.0	−95.0	13.3	−47.9	−25.2

Thus, the petrochemical industry produces solvents and chemicals of various grades or specifications which are used to produce industrial organic chemicals, including alcohols and aldehydes, butylene, butadiene, ethylene, propylene, toluene, styrene, acetylene, benzene, ethylene oxide, ethylene glycol, acrylonitrile, acetone, acetic acid, acetic anhydride, and ammonia. Petrochemicals are widely used in agriculture, in the manufacture of plastics, synthetic fibers, and explosives, and in the aircraft and automobile industries. The industrial organic chemicals produced from petrochemicals are employed in downstream industries including plastics and resins, synthetic fibers, elastomers, plasticizers, explosives, surface-active agents, dyes, surface coatings, pharmaceuticals, and pesticides.

2.4 Occurrence and Uses of Common Hydrocarbon Derivatives

Hydrocarbon derivatives, are (by definition) compounds containing carbon and hydrogen *only*, the hydrocarbon content may be as high as 97% by weight in a conventional (lighter) paraffinic crude oil or about 50% by weight in heavy crude oil and less than 30% by weight in tar sand bitumen. However, within the hydrocarbon constituents, there is also considerable variation of chemical type and molecular weight. Crude oil hydrocarbon derivatives may be paraffinic, alicyclic, or aromatic and occur in varying concentrations within the different fraction of a single crude oil. Thus, the constituents of crude oil that occur in varying amounts depend on the source and character of the oil (Speight, 2014).

Alkane derivatives

Alkane derivatives (*also called normal paraffins or n-paraffins*) are the major constituents of crude oil and the major constituents of most crude oil products. Alkane derivatives are also called saturates or saturated hydrocarbon derivatives. These constituents are characterized by unbranched (linear) or branched (non-linear) chains of carbon atoms with attached hydrogen atoms; alkane derivatives contain no carbon-carbon double bonds (hence the designation *saturated*), are generally insoluble in cold water—examples of alkane derivatives are pentane (C_5H_{12}) and heptane (C_7H_{16}). Linear and branched alkane derivatives are sometimes called paraffins. These compounds have an empirical formula of C_nH_{2n+2}.

Linear alkane derivatives (n-paraffin derivatives) are compounds in which the carbon atoms are arranged in a straight chain or row, whereas the branched-chain alkane derivative are often referred to as isoparaffins or iso-alkane derivatives. Examples of these alkane derivatives are methane, ethane, propane, butane, pentane, hexane, and heptane. Methane is the simplest hydrocarbon consisting of one carbon and four hydrogen atoms (CH_4) molecules and is a colorless, odorless gas at room temperature while ethane has two carbon molecules and six hydrogen atoms (C_2H_6) molecules. Like methane, ethane is a colorless, odorless gas at room temperature.

Propane (C_3H_8) and butane (C_4H_{10}) can be liquefied at fairly low pressures, and are used, for example, in the propane gas burner, or as propellants in aerosol sprays. The alkane derivatives from pentane to octane are highly volatile liquids and good solvents for nonpolar substances. They are used as fuels in internal combustion engines. Alkane derivatives from nonane (C_9H_{20}) to hexadecane ($C_{16}H_{34}$) are liquids of higher viscosity and are used in diesel fuel and in aviation fuel (kerosene). The alkane derivatives with 17 to 35 ($C_{17}H_{36}$ to ($C_{35}H_{72}$) are typically solids at room temperature and are the major constituents of waxes.

Cycloparaffin derivatives (also called naphthenes or naphthene derivatives)

The cycloalkane derivatives (also known as naphthene derivatives) are a special class of alkane derivatives with ring structure and have the formula C_nH_{2n} and all carbon-hydrogen (C-H) bonds are saturated. As such, naphthenic hydrocarbon derivatives in crude oil are also relatively stable compounds. The boiling point and density of the naphthene derivatives are higher than those of alkane derivatives having the same number of carbon atoms. Naphthene derivatives commonly present in crude oil are rings with five or six carbon atoms which typically carry alkyl substituents attached to them. Multi-ring naphthene derivatives are present in the higher-boiling (higher molecular weight) fractions of the crude oil (Speight, 2014).

The predominant mono-cycloalkane derivatives in crude oil are in the cyclopentane series, having five carbon atoms in the ring, and in the cyclohexane derivatives, having a six-membered ring. The monocyclic (single ring) derivatives and the bicyclic (two-ring) naphthene derivatives are generally the major types of cycloparaffin derivatives in the lower-boiling fractions of crude oil and the boiling point and molecular weight of these derivatives are increased by the presence of alkyl chains on the rings. The higher-boiling (higher molecular weight) fractions in crude oil, such as the lubricating oil fraction, may contain two to six rings per molecule (Speight, 2014). Also, as the molecular weight (or boiling point) of the crude oil fraction increases, there is an accompanying increase in the amount of cycloparaffin derivatives in the fraction.

In the asphaltic (naphthenic) crude oils, the gas oil fraction can contain considerable amounts of naphthenic ring systems. However, as the molecular weight of the fraction increases, the occurrence of condensed naphthene ring systems and alkyl-substituted naphthene ring systems also increase (Speight, 2014, 2017).

Olefin derivatives (also called alkene derivatives)

Olefin derivatives (i.e., alkene derivatives, RCH = CHR) are acyclic (branched or unbranched) hydrocarbon derivatives having one carbon-to-carbon double bond (> C = C <) with the general molecular formula C_nH_{2n}. The alkene derivatives can vary in the number of double bonds per molecule, making them mono-olefin derivatives, diolefin derivatives, tri-olefin derivatives and can also be categorized as either cyclic or acyclic. The cyclic olefin derivatives have a double bond between carbon atoms which makes up a closed ring of compounds while acyclic (non-cyclic) olefin derivatives form an open-chain molecular system.

The presence of olefin derivatives in crude oil has been under dispute for many years. Olefin derivatives are reactive hydrocarbon derivatives that are unlikely to survive unchanged in nature. However, there are investigators who claim that olefin derivatives are actually present but, in fact, these claims usually refer to distilled fractions, and it is very difficult to entirely avoid cracking during the distillation process (Speight, 2014, 2017). Nevertheless, evidence for the presence of considerable proportions of olefin derivatives in Pennsylvanian crude oils has been obtained; spectroscopic and chemical methods showed that the crude oils, as well as all distillate fractions, contained up to 3% w/w olefin derivatives. Hence, although the opinion that crude oil does not contain olefin derivatives requires some revision, it may be reasonable to assume that the Pennsylvania crude oils may hold an exceptional position and that olefin derivatives are present in crude oil in only a few special cases. The presence of diene derivatives (such as $RCH = CH = CHR'$) and acetylene derivatives (such as $RC \equiv CR'$) is considered to be extremely unlikely because of the tendency for these molecular species to react because of the susceptibility of these unsaturated compounds to oxidation over time (Speight, 2014, 2017).

Olefin derivatives are produced by using thermal hydrocarbon cracking at high heat to refine gaseous or liquid hydrocarbon feedstock, such as naphtha and natural gas condensates like ethane and propane, into smaller hydrocarbon chains. The olefin derivatives can be sold as-is or recombined into new configurations to form many high-value chemical compounds, including polyethylene and synthetic rubber (Speight, 2019b).

Single-ring aromatic derivatives

Aromatic constituents are characterized by the presence of an aromatic ring with six carbon atoms and are considered to be the most acutely toxic component of crude oil constituents because of their association with chronic and carcinogenic effects; the rings can carry alkyl of naphthene substituents further distinguishes the aromatic constituents—low molecular weight aromatic constituents may have a noticeable solubility in water, increasing the potential for exposure to aquatic resources.

Aromatic derivatives are compounds that contain at least one benzene ring. The benzene ring is very stable and does not crack to smaller components. Aromatic derivatives are not a preferred feedstock because few of the molecules will crack.

Multi-ring aromatic derivatives

Aromatic constituents with two or more condensed rings are referred to as polynuclear aromatic hydrocarbon derivatives (PNAs, sometime PAHs); the most abundant aromatic hydrocarbon families in crude oil and crude oil products have two and three fused rings with one to four carbon atom alkyl group substitutions—condensed aromatic constituents with more than two condensed rings (three-to-five) are also present in the higher boiling fractions of crude oil but higher condensed systems are unlikely (Speight, 2014).

The configuration of rings in the polynuclear aromatic hydrocarbon derivatives strongly influences the physical and chemical properties of these compounds. For example, the three-ring aromatic derivatives anthracene and phenanthrene have significantly different properties. Anthracene is used in the manufacture of certain dyes. Steroids, a large group of naturally occurring substances, contain the phenanthrene structure. In crude oil, the polynuclear aromatic hydrocarbon derivatives exist mostly as alkyl substituted ring systems such that the substituent alkyl groups (e.g., methyl, ethyl) replace (substitute for) the hydrogen atoms on the rings.

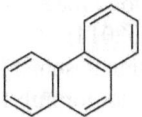

Phenanthrene

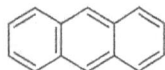

Anthracene

A typical crude oil contains approximately 1% polynuclear aromatic hydrocarbon derivatives. Concentrations of total carcinogenic polynuclear aromatic hydrocarbon derivatives [such as benzo(a)pyrene] range from 12 ppm to < 100 ppm.

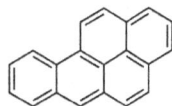

Benzo(a)pyrene

Fresh crude oil will contain a fraction of volatile hydrocarbon derivatives, some of which may pose a threat to human health such as benzene, toluene, xylenes, and other aromatic derivatives. However, the relative mass fraction of volatile hydrocarbon derivatives in crude oil is significantly less than that found in crude oil distillate products (such as naphtha, which is often used as a blend stock or precursor to gasoline). The percent by volume (% v/v) of benzene in gasoline may range up to 3% (30,000 ppm), while the benzene content of crude oil is usually on the order of 0.2% (2,000 ppm) or less. As a result of the lower percentage of volatile aromatic constituents in a product such as gasoline, vapor emissions from crude oil-contaminated soils are expected to be (but not always, depending upon the crude oil) much less than potential emissions from gasoline-contaminated soils.

If the only constituents of crude oil were the hydrocarbon derivatives, the complexity is further illustrated by the number of potential isomers, i.e., molecules having the same atomic formula, that can exist for a given number of paraffinic carbon atoms and that increases rapidly as molecular weight increases For example:

Butane (C_4H_{10})   :   2 possible isomers
Pentane (C_5H_{12})   :   3 possible isomers
Hexane (C_6H_{14})   :   5 possible isomers
Heptane (C_7H_{16})   :   8 possible isomers
Octane (C_8H_{18})   :   18 possible isomers
Nonane (C_9H_{20})   :   35 possible isomers
Decane ($C_{10}H_{22}$)   :   75 possible isomers
Undecane ($C_{11}H_{24}$) :   159 possible isomers
Dodecane ($C_{12}H_{26}$) :   355 possible isomers
Eicosane ($C_{20}H_{42}$) :   336,319 possible isomers

This same increase in number of isomers with molecular weight also applies to the other molecular types that are present in crude oil (Speight, 2014). Since the molecular weights of the molecules found in crude oil can vary from that of methane (CH_4; molecular weight: 16) to several thousand (Speight, 2001, 2002, 2005, 2014), it is clear that the higher boiling (higher molecular weight) nonvolatile fractions can contain virtually unlimited numbers of molecules. However, not all isomers of the above hydrocarbon derivatives occur in crude oil. In reality the number of molecules in any specified fraction is limited by the nature of the precursors of crude oil, their chemical structures, and the physical conditions that are prevalent during the maturation processes in which the organic precursors are converted to the constituents of crude oil (Speight, 2014).

By way of definition and clarity, a structural isomer (also referred to as a constitutional isomer) is an isomer one in which two or more hydrocarbon derivatives have the same molecular formulas but different structures. For example, the two isomers of pentane have the same chemical formula (C_5H_{12}), but are different insofar as the location of the methyl group is different:

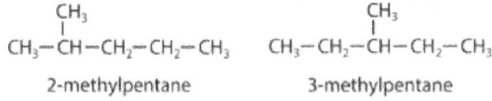

$$CH_3-\overset{\overset{\displaystyle CH_3}{|}}{CH}-CH_2-CH_2-CH_3 \qquad CH_3-CH_2-\overset{\overset{\displaystyle CH_3}{|}}{CH}-CH_2-CH_3$$

2-methylpentane   3-methylpentane

Similarly, olefin derivatives (also referred to as alkene derivatives) can also demonstrate structural isomerism. In the olefin series, there are multiple structural isomers based on where in the chain the double bond occurs as illustrated for the structural formula of 1-butene and the structural formula of 2-butene:

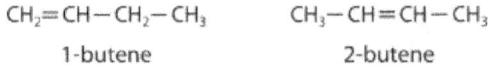

$$CH_2=CH-CH_2-CH_3 \qquad\qquad CH_3-CH=CH-CH_3$$

1-butene   2-butene

By definition, the number in the name of the alkene derivative refers to the lowest numbered carbon in the chain that is a part of the double bond.

However, olefin derivatives are not typical constituents of crude oil because of the reactivity of these hydrocarbon derivatives but are frequent constituents of the products of thermal (non-hydrotreating) processes.

Finally, a stereoisomer has the same connectivity of the constituent atoms in the molecule but has a different arrangement in three-dimensional space. Moreover, there are different classifications of stereoisomers depending on the manner in which arrangements differ from one another. Using 2-butene as the example:

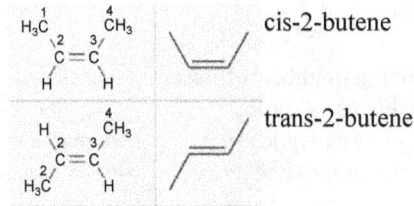

cis-2-butene

trans-2-butene

Stereoisomers will also exhibit different boiling points and melting points. Thus:

Isomer	Boiling point	Melting point
cis-2-butene	3.7°C	−139°C
trans-2-butene	0.9°C	−105°C

This differential in the boiling points and melting points can, in some case, influence the method selected for environmental cleanup.

2.5 Refinery Waste

The chemicals in crude oil and crude oil products vary from (chemically speaking) simple hydrocarbon derivatives of low-to-medium molecular weight to organic compounds containing sulfur, oxygen, and nitrogen, as well as compounds containing metallic constituents, particularly vanadium nickel, iron, and copper. Many of these latter compounds are of indeterminate molecular weight.

Crude oil refineries are complex, but integrated, unit process operations that produce a variety of products from various feedstock blends (Figure 4.1) (Meyers, 1997; Parkash, 2003; Gary et al., 2007; Speight, 2014; Hsu and Robinson, 2017; Speight, 2017). During crude oil refining, refineries use and generate an enormous amount of chemicals, some of which are present in air emissions, wastewater, or solid wastes. Emissions are also created through the combustion of fuels, and as byproducts of chemical reactions occurring when crude oil fractions are upgraded. A large source of air emissions is, generally, the process heaters and boilers that produce carbon monoxide, sulfur oxides, and nitrogen oxides, leading to pollution and the formation of acid rain.

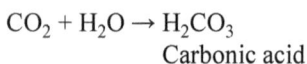

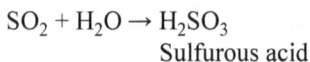

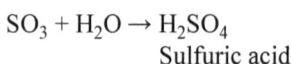

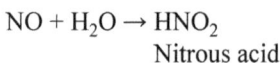

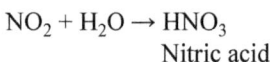

$$CO_2 + H_2O \rightarrow H_2CO_3$$
$$\text{Carbonic acid}$$

$$SO_2 + H_2O \rightarrow H_2SO_3$$
$$\text{Sulfurous acid}$$

$$2SO_2 + O_2 \rightarrow 2SO_3$$
$$SO_3 + H_2O \rightarrow H_2SO_4$$
$$\text{Sulfuric acid}$$

$$NO + H_2O \rightarrow HNO_2$$
$$\text{Nitrous acid}$$

$$2NO + O_2 \rightarrow NO_2$$
$$NO_2 + H_2O \rightarrow HNO_3$$
$$\text{Nitric acid}$$

Hence, there is the need for gas-cleaning operations on a refinery site so that such gases are cleaned from the gas stream prior to entry into the atmosphere.

In addition, some processes create considerable amounts of particulate matter and other emissions from catalyst regeneration or decoking processes. Volatile chemicals and hydrocarbon derivatives are also released from equipment leaks, storage tanks, and wastewaters. Other cleaning units such as the installation of filters, electrostatic precipitators and cyclones can mitigate part of the problem.

Process wastewater is also a significant effluent from a number of refinery processes. Atmospheric and vacuum distillation create the largest volumes of process wastewater, about 26 gallons per barrel of oil processed. Fluid catalytic cracking and catalytic reforming also generate considerable amounts of wastewater (15 and 6 gallons per barrel of feedstock, respectively). A large portion of wastewater from these three processes is contaminated with oil and other impurities and must be subjected to primary, secondary and sometimes tertiary water treatment processes, some of which also create hazardous waste.

Wastes, residua and by-products are produced by a number of processes. Residuals produced during refining can be but are not necessarily wastes. They can be recycled or regenerated, and in many cases do not become part of the waste stream but are useful products. For example, processes utilizing caustics for neutralization of acidic gases or solvent (e.g., alkylation, sweetening/chemical treating, and lubricating oil manufacture) create the largest source of residuals in the form of spent caustic solutions. However, nearly all of these caustics are recycled.

The treatment of oily wastewater from distillation, catalytic reforming and other processes generates the next largest source of residuals in the form of biomass sludge from biological treatment and pond sediments. Water treatment of oily wastewater also produces a number of sludge materials associated with oil-water separation processes. Such sludge is often recycled in the refining process and are not considered wastes.

Catalytic processes (fluid catalytic cracking, catalytic hydrocracking, hydrotreating, isomerization, ethers manufacture) also create some residuals in the form of spent catalysts and catalyst fines or particulates. The latter are sometimes separated from exiting gases by electrostatic precipitators or filters. These are collected and disposed of in landfills or may be recovered by off-site facilities.

Residua, that are produced by distillation that is a concentration process, contain significantly less hydrocarbon constituents that the original crude oil. The constituents of residua may, depending on the crude oil, may be molecular entities of which the majority contains at least one heteroatom.

Typical refinery products include (1) natural gas and liquefied petroleum gas (LPG), (2) solvent naphtha, (3) kerosene, (4) diesel fuel, (5) jet fuel, (6) lubricating oil, (7) various fuel oils, (8) wax, (9) residua, and (10) asphalt (Parkash, 2003; Gary et al., 2007; Speight, 2014; Hsu and Robinson, 2017; Speight, 2017). A single refinery does not necessarily produce all of these products. Some refineries are dedicated to particular products, e.g., the production of gasoline or the production of lubricating oil or the production of asphalt. However, the issue is that refineries also produce a variety of waste products (Table 4.1) (EPA, 1995) that must be disposed in an environmentally acceptable manner.

Waste treatment processes also account for a significant area of the refinery, particularly sulfur compounds in gaseous emissions together with various solid and liquid extracts and wastes generated during the refining process. The refinery is therefore composed of a complex system of stills, cracking units, processing and blending units and vessels in which the various reactions take place, as well as packaging units for products for immediate distribution to the retailer, e.g., lubricating oils. Bulk storage tanks usually grouped together in tank farms are used for storage of both crude and refined products. Other tanks are used in the processes outlined, e.g., treating, blending and mixing whilst others are used for spill and fire control systems. A boiler and electrical generating system usually operate for the refinery as a whole.

There are several hundred individual hydrocarbon chemicals defined as crude oil-based. Furthermore, each crude oil product has its own mix of constituents because crude oil varies in composition from one reservoir to another and this variation may be reflected in the finished product(s).

Crude oil hydrocarbon derivatives are environmental contaminants but they are not usually classified as hazardous wastes (Irwin, 1997). Soil and groundwater crude oil hydrocarbon contamination has long been of concern and has spurred various analytical and site remediation developments, e.g., risk-based corrective actions. In some instances, it may appear that such cleanup operations were initiated with an incomplete knowledge of the charter and behavior of the contaminants. The most appropriate first assumption is that the spilled constituents are toxic to the ecosystem. The second issue is an investigation of the products of the spilled material to determine an appropriate cleanup method. The third issue is whether or not the chemical nature of the constituents has changed during the time since the material was released into the environment. If it has, a determination must be made of the effect of any such changes on the potential cleanup method.

Despite the large number of hydrocarbon derivatives found in crude oil products and the widespread nature of crude oil use and contamination, many of the lower boiling constituents are well characterized in terms of physical properties but only a relatively small number of the compounds are well characterized for toxicity. The health effects of some fractions can be well characterized, based on their components or representative compounds (e.g., light aromatic fraction benzene-toluene-ethylbenzene-xylene isomers). However, higher molecular weight (higher boiling) fractions have far fewer well-characterized compounds.

Typical refinery products include (1) natural gas and liquefied petroleum gas (LPG), (2) solvent naphtha, (3) kerosene, (4) diesel fuel, (5) jet fuel, (6) lubricating oil, (7) various fuel oils, (8) wax, (9) residua, and (10) asphalt. A single refinery does not necessarily produce all of these products but any refinery may also produce a variety of waste chemicals that must be disposed in an environmentally acceptable manner. Example of such products (listed Alphabetically) are (1) acid sludge, (2) spent caustic, (3) spent acids, and (4) spent catalysts. In addition, excluding accidental spills, there are four main sources of pollutant emissions in the production of petrochemicals: (1) process vent discharges, (2) equipment leaks, (3) secondary sources, and (4) storage.

Process vent discharges can be from reactor vessels, recovery columns, and other process vessels. Equipment leaks include pump seals, process valves, compressors, safety relief valves, flanges, open-ended lines, and sampling connections. Secondary sources include process and other waste streams. However, little information is available regarding amounts of pollutant emissions from the entire petrochemical industry. Many petrochemical processes are located at or near crude oil refining operations; therefore, many of the air pollutants and hazardous wastes generated by the crude oil industry are also present at petrochemical facilities and most reports refer to facility-wide pollution.

More pertinent to the present text, the waste streams from the petrochemical industry are quite similar to those of the crude oil refining industry. Limited data are available, but almost a/l assume waste management operations and facilities are probably of the same degree of sophistication as those of the crude oil refining industry.

Waste water from crude oil refining consist of cooling water, process water, storm water, and sanitary sewage water (Speight, 2005, 2014, 2017). A large portion of water used in crude oil refining is used for cooling. Most cooling water is recycled over and over. Cooling water typically does not come into direct contact with process oil streams and therefore contains less contaminants than process wastewater. However, it may contain some oil contamination due to leaks in the process equipment.

Water used in processing operations accounts for a significant portion of the total wastewater. Process wastewater arises from desalting crude oil, steam stripping operations, pump gland cooling, product fractionator reflux drum drains and boiler blowdown. Because process water often comes into direct contact with oil, it is usually highly contaminated. Storm water (i.e., surface water runoff) is intermittent and will contain constituents from spills to the surface, leaks in equipment and any materials that may have collected in drains. Runoff surface water also includes water coming from crude and product storage tank roof drains.

Refinery effluent water contains various hydrocarbon components including gasoline blending stocks, kerosene, diesel fuel and heavier liquids (Speight, 2005). Also present may be suspended mineral solids, sand, salt, organic acids and sulfur compounds. The nature of the components depends on the constituents of the inlet crude oil as well as the processing scheme of the refinery. Most of these constituents would be undesirable in the effluent water, so it is necessary to treat the water to remove the contaminants.

Undesirable effects of hydrocarbon derivatives in water include taste and odor contamination in addition to toxicity. Crude oil hydrocarbon derivatives can impart a perceptible unpleasant taste and can adversely affect aquatic organisms. Naphthenic acid from refineries can have a toxic effect on plant and animal life and aromatic hydrocarbon derivatives are toxic and/or carcinogenic to humans and animals as well as to aquatic life.

Oil can also occur in a refinery wastewater stream may exist in one or more of three forms: (1) free oil, (2) emulsified oil, and (3) dissolved oil.

Free oil is oil in the form of separate oil globules of sufficient size that they can rise as a result of buoyancy force to the top of the water. Separators may readily be designed to remove this type of oil.

Emulsified oil is oil in the form of much smaller droplets or globules with a diameter of 20 microns or less which form a stable suspension in the water. A true emulsion will not separate by gravity regardless of how long the emulsion stands under quiescent conditions. The term *emulsified oil* may also be applied to emulsions where the droplets are so small that they will not rise at a rate that allows a practical size separation device. It is possible to design enhanced gravity separators to treat waters containing this type of oil, but generally it is only practical for small flow rates.

Dissolved oil cannot be removed by gravity separation and other methods must be adopted. Such means include (1) biological treatment (which is especially pertinent to this text), or (2) adsorption by activated carbon or other adsorbents or absorbents.

Waste waters are treated in onsite wastewater treatment facilities and then discharged to publicly owned treatment works (POTWs) or discharged to surfaces waters under National Pollution Discharge Elimination System (NPDES) permits.

Many types of separation methods have been used to remove oil from refinery wastewater with varying degrees of success (19). Some of the systems currently in use are: (1) API separators, (2) flocculation units, (3) dissolved and induced air flotation (DAF and IF) units, (4) coalescing plate separators, and (5) multiple angle separators.

Crude oil refineries typically utilize primary and secondary wastewater treatment. Primary wastewater treatment consists of the separation of oil, water and solids in two stages. During the first stage, an API separator, a corrugated plate interceptor, or other separator design is used. Wastewater moves very slowly through the separator allowing free oil to float to the surface and be skimmed off, and solids to settle to the bottom and be scraped off to a sludge collecting hopper. The second stage utilizes physical or chemical methods to separate emulsified oils from the wastewater. Physical methods may include the use of a series of settling ponds with a long retention time, or the use of dissolved air flotation (DAF).

In the dissolved air flotation method, air is bubbled through the wastewater, and both oil and suspended solids are skimmed off the top. Chemicals, such as ferric hydroxide [$Fe(OH)_3$] or aluminum hydroxide [$Al(OH)_3$], can be used to coagulate impurities into a froth or sludge that can be more easily skimmed off the top. Some wastes associated with the primary treatment of wastewater at crude oil refineries may be considered hazardous and include: (1) API separator sludge, (2) primary treatment sludge, (3) sludge from other gravitational separation techniques, (4) float from dissolved air flotation units, and (5) wastes from settling ponds.

After primary treatment, the wastewater can be discharged to a publicly owned treatment works (POTW) or undergo secondary treatment before being discharged directly to surface waters under a National Pollution Discharge Elimination System (NPDES) permit. In secondary treatment, dissolved oil and other organic pollutants may be consumed biologically by microorganisms. Biological treatment may require the addition of oxygen through a number of different techniques, including activated sludge units, trickling filters, and rotating biological contactors. Secondary treatment generates bio-mass waste that is typically treated anaerobically and then dewatered.

Some refineries employ an additional stage of wastewater treatment called polishing to meet discharge limits. The polishing step can involve the use of activated carbon, anthracite coal, or sand to filter out any remaining impurities, such as biomass, silt, trace metals and other inorganic chemicals, as well as any remaining organic chemicals.

Certain refinery wastewater streams are treated separately, prior to the wastewater treatment plant, to remove contaminants that would not easily be treated after mixing with other wastewater. One such waste stream is the sour water drained from distillation reflux drums. Sour water contains dissolved hydrogen sulfide and other organic sulfur compounds and ammonia which are stripped in a tower with gas or steam before being discharged to the wastewater treatment plant.

Wastewater treatment plants are a significant source of refinery air emissions and solid wastes. Air releases arise from fugitive emissions from the numerous tanks, ponds and sewer system drains. Solid wastes are generated in the form of sludge from a number of the treatment units.

The refining industry will certainly continue to improve operations to minimize the amounts of hydrocarbon derivatives in wastewater streams. Some hydrocarbon derivatives will still enter wastewater streams because of small spills and leaks, and it will be necessary to recover these products to further reduce the amount of hydrocarbon derivatives in refinery effluents.

Disposal of solid wastes is a significant problem for the petrochemical industry. Waste solids include water treatment sludge, ashes, fly ash and incinerator residue, plastics, ferrous and nonferrous metals, catalysts, organic chemicals, inorganic chemicals, filter cakes, and viscous solids.

2.5.1 Acid Sludge

Sludge produced during the use of sulfuric acid as a treating agent is mainly of two types: (1) sludge from light oils (gasoline and kerosene) and (2) sludge from lubricating stocks, medicinal oils, and the like. In the treatment of the latter oils it appears that the action of the acid causes precipitation of asphaltene constituents and resin constituents, as well as the solution of color-bearing and sulfur compounds. Sulfonation and oxidation-reduction reactions also occur but to a lesser extent since much of the acid can be recovered. In the desulfurization of cracked distillates, however, chemical interaction is more important, and polymerization, ester formation, aromatic-olefin condensation, and sulfonation also occur. Nitrogen bases are neutralized, and the acid dissolves naphthenic acids; thus the composition of the sludge is complex and depends largely on the oil treated, acid strength, and the temperature.

Sulfuric acid sludge from *iso*-paraffin alkylation and lubricating oil treatment are frequently decomposed thermally to produce sulfur dioxide (which is returned to the sulfuric acid plant) and *sludge acid coke*. The coke, in the form of small pellets, is used as a substitute for charcoal in the manufacture of carbon disulfide. Sulfuric acid coke is different from other crude oil coke in that it is pyrophoric in air and also reacts directly with sulfur vapors to form carbon disulfide

2.5.2 Spent Acid

In addition, to spent caustic solutions, spent acid solutions also occur in a refinery. By analogy to spent caustic, spent acid is an acidic solution that has become weakened, diluted or exhausted by constant use and the solution retains very little of the original acidic nature.

Naphthenic acids are complex carboxylic acids that are believed to have a cyclopentane ring or cyclohexane ring in the molecule, occur in crude oil. They seem to be of little consequence environmentally since thermal decarboxylation can occur during the distillation process. During this process, the temperature of the crude oil in the distillation column can reach as high as 395°C (740°F). Hence decarboxylation is possible.

$$RCO_2H \rightarrow RH + CO_2$$

However, inorganic acids are used in various processes to treat unfinished crude oil products such as gasoline, kerosene, and lubricating oil stocks are treated with sulfuric acid for improvement of color, odor, and other properties.

For example, treating of crude oil distillates with sulfuric acid is generally applied to dissolve unstable or colored substances and sulfur compounds, as well as to precipitate asphaltic materials. When drastic conditions are employed, as in the treatment of lubricating fractions with large amounts of concentrated acid or when fuming acid is used in the manufacture of white oils, considerable quantities of crude oil sulfonic acids are formed.

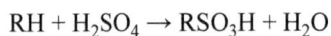

$$RH + H_2SO_4 \rightarrow RSO_3H + H_2O$$

Paraffin   sulfonic acid

Two general methods are applied for the recovery of sulfonic acids from sulfonated oils and their sludge. In one case the acids are selectively removed by adsorbents or by solvents (generally low-molecular-weight alcohols), and in the other case the acids are obtained by salting out with organic salts or bases.

Crude oil sulfonic acids may be roughly divided into those soluble in hydrocarbon derivatives and those soluble in water. These acids (typically hydrocarbon-soluble acids) are often referred to as mahogany acids because of the color and the water-soluble acids are referred to as green acids. The composition of each of acid varies with the nature of the oil sulfonated and the concentration of the acids produced. In general, those formed during light acid treatment are water-soluble; oil-soluble acids result from more drastic sulfonation.

Sulfonic acids are used as detergents made by the sulfonation of alkylated benzene. The number, size, and structure of the alkyl side chains are important in determining the performance of the finished detergent. The salts of mixed crude oil sulfonic acids have many other commercial applications. They find use as anticorrosion agents, leather softeners, and flotation agents and have been used in place of red oil (sulfonated castor oil) in the textile industry. Lead salts of the acids have been employed in greases as extreme pressure agents, and alkyl esters have been used as alkylating agents. The alkaline earth metal (magnesium, Mg, calcium, Ca, and barium, Ba) salts are used in detergent compositions for motor oils, and the alkali metal (K and Na) salts are used as detergents in aqueous systems.

The *sulfuric acid sludge* from sulfuric acid treatment is frequently used as a source (through thermal decomposition) to produce sulfur dioxide (SO_2) that is returned to the sulfuric acid plant) and *sludge acid coke*. The coke, in the form of small pellets, is used as a substitute for charcoal in the manufacture of carbon disulfide. Sulfuric acid coke is different from other crude oil coke in that it is pyrophoric in air and also reacts directly with sulfur vapor to form carbon disulfide.

In recent years, the crude oil refining industry has placed an increasing emphasis on the safety of the use of hydrogen fluoride (HF) in crude oil refineries. This acid is used in the alkylation process—the process is increasingly important in the production of a blend stock for high-quality gasoline. However, caution is advised because hydrofluoric acid is hazardous and corrosive and, if accidentally released, can form a vapor cloud.

Pure hydrogen fluoride is a clear, colorless, corrosive liquid that is readily soluble in water and can form a vapor cloud if released to the atmosphere. The acid has a sharp, penetrating odor that human beings can detect at very low concentrations (0.04 yo 0.13 ppm), in the air. To protect against adverse effects from exposure to hydrofluoric acid in the workplace, the Occupational Safety and Health Administration has established a permissible exposure limit (PEL) of 3 ppm averaged over an eight-hour work shift.

2.5.3 Spent Catalyst

The refining of crude oil fractions occurs over many types of catalytic materials and cracking catalysts can differ markedly in both activity to promote the cracking reaction and in the quality of the products obtained from cracking the feedstocks (Parkash, 2003; Gary et al., 2007; Speight, 2014; Hsu and Robinson, 2017; Speight, 2017).

Many processes involve the use of a catalyst, which is a material that aids or promotes a chemical reaction between other substances but does not react itself. Catalysts increase the rate of a reaction and can provide control by increasing the desired reaction(s) while decreasing the undesirable reactions.

Cracking of crude oil fractions occurs over many types of catalytic materials and cracking catalysts can differ markedly in both activity to promote the cracking reaction and in the quality of the products obtained from cracking the feedstocks (Gates et al., 1979; Wojciechowski and Corma, 1986; Stiles and Koch, 1995; Cybulski and Moulijn, 1998; Occelli and O'Connor, 1998). Activity can be related directly to the total number of active (acid) sites per unit weight of catalyst and also to the acidic strength of these sites. Differences in activity and acidity regulate the extent of various secondary reactions occurring and thus the product quality differences. The acidic sites are considered to be Lewis- or Brønsted-type acid sites, but there is much controversy as to which type of site predominates.

Commercial synthetic catalysts are amorphous and contain more silica than is called for by the preceding formulae; they are composed of 10 to 15% alumina (Al_2O_3) and 85 to 90% silica (SiO_2). The natural materials, montmorillonite, a non-swelling bentonite, and halloysite, are hydrosilicates of aluminum, with a well-defined crystal structure and approximate composition of $Al_2O_3.4Si_2O$. xH_2O. Some of the newer catalysts contain up to 25% alumina and are reputed to have a longer active life.

The catalysts are porous and highly adsorptive, and their performance is affected markedly by the method of preparation. However, it is worthy of note that two catalysts that are chemically identical but have pores of different size and distribution may have different activity, selectivity, temperature coefficient of reaction rate, and response to the various catalyst poisons.

Furthermore, the catalysts must be stable to physical impact loading and thermal shocks and must also be capable of withstanding the action of carbon dioxide, air, nitrogen compounds, and steam. The catalysts should also be resistant to sulfur compounds in which case synthetic catalysts and certain selected clays may be more efficient than the typical untreated natural catalysts.

2.5.4 Spent Caustic

Spent caustic solutions is an alkaline solution that has become weakened, diluted or exhausted by constant use and the solution retains very little of the original alkaline nature.

Several process use the principle of an alkali wash (also referred to as caustic wash), which is involves treatment of refinery distillates with aqueous sodium hydroxide to remove any acidic contaminants. Used caustic (spent caustic) in a refinery spent caustic comes from multiple sources (Parkash, 2003; Gary et al., 2007; Speight, 2014; Hsu and Robinson, 2017; Speight, 2017) in which sulfide derivatives and organic acid derivatives are removed from the product streams into the aqueous caustic phase. The sodium hydroxide is consumed and the resulting wastewaters are often mixed and called refinery spent caustic.

Thus, refinery wastewater may be highly alkaline and contain oil, sulfides, ammonia, and/or phenol. The potential exists in the coking process for exposure to burns when handling hot coke or in the event of a steam-line leak, or from steam, hot water, hot coke, or hot slurry that may be expelled when opening cokers.

The caustic streams used in various refining, petrochemical, and chemical operations often end up in a waste caustic tank for disposal. This waste caustic can produce a unique disposal challenge depending upon the components to be removed. A common method of remediation is a wet air oxidation process that is designed to convert the reactive sulfide components such as sodium sulfide and sodium mercaptide to sodium sulfate. Organic contaminants such as phenols can also be significantly reduced. Wet air oxidation treatment allows spent caustic to be routed to a standard biological wastewater treatment plant without creating operational or odor problems.

Light and heavy hydrocarbon derivatives (oils) will adversely affect the operation of a wet air oxidation process by one or more of the following: (1) foul the system heat exchangers causing frequent heat exchanger cleaning, system shutdown, and localized corrosion problems, (2) exceed the supply for of available oxygen, reducing system capacity or resulting in system shutdown due to oxygen deficiency, and (3) cause the system to be oversized in design that increases the capital and operating cost.

2.5.5 Sulfonic Acids

Sulfonic acids are produced by when crude oil is treated with sulfuric acid. Sulfuric acid treating of crude oil distillates is generally applied to dissolve unstable or colored substances and sulfur compounds, as well as to precipitate asphaltic materials. When drastic conditions are employed, as in the treatment of lubricating fractions with large amounts of concentrated acid or when fuming acid is used in the manufacture of white oils, considerable quantities of sulfonic acids are formed. Extensive side reactions, mainly oxidation, also occur and increase with the proportion of sulfur trioxide in the acid.

Many of the lower paraffins are physically absorbed by concentrated and fuming sulfuric acids; chemical activity increases with rise in molecular weight, and compounds containing tertiary carbons are especially responsive. *n*-Hexane, *n*-heptane, and *n*-octane are essentially inactive in cold fuming acid; but at the boiling point of the hydrocarbon derivatives rapid sulfonation takes place to give mono- and disulfonic acids:

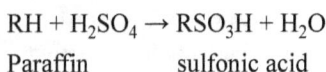

$$RH + H_2SO_4 \rightarrow RSO_3H + H_2O$$

Paraffin   sulfonic acid

The five- and six-membered ring lower naphthene derivatives are stable to cold concentrated sulfuric acid, but fuming sulfuric acid reacts with cyclohexane to give mono- and di-naphthene sulfonic acid derivatives and mono-aromatic sulfonic acid derivatives, along with products based on cyclic olefin derivatives formed through hydrogen-transfer reactions.

The action of sulfuric acid on hydrocarbon derivatives is indeed quite complex, but it is obvious that a reaction occurs readily with such compound types as aromatic derivatives and those tertiary carbon atoms in naphthenic rings that are both present in the lubricating fractions of crude oil. Ordinarily, a charge stock for sulfuric acid treatment will already have been refined by solvent extraction with, say, furfural (Chapter 20) to remove those more highly aromatic constituents. Thus the remaining hydrocarbon derivatives, which give higher yields of better sulfonated products, are those in which aromatic rings are entirely absent or are low in proportion relative to the naphthene rings and paraffinic chains, and hence the preferred sulfonic acids of commerce are probably naphthene sulfonic acids.

Sulfonic acids are also used as detergents made by the sulfonation of alkylated benzene. The number, size, and structure of the alkyl side chains are important in determining the performance of the finished detergent.

Two general methods are applied for the recovery of sulfonic acids from sulfonated oils and their sludge. (1) In one case the acids are selectively removed by adsorbents or by solvents (generally low-molecular-weight alcohols), and (2) in the other case the acids are obtained by salting out with organic salts or bases.

Sulfonic acids may be roughly divided into those soluble in hydrocarbon derivatives and those soluble in water. Because of their color, hydrocarbon-soluble acids are referred to as mahogany acids, and the water-soluble acids are referred to as green acids. The composition of each type varies with the nature of the oil sulfonated and the concentration of the acids produced. In general, those formed during light acid treatment are water soluble; oil-soluble acids result from more drastic sulfonation.

The salts of mixed sulfonic acids have many commercial applications. They find use as anticorrosion agents, leather softeners, and flotation agents and have been used in place of red oil (sulfonated castor oil) in the textile industry. Lead salts of the acids have been employed in greases as extreme pressure agents, and alkyl esters have been used as alkylating agents. The alkaline earth metal (Mg, Ca and Ba) salts are used in detergent compositions for motor oils, and the alkali metal (K and Na) salts are used as detergents in aqueous systems.

2.5.6 Product Blending

The modern crude oil refinery consists of a very complex mix of high technology processes which efficiently convert the wide array of crude oils into the hundreds of specification products we use daily. Each refinery has its own unique processing configuration as a result of the logistics and associated economics related to its specific crude oils and products markets. The refiner must continuously optimizing the mix of product volumes and this is accomplished through executing decisions regarding parameters as varied as (1) feedstock selection, (2) adjustments in product cut-points, and (3) reactor severities in individual processes because an imbalance in any two of these options can cause incompatibility of the blended product leading to environmental contamination. Additional options include changing the dispositions of intermediate product streams to alternative processing units, or alternative finished product blends.

In fact, many refinery products are typically the result of blending several component streams or blending stocks. In most cases, product blending is accomplished by controlling the volumes of blend stocks from individual component storage tanks that are mixed in the finished product storage tank. Samples of the finished blend are then analyzed by laboratory testing for all product specifications prior to shipping. Alternatively, *in-line blending* refers to pipeline shipments in which the finished product is actually blended directly into the product pipeline (as opposed to a standing product storage tank).

The most commonly recognized blending operations occur in the gasoline production section of the refinery. The various gasoline streams are so that specifications (dependent upon geographic location, environmental regulations and weather patterns) can be met.

Gasoline blending involves combining of the components that make up motor gasoline. The components include the various hydrocarbon streams produced by distillation, cracking, reforming, and polymerization, tetraethyl lead, and identifying color dye, as well as other special-purpose components, such as solvent oil and anti-icing compounds. The physical process of blending the components is simple, but determination of how much of each component to include in a blend is much more difficult. The physical operation is carried out by simultaneously pumping all the components of a gasoline blend into a pipeline that leads to the gasoline storage, but the pumps must be set to deliver automatically the proper proportion of each component. Baffles in the pipeline are often used to mix the components as they travel to the storage tank.

Selection of the components and their proportions in a blend is the most complex problem in a refinery. Many different hydrocarbon streams may need to be blended to produce quality gasoline. Each property of each stream is a variable, and the effect on the product gasoline is considerable. For example, the low octane number of straight-run naphtha limits its use as a gasoline component, although its other properties may make it desirable. The problem is further complicated by changes in the properties of the component streams due to processing changes. For example, an increase in cracking temperature produces a smaller volume of higher octane cracked naphtha, but before this cracked naphtha can be included in a blend, adjustments must be made in the proportions of the other hydrocarbon components. Similarly, the introduction of new processes and changes in the specifications of the finished gasoline dictate reevaluation of the components that make up the gasoline (Gibbs, 1989).

Gasoline blending is not the only blending operations and other product blending operations are also in operation in a refinery. The applicable specifications vary by product but typically include properties pertinent to the behavior of the product in use. Many product specifications do not blend linearly by component volumes. In these circumstances, the finished blend properties are predicted using experience-based algorithms for the applicable blending components.

2.5.7 Waste by Process

Crude oil refineries are complex, but integrated, unit process operations that produce a variety of products from various feedstocks and feedstock blends (Parkash, 2003; Gary et al., 2007; Speight, 2014; Hsu and Robinson, 2017; Speight, 2017). During crude oil refining, refineries use and generate an enormous amount of chemicals, some of which are present in air emissions, wastewater, or solid wastes (Table 2.6). Emissions are also created through the combustion of fuels, and as byproducts of chemical reactions occurring when crude oil fractions are upgraded. A large source of air emissions is, generally, the process heaters and boilers that produce carbon monoxide, sulfur oxides, and nitrogen oxides, leading to pollution and the formation of acid rain.

$CO_2 + H_2O \rightarrow H_2CO_3$ (carbonic acid)

$SO_2 + H_2O \rightarrow H_2SO_3$ (sulfurous acid)

$2SO_2 + O_2 \rightarrow 2SO_3$

$SO_3 + H_2O \rightarrow H_2SO_4$ (sulfuric acid)

$NO + H_2O \rightarrow HNO_2$ (nitrous acid)

$2NO + O_2 \rightarrow NO_2$

$NO_2 + H_2O \rightarrow HNO_3$ (nitric acid)

Table 2.6: Examples of Potential Emissions and Waste from Refinery Processes.

Process	Air emissions	Residual waste
Crude oil desalting	Heater gas (CO, SOx, NOx, hydrocarbons and particulate matter	Crude oil/desalter sludge composed of iron rust
	Fugitive emissions (hydrocarbons)	Clay, sand, water, emulsified oil and wax, metals
Atmospheric distillation	Heater stack gas (CO, SOx, NOx, hydrocarbons and particulate matter)	Typically, little or no residual waste generated
	Fugitive emissions (hydrocarbons)	
Vacuum Distillation	Steam ejector emissions, (hydrocarbons)	Typically, little or no residual waste generated
	Heater stack gas (CO, SOx, NOx, hydrocarbons, particulate matter	
	Fugitive emissions (hydrocarbons)	
Thermal cracking	Heater stack gas (CO, SOx, NOx, hydrocarbons and particulate matter)	Typically, little or no residual waste generated
Visbreaking	Heater stack gas (CO, SOx, NOx, hydrocarbons and particulate matter)	Typically, little or no residual waste generated
Coking	Heater stack gas (CO, SOx, NOx, hydrocarbons, particulate matter)	Coke dust (carbon particles and hydrocarbons)
	Fugitive emissions (hydrocarbons)	
	Decoking emissions (hydrocarbons and particulate matter)	
Catalytic cracking	Heater stack gas (CO, SOx, NOx, hydrocarbons, particulate matter)	Spent catalysts (metals, hydrocarbons)
	Fugitive emissions (hydrocarbons)	Spent catalyst fines
	Catalyst regeneration emissions (CO, NOx, SOx, particulate matter)	Aluminum silicate and metals
Hydrocracking	Heater stack gas (CO, SOx, NOx, hydrocarbons, particulate matter)	Spent catalysts fines
	Fugitive emissions (hydrocarbons)	
	Catalyst regeneration emissions (CO, NOx, S, catalyst dust)	

Hence, there is the need for gas-cleaning operations on a refinery site so that such gases are cleaned from the gas stream prior to entry into the atmosphere.

In addition, some processes create considerable amounts of particulate matter and other emissions from catalyst regeneration or decoking processes. Volatile chemicals and hydrocarbon derivatives are also released from equipment leaks, storage tanks, and wastewaters. Other cleaning units such as the installation of filters, electrostatic precipitators and cyclones can mitigate part of the problem.

Process wastewater is also a significant effluent from a number of refinery processes. Atmospheric distillation units and vacuum distillation units create the largest volumes of process wastewater, about 26 gallons per barrel of oil processed. Fluid catalytic cracking and catalytic reforming also generate considerable amounts of wastewater (15 and 6 gallons per barrel of feedstock, respectively). A large portion of wastewater from these three processes is contaminated with oil and other impurities and must be subjected to primary, secondary and sometimes tertiary water treatment processes, some of which also create hazardous waste.

Wastes, residua and by-products are produced by a number of processes. Residuals produced during refining can be but are not necessarily wastes. They can be recycled or regenerated, and in many cases do not become part of the waste stream but are useful products. For example, processes utilizing caustics for neutralization of acidic gases or solvent (e.g., alkylation, sweetening/chemical treating, lubricating oil manufacture) create the largest source of residuals in the form of spent caustic solutions. However, nearly all of these caustics are recycled.

The treatment of oily wastewater from distillation, catalytic reforming and other processes generates the next largest source of residuals in the form of biomass sludge from biological treatment and pond sediments. Water treatment of oily wastewater also produces a number of sludge materials associated with oil-water separation processes. Such sludge is often recycled in the refining process and is not always considered to be a waste product.

Catalytic processes (fluid catalytic cracking, catalytic hydrocracking, hydrotreating, isomerization, ethers manufacture) also create some residuals in the form of spent catalysts and catalyst fines or particulates. The latter are sometimes separated from exiting gases by electrostatic precipitators or filters. These are collected and disposed of in landfills or may be recovered by off-site facilities.

In terms of individual processes, the potential for waste generation and, hence, leakage of emissions is covered in the following sections.

2.5.7.1 Dewatering and Desalting

An aspect of crude oil treating (prior to the production of products) that must also be considered (after transportation problems) as a potential source of environmental pollution and that is the dewatering and desalting process. While this process is a first step of crude oil treatment in a refinery, it is not always considered to be the first step in crude oil refining (crude oil conversion).

When produced from the reservoir, crude oil often contains water, inorganic salts, suspended solids, and water-soluble trace metals. As a first step in the refining process (even as a preliminary step to transportation), in order to reduce corrosion, plugging, and fouling of equipment and to prevent poisoning the catalysts in processing units, these contaminants must be removed by desalting (dehydration) (Parkash, 2003; Gary et al., 2007; Speight, 2014a, 2015; Hsu and Robinson, 2017).

The two most typical methods of crude oil desalting, chemical separation and electrostatic separation, use hot water as the extraction agent. In chemical desalting, water and chemical surfactant (demulsifiers) are added to the crude oil, heated so that salts and other impurities dissolve into the water or attach to the water, and then held in a tank where they settle out. Electrical desalting is the application of high-voltage electrostatic charges to concentrate suspended water globules in the bottom of the settling tank. Surfactants are added only when the crude has a large amount of suspended solids. Both methods of desalting are continuous. A third and less-common process involves filtering heated crude oil using diatomaceous earth.

In the process, the feedstock crude oil is heated to between 65 and 177°C (150 and 350°F) to reduce viscosity and surface tension for easier mixing and separation of the water. The temperature is limited by the vapor pressure of the crude oil constituents. In both methods other chemicals may be added. Ammonia is often used to reduce corrosion. Caustic or acid may be added to adjust the pH of the water wash. Wastewater and contaminants are discharged from the bottom of the settling tank to the wastewater treatment facility. The desalted crude is continuously drawn from the top of the settling tanks and sent to the crude distillation (fractionating) tower.

Since desalting is a closed process, there is little potential for exposure to the feedstock unless a leak or release occurs. However, whenever elevated temperatures are used when desalting sour (sulfur-containing) crude oil, hydrogen sulfide will be present. And, depending on the crude feedstock and the treatment chemicals used, the wastewater will contain varying amounts of chlorides, sulfides, bicarbonates, ammonia, hydrocarbon derivatives, phenol, and suspended solids. If diatomaceous earth is used in filtration, exposures should be minimized or controlled.

The dewatering-desalting process (Parkash, 2003; Gary et al., 2007; Speight, 2014a, 2015; Hsu and Robinson, 2017; Speight, 2017) creates an oily desalter sludge that may be a hazardous waste and a high temperature salt wastewater stream (treated along with other refinery wastewater) (Chapter 2). The primary polluting constituents in desalter wastewater include hydrogen sulfide, ammonia, phenol, high levels of suspended solids, and dissolved solids, with a high biochemical oxygen demand (BOD). In some cases it is possible to recycle the desalter effluent water back into the desalting process, depending upon the type of crude being processed.

Wastewaters may be collected in separate drainage systems (for process, sanitary and storm water) although industrial and storm water systems may in some cases be combined. In addition, ballast water from bulk crude tankers may be pumped to receiving facilities at the refinery site prior to removal of floating oil in an interceptor and treatment as for other wastewater streams.

On-site treatment facilities may exist for wastewater or treatment may take place at a public wastewater treatment plant. Storm water/process water is generally passed to a separator or interceptor prior to leaving the site which takes out free-phase oil (i.e., floating product) from the water prior to discharge, or prior to further treatment, e.g., in settling lagoons).

Discharge from wastewater treatment plants is usually passed to a nearby watercourse. Other wastes that are typical of a refinery include (1) waste oils, process chemicals, still resides, (2) non-specification chemicals and/or products, (3) waste alkali (sodium hydroxide), (4) waste oil sludge (from interceptors, tanks, and lagoons), and (5) solid wastes (cartons, rags, catalysts, and coke).

Crude oil often contains water, inorganic salts, suspended solids, and water-soluble trace metals. As a first step in the refining process, to reduce corrosion, plugging, and fouling of equipment and to prevent poisoning the catalysts in processing units, these contaminants must be removed by desalting (dehydration).

The two most typical methods of crude oil desalting, chemical separation and electrostatic separation, use hot water as the extraction agent. In chemical desalting, water and chemical surfactant (demulsifiers) are added to the crude oil, heated so that salts and other impurities dissolve into the water or attach to the water, and then held in a tank where they settle out. Electrical desalting is the application of high-voltage electrostatic charges to concentrate suspended water globules in the bottom of the settling tank. Surfactants are added only when the crude has a large amount of suspended solids. Both methods of desalting are continuous. A third and less-common process involves filtering heated crude oil using diatomaceous earth.

The feedstock crude oil is heated to between 65 and 177°C (150° and 350°F) to reduce viscosity and surface tension for easier mixing and separation of the water. The temperature is limited by the vapor pressure of the crude oil constituents. In both methods other chemicals may be added. Ammonia is often used to reduce corrosion. Caustic or acid may be added to adjust the pH of the water wash. Wastewater and contaminants are discharged from the bottom of the settling tank to the wastewater treatment facility. The desalted crude is continuously drawn from the top of the settling tanks and sent to the crude distillation (fractionating) tower.

Since desalting is a closed process, there is little potential for exposure to the feedstock unless a leak or release occurs. However, whenever elevated temperatures are used when desalting sour (sulfur-containing) crude oil, hydrogen sulfide will be present. And, depending on the crude feedstock and the treatment chemicals used, the wastewater will contain varying amounts of chlorides, sulfides, bicarbonates, ammonia, hydrocarbon derivatives, phenol, and suspended solids. If diatomaceous earth is used in filtration, exposures should be minimized or controlled.

Desalting creates an oily desalter sludge that may be a hazardous waste and a high temperature salt wastewater stream (treated along with other refinery wastewaters). The primary polluting constituents in desalter wastewater include hydrogen sulfide, ammonia, phenol, high levels of suspended solids, and dissolved solids, with a high biochemical oxygen demand (BOD). In some cases it is possible to recycle the desalter effluent water back into the desalting process, depending upon the type of crude being processed.

2.5.7.2 Gas Processing

Gas processing either from field wells or within the refinery also, unfortunately, offers opportunities for pollution.

In the field, the gas from high-pressure wells is usually passed through field separators at the well to remove hydrocarbon condensate and water. Natural gasoline, butane, and propane are usually present in the gas, and gas processing plants are required for the recovery of these liquefiable constituents. In addition, hydrogen sulfide must be removed before the gas can be utilized. The gas is usually sweetened by absorption of the hydrogen sulfide in an amine (olamine) solution (Chapter 25). Other methods, such as carbonate processes, solid bed absorbents, and physical absorption, are employed in the other sweetening plants.

The major emission sources in the natural gas processing industry are compressor engines, acid gas wastes, fugitive emissions from leaking process equipment, and (if present) glycol dehydrator vent streams. Regeneration of the glycol solutions used for dehydrating natural gas can release significant quantities of benzene, toluene, ethylbenzene, and xylene, as well as a wide range of less toxic organics.

Many chemical processes are available for sweetening natural gas (Mokhatab et al., 2006; Speight, 2014, 2017, 2018). At present, the amine (olamine) process (also known as the Girdler process) is the most widely used method for hydrogen sulfide removal. The recovered hydrogen sulfide gas stream may be: utilized for the production of elemental sulfur or sulfuric acid.

Emissions will result from gas sweetening plants only if the acid waste gas from the amine process is flared or incinerated. Most often, the acid waste gas is used as a feedstock in nearby sulfur. If flaring or incineration is practiced, the major pollutant of concern is sulfur dioxide. Most plants employ elevated smokeless flares or tail gas incinerators for complete combustion of all waste gas constituents, including near quantitative conversion of hydrogen sulfide to sulfur dioxide. Little particulate, smoke, or hydrocarbon derivatives result from these devices, and because gas temperatures do not usually exceed 650°C (1200°F), significant quantities of nitrogen oxides are not formed.

2.5.7.3 Distillation

After desalting (Chapter 1), the crude oil is then heated in a heat exchanger and furnace to temperatures not exceeding 400°C (750°F) and fed to a vertical, distillation column at atmospheric pressure where most of the feedstock is vaporized and separated into its various fractions by condensing on 30 to 50 fractionation trays, each corresponding to a different condensation temperature. The lower boiling fractions condense and are collected towards the top of the column. Higher boiling fractions, which may not vaporize in the column, are further separated later by vacuum distillation.

Atmospheric and vacuum distillation units are closed processes and exposures are expected to be minimal. Both atmospheric distillation units and vacuum distillation units produce refinery fuel gas streams containing a mixture of light hydrocarbon derivatives, hydrogen sulfide and ammonia. These streams are processed through gas treatment and sulfur recovery units to recover fuel gas and sulfur. Sulfur recovery creates emissions of ammonia, hydrogen sulfide, sulfur oxides, and nitrogen oxides.

Within each atmospheric distillation tower, a number of side streams (at least four) of low-boiling point components are removed from the tower from different trays. These low-boiling point mixtures are in equilibrium with heavier components which must be removed. The side streams are each sent to a different small stripping tower containing four to 10 trays with steam injected under the bottom tray. The steam strips the light-end components from the heavier components and both the steam and light-ends are fed back to the atmospheric distillation tower above the corresponding side stream draw tray. Fractions obtained from atmospheric distillation include naphtha, gasoline, kerosene, light fuel oil, diesel oils, gas oil, lube distillate, and the residuum (bottoms).

Most of these fractions can be sold as finished products, or blended with products from downstream processes. Another product produced in atmospheric distillation, as well as many other refinery processes, is the light, non-condensable refinery fuel gas (mainly methane and ethane). Typically this gas also contains hydrogen sulfide and ammonia gases (*sour gas* or *acid gas*). The sour gas is sent to the gas processing section that separates the fuel gas so that it can be used as fuel in the refinery heating furnaces.

Vacuum distillation typically follows atmospheric distillation and is the distillation of crude oil fractions at low pressure (0.2 to 0.7 psi) to increase volatilization and separation. In most systems, the vacuum inside the fractionator is maintained with steam ejectors and vacuum pumps, barometric condensers or surface condensers. The injection of superheated steam at the base of the vacuum fractionator column further reduces the partial pressure of the hydrocarbon derivatives in the tower, facilitating vaporization and separation. The higher fractions from the vacuum distillation column are processed downstream into more valuable products through either cracking or coking operations.

Both atmospheric distillation units and vacuum distillation units produce refinery fuel gas streams containing a mixture of light hydrocarbon derivatives, hydrogen sulfide and ammonia. These streams are processed through gas treatment and sulfur recovery units to recover fuel gas and sulfur. Sulfur recovery creates emissions of ammonia, hydrogen sulfide, sulfur oxides, and nitrogen oxides.

When sour (high-sulfur) crude oil is processed, there is potential for exposure to hydrogen sulfide in the preheat exchanger and furnace, tower flash zone and overhead system, vacuum furnace and tower, and bottoms exchanger. Hydrogen chloride may be present in the preheat exchanger, tower top zones, and overheads. Wastewater may contain water-soluble sulfides in high concentrations and other water-soluble compounds such as ammonia, chlorides, phenol, mercaptans, etc., depending upon the crude feedstock and the treatment chemicals. Safe work practices and/or the use of appropriate personal protective equipment may be needed for exposures to chemicals and other hazards such as heat and noise, and during sampling, inspection, maintenance, and turnaround activities.

Air emissions from a distillation unit include emissions from the combustion of fuels in process heaters and boilers, fugitive emissions of volatile constituents in the crude oil and fractions, and emissions from process vents. The primary source of emissions is combustion of fuels in the crude pre-heat furnace and in boilers that produce steam for process heat and stripping. When operating in an optimum condition and burning cleaner fuels (e.g., natural gas, refinery gas), these heating units create relatively low emissions of sulfur oxides (SO_x), nitrogen oxides (NO_x), carbon monoxide (CO), hydrogen sulfide (H_2S), particulate matter, and volatile hydrocarbon derivatives. If fired with lower grade fuels (e.g., refinery fuel pitch, coke) or operated inefficiently (incomplete combustion), heaters can be a significant source of emissions.

Fugitive emissions of volatile hydrocarbon derivatives arise from leaks in valves, pumps, flanges, and other similar sources where crude and its fractions flow through the system. While individual leaks may be minor, the combination of fugitive emissions from various sources can be substantial. Those potentially released during crude distillation include ammonia, benzene, toluene, and xylenes, among others. These emissions are controlled primarily through leak detection and repair programs and occasionally through the use of special leak resistant equipment.

Distillation units generate considerable wastewater. The process water used in distillation often comes in direct contact with oil, and can be highly contaminated. Both atmospheric distillation and vacuum distillation produce an oily sour wastewater (condensed steam containing hydrogen sulfide and ammonia) from side stripping fractionators and reflux drums. Many refineries now use vacuum pumps and surface condensers in place of barometric condensers to eliminate the generation of the wastewater stream and reduce energy consumption. Reboiled side stripping towers rather than open steam stripping can also be utilized on the atmospheric tower to reduce the quantity of sour water condensate.

Typical constituents of sour wastewater streams from crude distillation include hydrogen sulfide, ammonia, suspended solids, chlorides, mercaptans, and phenol, characterized by a high pH.

2.5.7.4 Visbreaking and Thermal Cracking

Thermal cracking processes (of which visbreaking is used as the example) cause the decomposition of higher molecular weight fractions to lower boiling products. The process has been largely replaced by catalytic cracking and some refineries no longer employ thermal cracking but, because of the increase number of high-boiling feedstocks entering refineries and the propensity of these feedstocks to poison catalysts, there has been a re-emergence of interest in thermal cracking processes.

Thermal cracking (like visbreaking) reduces the production of less valuable products such as heavy fuel oil and cutter stock and increase the feedstock to the catalytic cracker and gasoline yields. In a thermal cracking process, heavy gas oils and the residuum from the vacuum distillation process are typical feedstocks. The feed stock is heated in a furnace or other thermal unit to up to 540°C (1,000°F) and then fed to a reaction chamber which is kept at a pressure of about 140 psig. The product is then fed to a flasher chamber, where pressure is reduced and lower boiling products vaporize and are drawn off as overhead to a fractionating tower where the various fractions are separated. The bottoms consist of heavy cracked residuum (*pitch*), part of which may be used for fuel or recycled for further cracking.

Visbreaking operates in a similar manner to thermal cracking (480°C, 895°F; outlet pressure: approximately 100 psi) except the product is quenched to mitigate coke forming reactions. The process is a mild thermal cracking operation that can be used to reduce the viscosity of residua to allow the products to meet fuel oil specifications. Alternatively, the visbroken residua could be blended with lighter product oils to produce fuel oils of acceptable viscosity. By reducing the viscosity of the residuum, visbreaking reduces the amount of light heating oil that is required for blending to meet the fuel oil specifications. In addition to the major product, fuel oil, material in the gas oil and gasoline boiling range is produced. The gas oil may be used as additional feed for catalytic cracking units, or as heating oil.

Thermal cracking and visbreaking tend to produce a relatively small amount of fugitive emissions and sour wastewater. Usually some wastewater is produced from steam strippers and the fractionator. Wastewater is also generated during unit cleanup and cooling operations and from the steam injection process to remove organic deposits from the soaker or from the coil. Combined wastewater flows from thermal cracking and coking processes are about 3.0 gallons per barrel of process feed.

2.5.7.5 Coking Processes

Coking is a cracking process used primarily to reduce refinery production of low-value residual fuel oils to transportation fuels, such as gasoline and diesel fuel. As part of the upgrading process, coking also produces crude oil coke, which is essentially solid carbon with varying amounts of impurities, and is used as a fuel for power plants if the sulfur content is low enough. Coke also has non-fuel applications as a raw material for many carbon and graphite products including anodes for the production of aluminum, and furnace electrodes for the production of elemental phosphorus, titanium dioxide, calcium carbide and silicon carbide.

In delayed coking operations, the same basic process as thermal cracking is used except feed streams are allowed to react longer without being cooled. The delayed coking feed stream of residual oils from various upstream processes is first introduced to a fractionating tower where residual lighter materials are drawn off and the heavy ends are condensed. The heavy ends are removed and heated in a furnace to about 480 to 540°C (900 to 1,000°F) and then fed to an insulated vessel called a coke drum where the coke is formed. When the coke drum is filled with product, the feed is switched to an empty parallel drum. Hot vapors from the coke drums, containing cracked lighter hydrocarbon products, hydrogen sulfide, and ammonia, are fed back to the fractionator where they can be treated in the sour gas treatment system or drawn off as intermediate products. Steam is then injected into the full coke drum to remove hydrocarbon vapors, water is injected to cool the coke, and the coke is removed. Typically, high pressure water jets are used to cut the coke from the drum.

Delayed coking and fluid coking produce a relatively small amount of sour wastewater from the associated steam strippers and fractionators. Wastewater is generated during coke removal and cooling operations and from the steam injection process to cut coke from the coke drums. Combined wastewater flows from thermal cracking and coking processes are about 3.0 gallons per barrel of process feed.

Like most separation processes in the refinery, the process water used in coker fractionators (as is also the case in other product fractionators) often comes in direct contact with oil, and can have a high oil content (much of that oil can be recovered through wastewater oil recovery processes). Thus, the main constituents of sour water from catalytic cracking include high levels of oil, suspended solids, phenols, cyanides, hydrogen sulfate, and ammonia. Typical wastewater flow from catalytic cracking is about 15.0 gallons per barrel of feed processed (more than one-third of a gallon of waste water for every gallon of feed processed), and represents the second largest source of wastewater in the refinery.

The need to upgrade heavy crude oil is increasing the use of coking technologies—especially delayed coking—which in turn is increasing the amount of particulate matter (*dust* or *fines*) produced. Particulate emissions from decoking can also be considerable. Coke-laden water from decoking operations in delayed cokers (hydrogen sulfide, ammonia, suspended solids), coke dust (carbon particles and hydrocarbon derivatives) occur. One aspect of particulate matter mitigation is to insure that the feedstock is free of particulate matter before entry into the coking unit. Other steps to mitigate the production of particulate matter include changes to (1) the fractionator bottoms design, (2) pump design, (3) heater design, and (4) fluid velocity (Sayles and Romero, 2013).

2.5.7.6 Fluid Catalytic Cracking

Fluid bed catalytic cracking processes use heat, pressure and a catalyst to produce lower boiling products from high boiling feedstocks. Catalytic cracking has largely replaced thermal cracking because it is able to produce more gasoline with a higher octane and less heavy fuel oils and light gases. Feedstocks are light and heavy oils from the crude oil distillation unit which are processed primarily into gasoline as well as some fuel oil and light gases. Most catalysts used in catalytic cracking consist of mixtures of crystalline zeolites and amorphous synthetic silica-alumina. The catalytic cracking processes, as well as most other refinery catalytic processes, produce coke which collects on the catalyst surface and diminishes its catalytic properties. The catalyst, therefore, needs to be regenerated continuously or periodically essentially by burning the coke off the catalyst at high temperatures. The method and frequency in which catalysts are regenerated are a major factor in the design of catalytic cracking units. A number of different catalytic cracking designs are currently in use including fixed-bed reactors, moving-bed reactors, fluidized-bed reactors, and once-through units. The fluidized- and moving-bed reactors are by far the most prevalent.

Fluid catalytic cracking is one of the largest sources of air emission in refineries. Air emissions are released in process heater flue gas, as fugitive emissions from leaking valves and pipes, and during regeneration of the cracking catalyst. If not controlled, catalytic cracking is one of the most substantial sources of carbon monoxide and particulate emissions in the refinery. In non-attainment areas where carbon monoxide and particulates are above acceptable levels, carbon monoxide waste heat boilers (CO boiler) and particulate controls are employed. Carbon monoxide produced during regeneration of the catalyst is converted to carbon dioxide either in the regenerator or further downstream in a carbon monoxide waste heat boiler (CO boiler). Catalytic crackers are also significant sources of sulfur oxides and nitrogen oxides. The nitrogen oxides produced by catalytic crackers is expected to be a major target of emissions reduction in the future.

Catalytic cracking units, like coking units, usually include some form of fractionation or steam stripping as part of the process configuration. These units all produce sour waters and sour gases containing some hydrogen sulfide and ammonia. Like crude oil distillation, some of the toxic releases reported by the refining industry are generated through sour water and gases, notably ammonia. Gaseous ammonia often leaves fractionating and treating processes in the sour gas along with hydrogen sulfide and fuel gases.

Catalytic cracking produces large volumes of wastewater and spent catalysts. Catalytic cracking (primarily fluid catalytic cracking) generates considerable sour wastewater from fractionators used for product separation, from steam strippers used to strip oil from catalysts, and in some cases from scrubber water. The steam stripping process used to purge and regenerate the catalysts can contain metal impurities from the feed in addition to oil and other contaminants. Sour wastewater from the fractionator/gas concentration units and steam strippers contain oil, suspended solids, phenols, cyanides, hydrogen sulfide, ammonia), spent catalysts (metals from crude oil and hydrocarbon derivatives).

Catalytic cracking generates significant quantities of spent process catalysts (containing metals from crude oils and hydrocarbon derivatives) that are often sent off-site for disposal or recovery or recycling. Management options can include land filling, treatment, or separation and recovery of the metals. Metals deposited on catalysts are often recovered by third-party recovery facilities. Spent catalyst fines (containing aluminum silicate and metals) from electro-static precipitators are also sent off-site for disposal and/or recovery options.

Catalytic crackers also produce a significant amount of fine catalyst dust that results from the constant movement of catalyst grains against each other. This dust contains primarily alumina (Al_2O_3) and small amounts of nickel (Ni) and vanadium (V), and is generally carried along with the carbon monoxide stream to the carbon monoxide waste heat boiler. The dust is separated from the carbon dioxide stream exiting the boiler through the use of cyclones, flue gas scrubbing or electrostatic precipitators, and may be disposed of at an offsite facility.

2.5.7.7 Hydrocracking and Hydrotreating

Hydrotreating and hydroprocessing are similar processes used to remove impurities such as sulfur, nitrogen, oxygen, halides and trace metal impurities that may deactivate process catalysts. Hydrotreating also upgrades the quality of fractions by converting olefin derivatives and diolefin derivatives to paraffins for the purpose of reducing gum formation in fuels. Hydroprocessing, which typically uses residua from the crude distillation units, also cracks these heavier molecules to lower boiling more saleable products. Both hydrotreating and hydroprocessing units are usually placed upstream of those processes in which sulfur and nitrogen could have adverse effects on the catalyst, such as catalytic reforming and hydrocracking units. The processes utilize catalysts in the presence of substantial amounts of hydrogen under high pressure and temperature to react the feedstocks and impurities with hydrogen. The reactors are usually fixed-bed with catalyst replacement or regeneration done after months or years of operation often at an off-site facility. In addition to the treated products, the process produces a stream of light fuel gases, hydrogen sulfide, and ammonia. The treated product and hydrogen-rich gas are cooled after they leave the reactor before being separated. The hydrogen is recycled to the reactor.

Hydrocracking typically utilizes a fixed-bed catalytic cracking reactor with cracking occurring under substantial pressure (1,200 to 2,000 psi) in the presence of hydrogen. Feedstocks to hydrocracking units are often those fractions that are the most difficult to crack and cannot be cracked effectively in catalytic cracking units. These include: middle distillate, cycle oil, residual fuel oil, and reduced crude oil. The hydrogen suppresses the formation of heavy residual material and increases the yield of gasoline by reacting with the cracked products. However, this process also breaks the heavy, sulfur and nitrogen bearing hydrocarbon derivatives and releases these impurities to where they could potentially foul the catalyst. For this reason, the feedstock is often first hydrotreated to remove impurities before being sent to the catalytic hydrocracker. Sometimes hydrotreating is accomplished by using the first reactor of the hydrocracking process to remove impurities. Water also has a detrimental effect on some hydrocracking catalysts and must be removed before being fed to the reactor. The water is removed by passing the feedstock through a silica gel or molecular sieve dryer. Depending on the products desired and the size of the unit, catalytic hydrocracking is conducted in either single stage or multi-stage reactor processes. Most catalysts consist of a crystalline mixture of silica-alumina with small amounts of rare earth metals.

Hydrocracking generates air emissions through process heater flue gas, vents, and fugitive emissions. Unlike fluid catalytic cracking catalysts, hydrocracking catalysts are usually regenerated off-site after months or years of operations, and little or no emissions or dust are generated. However, the use of heavy oil as feedstock to the unit can change this balance.

Hydrocracking produces less sour wastewater than catalytic cracking. Hydrocracking, like catalytic cracking, produces sour wastewater at the fractionator. These processes include processing in a separator (API separator, corrugated plate interceptor) that creates sludge. Physical or chemical methods are then used to separate the remaining emulsified oils from the wastewater. Treated wastewater may be discharged to public wastewater treatment, to a refinery secondary treatment plant for ultimate discharge to public wastewater treatment, or may be recycled and used as process water. The separation process permits recovery of usable oil, and also creates a sludge that may be recycled or treated as a hazardous waste.

In addition, oily sludge from the wastewater treatment facility that results from treating sour wastewaters may be hazardous wastes (unless they are recycled in the refining process). These include API separator sludge, primary treatment sludge, sludge from various gravitational separation units, and float from dissolved air flotation units.

Propylene, another source of toxic releases from refineries, is produced as a light end during cracking and coking processes. It is volatile as well as soluble in water, which increases its potential for release to both air and water during processing.

Like catalytic cracking, hydrocracking processes generate a toxic metal compounds, many of which are present in spent catalyst sludge and catalyst fines generated from catalytic cracking and hydrocracking. These include metals such as nickel (Ni), cobalt (Co), and molybdenum (Mo).

Hydrotreating generates air emissions through process heater flue gas, vents, and fugitive emissions. Unlike fluid catalytic cracking catalysts, hydrotreating catalysts are usually regenerated off-site after months or years of operations, and little or no emissions or dust are generated from the catalyst regeneration process at the refinery. Section 10 provides air emissions factors for emissions from process heaters and boilers used throughout the refinery.

The off-gas stream from hydrotreating is usually very rich in hydrogen sulfide and light fuel gas. This gas is usually sent to a sour gas treatment and sulfur recovery unit along with other refinery sour gases.

Fugitive air emissions of volatile components released during hydrotreating may also be toxic components. These include toluene, benzene, xylenes, and other volatiles that are reported as toxic chemical releases under the United States EPA Toxics Release Inventory.

Hydrotreating generates sour wastewater from fractionators used for product separation. Like most separation processes in the refinery, the process water used in fractionators often comes in direct contact with oil, and can be highly contaminated. It also contains hydrogen sulfide and ammonia and must be treated along with other refinery sour waters. In hydrotreating, sour wastewater from fractionators is produced at the rate of about 1.0 gallon per barrel of feed.

Oily sludge from the wastewater treatment facility that result from treating oily and/or sour wastewaters from hydrotreating and other refinery processes may be hazardous wastes, depending on how they are managed. These include API separator sludge, primary treatment sludge, sludge from various gravitational separation units, and float from dissolved air flotation units.

Hydrotreating also produces some residuals in the form of spent catalyst fines, usually consisting of aluminum silicate and some metals (e.g., cobalt, molybdenum, nickel, tungsten). Spent hydrotreating catalyst is now listed as a hazardous waste (K171) (except for most support material). Hazardous constituents of this waste include benzene and arsenia (arsenic oxide, As_2O_3). The support material for these catalysts is usually an inert ceramic (e.g., alumina, Al_2O_3).

2.5.7.8 Catalytic Reforming

Catalytic reforming uses catalytic reactions to process primarily low octane heavy straight run (from the crude distillation unit) gasolines and naphtha into high octane aromatic derivatives (including benzene). There are four major types of reactions which occur during reforming processes: (1) dehydrogenation of naphthenes to aromatic derivatives, (2) dehydrocyclization of paraffins to aromatic derivatives, (3) isomerization, and (4) hydrocracking. The dehydrogenation reactions are endothermic, requiring that the hydrocarbon stream be heated between each catalyst bed. All but the hydrocracking reaction release hydrogen which can be used in the hydrotreating or hydrocracking processes. Fixed-bed or moving bed processes are utilized in a series of three to six reactors.

Feedstocks to catalytic reforming processes are usually hydrotreated first to remove sulfur, nitrogen and metallic contaminants. In continuous reforming processes, catalysts can be regenerated one reactor at a time, once or twice per day, without disrupting the operation of the unit. In semi regenerative units, regeneration of all reactors can be carried out simultaneously after three to twenty four months of operation by first shutting down the process.

Emissions from catalytic reforming include fugitive emissions of volatile constituents in the feed, and emissions from process heaters and boilers. As with all process heaters in the refinery, combustion of fossil fuels produces emissions of sulfur oxides, nitrogen oxides, carbon monoxide, particulate matter, and volatile hydrocarbon derivatives.

Toluene, xylene, and benzene are toxic aromatic chemicals that are produced during the catalytic reforming process and used as feedstocks in chemical manufacturing. Due to their highly volatile nature, fugitive emissions of these chemicals are a source of their release to the environment during the reforming process. Point air sources may also arise during the process of separating these chemicals.

In a continuous reformer, some particulate and dust matter can be generated as the catalyst moves from reactor to reactor, and is subject to attrition. However, due to catalyst design little attrition occurs, and the only outlet to the atmosphere is the regeneration vent, which is most often scrubbed with a caustic to prevent emission of hydrochloric acid (this also removes particulate matter). Emissions of carbon monoxide and hydrogen sulfide may occur during regeneration of catalyst.

2.5.7.9 Alkylation

Alkylation is used to produce a high octane gasoline blending stock from the iso-butane formed primarily during catalytic cracking and coking operations, but also from catalytic reforming, crude distillation and natural gas processing.

Alkylation joins an olefin and an iso-paraffin compound using either a sulfuric acid or hydrofluoric acid catalyst. The products are alkylates including propane and butane liquids. When the concentration of acid becomes less than 88%, some of the acid must be removed and replaced with stronger acid. In the hydrofluoric acid process, the slip stream of acid is redistilled. Dissolved polymerization products are removed from the acid as thick dark oil. The concentrated hydrofluoric acid is recycled and the net consumption is about 0.3 pounds per barrel of alkylate produced.

Hydrofluoric acid alkylation units require special engineering design, operator training and safety equipment precautions to protect operators from accidental contact with hydrofluoric acid which is an extremely hazardous substance. In the sulfuric acid process, the sulfuric acid removed must be regenerated in a sulfuric acid plant which is generally not a part of the alkylation unit and may be located off-site. Spent sulfuric acid generation is substantial; typically in the range of 13 to 30 pounds per barrel of alkylate. Air emissions from the alkylation process may arise from process vents and fugitive emissions.

Alkylation combines low-molecular-weight olefin derivatives (primarily a mixture of propylene and butylene) with isobutene in the presence of a catalyst, either sulfuric acid or hydrofluoric acid. The product is called alkylate and is composed of a mixture of high-octane, branched-chain paraffinic hydrocarbon derivatives. Alkylate is a premium blending stock because it has exceptional antiknock properties and is clean burning. The octane number of alkylate depends mainly upon the kind of olefin derivatives used and upon operating conditions.

Emissions from alkylation processes and polymerization processes include fugitive emissions of volatile constituents in the feed, and emissions that arise from process vents during processing. These can take the form of acidic hydrocarbon gases, non-acidic hydrocarbon gases, and fumes that may have a strong odor (from sulfonated organic compounds and organic acids, even at low concentrations). To prevent releases of hydrofluoric acid, refineries install a variety of mitigation and control technologies (e.g., acid inventory reduction, hydrogen fluoride detection systems, isolation valves, rapid acid transfer systems, and water spray systems).

In hydrofluoric acid alkylation processes, acidic hydrocarbon gases can originate anywhere hydrogen fluoride is present (e.g., during a unit upset, unit shutdown, or maintenance). Hydrofluoric acid alkylation units are designed to pipe these gases from acid vents and valves to a separate closed-relief system where the acid is neutralized. The basins are tightly covered and equipped with a gas scrubbing system to remove odors, using either water or activated charcoal as the scrubbing agent. Another source of emissions is combustion of fuels in process boilers to produce steam for strippers. As with all process heaters in the refinery, these boilers produce significant emissions of sulfur oxides, nitrogen oxides, carbon monoxide, particulate matter, and volatile hydrocarbon derivatives.

Alkylation generates relatively low volumes of wastewater, primarily from water washing of the liquid reactor products. Wastewater is also generated from steam strippers, depropanizers and debutanizers, and can be contaminated with oil and other impurities. Liquid process waters (hydrocarbon derivatives and acid) originate from minor undesirable side reactions and from feed contaminants, and usually exit as a bottoms stream from the acid regeneration column. The bottoms product is an acid-water mixture that is sent to the neutralizing drum. The acid in this liquid eventually ends up as insoluble calcium fluoride.

Sulfuric acid alkylation generates considerable quantities of spent acid that must be removed and regenerated. Nearly all the spent acid generated at refineries is regenerated and recycled and, although technology for on-site regeneration of spent sulfuric acid is available, the supplier of the acid may perform this task off-site. If sulfuric acid production capacity is limited, acid regeneration is often done on-site. The development of internal acid regeneration for hydrofluoric acid units has virtually eliminated the need for external regeneration, although most operations retain one for start-ups or during periods of high feed contamination.

Both sulfuric acid and hydrofluoric acid alkylation units generate neutralization sludge from treatment of acid-laden streams with caustic solutions in neutralization or wash systems. Sludge from hydrofluoric acid alkylation neutralization systems consists largely of calcium fluoride and unreacted lime, and is usually disposed of in a landfill. It can also be directed to steel manufacturing facilities, where the calcium fluoride can be used as a neutral flux to lower the slag-melting temperature and improve slag fluidity. Calcium fluoride can also be routed back to a hydrofluoric acid manufacturer.

A basic step in hydrofluoric acid manufacture is the reaction of sulfuric acid with fluorspar (calcium fluoride) to produce hydrogen fluoride and calcium sulfate. Spent alumina is also generated by defluoridization of hydrofluoric acid alkylation products over alumina. It is disposed of or sent to the alumina supplier for recovery. Other solid residuals from hydrofluoric acid alkylation include any porous materials that may have come in contact with the hydrofluoric acid.

2.5.7.10 Isomerization and Polymerization

Isomerization is used to alter the arrangement of a molecule without adding or removing anything from the original molecule. Typically, paraffins (butane or pentane from the crude distillation unit) are converted to iso-paraffins having a much higher octane. Isomerization reactions take place at temperatures in the range of 95 to 205°C (200 to 400°F) in the presence of a catalyst that usually consists of platinum on a base material. Two types of catalysts are currently in use. One requires the continuous addition of small amounts of organic chlorides which are converted to hydrogen chloride in the reactor. In such a reactor, the feed must be free of oxygen sources including water to avoid deactivation and corrosion problems. The other type of catalyst uses a molecular sieve base and does not require a dry and oxygen free feedstock. Both types of isomerization catalysts require an atmosphere of hydrogen to minimize coke deposits; however, the consumption of hydrogen is negligible. Catalysts typically need to be replaced about every two to three years or longer. Platinum is then recovered from the used catalyst off-site.

Light ends are stripped from the product stream leaving the reactor and are then sent to the sour gas treatment unit. Some isomerization units utilize caustic treating of the light fuel gas stream to neutralize any entrained hydrochloric acid. This will result in a calcium chloride (or other salts) waste stream. Air emissions may arise from the process heater, vents and fugitive emissions. Wastewater streams include caustic wash and sour water.

Isomerization processes produce sour water and caustic wastewater. The ether manufacturing process utilizes a water wash to extract methanol or ethanol from the reactor effluent stream. After the alcohol is separated this water is recycled back to the system and is not released. In those cases where chloride catalyst activation agents are added, a caustic wash is used to neutralize any entrained hydrogen chloride. This process generates a caustic wash water that must be treated before being released. This process also produces a calcium chloride neutralization sludge that must be disposed of off-site.

Polymerization is occasionally used to convert propylene and butene to high octane gasoline blending components. The process is similar to alkylation in its feed and products, but is often used as a less expensive alternative to alkylation.

The reactions typically take place under high pressure in the presence of a phosphoric acid catalyst. The feed must be free of sulfur, which poisons the catalyst; basic materials, which neutralize the catalyst; and oxygen, which affects the reactions. The propylene and butene feedstock is washed first with caustic to remove mercaptans (molecules containing sulfur), then with an amine solution to remove hydrogen sulfide, then with water to remove caustics and amines, and finally dried by passing through a silica gel or molecular sieve dryer. Air emissions of sulfur dioxide may arise during the caustic washing operation. Spent catalyst, which typically is not regenerated, is occasionally disposed as a solid waste. Wastewater streams will contain caustic wash and sour water with amines and mercaptans.

2.5.7.11 Deasphalting

Propane deasphalting produces lubricating oil base stocks by extracting asphaltenes and resins from vacuum distillation residua. Propane is the usual solvent of choice due to its unique solvent properties. At lower temperatures (38 to 60°C, 100 to 140°F), paraffins are very soluble in propane and at higher temperatures (about 93°C, 200°F) hydrocarbon derivatives are almost insoluble in propane. The propane deasphalting process is similar to solvent extraction in that a packed or baffled extraction tower or rotating disc contactor is used to mix the oil feed stocks with the solvent. In the tower method, four to eight volumes of propane are fed to the bottom of the tower for every volume of feed flowing down from the top of the tower. The oil, which is more soluble in the propane, dissolves and flows to the top. The higher molecular weight polar asphalt constituents flow to the bottom of the tower where they are removed in a propane mix. Propane is recovered from the two streams through two-stage flash systems followed by steam stripping in which propane is condensed and removed by cooling at high pressure in the first stage and at low pressure in the second stage. The asphalt recovered can be blended with other asphalt, or heavy fuel oil, or can be used as feed to the coker. The propane recovery stage results in propane contaminated water that typically is sent to the wastewater treatment plant.

Air emissions may arise from fugitive propane emissions and process vents. These include heater stack gas (carbon monoxide, sulfur oxides, nitrogen oxides, and particulate matter) as well as hydrocarbon emission such as fugitive propane, and fugitive solvents. Steam stripping wastewater (oil and solvents), solvent recovery wastewater (oil and propane) are also produced.

2.5.7.12 Dewaxing

Dewaxing of lubricating oil base stocks is necessary to ensure that the oil will have the proper viscosity at lower ambient temperatures. Two types of dewaxing processes are used: selective hydrocracking and solvent dewaxing.

In selective hydrocracking, one or two zeolite catalysts are used to selectively crack the wax paraffins. Solvent dewaxing is more prevalent. In solvent dewaxing, the oil feed is diluted with solvent to lower the viscosity, chilled until the wax is crystallized, and then filtered to remove the wax. Solvents used for the process include propane and mixtures of methyl ethyl ketone (MEK) with methyl isobutyl ketone (MIBK) or MEK with toluene. Solvent is recovered from the oil and wax through heating, two-stage flashing, followed by steam stripping. The solvent recovery stage results in solvent contaminated water which typically is sent to the wastewater treatment plant. The wax is either used as feed to the catalytic cracker or is de-oiled and sold as industrial wax.

Dewaxing processes also produce heater stack gas (carbon monoxide, sulfur oxides, nitrogen oxides, and particulate matter) as well as hydrocarbon emission such as fugitive propane, and fugitive solvents. Steam stripping wastewater (oil and solvents), solvent recovery wastewater (oil and propane) are also produced. The fugitive solvent emissions may be toxic (toluene, methyl ethyl ketone, methyl isobutyl ketone).

2.5.8 Types of Waste

Pollution associated with crude oil refining typically includes VOCs (volatile organic compounds), carbon monoxide (CO), sulfur oxides (SO$_x$), nitrogen oxides (NO$_x$), particulates, ammonia (NH$_3$), hydrogen sulfide (H$_2$S), metals, spent acids, and numerous toxic organic compounds. Sulfur and metals result from the impurities in crude oil. The other wastes represent losses of inputs and final product. These pollutants may be discharged as air emissions, waste water, or solid waste. All of these wastes are treated. However, air emissions are more difficult to capture than waste water or solid waste. Thus, air emissions are the largest source of untreated wastes released to the environment.

More specifically to crude oil and crude oil products, the alkane derivatives in gasoline and some other crude oil products are CNS depressants. In fact, gasoline was once evaluated as an anesthetic agent. However, sudden deaths, possibly as a result of irregular heartbeats, have been attributed to those inhaling vapors of hydrocarbon derivatives such as those in gasoline.

Alkane derivatives of various types of crude oils and various crude oil products were biodegraded faster than the *unresolved fractions*. Different types of crude oils and products biodegraded at different rates in the same environments. An oil product is a complex mixture of organic chemicals and contains within it less persistent and more persistent fractions. The range between these two extremes is greatest for crude oils. Since the many different substances in crude oil have different physical and chemical properties, summarizing the fate of crude oil in general (or even a specific crude oil) is very difficult. Solubility-fate relationships must be considered.

The relative proportion of hazardous constituents present in crude oil is typically quite variable. Therefore, contamination will vary from one site to another. In addition, the farther one progresses from lighter towards heavier constituents (the general progression from lower molecular weight to higher molecular weight constituents) the greater the percentage of polynuclear aromatic hydrocarbon derivatives and other semi-volatile constituents or non-volatile constituents (many of which are not so immediately toxic as the volatiles but which can result in long-term/chronic impacts). These higher molecular weight constituents thus need to be analyzed for the semi volatile compounds that typically pose the greatest long-term risk.

In addition to large oil spills, crude oil hydrocarbon derivatives are released into the aquatic environments from natural seeps as well as non-point source urban runoffs. Acute impacts from massive one-time spills are obvious and substantial. The impacts from small spills and chronic releases are the subject of much speculation and continued research. Clearly, these inputs of crude oil hydrocarbon derivatives have the potential for significant environmental impacts, but the effects of chronic low-level discharges can be minimized by the net assimilative capacities of many ecosystems, resulting in little detectable environmental harm.

Short-term (acute) hazards of lighter, more volatile and water soluble aromatic compounds (such as benzenes, toluene, and xylenes) include potential acute toxicity to aquatic life in the water column (especially in relatively confined areas) as well as potential inhalation hazards. However, the compounds which pass through the water column often tend to do so in small concentrations and/or for short periods of time, and fish and other pelagic or generally mobile species can often swim away to avoid impacts from spilled oil in open waters. Most fish are mobile and it is not known whether or not they can sense, and thus avoid, toxic concentrations of oil.

However, there are some potential effects of spilled oil on fish. The impacts to fish are primarily to the eggs and larvae, with limited effects on the adults. The sensitivity varies by species; pink salmon fry are affected by exposure to water-soluble fractions of crude oil, while pink salmon pink salmon eggs are very tolerant to benzene and water-soluble crude oil. The general effects are difficult to assess and quantitatively document due to the seasonal and natural variability of the species. Fish rapidly metabolize aromatic hydrocarbon derivatives due to their enzyme system.

Long-term (chronic) potential hazards of lighter, more volatile and water soluble aromatic compounds include contamination of groundwater. Chronic effects of benzene, toluene, and xylene include changes in the liver and harmful effects on the kidneys, heart, lungs, and nervous system.

At the initial stages of a release, when the benzene-derived compounds are present at their highest concentrations, acute toxic effects are more common than later. These non-carcinogenic effects include subtle changes in detoxifying enzymes and liver damage. Generally, the relative aquatic acute toxicity of crude oil will be the result of the fractional toxicities of the different hydrocarbon derivatives present in the aqueous phase. Tests indicate that naphthalene-derived chemicals have a similar effect.

Except for short-term hazards from concentrated spills, BTEX compounds (benzene, toluene, ethyl benzene, and xylenes) have been more frequently associated with risk to humans than with risk to non-human species such as fish and wildlife. This is partly because plants, fish, and birds take up only very small amounts and because this volatile compound tends to evaporate into the atmosphere rather than persisting in surface waters or soils. However, volatiles such as this compound have can pose a drinking water hazard when they accumulate in ground water. See also, BTEX entry, and entries for benzene, toluene, ethyl benzene, and xylenes.

Crude oil is naturally weathered according to its physical and chemical properties, but during this process living species within the local environment may be affected via one or more routes of exposure, including ingestion, inhalation, dermal contact, and, to a much lesser extent, bioconcentration through the food chain. Aromatic compounds of concern include alkylbenzenes, toluene, naphthalenes, and polynuclear aromatic hydrocarbon derivatives (PNAs). Moreover, both atmospheric and hydrospheric impacts must be assessed when considering toxic implications from a crude oil release containing significant quantities of these single-ring aromatic compounds.

Gases and Lower Boiling Constituents

Gases and lower boiling constituents (refinery gases) contain one or more organic and inorganic constituents and are mixtures of individual compounds existing in the gaseous phase at normal environmental temperatures. These constituents typically have extremely low melting and boiling points. They also have high vapor pressures and low octanol/water partition coefficients. The aqueous solubility of these components varies, and can range from low parts per million (hydrogen gas) to several hundred thousand parts per million (ammonia). The environmental fate characteristics of refinery gases are governed by these physical-chemical attributes.

All components of these gases will partition to the air where interaction with hydroxyl radicals may be either an important fate process or have little influence, depending on the constituent. Many of the gases are chemically stable and may be lost to the atmosphere or simply become involved in the environmental recycling of their atoms. Some show substantial water solubility, but their volatility eventually causes these gases to enter the atmosphere (US EPA 2009).

The numerous process heaters used in refineries to heat process streams or to generate steam (boilers) for heating or steam stripping, can be potential sources of SO_x, NO_x, CO, particulates and hydrocarbon derivatives emissions. When operating properly and when burning cleaner fuels such as refinery fuel gas, fuel oil or natural gas, these emissions are relatively low. If, however, combustion is not complete, or heaters are fired with refinery fuel pitch or residuals, emissions can be significant.

The majority of gas streams exiting each refinery process contain varying amounts of refinery fuel gas, hydrogen sulfide and ammonia. These streams are collected and sent to the gas treatment and sulfur recovery units to recover the refinery fuel gas and sulfur. Emissions from the sulfur recovery unit typically contain some H_2S, SO_x, and NO_x. Other emissions sources from refinery processes arise from periodic regeneration of catalysts. These processes generate streams that may contain relatively high levels of carbon monoxide, particulates and volatile organic compounds. Before being discharged to the atmosphere, such off-gas streams may be treated first through a carbon monoxide boiler to burn carbon monoxide and any volatile organic compounds, and then through an electrostatic precipitator or cyclone separator to remove particulates.

Sulfur is removed from a number of refinery process off-gas streams (sour gas) in order to meet the SO_x emissions limits of the Clean Air Act and to recover saleable elemental sulfur. Process off-gas streams, or sour gas, from the coker, catalytic cracking unit, hydrotreating units and hydroprocessing units can contain high concentrations of hydrogen sulfide mixed with light refinery fuel gases. Before elemental sulfur can be recovered, the fuel gases (primarily methane and ethane) need to be separated from the hydrogen sulfide. This is typically accomplished by dissolving the hydrogen sulfide in a chemical solvent. Solvents most commonly used are amines, such as diethanolamine (DEA). Dry adsorbents such as molecular sieves, activated carbon, iron sponge and zinc oxide are also used. In the amine solvent processes, DEA solution or another amine solvent is pumped to an absorption tower where the gases are contacted and hydrogen sulfide is dissolved in the solution. The fuel gases are removed for use as fuel in process furnaces in other refinery operations. The amine-hydrogen sulfide solution is then heated and steam stripped to remove the hydrogen sulfide gas.

Current methods for removing sulfur from the hydrogen sulfide gas streams are typically a combination of two processes: the Claus Process followed by the Beavon Process, Scot Process, or the Wellman-Land Process. The Claus process consists of partial combustion of the hydrogen sulfide-rich gas stream (with one-third the stoichiometric quantity of air) and then reacting the resulting sulfur dioxide and unburned hydrogen sulfide in the presence of a bauxite catalyst to produce elemental sulfur.

Since the Claus process by itself removes only about 90 percent of the hydrogen sulfide in the gas stream, the Beavon, SCOT, or Wellman-Lord processes are often used to further recover sulfur. In the Beavon process, the hydrogen sulfide in the relatively low concentration gas stream from the Claus process can be almost completely removed by absorption in quinone solution. The dissolved hydrogen sulfide is oxidized to form a mixture of elemental sulfur and hydro-quinone. The solution is injected with air or oxygen to oxidize the hydroquinone back to quinone. The solution is then filtered or centrifuged to remove the sulfur and the quinone is then reused. The Beavon process is also effective in removing small amounts of sulfur dioxide, carbonyl sulfide, and carbon disulfide that are not affected by the Claus process. These compounds are first converted to hydrogen sulfide at elevated temperatures in a cobalt molybdate catalyst prior to being fed to the Beavon unit. Air emissions from sulfur recovery units will consist of hydrogen sulfide, SO_x and NO_x in the process tail gas as well as fugitive emissions and releases from vents.

The SCOT process is also widely used for removing sulfur from the Claus tail gas. The sulfur compounds in the Claus tail gas are converted to hydrogen sulfide by heating and passing it through a cobalt-molybdenum catalyst with the addition of a reducing gas. The gas is then cooled and contacted with a solution of diisopropanolamine (DIPA) which removes all but trace amounts of hydrogen sulfide. The sulfide-rich diisopropanolamine is sent to a stripper where hydrogen sulfide gas is removed and sent to the Claus plant and the cleaned diisopropanolamine is returned to the absorption column.

Most refinery process units and equipment are sent to a collection unit called the blowdown system. Blowdown systems provide for the safe handling and disposal of liquid and gases that are either automatically vented from the process units through pressure relief valves, or that are manually drawn from units. Recirculated process streams and cooling water streams are often manually purged to prevent the continued buildup of contaminants in the stream. Part or all of the contents of equipment can also be purged to the blowdown system prior to shut down before normal or emergency shutdowns. Blowdown systems utilize a series of flash drums and condensers to separate the blowdown into its vapor and liquid components. The liquid is typically composed of mixtures of water and hydrocarbon derivatives containing sulfides, ammonia, and other contaminants, which are sent to the wastewater treatment plant. The gaseous component typically contains hydrocarbon derivatives, hydrogen sulfide, ammonia, mercaptans, solvents, and other constituents, and is either discharged directly to the atmosphere or is combusted in a flare. The major air emissions from blowdown systems are hydrocarbon derivatives in the case of direct discharge to the atmosphere and sulfur oxides when flared.

Many of the gaseous and liquid constituents of the lower boiling fractions of crude oil and also in crude oil products fall into the class of chemicals which have the one or more of the following characteristics are considered to be hazardous by the Environmental Protection agency in terms of the following properties (1) ignitability-flammability, (2) corrosivity, (3) reactivity, and (4) hazardous.

An ignitable liquid is a liquid that has a flash point of less than 60°C (140°F). Examples are: benzene, hexane, heptane, benzene, pentane, low-boiling naphtha sometimes referred to as petroleum ether, toluene, and the xylene isomers. An aqueous solution that has a pH of less than or equal to 2, or greater than or equal to 12.5 is considered *corrosive*. Most crude oil constituents and crude oil products are not corrosive but many of the chemicals used in refineries are corrosive. Corrosive materials also include substances such as sodium hydroxide and some other acids or bases. Chemicals that react violently with air or water are considered *reactive*. Examples are sodium metal, potassium metal, phosphorus, etc. Reactive materials also include strong oxidizers such as perchloric acid ($HClO_4$), and chemicals capable of detonation when subjected to an initiating source, such as solid, dry < 10% H_2O picric acid, benzoyl peroxide or sodium borohydride ($NaBH_4$). Solutions of certain cyanide or sulfides that could generate toxic gases are also classified as reactive. *Hazardous chemicals* have toxic, carcinogenic, mutagenic or teratogenic effects on humans or other life forms and are designated either as *Acutely Hazardous Waste* or *Toxic Waste* by the Environmental Protection Agency. Substances containing any of the toxic constituents so listed are to be considered *hazardous* unless, after considering the following factors it can reasonably be concluded that the chemical (waste) is not capable of posing a substantial present or potential hazard to public health or the environment when improperly treated, stored, transported or disposed of, or otherwise managed.

The issues to be held in consideration are (1) the nature of the toxicity presented by the constituent, (2) the concentration of the constituent in the waste, (3) the potential of the constituent or any toxic degradation product of the constituent to migrate from the waste into the environment under the types of improper management, (4) the persistence of the constituent or any toxic degradation product of the constituent, (5) the potential for the constituent or any toxic degradation product of the constituent to degrade into non-harmful constituents and the rate of degradation, (6) the degree to which the constituent or any degradation product of the constituent accumulates in an ecosystem, (7) the plausible types of improper management to which the waste could be subjected, (8) the quantities of the waste generated at individual generation sites or on a regional or national basis, (9) the nature and severity of the public health threat and environmental damage that has occurred as a result of the improper management of wastes containing the constituent, and (10) actions taken by other governmental agencies or regulatory programs based on the health or environmental hazard posed by the waste or waste constituent.

Such lists of chemicals are not always complete and omission of a chemical from this list does not mean it is without toxic properties or any other hazard.

Higher Boiling Constituents

Naphthalene and its homologs are less acutely toxic than benzene but are more prevalent for a longer period during oil spills. The toxicity of different crude oils and refined oils depends on not only the total concentration of hydrocarbon derivatives but also the hydrocarbon composition in the water-soluble fraction (WSF) of crude oil, water solubility, concentrations of individual components, and toxicity of the components. The water-soluble fractions prepared from different oils will vary in these parameters. Water-soluble fractions (WSF) of refined oils (for example, No. 2 fuel oil and Bunker C oil) are more toxic than water-soluble fraction of crude oil to several species of fish (killifish and salmon). Compounds with either more rings or methyl substitutions are more toxic than less substituted compounds, but tend to be less water soluble and thus less plentiful in the water-soluble fraction.

Among the polynuclear aromatic hydrocarbon derivatives, the toxicity of crude oil is a function of its di- and tri-aromatic hydrocarbon content. Like the single aromatic ring variations, including benzene, toluene, and the xylenes, all are relatively volatile compounds with varying degrees of water solubility.

There are indications that pure naphthalene (a constituent of moth balls that are, by definition, toxic to moths) and alkyl naphthalene derivatives are from three-to-ten times more toxic to test animals than are benzene and alkyl benzene derivatives. In addition, and because of the low water-solubility of tricyclic and polycyclic (polynuclear) aromatic hydrocarbon derivatives (that is, those aromatic hydrocarbon derivatives heavier than naphthalene), these compounds are generally present at very low concentrations in the water-soluble fraction of oil. Therefore, the results of this study and others conclude that the soluble aromatic derivatives of crude oil (such as benzene, toluene, ethylbenzene, xylenes, and naphthalenes) produce the majority of its toxic effects in the environment.

Once the acutely toxic lighter compounds have been left the aquatic environment through volatilization or degradation, the main concern is chronic effects from heavier and more alkylated polynuclear aromatic hydrocarbon derivatives.

Bird species with water habitats are the species most commonly affected by oil spills and releases. Oil itself breaks down the protective waxes and oils in the feathers and fur of birds and animals and disrupts the fine strand structure of the feathers resulting in a loss of heat retention and buoyancy and possible hypothermia and death. Oiled birds often ingest crude oil while attempting to remove the crude oil from their feathers. The effects of ingested crude oil include anemia, pneumonia, kidney, and liver damage, decreased growth, altered blood chemistry, and decreased egg production and viability. Ingestion of the oil can also kill animals by interfering with their ability to digest food. Chicks may be exposed to crude oil by ingesting food regurgitated by impacted adults.

The dynamics of the oil-in-water dispersion (OWD) are complex and have relevance related to potential toxicity or hazard. In comparing the toxicities to marine animals of oil-in-water dispersions prepared from different oils, not only the amount of oil added but also the concentrations of oil in the aqueous phase and the composition and dispersion-forming characteristics of the parent oil must be taken into consideration. In comparing the potential impacts of spills of different oils on the marine biotic community, the amount of oil per unit water volume required to cause mortality is of greater importance than any other aspect of the crude oil behavior.

Several compounds in crude oil products are carcinogenic. The larger and higher molecular weight aromatic structures (with four to five aromatic rings), which are the more persistent in the environment, have the potential for chronic toxicological effects. Since these compounds are non-volatile and are relatively insoluble in water, their main routes of exposure are through ingestion and epidermal contact. Some of the compounds in this classification are considered possible human carcinogens; these include benzo(a and e)pyrene, benzo(a)anthracene, benzo(b, j, and k)fluorene, benzo(ghi)perylene, chrysene, dibenzo(ah)anthracene, and pyrene.

Mixtures of polynuclear aromatic hydrocarbon derivatives are often carcinogenic and possibly phototoxic. One way to approach site specific risk assessments would be to collect the complex mixture of polynuclear aromatic hydrocarbon derivatives and other lipophilic contaminants in a semi-permeable membrane device (SPMD, also known as a *fat bag*), then test the mixture for carcinogenicity, toxicity, and phototoxicity.

The solubility of hydrocarbon components in crude oil products is an important property when assessing toxicity. The water solubility of a substance determines the routes of exposure that are possible. Solubility is approximately inversely proportional to molecular weight; lighter hydrocarbon derivatives are more soluble in water than higher molecular weight compounds. Lower molecular weight hydrocarbon derivatives (C4 to C8, including the aromatic compounds) are relatively soluble, up to about 2,000 ppm, while the higher molecular weight hydrocarbon derivatives are nearly insoluble. Usually, the most soluble components are also the most toxic.

Finally, the toxicity of crude oil may be affected by factors such as "weathering" time or the addition of oil dispersants. *Weathered* crude oil and *fresh* crude oil may have different toxicities, depending on oil type and weathering time. *Wastewater*

A number of wastewater issues face the refining industry. These issues include chemicals in waste process waters. However, efforts by the industry are being continued to eliminate any water contamination that may occur, whether it be from inadvertent leakage of crude oil or crude oil products or leakage of contaminated water from one or more processes. In addition to monitoring organics in the water, metals concentration must be continually monitored since heavy metals tend to concentrate in the body tissues of fish and animals and increase in concentration as they go up the food chain. General sewage problems face every municipal sewage treatment facility regardless of size.

Primary treatment (solid settling and removal) is required and secondary treatment (use of bacteria and aeration to enhance organic degradation) is becoming more routine, tertiary treatment (filtration through activated carbon, applications of ozone, and chlorination) have been, or are being, implemented by all refineries.

Wastewater pretreaters that discharge water into sewer systems have new requirements. Pollutant standards for sewage sludge have been set. Toxics in the water must be identified and plans must be developed to alleviate any problems. In addition, regulators have established, and continue to establish, water-quality standards for priority toxic pollutants.

Waste water

From crude oil refining consist of cooling water, process water, storm water, and sanitary sewage water. A large portion of water used in crude oil refining is used for cooling and most cooling water is recycled. Cooling water typically does not come into direct contact with process oil streams and therefore contains a lesser amount of contaminants than process wastewater. However, it may contain some oil contamination due to leaks in the process equipment.

Refinery effluent water contains various hydrocarbon components including gasoline blending stocks, kerosene, diesel fuel and heavier liquids. Also present may be suspended mineral solids, sand, salt, organic acids and sulfur compounds. The nature of the components depends on the constituents of the inlet crude oil as well as the processing scheme of the refinery (Speight, 2005). Most of these constituents would be undesirable in the effluent water, so it is necessary to treat the water to remove the contaminants.

Water used in processing operations accounts for a significant portion of the total wastewater. Process wastewater arises from desalting crude oil, steam stripping operations, pump gland cooling, product fractionator reflux drum drains and boiler blowdown. Because process water often comes into direct contact with oil, it is usually highly contaminated. Storm water (i.e., surface water runoff) is intermittent and will contain constituents from spills to the surface, leaks in equipment and any materials that may have collected in drains. Runoff surface water also includes water coming from crude and product storage tank roof drains.

The issue face the refining industry includes chemicals in waste process waters. However, efforts by the industry are being continued to eliminate any water contamination that may occur, whether it be from inadvertent leakage of crude oil or crude oil products or leakage of contaminated water from one or more processes. In addition to monitoring organics in the water, metals concentration must be continually monitored since heavy metals tend to concentrate in the body tissues of fish and animals and increase in concentration as they go up the food chain. General sewage problems face every municipal sewage treatment facility regardless of size.

Primary treatment (solid settling and removal) is required and secondary treatment (use of bacteria and aeration to enhance organic degradation) is becoming more routine, tertiary treatment (filtration through activated carbon, applications of ozone, and chlorination) have been, or are being, implemented by all refineries.

After primary treatment, the wastewater can be discharged to a publicly owned treatment works (POTW) or undergo secondary treatment before being discharged directly to surface waters under a National Pollution Discharge Elimination System (NPDES) permit. In secondary treatment, dissolved oil and other organic pollutants may be consumed biologically by microorganisms. Biological treatment may require the addition of oxygen through a number of different techniques, including activated sludge units, trickling filters, and rotating biological contactors. Secondary treatment generates bio-mass waste which is typically treated anaerobically and then dewatered.

Some refineries employ an additional stage of wastewater treatment called polishing to meet discharge limits. The polishing step can involve the use of activated carbon, anthracite coal, or sand to filter out any remaining impurities, such as biomass, silt, trace metals and other inorganic chemicals, as well as any remaining organic chemicals.

Certain refinery wastewater streams are treated separately, prior to the wastewater treatment plant, to remove contaminants that would not easily be treated after mixing with other wastewater. One such waste stream is the sour water drained from distillation reflux drums. Sour water contains dissolved hydrogen sulfide and other organic sulfur compounds and ammonia which are stripped in a tower with gas or steam before being discharged to the wastewater treatment plant.

Wastewater treatment plants are a significant source of refinery air emissions and solid wastes. Air releases arise from fugitive emissions from the numerous tanks, ponds and sewer system drains. Solid wastes are generated in the form of sludge from a number of the treatment units.

If liquid hydrocarbon derivatives are released to ground water and surface waters, migration off-site can occur resulting in continuous *seeps* to surface waters. While the actual volume of hydrocarbon derivatives released in such a manner are relatively small, there is the potential to contaminate large volumes of ground water and surface water possibly posing a substantial risk to human health and the environment.

Wastewater pretreaters that discharge water into sewer systems have new requirements. Pollutant standards for sewage sludge have been set. Toxics in the water must be identified and plans must be developed to alleviate any problems. In addition, regulators have established, and continue to establish, water-quality standards for priority toxic pollutants.

Solid Waste

Solid wastes are generated from many of the refining processes, crude oil handling operations, as well as wastewater treatment. Both hazardous and non-hazardous wastes are generated, treated and disposed. Refinery wastes are typically in the form of sludge (including sludge from wastewater treatment), spent process catalysts, filter clay, and incinerator ash. Treatment of these wastes includes incineration, land treating off-site, land filling onsite, land filling off-site, chemical fixation, neutralization, and other treatment methods.

A significant portion of the non-crude oil product outputs of refineries is transported off-site and sold as byproducts. These outputs include sulfur, acetic acid, phosphoric acid, and recovered metals. Metals from catalysts and from the crude oil that have deposited on the catalyst during the production often are recovered by third party recovery facilities

Storage tanks are used throughout the refining process to store crude oil and intermediate process feeds for cooling and further processing. Finished crude oil products are also kept in storage tanks before transport off site. Storage tank bottoms are mixtures of iron rust from corrosion, sand, water, and emulsified oil and wax, which accumulate at the bottom of tanks. Liquid tank bottoms (primarily water and oil emulsions) are periodically drawn off to prevent their continued build up. Tank bottom liquids and sludge are also removed during periodic cleaning of tanks for inspection. Tank bottoms may contain amounts of tetraethyl or tetramethyl lead (although this is increasingly rare due to the phase-out of leaded products), other metals, and phenols. Solids generated from leaded gasoline storage tank bottoms are listed as a RCRA hazardous waste.

2.5.9 Waste Toxicity

With few exceptions, the constituents of crude oil, crude oil products, and the various emissions are hazardous to the health. There always exceptions that will be cited in opposition to such a statement, the most common exception being the liquid paraffin that is used medicinally to lubricate the alimentary tract. The use of such medication is common among miners who breathe and swallow coal dust every day during their work shifts.

Another approach is to consider crude oil constituents in terms of transportable materials, the character of which is determined by several chemical and physical properties (i.e., solubility, vapor pressure, and propensity to bind with soil and organic particles). These properties are the basis of measures of leachability and volatility of individual hydrocarbon derivatives. Thus, crude oil transport fractions can be considered by equivalent carbon number to be grouped into thirteen different fractions. The analytical fractions are then set to match these transport fractions, using specific *n*-alkane derivatives to mark the analytical results for aliphatic compounds and selected aromatic compounds to delineate hydrocarbon derivatives containing benzene rings.

Although chemicals grouped by transport fraction generally have similar toxicological properties, this is not always the case. For example, benzene is a carcinogen but many alkyl-substituted benzene derivatives do not fall under this classification. However, it is more appropriate to group benzene with compounds that have similar environmental transport properties than to group it with other carcinogens such as benzo(a)pyrene that have very different environmental transport properties.

Nevertheless consultation of any reference work that lists the properties of chemicals will show the properties and hazardous nature of the types of chemicals that are found in crude oil. In addition, crude oil is used to make crude oil products, which can contaminate the environment.

The range of chemicals in crude oil and crude oil products is so vast that summarizing the properties and/or the toxicity or general hazard of crude oil in general or even for a specific crude oil is a difficult task. However, crude oil and some crude oil products, because of the hydrocarbon content, are at least theoretically biodegradable but large-scale spills can overwhelm the ability of the ecosystem to break the oil down. The toxicological implications from crude oil occur primarily from exposure to or biological metabolism of aromatic structures. These implications change as an oil spill ages or is weathered.

2.6 Entry into the Environment

It is almost impossible to transport, store and refine crude oil without spills and losses. It is difficult to prevent spills resulting from failure or damage on pipelines. It is also impossible to install control devices for controlling the ecological properties of water and the soil along the length of all pipelines. The soil suffers the most ecological damage in the damage areas of pipelines. Crude oil spills from pipelines lead to irreversible changes of the soil properties. The most affected soil properties by crude oil losses from pipelines are filtration, physical and mechanical properties. These properties of the soil are important for maintaining the ecological equilibrium in the damaged area.

Principal sources of releases to air from refineries include: (1) combustion plants, emitting sulfur dioxide, oxides of nitrogen and particulate matter, (2) refining operations, emitting sulfur dioxide, oxides of nitrogen, carbon monoxide, particulate matter, volatile organic compounds, hydrogen sulfide, mercaptans and other sulfurous compounds, (3) bulk storage operations and handling of volatile organic compounds (various hydrocarbon derivatives). In light of this, it is necessary to consider (1) regulatory requirements—air emission permits stipulating limits for specific pollutants, and possibly health and hygiene permit requirements, (2) requirement for monitoring program, and (3) requirements to upgrade pollution abatement equipment.

2.6.1 Storage and Handling of Crude Oil and Crude Oil Products

Large quantities of environmentally-sensitive crude oil products are stored in (1) tank farms (multiple tanks), (2) single above-ground storage tanks (ASTs), (3) semi-underground storage tanks or underground storage tanks (USTs). Smaller quantities of materials may be stored in drums and containers of assorted compounds (such as lubricating oil, engine oil, and other products for domestic supply).

In light of this, it is also necessary to consider (1) secondary containment of tanks and other storage areas and integrity of hard standing (without cracks, impervious surface) to prevent spills reaching the wider environment: also secondary containment of pipelines where appropriate, (2) age, construction details and testing program of tanks, (3) labeling and environmentally secure storage of drums (including waste storage), (4) accident/fire precautions, emergency procedures, and (5) disposal/recycling of waste or "out of spec" oils and other materials.

There is a potential for significant soil and groundwater contamination to have arisen at crude oil refineries. Such contamination consists of (1) crude oil hydrocarbon derivatives including lower boiling, very mobile fractions (paraffins, cycloparaffins and volatile aromatics such as benzene, toluene, ethylbenzene and xylenes) typically associated with gasoline and similar boiling range distillates, (2) middle distillate fractions (paraffins, cycloparaffins and some polynuclear aromatics) associated with diesel, kerosene, and lower boiling fuel oil, which are also of significant mobility, (3) higher boiling distillates long-chain paraffins, cycloparaffins and polynuclear aromatics that are associated with lubricating oil and heavy fuel oil, (4) various organic compounds associated with crude oil hydrocarbon derivatives or produced during the refining process, e.g., phenols, amines, amides, alcohols, organic acids, nitrogen and sulfur containing compounds, (5) other organic additives, e.g., anti-freeze (glycols), alcohols, detergents and various proprietary compounds, (6) organic lead, associated with leaded gasoline and other heavy metals.

Key sources of such contamination at crude oil refineries are at (1) transfer and distribution points in tankage and process areas, also general loading and unloading areas, (2) land farm areas, (3) tank farms, (4) individual above-ground storage tanks and particularly individual underground storage tanks, (5) additive compounds, and (6) pipelines, drainage areas as well as on-site waste treatment facilities, impounding basins, lagoons, especially if unlined.

Whilst contamination may be associated with specific facilities the contaminants are relatively highly mobile in nature and have the potential to migrate significant distances from the source in soil and groundwater. Crude oil hydrocarbon contamination can take several forms: free-phase product, dissolved-phase, emulsified phase, or vapor phase. Each form will require different methods of remediation so that clean-up may be complex and expensive. In addition, crude oil hydrocarbon derivatives include a number of compounds of significant toxicity, e.g., benzene and some polyaromatics are known carcinogens. Vapor phase contamination can be of significance in terms of odor issues.

Due to the obvious risk of fire, refineries are equipped with sprinkler or spray systems that may draw upon the main supply of water, or water held in lagoons, or from reservoirs or neighboring water courses. Such water will be polluting and require containment.

Refining facilities require significant volumes of water for on-site processes (e.g., coolants, blow-downs, etc.) as well as for sanitary and potable use. Wastewater will derive from these sources (process water) and from storm water run-off. The latter could contain significant concentrations of crude oil products.

Crude oil hydrocarbon derivatives, either dissolved, emulsified or occurring as free-phase will be the key constituents although wastewater may also contain significant concentrations of phenols, amines, amides, alcohols, ammonia, sulfide, heavy metals and suspended solids.

Wastewaters may be collected in separate drainage systems (for process, sanitary and storm water) although industrial and storm water systems may in some cases be combined. In addition, ballast water from bulk crude tankers may be pumped to receiving facilities at the refinery site prior to removal of floating oil in an interceptor and treatment as for other wastewater streams.

On-site treatment facilities may exist for wastewater or treatment may take place at a public wastewater treatment plant. Storm water/process water is generally passed to a separator or interceptor prior to leaving the site which takes out free-phase oil (i.e., floating product) from the water prior to discharge, or prior to further treatment, e.g., in settling lagoons.

Discharge from wastewater treatment plants is usually passed to a nearby watercourse.

Other wastes that are typical of a refinery include (1) waste oils, process chemicals, still resides, (2) non-specification chemicals and/or products, (3) waste alkali (sodium hydroxide), (4) waste oil sludge (from interceptors, tanks, and lagoons), and (5) solid wastes (cartons, rags, catalysts, and coke).

2.6.2 *Release into the Environment*

Crude oil products released into the environment undergo weathering processes with time. These processes include evaporation, leaching (transfer to the aqueous phase) through solution and entrainment (physical transport along with the aqueous phase), chemical oxidation, and microbial degradation. The rate of weathering is highly dependent on environmental conditions. For example, gasoline, a volatile product, will evaporate readily in a surface spill, while gasoline released below 10 feet of clay topped with asphalt will tend to evaporate slowly (weathering processes may not be detectable for years).

An understanding of weathering processes is valuable to environmental test laboratories. Weathering changes product composition and may affect testing results, the ability to bio-remediate, and the toxicity of the spilled product. Unfortunately, the database available on the composition of weathered products is limited.

However, biodegradation processes, which influence the presence and the analysis of crude oil hydrocarbon at a particular site, can be very complex. The extent of biodegradation is dependent on many factors including the type of microorganisms present, environmental conditions (e.g., temperature, oxygen levels, and moisture), and the predominant hydrocarbon types. In fact, the primary factor controlling the extent of biodegradation is the molecular composition of the crude oil contaminant. Multiple ring cycloalkanes are hard to degrade, while polynuclear aromatic hydrocarbon derivatives display varying degrees of degradation. Straight-chain alkane derivatives biodegrade rapidly with branched alkanes and single saturated ring compounds degrading more slowly.

The primary processes determining the fate of crude oils and oil products after a spill are (1) dispersion, (2) dissolution, (3) emulsification, (4) evaporation, (5) leaching, (6) sedimentation, (7) spreading, and (8) wind. These processes are influenced by the spill characteristics, environmental conditions, and physicochemical properties of the spilled material.

2.6.2.1 *Dispersion*

The physical transport of oil droplets into the water column is referred to as dispersion. This is often a result of water surface turbulence, but also may result from the application of chemical agents (dispersants). These droplets may remain in the water column or coalesce with other droplets and gain enough buoyancy to resurface. Dispersed oil tends to biodegrade and dissolve more rapidly than floating slicks because of high surface area relative to volume. Most of this process occurs from about half an hour to half a day after the spill.

2.6.2.2 Dissolution

Dissolution is the loss of individual oil compounds into the water. Many of the acutely toxic components of oils such as benzene, toluene and xylene will readily dissolve into water. This process also occurs quickly after a discharge, but tends to be less important than evaporation. In a typical marine discharge, generally less than 5 percent of the benzene is lost to dissolution while greater than 95 percent is lost to evaporation. For alkylated polynuclear aromatic compounds, solubility is inversely proportional to the number of rings and extent of alkylation. The dissolution process is thought to be much more important in rivers because natural containment may prevent spreading, reducing the surface area of the slick and thus retarding evaporation. At the same time, river turbulence increases the potential for mixing and dissolution. Most of this process occurs within the first hour of the spill.

Aromatics, and especially BTEX, tend to be the most water-soluble fraction of crude oil. Crude oil contaminated groundwater tends to be enriched in aromatics relative to other crude oil constituents. Relatively insoluble hydrocarbon derivatives may be entrained in water through adsorption into kaolinite particles suspended in the water or as an agglomeration of oil droplets (microemulsion). In cases where groundwater contains only dissolved hydrocarbon derivatives, it may not be possible to identify the original crude oil product because only a portion of the free product will be present in the dissolved phase. As whole product floats on groundwater, the free product will gradually lose the water-soluble compounds. Groundwater containing entrained product will have a gas chromatographic fingerprint that is a combination of the free product chromatogram plus enhanced amounts of the soluble aromatics.

Generally, dissolved aromatics may be found quite far from the origin of a spill but entrained hydrocarbon derivatives may be found in water close to the crude oil source. Oxygenates, such as methyl-t-butyl ether (MTBE), are even more water soluble than aromatics and are highly mobile in the environment.

2.6.2.3 Emulsification

Certain oils tend to form water-in-oil emulsions (where water is incorporated into oil) or "mousse" as weathering occurs. This process is significant because, for example, the apparent volume of the oil may increase dramatically, and the emulsification will slow the other weathering processes, especially evaporation. Under certain conditions, these emulsions may separate and release relatively fresh oil. Most of this process occurs from about half a day to two days after the spill.

2.6.2.4 Evaporation

Evaporative processes are very important in the weathering of volatile crude oil products, and may be the dominant weathering process for gasoline. Automotive gasoline, aviation gasoline, and some grades of jet fuel (e.g., JP-4) contain 20% to 99% highly volatile constituents (i.e., constituents with less than nine carbon atoms).

Evaporative processes begin immediately after oil is discharged into the environment. Some light products (like 1- to 2-ring aromatic hydrocarbon derivatives and/or low molecular weight alkanes less than n-C15) may evaporate entirely; a significant fraction of heavy refined oils also may evaporate. For crude oils, the amount lost to evaporation can typically range from approximately 20 to 60 percent. The primary factors that control evaporation are the composition of the oil, slick thickness, temperature and solar radiation, wind speed and wave height. While evaporation rates increase with temperature, this process is not restricted to warm climates. For the Exxon Valdez incident, which occurred in cold conditions (March 1989), it has been estimated that appreciable evaporation occurred even before all the oil escaped from the ship, and that evaporation ultimately accounted for 20 percent of the oil. Most of this process occurs within the first few days after the spill.

It is not unusual for evaporative processes, however, to be working simultaneously with other processes to remove the volatile aromatics such as benzene and toluene.

2.6.2.5 Leaching

Leaching processes introduce hydrocarbon into the water phase by solubility and entrainment. Leaching processes of crude oil products in soils can have a variety of potential scenarios. Part of the aromatic fraction of a crude oil spill in soil may partition into water that has been in contact with the contamination.

2.6.2.6 Sedimentation or Adsorption

As mentioned above, most oils are buoyant in water. However, in areas with high levels of suspended sediment, crude oil constituents may be transported to the river, lake, or ocean floor through the process of sedimentation. Oil may adsorb to sediments and sink or be ingested by zooplankton and excreted in fecal pellets that may settle to the bottom. Oil stranded on shorelines also may pick up sediments, float with the tide, and then sink. Most of this process occurs from about two to seven days after the spill.

2.6.2.7 Spreading

As oil enters the environment, it begins to spread immediately. The viscosity of the oil, its pour point, and the ambient temperature will determine how rapidly the oil will spread, but light oils typically spread more rapidly than heavy oils. The rate of spreading and ultimate thickness of the oil slick will affect the rates of the other weathering processes. For example, discharges that occur in geographically contained areas (such as a pond or slow-moving stream) will evaporate more slowly than if the oil were allowed to spread. Most of this process occurs within the first week after the spill.

2.6.2.8 Wind

Wind (Aeolian) transport (relocation by wind) can also occur and is particular relevant when catalyst dust and coke dust are considered. Dust becomes airborne when winds traversing arid land with little vegetation cover pick up small particles such as catalyst dust, coke dust and other refinery debris and send them skyward. Wind transport may occur through *suspension*, *saltation*, or *creep* of the particles.

References

EPA. 1996. Study of Selected Crude oil Refining Residuals. Office of solid Waste Management, Environmental Protection Agency, Washington, DC.

Gary, J.G., Handwerk, G.E., and Kaiser, M.J. 2007. Petroleum Refining: Technology and Economics, 5th Edition. CRC Press, Taylor & Francis Group, Boca Raton, Florida.

Hsu, C.S., and Robinson, P.R. (Eds.). 2017. Handbook of Petroleum Technology. Springer International Publishing AG, Cham, Switzerland.

Irwin, R.J. 1997. Petroleum. In Environmental Contaminants Encyclopedia, National Park Service, Water Resources Divisions, Water Operations Branch, Fort Collins, Colorado.

Meyers, R.A. 1997. Handbook of Crude oil Refining Processes 2nd Edition. McGraw-Hill, New York.

Mokhatab, S., Poe, W.A., and Speight, J.G. 2006. Handbook of Natural Gas Transmission and Processing. Elsevier, Amsterdam, Netherlands.

Parkash, S. 2003. Refining Processes Handbook. Gulf Professional Publishing, Elsevier, Amsterdam, Netherlands.

Speight, J.G. 2005. Environmental Analysis and Technology for the Refining Industry. John Wiley & Sons Inc., Hoboken, New Jersey.

Speight, J.G., and Arjoon, K.K. 2012. Bioremediation of Crude oil and Crude oil Products. Scrivener Publishing, Beverly, Massachusetts.

Speight, J.G. 2014. The Chemistry and Technology of Petroleum 5th Edition. CRC Press, Taylor and Francis Group, Boca Raton, Florida.

Speight, J.G., and Exall, D.I. 2014. Refining Used Lubricating Oils. CRC Press, Taylor and Francis Group, Boca Raton, Florida.

Speight, J.G. 2015. Asphalt Materials Science and Technology. Butterworth-Heinemann, Elsevier, Oxford, United Kingdom.

Speight, J.G. 2017. Handbook of Petroleum Refining. CRC Press, Taylor and Francis Group, Boca Raton, Florida.

Speight, J.G. 2018. Handbook of Natural Gas Analysis. John Wiley & Sons Inc., Hoboken, New Jersey.

Speight, J.G. 2019. Handbook of Petrochemical Processes. CRC Press, Taylor & Francis Group, Boca Raton, Florida.

Ware, G.W., and Whitacre, D.M. 2004. An Introduction to Insecticides. In The Pesticide Book 6th Edition. MeisterPro Information Resources, Meister Media Worldwide, Willoughby, Ohio.

CHAPTER 3

Test Methods for Crude Oil and Crude Oil Products

An essential aspect of the investigation of any spill involving crude oil and crude oil products must involve a description of the methods by which the crude oil and the crude oil products are identified as well as the methods which can be used to determine any potential chemical and physical changes that the spilled constituents undergo after the spill has occurred (Daling et al., 1990). This is especially true insofar as the changes that occur to the spilled material give an indication of the chemical mechanisms and the physical mechanisms of the biodegradation of the material leading to conclusions about the effects of the spill on the environment.

However, just as the chemical and physical properties of crude oil have offered challenges in selecting and designing optimal upgrading schemes (Parkash, 2003; Gary et al., 2007; Speight, 2014; Hsu and Robinson, 2017; Speight, 2017), these properties also introduce challenges when determining the effects of crude oil and its product on the environment. In particular, predicting the fate of the polynuclear aromatic systems, the heteroatom systems (principally compounds containing nitrogen and sulfur), and the metal-containing systems principally compounds of vanadium, nickel, and iron in the feedstocks is the subject of many studies and migration models. These constituents generally cause processing problems and knowledge of the behavior of these elements is essential for process improvements, process flexibility, and environmental compliance.

Because of the variation in the amounts of chemical types and bulk fractions in crude oil and the chemical types in crude oil products, it should not be surprising that crude oil and its products exhibit a wide range of physical properties and several relationships can be made between various physical properties. Whereas the properties such as viscosity, density, boiling point, and color of crude oil may vary widely, the ultimate or elemental analysis varies, as already noted, over a narrow range for a large number of crude oil samples. The carbon content is relatively constant, while the hydrogen and heteroatom contents are responsible for the major differences between crude oil. The nitrogen, oxygen, and sulfur can be present in only trace amounts in some crude oil, which as a result consists primarily of hydrocarbon derivatives. On the other hand, crude oil containing 9.5% w/w heteroatoms may contain essentially no true hydrocarbon constituents insofar as the constituents contain at least one or more heteroatoms, i.e., compounds that contain at least one nitrogen, oxygen, sulfur, and/or metals containing constituents within the molecular structure. Coupled with the changes brought about to the crude oil constituents by refinery operations and the multitude of chemical reactions that can occur during refining, it is not surprising that crude oil characterization is a monumental task.

In fact, the analysis of crude oil and crude oil products as they affect the environment into which one or the other is spilled requires not only knowledge of the chemical and physical properties but also knowledge of the chemical and physical reactivity of the constituents of crude oil and crude products. In addition, crude oil (and to some extent crude oil products) varies markedly in properties and composition according to the source as well as exhibiting variations in chemical and physical reactivity. Thus, knowledge of the reactivity of the constituents of crude oil and crude oil products is required for optimization of cleanup processes as well as for the development and design of new cleanup processes.

Furthermore, the behavior of crude oils and crude oil products is determined by the chemistry of the constituents which can be grouped into several broad classes of compounds which are: (1) saturated constituents, including waxes, (2) aromatic constituents, (3) the constituents of the resin fraction, and (4) the constituents of the asphaltene fraction.

The saturate constituents are alkane derivatives with structures of C_nH_{2n+2} (aliphatic compounds) or C_nH_{2n} in the case of cyclic saturates (alicyclic compounds). Lower molecular weight saturated compounds (< C18) are the most dispersible components of oils whereas higher molecular weight saturated compounds (C18+, waxes which are predominantly straight-chain saturates with melting points above 20°C, 68°F) can produce anomalous evaporation, dispersion, emulsification, and flow behavior.

Aromatic compounds have at least one benzene ring as part of their chemical structure. The lower molecular weight aromatic compounds (compounds that have one and two aromatic rings) do have a measurable solubility in water and will also evaporate rapidly from spilled crude oil. Higher molecular weight aromatic compounds exhibit neither the water solubility nor the tendency to evaporate that is exhibited by the lower molecular weight compounds.

The constituents of the resin fraction and the asphaltene fraction are similar insofar as the constituents of both fractions are composed of condensed aromatic nuclei which may carry alkyl and alicyclic systems containing heteroatoms such as nitrogen, oxygen, sulfur, and metals such as nickel, vanadium, and iron. The constituents of both of these fraction do not evaporate, disperse, or degrade but do tend to stabilize water-in-oil emulsions.

Also, the initial inspection of the nature of the crude oil will provide deductions about the most logical means of clean up and any subsequent environmental effects. Indeed, careful evaluation of crude oil from physical property data is a major part of the initial study of any crude oil that has been released to the environment. Proper interpretation of the data resulting from the inspection of crude oil requires an understanding of their significance.

Through the judicious use of resources and the application of the principles of environmental analysis, environmental science, and environmental engineering, it is possible to reach a state where pollution in minimal and not a threat to the future (Speight, 1996; Manahan, 1999; Woodside, 1999; Speight, 2005). Such a program will not only involve well-appointed suites of analytical tests but also subsequent studies that cover the effects of changes in the environmental conditions on the flora and fauna of a region to the more esoteric studies of animals in laboratories. These studies can include aspects of chemistry, chemical engineering, microbiology, and hydrology as they can be applied to solve environmental problems (Pickering and Owen, 1994; Speight, 1996; Schwarzenbach et al., 2003; Tinsley, 2004). As an historical aside, environmental engineering (formerly known as sanitary engineering) originally developed as a sub-discipline of civil engineering.

As already noted (Chapter 1, Chapter 2), the chemical composition of crude oil and crude oil products is complex and to make it even more complicated, the composition of a crude oil-related spilled material will change over time following release into the environment (Speight, 2005).

There is a significant number of crude oil hydrocarbon derivatives impacted sites and evaluation and remediation of these sites may be difficult arise from the complexity of the issues (analytical, scientific, and regulatory not to mention economic) regarding impacted water and soil. Such factors make it essential that the most appropriate analytical methods are selected from a comprehensive list of methods and techniques that are used for the analysis of environmental samples (Dean, 1998; Miller, 2000; Budde, 2001; Sunahara et al., 2002; Nelson, 2003; Smith and Cresser, 2003; Speight, 2005, 2015). But once a method is selected, it may not be the ultimate answer to solving the problem of identification and, hence, behavior (Patnaik, 2004). There is a significant number of crude oil hydrocarbon impacted sites and evaluation and remediation of these sites may be difficult arise from the complexity of the issues (analytical, scientific, and regulatory not to mention economic) regarding impacted water and soil.

3.1　The Need for Test Methods

Considering the complexity of the composition of crude oil and crude oil products (Speight, 2014), it is not surprising that crude oil and crude oil-derived chemicals are environmental pollutants (Speight, 2005, 2014). The economy of many countries is highly dependent on crude oil for energy production and widespread use has led to enormous releases to the environment of crude oil, crude oil products, exhaust from internal combustion engines, emissions from oil-fired power plants, and industrial emissions where fuel oil is employed.

An important step in assessing the effects of crude oil and crude oil products that have been released into the environment is to evaluate the nature of the particular mixture and eventually select an optimum remediation technology for that mixture. As a general rule, unrefined crude oil is complex mixtures composed of the same compounds, but the quantities of the individual compounds differ in crude oils from different locations. This rule of thumb implies that the quantities of some compounds can be zero in a given mixture of compounds that comprise a crude oil from a specific location.

In very general terms, crude oil is a mixture of: (a) hydrocarbon types, (b) nitrogen compounds, (c) oxygen compounds, (d) sulfur compounds, and (e) metallic constituents. Crude oil products are less well defined in terms of heteroatom compounds and are better defined in terms of the hydrocarbon types present (Table 3.1). However, this general definition is not adequate to describe the true composition as it relates to the behavior of the crude oil, and its products, in the environment. For example, the occurrence of amphoteric species (i.e., compounds having a mixed acid/base nature) is not always addressed nor is the phenomenon of molecular size or the occurrence of specific functional types that can play a major role.

In the present context, the most common physical properties used to describe crude oil are (alphabetically rather than by preference) are (1) boiling point range, (2) density, (3) viscosity while crude oil products may require other properties to be included as part of the product specifications. Other properties include the compositions of the crude oil composition or crude oil product composition which is defined in terms of (1) the elemental composition, (2) the chemical composition, and (3) the fractional composition. All three are interrelated although the closeness of the elemental composition makes it difficult to relate precisely to the chemical composition and the fractional composition. The chemical composition and the fractional composition are somewhat easier to relate because of the quantities of heteroatoms (nitrogen, oxygen, sulfur, and metals) that occur in the higher boiling fractions (Speight, 2014, 2015; Rullkötter and Farrington, 2021).

Furthermore, the toxicity of polynuclear aromatic hydrocarbon derivatives is perhaps one of the most serious long-term problems associated with the use of crude oil. They comprise a large class of crude oil compounds containing two or more benzene rings. Polynuclear aromatic hydrocarbon derivatives are formed in nature by long-term, low-temperature chemical reactions in sedimentary deposits of organic materials and in high-temperature events such as volcanoes and forest fires. Polynuclear aromatic hydrocarbon derivatives accumulate in soil, sediment, and biota. At high concentrations, they can be acutely toxic by disrupting membrane function. Many cause sunlight-induced toxicity in humans and fish and other aquatic organisms. In addition, long-term chronic toxicity has been demonstrated in a wide variety of organisms. Through metabolic activation, some polynuclear aromatic hydrocarbon derivatives *form* reactive intermediates that bind to deoxyribonucleic acid (DNA). For this reason, many of these hydrocarbon derivatives are *mutagenic* (tending to cause mutations), *teratogenic* (tending to cause developmental malformations), or *carcinogenic* (tending to cause cancer).

Table 3.1: Compound Types in Crude Oil and Crude Oil Products.

Class	Compound Types
Saturated hydrocarbons	n-Paraffins
	iso-Paraffins and other branched paraffins
	Cycloparaffins (naphthenes)
	Condensed cycloparaffins (including steranes, hopanes)
	Alkyl side chains on ring systems
Unsaturated hydrocarbons	Olefins not indigenous to petroleum; present in products of thermal processes
Aromatic hydrocarbons	Benzene systems
	Condensed aromatic systems
	Condensed aromatic-cycloalkyl systems
	Alkyl side chains on ring systems
Saturated heteroatomic systems	Alkyl sulfides
	Cycloalkyl sulfides
	Alkyl side chains on ring systems
Aromatic heteroatomic systems	Furans (single-ring and multi-ring systems)
	Thiophenes (single-ring and multi-ring systems)
	Pyrroles (single-ring and multi-ring systems)
	Pyridines (single-ring and multi-ring systems)
	Mixed heteroatomic systems
	Amphoteric (acid-base) systems
	Alkyl side chains on ring systems

Preliminary tests to assess feasibility of biotreatment of the contamination by hydrocarbon derivatives (ASTM, 2021) are typically performed in the laboratory (for example degradation experiments) or in the field (chiefly in situ respiration tests). In general, the quantity of soil used in laboratory experiments is relatively small compared to the quantity of soil that has to be treated on-site and thus spatial heterogeneity is rarely considered in laboratory experiments (Aichberger et al., 2005). This might in particular be true for small scale experiments and/or for inhomogeneous sites.

In fact, data for biodegradation rates determined in the laboratory experiments might be higher than would be expected, thus favoring higher degradation rates compared to field data (Davis et al. 2003; Höhener et al., 2003). Laboratory tests to determine biodegradability potential indirectly via oxygen consumption or carbon dioxide may require corrections to the rates then have to be corrected by background respiration result from biodegradation of organic substances and other oxygen consuming processes (Balba et al., 1998; Baker et al., 2000). However, the manner in which the corrections are applied must be beyond reproach or claims of falsification of the data will be the most likely result (Speight and Foote, 2011).

Field tests must take into consideration the various site conditions comprising local heterogeneity and changing environmental conditions. By the application of preliminary field tests, relevant information can be deducted on the definite system design including actual flow rates, radii of influence, blower and well layout (Leeson and Hinchee, 1997; Baker et al., 2000; Höhener et al., 2003).

Preliminary site assessments account for three critical prerequisites for bioremediation namely for (1) the biodegradability of the predominating compounds, (2) contaminant bioavailability to indigenous microbial populations, and (3) environmental conditions present at a site.

First, *hydrocarbon biodegradation* under specific environmental conditions, the degree of is mainly affected by the type of hydrocarbon derivatives in the contaminant matrix (Huesemann, 1995).

Of the various crude oil fractions, n-alkanes and branched alkanes of intermediate length (C_{10} to C_{20}) are the preferred substrates to microorganisms and tend to be most readily degradable. Cycloalkanes are degraded more slowly than corresponding n-alkanes (unbranched) and branched alkanes.

Second, *contaminant bioavailability* may be determined by a series of preliminary tests. Strong interactions between soil matrix and hydrophobic pollutants can evolve causing pollutant retention or even irreversible binding to sorbents. This phenomenon known as ageing increases with time and significantly reduces bioavailability of hydrophobic contaminants in soil (Hatzinger and Alexander, 1995). Pollutant retention over time is governed by physical-chemical characteristics of the pollutant and by soil characteristics. Strong or even irreversible sorption onto soil is in general attributed to the soil organic matter (Luthy et al., 1997; Huang et al., 2003). However, the degree of degradation of the hydrocarbon derivatives is affected mainly by the type of hydrocarbon contaminants matrix and only to a lesser extent by soil characteristics (Huesemann, 1995; Nocentini et al., 2000; Breedveld and Sparrevik, 2001). This might be true in particular for soils derived from further depths in the subsoil, where relatively low amounts of soil organic matter are present (Nierop and Verstraten 2003).

Third, *environmental conditions* have to be regarded and include factors such as: temperature, pH, moisture content, availability of mineral nutrients, and contaminant concentration. Most microorganisms can degrade specific types of hydrocarbon derivatives under extreme environmental conditions even if rates might be lower and/or degradation might be incomplete (Margesin et al., 1997; Mohn and Stewart, 2000). Most crude oil-related hydrocarbon derivatives are readily degraded by means of aerobic microorganisms although hydrocarbon derivatives can be degraded in the absence of oxygen (anaerobic conditions) if alternate electron acceptors such as nitrate, manganese (IV), iron (III), sulphate and carbon dioxide hydrocarbon derivatives are present—however, the rate anaerobic degradation may be substantially lower than the rate of aerobic degradation (Holliger and Zehnder, 1996; Heider et al., 1999; Grishchenkov et al., 2000; Boopathy, 2002; Massias et al., 2003).

The addition of nutrients has been reported to have a beneficiary effect on hydrocarbon degradation in soil (Dott et al., 1995; Breedveld and Sparrevik, 2001; Chaîneau et al., 2003)—typically a carbon/nitrogen/phosphorous (C/N/P) ratio of 100/10/1 is commonly proposed (Oudot and Dutrieux, 1989; Atagana et al., 2003) although water is often needed to promote optimal microbial activity (Margesin et al. 2000).

Above all, the long-term fate of crude oil hydrocarbon derivatives in areas where spills have occurred needs to be determined. This is only possible through knowledge of the constituents of crude oil and crude oil products as well as application of the relevant text methods at the time of the spill. The data can then be used to determine whether or not contamination by the hydrocarbon derivatives will persist indefinitely.

The methods that are employed measure the concentration of total crude oil hydrocarbon derivatives generate a single number that represents the combined concentration of all crude oil hydrocarbon derivatives in a sample that are measurable by the particular method (Speight, 2005). Therefore, the determination of the total crude oil hydrocarbon derivatives in a sample is method dependent. On the other hand, methods that measure a crude oil group type concentration separate are used to quantify different categories of hydrocarbon derivatives (e.g., saturates, aromatics, and polars/resins) (Speight, 2007). The results of crude oil group type analyses can be useful for product identification because products such as, for example, gasoline, diesel fuel, and fuel oil have characteristic levels of various structural moieties of the hydrocarbon derivatives. Thus, the methods that measure identifiable crude oil fractions can be used to indicate and/or quantify the changes that have occurred through weathering of the sample. More specifically, weathering is a series of chemical and physical changes that cause the spilled material to degrade (break down) and become heavier (more dense) than water. The processes that constitute weathering processes occur at different rates but typically begin as soon as the crude oil or the crude oil product is spilled and usually proceed most rapidly immediately after the spill. Most weathering processes are highly temperature dependent and will slow to insignificant rates as temperatures approach freezing.

Also, if mechanical recovery systems (Chapter 5) are deployed for cleanup, the type of skimmers and pumps used may need to be changed as the constituents of the crude oil and crude oil product undergo weathering with a rise in viscosity rises and the formation of emulsions. For example, oleophilic (oil attracting) disc skimmers rely on the spilled material adhering to the disc for recovery. However, an emulsion acts as a 'shear-thinning' fluid such that when a twisting movement is applied, for example by a spinning disc, the water droplets in the emulsion align all in one direction, reducing the viscosity and causing the emulsion to be sliced through rather than adhering to the disc. The same effect occurs with centrifugal pumps, where the pump impeller may spin without efficient movement of the emulsion through the pump. For this reason, positive displacement pumps are recommended for the transfer of emulsions.

Although these methods measure different categories of crude oil hydrocarbon derivatives, there are several basic steps that are common to the analytical processes for all methods, no matter the method type or the environmental matrix. In general, these steps are: (1) sample collection and preservation—requirements specific to environmental matrix and analytes of interest, (2) extraction—separation of the analytes of interest from the sample matrix, (3) concentration—enhances the ability to detect analytes of interest, (4) cleanup, dependent upon the need to remove interfering compounds, and (5) measurement, or quantification, of the analytes (Dean, 1998). Each step affects the final result, and a basic understanding of the steps is vital to data interpretation.

3.2 Chemical and Physical Properties of Crude Oil and Crude Oil Products

Crude oils of different origin vary widely in their physical and chemical properties (Chapter 1) and, in addition, whereas many refined products tend to have well-defined properties irrespective of the crude oil from which they are derived. However, it is worth remembering that crude oil products (Chapter 2), while meeting the physical specification for the designated product, may also vary in composition. For example, the fuel oil products, which contain varying proportions of the residues of the refining process blended with lighter refined products, also vary considerably in their properties.

The main physical properties that are considered to affect the behavior of crude oils and crude oil products and the persistence of an oil spilled at on the land or a sea are often cited as (1) specific gravity, (2) distillation characteristics, (3) vapor pressure, (4) viscosity, and (5) pour point. These properties are dependent on chemical composition of the spilled material, such as the proportion of volatile components and the content of waxes, resin constituents, and asphaltene constituents (Parkash, 2003; Gary et al., 2007; Speight, 2014; Hsu and Robinson, 2017; Speight, 2017). But these properties of properties is not the end of the story.

By way of definition, the vapor pressure provides an indication of the volatility of a crude oil or a crude oil product (typically presented as the Reid Vapor Pressure) measured at 37.8°C (100°F). A vapor pressure greater than 23 mm Hg (3 kPa) is the criterion for evaporation to occur under most conditions whereas a vapor pressure greater than 760 mmHg (100 kPa) indicate that the substance is a gas. AS an examples, naphtha, has a vapor pressure on the order of between 300 to 600 mm Hg (40 to 80 kPa) and is highly volatile with a high proportion of the constituents boiling at low temperature (i.e., 200°C, < 390°F).

However, there is a host of properties that can influence the behavior of crude oil and crude oil products when spilled into an ecosystem and rather than show any preference for the major drivers in spoil behavior, the properties that can influence the behavior of the constituents of the spill are presented in the following sub-sections without any preference other than alphabetical order by property name. In fact, the multicomponent composition and corresponding physical properties data of crude oil and crude oil products are needed as input to the determination of the environmental fate of the spilled constituents. Examples of physical properties of crude oil and crude oil products are presented (alphabetically and not in any order of preference) in the following sub-sections.

The importance of chemical and physical properties is dependent upon the purity of the crude oil or the crude oil product. In the strictest sense, crude oil and crude oil products are complex chemical mixtures which contain a variety of hydrocarbon derivatives as well as non-hydrocarbon compounds (i.e., compounds that contain a heteroatom such as nitrogen and/or oxygen and/or sulfur and/or metals) that confer properties on the mixture that may not be reflected in the composition. Therefore, it is necessary to apply various text methods to crude oil and crude oil products to determine if the material is suitable for processing and (in the case of the products) for sale with a designated use in mind. The more common tests are introduced in the following sub-sections—these tests are presented in alphabetical order with no preference given to any particular test method.

Crude oil producers and refiners typically do not have precise information about the extent to which the crude oil (or crude oil products) (1) will evaporate and how quickly, (2) the ability of the oil to disperse naturally, (3) the ability of the oil to disperse with the aid of dispersants, (4) the tendency of the oil to form emulsions, (5) the tendency of the oil (or its emulsions) to sink or submerge, (6) the viscosity of the oil at ambient temperatures, (7) the changing viscosity of the oil as the more volatile constituents evaporate, and (8) health hazard to on-site personnel as well as the toxicity of the crude oil (crude oil product) constituents land- based or aquatic flora and non-human fauna.

The importance of each of these properties is dependent upon the purity of the crude oil or crude oil product spilled as well as of the properties of the site (for example, soil composition, which amongst other properties is site specific) and the prevailing climatic conditions (arctic, sub-arctic, temperate, and tropical—which are also site specific). Because of this site specificity and the complex nature of crude oil and crude oil products (the latter is often refinery specific), generalizations of the interactions of crude oil and crude oil products on various sites are to be avoided.

On order to estimate the impact of a crude oil or crude oil product spill several non-conventional (non-typical) properties must be assessed and these properties are presented below in alphabetical order—and not necessarily in the order of importance.

3.2.1 Adhesion

Adhesion of the constituents of the spilled material are an important aspect of the release and introduction of crude oil and crude oil products into the environment. For example, the biodegradation of poorly water-soluble liquid hydrocarbons is often limited by low availability

The basic features of the structure, properties, and modification of chemicals spilled into the environment affect the adhesive behavior after the spill. Adhesive properties and bond performance, including durability, are derived from chemical composition (functional groups), molecular organization (branching, molecular weight distribution, cross-linking) and physical state (elastomer, thermoplastic, thermoset, crystalline). The various types of adhesives include those derived from interaction of the spilled chemicals with the soil and with minerals. The transformation of the chemicals and the properties of the soil and mineral surfaces play a critical role in the bonding of the chemicals to a mineral (including soil) as do modifications of interfacial interactions between the chemicals and water. Resistance to environmental effects such as (1) humidity, (2) temperature, and (3) microbial attack determines the durability of bonds and the retention of the spilled materials in the environment.

The adhesion (sometimes referred to as the "stickiness" of a crude oil of a crude oil product) may be noted as one of the issues that accompanies of a spill. The adhesion of the constituents of a crude oil (or a crude oil product) to the surfaces of soil, mineral matter, and vegetation can greatly impede cleanup. Although important in the context of oil spill response, adhesion is a property that is not typically determined during the typical analysis of crude oils and crude oil products.

Also, adhesion of microorganisms to an oil-water interface can enhance this availability, whereas detaching cells from the interface can reduce the rate of biodegradation. The capability of microbes to adhere to the interface is not limited to hydrocarbon degraders, nor is it the only mechanism to enable rapid uptake of hydrocarbons, but it represents a common strategy. In fact, microbial adhesion can benefit growth on and biodegradation derivatives of difficultly water-soluble hydrocarbons such as n-alkane derivatives and large polycyclic aromatic hydrocarbon derivatives dissolved in a non-aqueous phase. Adhesion is particularly important when the hydrocarbon derivatives are not emulsified, giving limited interfacial area between the two liquid phases. When mixed communities of microbial species are involved in biodegradation, the ability of the microbial cells to adhere to the interface can enable selective growth and enhance bioremediation with time (Abbasnezhad et al., 2011).

More generally, the forces that cause adhesion and cohesion can be divided into several types: (1) chemical adhesion, (2) dispersive adhesion, and (3) diffusive adhesion.

Accordingly, a test has been developed, which uses a standard surface that gives a semi-quantitative measure of this adhesive property. Test parameters that can be evaluated include temperature, oil viscosity, time, and test-surface area. The usefulness of the test lies in the produced data which allow the investigator to make reasonable deductions on the behavior of crude oil and its products under standard conditions thereby removing mush of the guesswork from the behavior of the spilled constituents. Such data, with a knowledge of the composition of the crude oil and/or the crude oil product will assist the investigator to identify the principal constituents that are responsible for such behavior.

3.2.2 Biological Oxygen Demand

The biological oxygen demand (BOD) is a measure of the amount of oxygen required to remove waste organic matter from the environment in the process of decomposition by aerobic bacteria (those bacteria that live only in an environment containing oxygen). The data can also be used to give an indication of the potential destruction rate of the spilled material. The biological oxygen demand is commonly expressed in milligrams of oxygen consumed per liter of sample during 5 days of incubation at 20°C (68°F) and is often used as a robust surrogate of the degree of contamination. Alternatively, the unit lb/lb indicates the pounds of oxygen consumed by each pound of crude oil or crude oil product constituents per unit of time.

There are two commonly recognized methods for the measurement of the biological oxygen demand: (1) the dilution method and (2) the manometric method.

In the *dilution method*, a small but measured amount of micro-organism is added to each sample being tested after which the sample is diluted with oxygen-saturated water, followed by inoculating with a fixed aliquot of the micro-organism, measuring the dissolved oxygen, and then sealing the sample to prevent further oxygen dissolution. The crude oil or crude oil product sample is kept at 20°C (68°F) in the dark to prevent photosynthesis (and thereby the addition of oxygen) for five days, and the dissolved oxygen is measured again. The difference between the final dissolved oxygen and the initial dissolved oxygen is the biological oxygen demand.

A substance that absorbs carbon dioxide (typically lithium hydroxide) is added in the container above the sample level. The crude oil or crude oil product sample is stored in conditions identical to the dilution method. Oxygen is consumed and, as ammonia oxidation is inhibited, carbon dioxide is released.

3.2.3 Boiling Point Distribution

In the crude oil refining industry, boiling range distribution data are used (1) to assess crude oil crude quality before purchase, (2) to monitor crude oil quality during transportation, (3) to evaluate crude oil for refining, (4) to provide information for the optimization of refinery processes, and

Table 3.2: General Summary of the Bulk Products from Crude Oil and the Distillation Range.*

Product	Lower carbon limit	Upper carbon limit	Lower boiling point °C	Upper boiling point °C	Lower boiling point °F	Upper boiling point °F
Refinery gas	C1	C4	−161	−1	−259	31
Liquefied petroleum gas	C3	C4	−42	−1	−44	31
Naphtha	C5	C16	36	287	97	549
Kerosene	C8	C18	126	258	302	575
Fuel oil	C12	> C20	216	421	> 343	> 649
Lubricating oil	> C20		> 343		> 649	
Wax	C17	> C20	302	> 343	575	> 649
Residue	> C20		> 343		> 649	

* The carbon numbers and boiling points are estimates because of the variations between refinery processes and are inserted for illustrative purposes only.

(5) to provide information for the investigation of the changes that can (and will) occur in crude oil or a crude oil product as a result of release to the environment (Table 3.1, Table 3.2).

In the environment, volatility is an important aspect of the spilled material because volatile organic compounds (VOCs) are an important aspect of environmental pollution because of the contribution of these chemicals to the formation of ground-level ozone. Volatile organic compounds (VOC) means any compound of carbon, excluding carbon monoxide, carbon dioxide, carbonic acid, metallic carbides or carbonates and ammonium carbonate, which participates in atmospheric photochemical reactions as, for example, when volatile organic compounds are released into the atmosphere, they react with nitrogen oxides (NOx) to create ozone (O_3).

3.2.4 Chemical Dispersability

The term dispersion (in the current context) refers to the dispersion of pollutant chemicals by the transportation of the chemicals into the atmosphere, into an aqueous ecosystem, or into a land ecosystem after being emitted from the source. Thus, the dispersability of crude oil or crude oil products is a measure or indication of the crude oil to spread on land or water. In some spill situations and under appropriate conditions, dispersants may be an effective countermeasure for minimizing contamination of land-based sites and/or shorelines.

The factors that affect the dispersability of released chemicals are (1) the meteorological conditions, such as the wind speed, the wind direction, and the atmospheric stability, (2) the emission height, such as ground level sources such as road traffic or high level sources such as tall chimneys, (3) the local and regional geographical features, and (4) the source, such as a fixed point that can be chimney or a diffuse number of sources such as cars and solvents.

During dispersion the constituents of crude oil and crude oil products may undergo a wide array of changes. For examples, dilution occurs owing to mixing of the volatile chemicals with the air. Also, separation or accumulation of pollutant chemicals can occur on the basis of physical characteristics of the pollutant. In addition, chemical reactions can occur, breaking down the original pollutant or converting it into new compounds. Some chemicals can also be removed from the transporting medium through deposition, for example, by settling out under the effects of gravity, by rain-wash or by interception (scavenging) by plants and other obstructions. Many pollutant chemicals can, therefore show extremely complex dispersion patterns, especially in environments (such as cities and towns) where there are a large number of emission sources and major variations in environmental conditions. This complexity means that it is often very difficult to model or measure pollutant patterns and trends, and thus to predict levels of human exposure.

In terms of testing for the dispersability of released chemicals, the *swirling flask test* (SFT) was developed for determining the effectiveness of various dispersants with different crude oils and products (Fingas et al., 1989a). Quantitation is accomplished through the use of gas chromatograph with flame ionization detection (GC/FID) (Fingas et al., 1995b). In the test, a sample of the crude oil (or crude oil product) and dispersant are pre-mixed in a ratio of 25:1 and applied to salt water (3.3% w/w NaCl in water) in a ratio of 1:1200. The test method can be employed to determine the effects of oil-to-water ratios and settling times (Fingas et al., 1989b) and energy (Fingas et al., 1991; Fingas et al., 1992; Fingas et al., 1993).

While this test method is useful for assessing the tendency of the constituents of crude oil or crude oil products to disperse on water, unless specific experiment are also carried out to determine the effects of oxidation on the constituents of the crude oil (or the constituents of the crude oil product) dispersibility the deductions of crude oil behavior will be missing an important parameter.

Briefly, in a water system, oxidation occurs when oil contacts the water and oxygen combines with the hydrocarbon derivatives in the spilled material to produce water-soluble compounds. This process affects oil slicks—mostly around the edges of the slick—and if the slick is sufficiently thick—may only partially oxidize, forming tar balls which may linger in the environment.

3.2.5 Density, Specific Gravity, and API Gravity

Density is the mass per unit volume of a substance Density is temperature-dependent. It is an important property of crude oil and crude oil products as it gives the investigator(s) indications of whether or not the contaminant(s) will float on water. For environmental applications, knowledge of the liquid density is required to ascertain emissions into air from a liquid spill.

The specific gravity or relative density of an oil is the density of the oil in relation to pure water, which has a specific gravity of 1. Most crude oils and crude oil products are less dense than water, which typically has a specific gravity on the order of 1.00 and sea water has a density on the order of 1.025. The American Petroleum Institute gravity scale (presented as °API) is commonly used to describe the specific gravity of crude oils and petroleum products and is derived as follows:

°API = [141.4/(specific gravity)] – 131.5

In addition to determining whether or not the oil will float, the specific gravity can also give a general indication of other properties of the crude oil or the crude oil product. For example, crude oil and crude oil products oils with a low specific gravity (high °API) tend to contain a high proportion of volatile components and to be of low viscosity.

Thus, the crude oil or the crude oil product will float on water if the density of the crude oil or the crude oil product is less than that of the water. However, heavy crude oil (as well as extra heavy crude oil and tar sand bitumen) may have densities greater than 1.0 g/mL and their buoyancy behavior will vary depending on the salinity and temperature of the water. The density of spilled crude oils and crude oil products oil will also increase with time, as the more volatile (and less dense) components are lost. After considerable evaporation, the density of some crude oils may increase enough for the oils to submerge below the water surface.

The density of a chemical is one of its most important and easily-measured physical properties. The density of a chemical is widely used property that helps to identify pure (and impure) chemicals and is used to characterize and estimate the composition of many kinds of mixtures.

In general, gases have the lowest densities, but these densities are highly dependent on the pressure and temperature which must always be specified. To the extent that a gas exhibits ideal behavior (low pressure, high temperature), the density of a gas is directly proportional to the masses of its component atoms, and thus to its molecular weight. Measurement of the density of a gas is a simple experimental way of estimating its molecular weight. Liquids encompass an intermediate range of densities. Mercury, being a liquid metal, is something of an outlier. Liquid densities are largely independent of pressure, but they are somewhat temperature-sensitive. The density of solids shows a wide variation—metals have the highest densities

All substances tend to expand as they are heated, causing the same mass to occupy a greater volume, and thus lowering the density. For most solids, this expansion is relatively small, but it is far from negligible; for liquids, it is greater.

In the context of crude oil and crude oil products, two density-related properties of crude oil and crude oil products are often used: (1) specific gravity and (2) American Petroleum Institute (API) gravity. The API gravity scale (presented as °API) arbitrarily assigns an API gravity of 10° to pure water. Thus:

°API = [141.5/(specific gravity at 15.6°C) – 131.5]

The scale is commercially important for ranking the quality of crude oil—heavy crude oil typically (but arbitrarily) has a gravity < 20°API; medium oils are 20 to 35°API; light oils are 35 to 45°API while liquid crude oil product can have an API gravity up to 65°. Tar sand bitumen typically has an API gravity < 10°.

Crude oil (unless it is a specific heavy oil or tar sand bitumen) and crude oil products (unless it is a residual fuel oil or asphalt) will float on water if the density of the crude oil or crude oil product is less than that of the water. This behavior is typical of all crude oils and distillate products for both salt and fresh water. Some heavy oils, tar sand Bitumens, and residual fuel oils may have densities greater than 1.0 g/mL and their buoyancy behavior varies depending on the salinity and temperature of the water (Speight, 2009). Crude oil and crude oil products float on water because these materials are less dense than water and although oil spills are detrimental to the environment, the ability of crude oil (and crude oil products) to float aids cleanup.

As stated above, density (which is temperature dependent) is an important property of crude oil and crude oil products as it gives the investigator(s) indications of whether or not the contaminant(s) will float on water. However, such effects are not lasting.

Density is also influence by the oxidation of crude oil constituents. The inclusion of polar functions such as hydroxyl groups (-OH) or carbonyl groups (> C=O) (a result of the oxidation process) causes an increase in the density of the crude oil or crude oil product (relative to the original material). Thus, as the density of a weathering oil approaches the density of water, contact with even small amounts of sand, clay, or other suspended sediment can catalyze the oxidation process that can lead to submergence of the oxidized crude oil or the oxidized crude oil product.

Thus, given that the density of fresh water is 1.00 g/cm^3 at environmental temperatures and the density of crude oils and crude oil products commonly range from 0.7 to 0.99 g/cm^3, most silled crude oils or crude oil products will float on water. Because the density of seawater is 1.03 g/cm^3, even the heaviest crude oil and many of the more viscous products will also oils will usually float on seawater. But evaporative losses of the low molecular weight volatile constituents can lead to significant increases in density of the non-volatile constituents. Accordingly, those materials can submerge and may sink to the bottom of the lake or ocean.

3.2.6 Emulsion Formation

An emulsion is a mixture of two or more liquids that are usually immiscible. Examples include crude oil and water which can form an oil-in-water emulsion, wherein the oil is the dispersed phase, and water is the dispersion medium. In addition, crude oil (as well as crude oil products) and water can also form a water-in-oil emulsion, wherein water is the dispersed phase and crude oil is the external phase. Whether an emulsion of oil and water exists as a water-in-oil emulsion or an oil-in-water emulsion depends on the volume fraction of both phases and the type of emulsifier (surfactant) present. Multiple emulsions are also possible, including a water-in-oil-in-water emulsion and an oil-in-water-in-oil emulsion. Emulsification is the process that forms emulsions, which are mixtures of small droplets of oil and water.

When formed from crude oils or crude oil products that have been spilled at sea, emulsions can have very different characteristics from their parent crude oils or the crude oil products. This has important implications for the fate and behavior of the spilled material and the subsequent cleanup operation. It is desirable, therefore, to determine if a crude oil the crude oil product is likely to form an emulsion, and if so, whether that emulsion is stable the physical characteristics of the emulsion.

The stability of an emulsion refers to the ability of the emulsion to resist change in its properties over time. There are three types of instability in emulsions which are (1) flocculation, (2) creaming, and (3) coalescence. Flocculation occurs when there is an attractive force between the droplets, so they form flocs. Coalescence occurs when droplets bump into each other and combine to form a larger droplet, so the average droplet size increases over time. Creaming occurs when the droplets rise to the top of the emulsion under the influence of buoyancy.

Thus, an emulsion is a type of colloid formed by combining two liquids that normally do not mix and typically one liquid contains a dispersion of the other liquid. A water-in-oil emulsion is a stable dispersion of small droplets of water in oil. When formed from crude oils spilled at sea, these emulsions can have very different characteristics from their parent crude oils. This has important implications for the fate and behavior of the oil and its subsequent cleanup. In fact, stable water-in-oil emulsions, often formed after oil spills, contribute to the difficulties of cleanup due to their persistence and physical properties. It is desirable, therefore, to determine (1) if oil is likely to form an emulsion, (2) the physical characteristics of the emulsion, and (3) whether any such emulsion is stable.

In an older test method, the tendency for a crude oil to form a water-in-oil emulsion was measured using a method based on the rotating flask apparatus (Mackay and Zagorski, 1982). In a newer variation, the reproducibility is considerably improved and several parameters (1) the water-to-oil ratio, (2) the fill volume, and (3) the orientation of the vessels were found to be important parameters affecting emulsion formation.

Emulsification of oil is caused by the uptake of water by the oil which results in a substance with increased viscosity. This oil/water substance can severely hamper recovery capabilities of skimmers; reduce pumping volumes; and increase handling, oily waste disposal, segregation, and storage problems. In addition, non-mechanical response techniques, such as the use of dispersants and in situ burning, are rendered less effective or ineffective by formation of high-viscosity emulsions.

Emulsion formation and behavior is influenced by the oxidation of crude oil constituents (Speight, 2014, 2015). The inclusion of polar functions such as hydroxyl groups (-OH) or carbonyl groups ($> C=O$) (a result of the oxidation process) causes an increase in the density of the emulsion (relative to the original unoxidized crude oil) and with an increased propensity to form emulsions. As a result, the emulsion and sinks to various depths or even to the seabed, depending on the extent of the oxidation and the resulting density. This may give the erroneous appearance (leading to erroneous deductions with catastrophic consequences) that the crude oil spill (as evidenced from the crude oil remaining on the surface of the water) is less than it actually was. The so-called *missing* oil will undergo further chemical changes and eventually reappear on the water surface or on a distant beach.

3.2.7 Evaporation

Evaporation is the removal of the lower-boiling constituents from crude oil or a crude oil product usually under ambient conditions or, in the current context, under the conditions prevalent at the spill site.

Evaporation occurs when the lower boiling constituents of crude oil or crude oil products vapors and leave behind the high boiling constituents which may undergo weathering or, in the case of a water spill, sink to the bottom of the ware system. Spills of lower boiling products, such as naphtha and kerosene, contain a high proportion of flammable constituents (often referred to as light ends) which may evaporate within a relatively short time. On the other hand, the constituents of the higher boiling more viscous materials are less likely to evaporate.

Evaporation rate and loss are of importance for all volatile constituents of crude oil and crude oil products. With condensation and precipitation, evaporation is one of the three main steps in the Earth's water cycle. Evaporation accounts for approximately 90 percent of the moisture in the atmosphere of the Earth and the remainder is due to plant transpiration. In an enclosed environment, a liquid will evaporate until the surrounding air is saturated. Evaporation of water occurs when the surface of the liquid is exposed, allowing molecules to escape and form water vapor; this vapor can then rise up and form clouds. With sufficient energy, the liquid will turn into vapor. Evaporation is an important aspect of the water cycle. Heat from the sun, or solar energy, powers the evaporation process. In the process, moisture from soil is evaporated as well as from lakes and oceans.

While pure compounds evaporate at constant rates, crude oil and crude oil products, which are composed of thousands of compounds, do not. Rapid initial loss of the more volatile fractions is followed by progressively slower loss of less volatile constituents.

Crude oil and crude oil products evaporate at a logarithmic rate with respect to time (Fingas, 1995, 1998). Crude oil products with fewer constituents (such as diesel fuel) evaporate at a rate which is square root with respect to time, which is a result of the number of components evaporating. Such changes could be reflected in an increase in adhesion of the contaminant constituents to the soil or rock.

3.2.8 Flash Point and Fire Point

The *flash point* of crude oil or a crude oil product is the temperature to which the sample must be heated to produce a vapor/air mixture above the liquid fuel that is ignitable when exposed to an open flame under specified test conditions. Flash point is an important property in relation to the safety of spill cleanup operations. Low boiling products such as naphtha and gasoline can be ignited under most ambient conditions and therefore pose a serious hazard when spilled. Many freshly spilled crude oils also have low flash points until the lighter components have evaporated or dispersed.

As the temperature increases, vapor pressure increases and as vapor pressure increases, the concentration of vapor of a flammable or combustible liquid in the air increases. Hence, temperature determines the concentration of vapor of the flammable liquid in the air. Below the flash point for a liquid, burning does not occur because the liquid is unable to give off sufficient vapor to form a burnable mixture with air. Hence, flash point is the minimum temperature at which the liquid gives off enough vapor to form a combustible mixture with air.

There is a broad range of flash points for crude oils and crude oil products, many of which are considered flammable. A liquid, such as naphtha, is considered to be flammable if its flash point is less than 60°C (140°F) and, which is flammable under all ambient conditions, poses a serious hazard when spilled. Many crude oils or crude or products may be flammable for a day or longer after being spilled, depending on the rate at which highly volatile components are lost by evaporation.

The *fire point* is the lowest temperature, corrected to one atmosphere pressure (14.7 psi), at which the application of a test flame to the crude oil or to the crude oil product sample surface causes the vapor of the oil to ignite and burn for at least five seconds. The fire point plays a key role in shaping ecosystems by serving as an agent of renewal and change. But fire can be deadly by destroying wildlife habitat and timber as well as polluting the air with emissions harmful to human health. Fire also releases carbon dioxide—a key greenhouse gas—into the atmosphere.

Both terms (i.e., flash point and fire point)—are used to describe liquids such as crude oil and crude oil products liquids. All liquids vaporize insofar as the some of the constituents of crude oil and crude oil products evaporate, to form a layer of vapor molecules above (and in contact with) the liquid surface and the vapor, when mixed with the surrounding air, forms a combustible mixture.

At any time after a spill of crude oil or a crude oil product, fire should always be considered an imminent hazard. Related to fire point, the flash point is a measure of the tendency of the crude oil or a crude oil product to form a flammable mixture with air under controlled laboratory conditions.

The ignition temperature (sometimes called the *autoignition temperature*) is the minimum temperature at which the material will ignite without a spark or flame being present. The method of measurement is given by ASTM E659 (Standard Test Method for Autoignition Temperature of Liquid Chemicals).

Also related to fire point, the flammability limits of vapor in air is an expression of the percent concentration in air (by volume) is given for the lower and upper limit. These values give an indication of relative flammability. The limits are sometimes referred to as *lower explosive limit* (LEL) and *upper explosive limit* (UEL).

3.2.9 Fractionation

The fractionation methods for crude oil and crude oil products can be used to measure both the volatile constituents and the extractable constituents. The available methods for identifying the constituents of crude oil and crude oil products are typically based on gas chromatography and are thus sensitive to a broad range of hydrocarbon derivatives. These fraction data also can be used in risk assessment.

One particular method is designed to characterize C_6 to C_{28+} crude oil hydrocarbon derivatives in soil as a series of aliphatic and aromatic carbon range fractions. The extraction methodology differs from other crude oil hydrocarbon methods because it uses *n*-pentane and not methylene chloride as the extraction solvent.

n-Pentane extracts crude oil hydrocarbon derivatives in this range most efficiently. The whole extract is separated into aliphatic and aromatic crude oil-derived fractions (EPA SW-846 3611, EPA SW-846 3630). Gas chromatographic parameters allow the measurement of a hydrocarbon range of n-hexane (C_6) to n-octacosane (C_{28+}), a boiling point range of approximately 65°C to 450°C (150 to 840°F).

Also, crude oil and crude oil products are generally considered a major source of pollutants in areas where they are located and are regulated by a number of environmental laws related to (1) air, (2) land, and (3) water.

Crude oil and crude oil products are a sources of air pollutants such as BTEX compounds (benzene, toluene, ethylbenzene, and xylene) as well as sources of particulate matter (PM), nitrogen oxides (NOx), carbon monoxide (CO), hydrogen sulfide (H2S), and sulfur dioxide (SO2). Hydrocarbon derivatives such as natural gas (methane) and other low boiling hydrocarbon derivatives are also released. The combination of volatile hydrocarbons and oxides of nitrogen also contribute to ozone formation, one of the most important air pollution problems in the United States.

Crude oil and crude oil products are also potential major contributors to the contamination of ground water and surface water. Wastewater in refineries may be highly contaminated given the number of sources it can come into contact with during the refinery process (such as equipment leaks and spills and the desalting of crude oil). This contaminated water may be process wastewaters from desalting, water from cooling towers, storm water, distillation, or cracking. It may contain oil residuals and many other hazardous wastes. This water is recycled through many stages during the refining process and goes through several treatment processes, including a wastewater treatment plant, before being released into surface waters.

Contamination of soils from crude oil and crude oil products is generally a less significant problem when compared to contamination of air and water. Production practices may have led to spills on the refinery property that now need to be cleaned up. Natural bacteria that may use the crude oil products as food are often effective at cleaning up crude oil spills and leaks compared to many other pollutants. Soil contamination including some hazardous wastes, spent catalysts or coke dust, tank bottoms, and sludge from the treatment processes can occur from leaks as well as accidents or spills on or off site during the transportation.

3.2.10 Leachability and Toxicity

Leachability is the movement of constituents (often generally regarded as contaminants) from crude oil and crude oil products carried by water downward through permeable soils. The factors that influence the liquid-solid partitioning (LSP) of a constituent, include solution pH, redox, the presence of dissolved organic matter, and biological activity.

As a start and for regulatory and remediation purposes, a standard test is needed to measure the likelihood of toxic substances getting into the environment and causing harm to organisms. The test (required by the United States Environmental Protection Agency) is the *toxicity characteristic leaching procedure* (TCLP, EPA SW-846 Method 1311), designed to determine the mobility of both organic and inorganic contaminants present in liquid, solid, and multiphase wastes.

The extraction is performed using acetic as the extraction fluid. The pH of the acetic acid/ sodium acetate buffer solution is maintained at 4.93. This sample/acetic acid mixture is subjected to rotary extraction, designed to accelerate years of material exposure in the shortest possible time. After extraction, the resulting liquid is subjected to analysis utilizing a list of contaminants that includes metals, volatile organic compounds, semi-volatile organic compounds, pesticides, and herbicides.

The toxicity characteristic leaching procedure may be subject to misinterpretation if the compounds under investigation are not included in the methods development or the list of contaminants leading to the potential for technically invalid results. However, an alternate procedure, the synthetic precipitation leaching procedure (SPLP, EPA SW-846 Method 1312) may be appropriate. This procedure is applicable for materials where the leaching potential due to normal rainfall is to be determined. Instead of the leachate simulating acetic acid mixture, nitric and sulfuric acids are utilized in an effort to simulate the acid rains resulting from airborne nitric and sulfuric oxides.

An alternate procedure, the synthetic precipitation leaching procedure (SPLP, EPA SW-846 Method 1312) may be appropriate. This procedure is applicable for materials where the leaching potential due to normal rainfall is to be determined. Instead of the leachate simulating acetic acid mixture, nitric and sulfuric acids are utilized in an effort to simulate the acid rains resulting from airborne nitric and sulfuric oxides.

Toxicity values are often reported as the amount of chemical that is lethal to indigenous floral and faunal populations. Thus:

LC_{50}: the median lethal concentration is the estimated concentration of a compound that will cause death to 50 percent of the test population in a specified time after exposure. In most instances, LC_{50} is statistically derived by analysis of mortalities in various test concentrations after a fixed period of exposure.

EC_{50}: the median effective concentration is used when an effect other than death is the observed endpoint. EC_{50} is the estimated concentration of the compound in water that will have a specific effect on 50 percent of the test population in a specified time after exposure. As with LC_{50}, the EC_{50} is generally derived statistically.

TLm: the median tolerance limit for a floral and faunal species for the spilled chemical; a term sometimes used instead of EC_{50}.

3.2.11 Metals Content

Metals content in crude oils can provide valuable information about the origin of those oils and can be used as a means of identifying the source of the crude oil that has been spilled.

A variety of tests (ASTM D1026, ASTM D1262, ASTM D1318, ASTM D1368, ASTM D1548, ASTM D1549, ASTM D2547, ASTM D2599, ASTM D2788, ASTM D3340, ASTM D3341, and ASTM D3605) have been designated for the determination of metals in crude oil and crude oil products. At the time of writing, the specific test for the determination of metals in whole feeds has not been designated. However, this task can be accomplished by combustion of the sample so that only inorganic ash remains. The ash can then be digested with an acid and the solution examined for metal species by atomic absorption (AA) spectroscopy or by inductively coupled argon plasma (ICP,) spectrometry.

3.2.12 Pour Point and Cloud Point

The pour point of a crude oil or a crude oil product is the temperature below which the liquid loses its flow characteristics. By definition, the pour point is the minimum temperature in which the oil has the ability to pour down from a beaker. Any change in the pour point of a crude oil or a crude oil product depends on both (1) the composition of the oil, (2) the properties of the oil, as well as (3) the relative molecular weight of the oil.

From a spill response point of view, it must be emphasized that the tendency of the oil to flow will be influenced by the size and shape of the container, the head of the oil, and the physical structure of the solidified oil. The pour point is particularly useful information in colder climates, where knowing whether oil is fluid enough to be pumped without special heating equipment, for example, would certainly affect cleanup and salvage decisions.

The cloud point, which is often used with the pour point, is the temperature below which a crude oil or a crude oil product no longer flows and is a function of the wax content and the asphaltene content.

When cooled, a crude oil or a crude oil product will reach a temperature (the cloud point) when the wax components (or in the case of asphaltenes, the asphaltene constituents) begin to form solid structures as a separate phase. This solid phase hinders the flow of the oil until on further cooling the pour point is reached, flow ceases and the oil changes from a liquid to a semi-solid.

3.2.13 Solubility in Aqueous Media

The solubility of crude oil or crude oil products in water (which should not be confused with the concept of emulsion formation) can be determined by bringing to equilibrium a volume of oil and water, and then analyzing the water phase. Typically, the solubility of crude oil (or crude oil products) in water is low, generally less than 100 ppm. However, solubility is an important issue after oil spill accidents because dissolved oil components are often acutely toxic to aquatic life.

Thus, solubility of crude oil or a crude oil product in water is the measure of how much of an oil will dissolve in the water column at a known temperature and pressure. The more polar the compound, the more soluble it is in water. For example, benzene (C_6H_6), toluene (C6H5CH3), ethyl benzene C6H5CH2CH3, and the xylene isomers ($H_3CC_6H_4CH_3$) (i.e., the BTEX compounds) are so frequently encountered in groundwater in part due to their high water solubility. However, the solubility in water is dependent on the nature of the multi-component mixture, such as naphtha, gasoline, kerosene, diesel, or crude oil or any crude oil product. The solubility of a constituent within a multi-component mixture may be orders of magnitude lower than the aqueous solubility of the pure constituent in water. Crude oil and crude oil products are complex mixtures of compounds, each of which partitions uniquely between oil and water; therefore, the water solubility varies between oils.

3.2.14 Sulfur Content

Removing sulfur from crude oil and crude oil products is an essential task for crude oil refineries and product upgrading in which the steric hindrance effect is the most important factor affecting the desulfurization activity of sulfur compounds via hydrotreating. Some polar sulfur compounds are difficult to hydrogenate to remove sulfur; however, the chemical structures of these compounds are unclear. However, understanding of their molecular structure is limited there is the need to focus on the formation and transformation mechanism of these polar sulfur compounds.

In the current context, during an oil spill, the sulfur content of the spilled material becomes a health and safety concern for cleanup personnel. In addition, if high sulfur oils are burning, the sulfur can lead to dangerous levels of sulfur dioxide.

The total sulfur content of oil can be determined by numerous standard techniques (ASTM D129, Standard Test Method for Sulfur in Crude oil Products) is applicable to crude oil products of low volatility and containing at least 0.1 mass percent sulfur. Sulfur contents are also determined in accordance with ASTM D4294 (Standard Test Method for Sulfur in Crude oil Products by Energy-Dispersive X-Ray Fluorescence Spectroscopy. This method is applicable to both volatile and non-volatile crude oil products with sulphur concentrations ranging from 0.05 to 5 mass percent.

When sulfur dioxide combines with water and air, it forms sulfurous acid and sulfuric acid, which are components of acid rain:

$$SO_2 + H_2O \rightarrow H_2SO_3$$

$$2SO_2 + 2H_2O + O_2 \rightarrow H_2SO_4$$

Acid rain can: (1) cause deforestation and (2) acidify waterways to the detriment of aquatic life.

3.2.15 Surface Tension and Interfacial Tension

The surface tension of crude oil (or a crude oil product), together with its viscosity, affects the rate at which a crude oil spill cured oil products spill will spread. Air/oil and oil/water interfacial tensions can be used to calculate a spreading coefficient which gives an indication of the tendency for the oil to spread. It is defined as:

Spreading coefficient = S_{WA} - S_{OA} - S_{WO}

S_{WA} is water/air interfacial tension, S_{OA} is oil/air interfacial tension, and S_{WO} is water/oil interfacial tension.

3.2.16 Total Petroleum Hydrocarbons

In the method, the term *total petroleum hydrocarbons* (total petroleum hydrocarbon derivatives) includes any crude oil constituent that falls within the measurable amount of crude oil-based hydrocarbon derivatives in the environment; the information obtained for total crude oil hydrocarbon derivatives depends on the analytical method used. Therefore, the difficulty associated with measurement of the total crude oil hydrocarbon derivatives is that the scope of the methods varies greatly (Speight, 2005). Interpretation of analytical results requires an understanding of how the determination was made (Speight, 2005, 2014, 2015).

There are many analytical techniques available that measure total crude oil hydrocarbon concentrations in the environment but no single method is satisfactory for the measurement of the entire range of crude oil-derived hydrocarbon derivatives. In addition, and because the techniques vary in the manner in which hydrocarbon derivatives are extracted and detected, each method may be applicable to the measurement of different subsets of the crude oil-derived hydrocarbon derivatives present in a sample. The four most commonly used total crude oil hydrocarbon analytical methods include (1) gas chromatography (GC), (2) infrared spectrometry (IR), (3) gravimetric analysis, and (4) immunoassay (Miller, 2000).

The total petroleum hydrocarbon (TPH) analyses are conducted to determine the total amount of hydrocarbon derivatives present in the environment (Rhodes et al., 1994; Pavlova and Ivanova,

2003; Speight, 2005, 2015). There are a wide variety of methods for measurement of the total crude oil hydrocarbon derivatives in a sample but analytical inconsistencies must be recognized because of the definition of total crude oil hydrocarbon derivatives and the methods employed for analysis (Rhodes et al., 1994; Speight, 2014, 2015). Thus, in practice, the term total crude oil hydrocarbon is defined by the analytical method since different methods often give different results because they are designed to extract and measure slightly different subsets of crude oil hydrocarbon derivatives.

The analysis for the total crude oil hydrocarbon derivatives (TPH) in a sample as a means of evaluating crude oil-contaminated sites is also an analytical method in common use. The data are used to establish target cleanup levels for soil or water is a common approach implemented by regulatory agencies in the United States, and in many other countries.

Thus, as often occurs in crude oil science (Speight, 2007), the definition of total crude oil hydrocarbon derivatives depends on the analytical method used because the total crude oil hydrocarbon derivatives measurement is the total concentration of the hydrocarbon derivatives extracted and measured by a particular method. The same sample analyzed by different methods may produce different values. For this reason, it is important to know exactly how each determination is made since interpretation of the results depends on understanding the capabilities and limitations of the method. If used indiscriminately, measurement of the total crude oil hydrocarbon derivatives in a sample can be misleading, leading to an inaccurate assessment of risk.

Thus, the choice of a specific method should be based on compatibility with the particular type of hydrocarbon contamination to be measured and, furthermore, the choice may depend upon local or regional regulatory requirements for the type of hydrocarbon contamination that is known, or suspected, to be present.

3.2.17 Viscosity

Viscosity is a measure of the resistance of a fluid to deformation in shear, i.e., the resistance of the fluid (such as crude oil or crude oil products) to flow or pouring. For example, high viscosity crude oil and crude oil products flow less easily than those of lower viscosity. All crude oils and crude oil products become more viscous (i.e., flow less readily) as the temperature falls, some more than others depending on the composition of the oil. Thus, the viscosity of the crude oil or the crude oil products is possibly the most important factor influencing oil dispersability. Tare indications of an upper viscosity limit, above which oil becomes non-dispersible which is crude oil (or crude oil product) dependent (some oils have a higher viscosity limit than others), but, for all oils, dispersability decreases with increasing viscosity.

As an example in the current context of spills, the viscosity of the spilled material usually determines the effectiveness of the dispersion of the chemical(s). Dispersants are most effective immediately after a spill and become less effective as oil weathering alters the properties of the oil, decreasing its dispersability. Operational parameters (e.g., dispersant droplet size, dispersant concentration, and mixing energy) also factor into the effectiveness of chemical dispersion.

More specifically, the viscosity of crude oil and crude oil products is largely determined by the content of high molecular weight polar molecules such as the constituents of the resin fraction and the asphaltene fraction. The greater the percentage of lower molecular weight constituents such as the constituents of the saturates fraction and the lower molecular weight aromatics fraction, it follows that the lower the amount of resin constituents and asphaltene constituents leading to lower viscosity. Temperature also affects viscosity, with a lower temperature resulting in a higher viscosity. The variations with temperature are commonly large. For example, a crude oil (or a crude oil product) that flows readily at 40°C (104°F) will be a much slower moving more viscus material at 10°C (50°F). Evaporative losses selectively remove lighter components and, consequently, increase the viscosity of the residual oil.

3.2.18 Volatility

The distillation characteristics of a crude oil or a crude oil product is influenced by the volatility. In the distillation process, as the temperature of crude oil or the crude oil product raised, different components reach the boiling point in succession, evaporate, and are then cooled and condensed. The distillation characteristics are expressed as the proportions of the parent oil that distil within specific ranges of temperature ranges. Some crude oil and crude oil products oils contain waxy or asphaltenic residues, which do not readily distil even at high temperatures and are also likely to persist in the environment.

On the other hand, at the lower boiling end of the volatility scale, benzene, toluene, ethylbenzene, and the xylene isomers (BTEX), and substituted benzene derivatives are the most common aromatic compounds in crude oil (often referred to as volatile organic compounds (VOCs), making up to a few percent of the total mass of some crude oils.

Benzene, toluene, ethylbenzene, and the xylene isomers (BTEX), and substituted benzenes are the most common aromatic compounds in petroleum, making up to a few percent of the total mass of some crude oils. They are the most soluble and mobile fraction of crude oil and many crude oil products, and as such, frequently enter soil, sediments, and ground water because of accidental spills, leakage of petroleum fuels from storage tanks and pipelines, and improper oil waste disposal practices.

A rapid, reliable, and effective method for direct determination of BTEX plus C_3-substituted benzenes has been developed using gas chromatography with mass spectrometric detection (GC/MS) (Wang et al., 1995). Vapor pressure, which is related to the amount of volatile organic compounds in the crude oil or crude oil product, is an important physical property of volatile liquids.

During weathering, oil becomes more viscous through evaporative loss of volatile compounds and by collecting water (emulsification). The degree and rate at which oil properties change as a result of weathering depend on the type of crude oil (or the crude oil product) and on the conditions surrounding the oil spill (weather, sea state, and location of the spill).

The vapor pressure is an important physical property of the volatile constituents of crude oil and crude oil products. It is the pressure that a vapor exerts on the surroundings. For volatile crude oils and crude oil products, the vapor pressure is used as an indirect measure of evaporation rate. The most commonly used method for determining the volatility of crude oil and crude oil products is the Reid vapor pressure (ASTM D323—Standard Test Method for Vapor Pressure of Crude oil Products). This test method determines vapor pressure at 37.8°C (100°F) of crude oil and crude oil products with an initial boiling point above 0°C (32°F). It is measured by saturating a known volume of oil in an air chamber of known volume and measuring the equilibrium pressure which is then corrected to one atmosphere (14.7 psi).

3.2.19 Water Content

Some crude oils and crude oil products sampled contain substantial amounts of water. Because any process that would separate the oil and water would also change the composition of the oil, most properties are often determined on the oils as received. Therefore, for those oils with significant water contents (> 5% w/w), many of the properties measured will not represent the properties of the dry oil.

Generally, water content is determined by the Karl Fischer titration (ASTM D6304). Water content could has not been identified as a direct governing environmental factor in the biodegradation of crude oil and crude oil products (Frijer et al., 1996). The rate of production of carbon dioxide is a means of expressing the mineralization rate of hydrocarbon derivatives.

3.2.20 Weathering Processes

The individual processes discussed in this subsection, although presented to some extent in the above sub-sections are included here as a group because of the participation of these processes in what is commonly referred to as the weathering process. However, the relative contribution of each process to the weathering of crude oils and crude oil products varies with time. In addition to these processes, a spill of crude oil or a crude oil product will move according to climatic conditions.

The term *spreading* refers to the movement of the spilled material immediately after the spill. The speed at which this movement occurs depends to a great extent on the viscosity of the oil and the volume spilled. Fluid, low viscosity oils spread much faster than those with high viscosity. As the oil spreads and the thickness of the material is reduced, the appearance of the material changes and rather than spreading as thin layers, semi-solid or highly viscous oils fragment into patches which move apart.

As an example, in open water, wind circulation patterns tend to cause oil to form narrow bands or 'windrows' parallel to the wind direction and, over time, the properties of the crude oil or the crude oil product become less important in determining the movement of the spill.

Evaporation occurs when the more volatile components of the crude oil or the crude oil product will evaporate to the atmosphere. The rate of evaporation depends on ambient temperatures and wind speed. In general those oil components, such as naphtha fractions, with a boiling point below 200°C (390°F) will evaporate quickly in temperate conditions. The higher the proportion of low-boiling constituents (as deduced from the distillation characteristics) the greater the degree of evaporation.

The initial spreading rate of the oil also affects the rate of evaporation since the larger the surface area, the faster the evaporation rate of the low boiling constituent. The non-volatile constituents that remain after evaporation of the lower boiling constituents have an increased density and an increased viscosity, which affects subsequent weathering processes as well as the effectively of the clean-up technology.

Spills of crude oil products, such as naphtha and kerosene, may evaporate completely within a few hours of the spill. When such extremely volatile crude oils and crude oil products are spilled in confined areas, there may be a risk of fire and explosion. In contrast, heavy fuel oil undergoes little, if any, evaporation and poses a minimal risk of explosion but the fire risk from such a spill remain a reality. If debris is ignited in a pool of oil in calm conditions, it can sustain a vigorous fuel oil fire.

The rate of *dispersion* is largely dependent upon the nature of the crude oil or the crude oil product and, in the case of a spill on land, can be influence by the adhesion of the constituents to the soil or minerals or in the case of a spill at sea can proceed rapidly with low viscosity oils in the presence of breaking waves.

Waves and turbulence at the sea surface can cause all or part of the spilled material to break up into droplets of varying sizes which become mixed into the upper layers of the water column. Smaller droplets remain in suspension while the larger ones rise back to the surface where they either coalesce with other droplets to reform a thick coating of oil or spread out in a very thin film.

An increased surface area presented by dispersed oil also promotes processes such as biodegradation, dissolution and sedimentation. Crude oil and crude oil products that remain fluid and spread unhindered by other weathering processes may disperse completely within days in moderate sea conditions. Conversely, viscous oils tend to form thick fragments on the water surface that show little tendency to disperse even with the addition of dispersants.

Emulsification is the process in which a crude oil or a crude oil product takes up water and form water-in-oil emulsions which can increase the volume of pollutant one or more order of magnitude. Viscous oils, such as heavy crude oils and the heavier fuel oils, tend to take up water more slowly than more fluid oils. As the emulsion develops, the asphaltene constituents can precipitate (i.e., phase separate) from the oil to coat the water droplets thereby increasing the stability of the emulsion.

Stable emulsions may contain as much as 70 to 80% v/v water are highly persistent in the environment. Less stable emulsions may separate into oil and water if heated by sunlight under calm conditions or when stranded on shorelines. The formation of water-in-oil emulsions reduces the rate of other weathering processes and is the main reason for the persistence of light and medium crude oils in an aqueous ecosystem.

The term *dissolution* refers to the rate and extent to which the constituents of a crude oil or crude oil product dissolves in water and depends upon the composition of the crude oil (or the crude oil product), spreading, the water temperature, turbulence of the water in the ecosystem, and degree of dispersion. The high molecular weight constituents of crude oil and crude oil products are (typically) insoluble in water whereas the lower molecular weight constituents (particularly the aromatic hydrocarbon derivatives such as benzene and toluene) are slightly soluble in water. However, these compounds are also the most volatile and are lost by evaporation.

The term *photo-oxidation* refers to the fraction of hydrocarbon derivatives with oxygen, which may either lead to the formation of soluble products or tars. Oxidation is promoted by sunlight and, although it occurs for the entire duration of the spill, its overall effect on dissipation is minor compared to that of other weathering processes. The overall result (in the case of a spill at sea) is the formation of tar balls (which usually consist of a solid outer crust of oxidized oil and sediment particles, surrounding a softer, less weathered interior) on a shoreline.

The terms *sedimentation* and *sinking* occur when dispersed oil droplets interact with sediment particles and organic matter suspended in the water column so that the droplets become dense enough to sink slowly to the base of the water system. For example, shallow coastal areas and the waters of river mouths and rover estuaries are often laden with suspended solids that can bind with dispersed oil droplets, thereby providing favorable conditions for sedimentation of oily particles.

Most crude oils and crude oil products have sufficiently low specific gravity to remain afloat unless they interact with and attach to more dense materials. However, some heavy crude oils, most heavy fuel oils, and water-in-oil emulsions have a specific gravity close to that of sea water and even minimal interaction with sediment can be sufficient to cause sinking. Some residual oils have a specific gravity greater than the specific gravity of sea water (> 1.025), thereby causing them to sink once spilled. In addition, some oils can sink following a fire, which not only consumes the lower boiling constituents of the oil but also results in the formation of higher molecular weight products as a consequence of the associated high temperatures. This must be a consideration if deliberate in-situ burning is contemplated as a response technique.

Sedimentation is one of the key long term processes leading to the accumulation of spilled oil in the marine environment. However, sinking of bulk oil is only rarely observed other than in shallow water, close to shore, primarily as a result of interaction with the shoreline.

Thus, an understanding of the way in which weathering processes interact is important when attempting to forecast the changing characteristics of an oil during the lifetime of the spilled material. Predictions of potential changes in the characteristics of the spilled material over time allow an assessment to be made of the likely persistence of constituents of the spilled material leading to (hopefully) the most option for spill response option. In this latter aspect of a spill response, a distinction is frequently made between (1) non-persistent oils, which because of their volatile nature and low viscosity tend to disappear rapidly from the sea surface, and (2) persistent oils, which dissipate more slowly and usually require a clean-up response. Examples of the former (i.e., then non-persistent oils) are naphtha (gasoline) and kerosene (diesel), whereas most crude oils, intermediate and heavy fuel oils, and bitumen are classed as persistent.

3.3 Petroleum Group Analysis

Crude oil group analyses are conducted to determine amounts of the crude oil compound classes (e.g., saturates, aromatics, and polars/resins) present in crude oil-contaminated samples. This type of measurement is sometimes used to identify fuel type or to track plumes. It may be particularly useful for higher boiling products, such as asphalt. Group type test methods include multidimensional gas chromatography (not often used for environmental samples), high performance liquid chromatography (HPLC), and thin layer chromatography (TLC) (Miller, 2000; Patnaik, 2004).

Test methods that analyze individual compounds (e.g., benzene-toluene-ethylbenzene-xylene mixtures and polynuclear aromatic hydrocarbon derivatives) are generally applied to detect the presence of an additive or to provide concentration data needed to estimate environmental and health risks that are associated with individual compounds. Common constituent measurement techniques include gas chromatography with second column confirmation, gas chromatography with multiple selective detectors and gas chromatography with mass spectrometry detection (GC/MS) (EPA 8240).

Many common environmental methods measure individual crude oil constituents or *target compound* rather than the whole signal from the total crude oil hydrocarbon derivatives. Each method measures a suite of compounds selected because of their toxicity and common use in industry.

For organic compounds, there are three series of target compound methods that must be used for regulatory purposes:

1. EPA 500 series: Organic Compounds in Drinking Water, as regulated under the Safe Drinking Water Act.

2. EPA 600 series: Methods for Organic Chemical Analysis of Municipal and Industrial Wastewater, as regulated under the Clean Water Act.

3. SW-846 series: Test Methods for Evaluating Solid Waste: Physical/Chemical Methods, as promulgated by the US EPA, Office of Solid Waste and Emergency Response.

The 500 and 600 series methods provide parameters and conditions for the analysis of drinking water and wastewater, respectively. One method (EPA SW-846) is focused on the analysis of nearly all matrices including industrial waste, soil, sludge, sediment, and water miscible and non-water miscible wastes. It also provides for the analysis of groundwater and wastewater but is not used to evaluate compliance of public drinking water systems.

Selection of one method over another is often dictated by the nature of the sample and the particular compliance or cleanup program for which the sample is being analyzed. It is essential to recognize that capabilities and requirements vary between methods when requesting any analytical method or suite of methods. Most compound-specific methods use a gas chromatographic selective detector, high performance liquid chromatography, or gas chromatography/mass spectrometry.

More correctly, group analytical methods are designed to separate hydrocarbon derivatives into categories, such as saturates, aromatics, resins and asphaltenes (SARA) or paraffins, iso-paraffins, naphthenes, aromatics, and olefins (PIANO). These chromatographic, gas chromatographic, and high performance liquid chromatographic methods (HPLC) were developed for monitoring refinery processes or evaluating organic synthesis products. Column chromatographic methods that separate saturates from aromatics are often used as preparative steps for further analysis by gas chromatography/mass spectrometry. Thin layer chromatography is sometimes used as a screening technique for crude oil product identification.

3.3.1 Thin Layer Chromatography

In the environmental field, thin layer chromatography (TLC) is best used for screening analyses and characterization of semi-volatile and non-volatile crude oil products. The precision and accuracy of the technique is inferior to other methods (EPA 8015, EPA 418.1) but when speed and simplicity are desired, thin layer chromatography may be a suitable alternative. For characterizations of crude oil products such as asphalt, the method has the advantage of separating compounds that are too high boiling to pass through a gas chromatograph. While thin layer chromatography does not have the resolving power of a gas chromatograph, it is able to separate different classes of compounds. Thin layer chromatography analysis is fairly simple and, since the method does not give highly accurate or precise results, there is no need to perform the highest quality extractions.

When the aromatic content of a sample is high, as with bunker C fuel oil, the detection limit can be near 100 ppm. It is often not possible to distinguish between similar products such as diesel and jet fuel. As with all chemical analyses, quality assurance tests should be run to verify the accuracy and precision of the method.

3.3.2 Immunoassay

Immunoassay methods correlate total crude oil hydrocarbon derivatives with the response of antibodies to specific crude oil constituents. Many of the methods measure only aromatics that have an affinity for the antibody, benzene-toluene-ethylbenzene-xylene, and polynuclear aromatic hydrocarbon analysis.

Synthetic antibodies have been developed to complex with crude oil constituents. Because the coloring agent reacts with the labeled enzyme, samples with high optical density contain low concentrations of analytes. Concentration is inversely proportional to optical density.

The antibodies used in immunoassay kits are generally designed to bond with selected compounds. A correction factor supplied by the manufacturer must be used to calculate the concentration of total crude oil hydrocarbon derivatives. The correction factor can vary depending on product type because it attempts to correlate total crude oil hydrocarbon derivatives with the measured surrogates.

Immunoassay tests do not identify specific fuel types and are best used as screening tools. The tests are dependent on soil type and homogeneity. In particular, for clay and other cohesive soils, the tests are limited by a low capacity to extract hydrocarbon derivatives from the sample.

A number of different testing kits based on immunoassay technology are available for rapid field determination of certain groups of compounds such as benzene-toluene-ethylbenzene-xylene (EPA 4030) or polynuclear aromatic hydrocarbon derivatives (EPA 4035, Polycyclic Aromatic Hydrocarbon derivatives by Immunoassay). The immunoassay screening kits are self-contained portable field kits that include components for sample preparation, instrumentation to read assay results, and immunoassay reagents.

Unless the immunoassay kit is benzene sensitive, the kit may display strong biases such as the low affinity for benzene relative to toluene, ethylbenzene, xylenes, and other aromatic compounds. This will cause an underestimation of the actual benzene levels in a sample. And, since benzene is often the dominant compound in leachates due to its high solubility, a low sensitivity for benzene is undesirable.

The quality of the analysis of polynuclear aromatic hydrocarbon derivatives is often dependent on the extraction efficiency. Clay and other cohesive soils lower the ability to extract polynuclear aromatic hydrocarbon derivatives. Another potential problem with polynuclear aromatic hydrocarbon analysis is that the test kits may have different responses for different compounds.

3.3.3 Gas Chromatography

Briefly, gas chromatography (also referred to as *gas-liquid chromatography*, GLC) is a method for separating the volatile components of various mixtures (Altgelt and Gouw, 1975; Fowlis, 1995; Grob, 1995). Gas-liquid chromatography is also used extensively for individual component identification, as well as percentage composition, in the gasoline boiling range.

The mobile phase is the carrier gas, and the gas selected has a bearing on the resolution. Nitrogen has very poor resolution ability, while helium or hydrogen are better choices with hydrogen being the best carrier gas for resolution. However, hydrogen is reactive and may not be compatible with all sets of target analytes. There is an optimum flow rate for each carrier gas to achieve maximum resolution. As the temperature of the oven increases, the flow rate of the gas changes due to thermal expansion of the gas. Most modern gas chromatographs are equipped with constant flow devices that change the gas valve settings as the temperature in the oven changes, so changing flow rates are no longer a concern. Once the flow is optimized at one temperature it is optimized for all temperatures.

Gas chromatography uses the principle of a stationary phase and a mobile phase. Much attention has been paid to the various stationary phases and books have been written on the subject as it pertains to crude oil chemistry.

Gas chromatographic methods are currently the preferred laboratory methods for measurement of total crude oil hydrocarbon measurement because they detect a broad range of hydrocarbon derivatives and provide both sensitivity and selectivity. In addition, identification and quantification of individual constituents of the total crude oil hydrocarbon mix is possible.

Methods based on gravimetric analysis are also simple and rapid but they suffer from the same limitations as infrared spectrometric methods (Speight, 2005, 2015). Gravimetric-based methods may be useful for oily sludge and wastewaters, which will present analytical difficulties for other more sensitive methods. Immunoassay methods for the measurement of total crude oil hydrocarbon are also popular for field testing because they offer a simple, quick technique for in-situ quantification of the total crude oil hydrocarbon derivatives.

For methods based on gas chromatography, the total crude oil hydrocarbon derivatives fraction is defined as any chemicals extractable by a solvent or purge gas and detectable by gas chromatography/flame ionization detection (GC/FID) within a specified carbon range. The primary advantage of such methods is that they provide information about the type of crude oil in the sample in addition to measuring the amount. Identification of product type(s) is not always straightforward, however, and requires an experienced analyst of crude oil products (Sullivan and Johnson, 1993; Speight, 2014, 2015). Detection limits are method-dependent as well as matrix-dependent and can be as low as 0.5 mg/L in water or 10 mg/kg in soil.

Chromatographic columns are commonly used to determine total crude oil hydrocarbon compounds approximately in the order of their boiling points. Compounds are detected by means of a flame ionization detector that responds to virtually all compounds that can burn. The sum of all responses within a specified range is equated to a hydrocarbon concentration by reference to standards of known concentration.

Two methods (EPA SW-846 8015 and 8015A) were, in the past, often quoted as the source of gas chromatography-based methods for the measurement of the total crude oil hydrocarbon derivatives in a sample. However, the original methods were developed for non-halogenated volatile organic compounds and were designed to measure a short target list of chemical solvents rather than crude oil hydrocarbon derivatives. Thus, because there was no universal method for total crude oil hydrocarbon derivatives, there were many variations of these methods. Recently, an updated method (EPA 8015B) provides guidance for the analysis of gasoline and diesel range organic compounds.

Gas chromatography-based methods can be broadly used for different kinds of crude oil contamination but are most appropriate for detecting nonpolar hydrocarbon derivatives with carbon numbers between C6 and C25 or C36. Many lubricating oils contain molecules with more than 40 carbon atoms. In fact, crude oil itself contains molecules having more than 100 carbons or more. These high molecular weight hydrocarbon derivatives are outside the detection range of the more common gas chromatographic methods, but specialized gas chromatographs are capable of analyzing such high molecular weight constituents.

Accurate quantification depends on adjusting the chromatograph to reach as high a carbon number as possible, then running a calibration standard with the same carbon range as the sample. There should also be a check for mass discrimination, a tendency for higher molecular weight hydrocarbon derivatives to be retained in the injection port. If a sample is suspected to be heavy oil or to contain a mixture of light oil and heavy oil, the most appropriate method must be used.

Gravimetric or infrared methods are often preferred for high molecular weight samples. These methods can even be used as a check on gas chromatographic data if it is suspected that high molecular weight hydrocarbon derivatives are present but are not being detected.

Calibration standards vary. Most methods specify a gasoline calibration standard for volatile range total crude oil hydrocarbon derivatives and a diesel fuel #2 standard for extractable range total crude oil hydrocarbon derivatives. Some methods use synthetic mixtures for calibration. Because most methods are written for gasoline or diesel fuel, total crude oil hydrocarbon derivatives methods may have to be adjusted to measure contamination by heavier hydrocarbon derivatives (e.g., heavy fuel oil, lubricating oil, or crude oil). Such adjustments may entail use of a more aggressive solvent, a wider gas chromatographic window that allows detection of molecules containing up to C36 or more, and a different calibration standard that more closely resembles the constituents of the sample under investigation.

Gas chromatographic methods can be modified and fine-tuned so that they are suitable for measurement of specific crude oil products or group types. These modified methods can be particularly useful when there is information on the source of contamination, but method results should be interpreted with the clear understanding that a modified method was used for detection of a specific carbon range.

Interpretation of gas chromatographic data is often complicated and the analytical method should always be considered when interpreting concentration data. For example, a volatile range analysis may be very useful for quantifying total crude oil hydrocarbon derivatives at a gasoline release site, but a volatile range analysis will not detect the presence of lube oil constituents. In addition, a modified method that has been specifically selected for detection of gasoline-range organics at a gasoline-contaminated site may also detect hydrocarbon derivatives from other crude oil releases because fuel carbon ranges frequently overlap. Gasoline is found primarily in the volatile range. Diesel fuel falls primarily in an extractable range. Jet fuel overlaps both the volatile and semi-volatile ranges. However, the detection of different kinds of crude oil constituents does not necessarily indicate that there have been multiple releases at a site. Analyses of spilled waste oil will frequently detect the presence of gasoline, and sometimes diesel. This does not necessarily indicate multiple spills since all waste oil contain some fuel. As much as 10% of used motor oil can consist of gasoline.

If the type of contaminant is unknown, a *fingerprint* analysis can help in the identification procedure. A *fingerprint* or *pattern recognition* analysis is a direct injection analysis where the chromatogram is compared to chromatograms of reference materials.

Furthermore, as a fuel evaporates or biodegrades, its pattern can change so radically that identification becomes difficult (Bartha, 1986). Consequently, a gas chromatographic fingerprint is not a conclusive diagnostic tool. The methods must for total crude oil hydrocarbon analysis must stress calibration and quality control, while pattern recognition methods stress detail and comparability.

The gas chromatographic methods usually cannot quantitatively detect compounds below C6 because these compounds are highly volatile and interference can occur from the solvent peak. As much as 25% of fresh gasoline can be below C6 but the problem is reduced for weathered gasoline and/or diesel range contamination because most of the very volatile hydrocarbon derivatives ($< C6$) may no longer be present in the sample. Gas chromatographic methods may also be inefficient for quantification of polar constituents (nitrogen, oxygen, and sulfur containing molecules). Some of the polar constituents are too reactive to pass through a gas chromatograph and thus will not reach the detector for measurement.

Oxygenated gasoline is sometimes analyzed by GC-based methods but it should be noted that the efficiency of purge methods is lower for oxygenates such as ethers and alcohols because detector response to oxygenates is lower relative to hydrocarbon derivatives. Therefore the data will be biased slightly low for ether-containing fuels compared to equivalent amounts of traditional gasoline. Methanol and ethanol elute before hexane and, consequently, they are not quantified and may not even be detected due to co-elution with the solvent.

In addition, cleanup steps do not perfectly separate crude oil hydrocarbon derivatives from biogenic material such as plant oils and waxes that are sometimes extracted from vegetation-rich soil.

Because crude oil is made up of so many isomers, many compounds, especially those with more than eight carbon atoms, co-elute with isomers of nearly the same boiling point. These unresolved compounds are referred to as the unresolved complex mixture. They are legitimately part of the crude oil signal, and unless otherwise specified, should be quantified. Quantifying such a mixture requires a baseline-to-baseline integration mode rather than a peak-to-peak integration mode. The baseline-to-baseline integration quantifies all of the crude oil constituents in the sample but the peak-to-peak integration only the individual resolved hydrocarbon derivatives (not including the unresolved complex mixture) are quantified.

For environmental analysis (Bruner, 1993), particularly the volatile samples such found in total crude oil hydrocarbon derivatives, the gas chromatograph is generally interfaced with a purge and trap system as described in the section on gas chromatographic methods. The photoionization detector works by bombarding compounds with ultraviolet (UV) light, generating a current of ions. Compounds with double carbon bonds, conjugated systems (multiple carbon double bonds arranged in a specific manner), and aromatic rings are easily ionized with the ultraviolet light generated by the photoionization detector lamp, while most saturated compounds require higher energy radiation.

One method (EPA 8020) that is suitable for volatile aromatic compounds is often referred to as benzene-toluene-ethylbenzene-xylene analysis, though the method includes other volatile aromatics.

Certain false positives are common (EPA 8020). For example, trimethylbenzenes, and gasoline constituents are frequently identified as chlorobenzenes (EPA 602, EPA 8020) because these compounds elute with nearly the same retention times from nonpolar columns. Cyclohexane is often mistaken for benzene (EPA 8015/8020) because both compounds are detected by a 10.2 eV photoionization detector and have nearly the same elution time from a nonpolar column (EPA 8015). The two compounds have very different retention times on a more polar column (EPA 8020) but a more polar column skews the carbon ranges (EPA 8015). False positives for oxygenates in gasoline are common, especially in highly contaminated samples.

For semi-volatile constituents of crude oil, the gas chromatograph is generally equipped with either a packed or capillary column. Either neat or diluted organic liquids can be analyzed via direct injection, and compounds are separated during movement down the column. The flame ionization detector uses a hydrogen-fueled flame to ionize compounds that reach the detector.

For polynuclear aromatic hydrocarbon derivatives, a method is available (EPA 8100) in which injection of sample extracts directly onto the column is the preferred method for sample introduction for this packed-column method.

Semi-volatile constituent are among the analytes that can be readily resolved and detected using the system. If a packed column is used, four pairs of compounds may not be adequately resolved and are reported as a quantitative sum: anthracene and phenanthrene, chrysene and benzo(a)anthracene, benzo(b) fluoranthene and benzo(k)fluoranthene, and dibenzo(a,h)anthracene and indeno(1,2,3-cd) pyrene. This issue can be resolved through the use of a capillary column in place of a packed column.

3.3.4 High Performance Liquid Chromatography

A high performance liquid chromatography system can be used to measure concentrations of target semi-volatile and non-volatile crude oil constituents. The system only requires that the sample be dissolved in a solvent compatible with those used in the separation.

In the method, polynuclear aromatic hydrocarbon derivatives are extracted from the sample matrix with a suitable solvent, which is then injected into the chromatographic system. Usually the extract must be filtered because fine particulate matter can collect at the inlet of the column, resulting in high back-pressures and eventual plugging of the column. For most hydrocarbon analyses, reverse phase high performance liquid chromatography (i.e., using a nonpolar column packing with a more polar mobile phase) is used. The most common bonded phase is the octadecyl (C18) phase. The mobile phase is commonly aqueous mixtures of either acetonitrile or methanol.

After the chromatographic separation, the analytes flow through the cell of the detector. A fluorescence detector shines light of a particular wavelength (the excitation wavelength) into the cell. Fluorescent compounds absorb light and reemit light of other, higher wavelengths (emission wavelengths). The emission wavelengths of a molecule are mainly determined by its structure. For polynuclear aromatic hydrocarbon derivatives, the emission wavelengths are mainly determined by the arrangement of the rings and vary greatly between isomers.

Some of the polynuclear aromatic hydrocarbon derivatives such as phenanthrene, pyrene, and benzo(g,h,i)perylene, are commonly seen in products boiling in the middle to heavy distillate range. In a method for their detection and analysis (EPA 8310) an octadecyl column and an aqueous acetonitrile mobile phase are used. Analytes are excited at 280 nm and detected at emission wavelengths of > 389 nm. Naphthalene, acenaphthene, and fluorene must be detected by a less-sensitive UV detector because they emit light at wavelengths below 389 nm. Acenaphthylene is also detected by UV detector.

The methods using fluorescence detection will measure any compounds that elute in the appropriate retention time range and which fluoresce at the targeted emission wavelength(s) (Falla Sotelo et al., 2008). In the case of one method (EPA 8310), the excitation wavelength excites most aromatic compounds. These include the target compounds and also many derivatized aromatics, such as alkyl aromatics, phenols, anilines, and heterocyclic aromatic compounds containing the pyrrole (indole, carbazole, etc.), pyridine (quinoline, acridine, etc.), furan (benzofuran, naphthofuran, etc.), and thiophene (benzothiophene, naphthothiophene, etc.) structures. In crude oil samples, alkyl polynuclear aromatic hydrocarbon derivatives are strong interfering compounds. For example, there are five methyl phenanthrene derivatives and over 20 dimethyl phenanthrene derivatives. The alkyl substitution does not significantly affect either the wavelengths or intensity of the phenanthrene fluorescence. For a very long time after the retention time of phenanthrene, the alkyl phenanthrene derivatives will interfere, affecting the measurements of all later-eluting target polynuclear aromatic hydrocarbon derivatives.

Interfering compounds will vary considerably from source to source and samples may require a variety of cleanup steps to reach required method detection limits. The emission wavelengths used (EPA 8310) are not optimal for sensitivity of the small ring compounds. With modern electronically-controlled monochromator, wavelength programs can be used which tune excitation and emission wavelengths to maximize sensitivity and/or selectivity for a specific analyte in its retention time window.

3.3.5 Gas Chromatography-Mass Spectrometry

A gas chromatography-mass spectrometry system is used to measure concentrations of target volatile and semi-volatile crude oil constituents. It is not typically used to measure the amount of total crude oil hydrocarbon derivatives.

The current method (EPA SW-846 8260) for the analysis of volatile compounds reveals that most of the compounds listed in these methods are not typically found in crude oil products. However, a method that uses selected ion monitoring (SIM) involves system set up to measure only selected target masses rather than scanning the full mass range. This technique yields lower detection limits for specific compounds. At the same time, it gives the more complete information available from the total ion chromatogram and the full-mass-range spectrum of each compound. The technique is sometimes used to quantify compounds present at very low concentrations in a complex hydrocarbon matrix. It can be used if the target compound's spectrum has a prominent fragment ion at a mass that distinguishes it from the rest of the hydrocarbon compounds.

The most common method for GC/MS analysis of semi-volatile compounds (EPA SW-846 8270) includes 16 polycyclic aromatic compounds, some of which commonly occur in middle distillate to heavy crude oil products. The method also quantifies phenols and cresols, compounds that are not hydrocarbon derivatives but may occur in crude oil products. Phenols and cresols are more likely found in crude oils and weathered crude oil products.

To reduce the possibility of false positives, the intensities of one to three selected ions are compared to the intensity of a unique target ion of the same spectrum. The sample ratios are compared to the ratios of a standard.

3.4 Other Analytical Methods

The are several alternate or complementary methods that measure different crude oil hydrocarbon categories, there are several basic steps that are common to the analytical processes for all methods, no matter the method type or the environmental matrix. In general, these steps are: (1) collection and preservation—requirements specific to environmental matrix and analytes of interest, (2) extraction so that separations of the analytes of interest from the sample matrix can be achieved, (3) concentration—enhances the ability to detect analytes of interest, (4) cleanup, dependent upon the need to remove interfering compounds, and (5) measurement, or quantification, of the analytes. Each step affects the final result, and a basic understanding of the steps is vital to data interpretation and, hence, environmental cleanup.

3.4.1 Infrared Spectroscopy

Infrared methods measure the absorbance of the C-H bond and most methods typically measure the absorbance at a single frequency (usually 2930 cm^{-1}) that corresponds to the stretching of aliphatic methylene (CH$_2$) groups. Some methods use multiple frequencies including 2960 cm^{-1} (CH$_3$ groups) and 2900 to 3000 cm^{-1} (aromatic C-H bonds).

Therefore, for infrared spectroscopic methods, the total crude oil hydrocarbon derivatives is any chemicals extracted by a solvent that is not removed by silica gel and can be detected by infrared spectroscopy at a specified wavelength. The primary advantage of the infrared-based methods is that they are simple and rapid. Detection limits (e.g., for EPA 418.1) are approximately 1 mg/L in water and 10 mg/kg in soil. However, the infrared method(s) can often suffers from poor accuracy and precision, especially for heterogeneous soil samples. Also, the infrared methods give no information on the type of fuel present in the sample and there is little, often no, information about the presence or absence of toxic molecules, and no specific information about potential risk associated with the contamination.

Samples are extracted with a suitable solvent (i.e., a solvent with no C-H bonds) and biogenic polar materials are removed with silica gel. Some polar crude oil constituents may be removed as part of the silica gel cleanup. The absorbance of the silica gel eluate is measured at the specified frequency and compared to the absorbance of a standard or standards of known crude oil hydrocarbon concentration. The absorbance is a measurement of the sum of all the compounds contributing to the result. However, infrared methods cannot provide information on the type of hydrocarbon contamination.

For all IR-based TPH methods, the C-H absorbance is quantified by comparing it to the absorbance of standards of known concentration. An assumption is made that the standard has an aliphatic-to-aromatic ratio and an infrared response similar to that of the sample. Consequently, it is important to use a calibration standard as similar to the type of contamination as possible (EPA 418.1).

The infrared method that has been most frequently used (EPA 418.1) is appropriate only for water samples. A separatory funnel liquid/liquid extraction technique is used to extract the hydrocarbon derivatives from the water. A method (EPA 5520D) using a Soxhlet extraction technique is suitable for sludge. This extraction is frequently used to adapt the method (EPA 418.1) to soil samples. An infrared-based supercritical fluid extraction method for diesel range contamination (EPA 3560) is available.

Similar to gas chromatographic methods, the data from infrared methods must be interpreted after considering certain limitations and interferences that can affect data quality. For example, the C-H absorbance is not always measured in exactly the same way. Within the set of methods that specify a single infrared measurement, some methods call for the measurement at precisely 2930 cm^{-1} while others (including EPA 418.1) require measurement at the absorbance maximum nearest 2930 cm^{-1}. This variation can make a significant difference in the magnitude of the result, and can lead to confusion when comparing duplicate sample results. If only C-H absorbance is measured, infrared methods will potentially underestimate the concentration of total crude oil hydrocarbon derivatives in samples that contain crude oil constituents such as benzene and naphthalene that do not contain alkyl C-H groups.

Because an infrared result is calculated as if the aromatics in the sample were present in the same ratio as in the calibration standard, accuracy depends upon use of a calibration standard as similar to the type of contamination as possible. Use of a dissimilar standard will tend to create a positive bias in highly aliphatic samples and a negative bias in highly aromatic samples.

In summary, infrared methods are prone to interferences (positive bias) from non-crude oil sources since many organic compounds have some type of alkyl group associated with them whether crude oil-derived or not.

3.4.2 Gravimetry

Gravimetric methods measure all chemicals that are extractable by a solvent, not removed during solvent evaporation, and capable of being weighed. Some gravimetric methods include a cleanup step to remove biogenic material. The advantage of gravimetric methods is that they are simple and rapid. Detection limits are approximately 5–10 mg/L in water and 50 mg/kg in soil.

However, gravimetric methods are not suitable for measurement of low boiling hydrocarbon derivatives that volatilize at temperatures below 70 to 85°C (158 to 185°F). They are recommended for use with (1) oily sludge, (2) for samples containing heavy molecular weight hydrocarbon derivatives, or (3) for aqueous samples when hexane is preferred as the solvent.

Gravimetric methods give no information on the type of fuel present, no information about the presence or absence of toxic compounds, and no specific information about potential risk associated with the contamination.

There are a variety of gravimetric oil and grease methods suitable for testing water and soil samples (e.g., EPA SW-846 9070, EPA 413.1, EPA 9071). Technically the result is an oil and grease result because no cleanup step is used. One method (EPA 9071) is used to recover low levels of oil and grease by chemically drying a wet sludge sample and then extracting it using Soxhlet apparatus. Results are reported on a dry-weight basis. The method is also used when relatively polar high molecular weight crude oil fractions are present, or when the levels of non-volatile grease challenge the solubility limit of the solvent. Specifically, the method (EPA SW-846 9071) is suitable for biological lipids, mineral hydrocarbon derivatives, and some industrial wastewater.

Gravimetric methods for oil and grease e.g., EPA SW-846 9071 measure anything that dissolves in the solvent and remains after solvent evaporation. These substances include hydrocarbon derivatives, vegetable oils, animal fats, waxes, soaps, greases and related biogenic material. Gravimetric methods for total crude oil hydrocarbon derivatives (EPA 1664) measure anything that dissolves in the solvent and remains after silica gel treatment and solvent evaporation.

This method (EPA 1664) is a liquid/liquid extraction gravimetric procedure that employs n-hexane as the extraction solvent, in place of 1,1,2-trichloroethane (CFC-113) and/or 1,2,2-trifluoroethane (Freon-113), for determination of the conventional pollutant oil and grease. However, n-hexane is a poor solvent for high molecular weight crude oil constituents (Speight, 2014, 2015).

All gravimetric methods measure any suspended solids that are not filtered from solution, including bacterial degradation products and clay fines. Method 9071 specifies using cotton or glass wool as a filter.

Because extracts are heated to remove solvent, these methods are not suitable for measurement of low boiling low molecular weight hydrocarbon derivatives (i.e., hydrocarbon derivatives having less than fifteen carbon atoms) that volatilize at temperatures below 70 to 85°C (158 to 185°F). Liquid fuels, from gasoline through No. 2 fuel oil, lose volatile constituents are during solvent removal. In addition, soil results that are reported on a dry-weight basis suffer from potential losses of lower boiling hydrocarbon constituents during moisture determination where the matrix is dried at approximately 103 to 105°C (217 to 221°F) for several hours in an oven.

3.5 Properties and Analysis of Crude Oil Products

Crude oil products are in the forms of gases, liquids, and solids. In order to protect the environment, analysts need consistent, reliable, and credible methodologies to produce analytical data about gaseous emissions (Patnaik, 2004). To fulfill this need in this text, descriptions are given of this chapter is devoted to descriptions and the various analytical methods that can be applied to identify gaseous emissions from a refinery. Each gas is, in turn, and referenced by its name rather than the generic term *petroleum gas* (ASTM D4150). However, the composition of each gas varies and recognition of this is essential before testing protocols are applied.

3.5.1 Gaseous Products

Crude oil is capable of producing gaseous pollutant chemicals (Guthrie, 1967; Rawlinson and Ward, 1973; Francis and Peters, 1980; Hoffman, 1983; Loehr, 1992; Olschewsky and Megna, 1992; Moustafa, 1996; Speight, 1996, 1999). The gaseous emissions are often characterized by chemical species identification, e.g., *inorganic gases* such as sulfur dioxide (SO_2), nitrogen oxides (NO_x), and carbon monoxide (CO) or *organic gases* such as chloroform ($CHCl_3$) and formaldehyde (HCHO). The rate of release or concentrating in the exhaust air stream (in parts per million or comparable units) along with the type of gaseous emission greatly predetermines the applicable control technology.

The three main greenhouse gases that are products of refining are carbon dioxide, nitrous oxide, and methane (Fogg and Sangster, 2003). Carbon dioxide is the main contributor to climate change. Methane is generally not as abundant as carbon dioxide but is produced during refining and, if emitted into the atmosphere is a powerful greenhouse gas and more effective at trapping heat. However, gaseous emissions associated with crude oil refining are more extensive that carbon dioxide and methane and typically include process gases, petrochemical gases, volatile organic compounds (VOCs), carbon monoxide (CO), sulfur oxides (SO_x), nitrogen oxides (NO_x), particulates, ammonia (NH_3), and hydrogen sulfide (H_2S). These effluents may be discharged as air emissions and must be treated. However, gaseous emissions are more difficult to capture than wastewater or solid waste and, thus, are the largest source of untreated wastes released to the environment.

In addition to the corrosion of equipment of acid gases, the escape into the atmosphere of sulfur-containing gases can eventually lead to the formation of the constituents of acid rain, i.e., the oxides of sulfur (SO_2 and SO_3). Similarly, the nitrogen-containing gases can also lead to nitrous and nitric acids (through the formation of the oxides NO_x, where x = 1 or 2) which are the other major contributors to acid rain. The release of carbon dioxide and hydrocarbon derivatives as constituents of refinery effluents can also influence the behavior and integrity of the ozone layer.

The processes that have been developed to accomplish gas purification vary from a simple once-through wash operation to complex multi-step recycling systems. In many cases, the process complexities arise because of the need for recovery of the materials used to remove the contaminants or even recovery of the contaminants in the original, or altered, form (Kohl and Riesenfeld, 1979; Newman, 1985).

The majority of gas streams exiting each refinery process collected and sent to the gas treatment and sulfur recovery units to recover the refinery fuel gas and sulfur. Emissions from the sulfur recovery unit typically contain some hydrogen sulfide (H_2S), sulfur oxides, and nitrogen oxides. Other emissions sources from refinery processes arise from periodic regeneration of catalysts. These processes generate streams that may contain relatively high levels of carbon monoxide, particulates and volatile organic compounds (VOCs). Before being discharged to the atmosphere, such off-gas streams may be treated first through a carbon monoxide boiler to burn carbon monoxide and any volatile organic compounds, and then through an electrostatic precipitator or cyclone separator to remove particulates.

Analysts need consistent, reliable, and credible methodologies to produce analytical data about gaseous emissions (Patnaik, 2004). To fulfill this need in this text, descriptions are given of this chapter is devoted to descriptions and the various analytical methods that can be applied to identify gaseous emissions from a refinery. Each gas is, in turn, and referenced by its name rather than the generic term *petroleum gas* (ASTM D4150). However, the composition of each gas varies and recognition of this is essential before testing protocols are applied.

One of the more critical aspects for the analysis of gaseous (or low boiling) hydrocarbon derivatives is the question of volumetric measurement (ASTM D1071) and sampling (ASTM D1145, ASTM D1247, ASTM D1265). However, sampling liquefied petroleum gas from a liquid storage system is complicated by existence of two phases (gas and liquid) and the composition of the supernatant vapor phase will, most probably, differ from the composition of the liquid phase. Furthermore, the compositions of both phases will vary as a sample (or sample) is removed from one or both phases. An accurate check of composition can only be made if samples are taken during filling of the tank or from a fully charged tank.

In general, the sampling of gaseous constituents and of liquefied gases is the subject of a variety of sampling methods (ASTM D5503), such as the manual method (ASTM D1265, ASTM D4057), the floating piston cylinder method (ASTM D3700), and the automatic sampling method (ASTM D4177, ASTM D5287). Methods for the preparation of gaseous and liquid blends are also available (ASTM D4051, ASTM D4307) including the sampling and handling of fuels for volatility measurements (ASTM D5842).

Sampling methane (CH_4) and ethane (C_2H_6) hydrocarbon derivatives is usually achieved using stainless steel cylinders, either lined or unlined. However, other containers may also be employed dependent upon particular situations. For example, glass cylinder containers or polyvinyl fluoride (PVF) sampling bags may also be used but, obviously, cannot be subjected to pressures that are far in excess of ambient pressure. The preferred method for sampling propane (C_3H_8) and butane (C_4H_{10}) hydrocarbon derivatives is by the use of piston cylinders (ASTM D3700) although sampling these materials as gases is also acceptable in many cases. The sampling of propane and higher boiling hydrocarbon derivatives is dependent upon the vapor pressure of the sample. Piston cylinders or pressurized steel cylinders are recommended for high-vapor pressure sampling where significant amounts of low boiling gases are present while atmospheric sampling may be used for samples having a low-vapor pressure.

To monitor a process, measurement of gaseous emissions is typically performed over the time of a process cycle or over the time of use of a particular product. Emission test cycles are repeatable sequences of operating conditions. Such timely analyses allows process monitoring as well as identification of any changes that can lead to potential leakage of the gas.

Hydrocarbon gases are amenable to analytical techniques and there has been the tendency, and it remains, for the determination of both major constituents and trace constituents than is the case with the heavier hydrocarbon derivatives. The complexity of the mixtures that are evident as the boiling point of crude oil fractions and crude oil products increases make identification of many of the individual constituents difficult, if not impossible. In addition, methods have been developed for the determination of physical characteristics such as calorific value, specific gravity, and enthalpy from the analyses of mixed hydrocarbon gases, but the accuracy does suffer when compared to the data produced by methods for the direct determination of these properties.

Bulk physical property tests, such as density and heating value, as well as some compositional tests, such as the Orsat analysis and the mercuric nitrate method for the determination of unsaturation, are still used. However, the choice of a particular test is dictated by (1) the requirements of the legislation, (2) the properties of the gas under study, and (3) the selection of a suitable suite of tests by the analyst to meet the various requirements. For example, judgment by the analyst is necessary whether or not a test that is applied to liquefied petroleum gas is suitable for process gas or natural gas insofar as inference from the non-hydrocarbon constituents will be minimal.

The first and most important aspect of gaseous testing is the measurement of the volume of gas (ASTM D1071). In this test method, several techniques are described and may be employed for any purpose where it is necessary to know the quantity of gaseous fuel. In addition, the thermophysical properties of methane (ASTM D3956), ethane (ASTM D3984), propane (ASTM D4362), *n*-butane (ASTM D4650), *iso*-butane (ASTM D4651) should be available for use and consultation (see also Stephenson and Malanowski, 1987).

3.5.2 Liquid Products

A wide variety of liquid products are produced from crude oil that varies from the high-volatile naphtha to the low-volatile lubricating oil (Guthrie, 1967; Speight, 2007). The liquid products are often characterized by a variety of techniques that include measurement of physical properties and fractionation into group types.

The purpose of this chapter section is to present examples of the various methods of analysis that can be applied to the analysis of various liquid effluents from refinery processes. The example chosen are those at the lower end of the crude oil product boiling range (naphtha) and at the higher end of the crude oil product boiling range (residual fuel oil). The environmental behavior of the products that are intermediate in boiling range between these two can be predicted, with some degree of caution since behavior is very much dependent upon composition. Nevertheless, similar tests can be applied from which reasonable deduction about behavior can be made.

In terms of waste definition, there are three basic approaches (as it pertains to crude oil, crude oil products, and non-crude oil chemicals) to defining crude oil or a crude oil product as hazardous: (1) a qualitative description of the waste by origin, type, and constituents; (2) classification by characteristics based upon testing procedures; and (3) classification as a result of the concentration of specific chemical substances (Chapter 1).

A wide variety of liquid products are produced from crude oil that varies from the high-volatile naphtha to the low-volatile lubricating oil (Guthrie, 1967; Speight, 1999). The liquid products are often characterized by a variety of techniques that include measurement of physical properties and fractionation into group types (Speight, 2014, 2015).

The impact of the release of liquid products on the environment can, in part be predicted from knowledge of the properties of the released liquid. Each part ocular liquid product from crude oil has its own set of unique analytical characteristics (Speight, 2014, 2015). Since these are well documented, there is no need for repetition here. The decision is to include the properties of the lowest boiling liquid product (naphtha) and a high boiling liquid product (fuel oil). From the properties of each product (as determined by analysis) a reasonable estimate can be made of other liquid products but the relationship may not be linear and is subject to the type of crude oil and the distillation range of the product.

Consequently, for the purposes of this chapter, naphtha and fuel oil have been selected as the liquids that are representative of the extremes of the boiling range of crude oil liquids. Naphtha is the fraction that commences boiling at room temperature while fuel boils at the higher extreme of atmospheric distillation and may even (especially residual fuel oil) contain non-volatile constituents that are not found in the equivalent-boiling range lubricating oil.

3.5.3 Solid Products

In terms of waste definition, there are three basic approaches (as it pertains to crude oil, crude oil products, and non-crude oil chemicals) to defining crude oil or a crude oil product as hazardous: (1) a qualitative description of the waste by origin, type, and constituents; (2) classification by characteristics based upon testing procedures; and (3) classification as a result of the concentration of specific chemical substances (Chapter 1).

Solid effluents are generated from many of the refining processes, crude oil handling operations, as well as wastewater treatment. Both hazardous and non-hazardous wastes are generated, treated and disposed. Refinery wastes are typically in the form of sludge (including sludge from wastewater treatment), spent process catalysts, filter clay, and incinerator ash. Treatment of these wastes includes incineration, land treating off-site, land filling onsite, land filling off-site, chemical fixation, neutralization, and other treatment methods.

A significant portion of the non-crude oil product outputs of refineries is transported off-site and sold as byproducts. These outputs include sulfur, acetic acid, phosphoric acid, and recovered metals. Metals from catalysts and from the crude oil that have deposited on the catalyst during the production often are recovered by third party recovery facilities.

Storage tanks are used throughout the refining process to store crude oil and intermediate process feeds for cooling and further processing. Finished crude oil products are also kept in storage tanks before transport off site. Storage tank bottoms are mixtures of iron rust from corrosion, sand, water, and emulsified oil and wax, which accumulate at the bottom of tanks. Liquid tank bottoms (primarily water and oil emulsions) are periodically drawn off to prevent their continued build up. Tank bottom liquids and sludge are also removed during periodic cleaning of tanks for inspection. Tank bottoms may contain amounts of tetraethyl or tetramethyl lead (although this is increasingly rare due to the phase-out of leaded products), other metals, and phenols. Solids generated from leaded gasoline storage tank bottoms are listed as a RCRA hazardous waste.

The importance of residua and asphalt to the environmental analyst arises from spillage or leakage in the refinery or on the road. In either case, the properties of these materials are detrimental to the ecosystem in which the release occurred. As with other crude oil products, knowledge of the properties of residua and asphalt can help determine the potential cleanup methods and may even allow regulators trace the product to the refinery where it was produced. In addition, the character of residua and asphalt render the usual test methods for *total petroleum hydrocarbons* are ineffective since high proportions of asphalt and residua are insoluble in the usual solvents employed for the test. Application of the test methods for *total petroleum hydrocarbons* to fuel oil is also subject to similar limitations.

Residua are the dark colored near solid or solid products of crude oil refining that are produced by atmospheric and vacuum distillation (Parkash, 2003; Gary et al., 2007; Speight, 2014; Hsu and Robinson, 2017; Speight, 2017). Asphalt is usually produced from a residuum and is a dark brown to black cementitious material obtained from crude oil processing and which contains very high molecular weight molecular polar species called asphaltenes that are soluble in carbon disulfide, pyridine, aromatic hydrocarbon derivatives, and chlorinated hydrocarbon derivatives (Gruse and Stevens, 1960; Guthrie, 1967; Broome and Wadelin, 1973; Weissermel and Arpe, 1978; Hoffman, 1983; Austin, 1984; Chenier, 1992; Hoffman and McKetta, 1993; Warne, 1998; Speight, 2014, 2015).

Residua and asphalt derive their characteristics from the nature of their crude oil precursor, the distillation process being a concentration process in which most of the heteroatoms and polynuclear aromatic constituents of the feedstock are concentrated in the residuum (Speight, 2015). Asphalt may be similar to its parent residuum but with some variation possible by choice of manufacturing process. In general terms, residua and asphalt are a hydrocarbonaceous material that consist constituents (containing carbon, hydrogen, nitrogen, oxygen, and sulfur) that are completely soluble in carbon disulfide (ASTM D4). Trichloroethylene or 1,1,1-trichloroethane has been used in recent years as solvents for the determination of asphalt (and residua) solubility (ASTM D2042).

An *asphalt emulsion* is a mixture of asphalt and an anionic agent such as the sodium or potassium salt of a fatty acid. The fatty acid is usually a mixture and may contain palmitic, stearic, linoleic, and abietic acids and/or high molecular weight phenols. Sodium lignate is often added to alkaline emulsions to effect better emulsion stability. Nonionic cellulose derivatives are also used to increase the viscosity of the emulsion if needed. The acid number is an indicator of its asphalt emulsification properties and reflects the presence of high molecular weight asphaltic or naphthenic acids. Diamines, frequently used as cationic agents, are made from the reaction of tallow acid amines with acrylonitrile, followed by hydrogenation. The properties of asphalt emulsions (ASTM D977, ASTM D2397) allow a variety of uses. As with other crude oil products, sampling is an important precursor to asphalt analysis and a standard method (ASTM D140) is available that provides guidance for the sampling of asphalts, liquid and semisolid, at point of manufacture, storage, or delivery.

The properties of residua and asphalt are defined by a variety of standard tests that can be used to define quality and remembering that the properties of residua vary with cut-point (Speight, 2015), i.e., the volume % of the crude oil helps the refiner produce asphalt of a specific type or property (ASTM D496). Roofing and industrial asphalts are also generally specified in various grades of hardness usually with a combination of softening point (ASTM D61, ASTM D2319, ASTM D3104, ASTM D3461) and penetration to distinguish grades (ASTM D312, ASTM D449).

Coke does not offer the same potential environmental issues as other crude oil products (Chapter 2). It is used predominantly as a refinery fuel unless other used for the production of a high-grade coke or carbon are desired. In the former case, the constituents of the coke that will release environmentally harmful gases such as nitrogen oxides, sulfur oxides, and particulate matter should be known. In addition, stockpiling coke on a site where it awaits use or transportation can lead to leachates that the result of rainfall (or acid rainfall) that are highly detrimental. In such a case, application of the toxicity characteristic leaching procedure to the coke (TCLP, EPA SW-846 Method 1311), that is designed to determine the mobility of both organic and inorganic contaminants present in materials such as coke, is warranted before stockpiling the coke in the open is warranted.

Crude oil coke is the residue left by the destructive distillation of crude oil residua in processes such as the delayed coking process (Parkash, 2003; Gary et al., 2007; Speight, 2014; Hsu and Robinson, 2017; Speight, 2017). That formed in catalytic cracking operations is usually non-recoverable, as it is often employed as fuel for the process.

Coke is a gray to black solid carbonaceous residue that is produced from crude oil during thermal processing; characterized by having a high carbon content (95% + by weight) and a honeycomb type of appearance and is insoluble in organic solvents. (ASTM D121) (Chapter 2) (Gruse and Stevens, 1960; Guthrie, 1967; Weissermel and Arpe, 1978; Hoffman, 1983; Austin, 1984; Chenier, 1992; Hoffman and McKetta, 1993; Speight, 2014, 2017).

Coke occurs in various forms and the terminology reflects the type of coke that can influence behavior in the environment. But no matter what the form, coke usually consists mainly of carbon (greater than 90 percent but usually greater than 95 per cent) and has a low mineral matter content (determined as ash residue). Coke is used as a feedstock in coke ovens for the steel industry, for heating purposes, for electrode manufacture and for production of chemicals. The two most important qualities are *green coke* and *calcined coke*. This latter category also includes *catalyst coke* deposited on the catalyst during refining processes: this coke is not recoverable and is usually burned as refinery fuel.

The test methods for coke are necessary for defining the coke as a fuel (for internal use in a refinery) or for other uses, particularly those test methods where prior sale of the coke is involved. Specifications are often dictated by environmental regulations, if not by the purchaser of the coke.

The test methods are the methods that are usually applied to crude oil coke but should not be thought of as the only test methods. In fact there are many test methods for coke (ASTM, 2021) and these test method should be consulted either when more detail is required or a fuller review is required.

Because crude oil coke is a solid, fairly inert material there has not been much concern about the related environmental effects. The typical battery of tests used to measure a chemical's impact on the environment, such as breakdown by sunlight, stability in water, breakdown in the soil and volatility, cannot be measured for crude oil coke. In addition, the lack of significant adverse effects in animal studies suggests that the residual oil associated with crude oil coke either does not come off, or is present in amounts too low to cause harmful effects.

The specific chemical composition of any given batch of crude oil coke is determined by the composition of the feedstocks used in the coking process, which in turn are dependent upon the composition of the crude oil and refinery processing from which the feedstock is derived. Coke produced from feedstocks with high proportions of asphaltene constituents will contain higher concentrations of sulfur and metals than cokes produced from low-asphaltene feedstocks (Speight, 2007).

Most of the sulfur in coke exists as organic sulfur bound to the carbon matrix. However, the structure of organic sulfur compounds in crude oil coke is largely unknown. Metals, mainly vanadium and nickel, occur as metal chelates or porphyrins in the asphaltene fraction (Speight, 2007). Some metals are intercalated in the coke structure and are not chemically bonded, so they become part of the combustion ash and particulate matter. Metal concentrations in coke normally increase upon calcining due to the weight loss from evolution of the volatile matter.

Typical parameters measured to define the chemical composition of crude oil coke are: (1) ash % w/w, (2) sulfur % w/w, (3) solvent-extractable material % w/w, (4) nickel % w/w (usually in parts per million—ppm, and (5) vanadium % w/w/(usually in parts per million—ppm. Green coke (because of the lower temperature used in its production) contains higher levels of residual hydrocarbon than other grades of coke. Calcining (approximately 1200 to 1350°C, 2190 to 2460°F) removes essentially all of the residual solvent-extractable material.

If released to the environment, both forms of crude oil coke would not be expected to undergo many of the environmental fate pathways. Because crude oil coke is predominantly elemental carbon, it would not be subject to photolytic processes not would it be susceptible to biodegradation by microorganisms. Depending on the particle size and density of the material, terrestrial releases will become incorporated into the soil or transported via wind or surface water flow. If released to the aquatic environment, crude oil coke will either incorporate into sediment or float on the surface, depending on the particle size and density in relation to water.

However, in order to reduce the environmental impact of crude oil coke (through combustion and/or storage in piles which are susceptible to run-off during period of rain and snow melt), gasification (which has a long history of commercialization) has been proposed as a means of converting crude oil coke (and other refinery waste streams) into power, steam and hydrogen for use in the production of clean fuels. Gasification units are already in operation in number of refineries and it is expected that other refineries will add these units in the future (Marano, 2003; Speight, 2011a).

At the high temperatures that are employed in gasification applications, many gasification reactions are equilibrium controlled. Methanation is favored at lower temperatures; thus, little methane is produced. Since the gasifier is operated at adiabatic conditions, the heat liberated by exothermic reactions must balance with the heat required by the endothermic reactions and the heat required to heat the feed streams.

Gasification temperature is controlled by the addition of water or steam—for slurried feedstocks, the slurry water accomplishes this control but for dry feedstocks such as crude oil coke steam must be injected with the feedstock to control temperature. Steam injection may also be used to adjust the composition of the product syngas.

Solid products from refining operations often receive less attention than the liquid and gaseous products for a variety of justifiable or unjustifiable reasons. In fact, the term *solid product* is a generic term that is often confusing and meant (incorrectly) to convey the idea of a *waste product*. However, in terms of waste definition, there are three basic approaches (as it pertains to crude oil, crude oil products, and non-crude oil chemicals) to defining crude oil or a crude oil product as hazardous: (1) a qualitative description of the waste by origin, type, and constituents; (2) classification by characteristics based upon testing procedures; and (3) classification as a result of the concentration of specific chemical substances (Chapter 1) (Speight, 2005).

Solid effluents are generated from many of the refining processes, crude oil handling operations, as well as wastewater treatment (Speight and Ozum, 2002; Speight, 2005, 2014, 2015). Both hazardous and non-hazardous wastes are generated, treated and disposed. Refinery wastes are typically in the form of sludge (including sludge from wastewater treatment), spent process catalysts, filter clay, and incinerator ash. Treatment of these wastes includes incineration, land treating off-site, land filling onsite, land filling off-site, chemical fixation, neutralization, and other treatment methods (Speight, 2005).

As with any complex mixture, each component of the sludge can have its own environmental impact or there is always the potential for associated reactions between the sludge components that change the potential for the environmental impact—usually to a higher potential.

Generally, sludge from crude oil refining operations has high pollution potentials and an aerobic biological method is an efficient method of its treatment. The choice of aerobic method over anaerobic system lies in the fact that crude oil industry produces a large volume of sludge coupled with a high concentration of biological oxygen demand (Asia et al., 2006).

A significant portion of the non-crude oil product outputs of refineries is transported off-site and sold as byproducts. These outputs include sulfur, acetic acid, phosphoric acid, and recovered metals. Metals from catalysts and from the crude oil that have deposited on the catalyst during the production often are recovered by third party recovery facilities.

Storage tanks are used throughout the refining process to store crude oil and intermediate process feeds for cooling and further processing. Finished crude oil products are also kept in storage tanks before transport off site. Storage tank bottoms are mixtures of sludge (typically emulsified oil and incompatible organic constituents as well as iron rust from corrosion) which accumulates at the bottom of tanks. Liquid tank bottoms are periodically drawn off to prevent their continued build up—sludge is also removed during periodic cleaning of tanks for inspection. Tank bottoms may contain amounts of tetraethyl or tetramethyl lead (although this is increasingly rare due to the phase-out of leaded products), other metals, and phenols. Solids generated from leaded gasoline storage tank bottoms are listed as a RCRA hazardous waste.

Disposal of the solid waste can be achieved by installation of a gasifier (Marano, 2003). In fact, a wide variety of feedstocks can be considered for gasification, ranging from solids to liquids to gaseous streams. Although when the feed is a gas or liquid, the operation is frequently referred to as partial oxidation (POX). From a process perspective, partial oxidation of gases and liquids is very similar to the gasification of solids.

The major requirement for a suitable feedstock is that it contains a significant content of carbon and hydrogen. Solid feedstocks include solid waste, residua, visbreaker bottoms, crude oil coke, and even biomass—the streams most commonly employed are generally low-value byproducts or waste streams generated by other processes (Speight, 2008, 2011a, 2011b).

References

Abbasnezhad, H., Gray, M.R., and Foght, J.M. 2011. Influence of adhesion on Aerobic Biodegradation and bioremediation of liquid hydrocarbons. Appl. Microbiol. Biotechnol., 92: 653–675.

Aichberger, H., Hasinger, M., Braun, R., and Loibner, A.P. 2005. Potential of preliminary test methods to predict biodegradation performance of petroleum hydrocarbons in soil. Biodegradation, 16: 115–125.

Altgelt, K.H., and Gouw, T.H. 1975. In Advances in Chromatograph. Giddings, J.C., Grushka, E., Keller, R.A. and Cazes, J. (Eds.). Marcel Dekker Inc., New York.

ASTM. 2021. Annual Book of ASTM Standards. American Society for Testing and Materials, West Conshohocken, Pennsylvania.

Atagana, H.I., Haynes, R.J., and Wallis, F.M. 2003. Optimization of soil physical and chemical conditions for the bioremediation of creosote-contaminated soil. Biodegradation, 14: 297–307.

Austin, G.T. 1984. Shreve's Chemical Process Industries. 5th Edition. McGraw-Hill, New York. Chapter 37.

Baker, R.J., Baehr, A.L., and Lahvis, M.A. 2000. Estimation of hydrocarbon biodegradation rates in gasoline-contaminated sediment from measured respiration rates. J. Contam. Hydrol., 41: 175–192.

Balba, M.T., Al-Awadhi, N., and Al-Daher, R. 1998. Bioremediation of oil-contaminated soil: microbiological methods for feasibility assessment and field evaluation. J. Microbiol. Methods, 32: 155–164.

Bartha, R. 1986. Biotechnology of petroleum pollutant biodegradation. Microb. Ecol., 12: 155–172.

Boopathy, R. 2002. Use of anaerobic soil slurry reactors for the removal of petroleum hydrocarbons in soil. Int. Biodeterioration Biodegradation, 52: 161–166.

Breedveld, G.D., and Sparrevik, M. 2001. Nutrient-limited biodegradation of PAH in various soil strata at a creosote contaminated site. Biodegradation, 11: 391–399.

Bruner, F. 1993. Gas Chromatographic Environmental Analysis: Principles, Techniques, and Instrumentation. John Wiley & Sons Inc., Hoboken, New Jersey.

Budde, W.L. 2001. The Manual of Manuals. Office of Research and Development, Environmental Protection Agency, Washington, DC.

Cao, J.R. 1992. Microwave Digestion of Crude Oils and Oil Products for the Determination of Trace Metals and Sulphur by Inductively-Coupled Plasma Atomic Emission Spectroscopy, Environment Canada Manuscript Report Number EE-140, Ottawa, Ontario, Canada.

Chaîneau, C.H., Morel, J.L., and Oudot, J. 1995. Microbial degradation in soil microcosms of fuel oil hydrocarbons from drilling cuttings. Environ. Sci. Technol., 29: 1615–1621.

Chaîneau, C.H., Yepremian, C., Vidalie, J.F., Ducreux, J., and Ballerini, D. 2003. Bioremediation of a crude oil-polluted soil: biodegradation, leaching and toxicity assessments. Water, Air, Soil Pollut., 144: 419–440.

Chenier, P. 1992. Survey of Industrial Chemistry. 2nd Revised Edition. VCH Publishers, New York. Chapters 7 and 8.

CHRIS. 1991. Chemical Hazards Response Information System (CHRIS). United States Coast Guard, Department of Transportation, Washington, DC.

Daling, P.S., Brandvik, P.J., Mackay, D., and Johansen, Ø. 1990. Characterization of crude oils for environmental purposes. Oil & Chemical Pollution, 7: 199–224.

Davis, C., Cort, T., Dai, D., Illangasekare, T.H., and Munakata-Marr, J. 2003. Effects of heterogeneity and experimental scale on biodegradation of diesel. Biodegradation, 14: 373–384.

Dean, J.R. 1998. Extraction Methods for Environmental Analysis. John Wiley & Sons, Inc., New York.

Dott, W., Feidieker, D., Steiof, M., Becker, P.M., and Kämpfer, P. 1995. Comparison of *ex situ* and *in situ* techniques for bioremediation of hydrocarbon-polluted soils. Int. Biodeterioration & Biodegradation, 301–316.

Dyroff, G.V. (Ed.). 1993. Manual on Significance of Tests for Petroleum Products: 6th Edition, American Society for Testing and Materials, West Conshocken, Pennsylvania.

Falla Sotelo, F., Araujo Pantoja, P., López-Gejo, J., Le Roux, J.G.A.C., Quina, F.H., and Nascimento, C.A.O. 2008. Application of fluorescence spectroscopy for spectral discrimination of crude oil samples. Brazilian Journal of Petroleum and Gas, 2(2): 63–71.

Fingas, M.F., Duval, W.S., and Stevenson, G.B. 1979. The Basics of Oil Spill Cleanup, Environment Canada, Ottawa, Ontario, Canada.

Fingas, M.F., Dufort, V.M., Hughes, K.A., Bobra, M.A., and Duggan, L.V. 1989a. Laboratory studies on oil spill dispersants. pp. 207–219. In: Flaherty, M. (Ed.). Chemical Dispersants - New Ecological Approaches. Publication No. STP 1084, American Society for Testing and Materials, West Conshohocken, Pennsylvania.

Fingas, M.F., White, B., Stoodley, R.G., and Crerar, I.D. 1989b. Laboratory Testing of Dispersant Effectiveness. Proceedings. 1989 Oil Spill Conference, American Petroleum Institute, Washington, D.C., pp. 365–373, 1989b.

Fingas, M.F., Kyle, D.A., Bier, I.E., Lukose, A., and Tennyson, E.J. 1991. Physical and chemical studies on oil spill dispersants: the effect of energy. Proceedings. 14th Arctic and Marine Oil Spill Program Technical Seminar, Environment Canada, Ottawa, Ontario, Canada. Page 87–106.

Fingas, M.F., Kyle, D.A., and Tennyson, E.J. 1992. Physical and chemical studies on oil spill dispersants: effectiveness variation with energy. Proceedings. 15th Arctic and Marine Oil Spill Program Technical Seminar, Environment Canada. Ottawa, Ontario, Canada Page 135–142.

Fingas, M.F., Kyle, D.A., and Tennyson, E.J. 1993. Physical and Chemical Studies on Dispersants: The Effect of Dispersant Amount on Energy. Proceedings. 16th Arctic and Marine Oil Spill Program Technical Seminar, Environment Canada, Ottawa, Ontario, Canada. Page 861–876.

Fingas, M.F., Fieldhouse, B., Gamble, L., and Mullin, J. 1995a. Studies of water-in-oil emulsions: stability classes and measurement. Proceedings. 18th Arctic and Marine Oil Spill Program Technical Seminar, Environment Canada, Ottawa, Ontario, Canada. Page 21–42.

Fingas, M.F., Kyle, D.A., Lambert, P., Wang, Z., and Mulling, J. 1995b. Analytical procedures for measuring oil spill dispersant effectiveness in the laboratory. Proceedings. 18th Arctic and Marine Oil Spill Program Technical Seminar, Environment Canada, Ottawa, Ontario, Canada. Page 339–354.

Fingas, M.F., Ackerman, F., Lambert, P., Li, K., Wang, Z., Mullin, J., Hannon, L., Wang, D., Steenkammer, A., Hiltabrand, R. Turpin, R., and Campagna, P. 1995c. The newfoundland offshore burn experiment: further results of emissions measurement. Proceedings. 18th Arctic and Marine Oil Spill Program Technical Seminar, Environment Canada, Ottawa, Ontario, pp. 915–995.

Fingas, M.F. 1995. The evaporation of oil spills. Proceedings. 18th Arctic and Marine Oil Spill Program Technical Seminar, Environment Canada, Ottawa, Ontario, Canada. Page 43–60.

Fingas, M.F. 1998. Studies on the evaporation of crude oil and petroleum products. II. Boundary Layer Regulation. J. Hazardous Materials. 57(1-3): 41–58.

Fowlis, I.A. 1995. Gas Chromatography. 2nd Edition. John Wiley & Sons Inc., New York.

Francis, W., and Peters, M.C. 1980. Fuels and Fuel Technology: A Summarized Manual. Pergamon Press, New York. Section B.

Frijer, J.I., De Jonge, H., Bounten, W., and Verstraten, J.M. 1996. Assessing mineralization rates of petroleum hydrocarbons in soils in relation to environmental factors and experimental scale. Biodegradation 7: 487–500.

Gary, J.G., Handwerk, G.E., and Kaiser, M.J. 2007. Petroleum Refining: Technology and Economics, 5th Edition. CRC Press, Taylor & Francis Group, Boca Raton, Florida.

Grishchenkov, V.G., Townsend, R.T., McDonald, T.J., Autenrieth, R.L., Bonner, J.S., and Boronin, A.M. 2000. Degradation of petroleum hydrocarbons by facultative anaerobic bacteria under aerobic and anaerobic conditions. Process Biochem., 35: 889–896.

Grob, R.L. 1995. Modern Practice of Gas Chromatography. 3rd Edition. John Wiley & Sons Inc., New York.

Gruse, W.A., and Stevens, D.R. 1960. Chemical Technology of Petroleum. McGraw-Hill, New York. Chapter 11.

Guthrie, V.B. 1967. In Petroleum Processing Handbook. Bland,W.F., and Davidson, R.L. (Eds.). McGraw-Hill, New York. Section 11.

Hatzinger, P.B., and Alexander, M. 1995. Effect of ageing chemicals in soil upon their biodegradability and extractability. Environ. Sci. Technol., 29: 537–545.

Heider, J., Spormann, A.M., Beller, H.R., and Widdel, F. 1999. Anaerobic bacterial metabolism of hydrocarbons. FEMS Microbiol. Rev., 22: 459–473.

Hoffman, H.L. 1983. In Riegel's Handbook of Industrial Chemistry. 8th Edition. Kent, J.A. (Ed.). Van Nostrand Reinhold Company, New York. Chapter 14.

Hoffman, H.L., and McKetta, J.J. 1993. Petroleum processing. *In*: McKetta, J.J. (Ed.). Chemical Processing Handbook. Marcel Dekker Inc., New York. Page 851.

Höhener, P., Duwig, C., Pasteris, G., Kaufmann, K., Dakhel, N., and Harms, H. 2003. Biodegradation of petroleum hydrocarbon vapors: laboratory studies on rates and kinetics in unsaturated alluvial sand. J. Contam. Hydrol., 1917: 1–23.

Holliger, C., and Zehnder, A.J.B. 1996. Anaerobic biodegradation of hydrocarbons. Curr. Opin. Biotechnol., 7: 326–330.

Hsu, C.S., and Robinson, P.R. (Eds.). 2017. Handbook of Petroleum Technology. Springer International Publishing AG, Cham, Switzerland.

Huang, W., Pent, P., Yu, Z., and Fu, J. 2003. Effects of organic matter heterogeneity on sorption and desorption of organic contaminants by soils and sediments. Appl. Geochem., 18: 955–972.

Huesemann, M.H. 1995. Predictive model for estimating the extent of petroleum hydrocarbon biodegradation in contaminated soils. Environ. Sci. Technol., 29: 7–18.

Jokuty, P., Fingas, M.F., Whiticar, S., and Fieldhouse, B. 1995. A Study of Viscosity and Interfacial Tension of Oils and Emulsions. Report No. EE-153. Environment Canada, Ottawa, Ontario, Canada.

Jokuty, P., Whiticar, S., McRoberts, K., and Mullin, J. 1996. Oil adhesion testing—recent results. Proceedings. 19th Arctic and Marine Oil Spill Program Technical Seminar, Environment Canada, Ottawa, Ontario, Canada. Page 9–27.

Leeson, A., and Hinchee, R.E. 1997. Soil Bioventing, Principles and Practice. CRC, Lewis Publishers, Boca Raton, Florida.

Loeher, R.C. 1992. In Petroleum Processing Handbook. McKetta, J.J. (Ed.). Marcel Dekker Inc., New York. Page 190.

Luthy, R.G., Aiken, G.R., Brusseau, M.L., Cunningham, S.D., Gschwend, P.M., Pignatello, J.J., Reinhard, M., Traina, S.J., Weber, W.J., and Westall, J.C. 1997. Sequestration of hydrophobic organic contaminants by geosorbents. Environ. Sci. Technol., 31: 3341–3347.

Mackay, D., and Zagorski, W. 1982. Studies of Water-in-Oil Emulsions. Report No. EE-34. Environment Canada, Ottawa, Ontario, Canada.

Manahan, S.E. 1999. Environmental Chemistry. 7th Edition. Lewis Publishers, Chelsea, Michigan.

Margesin, R., Zimmerbauer, A., and Schinner, F. 1997. Efficiency of indigenous and inoculated cold-adapted soil microorganisms for biodegradation of diesel oil in alpine soils. Appl. and Environ. Microbiol., 63: 2660–2664.

Margesin, R., Zimmerbauer, A., and Schinner, F. 2000. Monitoring of bioremediation by soil biological activities. Chemosphere, 40: 339–346.

Massias, D., Grossi, V., and Bertrand, J.C. 2003. *In situ* anaerobic degradation of petroleum alkanes in marine sediments: preliminary results. Geoscience, 335: 435–439.

Miller, M. (Ed.). 2000. Encyclopedia of Analytical Chemistry. John Wiley & Sons Inc., Hoboken, New Jersey.

Mohn, W.M., and Stewart, G.R. 2000. Limiting factors for hydrocarbon biodegradation at low temperature in arctic soils. Soil Biol. Biochem., 32: 1161–1172.

Nelson, P. 2003. Index to EPS Test Methods. US EPA New England Region, Boston MA.

Nierop, K.G.J., and Verstraten, J.M. 2003. Organic matter formation in sandy subsurface horizons of dutch coastal dunes in relation to soil acidification. Org. Geochem., 34: 499–513.

Nocentini, M., Pinelli, D., and Fava, F. 2000. Bioremediation of a soil contaminated by hydrocarbon mixtures: the residual concentration problem. Chemosphere, 41: 1115–1123.

Oudot, J., and Dutrieux, E. 1989. Hydrocarbon weathering and biodegradation in a tropical estuarine ecosystem. Mar. Environ. Res., 27: 195–213.

Parkash, S. 2003. Refining Processes Handbook. Gulf Professional Publishing, Elsevier, Amsterdam, Netherlands.

Patnaik, P. (Ed.). 2004. Dean's Analytical Chemistry Handbook. 2nd Edition. McGraw-Hill, New York.

Pavlova, A., and Ivanova, R. 2003. Determination of petroleum hydrocarbons and polycyclic aromatic hydrocarbons in sludge from wastewater treatment basins. J. Environ. Monit., 5: 319–323.

Pickering, K.T., and Owen, L.A. 1994. Global Environmental Issues. Routledge Publishers, New York.

Rhodes, I.A., Hinojas, E.M., Barker, D.A., and Poole, R.A. 1994. Pitfalls Using Conventional TPH Methods for Source Identification. Proceedings. Seventh Annual Conference: EPA Analysis of Pollutants in the Environment. Norfolk, VA. Environmental Protection agency, Washington, DC.

Rullkötter, J., and Farrington, J.W. 2021. What Was Released? Assessing the physical properties and chemical composition of petroleum and products of burned oil. Oceanography, 34(1): 44–57. https://tos.org/oceanography/article/what-was-released-assessing-the-physical-properties-and-chemical-composition-of-petroleum-and-products-of-burned-oil.

Schramm, L.L. (Ed.). 1992. Emulsions. Fundamentals and Applications in the Petroleum Industry. American Chemical Society, Washington, DC.

Schwarzenbach, R.P., Gschwend, P.M., and Imboden, D.M. 2003. Environmental Organic Chemistry. 2nd Edition. John Wiley & Sons Inc., New York.

Smith, K.A., and Cresser, M. 2003. Soil & Environmental Analysis: Modern Instrumental Techniques. Marcel Dekker Inc., New York. 2003.

Speight, J.G. 1996. Handbook of Environmental Technology. Taylor & Francis, Washington, DC.

Speight, J.G. 2015. Handbook of Petroleum Product Analysis 2nd Edition. John Wiley & Sons Inc., Hoboken, New Jersey.

Speight, J.G. 2005. Environmental Analysis and Technology for the Refining Industry John Wiley & Sons Inc., Hoboken, New Jersey.

Speight, J.G. 2009. Enhanced Recovery Methods for Heavy Oil and Tar Sands. Gulf Publishing Company, Houston, Texas.

Speight, J.G. 2011. An Introduction to Petroleum Technology, Economics, and Politics. Scrivener Publishing, Salem, Massachusetts.

Speight, J.G., and Foote, R. 2011. Ethics in Science and Engineering. Scrivener Publishing, Salem, Massachusetts.

Speight, J.G. 2014. The Chemistry and Technology of Petroleum 5th Edition. CRC Press, Taylor & Francis Group, Boca Raton, Florida.

Speight, J.G. 2015. Handbook of Petroleum Product Analysis 2nd Edition. John Wiley & Sons Inc., New York.

Speight, J.G. 2017. Handbook of Petroleum Refining. CRC Press, Taylor and Francis Group, Boca Raton, Florida.

Sullivan and Johnson. 1993. 'Oil' You Need to know about Crude—Implications of TPH Data for Common Petroleum Products. Soil. May, page 8.

Sunahara, G.I., Renoux, A.Y., Thellen, C., Gaudet, C.L., and Pilon, A. (Eds.). 2002. Environmental Analysis of Contaminated Sites. John Wiley & Sons Inc. New York.

Tinsley, I.J. 2004. Chemical Concepts in Pollutant Behavior 2nd Edition. John Wiley & Sons Inc., Hoboken, New Jersey.

Twardus, E.M. 1980. A Study to Evaluate the Combustibility and Other Physical and Chemical Properties of Aged Oils and Emulsions. Report Number EE-5. Environment Canada, Ottawa, Ontario, Canada.

US EPA. 2004. Environmental Protection Agency, Washington, DC. Web site: http://www.epa.gov.

Wang, Z., Fingas, M.F., Landriault, M., Sigouin, L., and Xu, N. 1995. Identification of Alkyl benzenes and direct determination of BTEX and (BTEX + C3-Benzenes) in Oils by GC/MS. Proceedings. 18th Arctic and Marine Oil Spill Program Technical Seminar, Environment Canada, Ottawa, Ontario, Canada. Page 141–164.

Warne, T.M. 1998. In manual on hydrocarbon analysis. 6th Edition. Drews, A.W. (Ed.). American Society for Testing and Materials, West Conshohocken, PA. Chapter 3.

Weissermel, K., and Arpe, H.-J. 1978. Industrial Organic Chemistry. Verlag Chemie. New York. Chapter 13.

Woodside, G. 1999. Hazardous Materials and Hazardous Waste Management 2nd Edition. John Wiley & Sons Inc., Hoboken, New Jersey.

CHAPTER 4
The Nature of Oil Spills

The environment represents continuation of life for all floral and faunal species and the entire life support system of humans depends on the well-being of all these species living on the Earth. The relationship between the environment and humanity is a relationship of interdependence insofar as one affects the other leading to a changing environment. In fact, human actions are destroying habitats and endangering the lives of floral and faunal species and more needs to be done to remedy these issues.

The capacity of the environment to absorb the effluents and other impacts of process technologies is not unlimited, as some would have us believe. The environment should be considered to be an extremely limited resource, and discharge of chemicals into it should be subject to severe constraints. Indeed, the declining quality of raw materials for industrial manufacturing processes dictates that more material must be processed to provide the needed fuels. And the growing magnitude of the effluents from various processes has moved above the line where the environment has the capability to absorb such process effluents without disruption.

In fact, the pollution of the environment is an important issue even when society is faced with economic and social crises because Earth is the only home that provides air, food, and other needs to humans. More generally, environmental pollution is the causative changes in the physical, chemical, and biological characteristics of the atmosphere (the air), the aquasphere (the water or water systems), and the geosphere (the land). Thus, environmental pollution can be broadly classified according to the components of environment that are polluted, for example (1) air pollution, (2) water pollution, and (3) land pollution, particularly the pollution of the soil.

Humanity faces serious challenges with pollution of the environment and its natural resources such as air, water, or land with different pollutants. The most harmful effect of pollution on the environment is that it causes both physical and biological effects which vary from mild to severe. Environmental pollution has reached worrying proportions worldwide and must be taken seriously for life to continue to exist on Earth.

The production of crude oil products generates substantial quantities of waste materials in different forms which are (1) gases, (2) liquids, such as low boiling constituents, high boiling constituents, waste water, spent caustic, and solids, such as filter clay.

4.1 Environmental Effects of Crude Oil Refining

Like any other raw material, crude oil is capable of producing chemical waste. By 1960 the crude oil-refining industry had become well established throughout the world. Demand for refined crude oil products had reached almost millions of barrels per day, with major concentrations of refineries in most developed countries. However, as the world became aware of the impact of industrial chemical waste on the environment, the crude oil-refining industry was a primary focus for change. Refiners added hydrotreating units to extract sulfur compounds from their products and began to generate large quantities of elemental sulfur. Effluent water, atmospheric emissions and combustion products also became a focus of increased technical attention (Carson and Mumford, 1988; Renzoni et al., 1994; Carson and Mumford, 1995; Edwards, 1995; Thibodeaux, 1995; Speight, 1996).

The waste products from any crude oil refinery or production process which has been dewatered is commonly referred to as slop oil. Residues, clay-treating filter wash A complex residuum from the solvent washing of clay-treating filters. However, during the crude oil refining and the production of crude oil products, a refinery produces other products such as (1) gases, (2) liquids, and (3) solids, which can cause harm to the environment if released or spilled. Each of these products must be treated to reduce any such effects.

4.1.1 Gases

Refineries (and gas processing plants) create and process a variety of gaseous product streams and waste streams which contain inorganic constituents such as hydrogen sulfide, hydrogen, carbon monoxide, and ammonia. Exposure to these substances is limited by exposure standards and their inherent flammability hazard.

Potential releases of crude oil and refinery gases from crude oil facilities can be characterized as either controlled or unintentional releases. Controlled releases are planned releases from pressure relief valves and venting valves, for safety purposes or maintenance, and are considered part of routine operations and occur under controlled conditions. Unintentional releases are typically characterized as unplanned releases due to spills or leaks from various equipment, valves, piping, flanges, etc., and may result from equipment failure, poor maintenance, lack of proper operating practices, adverse weather conditions or other unforeseen factors. Crude oil facilities are highly regulated and regulatory requirements established under various jurisdictions, as well as voluntary non-regulatory measures implemented by the crude oil industry, are in place to manage potential releases.

More generally, these gaseous products or waste materials rarely, if ever, leave the plant without further processing to recover the valuable hydrocarbon derivatives or inorganic constituents. Some are used within the plant for process heating (burned). Because they are piped and tanked under elevated pressure and present an extreme explosion hazard should there be a release, control technologies to prevent exposure have been in place since the earliest days of the crude oil industry.

The refinery gases are made up of predominantly one to four carbon atom hydrocarbon derivatives and inorganic components such as ammonia, hydrogen, nitrogen, hydrogen sulfide, mercaptans, carbon monoxide and carbon dioxide. Several refinery gases (process gases) also contain benzene and/or 1,3-butadiene. These gases can be used within the refinery as fuel gases to provide energy for other refinery processes. Alternatively, they can also undergo further refining to separate components and make them into commercially salable products.

All refinery gases contain one or more inorganic compounds in addition to hydrocarbon derivatives, which are (with the exception of asphyxiant gases such as hydrogen and nitrogen) typically more toxic than the C_1 to C_6 to both fauna and flora. Unlike other crude oil product categories (such as gasoline, diesel fuel, and lubricating oil), the inorganic and hydrocarbon constituents of refinery gases can be evaluated for hazardous properties individually. The predominant categories of refinery gases are (1) inorganic gases, (2) hydrocarbon gases, and (3) asphyxiant gases.

The *inorganic gases* are sub-categorized as (1) ammonia, (2) carbon Monoxide, (3) volatile mercaptans, and (4) hydrogen sulfide. The *hydrocarbon gases* include (1) C1 to C4 hydrocarbon derivatives, (2) C5 and C6 hydrocarbon derivatives, and (3) benzene. Finally the *asphyxiant gases* include (1) carbon dioxide, (2) hydrogen, and (3) nitrogen.

Inorganic compounds (with the exception of asphyxiant gases such as hydrogen and nitrogen) are typically more toxic than the majority of hydrocarbon derivatives in the C_1 to C_4 range. In contrast, hydrogen sulfide, ammonia, methyl mercaptan, and carbon monoxide are acutely toxic (API, 2009).

Air pollutants are responsible for a number of adverse environmental effects, such as photochemical smog, acid rain, death of forests, or reduced atmospheric visibility. Emissions of greenhouse gases are associated with the global warming. Certain air pollutants, including black carbon, not only contribute to global warming, but are also suspected of having immediate effect on regional climates.

Sulfur oxides, nitrogen oxides, hydrogen sulfide, and carbon dioxide are commonly produced during refining operations or during use of the refined products. For example, the most common toxic gases present in diesel exhaust include carbon monoxide, sulfur dioxide, nitric oxide, and nitrogen dioxide.

These gases are also classed as primary pollutants because they are emitted directly from the source and then react to produce secondary pollutant, such as acid rain (Speight, 1996). The emissions may include a number of biologically active substances that can pose a major health concern. These gases are classed as pollutants because (1) they may not be indigenous to the location or (2) they are ejected into the atmosphere in a greater-than natural concentration and are, in the current context, the product of human activity. Thus, they can have a detrimental effect on the environment in part or *in toto*.

For these pollutants, the atmosphere has the ability to cleanse itself within hours especially when the effects of the pollutant is minimized by the natural constituents of the atmosphere. For example, the atmosphere might be considered to be cleaning as a result of rain. However, removal of some pollutants from the atmosphere (e.g., sulfates and nitrates) by rainfall results in acid rain that can/will cause serious environmental damage to ecosystems within the water and land systems.

Several methods have been developed to estimate the exposure to such emissions. Most methods are based on either ambient air quality surveys or emission modeling. Exposure to other components of diesel emissions, such as polynuclear aromatic hydrocarbon derivatives, is also higher in occupational settings than it is in ambient environments. The principles of the techniques most often used in exhaust gas analysis include infrared (NDIR and FTIR), chemiluminescence, flame ionization detector (FID and fast FID), and paramagnetic methods.

Sulfur is removed from a number of refinery process off-gas streams (sour gas) in order to meet the sulfur oxide emissions limits of the Clean Air Act and to recover saleable elemental sulfur. Process off-gas streams, or sour gas, from the coker, catalytic cracking unit, hydrotreating units and hydroprocessing units can contain high concentrations of hydrogen sulfide mixed with light refinery fuel gases. Before elemental sulfur can be recovered, the fuel gases (primarily methane and ethane) need to be separated from the hydrogen sulfide. This is typically accomplished by dissolving the hydrogen sulfide in a chemical solvent. Solvents most commonly used are amines, such as diethanolamine (DEA). Dry adsorbents such as molecular sieves, activated carbon, iron sponge and zinc oxide are also used. In the amine solvent processes, DEA solution or another amine solvent is pumped to an absorption tower where the gases are contacted and hydrogen sulfide is dissolved in the solution. The fuel gases are removed for use as fuel in process furnaces in other refinery operations. The amine-hydrogen sulfide solution is then heated and steam stripped to remove the hydrogen sulfide gas.

Since the Claus process by itself removes only about 90 percent of the hydrogen sulfide in the gas stream, the Beavon process (Speight, 1993, page 268) SCOT (Shell Claus Off-gas Treating) process (Speight, 1993, page 316; Hydrocarbon Processing, 2002), or the Wellman-Lord process (Speight, 1993, page 327) are often used to further recover sulfur. The Claus process consists of partial combustion of the hydrogen sulfide-rich gas stream (with one-third the stoichiometric quantity of air) and then reacting the resulting sulfur dioxide and unburned hydrogen sulfide in the presence of a bauxite catalyst to produce elemental sulfur.

In the Beavon process, the hydrogen sulfide in the relatively low concentration gas stream from the Claus process can be almost completely removed by absorption in a quinone solution. The dissolved hydrogen sulfide is oxidized to form a mixture of elemental sulfur and hydroquinone. The solution is injected with air or oxygen to oxidize the hydroquinone back to quinone. The solution is then filtered or centrifuged to remove the sulfur and the quinone is then reused. The Beavon process is also effective in removing small amounts of sulfur dioxide, carbonyl sulfide, and carbon disulfide that are not affected by the Claus process. These compounds are first converted to hydrogen sulfide at elevated temperatures in a cobalt molybdate catalyst prior to being fed to the Beavon unit. Air emissions from sulfur recovery units will consist of hydrogen sulfide, sulfur oxides, and nitrogen oxides in the process tail gas as well as fugitive emissions and releases from vents.

In the SCOT process, the sulfur compounds in the Claus tail gas are converted to hydrogen sulfide by heating and passing it through a cobalt-molybdenum catalyst with the addition of a reducing gas. The gas is then cooled and contacted with a solution of di-isopropanolamine (DIPA) that removes all but trace amounts of hydrogen sulfide. The sulfide-rich di-isopropanolamine is sent to a stripper where hydrogen sulfide gas is removed and sent to the Claus plant. The di-isopropanolamine is returned to the absorption column.

In the Wellman-Lord process, sodium sulfite is used to capture the sulfur dioxide. The sodium bisulfite thus formed is later heated to evolve sulfur dioxide and regenerate the sulfite scrubbing material. The sulfur dioxide-rich product stream can be compressed or liquefied and oxidized to sulfuric acid, or reduced to sulfur.

Most refinery process units and equipment are manifolded into a collection unit, called the blowdown system. Blowdown systems provide for the safe handling and disposal of liquid and gases that are either automatically vented from the process units through pressure relief valves, or that are manually drawn from units. Recirculated process streams and cooling water streams are often manually purged to prevent the continued buildup of contaminants in the stream. Part or all of the contents of equipment can also be purged to the blowdown system prior to shut down before normal or emergency shutdowns. Blowdown systems utilize a series of flash drums and condensers to separate the blowdown into its vapor and liquid components. The liquid is typically composed of mixtures of water and hydrocarbon derivatives containing sulfides, ammonia, and other contaminants, which are sent to the wastewater treatment plant. The gaseous component typically contains hydrocarbon derivatives, hydrogen sulfide, ammonia, mercaptans, solvents, and other constituents, and is either discharged directly to the atmosphere or is combusted in a flare. The major air emissions from blowdown systems are hydrocarbon derivatives in the case of direct discharge to the atmosphere and sulfur oxides when flared.

Lowering the sulfur content in fuel oil will significantly reduce the threats to public health and sensitive ecosystems posed by sulfur dioxide emissions. Emissions of nitrogen oxides (NOx), which contribute to a number of public health and environmental problems, will also decrease with lower sulfur heating oil. The use of cleaner (environmentally friendly) fuel oil has the potential to improve combustion efficiency by reducing fouling rates of boiler and furnace heat exchangers and other components. Further, the availability of low sulfur fuel oil will enable the introduction of highly efficient condensing furnace technology. Both outcomes will lower emissions of CO_2 and other pollutants from this source sector by reducing fuel use.

Upon release into the environment, heavy fuel oil will break into small masses and will not spread as rapidly as less viscous oil. The density of some heavy fuel oils means that they may sink on release to water, rather than float on the surface like other crude oil fuels. Loss of the lower molecular weight components due to volatility and dissolution will increase the density of the floating oil, causing it to sink. This heavy fraction will assume a tar-like consistency and stick to exposed substrates or become adsorbed to particulates.

Weather conditions, temperature, and the location into which the crude oil or the crude oil product will gave a significant effect on the rate of dispersion while higher temperatures will increase the rate of evaporation of the lower boiling hydrocarbon derivatives (Figure 4.1). Also the

Discharge   Oil layer   Evaporation
                        Atmospheric pollution
              Dissolution
              Remaining oil layer   Evaporation
                                    Atmospheric pollution
                        Dissolution
                        Soluble constituents   Water-in-oil emulsion   Microbial degradation
                                                                        Oxidation
                                                                        Reaction with minerals
                                                                        Adsorption on minerals
                                               Oil-in-water emulsion   Microbial degradation
                                                                        Oxidation
                                                                        Reaction with minerals
                                                                        Adsorption on minerals
                                               Tar   Adsorption on minerals

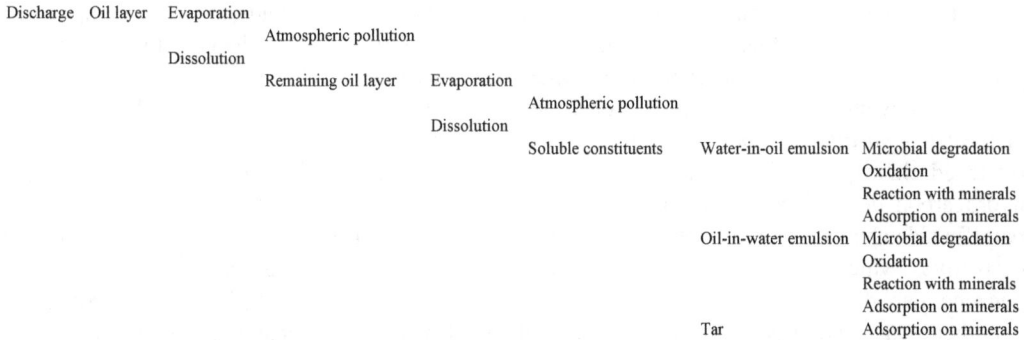

Figure 4.1: Examples of the Possible Modes of Dispersion of Spilled Crude Oil and Crude Oil Products.

temperature of the water temperature is a major factor in determining the extent of the environmental impact following a heavy fuel oil spill since higher temperatures will enhance loss of lower boiling constituents by evaporation as well as degradation processes.

Some of the non-hydrocarbon fraction of the refinery gases would not be expected to biologically degrade as these substances do not contain the chemical linkages necessary for microbial metabolism. For this reason, hydrogen, nitrogen, and carbon dioxide would not be susceptible to biodegradation. Furthermore, carbon dioxide is the final product in the biological mineralization of organic compounds. In contrast, ammonia can be readily oxidized to nitrite under aerobic conditions by autotrophic nitrifying bacteria (API, 2009).

Carbon monoxide has been reported to be microbially oxidized to carbon dioxide in pure cultures by a number of microbial species. It was also shown to be rapidly converted to carbon dioxide by indigenous soil microbial communities. Methanethiol can be both evolved and consumed in nature. It is produced by a variety of organisms through the decay of sulfur-containing organic matter under anoxic conditions. Methanethiol is known to undergo both aerobic and anaerobic biodegradation, but hydrogen sulfide does not usually biodegrade. The reduction of sulfate to hydrogen sulfide occurs in anoxic environments by anaerobic bacteria. Conversely, hydrogen sulfide can be oxidized to elemental sulfur and sulfate by a number of bacteria. Much of this cycling of sulfur occurs in sediments at the boundary layer between oxic and anoxic conditions.

Biodegradation of the hydrocarbon components in refinery gases may occur in soil and water. Gaseous hydrocarbon derivatives are widespread in nature and numerous types of microbes have evolved which are capable of oxidizing these substances as their sole energy source. Although volatilization is the predominant behavior for these gases, sufficient aqueous solubility and bioavailability is exhibited by these compounds. The use of gaseous carbon sources for cell growth is common among autotrophic organisms (Chapter 1). Higher chain length hydrocarbon derivatives typical of naphtha streams also are known to inherently biodegrade in the environment (Chapter 11).

Several of the constituents in refinery gases were shown to be highly hazardous to aquatic organisms in laboratory toxicity tests where exposure concentrations can be maintained over time (API, 2009). Hydrogen sulfide has been shown to be the most toxic constituent to fish and invertebrates. Given the physical-chemical characteristics of the refinery gases and the confined production and use within refineries, potential exposures to aquatic organisms would be greatest from accidental catastrophic releases. Fugitive emissions in refineries would not be expected to impact aquatic systems. Based on a simple conceptual exposure model analysis, emissions of refinery gases to the atmosphere would not likely result in acutely toxic concentrations in adjacent water bodies because such emitted gases will tend to remain in the atmosphere.

4.1.2 Liquids

Crude oil, as a mixture of hydrocarbon derivatives, is (theoretically) a biodegradable material. However, in very general terms (and as observed from elemental analyses), crude oil is a mixture of

(1) hydrocarbon derivatives, (2) nitrogen compounds, (3) oxygen compounds, (4) sulfur compounds, and (5) metallic constituents. However, this general definition is not adequate to describe the composition of crude oil as it relates to the behavior of these feedstocks.

It is convenient to divide the hydrocarbon components of crude oil into the following three classes: (1) *paraffin compounds*, which are saturated hydrocarbon derivatives with straight or branched chains, but without any ring structure, (2) naphthene derivatives, which are saturated hydrocarbon derivatives containing one or more rings, each of which may have one or more paraffin side chains—more correctly known as alicyclic hydrocarbon derivatives, and (3) aromatic derivatives, which are hydrocarbon derivatives that contain one or more aromatic nuclei, such as benzene, naphthalene, and phenanthrene ring systems, which may be linked up with (substituted) naphthene rings and/or paraffin side chains. Crude oil also contains appreciable amounts of organic non-hydrocarbon constituents, mainly sulfur-, nitrogen-, and oxygen-containing compounds and, in smaller amounts, organometallic compounds in solution and inorganic salts in colloidal suspension. These constituents appear throughout the entire boiling range of the crude oil but tend to concentrate mainly in the heavier fractions and in the nonvolatile residues (Speight, 2014).

Although their concentration in certain fractions may be quite small, their influence is important. For example, the thermal decomposition of deposited inorganic chlorides with evolution of free hydrochloric acid can give rise to serious corrosion problems in the distillation equipment. The presence of organic acid components, such as mercaptans (R-SH) and acids (R-CO$_2$H), can also promote environmental damage. In catalytic operations, passivation and/or poisoning of the catalyst can be caused by deposition of traces of metals (vanadium and nickel) or by chemisorption of nitrogen-containing compounds on the catalyst, thus necessitating the frequent regeneration of the catalyst or its expensive replacement. This carries with it the issues related to catalyst disposal.

Thermal processing can significantly increase the concentration of polynuclear aromatic hydrocarbon derivatives in the product liquid because the low-pressure hydrogen deficient conditions favor aromatization of naphthene constituents and condensation of aromatics to form larger ring systems. To the extent that more compounds like benzo(a)pyrene are produced, the liquids from thermal processes will be more carcinogenic than asphalt. This biological activity was consistent with the higher concentration of polynuclear aromatic hydrocarbon derivatives at 38.8 mg/g in the pitch compared to only 0.22 mg/g in the asphalt. Similarly, one would expect coker gas oils to contain more polynuclear aromatic hydrocarbon derivatives than unprocessed or hydroprocessed distillates, and thereby give a higher potential for carcinogenic or mutagenic effects.

In all cases careful separation of reaction products is important to the recovery of well-refined materials. This may not be easy if the temperature has risen as a consequence of chemical reaction. This will result in a persistent dark color traceable to reaction products that are redistributed as colloids. Separation may also be difficult at low temperature because of high viscosity of the stock, but this problem can be overcome by dilution with light naphtha or with propane.

In addition, delayed coking also requires the use of large volumes of water for hydraulic cleaning of the coke drum. However, the process water can be recycled if the oil is removed by skimming and suspended coke particles are removed by filtration. If this water is used in a closed cycle and treated to produce useable water, the impact of delayed coking on water treatment facilities and the environment is minimized. The flexicoking process offers one alternative to direct combustion of coke for process fuel. The gasification section is used to process excess coke to mixture of carbon monoxide (CO), carbon dioxide (CO$_2$), hydrogen (H$_2$), and hydrogen sulfide (H$_2$S) followed by treatment to remove the hydrogen sulfide. Maximizing the residue conversion and desulfurization of the residue in upstream hydroconversion units also maximizes the yield of hydrogen sulfide relative to sulfur in the coke. Currently, maximum residue conversion with minimum coke production is favored over gasification of coke (Menon and Mink, 1992).

The main uses of crude oil naphtha fall into the general areas of (1) precursor to gasoline and other liquid fuels, (2) solvents (diluents) for paints, (3) dry-cleaning solvents, (4) solvents for cutback asphalts, (5) solvents in rubber industry, and (6) solvents for industrial extraction processes. Turpentine, the older and more conventional solvent for paints has now been almost completely replaced by the cheaper and more abundant crude oil naphtha.

The term aliphatic naphtha refers to naphtha containing less than 0.1% v/v benzene and with carbon numbers from C_5 through C_{16}. Aromatic naphtha has carbon numbers from C_6 through C_{16} and contains significant quantities of aromatic hydrocarbon derivatives such as benzene (> 0.1% v/v), toluene, and xylene isomers. The final gasoline product as a transport fuel is a carefully blended mixture having a predetermined octane value. Thus, gasoline is a complex mixture of hydrocarbon derivatives that boils below 200°C (390°F). The hydrocarbon constituents in this boiling range are those that have four to twelve carbon atoms in their molecular structure.

Potential releases of naphtha from refineries and upgraders can be characterized as either controlled or unintentional releases. Controlled releases are planned releases from pressure relief valves, venting valves and drain systems that occur for safety purposes or maintenance, are considered part of routine operations and occur under controlled conditions. Unintentional releases are typically characterized as unplanned releases due to spills or leaks from various equipment, valves, piping, and flanges which result from equipment failure, poor maintenance, lack of proper operating practices, adverse weather conditions or other unforeseen factors. Refinery and upgrader operations are highly regulated and regulatory requirements established under various jurisdictions, as well as voluntary non-regulatory measures implemented by the crude oil industry, are in place to manage these releases.

If spilled or discharged in the environment, naphtha represents a threat of the toxicity of the constituents to land and/or to aquatic organisms. A significant spill may cause long-term adverse effects in the aquatic environment. The constituents of naphtha predominantly fall in the C_5–C_{16} carbon range: alkane derivatives, cycloalkane derivatives, aromatic derivatives and, if they are subject to a cracking process, alkene derivatives as well. Naphtha may also contain a preponderance of aromatic constituents (up to 65% v/v), others contain up to 40% v/v alkene derivatives, while all of the others are aliphatic in composition, up to 100% v/v.

Water solubility ranges from very low for the longest-chain alkane derivatives to high solubility for the simplest mono-aromatic constituents. Generally, the aromatic compounds are more soluble than the same-sized alkane derivatives, iso-alkane derivatives and cycloalkane derivatives. This indicates that the components likely to remain in water are the one- and two-ring aromatics (C_6–C_{12}). The C_9–C_{16} alkane derivatives, iso-alkane derivatives and one- and two-ring cycloalkane derivatives are likely to be attracted to sediments based on their low water solubilities and moderate to high octanol-water partition coefficient (log K_{ow}) and organic carbon-water partition coefficient (log K_{oc}) values.

Naphtha (especially the low-boiling naphtha) contains volatile organic compounds (VOCs) that are rapidly degraded in air, water, and soil. Considerable measures must be taken to prevent release of naphtha (constituents) to the atmosphere and minimize any exposure to the environment from activities in which naphtha is manufactured and used.

Constituents of naphtha can be carcinogenic, and frequently products sold as naphtha contain some impurities which may also have harmful properties of their own. The method of manufacture means that there is a range of distinct chemicals in naphtha which makes rigorous comparisons and identification of specific carcinogens difficult and is further complicated by exposure to a significant range of other known and potential carcinogens.

The most toxic components of fuel oils are the aromatics, such as benzene, toluene, xylene isomers, naphthalene derivative, and others. These aromatics are relatively highly soluble in water. After the aromatic fraction, toxicity decreases from olefins through naphthenes to paraffins. Within each of these groups, the lower molecular weight hydrocarbon tends to be more acutely toxic.

More pertinent to the present text, because of the publicity received over the years, heavy fuel oil is a blended product based on the residues from various refinery distillation and cracking processes. Heavy fuel oil is a viscous liquid with a characteristic odor and require heating for storage and combustion. Heavy fuel oil is used in medium to large industrial plants, marine applications and power stations in combustion equipment such as boilers, furnaces and diesel engines.

Heavy fuel oil is a general term and other names commonly used to describe this range of products include: residual fuel oil, bunker fuel, bunker C, fuel oil No. 6, industrial fuel oil, marine fuel oil and black oil. In addition, terms such as heavy fuel oil, medium fuel oil and light fuel oil are used to describe products for industrial applications to give a general indication of the viscosity and density of the product.

Short-term toxicity decreases as the type of fuel oil becomes less volatile (that is, No. 1 and No. 2 are moderately toxic, while toxicity decreases through No. 4, No. 5, and No. 6). Fuel oil No. 1 and No. 2 possesses moderate to high acute toxicity to biota with product-specific toxicity related to the type and concentration of aromatic compounds, while fuel oil No. 5 and No. 6 are considered to be less acutely toxic relative to other oil types. Fuel oil No. 4 has variable acute toxicity, depending on the amount of light fraction.

Like naphtha, potential releases of fuel oil from refineries and upgraders can be characterized as either controlled or unintentional releases. Controlled releases are planned releases from pressure relief valves, venting valves and drain systems that occur for safety purposes or maintenance, are considered part of routine operations and occur under controlled conditions. Unintentional releases are typically characterized as unplanned releases due to spills or leaks from various equipment, valves, piping, and flanges which result from equipment failure, poor maintenance, lack of proper operating practices, adverse weather conditions or other unforeseen factors. Refinery and upgrader operations are highly regulated and regulatory requirements established under various jurisdictions, as well as voluntary non-regulatory measures implemented by the crude oil industry, are in place to manage these releases.

Most fuel oil entering the environment comes from spills or leaking storage tanks. When spilled on soil, some components of fuel oil attach to soil. Furthermore, fuel spilled on water or soil, evaporates into the air. Fuel oil can also contaminate soil sediments and private drinking water supplies. Other environmental impacts associated with oil production include blowouts, spills, brine disposal, and the production of hydrogen sulfide. Transportation of crude oil and crude oil products oil (whether by sea-going vessel, land-going vessel, or pipelines) involves spill and leak hazards. Oil refining includes environmental effects such as explosions, fires, air emissions, noise, odor, and water runoff.

4.1.3 Solids

Catalyst disposal is a major concern in all refineries. In many cases the catalysts are regenerated at the refinery for repeated use. Disposal of spent catalysts is usually part of an agreement with the catalysts manufacturer whereby the spent catalyst is returned for treatment and re-manufacture.

The formation of considerable quantities of coke in the *coking* processes is a cause for concern since it not only reduces the yield of liquid products but also initiates the necessity for disposal of the coke. Stockpiling to coke may be a partial answer unless the coke contains leachable materials that will endanger the ecosystem as a result of rain or snow melt. In addition, the generation and emission of sulfur oxides (particularly sulfur dioxide) from combustion of sulfur-containing coke as plant fuel. Sulfur dioxide (SO_2) has a wide range of effects on health and on the environment. These effects vary from bronchial irritation upon short-term exposure to contributing to the acidification of lakes. Emissions of sulfur dioxide therefore, are regulated in many countries.

During the process of crude oil refining, large amounts of oily sludge, which contains oil, benzene derivatives, phenol derivatives, and other odorous and toxic substances, can be produced in refineries. Thus, a common form of solid waste (typically which is in a semi-solid form) in a refinery is the sludge generated from various processes. Generally, refinery sludge contains oil content (> 40% w/w) and several methods (such as centrifuging) are used to separate the oil, water and solids. The recovered oil is pumped back into the refinery process, while the solids and water are treated before disposal.

More specifically, solids in crude oil fall into two classes which are (1) basic sediment, which typically consist particles above 20 microns that centrifuge out of crude oil during standard BSW testing and (2) filterable solids, which consists of smaller particles typically are measured by filtration methods. As the domestic refining industry continues to move to heavier, dirtier crudes, the economic impact of these filterable solids continues to increase. The filterable solids can cause fouling, foaming, erosion, corrosion, catalyst contamination, end-product contamination, and oil contamination in effluent waters.

The sludge produced on acid treatment of crude oil distillates is complex in nature. Ester derivatives (RCO_2R^1, where R and R1 are the same or different aliphatic or aromatic groups) and alcohol derivatives (ROH, where R is an aliphatic or aromatic groups) are present from reactions with olefins. Sulfonation products from reactions with aromatic compounds, naphthene compounds, and phenols; and salts from reactions with nitrogen bases. In addition, such materials as naphthenic acids, sulfur compounds, and asphalt (residua constituents) material are all retained by direct solution. To these constituents must be added the various products of oxidation-reduction reactions: coagulated resins, soluble hydrocarbon derivatives, water, and free acid.

The disposal of the sludge must be handles with caution—the sludge contains unused free acid that must be removed by dilution and settling. The disposal is a comparatively simple process for the sludge resulting from treating gasoline and kerosene, the so-called light oils. The insoluble oil phase separates out as a mobile tar, which can be mixed and burned without too much difficulty. Sludge from heavy oil, however, separates out granular semisolids, which offer considerable difficulty in handling.

Many technologies prevent the harmful effects of solids at each phase of the refining process without removing the solids from the process stream, so problems pass on to downstream operations. The most economical approach is to remove solids as soon as possible at the desalter.

Particulate matter is a complex emission that is classified as either suspended particulate matter, total suspended particulate matter, or simply particulate matter. For human health purposes, the fraction of particulate matter that has been shown to contribute to respiratory diseases is termed PM_{10}, i.e., particulate matter with sizes less than 10 microns. From a control standpoint, particulate matter can be characterized as follows: (1) particle size distribution; and (2) particulate matter concentration in the emission (mg/m^3). On occasion, physical property descriptions may also be employed when there are specific control applications.

Both solid particles and condensable liquid droplets are generated from most combustion sources including fuel oil burners. Most of the particulate matter emitted by combustion sources is classified as fine particulate matter with diameters less than 2.5 microns (PM2.5). Primary particulates include unburned carbonaceous materials (soot) that are directly emitted into the air. Secondary particulates, such as sulfates, are formed after sulfur dioxide is emitted into the air from combustion sources burning sulfur-containing fuels. Particulate matter less than 10 microns in size (PM10) is linked to a number of adverse health outcomes including asthma, bronchitis, cardiac arrhythmia, and heart attacks. Sulfates are also the primary cause of regional haze and acid deposition (acid rain). Direct PM emissions from residential and small commercial oil burners in the form of soot have decreased by approximately 95 percent over the past three decades (as will be discussed later in this section). Sulfates that condense in the outdoor air after being emitted by oil heating equipment are now the predominate form of particulate associated with emissions from fuel oil burners. Reducing the sulfur content of the fuel can lower sulfate emissions.

In collecting methods, particulate matter emissions are determined through gravimetric analysis of the particulates collected on a sampling filter. Alternatively, the sample can be analyzed using thermal mass analysis (e.g., coulometric analysis). A number of other properties, for instance surface area or biological activity, can be also analyzed. Collecting methods, and especially gravimetric analysis, are well established as the most common method of particulate matter emission determination.

The health effects of particulate matter (a complex mixture of solids and liquids) emissions are not yet well understood but are recognized as major contributors to health problems. Biological activity of particulate matter may be related to particle sizes and/or particle composition. Furthermore, it has generally been concluded that exposure to particulate matter may cause increased morbidity and mortality, such as from cardiovascular disease. Long-term exposure to particulate emissions is also associated with small increase in the relative risk of lung cancer.

4.2 Understanding Spills of Crude Oil and Crude Oil Products

In order to combat any threat to the environment, it is necessary to understand the nature and magnitude of the problems involved (Ray and Guzzo, 1990; Speight, 2005; Speight and Arjoon, 2012). It is in such situations that environmental technology has a major role to play. Environmental issues even arise when outdated laws are taken to task. Thus the concept of what seemed to be a good idea at the time the action occurred no longer holds when the law influences the environment.

In the remediation of the spill of crude oil or a crude oil products, it is critically important to understand the options for cleanup. After a spill, timely decisions must be made about the most effective response to mitigate the environmental impacts of the spill. Multiple technologies are typically used in a spill response, including physical methods (skimmers, sorbents, or solidifying agents) and/or chemical methods (dispersants or *in situ* burning). Each technology depends on the specific spill conditions and properties of the spilled oil. A major requirement in the response to a spill involves the analysis and characterization of the crude oil or the crude oil product (Chapter 3). These materials are heterogeneous mixtures of hydrocarbons with physical properties that change depending on the exact oil composition, and on the environmental conditions.

Both the chemical composition and the physical properties of oils have been carefully analyzed and documented for thousands of crude oils and refined oil products. Physical properties influencing spill cleanup include viscosity, density, API gravity, pour point, flash point, and the chemical composition of the oil (i.e., wax content and SARA (saturated hydrocarbon, aromatic hydrocarbon, resin, asphaltene) content). Although there is a comprehensive list of the properties that may (or may not) need to be investigated (Chapter 3), the common oil properties used in spill remediation research are: (1) the viscosity, which is the resistance to flow of a liquid, (2) the density, which is the mass per unit volume, (3) the API gravity, which is a dimensionless form of density, describing a fluid's density relative to water, (4) the flash point, which is the lowest temperature at which a substance can vaporize to form an ignitable mixture, (5) the chemical composition, which is the content of the chemical groups in the original crude oil or the crude oil product, (6) the SARA analysis, which is the content of saturates, aromatics, resins, and asphaltenes in original and is often correlated with the bulk physical properties of crude oil or the crude oil product, (7) the sulfur content, and (8) in some cases, the wax content of the crude oil or the crude oil product may also be important since waxes are malleable near ambient temperature and are typically insoluble in water, and soluble in organic solvents and the wax content is often correlated with viscosity.

Oil spills are any uncontrolled release of crude oil and crude oil products which can be either (1) an accidental or (2) intentional release of crude oil or a crude oil product into the environment as a result of human activity (drilling, manufacturing, storing, transporting, waste management). The spills may be due to release of crude oil from tankers, pipelines, railcars, offshore platforms, drilling rigs and wells, as well as spills of refined crude oil products and their byproducts, heavy fuels used by large ships such as bunker fuel (often called Bunker C oil), or the spill of any oily refuse or waste oil. Such spills can pollute air, water, or land. Spills of crude oil and crude oil products have huge and immediate economic, social, and environmental impacts, it endanger public health, imperil drinking water, devastate natural resources, and disrupt the economy (Table 4.1, Table 4.2).

Table 4.1: Selection of Reported Oil Spills in the Recent Past.*

Spill/Tanker	Location	Date**	Barrels***
ABT Summer	Angola	May, 1991	1,907
Amoco Cadiz	France, Brittany	March, 1978	1,635
Atlantic Empress/Aegean Captain	Trinidad and Tobago	July, 1979	2,105
Castillo de Bellver	South Africa, Saldanha Bay	August, 1983	1,848
Deepwater Horizon	Gulf of Mexico	April, 2010	4,100–4,900
Fergana Valley	Uzbekistan	March, 1992	2,090
Gulf War oil spill	Kuwait, Iraq	January 1991	6,000–8,000
Ixtoc I	Mexico, Gulf of Mexico	June, 1979	3,329–3,520
Kuwaiti Oil Fires	Kuwait	January, 1991	1,000,000
Kuwaiti Oil Lakes	Kuwait	January 1991	25,000–50,000
MT Haven	Genoa, Italy	April, 1991	1,400,000
Nowruz Field Platform	Iran, Persian Gulf	February, 1983	1,900
Odyssey	Nova Scotia, Canada	November, 1988	968
Torrey Canyon	Cornwall, England	March, 1967	872

Source: https://en.wikipedia.org/wiki/Oil_spill
See also: https://en.wikipedia.org/wiki/List_of_oil_spills
* Listed alphabetically rather than by any other parameter.
** Month and year first reported.
*** Thousands (approximate); 1 barrel (bbl) of oil is equivalent to 35 imperial gallons or 42 US gallons.

Spills of crude oil and crude oil products can occur as a result of natural disasters, such as hurricanes and earthquakes, disrupting wells, oil tankers, and storage facilities, oil tankers colliding with each other or other maritime objects or destruction of oil wells or storage facilities during war or as acts of terrorism. Crude oil can also escape from the reservoir to the ground surface or the ocean floor naturally and is caused by the escape of the oil (and gas) through fractures in rock formations or directly from rock outcrops.

Furthermore, using terminology common to environmental technology, spills from tankers, pipelines, and oil wells are examples of *point sources* of pollution, where the origin of the contaminants is a single identifiable point 2005 (Speight, 1996; Speight and Lee, 2000; Speight, 2005). They also represent catastrophic releases of a large volume of pollutants in a short period of time. But the majority of pollution from oil is from nonpoint sources, where small amounts coming from many different places over a long period of time add up to large-scale effects. For example, the majority of the crude oil products released by human activity into oceans worldwide is a result of small spills during crude oil consumption. These minor unreported spills can include routine discharges of fuel from commercial vessels or leakage from recreational boats.

Spills of crude oil and crude oil products tend to collect in hazardous concentrations in the soil or wastewater coming out of cities and other populated areas. Runoff from asphalt-covered roads and parking lots enters storm drains, streams, and lakes and eventually travels to the ocean, affecting all of the ecosystems through which it passes. As cities grow, more and more crude oil products—gasoline, solvents, and lubricants—are often improperly disposed into drains and sewage pipes.

When assessing a situation for oil spill cleanup, there are several elements to consider and include (1) the type and character of the crude oil, (2) the quantity of the crude oil, (3) the source of the spill, (4) the ecosystem into which the oil or the product is spilled, (5) the effect of the spill on the flora and fauna of the ecosystem, and (6) the proximity to populated areas are all necessary considerations. Other key factors or variables that influence oil-spill impacts can be identified in several domains: the oil spill itself, disaster management, marine physical environment, marine biology, human health and society, economy, and policy.

Table 4.2: Emissions and waste from refinery processes.

Process	Air Emissions	Residual Wastes Generated
Crude oil desalting	Heater stack gas (CO, SOx, NOx, hydrocarbons and particulates), fugitive emissions (hydrocarbons)	Crude oil/desalter sludge (iron rust, clay, sand, water, emulsified oil and wax, metals)
Atmospheric distillation Vacuum Distillation	Heater stack gas (CO, SOx, NOx, hydrocarbons and particulates), vents and fugitive emissions (hydrocarbons) Steam ejector emissions (hydrocarbons), heater stack gas (CO, SOx, NOx, hydrocarbons and particulates), vents and fugitive emissions (hydrocarbons)	Typically, little or no residual waste generated
Thermal Cracking/ Visbreaking	Heater stack gas (CO, SOx, NOx, hydrocarbons and particulates), vents and fugitive emissions (hydrocarbons)	Typically, little or no residual waste generated
Coking	Heater stack gas (CO, SOx, NOx, hydrocarbons and particulates), vents and fugitive emissions (hydrocarbons) and decoking emissions (hydrocarbons and particulates)	Coke dust (carbon particles and hydrocarbons)
Catalytic Cracking	Heater stack gas (CO, SOx, NOx, hydrocarbons and particulates), fugitive emissions (hydrocarbons) and catalyst regeneration (CO, NOx, SOx, and particulates)	Spent catalysts (metals from crude oil and hydrocarbons), spent catalyst fines from electrostatic precipitators (aluminum silicate and metals)
Catalytic Hydro-cracking	Heater stack gas (CO, SOx, NOx, hydrocarbons and particulates), fugitive emissions (hydrocarbons) and catalyst regeneration (CO, NOx, SOx, and catalyst dust)	Spent catalysts fines
Hydrotreating/ Hydroprocessing	Heater stack gas (CO, SOx, NOx, hydrocarbons and particulates), vents and fugitive emissions (hydrocarbons) and catalyst regeneration (CO, NOx, SOx)	Spent catalyst fines (aluminum silicate and metals)
Alkylation	Heater stack gas (CO, SOx, NOx, hydrocarbons and particulates), vents and fugitive emissions (hydrocarbons)	Neutralized alkylation sludge (sulfuric acid or calcium fluoride, hydrocarbons)
Isomerization	Heater stack gas (CO, SOx, NOx, hydrocarbons and particulates), HCl (potentially in light ends), vents and fugitive emissions (hydrocarbons)	Calcium chloride sludge from neutralized HCl gas
Polymerization	H2S from caustic washing	Spent catalyst containing phosphoric acid
Catalytic Reforming	Heater stack gas (CO, SOx, NOx, hydrocarbons and particulates), fugitive emissions (hydrocarbons) and catalyst regeneration (CO, NOx, SOx)	Spent catalyst fines from electrostatic precipitators (alumina silicate and metals)
Solvent Extraction	Fugitive solvents	Little or no residual wastes generated
Dewaxing	Fugitive solvents, heaters	Little or no residual wastes generated
Propane Deasphalting	Heater stack gas (CO, SOx, NOx, hydrocarbons and particulates), fugitive propane	Little or no residual wastes generated
Wastewater treatment	Fugitive emissions (H2S, NH3, and hydrocarbons)	API separator sludge (phenols, metals and oil), chemical precipitation sludge (chemical coagulants, oil), DAF floats, biological sludge (metals, oil, suspended solids), spent lime

The difficulty of oil spill cleanup jobs vary greatly depending on complications that may arise from environmental factors. The *International Petroleum Industry Environmental Conservation Association* (IPIECA) classification system includes three tiers of oil spills that are differentiated according to the severity of the spill as well as the response systems required to effectively remedy (remediate) the spilled material. The tiered structure provides the crude oil industry and governmental authorities with a system to prepare for three levels of environmental emergency.

A Tier 1 spill is a low-severity spill which tends to be operational incident and occurs at or near the operator's facility and may require only a local response team. A Tier 2 spill is the result of an accident and which may require national or regional response teams with specialized knowledge to intervene and such spills extend outside the operational area of the oil or gas facility. A Tier 3 spill is an accident that is global in terms of the need for the necessary large-scale response.

As part of the practice of bioremediation, it is necessary to formulate an oil spill response plan. The best way to do this is to assume that a spill will occur at some time during the use and handling of the crude oil and/or the crude oil product. Thus, in order to prevent the contamination of the environment (should, in the worst case scenario) a spill occur, requires a set of preparedness steps ensure that a worksite is ready to respond to an oil spill. The oil spill response requires (1) a viable reconciliation of the physical and substance properties, (2) transport of the spilled oil, (3) weathering of the spilled oil, (4) the choice of cleanup strategies, and (5) climatic conditions on land and over the ocean.

In order to apply bioremediation technologies to crude oil or crude oil products (or, for that matter, any collection of organic compounds) it is essential to know the origin of the spilled material as well as the methods of manufacture and the composition of the material and the role played by the composition in biodegradation (Chapter 10). Without such background knowledge and information the cleanup technology could be more difficult to apply or even register as a failed technology.

Crude oil and crude oil products are no exception to this and even requires, on the basis of its complexity, additional efforts are required to assure cleanup in the form of complete removal of any such contaminants from a spill area. The fate of spilled of crude oil or crude oil products in water systems is exceptionally difficult to predict—crude oil and crude oil products oxidize on the surface after which the form oil-in-water emulsions.

Crude oil and crude oil products oxidize on the surface after which the form oil-in-water emulsions. The inclusion of polar functions such as hydroxyl groups (-OH) or carbonyl groups ($> C=O$) (a result of the oxidation process) causes an increase in the density of the emulsion (relative to the original unoxidized crude oil) and with an increased propensity to form emulsions. As a result, the emulsion and sinks to various depths or even to the seabed, depends on the extent of the oxidation and the resulting density. This may give the erroneous appearance (leading to erroneous deductions with catastrophic consequences) that the crude oil spill (as evidenced from the crude oil remaining on the surface of the water) is less than it actually was. The so-called *missing* oil will undergo further chemical changes and eventually reappear on the water surface or on a distant beach. Such behavior can confuse (and has confused) many cleanup (or oil spill) *experts* to the detriment of the environment. Thus understanding the chemistry of crude oil and its products is not merely a chemist's pipe dream but is an essential part of Oil spill response planning allows responders and supplies in and out of the area with as minimal disruption to the natural surroundings as possible.

All facilities that handle, store, or transport crude oil and crude oil product should have an approved oil spill response plan (OSRP) before operations begin, and must operate their facilities in accordance with that OSRP to ensure a quick and effective response to a worst-case discharge from a facility to the maximum extent practicable.

Oil spill response procedures, differ depending or where they occur but a good spillage response plan usually have four major elements that are (1) hazard identification, (2) vulnerability analysis, (3) risk assessment, and (4) response actions. Hazard identification and vulnerability analysis are used to develop a risk assessment which is then used as the basis for planning specific response actions.

Finally, various methods exist for the testing of biodegradability of substances which can be assessed, for example, by (1) measurement of the amount of carbon dioxide or methane, for anaerobic cases, produced during a specified period, (2) measurement of the loss of dissolved organic carbon for substances which are water soluble, (3) measurement of the loss of hydrocarbon infrared bands, as well and there are yet others which measure the uptake of oxygen by the activities of microorganisms (biochemical oxygen demand, BOD). These test methods are presented elsewhere in this book (Chapter 3).

4.2.1 Types of Spills

A spill is defined as an uncontrolled release of a chemical and can be categorized into two types known as (1) a minor spill, which can be considered as simple spills as they are small, confined, and present minimal hazards, and (2) a major spill, which could represent a significant environmental risk or serious human health risk as a result of release or exposure and may also be a larger-volume spill. Major spills require an external emergency response and can result in dangerous disasters. Consequences to ecosystems and economies can be felt for decades following a major oil spill. A minor chemical spill is one that the individual or a company can clean up with limited or no environmental impact and health risks to living creatures. Even a minor spill can create a dangerous situation for the health and safety of the workers.

There are many different kinds of crude oil and crude oil products with varying viscosity, volatility, and toxicity (Speight, 2014, 2017). Therefore when spilled, the various types of crude oil (or the crude oil product) oil can have different effects on the environment differently. They are categorized as follows: Group 1 (Non-Persistent Light Oils), Group 2 (Persistent Light Oils), Group 3 (Medium Oils), Group 4 (Heavy Oils) and Group 5 (Sinking Oils). These different groups of oil form the different types of oil spills found in the environment.

A Group 1 oil spill will usually evaporate within 1 to 2 days after the spill since the spilled material is highly volatile and do not leave a residue behind after evaporation. The risk when handling this type of spills is that it is high flammability and the spill also produces a toxic air hazard. Group 2 which comprises of diesel, No. 2 fuel oil, and light crude oil, which are moderately volatile with moderate concentrations of toxic (soluble) compounds therefore cleanup can be very effective. Groups 3, 4 and 5 have severe and long-term effects on the environment. Also, the oils included in these categories have the tendency to smother organisms. Group 3 oils can be effectively cleaned up if the cleanup operations are conducted quickly while Group 4 cleanup is difficult under all conditions and there is little or no evaporation or dissolution for Group 4 and 5 types of crude oil or crude oil products and there is often long-term contamination of sediments.

4.2.2 Composition of a Spill

Transportation of crude oil and crude oil products from the point of production to the point of processing or sales often times causes spillage with unfavorable results. At the point when the material spills onshore, there is an impact on the indigenous microorganisms and other land (and water) properties.

Thus, there is the need to know (1) the rate of evaporation of the crude oil or the crude oil product, (2) the detailed chemical composition of the material oil, (3) the differential compositional changes over time and the way in which these changes affect the behavior of the spill and fate of the spill constituents in the environment, (4) the viscosity of the oil at ambient temperature as it evaporates, (5) whether or not the spilled material is likely to sink or submerge, (6) whether or not the use of chemical dispersants can enhance its dispersion, (7) whether or not emulsions will form, (8) the toxicity of the spilled material to marine or aquatic organisms, and, last but by no means least, (9) the hazard to on-site personnel during cleanup (Wang et al., 2003).

In general, the smaller and lower molecular weight constituents of crude oil and crude oil products are more susceptible to processes such as evaporation, dissolution, and biodegradation, while the higher molecular weight more hydrophobic constituents tend to adhere to living organisms or particulates (i.e., soil constituents) and persist in the ecosystem. The presence of certain compounds, also determines the acute and chronic toxicity of the spilled oil (Overton et al., 2016).

Crude oil hydrocarbon derivatives are predominantly one of two types (1) alkane derivatives and (2) aromatic derivatives. The alkane constituents tend to be less toxic than the aromatic constituents and are much more readily biodegraded naturally; most can be ingested as food by some microorganisms. On the other hand, the aromatic constituents which are based on a 6-carbon ring, tend to be the molecular compounds in oil that are the most toxic to flora and fauna. A notable case is polynuclear aromatic hydrocarbon derivatives (PNAs, also called polyaromatic hydrocarbon derivatives, PAHs), which have multiple rings composed of aromatic carbon atoms and which can also be persistent in the environment.

The overall quantitative distribution of hydrocarbon and non-hydrocarbon compounds in crude oils and crude oil products, the molecular size, structure, and polarity of the constituents determine the viscosity of the spilled material at any temperature and pressure (under both subsurface and degassed surface conditions). The molecular size, structure, and polarity of individual compounds and the overall bulk oil composition determine the oil/water partition behavior of individual compounds based on their respective water solubilities and volatilities, and also determine rates of degradation by natural microorganisms under aerobic and anaerobic conditions and their interaction with the environment especially in oil spills.

In fact, the spill of crude oil or a crude oil product demonstrates the importance of interwoven chemical, physical, and biological processes in regulating the transport and fate of hydrocarbon derivatives in the marine environment. Thus, compositional information for crude oil (gas and oil) released by the well at the seafloor is essential for evaluating the fate of the spilled constituents (Kessler et al., 2011).

Lighter (i.e., low density, low viscosity) crude oils (and their related products) tend to evaporate and degrade (break down) very quickly. Higher viscosity materials, however, tend to form a thick oil-and-water mixture (often referred to as a mousse) which clings to rocks and sand. Heavier oils exposed to sunlight and wave action also tend to form dense, highly viscous products (often referred to as tar balls) and asphalt that are very difficult to remove from rocks and sediments. Such deposits typically require more aggressive cleanup than those from lighter ones.

In summary, the nature of the environmental damage caused by the spill of crude oil or a crude oil product will also vary according to the type of crude oil or the crude oil product. For example light (volatile) crude oil and light refined (volatile) crude oil products may constitute a fire and explosion hazard when spilled. Also, emulsified crude and heavy fuel oils, have the tendency to coat and smother the environment and living things present and, in addition, further problems can arise when the residues sink. Between the two extreme of gasolines and heavy fuel oil, there are many intermediate crude oils and refined products that are transported by tankers that require varying cleanup plans.

4.2.3 Dangers of a Spill in the Workplace

In addition to the environmental implications of the spillage of crude oil or a crude oil product, there are always implications related to the health of the staff. The key is to recognize the risks from all sources and to be prepared to act accordingly.

How dangerous a spill really is, can only be assessed by the properties of the actual material itself, how much material was spilled, where the spill occurred and what surface received the spill, the amount of ventilation in the area, and the temperature of the surface, immediate area, and the chemical itself. Regardless of the level of hazard involved, it is essential that any company that transports, stores or handles hazardous materials ensure spills are properly cleaned up to minimize environmental impacts and workers are not injured. The Environment Protection Agency (EPA) states that site-specific scenarios and response resources must be addressed for small, medium, and worst-case spills, whether chemical, biological or crude oil.

The protection of human health and the environment are some of the most important aspects of assessment and cleanup at an oil spill site. Protection includes using safe work practices to minimize safety and health risks to the hazmat personnel, support contractors, regulators, industry representatives, and the community. Proper safety procedures for oil spill workers and volunteers should be addressed before cleanup activities begin. This is because oil spill cleanup workers can face potential hazards from oil byproducts, dispersants, detergents and degreasers. Drowning, heat illness and falls also pose hazards, as can encounters with insects, snakes and other wild species native to the impacted areas. They can also suffer from a multitude of illnesses and injuries, such as ataxia, migraines, and various lung diseases. It's the small steps such as doing some maintenance, being prepared and acting fast that go a long ways in creating a safe, employee-friendly workplace.

4.2.4 Causes of Crude Oil Spills

The risks of spillage are always present, they take place when there is failure of the oil drilling machinery, due to human error, carelessness, deliberate acts or mistakes, or because of natural disasters or marine accidents especially for refineries or tankers shipping any form of crude oil product. Such spills tend to happen when there are accidents during transportation of the crude oil or crude oil products (whether by sea-going vessel, land-going vessel, or pipelines) is equipment failure or pipeline rupture, a crude oil tanker ship sinks, or drilling operations go wrong. Equipment failure and human error were the major causes of spills.

4.2.4.1 Transportation

In addition to the conventional meaning of the term *process*, the *transportation* of crude oil also needs to be considered here.

Oil spills during crude oil *transportation* have been the most visible problem. There have also been instances of oil wells at sea "blowing out", or flowing uncontrollably, although the amounts from blowouts tend to be smaller than from tanker accidents. The 1979 Ixtoc I blowout in the Gulf of Mexico was an exception, as it flowed an estimated 3 million barrels over many months.

Tanker accidents typically have a severe impact on ecosystems because of the rapid release of hundreds of thousands of barrels of crude oil (or crude oil products) into a small area. The largest single spill to date is believed to have occurred during the 1991 Gulf War, when as much as ten million barrels were dumped in the Persian Gulf by Iraq, apparently intentionally. More typical was the 1989 spill from the tanker Exxon Valdez, where two hundred fifty thousand barrels were lost in Alaskan coastal waters.

While oil, as a hydrocarbon, is at least theoretically biodegradable, large-scale spills can overwhelm the ability of the ecosystem to break the oil down (Speight and Arjoon, 2012). Over time the lighter portions of crude oil evaporate, leaving the non-volatile portion. Oil itself breaks down the protective waxes and oils in the feathers and fur of birds and animals, resulting in a loss of heat retention and causing death by freezing. Ingestion of the oil can also kill animals by interfering with their ability to digest food. Some crude oils contain toxic metals as well. The impact of any given oil spill is determined by the size of the spill, the degree of dispersal, and the chemistry of the oil. Spills at sea are thought to have a less detrimental effect than spills in shallow waters.

4.2.4.2 Errors by Personnel

Reducing accidents is the obvious answer. But the causes are widely misunderstood. Shipping records often list consequences—collisions, groundings and explosions—rather than reasons, such as poor navigation, lack of maintenance, miscommunication and other human errors. There are three categories that are known to contribute to accidents in the offshore oil industry, including tanker operations based on the human influence, these are individual factors, group factors, and organizational factors. At the organizational level, various factors may contribute to an increase in incidents and accidents, including cost-cutting programs and the level of communication between work-sites. In relation to the maritime industry, two of the most recognized and studied individual factors are (1) inadequate knowledge and (2) fatigue (Gordon, 1998).

Policies would be more effective if they acknowledged the role of human error. However, human factors can be difficult to eliminate but by studying past accidents or spills and drawing lessons from the various industries in general, an understanding of the dominance of the human-technology interface to guide and enhance oil spill prevention efforts.

4.2.4.3 Equipment Breakdown

Equipment failure alone is more likely to cause small and medium oil spills than large spills, but equipment failure may act as a contributing factor to larger spills (ABSG, 2016). Recently, equipment failure and other factors were cited as the causes for the 2010 Deepwater Horizon oil-spill incident. The investigation report identified a well blowout as one of the contributing factors for the incident (USCG, 2011). In the Arctic OCS, equipment failure is likely to be a more influential causal factor due to the additional stresses and pressures on equipment from the harsh environment.

Mechanical failure of a pipeline may include connection failures or material failures. Connection failures most likely occur at the source or end of the pipeline where the pipeline meets the platform or onshore facility, and material failures may occur throughout the entirety of the pipeline. Mechanical failure may include failure of bends, bolts, connectors, fittings, clamps, joins, valves, or other features of the pipeline.

4.2.4.4 Natural Disasters

Natural Hazards (and the resulting disasters) are the result of naturally occurring processes that have operated throughout history of the Earth that will have a negative effect on humans. Very heavy storms in the oceans, shaking of the sea floor due to earthquakes, and hurricanes have contributed to oil tanker ship accidents or breakage/leakage of underground pipelines thereby causing colossal oil spills.

Weather is the most common causal factor for medium and large oil spills on platforms. Weather as a factor identifies spills caused by standard weather events, such as strong tides, rough seas, and waves but since 1972, hurricanes have caused the most oil spills of 50 bbls or more. Hurricanes not only can have dramatic effects on the number of spills occurring, but they can also cause spills that are large and long lasting if wells are not shut in properly or timely.

Also, many natural hazards can impact pipelines, but mud slides and hurricanes are most notable. The strong winds and currents of hurricanes and storms can threaten the structural integrity of pipelines while mudslides threaten pipelines as they remove supporting sediment and pose external forces from moving sediment.

4.2.4.5 Acts of Terrorism, War, Vandalism, or Illegal Dumping

Deliberate acts by terrorists, vandals, or countries at war have contributed to a number of oil spillages. The illegal dumping of oil waste along the coastline has been and still is a problem for many Governments in the world. Marine habitats and inland water systems have been polluted by oil spills especially by acts of vandalism, sabotage or terror activities with an aim of completely destroying the perceived enemy's economic wealth base. Most of the criminals target major pipelines. Such incidences are rampant in war and crude oil resource zones, as what resulted in the Gulf War Oil Spill.

When crude oil being the major source of national income, as well as a key product for the export market, it becomes a target to derail the economy of a country. Interfering with oil pipelines and installations has assumed huge dimensions and a variety of forms such as oil pipeline vandalism and scooping, and oil terrorism. In fact, recurrent issues over pipeline vandalism in different parts of the country can translate into financial losses sometimes running into a billion dollars, or more.

These actions can results in the spillage of crude oil (or crude oil products) because the taking away of oil by whatever means and diverting for personal benefit, therefore intentionally destroying pipelines, platforms, loading bays and other facilities.

4.3 Entry into the Environment and Toxicity

Environmental contaminants are chemicals (of which the constituents of crude oil and crude oil products are example) that accidentally or deliberately enter the environment and are often, but not always, as a result of human activities. Some of these contaminants may have been manufactured for industrial use and because they are very stable, they do not break down easily.

The main pathway for contaminants to enter the environment from is, as for many other sources, through release into the an ecosystem. A large number of substances have been detected—most at low concentrations—in the form of gases, particulate matter, liquids, semi-solids, and solids. Among the emitted contaminants are compounds containing metals and a large number of organic compounds as well as oxides of nitrogen (NO_x, where x = 1 or 2), sulfur (SO_x, where x = 2 or 3), and carbon (CO_x, were x = 1 or 2). As the (gas, liquid, or solid) which move through the environment to points of contact with the flora and fauna of an ecosystem.

In the current context, key sources of such contamination at crude oil refineries are at (1) transfer and distribution points in tankage and process areas, also general loading and unloading areas, (2) land farm areas, (3) tank farms, (4) individual above-ground storage tanks and particularly individual underground storage tanks, (5) additive compounds, and (6) pipelines, drainage areas as well as on-site waste treatment facilities, impounding basins, lagoons, especially if the basins and lagoons are unlined.

Spills of crude oil from tankers, pipelines, and oil wells represent catastrophic releases of a large volume of pollutants in a short period of time. There are also lesser amounts of crude oil-based pollutants that come from many different places over a long period of time that add up to large-scale effects. The spilled material tends to collect in hazardous concentrations in the soil or in groundwater. The effects spill must be mitigated in order to preserve the environment.

Large quantities of environmentally-sensitive crude oil products are stored in (1) tank farms (multiple tanks), (2) single above-ground storage tanks (ASTs), (3) semi-underground storage tanks or underground storage tanks (USTs). Smaller quantities of materials may be stored in drums and containers of assorted compounds (such as lubricating oil, engine oil, and other products for domestic supply).

Leaking underground storage tanks have been and, in some countries, continue to be a serious threat to groundwater supplies, especially if these tanks are located in sensitive ground water areas. Sensitive ground water areas are areas near public or private water supply wells and areas of aquifer recharge.

Problems resulting from leaking tanks can include: surface water contamination, surface and subsurface soil contamination and property damage. Property damage can include seepage into buildings and damage to buried telephone conduits. Vapor leakage from underground tanks depends upon soil and weather conditions. Leaking vapors can cause health problems ranging from nausea to respiratory distress. The potential for combustion and fire is a concern for tanks located next to buildings since vapors may leak into the building. Other less frequent problems associated with underground storage tanks include spills from overfilling the tank. This problem, if caught quickly, may involve only minor cleanup.

Leaks from underground storage tanks result from defects in tank material, improper fittings, improper installation, and damage during installation, corrosion or mechanical failure of the pipes and fittings or tanks which are improperly abandoned or removed. Older tanks, especially those constructed of bare steel and those tanks located in highly corrosive areas, are more susceptible to leakage problems.

Thus, it is necessary to consider (1) secondary containment of tanks and other storage areas and integrity of hard standing (without cracks, impervious surface) to prevent spills reaching the wider environment: also secondary containment of pipelines where appropriate, (2) age, construction details and testing program of tanks, (3) labeling and environmentally secure storage of drums (including waste storage), (4) accident/fire precautions, emergency procedures, and (5) disposal/ recycling of waste or "out of spec" oils and other materials.

There is a potential for significant soil and groundwater contamination to have arisen at crude oil refineries. Such contamination consists of (1) crude oil hydrocarbon derivatives including lower boiling, very mobile fractions (paraffins, cycloparaffins and volatile aromatics such as benzene, toluene, ethylbenzene, and the isomers of xylene) typically associated with gasoline and similar boiling range distillates, (2) middle distillate fractions (paraffins, cycloparaffins and some polynuclear aromatics) associated with diesel, kerosene, and lower boiling fuel oil, which are also of significant mobility, (3) higher boiling distillates (long-chain paraffins, cycloparaffins and polynuclear aromatics that are associated with lubricating oil and heavy fuel oil, (4) various organic compounds associated with crude oil hydrocarbon derivatives or produced during the refining process, e.g., phenols, amines, amides, alcohols, organic acids, nitrogen and sulfur containing compounds, (5) other organic additives, e.g., anti-freeze (glycols), alcohols, detergents and various proprietary compounds, (6) organic lead, associated with leaded gasoline and other heavy metals.

Wastewaters may be collected in separate drainage systems (for process, sanitary and storm water) although industrial and storm water systems may in some cases be combined. In addition, ballast water from bulk crude tankers may be pumped to receiving facilities at the refinery site prior to removal of floating oil in an interceptor and treatment as for other wastewater streams.

On-site treatment facilities may exist for wastewater or treatment may take place at a public wastewater treatment plant. Storm water/process water is generally passed to a separator or interceptor prior to leaving the site which takes out free-phase oil (i.e., floating product) from the water prior to discharge, or prior to further treatment, e.g. in settling lagoons).

Discharge from wastewater treatment plants is usually passed to a nearby watercourse.

Other wastes that are typical of a refinery include (1) waste oils, process chemicals, still resides, (2) non-specification chemicals and/or products, (3) waste alkali (sodium hydroxide), (4) waste oil sludge (from interceptors, tanks, and lagoons), and (5) solid wastes (cartons, rags, catalysts, and coke).

Key sources of such contamination at crude oil refineries are at (1) transfer and distribution points in tankage and process areas, also general loading and unloading areas, (2) land farm areas, (3) tank farms, (4) individual above-ground storage tanks and particularly individual underground storage tanks, (5) additive compounds, and (6) pipelines, drainage areas as well as on-site waste treatment facilities, impounding basins, lagoons, especially if unlined.

Whilst contamination may be associated with specific facilities the contaminants are relatively highly mobile in nature and have the potential to migrate significant distances from the source in soil and groundwater. Crude oil hydrocarbon contamination can take several forms such as (1) a free-phase product, (2) a dissolved-phase product, (3) an emulsified phase product, or (4) a vapor phase product. Each form will require different methods of remediation so that clean-up may be complex and expensive. In addition, crude oil hydrocarbon derivatives include a number of compounds of significant toxicity, e.g., benzene and some polynuclear aromatic derivatives are known carcinogens. Vapor phase contamination can be of significance in terms of odor issues.

Due to the obvious risk of fire, refineries are equipped with sprinkler or spray systems that may draw upon the main supply of water, or water held in lagoons, or from reservoirs or neighboring water courses. Such water will be polluting and require containment.

Refining facilities require significant volumes of water for on-site processes (e.g., coolants, blow-downs, etc.) as well as for sanitary and potable use. Wastewater will derive from these sources (process water) and from storm water run-off. The latter could contain significant concentrations of crude oil product.

Crude oil hydrocarbon derivatives, either dissolved, emulsified or occurring as free-phase will be the key constituents although wastewater may also contain significant concentrations of phenols, amines, amides, alcohols, ammonia, sulfide, heavy metals and suspended solids.

Crude oil and crude oil products released into the environment undergo weathering processes with time. These processes include evaporation, leaching (transfer to the aqueous phase) through solution and entrainment (physical transport along with the aqueous phase), chemical oxidation, and microbial degradation. The rate of weathering is highly dependent on environmental conditions. For example, gasoline, a volatile product, will evaporate readily in a surface spill, while gasoline released below 10 feet of clay topped with asphalt will tend to evaporate slowly (weathering processes may not be detectable for years).

An understanding of weathering processes is valuable to environmental test laboratories. Weathering changes product composition and may affect testing results, the ability to bio-remediate, and the toxicity of the spilled product. Unfortunately, the database available on the composition of weathered products is limited.

However, biodegradation processes, which influence the presence and the analysis of crude oil hydrocarbon at a particular site, can be very complex. The extent of biodegradation is dependent on many factors including the type of microorganisms present, environmental conditions (e.g., temperature, oxygen levels, and moisture), and the predominant hydrocarbon types. In fact, the primary factor controlling the extent of biodegradation is the molecular composition of the crude oil contaminant. Multiple ring cycloalkane derivatives are hard to degrade, while polynuclear aromatic hydrocarbon derivatives display varying degrees of degradation. Straight-chain alkane derivatives biodegrade rapidly with branched alkane derivatives and single saturated ring compounds degrading more slowly.

4.3.1. Entry into the Environment

The primary processes determining the fate of crude oils and oil products after a spill are (1) dispersion, (2) dissolution, (3) emulsification, (4) evaporation, (5) leaching, (6) sedimentation, (7) spreading, and (8) wind. These processes are influenced by the spill characteristics, environmental conditions, and physicochemical properties of the spilled material.

4.3.1.1 Dispersion

The physical transport of oil droplets into the water column is referred to as dispersion. This is often a result of water surface turbulence, but also may result from the application of chemical agents (dispersants). These droplets may remain in the water column or coalesce with other droplets and gain enough buoyancy to resurface. Dispersed oil tends to biodegrade and dissolve more rapidly than floating slicks because of high surface area relative to volume. Most of this process occurs from about half an hour to half a day after the spill.

4.3.1.2 Dissolution

Dissolution is the loss of individual oil compounds into the water. Many of the acutely toxic components of oils such as benzene, toluene, and the xylene isomers will readily dissolve into water. This process also occurs quickly after a discharge, but tends to be less important than evaporation. In a typical marine discharge, generally less than 5 percent of the benzene is lost to dissolution while greater than 95 percent is lost to evaporation. For alkylated polynuclear aromatic compounds, solubility is inversely proportional to the number of rings and extent of alkylation. The dissolution process is thought to be much more important in rivers because natural containment may prevent spreading, reducing the surface area of the slick and thus retarding evaporation. At the same time, river turbulence increases the potential for mixing and dissolution. Most of this process occurs within the first hour of the spill.

Aromatics, and especially BTEX, tend to be the most water-soluble fraction of crude oil. Crude oil contaminated groundwater tends to be enriched in aromatics relative to other crude oil constituents. Relatively insoluble hydrocarbon derivatives may be entrained in water through adsorption into kaolinite particles suspended in the water or as an agglomeration of oil droplets (microemulsion). In cases where groundwater contains only dissolved hydrocarbon derivatives, it may not be possible to identify the original crude oil product because only a portion of the free product will be present in the dissolved phase. As whole product floats on groundwater, the free product will gradually lose the water-soluble compounds. Groundwater containing entrained product will have a gas chromatographic fingerprint that is a combination of the free product chromatogram plus enhanced amounts of the soluble aromatics.

Generally, dissolved aromatics may be found quite far from the origin of a spill but entrained hydrocarbon derivatives may be found in water close to the crude oil source. Oxygenates, such as methyl-t-butyl ether (MTBE), are even more water soluble than aromatics and are highly mobile in the environment.

4.3.1.3 Emulsification

Certain oils tend to form water-in-oil emulsions (where water is incorporated into oil) or "mousse" as weathering occurs. This process is significant because, for example, the apparent volume of the oil may increase dramatically, and the emulsification will slow the other weathering processes, especially evaporation. Under certain conditions, these emulsions may separate and release relatively fresh oil. Most of this process occurs from about half a day to two days after the spill.

4.3.1.4 Evaporation

Evaporative processes are very important in the weathering of volatile crude oil products, and may be the dominant weathering process for gasoline. Automotive gasoline, aviation gasoline, and some grades of jet fuel (e.g., JP-4) contain 20% to 99% v/v highly volatile constituents (i.e., constituents with less than nine carbon atoms).

Evaporative processes begin immediately after oil is discharged into the environment. Some light products (like 1- to 2-ring aromatic hydrocarbon derivatives and/or low molecular weight alkane derivatives less than n-C15) may evaporate entirely; a significant fraction of heavy refined oils also may evaporate. For crude oils, the amount lost to evaporation can typically range from approximately 20 to 60 percent. The primary factors that control evaporation are the composition of the oil, slick thickness, temperature and solar radiation, wind speed and wave height. While evaporation rates increase with temperature, this process is not restricted to warm climates. For the Exxon Valdez incident, which occurred in cold conditions (March, 1989), it has been estimated that appreciable evaporation occurred even before all the oil escaped from the ship, and that evaporation ultimately accounted for 20 percent of the oil. Most of this process occurs within the first few days after the spill.

It is not unusual for evaporative processes, however, to be working simultaneously with other processes to remove the volatile aromatics such as benzene and toluene.

4.3.1.5 Leaching

Leaching processes introduce hydrocarbon into the water phase by solubility and entrainment. Leaching processes of crude oil products in soils can have a variety of potential scenarios. Part of the aromatic fraction of a crude oil spill in soil may partition into water that has been in contact with the contamination.

4.3.1.6 Sedimentation or Adsorption

As mentioned above, most oils are buoyant in water. However, in areas with high levels of suspended sediment, crude oil constituents may be transported to the river, lake, or ocean floor through the process of sedimentation. Oil may adsorb to sediments and sink or be ingested by zooplankton and excreted in fecal pellets that may settle to the bottom. Oil stranded on shorelines also may pick up sediments, float with the tide, and then sink. Most of this process occurs from about two to seven days after the spill.

4.3.1.7 Spreading

As oil enters the environment, it begins to spread immediately. The viscosity of the oil, its pour point, and the ambient temperature will determine how rapidly the oil will spread, but light oils typically spread more rapidly than heavy oils. The rate of spreading and ultimate thickness of the oil slick will affect the rates of the other weathering processes. For example, discharges that occur in geographically contained areas (such as a pond or slow-moving stream) will evaporate more slowly than if the oil were allowed to spread. Most of this process occurs within the first week after the spill.

4.3.1.8 Wind

Wind (Aeolian) transport (relocation by wind) can also occur and is particular relevant when catalyst dust and coke dust are considered. Dust becomes airborne when winds traversing arid land with little vegetation cover pick up small particles such as catalyst dust, coke dust and other refinery debris and send them skyward. Wind transport may occur through *suspension, saltation,* or *creep* of the particles.

4.3.2 Toxicity

With few exceptions, the constituents of crude oil, crude oil products, and the various emissions are hazardous to the health. There always exceptions that will be cited in opposition to such a statement, the most common exception being the liquid paraffin that is used medicinally to lubricate the alimentary tract. The use of such medication is common among miners who breathe and swallow coal dust every day during their work shifts.

Another approach is to consider crude oil constituents in terms of transportable materials, the character of which is determined by several chemical and physical properties (i.e., solubility, vapor pressure, and propensity to bind with soil and organic particles). These properties are the basis of measures of leachability and volatility of individual hydrocarbon derivatives. Thus, crude oil transport fractions can be considered by equivalent carbon number to be grouped into thirteen different fractions. The analytical fractions are then set to match these transport fractions, using specific *n*-alkane derivatives to mark the analytical results for aliphatic compounds and selected aromatic compounds to delineate hydrocarbon derivatives containing benzene rings.

Although chemicals grouped by transport fraction generally have similar toxicological properties, this is not always the case. For example, benzene is a carcinogen but many alkyl-substituted benzenes do not fall under this classification. However, it is more appropriate to group benzene with compounds that have similar environmental transport properties than to group it with other carcinogens such as benzo(a)pyrene that have very different environmental transport properties.

Nevertheless consultation of any reference work that lists the properties of chemicals will show the properties and hazardous nature of the types of chemicals that are found in crude oil. In addition, crude oil is used to make crude oil products, which can contaminate the environment.

The range of chemicals in crude oil and crude oil products is so vast that summarizing the properties and/or the toxicity or general hazard of crude oil in general or even for a specific crude oil is a difficult task. However, crude oil and some crude oil products, because of the hydrocarbon content, are at least theoretically biodegradable but large-scale spills can overwhelm the ability of the ecosystem to break the oil down. The toxicological implications from crude oil occur primarily from exposure to or biological metabolism of aromatic structures. These implications change as an oil spill ages or is weathered.

4.3.2.1 Lower Boiling Constituents

Many of the gaseous and liquid constituents of the lower boiling fractions of crude oil and also in crude oil products fall into the class of chemicals which have the one or more of the following characteristics are considered to be hazardous by the Environmental Protection agency.

Ignitability-Flammability

A liquid that has a flash point of less than 60°C (140°F) is considered ignitable. Some examples are: benzene, hexane, heptane, benzene, pentane, crude oil ether (low boiling), toluene, and the xylene isomers.

Corrosivity

An aqueous solution that has a pH of less than or equal to 2, or greater than or equal to 12.5 is considered corrosive. Most crude oil constituents and crude oil products are not corrosive but many of the chemicals used in refineries are corrosive. Corrosive materials also include substances such as sodium hydroxide and some other acids or bases.

Reactivity

Chemicals that react violently with air or water are considered hazardous. Examples are sodium metal, potassium metal, phosphorus, etc. Reactive materials also include strong oxidizers such as perchloric acid, and chemicals capable of detonation when subjected to an initiating source, such as solid, dry ($< 10\% H_2O$ v/v) picric acid, benzoyl peroxide or sodium borohydride. Solutions of certain cyanide or sulfides that could generate toxic gases are also classified as reactive.

The potential for finding such chemicals in a refinery is subject to the function and product slate of the refinery and/or the petrochemical complex.

Hazardous Chemicals

Many chemicals have been shown in scientific studies to have toxic, carcinogenic, mutagenic or teratogenic effects on humans or other life forms and are designated either as *Acutely Hazardous Waste* or *Toxic Waste* by the Environmental Protection Agency. Substances found to be fatal to humans in low doses or, in the absence of data on human toxicity, have been shown in studies to have an oral LD_{50} toxicity (rat) of less than 2 milligrams per liter, or a dermal LD_{50} toxicity (rabbit) of less than 200 milligrams per kilogram or is otherwise capable of causing or significantly contributing to an increase in serious irreversible, or incapacitating reversible illness are designated as Acute Hazardous Waste. Materials containing any of the toxic constituents so listed are to be considered hazardous waste, unless, after considering the following factors it can reasonably be concluded (by the Department of Environmental Health and Safety) that the waste is not capable of posing a substantial present or potential hazard to public health or the environment when improperly treated, stored, transported or disposed of, or otherwise managed.

The issues to be held in consideration are (1) the nature of the toxicity presented by the constituent, (2) the concentration of the constituent in the waste, (3) the potential of the constituent or any toxic degradation product of the constituent to migrate from the waste into the environment under various types of improper management, (4) the persistence of the constituent or any toxic degradation product of the constituent, (5) the potential for the constituent or any toxic degradation product of the constituent to degrade into non-harmful constituents and the rate of degradation, (6) the degree to which the constituent or any degradation product of the constituent accumulates in an ecosystem, (7) the plausible types of improper management to which the waste could be subjected, (8) the quantities of the waste generated at individual generation sites or on a regional or national basis, (9) the nature and severity of the public health threat and environmental damage that has occurred as a result of the improper management of wastes containing the constituent, and (10) actions taken by other governmental agencies or regulatory programs based on the health or environmental hazard posed by the waste or waste constituent. Other factors that may be appropriate may also be considered.

For the analysts, laboratories wishing to dispose of materials containing dilute concentrations of these constituents should contact the Department of Environmental Health and Safety for advice regarding the proper disposition of the materials. In addition, the list of such materials is not included here as it is subject to periodic updates. Furthermore, the list is not meant to be complete and may not include substances that have the hazardous characteristics as defined above. Omission of a chemical from this list does not mean it is without toxic properties or any other hazard.

More specifically to crude oil and crude oil products, the alkane derivatives in gasoline and some other crude oil products are CNS depressants. In fact, gasoline was once evaluated as an anesthetic agent. However, sudden deaths, possibly as a result of irregular heartbeats, have been attributed to those inhaling vapors of hydrocarbon derivatives such as those in gasoline.

Alkane derivatives of various types of crude oils and various crude oil products were biodegraded faster than the *unresolved fractions*. Different types of crude oils and products biodegraded at different rates in the same environments. An oil product is a complex mixture of organic chemicals and contains within it less persistent and more persistent fractions. The range between these two extremes is greatest for crude oils. Since the many different substances in crude oil have different physical and chemical properties, summarizing the fate of crude oil in general (or even a particular oil) is very difficult. Solubility-fate relationships must be considered.

The relative proportion of hazardous constituents present in crude oil is typically quite variable. Therefore, contamination will vary from one site to another. In addition, the farther one progresses from lighter towards heavier constituents (the general progression from lower molecular weight to higher molecular weight constituents) the greater the percentage of polynuclear aromatic hydrocarbon derivatives and other semi-volatile constituents or non-volatile constituents (many of which are not so immediately toxic as the volatiles but which can result in long-term/chronic impacts). These higher molecular weight constituents thus need to be analyzed for the semi volatile compounds that typically pose the greatest long-term risk.

In addition to large oil spills, crude oil hydrocarbon derivatives are released into the aquatic environments from natural seeps as well as non-point source urban runoffs. Acute impacts from massive one-time spills are obvious and substantial. The impacts from small spills and chronic releases are the subject of much speculation and continued research. Clearly, these inputs of crude oil hydrocarbon derivatives have the potential for significant environmental impacts, but the effects of chronic low-level discharges can be minimized by the net assimilative capacities of many ecosystems, resulting in little detectable environmental harm.

Short-term (acute) hazards of lighter, more volatile and water soluble aromatic compounds (such as benzenes, toluene, and the isomers of xylene) include potential acute toxicity to aquatic life in the water column (especially in relatively confined areas) as well as potential inhalation hazards. However, the compounds which pass through the water column often tend to do so in small concentrations and/or for short periods of time, and fish and other pelagic or generally mobile species can often swim away to avoid impacts from spilled oil in open waters. Most fish are mobile and it is not known whether or not they can sense, and thus avoid, toxic concentrations of oil.

However, there are some potential effects of spilled oil on fish. The impacts to fish are primarily to the eggs and larvae, with limited effects on the adults. The sensitivity varies by species; pink salmon fry are affected by exposure to water-soluble fractions of crude oil, while pink salmon pink salmon eggs are very tolerant to benzene and water-soluble crude oil. The general effects are difficult to assess and quantitatively document due to the seasonal and natural variability of the species. Fish rapidly metabolize aromatic hydrocarbon derivatives due to their enzyme system.

Long-term (chronic) potential hazards of lighter, more volatile and water soluble aromatic compounds include contamination of groundwater. Chronic effects of benzene, toluene, and the isomers of xylene include changes in the liver and harmful effects on the kidneys, heart, lungs, and nervous system.

At the initial stages of a release, when the benzene-derived compounds are present at their highest concentrations, acute toxic effects are more common than later. These non-carcinogenic effects include subtle changes in detoxifying enzymes and liver damage. Generally, the relative aquatic acute toxicity of crude oil will be the result of the fractional toxicities of the different hydrocarbon derivatives present in the aqueous phase. Tests indicate that naphthalene-derived chemicals have a similar effect.

Except for short-term hazards from concentrated spills, BTEX compounds (benzene, toluene, ethyl benzene, and the isomers of xylene) have been more frequently associated with risk to humans than with risk to non-human species such as fish and wildlife. This is partly because plants, fish, and birds take up only very small amounts and because this volatile compound tends to evaporate into the atmosphere rather than persisting in surface waters or soils. However, volatiles such as this compound have can pose a drinking water hazard when they accumulate in ground water. See also, BTEX entry, and entries for benzene, toluene, ethyl benzene, and the isomers of xylene.

Crude oil is naturally weathered according to its physical and chemical properties, but during this process living species within the local environment may be affected via one or more routes of exposure, including ingestion, inhalation, dermal contact, and, to a much lesser extent, bioconcentration through the food chain. Aromatic compounds of concern include alkylbenzene derivatives, toluene, naphthalene derivatives, and polynuclear aromatic hydrocarbon derivatives (PNAs). Moreover, both atmospheric and hydrospheric impacts must be assessed when considering toxic implications from a crude oil release containing significant quantities of these single-ring aromatic compounds.

4.3.2.2 Higher Boiling Constituents

Naphthalene and its homologs are less acutely toxic than benzene but are more prevalent for a longer period during oil spills. The toxicity of different crude oils and refined oils depends on not only the total concentration of hydrocarbon derivatives but also the hydrocarbon composition in the water-soluble fraction (WSF) of crude oil, water solubility, concentrations of individual components, and toxicity of the components. The water-soluble fractions prepared from different oils will vary in these parameters. Water-soluble fractions (WSF) of refined oils (for example, No. 2 fuel oil and Bunker C oil) are more toxic than water-soluble fraction of crude oil to several species of fish (killifish and salmon). Compounds with either more rings or methyl substitutions are more toxic than less substituted compounds, but tend to be less water soluble and thus less plentiful in the water-soluble fraction.

Among the polynuclear aromatic hydrocarbon derivatives, the toxicity of crude oil is a function of its di- and tri-aromatic hydrocarbon content. Like the single aromatic ring variations, including benzene, toluene, and the xylene isomers, all are relatively volatile compounds with varying degrees of water solubility.

There are indications that pure naphthalene (a constituent of moth balls that are, by definition, toxic to moths) and alkyl naphthalene derivatives are from three-to-ten times more toxic to test animals than are benzene and alkyl benzene derivatives. In addition, and because of the low water-solubility of tricyclic and polycyclic (polynuclear) aromatic hydrocarbon derivatives (that is, those aromatic hydrocarbon derivatives heavier than naphthalene), these compounds are generally present at very low concentrations in the water-soluble fraction of oil. Therefore, the results of this study and others conclude that the soluble aromatics of crude oil (such as benzene, toluene, ethylbenzene, xylene isomers and derivatives, and naphthalene derivatives) produce the majority of the toxic effects in the environment.

Once the acutely toxic lighter compounds have been left the aquatic environment through volatilization or degradation, the main concern is chronic effects from heavier and more alkylated polynuclear aromatic hydrocarbon derivatives.

Bird species with water habitats are the species most commonly affected by oil spills and releases. Oil itself breaks down the protective waxes and oils in the feathers and fur of birds and animals and disrupts the fine strand structure of the feathers resulting in a loss of heat retention and buoyancy and possible hypothermia and death. Oiled birds often ingest crude oil while attempting to remove the crude oil from their feathers. The effects of ingested crude oil include anemia, pneumonia, kidney, and liver damage, decreased growth, altered blood chemistry, and decreased egg production and viability. Ingestion of the oil can also kill animals by interfering with their ability to digest food. Chicks may be exposed to crude oil by ingesting food regurgitated by impacted adults.

The dynamics of the oil-in-water dispersion (OWD) are complex and have relevance related to potential toxicity or hazard. In comparing the toxicities to marine animals of oil-in-water dispersions prepared from different oils, not only the amount of oil added but also the concentrations of oil in the aqueous phase and the composition and dispersion-forming characteristics of the parent oil must be taken into consideration. In comparing the potential impacts of spills of different oils on the marine biotic community, the amount of oil per unit water volume required to cause mortality is of greater importance than any other aspect of the crude oil behavior.

Several compounds in crude oil products are carcinogenic. The larger and higher molecular weight aromatic structures (with four to five aromatic rings), which are the more persistent in the environment, have the potential for chronic toxicological effects. Since these compounds are non-volatile and are relatively insoluble in water, their main routes of exposure are through ingestion and epidermal contact. Some of the compounds in this classification are considered possible human carcinogens; these include benzo(a and e)pyrene, benzo(a)anthracene, benzo(b, j, and k)fluorene, benzo(ghi)perylene, chrysene, dibenzo(ah)anthracene, and pyrene.

Mixtures of polynuclear aromatic hydrocarbon derivatives are often carcinogenic and possibly phototoxic. One way to approach site specific risk assessments would be to collect the complex mixture of polynuclear aromatic hydrocarbon derivatives and other lipophilic contaminants in a semi-permeable membrane device (SPMD, also known as a *fat bag*), then test the mixture for carcinogenicity, toxicity, and phototoxicity.

The solubility of hydrocarbon components in crude oil products is an important property when assessing toxicity. The water solubility of a substance determines the routes of exposure that are possible. Solubility is approximately inversely proportional to molecular weight—lower molecular weight hydrocarbon derivatives are more soluble in water than higher molecular weight compounds. Lower molecular weight hydrocarbon derivatives (C4 to C8, including the aromatic compounds) are relatively soluble, up to about 2,000 ppm, while the higher molecular weight hydrocarbon derivatives are nearly insoluble. Usually, the most soluble components are also the most toxic.

Finally, the toxicity of crude oil may be affected by factors such as "weathering" time or the addition of oil dispersants. *Weathered* crude oil and *fresh* crude oil may have different toxicities, depending on oil type and weathering time.

4.3.2.3 Wastewater

A number of wastewater issues face the refining industry. These issues include chemicals in waste process waters. However, efforts by the industry are being continued to eliminate any water contamination that may occur, whether it be from inadvertent leakage of crude oil or crude oil products or leakage of contaminated water from one or more processes. In addition to monitoring organics in the water, metals concentration must be continually monitored since heavy metals tend to concentrate in the body tissues of fish and animals and increase in concentration as they go up the food chain. General sewage problems face every municipal sewage treatment facility regardless of size.

Primary treatment (solid settling and removal) is required and secondary treatment (use of bacteria and aeration to enhance organic degradation) is becoming more routine, tertiary treatment (filtration through activated carbon, applications of ozone, and chlorination) have been, or are being, implemented by all refineries.

Wastewater pretreating units that discharge water into sewer systems have new requirements. Pollutant standards for sewage sludge have been set. Toxics in the water must be identified and plans must be developed to alleviate any problems. In addition, regulators have established, and continue to establish, water-quality standards for priority toxic pollutants.

4.4 General Methods for Soil and Groundwater Remediation

By definition, environmental remediation deals with the removal of pollutants or contaminants from groundwater, surface water, soil, and sediment (Chapter 4). The term water remediation refers to the processes by which contaminants are biologically removed from surface water and from ground water. In a similar manner, the term soil remediation refers to processes by which contaminants are biologically removed from surface water and from soils (which includes topsoil, subsoil, and sediment).

Environmental remediation is necessary to safeguard people's social, economic and physical health and wellbeing while balancing risk management and cost control to provide the best path forward in site rehabilitation. Moreover, human economic systems are not always sufficiently adequate to deliver an acceptable balance of environmental goals and social goals and remediation actions need to be justified and optimized. In addition, remedial action is generally subjected to a variety of regulatory requirements which are based on assessments of the risks to the environment and to human health.

The goal of remediation is to reduce the quantity, concentration, mobility and toxicity of the contaminants that are present at a site and typically begins with an assessment of the site to determine the best applicable (or available) technology (either of which are often shortened to the acronym BAT) that would be the most appropriate for the site. The level of effort and resources expended to gather and integrate the information for remediation and/management decision-making need to be commensurate with (1) the risk assessment, which is the likelihood and level of the risk, (2) the severity of the consequences taking into consideration the sensitivity of the site use, (3) the environmental setting, extent, mobility and nature or complexity of the contamination, and (4) any regulatory requirements.

In addition to effects on human health and the environment, the following issues may need to be considered prior to deciding the remediation objectives (1) the effects of contamination, including toxicity, (2) the potential for bioaccumulation and persistence, (3) the potential risk posed by residual contamination, (4) the effectiveness and acceptability of any controls that might be involved, (5) the expected effectiveness, practicability and outcome of the proposed remediation, and (6) the management strategy for site continuance.

In order for the remediation technology to be successful, the technology must have satisfactory technical performance, commercial viability and must appeal to the law and environmental officials and also to the public. Thus, the site is assessed to determine its existing state and detailed data about the area is collected and, once the site assessment is complete, the next step is to categorize the collected information and begin to strategize a course of action that will bring the desired results. Once data has been gathered, a plan of the estimated time and budget required for the entire project can be developed.

Documentation of the entire process is undertaken in order to make available a clear understanding of the whole project as well to provide a permanent record of the entire site rehabilitation project as well as the actual work of remediation and mitigation to be performed. The documentation must as include any post-project monitoring for factors such as the presence of contaminants. Once the site has been mitigated and remediated it can be redeveloped, often referred to as a brownfield site.

Environmental remediation may also be subdivided into (1) *in situ* processes and (2) *ex situ* processes. Using soil or sediment as the example, the *in situ* processes for soil remediation or sediment remediation are used to treat the contaminant(s) in place without removing the soil from its position on the site. On the other hand, *ex situ* processes involve (1) removal of the soil or sediment by excavation, (2) treating the soil or sediment, and (3) returning the soil or sediment to the original site.

4.4.1 Bioremediation

Bioremediation is a process that intensifies the actions of natural biological species to remediate polluted groundwater and contaminated soil environment consequently restoring the original natural surroundings and preventing further pollution. The process uses biological microbes to do the cleanup work.

In terms of a spillage of crude oil or a crude oil product, the microbes attack the spill thereby converting the crude oil (or the crude oil product) into water and gases which are then released (by the microbes) into the soil or into the ground water. Thus:

Crude oil (or product) –microbial attack → water + gases

Water + gases–microbial release → into soil

Water + gases–microbial release → into groundwater

However, an important factor in selecting bioremediation as the appropriate process for site remediation relates to the potential for the contaminants to be susceptible to biodegradation by the biological organisms (microbes) at the site. While the process is useful for the degradation of many types of organic compounds and is a natural process, if the process is not controlled it is possible that the organic contaminants may not be suitably degraded and, in such a case, the result could be the production of toxic by-products that have greater mobility than the initial contaminants.

There are two main classifications of bioremediation and are (1) the *in situ* method, which does not require any excavation soils so may be less expensive and cause less release of contaminants, and (2) the *ex situ* method (Figure 4.2) (Azubuike et al., 2016). The *in situ* bioremediation process relies on the indigenous microbial flora of subsurface soils and groundwater which have adapted to the organic chemical wastes found there and are able to degrade some or all the components of these wastes.

On the other hand, *ex situ* remediation techniques involve the physical removal of the contaminated material for the treatment process and are typically based on (1) the cost of treatment, (2) the depth of the pollutant, (3) the type of the pollutant, (4) the degree and the extent of the of pollution, (5) the geographical location, (6) the geology of the polluted site, and (7) the performance standards of the method. *Ex situ* remediation includes techniques include methods such as land

Method	Method	Sub-method
In situ	Natural	
	Enhanced	Bioslurping
		Bioventing
		Biosparging
		Phytoremediation
Ex situ	Biopile	
	Windrow	
	Bioreactor	

Figure 4.2: Schematic Illustration of the Various Bioremediation Methods.

farming, biopiling, and processing by bioreactors during chemical, physical and thermal processes. Typically, groundwater contamination is treated in-situ but if the necessary soil conditions in the soil cannot be achieved, *in situ* treatment may not be possible and the soil may have to be excavated and treated *ex situ*.

Since, the feasibility of bioremediation depends on the location of contaminants; there are two basic bioremediation techniques that can be applied which are (1) biostimulation and (2) bioaugmentation. Biostimulation provides nutrients and suitable physiological conditions for the growth of the indigenous microbial populations while, on the other hand, bioaugmentation allows for the addition of oil-degrading micro-organisms to supplement the indigenous populations present in the contaminated site.

Certain micro-organisms are capable of removing toxic chemicals and pathogens by changing the composition of the contaminants (such as crude oil and crude oil products) into gases, such as like methane and carbon dioxide that can be represented by the simple equations (Sharma and Reddy, 2004):

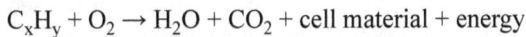

$$C_xH_y + O_2 \rightarrow CH_4 + CO_2$$

$$C_xH_y + O_2 \rightarrow H_2O + CO_2 + \text{cell material} + \text{energy}$$

However, it must be recognized that both methane and carbon dioxide are members of the group referred to as greenhouse gases and methane is a much more potent gas than carbon dioxide. What is not often recognized is that methane is emitted by human activities as well as by natural sources such as from natural wetlands and by the raising of livestock (such as by the flatulence of cattle which produces copious amounts of methane) is often ignored as a contributor to global warming and to global climate change (Speight, 2020). In fact, on a unit weight basis, the comparative impact of methane is on the order of 25 times greater than the impact of carbon dioxide over a 100-year period (IPCC, 2007, 2013) and steps must be taken to trap these gases rather than allow deliberate or inadvertent release to the atmosphere.

The factors that can affect bioremediation are as (1) the concentration of the contaminants, (2) the contaminant bioavailability, (3) the site characteristics, (4) the acidity or the alkalinity, i.e., pH, (5) the redox conditions, (6) the potential nutrients, and (7) the temperature.

In aerobic conditions, micro-organisms (such as Pseudomonas, Alcaligenes, Sphingomonas, Rhodococcus, and Mycobacterium) use the available oxygen to convert organic contaminants to carbon dioxide and water. On the other hand, anaerobic conditions (i.e., conditions in which there is a deficiency or absence of oxygen) support biological activity in which little or no oxygen is present.

In the process, the remediation techniques pathways include transfer of contaminants either (1) alone or (2) with contaminated soils or groundwater to other place for final treatment or disposal, confinement, and destruction of the contaminants in place (Nyer, 1998; Mulligan et al., 2001). The technical principles for remediation can be divided into chemical processes, physical processes, and biological processes. These technologies are contained within five categories of general approaches to remediation which are (1) contaminant isolation, (2) contaminant immobilization, (3) toxicity reduction of the contaminant, (4) physical separation of the contaminant, and (5) extraction of the contaminant. In many cases, each of these approaches are inter-related and may be accomplished as one step. However, in order to select an appropriate soil remediation approach, it is necessary to recognize that the method is dependent on the nature of soil, feasibility of contaminant isolation, soil microbiome, cost and contaminant nature.

Remediation techniques which have been commonly used include (alphabetically rather than in any order of preference) the following (1) bioremediation, (2) containment, (3) dredging or excavation, (4) *in situ* oxidation, (5) nano-remediation, (6) pump and treat, (7) soil vapor extraction, (8) solidification and stabilization, (9) surfactant enhanced aquifer remediation, and (10) thermal desorption plus an extra section entitled other aspect of remediation, which focuses on the removal of metals from a spill site. A brief description of each of these processes is presented in the following sections.

4.4.2 Containment

Containment is a key technology for controlling the spread of a contaminant (or a mixture of contaminants as is often the case when crude oil or a crude oil product is spilled into the environment). The technology should be employed as quickly as possible following a spill in order to prevent or at least, significantly reduce the migration of contaminants in soils or groundwater. This can be achieved by controlling the flow of the fluid that carries the contaminant, or by directly immobilizing the contaminant with the use of a physical barrier. The type of barrier is site dependent and contaminant dependent—to prevent uncontrolled exposure to the contaminated soil, separation at the base of the containment cell to prevent infiltration or the loss of leachate and impacts to underlying soils and/or groundwater.

There are four major types of physical barriers which are (1) a slurry wall, which is a technique used to build an impermeable barrier in areas of soft (permeable) earth, (2) a grout curtain, which is constructed by drilling and grouting a linear sequence of holes to form a barrier that is used up-gradient of the contaminated area to prevent clean water from migrating through waste, or down-gradient, to limit migration of contaminants, (3) sheet piling, which is a type of type of retaining wall in which segments with indented profiles interlock to form a wall with alternating indents and outdents, and (4) booms, which act as physical barriers that enclose floating oil and prevent it from spreading.

In general, environmental conditions may limit effectiveness, as it is difficult to obtain uniform immobilization due to the natural heterogeneity of the subsurface. Therefore, various techniques (such as solidification, stabilization, encapsulation, and immobilization), which are based on injecting a solution containing a compound that will cause immobilization or encapsulation of the contaminant(s), can be used to immobilize subsurface contaminants by fixing them in an impermeable, immobile solid matrix.

The cost factors for containment relates to the types and quantities of construction necessary. Monitoring systems are necessary. For containment to be successful, there are some key challenges that must be addressed. These include uncertainties related to (1) the long-term durability of synthetic lining systems, (2) any unpredictable post-closure maintenance and monitoring costs, (3) any delay in the degradation of the waste where waste composition can vary over time, and (4) any excessive costs in construction and operation of containment landfills.

In terms of landfill sites, most management strategies take advantage of the natural hydrogeological characteristics and properties of the subsurface. Containments methods have been referred to alternatively as (1) solidification, (2) stabilization, (3) encapsulation, and (4) immobilization which are based on the injection of a solution containing a compound that will cause immobilization or encapsulation of the contaminant(s). However, it must be noted that the encapsulation process does not filter the contaminant(s) from the soil so much as it separates the contaminant(s).

4.4.3 Dredging or Excavation

The discharge of contaminants into the various waterways (ponds, lakes, streams, rivers, and oceans) is a threat to the floral systems and the faunal systems (including humans) that depend upon these waterways for their existence. In fact, some of the waterways in various countries are so polluted by contaminants that there is a constant danger to life in the water and an effective cleanup method is essential to remove the contaminant threat.

Dredging or excavation are methods that can be used to remove contaminated sediments from freshwater or from marine water bodies in order to reduce risks to human health and the environment.

The two most common methods of removing contaminated soil or sediment for a body of water are (1) excavation and (2) dredging either while the soil or sediment is submerged are the two most common means of removing contaminated sediment from a water body.

Excavation refers to sediment removal conducted after the water above the sediment has been removed. In an excavation process, a segment of the sediment and water column is isolated in an enclosure after which the enclosure is dewatered and the exposed sediment is removed using conventional land-based excavation equipment. Dredging is the form of excavation carried out underwater or partially underwater, in shallow waters or ocean waters and is a method in which sediment and other materials from the bottom of the water are removed by pumping them through pipelines and into a processing facility.

The nature of the material to be removed from a waterway by, for example, dredging decides the type of equipment or method to be adopted to carry out the work. In addition, other considerations that need to be evaluated include sediment removal, transport, staging, treatment (pretreatment, treatment of water and sediment, if necessary), and disposal (liquids and solids).

The selection of sediment removal from the water as a remedial approach should be based on an overall assessment using criteria appropriate for the specific site being investigated. Sometimes, a single site may use hydraulic dredging in some segments and mechanical dredging in another segment, in order to leverage the advantages of each.

The mechanical dredge equipment typically consists of (1) a bucket equipped with a cutting and grabbing edge, (2) a crane or other means of lowering, manipulating, and retrieving the bucket (with the dredge material) through the water column, (3) and transportation of the sediment (usually by a barge) from the dredging site to a sediment handling and processing facility or to a disposal facility.

The application of a dredging process or an excavation process relates to the advantage of removing contaminated sediment from the aquatic environment which achieves a designated cleanup level for the site. In fact, removing contaminated sediment allows for some flexibility regarding future use of the water body.

4.4.4 In situ Oxidation

The *in situ* process for chemical oxidation is a redox process that typically involves one or more reduction/oxidation (redox) reactions that chemically convert hazardous compounds to nonhazardous compounds or to less toxic compounds that are more stable, less mobile, or inert than the original contaminant(s).

By definition, a redox reaction (i.e., an oxidation-reduction) reaction is a type of chemical reaction that involves a transfer of electrons between two species in which there is a change in the oxidation number between the reactants and the products.

The oxidizing agents most commonly used for treatment of hazardous contaminants in soil and groundwater are chemicals such as hydrogen peroxide (H_2O_2), potassium permanganate ($KMnO_4$), sodium persulfate ($Na_2S_2O_8$), and ozone (O_3). While applicable to soil contamination and some source zone contamination, these oxidants have been applied to the remediation of groundwater. However, the permanganate-based *in situ* oxidation is more fully developed than process which use the other oxidants. In the process using the permanganate oxidant as the example, the oxidant causes the oxidation of another substance while it (the oxidant) is reduced in the process. Thus:

Reduction formula:

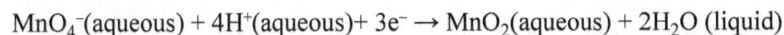

$$MnO_4^-(\text{aqueous}) + 4H^+(\text{aqueous}) + 3e^- \rightarrow MnO_2(\text{aqueous}) + 2H_2O \text{ (liquid)}$$

Hydrogen peroxide is one of the most powerful oxidizers known—stronger than chlorine (Cl_2), chlorine dioxide (ClO_2), and potassium permanganate ($KMnO_4$). By adjusting the reaction parameters (e.g., pH, temperature, dose, reaction time, and/or catalyst addition), hydrogen peroxide can often be made to oxidize one pollutant over another, or even to favor different oxidation products from the same pollutant.

Equation for reduction:

$$H_2O_2 + 2H^+ + 2e^- \rightarrow 2H_2O$$

The use of hydrogen peroxide to chemically oxidize soils contaminated with chemicals (such as crude oil residues, solvents, pesticides, and wood preservatives) present one of the more difficult challenges for remediation specialists. This method offers several advantages over these current treatment methods on the basis that the process involves Fenton's reagent (a solution of hydrogen peroxide, H_2O_2, with ferrous iron, typically iron (Fe^{2+}) sulfate, $FeSO_4$) whereby the transition metal catalyst is either provided by iron oxides within soil or added separately as a solubilized iron salt. The oxidation occurs both at the soil interface (in the case of dense non-aqueous phase liquids, DNAPLs) and within the interstitial groundwater (in the case of desorbed organics). Variables that the application of this technology are (1) the hydrogen peroxide/contaminant ratio, (2) the hydrogen peroxide concentration, (3) the soil type, (4) the reaction time, and (5) the contaminant type.

Ozone (O_3) has an oxidation potential much greater than hydrogen peroxide which can be used to destroy contamination *in situ* by crude oil constituents and by the constituents of crude oil products. Typically, the ozone needs to be generated near the treatment area as it is also unstable at high concentrations. It is generated artificially by electric generators that produce a high-voltage discharge in air or oxygen and can be delivered to the subsurface through sparge wells. Thus:

$$O_2 + \text{Electric Current} \rightarrow O_3$$

$$O_3 + 2H^+ + 2e^- \rightarrow O_2 + H_2O$$

The selection of ozone in comparison to potassium permanganate or hydrogen peroxide is based primarily on the fact that ozone is applied as a gas and is Thus, Ozone is also effective in delivering oxygen to enhance subsurface bioremediation of areas hat bare impacted by the spillage of crude oil or crude oil products.

Sodium persulfate ($Na_2S_2O_8$) is a chemical oxidant that is suitable for the treatment of soil and groundwater when contaminated with chemicals such as chlorinated alkene derivatives, chlorinated alkane derivatives, mixtures of benzene, toluene, ethylbenzene and the xylene isomers (BTEX), polynuclear hydrocarbon derivatives (PNAs or PAHs), polychlorobiphenyl derivatives (PCBs), as well as dichlorobenzene and tri-chlorobenzene. The persulfate is prepared by the electrolytic oxidation of sodium hydrogen sulfate

$$2NaHSO_4 \rightarrow Na_2S_2O_8 + H_2$$

The persulfate anion is one of the strongest oxidants used in remediation and the standard oxidation reduction equation is

$$S_2O_8^{2-} + 2e^- \rightarrow 2SO_4^{2-}$$

Chemical oxidation is typically conducted *in situ* in place and therefore does not require excavation of the soil or groundwater pump and treat. Also, the method is applicable to a wide range of contaminants which can be destroyed in-situ and is a relatively fast treatment that enhances post-oxidation microbial activity and natural attenuation.

4.4.5 Metals Removal

One aspect of remediation that is often omitted when dealing with chemical contaminants is the removal of metals from a contaminated site. Even though several processes might be necessary for the removal of metal contaminants from a site, the concept is worthy of mention here because of the importance of removing such contaminants. In fact, metals do occur in crude oil and although the predominant metals are nickel and vanadium (Speight, 2014, 2017) the presence of other metals is also a factor after a crude oil spill.

Another method is the solidification/stabilization method which contains the contaminants in an area by mixing or injecting agents, for example: (1) the process of solidification results in the encapsulation of contaminants in a solid matrix while (2) the stabilization process involves the formation of chemical bonds to reduce contaminant mobility. Another approach is size selection processes for removal of the larger, cleaner particles from the smaller more polluted ones. To accomplish this, several processes are used and include the use of (1) hydrocyclones, (2) fluidized bed separation, and (3) and flotation. Other process use the principle of electrokinetic which involve passing a low intensity electric current between a cathode and an anode imbedded in the contaminated soil and, as a result, ions and small charged particles, in addition to water, are transported between the electrodes.

Another method that can be applied to the recovery of metals from contaminated soil is the leaching method in which the soil is washed (or flushed) with water which involves the addition of water with or without additives including organic and inorganic acids, sodium hydroxide which can dissolve organic soil matter, water soluble solvents such as methanol, nontoxic cations, complexing agents such as ethylenediaminetetraacetic acid (EDTA), acids in combination with complexation agents or oxidizing/reducing agents. Biosurfactants (i.e., biologically produced surfactants) may also be promising agents for enhancing removal of metals from contaminated soils and sediments.

Techniques for the extraction of metals by biological means include processes such as (1) bioleaching, which is a natural process that involves use of the interactions of micro-organisms such as bacteria or archaea (single-celled organisms that lack cell nuclei) to extract metals from the soil, and (2) phytoremediation, which involves the use of living plants and associated soil microbes to reduce the concentrations or toxic effects of contaminants in the environment. Bioleaching involves *Thiobacillus* sp. bacteria which can reduce sulphur compounds under aerobic and acidic conditions (pH 4) at temperatures on the order of 15 to 55°C (59 to 131°F).

In summary, the main techniques that have been used for the removal of metal contaminants from soil are (1) solidification/stabilization, (2) *in situ* extraction, and (3) electrokinetics. Site characteristics are of paramount importance in choosing the most appropriate remediation method (Mulligan et al., 2001).

4.4.6 Nanoremediation

The term nanoremediation (nano-remediation) refers to the use of nanoparticles (micrometer-scale particles, i.e., particle that are 1×10^{-6} meter in size) for environmental remediation which involves the use of nanoparticles (NPs) for the treatment of contaminated groundwater and soil. Depending on the properties of particles, a nano remediation process typically involves reduction, oxidation, sorption, or a combination of these sub-processes (Lee et al., 2014). The processes involve the application of reactive nanomaterials for transformation and detoxification of pollutants.

By way of clarification, nanomaterials exhibit unique physical and chemical properties and, hence, have properties that are adaptable to both chemical reduction and catalysis to mitigate the effects of pollutant. Examples of these nanoscale materials that are suitable for remediation processes include nanoscale zeolites, metal oxides, carbon nanotubes and fibers, enzymes, various noble metals (as bimetallic nanoparticles, BNPs), and titanium dioxide. In addition, a nano-remediation process can provide (in some instances, depending upon site conditions and properties) can provide a faster, potentially more cost-effective manner to clean up contaminated soils, sediments, and groundwater (Sridevi and Lakshmi 2013).

For nano-remediation *in situ*, there is not the need to pump groundwater from the formation for above-ground treatment and there is no need to transport the soil to other sites for treatment and disposal (Otto et al., 2008). Also, in the absence of significant surface electrostatic forces, the nano-size particles can be suspended in water thereby allowing direct injection into the subsurface where the contaminants are present. However, the properties of the nano-particles must be well understood—as an examples, there is the issues of the potential toxicity of the particles to floral and faunal species as well as the tendency of nano-particles to aggregate, rendering them less effective in the environment.

4.4.7 Pump and Treat

The pump and treat method, which is a widely used remediation method for cleanup contaminated groundwater offers a way to contaminants such as dissolved chemicals, including industrial solvents, metals, and various types of crude oil products.

Pump and treat systems can be designed for two very different goals which are (1) containment, which prevents the contaminant(s) from spreading, and (2) restoration, which is the removal of the contaminant(s). In pump-and-treat systems designed for containment, the extraction rate is established as the minimum rate that is sufficient to prevent enlargement of the contaminated zone while in the system that is designed for restoration, the pumping rate is typically much higher.

At some sites with simple geology and dissolved contaminants, pump-and-treat systems appear are capable for use in cleaning groundwater to health-based standards in a relatively short time. However, the chemical nature of contaminants and the geologic conditions of the can prevent pump-and-treat systems from restoring aquifers to health-based standards in a relatively short time. An effective pump and treat system depend on careful design of the pumping and treatment components based on the hydrogeological information gathered at the site.

There are four types of contaminant characteristics that can complicate ground water cleanup and are (1) the tendency of the contaminant—which may be organic or inorganic—to adsorb or absorb to solid materials and (2) the tendency for certain organic contaminants to remain undissolved as a nonaqueous phase, (3) the mass of contaminant released, and (4) the length of time the contaminant remained in the subsurface before cleanup.

Since the length of time required for remediation would increases with the amount of contaminant mass and the size of the source area, monitoring would provide the necessary information to optimize the system performance for it to reach its clean up goals. The parameters to be monitored and the necessary frequency of monitoring vary from one spill sites to another but these may also be due to the water level at (1) the numerous sampling points throughout and around the contaminated zone, (2) the contaminant concentrations in groundwater at numerous sampling points throughout and around the contaminated zone, (3) the contaminant concentration of the contaminant(s) in the extracted ground water, (4) contaminant concentrations of the contaminant(s) in the treatment system effluent, (5) the flow rates from the extraction wells and through the treatment system as well as other operational parameters.

4.4.8 Soil Vapor Extraction

The soil vapor extraction (SVE) method is a physical treatment process is a common and effective vacuum extraction process which utilizes the volatility of the crude oil constituents as a primary method of remediation to remove contaminant vapors from below ground for treatment above ground. The method is often used as a stand-alone remedy when volatile organic compounds are only impacting unsaturated soil.

The process uses the volatility of the contaminants to allow mass transfer from the adsorbed phase (in the soil) and the dissolved phases (in the groundwater) to the vapor phase followed by removal of the contaminants under vacuum. Thus, the purpose of the process is to the vapors by inducing air flow within the vadose zone. In the process, air is drawn (or pumped) through the contaminated soil where it mixes with contaminant vapors and the air-contaminant mixture is drawn out through an exit (or production) well to the surface for treatment and disposal. In this manner, the process can be used to clean up contamination under buildings and cause little disruption to nearby activities when in full operation and to remove volatile organic compounds above and below the water table.

The cleanup time depends on several factors vary from site to site such as (1) the concentration of the contaminant(s), (2) the size—i.e., the areal extent—and depth of the contaminated area, (3) the density of the soil, and (4) the moisture content of the soil, any of which can slow the movement of the contaminant vapor. The treatment of the vapors involves no harmful chemicals that must be transported to or from the site insofar as (in the latter case) the vapors are contained from extraction to treatment.

The process also provides oxygen to the subsurface which enhances the biodegradation of the volatile and nonvolatile hydrocarbon contaminants and can also be combined with *in situ* air sparging (IAS) to quickly removing volatile organic contamination from soils and groundwater.

4.4.9 Solidification and Stabilization

Solidification and stabilization (also referred to as called immobilization, fixation, or encapsulation) can be defined separately but they are usually implemented simultaneously through a single treatment process which is a soil remediation process by which the contaminants are rendered immobile through reactions with additives. The process reduces the mobility of contaminants in the environment through the use of both physical methods and chemical methods. Thus, the solidification and stabilization process is a method whereby contaminants may be chemically bound or encapsulated into a matrix. The process does not destroy the contaminants but instead traps or immobilizes the contaminants within a host medium (such as soil, sand, and/or building materials that contain them) instead of removing them by means of chemical principles or physical principles.

The solidification procedure involves mixing a waste with a binding agent, which is used as a materials to coagulate loose materials. On the other hand, stabilization refers to processes that involve chemical reactions, which reduce the leachability of the product and, hence, stabilize the contaminant(s). When the two terms are combined, the result is a process that accomplishes one or more of the following aspects (1) improved physical characteristics and handling of the contaminant(s), and (2) a limit in the solubility of hazardous constituents in the waste (EPA, 1989).

The stabilization and solidification phenomenon can be classified into three types of processes which are (1) aqueous stabilization/solidification, (2) polymer stabilization/solidification, and (3) vitrification, which involves the use of a high temperature. Also, process efficiency is improved by the use of a binder of which examples are cement, asphalt (which is a thermoplastic material which is used to microencapsulate waste), fly ash, and clay minerals. Organic polymer binders are also used to encapsulate both inorganic and organic wastes. In fact, an asphalt binder can be used directly to microencapsulate large quantities of waste or it can be used in an emulsion. In this process (i.e., the emulsion process), polymers such as urea formaldehyde, polybutadiene, polyurethane, polyethylene or polypropylene can be used to immobilize wastes by way of microencapsulation and/or macro encapsulation. Whereas vitrification uses electrical energy to heat a broad spectrum of wastes and soil which melt and transform into a glass-like material.

Furthermore, as might be anticipated, the physical and chemical characteristics of contaminated soil play a vital role in predicting the conditions of reaction and have impact on process performance. It is therefore essential that site conditions are well understood to enable treatment to be applied in a thorough and timely manner and the ability to accurately predict long-term effectiveness of the process is necessary for accurate design and selection of optimal binders and stabilization materials (Sharma and Reddy, 2004).

Finally, the process has been used for cleanup of sites containing the following types of contaminants (1) halogenated semi-volatile contaminants, (2) non-halogenated semi-volatile contaminants, (3) non-volatile contaminants (4) non-volatile metals, (5) radioactive materials,

(6) polychorinated biphenyl derivatives, (7) pesticides, and (8) dioxin derivatives and furan derivatives. With the development of effective binders, the process can be used to treat a wide range of organic and inorganic contaminants in soil and also in sludge.

4.4.10 Surfactant Enhanced Aquifer Remediation

Surfactant enhanced aquifer remediation (SEAR) involves the use of a surface active agent (a surfactant) which is a chemical agent that enhances the solubility of organic contaminants in aqueous media. The surfactant typically contains two distinct chemically active moieties which are (1) a hydrophilic moiety, which is a water-attracting group, such as an acid anion (a carboxylate ion -CO_2^- a or a sulfite or sulfate ion or SO_3^-, SO_4^-) and (2) a hydrophobic (water-repellent) group, such as an alkyl chain. Thus, the surfactant can be soluble in both water and an organic liquid (such as crude oil) thereby reducing the interfacial tension (IFT), which is the force that exists between two fluids but maintains the fluids as separate phases (i.e., between the aqueous and organic phases).

Surfactants are amphiphilic molecules (molecules possessing both hydrophilic and lipophilic properties) and are absorbed in the air-water interface (Figure 4.3) and it is this property that allows these chemicals to increase the solubility of nonaqueous-phase liquids (NAPLs) in water. The Surfactant enhanced aquifer remediation process involves the injection of a surfactant solution consisting of surfactant, electrolyte, cosolvent (such as ethyl alcohol), and water.

This remediation technology addresses the removal of immiscible-phase liquid contaminants, also termed nonaqueous phase liquids (NAPLs), from the saturated zone.

In the process, a chemical solution is pumped across a contaminated zone by introduction at an injection point and removal from an extraction point—which is analogous to the recovery of crude oil from an oil-bearing formation (i.e., the reservoir).

When designing a surfactant flood, surfactants are screened for acceptable toxicity and biodegradation characteristics, and minimal sorption to the aquifer mineral surfaces. In most aquifers, mineral surfaces are negatively charged; thus, anionic surfactants are selected because they are composed of negatively charged water-soluble heads, which exhibit minimal sorption. The solubilization as well as the reduction of the interfacial tension of most anionic surfactants are sensitive to the electrolyte concentration.

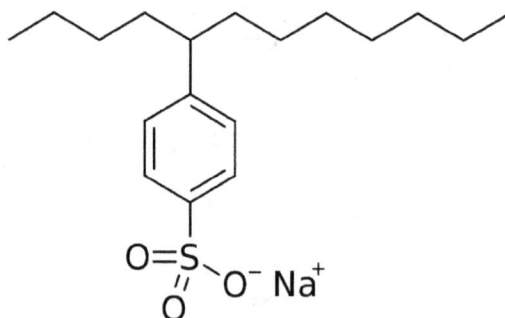

Figure 4.3: Representation of a Surfactant Using 4-(5-Dodecyl) Benzenesulfonate which is a Common Surfactants.

4.4.11 Thermal Desorption

The thermal desorption is typically used to clean up soil that is contaminated with volatile organic compounds (sometimes referred to as VOCs) and semi-volatile organic compounds (sometimes referred to as SVOCs) at depths shallow enough to reach through excavation.

Thus, thermal desorption is one of the methods commonly used to remediate contaminated soil and involves a thermally-induced physical separation process that separates contaminants from

Treatment stage	Product
Contaminated material	Oversized objects (removed)
	Crude feedstock
	Thermal desorption
	Vapors cleaning
	Contaminants disposal
	Treated material

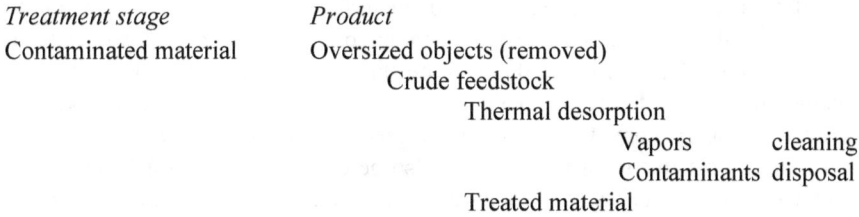

Figure 4.4: Schematic Illustration of the Thermal Desorption Process.

soil by heating contaminated material to evaporate hydrocarbon impurities and water (Figure 4.4). Thermal desorption is a different process from incineration—which uses heat to destroy the contaminant(s)—insofar as the thermal desorption process uses heat to physically separate the contaminant(s) from the soil and, therefore, require further treatment of the contaminant(s).

The process removes organic contaminants from soil, sludge or sediment by use of a thermal desorption unit to evaporate the contaminants, which include volatile organic compounds and some semi-volatile organic compounds (sometimes referred to as SVOCs).

Volatile organic compounds such as solvents and gasoline evaporate easily when heated whereas semi-volatile organic compounds require higher temperatures to evaporate and include diesel fuel, creosote (a wood preservative), coal tar, and several pesticide derivatives.

Thermal desorption (Figure 4.4) involves excavating the soil (or other contaminated material) for treatment in a thermal desorption unit which may be assembled at the site or the material may be loaded into trucks and transported to an offsite thermal desorption facility. Low-temperature thermal desorption is used to heat the solid material to 93 to 315°C (200 to 600°F) to treat volatile organic compounds. If semi-volatile organic compounds are present, then high-temperature thermal desorption is used to heat the soil to 315 to 535°C (600 to 1000°F). Gas collection equipment captures the contaminated vapors.

The gases (or vapors) from the thermal desorption unit often require further treatment, such as removing dust particles (Speight, 2019). At some sites, when there is a high concentrations of organic vapors, cooling followed by condensing will cause the vapor to revert to a liquid form and the liquid chemicals may be recycled for reuse (in a petrochemical plant) or destroyed by incineration. Often, treated soil can be used to fill in the excavation at the site but if the treated soil still contains contaminants (such as metal constituents that do not evaporate), the still-contaminated soil must be sent for disposal at an appropriate (i.e., regulated) landfill. The three major principles that control this phenomenon are (1) volatilization, (2) adsorption-desorption, and (3) diffusion (Sharma et al., 2004).

The factors that affect the performance of thermal desorption are (1) the temperature, (2) the heating time, (3) the heating rate, (4) the carrier gas, (5) the particle size of the soil, (6) the moisture content, and (7) the initial concentration of contaminants. Furthermore, based on the operating temperature of the desorption unit, the thermal desorption process can be sub-categorized into two groups which are (1) high temperature thermal desorption, HTTD, and (2) low temperature thermal desorption, LTTD. Also, thermal desorption systems fall into two general classes which are (1) stationary units which require transportation of the contaminated material (i.e., soil) from the site to the facility and (2) mobile units, which can be transported to the site and operated on-site.

In terms of the process, there are two main steps which are (1) the polluted materials are treated at temperatures on the order of 345 to 480°C (650 to 900°F) and (2) then discharged into a cooling unit after which the treated material is discharged from the cooling unit and is ready for testing and (subject to the test results) subsequent recycling.

The design of the thermal desorption system is focused on the volatilization of the contaminants while attempting to mitigate the potential for oxidation. Also, pretreatment of contaminated material involves a sifting step to remove large pieces of material and any other foreign objects. If the contaminated material is wet or has a substantial amount of contamination, it may need to be mixed with sand or dried to render a feedstock that is more enable to the desorption treatment.

Two common thermal desorption designs are the rotary dryer and thermal screw. For the thermal screw units, screw conveyors or hollow augers are used to transport the medium through an enclosed trough. Post treatment testing must also be carried out to ensure quality control and consistent remediation as well as the efficiency of the method used as well as the effectiveness (Figure 4.4) (Sharma and Reddy, 2004).

One major advantage of thermal desorption is that it can be extremely fast and is, therefore, often used for time-sensitive projects, although it does require that the soil has to be excavated. Also, the treated soil may no longer be able to support microbiological activity that breaks down contaminants which, if the soil is returned to a previously or partially contaminated site, this may be of some concern for future bioremediation efforts.

4.5 Remediation Management

Remediation management is the process of managing remediation activities by looking at remediation activity holistically, from end to end, beginning with identifying risks, and moving to mitigating risks, through to monitoring the effectiveness of solutions and keeping watch for additional risks as they occur.

Remediation management looks at remediation activity holistically, from end to end and commences with the identification of risks, and moving to mitigating risks, monitoring the effectiveness of solutions and keeping watch for additional risks as they occur. The management of remediation activities should be a cycle insofar as the remediation activities should start with the preparation of a remedial action plan (often referred to as RAP) and an environmental management plan (EMP).

The issues associated with environmental remediation are an often-changing mix of regulations and liabilities factors combined with an unpredictable subset of potential outcomes, pitfalls, and costs, environmental remediation can make assessment, planning, and response nearly impossible to quantify. Thus, the organizations with well-established remediation management systems and processes frequently experience increased costs due to unanticipated issues.

Remediation programs must be both driven by a customer-centric outlook and maintain commercial outcomes. A remediation program also requires communication and having a clear communication strategy is a necessary aspect of business. Where necessary, processes can be developed to enable the remediation of any adverse impacts attributable to business operations. Remediation is becoming one of the most important steps in management insofar as widespread remediation has mitigated the potential of world-wide environmental contamination through the prevention of occurrence of accidents.

A remedial action plan (often referred to as the RAP) is a plan that outlines the means by which a company will follow the accepted remedial actions. The plan will explain the type of remedial action that the said company needs to take and the variations in the remedial action can include (1) actions on policies, (2) recall, (3) repair, and (4) any environmental issues.

Thus, the plan will ensure that the company can successfully remedy any environmental issues (problems) that occur. However, as might be anticipated, the remedial action that a company will enact will depend on the nature of the problem. The main requirements for a remediation plan are to outline which remediation methods are necessary and the action plan must include the following components (1) background, (2) previous investigations, (3) remediation method, (4) goals, (5) monitoring, (6) health and safety aspects, and (7) regulatory oversight.

A remedial action plan typically contains the following elements, based on the nature, extent, type, volume, or complexity of the release: (1) a brief summary of the site characterization report conclusions, (2) a copy of the plans relating to worker health and safety, (3) management of wastes generated and quality assurance/quality control procedures, as they relate to the remedial action, and (4) a list of required (using the United States as the example) federal, state, and local permits or approvals to conduct the remedial action. Also, a presentation must be included of the means by which the remedial action will attain the selected remediation standard for the site, the results of treatability, bench scale or pilot scale studies or other data collected to support the remedial action.

The plan may also include (and describe) the descriptions such as (1) the operation and maintenance details for the remedial action, (2) the schedule for monitoring, sampling and site inspections, (3) a site map showing information pertinent to the remedial action, (4) a description of the media and parameters to be monitored or sampled during the remedial action, (5) the analytical methods to be utilized, and (6) the proposed post-remediation care requirements.

If no remedial action is performed, done when a flaw is discovered the company can risk facing serous negative consequences which can range from bad (1) publicity, (2) lawsuits, and/or fines— especially a combination of the latter two consequences.

References

Azubuike, C.C., Chikere, C.B., and Okpokwasili, G.C. 2016. Bioremediation techniques–classification based on site of application: principles, advantages, limitations and prospects. World Journal of Microbiology and Biotechnology, 32: 180–197.

Carson, C.J., and Mumford, P.A. 1995. The Safe Handling of Chemicals in Industry. Volume 3. John Wiley & Sons Inc., New York.

Carson, C.J., and Mumford, P.A. 1988. The Safe Handling of Chemicals in Industry. Volumes 1 and 2. John Wiley & Sons Inc., New York.

Edwards, J.D. 1995. Industrial Wastewater Treatment: A Guidebook. CRC Press Inc., Boca Raton, FL.

EPA. 2015. A Citizen's Guide to Thermal Desorption. United States Environmental Protection Agency, Washington, DC. https://www.epa.gov/sites/default/files/2015-04/documents/a_citizens_guide_to_thermal_desorption.pdf.

IPCC. 2007. Climate Change 2007: The physical science basis. *In*: Solomon, S., Qin, D., Manning, M., Chen Z., Marquis, M., Averyt, K.B., Tignor, M., and Miller, H.L. (Eds.). Contribution of Working Group I to the Fourth Assessment Report of the Intergovernmental Panel on Climate Change. Cambridge University Press. Cambridge, United Kingdom.

IPCC. 2013. Climate Change 2013: The physical science basis. *In*: Stocker, T.F., Qin, D., Plattner, G.-K., Tignor, M., Allen, S.K., Boschung, J., Nauels, A., Xia, Y., Bex, V., and Midgley, P.M. (Eds.). Contribution of Working Group I to the Fifth Assessment Report of the Intergovernmental Panel on Climate Change. Cambridge University Press, Cambridge, United Kingdom.

Lee, C.C., Lien, H.L., Wu, S.C., Doong, R.A., and Chao, C.C. 2014. Reduction of priority pollutants by nanoscale zerovalent iron in subsurface environments. pp. 63–96. *In*: Reisner, D.E., and Pradeep, T. (Eds.). Aqua-nanotechnology: Global prospects. CRC Press, Taylor & Francis Group, Boca Raton, Florida. .

Mulligan, C.N., Yong, R.N., and Gibbs, B.F. 2001. Remediation technologies for metal-contaminated soils and groundwater: an evaluation. Engineering Geology, 60(1-4): 193–207.

Nyer, E.K. 1998. Groundwater and Soil Remediation: Practical Methods and Strategies, Volume II. CRC Press, Taylor & Francis Group, Boca Raton, Florida.

Otto, M., Floyd, M., and Bajpai, S. 2008. Nanotechnology for site remediation. Remediation, 19(1): 99–108.

Ray, D.L., and Guzzo, L. 1990. Trashing The Planet: How Science Can Help Us Deal With Acid Rain, Depletion of The Ozone, and Nuclear Waste (Among Other Things). Regnery Gateway, Washington, DC.

Renzoni, A., Fossi, M.C., Lari, L., and Mattei, N. 1994. Contaminants in the Environment: A Multidisciplinary Assessment of risks to Man and Other Organisms. CRC Press Inc., Boca Raton, FL.

Sharma, H.D., and Reddy, K.R. 2004. Geoenvironmental Engineering: Site Remediation, Waste Containment, and Emerging Waste Management Technologies. John Wiley & Sons Inc., Hoboken, New Jersey.

Speight, J.G. 1996. Environmental Technology Handbook. Taylor & Francis Publishers, Washington, DC.

Speight, J.G., and Lee, S. 2000. Environmental Technology Handbook 2nd Edition. Taylor & Francis, New York.

Speight. J.G. 2005. Environmental Analysis and Technology for the Refining Industry. John Wiley & Sons Inc., Hoboken, New Jersey.

Speight, J.G., and Arjoon, K.K. 2012. Bioremediation of Petroleum and Petroleum Products. Scrivener Publishing, Salem, Massachusetts.

Speight, J.G. 2014. The Chemistry and Technology of Petroleum 5th Edition. CRC Press, Taylor & Francis Group, Boca Raton, Florida.

Speight, J.G. 2017. Handbook of Petroleum Refining. CRC Press, Taylor & Francis Group, Boca Raton, Florida.

Speight, J.G. 2019. Natural Gas: A Basic Handbook 2nd Edition. Gulf Publishing Company, Elsevier, Cambridge, Massachusetts. 2019.

Speight, J.G. 2020. Global Climate Change Demystified. Scrivener Publishing, Beverly, Massachusetts.

Sridevi, V., and Lakshmi, P.K.S. 2013. Role of nZVI, Metal Oxide and Carbon Nanotube for Ground Water Cleanup and Their Comparison: An Overview. Advanced Nanomaterials and Emerging Engineering Technologies (CANMET), 2013 International Conference, 737,742, 24–26. July.

Thibodeaux, L.J. 1995. Environmental Chemodynamics. John Wiley & sons Inc., New York. US EPA. 2012. A Citizen's Guide to Thermal Desorption. EPA 542-F-12-020. United States Environmental Agency, Washington DC. September.

CHAPTER 5
Overview of Oil Spill Clean Up Methods

As, the global population continues to rise at an astonishing rate, the environmental consequences have led to an increase in the number of incidents on land, air, and water resources that has resulted in the contamination of the environment by toxic materials and other pollutants, threatening humans and ecosystems with serious health risks. Although a better and cheaper option than cleaning up or mitigating the impact of pollution would have been to avoid it there are cleanup programs that use natural resources and energy efficiently.

Since the early 1990s, governments and the business communities have invested significant resources in supporting the development, testing, and deployment of innovative technologies. These remediation technologies have evolved rapidly in the past decade and even though there is no such a single universal method for environmental cleanup, there have been many alternatives of remediation techniques. Cleanup technologies may be specific to the contaminant (or contaminant class) and to the site and there are at least three different options: biological, physical, and chemical remediation that are available.

As a result, evolution of cleanup technologies has yielded four general categories of remediation approaches: (1) physical removal, with or without treatment of the spilled material, (2) *in situ* conversion of the spilled material by physical or chemical means to less toxic or less mobile forms, (3) containment of the spilled material, and (4) passive cleanup or natural attenuation of the spilled material. A combination of two or more technologies may be necessary for use at some sites.

There are laws that applies to every business and public body referred to as environmental cleanup laws, that are designed to respond after-the-fact to environmental contamination and govern the removal of pollution or contaminants from environmental media such as soil, sediment, surface water, or groundwater. They are a complex combination of state, federal, and international treaty law pertaining to issues of concern to the environment and protecting natural resources. It also attempts to minimize the impact that humans have on our planet with the share the goal of protecting the environment. In fact, around the world, a draft of new legislation to help protect the environment has just come into force.

Hydrocarbon species can enter the soil environment from a number of sources. The origin of the contaminants has a significant bearing upon the species present and hence the analytical methodology to be used (Driscoll et al., 1992). Unlike other chemicals (notably pesticides), hydrocarbon derivatives were generally not applied to soils for a purpose and thus hydrocarbon contamination results almost entirely from misadventure. The source that is probably most familiar to persons involved in the study of contaminated sites is leakage from underground storage tanks. This is particularly important at the site of former service stations and the hydrocarbon derivatives involved are generally in the gasoline or diesel range. Other major sources include spillage during refueling and lubrication, the hydrocarbon derivatives being within the diesel and heavy oil range. Places in which transfer and handling of crude oils takes place (such as tanker terminals and oil refineries) are also potential places of contamination, the oil being largely of the heavier hydrocarbon type.

Since the group of chemicals generally referred to as *total petroleum hydrocarbons* have widely differing properties, they are likely to present a significant analytical challenges. Additionally, the hydrocarbon derivatives will be associated with the soil in different ways and hence the strength of the hydrocarbon interaction (usually sorption) will vary according to the nature of the hydrocarbon as well as with the nature of any other organic matter present in the soil.

Thus, the relevant chemistry of hydrocarbon derivatives likely to be encountered at contaminated sites is briefly reviewed and the importance of hydrocarbon speciation noted in terms of a toxicological basis for risk assessment. Hydrocarbon interaction with soil contaminants is important both in terms of their toxicology and also their accessibility by analytical methods. There is no simple procedure that will give an overall picture of hydrocarbon derivatives present at contaminated sites. This is largely because the molecules are present in two separate categories— viz. volatile and semi- or non-volatile. These two categories require significantly different sample collection, handling and management techniques. Volatile hydrocarbon derivatives may be collected by zero headspace procedures or by immediate immersion of the soil into methanol.

The analysis involves gas chromatographic methods such as purge and trap, vacuum distillation, and headspace. On the other hand, samples for the determination of semi and non-volatile hydrocarbon derivatives need not be collected in such a rigorous manner. They require extraction by techniques such as solvent or supercritical fluid on arrival at the laboratory. Some cleanup of extracts is also necessary in most cases and the analytical finish is again by gas chromatography. Detectors used range from flame ionization to Fourier transform infrared and mass spectrometric, the latter types being necessary to achieve speciation of the component hydrocarbon derivatives.

The determination of hydrocarbon contaminants in soil is one of the most frequently performed analyses in the study of contaminated sites and is also one of the least standardized. Given the wide variety of hydrocarbon contaminants that can potentially enter and exist in the soil environment, a need exists for methods that satisfactorily quantify these chemicals. Formerly, the idea of total hydrocarbon determination in soil was seen as providing a satisfactory tool for assessing contaminated sites but the nature of the method and the site specificity dictates a risk-based approach in data assessment. Quantitation of particular hydrocarbon species may be required.

Currently, many regulatory agencies recommend the common methods (EPA 418.1, EPA 801.5 Modified) or similar methods for analysis during remediation of contaminated sites. In reality, there is no standard for the measurement of total petroleum hydrocarbon derivatives since each method may need to be chosen or adapted on the basis of site specificity.

There is a trend toward use of GC techniques in analysis of soils and sediments. One aspect of these methods is that *volatiles* and *semi-volatiles* are determined separately. The volatile or gasoline range organic constituents are recovered using purge-and-trap or other stripping techniques. Semi-volatiles are separated from the solid matrix by solvent extraction. Other extraction techniques have been developed to reduce the hazards and the cost of solvent use and to automate the process and techniques include supercritical fluid extraction (SFE), microwave extraction, Soxhlet extraction, sonication extraction, and solid phase extraction (SPE) (EPA 3540C). Capillary column techniques have largely replaced the use of packed columns for analysis, as they provide resolution of a greater number of hydrocarbon compounds.

Because of the overall complexity of the problem and of the spectrum of hydrocarbon derivatives likely to be encountered, it is impossible to view the total; petroleum hydrocarbons as a single entity. There have been many approaches to the problem, but the simplest and one most frequently used is the one based on the vapor pressure ranges of the relevant organic constituents. This also relates to the sampling methodology employed and the approach consists of sub-dividing the hydrocarbon derivatives into the most volatile fraction referred to as gasoline range organics (GRO) and the less volatile fraction. In the case of monitoring of storage tanks, a sub-fraction (known as diesel range organics or DRO) is often distinguished amongst the semi-volatile fraction.

As regards a contaminated soil, this type of analysis may not possible because the various hydrocarbon derivatives cannot be extracted from the sample with equal efficiency. Volatile organic compounds require special procedures to achieve satisfactory recovery from the soil matrix. It thus becomes important to distinguish between those compounds that are considered to be volatile and those that rank as semi-volatile compounds or non-volatile compounds.

The most obvious method of cleanup involves the natural recovery of the oil. This is a method that allows a site to recover without intervention or intrusion and is considered to be the simplest method of dealing with the oil spill cleanup operation is to make use of the diversity of nature like the sun, the wind, the weather, tides, or naturally occurring microbes. Depending upon the type, magnitude, extent, and persistence of the natural or anthropogenic disturbance, the natural recovery process generally follows one of two basic patterns. In one pattern, that survive the effects of the oil spill are released from the competition for space, and may even be temporarily relieved of a full complement of predators. These surviving organisms may rapidly reoccupy the habitat.

Natural recovery is considered an appropriate clean up method where oiling has occurred on high energy beaches where boulders and rocks are found. The wave action will remove most of the oil in a relatively short time. This method is utilized when the shorelines are remote or inaccessible. It can be implemented also when either treatment or cleaning of the stranded oil may cause more damage than leaving the shore to recover naturally or other response techniques either cannot accelerate natural recovery or may not be practical.

Natural recovery processes comprises of weathering which includes evaporation, oxidation, biodegradation, and emulsification. Weathering is a series of chemical and physical changes that cause spilled oil to break down and become (through oxidation and the incorporation of oxygen-contain functions) heavier than water thereby sinking to the bottom of the water system. Thus, weathering changes the physical properties, chemical reactivity, and toxic chemical content of the oil across a wide range of time scales and can have a significant impact on the properties of a slick and affect dispersant performance.

Evaporation is an aspect of oil spill insofar as the volume of a light crude oil can be reduced considerably. However, heavy crude oil or residual oil will only lose approximately 5% of the volume (often less) in the first few days following a spill.

Oxidation occurs when oil contacts the water and oxygen combines with the oil to produce water-soluble compounds. This process is promoted by sunlight and the extent to which it occurs depends on the type of oil and the form in which it is exposed to sunlight. For example, thick oil slicks may only partially oxidize, forming *tar balls*, which are dense, sticky, black spheres may linger in the environment, and can collect in the sediments of slow moving streams or lakes or wash up on shorelines long after a spill.

Although environmental pollution is the major source of problems in human health and the sustainable development of society and economy in the world, there are nature-based solutions. Ecosystems and some specific plant and microbes can be of assistance to break down pollutants. There are six main groups of microbes and the bacteria and fungi provide the environment with a self-cleaning capability, which is a holistic approach of leveraging the existing resources of nature (Table 5.1) (Kohli, 2019).

Wind and waves will naturally disperse the oil over time—part of the spiled material will evaporate and naturally occurring microbes will also degrade the oil. This nature-based solution, enable us to make use of the planet's intrinsic restorative capacity.

Although there is no single definition of green in terms of the environment. A broad definition is being environment friendly or beneficial to the environment (Vilsack, 2009) and the so-called green solutions generally improve the environment and/or use fewer natural resources. In general, the production or performance of green processes, products, technologies, and procedures are

Table 5.1: The Six Main Groups of Microbes.

Archaea	A group of unicellular prokaryotic cells that can produce methane during their metabolism. They are specifically adapted to a wide variety of environmental conditions by means of membranes and metabolism.
Bacteria	Unicellular prokaryotic organisms have a unique type of cell wall and cell membrane that distinguishes them from Archaea
Fungi	Non-photosynthetic eucaryotes that absorb their nutrients directly from the environment. This group includes mushrooms, molds, and yeast.
Protista	Animal-like, non-photosynthetic eukaryotes common in moist environments.
Viruses	Composed of nucleic acid (DNA or RNA) and protein and have some of the characteristics of life.
Microbial mergers	Combinations and collaborations between different microbe species.

favorable compared to other options; that is, they have a more positive impact on their local or global surroundings, at a specific time, than other possible alternatives.

It is the practice of considering all environmental effects of remedy implementation and incorporating options to maximize net environmental benefit of cleanup actions (US EPA, 2010b). However, the conventional techniques are used often, but they can only transform their form; therefore bioremediation over the conventional methods is most promising technology for cleaning the environment (Speight and Arjoon, 2012; Singh et al., 2017).

Bioremediation refers to the use of natural micro-organisms (such as bacteria, yeasts, and fungi) for remediation of polluted air, water, soil, or sediment. The process occurs naturally. Its application in oil spill cleanup involves the artificial introduction of biological agents such as fertilizers and nutrients to native micro-organisms in the contaminated site so they proliferate (bio-stimulation) or the introduction of non-native micro-organisms (bioaugmentation) to speed up the natural process of biodegradation so as to protect shorelines, wetlands, and other marshy areas affected by spills from further damage.

However, natural recovery is not always (if ever) possible and other applied methods are used either alone or in concert with another method, including bioremediation. To give a the reader a better understanding of oil spill cleanup methods, The alternate methods are introduced in the following sub-sections which cover non-bioremediation methods of cleanup of oil spills at sea and on the land (Speight and Arjoon, 2012). But first, it is necessary to understand the types of effluents and the potential for these effluents to interact with the environment.

5.1 Types of Effluents

The refining industry, as well as other industries, will increasingly feel the effects of the land bans on their hazardous waste management practices. Current practices of land disposal must change along with management attitudes for waste handling. The way refineries handle their waste in the future depends largely on the ever-changing regulations (CFR, 2004). Waste management is the focus and reuse/recycle options must be explored to maintain a balanced waste management program. This requires that a waste be recognized as either *non-hazardous* or *hazardous*.

However, before a refinery can determine if its waste is hazardous, it must first determine that the waste is indeed a solid waste. In 40 CFR 261.2, the definition of solid waste can be found. If a waste material is considered a solids waste, it may be a hazardous waste in accordance with 40 CFR 261.3. There are two ways to determine whether a waste is hazardous. These are to see if the waste is listed in the regulations or to test the waste to see if it exhibits one of the characteristics (40 CFR 261).

The first step to be taken by a generator of waste is to determine whether that waste is hazardous. Waste may be hazardous by being listed in the regulations, or by meeting any of the four characteristics: ignitability, corrosivity, reactivity, and extraction procedure (EP) toxicity.

Generally: (1) if the material has a flash point less than 140°F it is considered ignitable; (2) if the waste has a pH less than 2.0 or above 12.5, it is considered corrosive. It may also be considered corrosive if it corrodes stainless steel at a certain rate; (3) a waste is considered reactive if it is unstable and produces toxic materials, or it is a cyanide or sulfide-bearing waste which generates toxic gases or fumes; (4) a waste which is analyzed for EP toxicity and fails is also considered a hazardous waste. This procedure subjects a sample of the waste to an acidic environment. After an appropriate time has elapsed, the liquid portion of the sample (or the sample itself if the waste is liquid) is analyzed for certain metals and pesticides. Limits for allowable concentrations are given in the regulations.

In terms of waste definition, there are three basic approaches (as it pertains to crude oil, crude oil products, and non-crude oil chemicals) to defining crude oil or a crude oil product as hazardous: (1) a qualitative description of the waste by origin, type, and constituents; (2) classification by characteristics based upon testing procedures; and (3) classification as a result of the concentration of specific chemical substances (Chapter 3). In addition, there are recommended protocols that must occur as a precluded to cleanup of emissions and mitigating future releases.

There are four lists of hazardous wastes in the regulations. These are wastes from nonspecific sources (F list), wastes from specific sources (K list), acutely toxic wastes (P list), and toxic wastes (U list). And there are the four characteristics mentioned before: ignitability, corrosivity, reactivity, and extraction procedure toxicity. Certain waste materials are excluded from regulation under RCRA. The various definitions and situations that allow waste to be exempted can be confusing and difficult to interpret. One such case is the interpretation of the *mixture* and *derived-from* rules. According to the mixture rule, mixtures of solid waste and listed hazardous wastes are, by definition, considered hazardous. Likewise, the *derived-from* rule defines solid waste resulting from the management of hazardous waste to be hazardous (40 CFR 261.3a and 40 CFR. 261.1c).

There are five specific listed hazardous wastes (K list) generated in refineries. These are K048-K052. Additional listed wastes, those from nonspecific sources (F list) and those from the commercial chemical product lists (P and U), may also be generated at refineries. Because of the mixture and derived-from rules, special care must be taken to ensure that hazardous wastes do not *contaminate* non-hazardous waste. Under the mixture rule, adding one drop of hazardous waste in a container of non-hazardous materials makes the entire container contents a hazardous waste.

As an example of the problems such mixing can cause, consider the case with API separator sludge that is a listed hazardous waste (K051). The wastewater from a properly operating API separator is not hazardous unless it exhibits one of the characteristics of a hazardous waste. That is, the derived-from rule does not apply to the wastewater. However, if the API separator is not functioning properly, solids carry over in the wastewater can occur. In this case, the wastewater contains a listed hazardous waste, the solids from the API sludge, and the wastewater would be considered a hazardous waste because it is a mixture of a non-hazardous waste and a hazardous waste.

This wastewater is often further cleaned by other treatment systems (filters, impoundments, etc.). The solids separating in these systems continue to be API separator sludge, a listed hazardous waste. Therefore, all downstream wastewater treatment systems are receiving and treating a hazardous waste and are considered hazardous waste management units subject to regulation.

Oily wastewater is often treated or stored in unlined wastewater treatment ponds in refineries. These wastes appear to be similar to API separator waste.

5.1.1 Gaseous Effluents

Crude oil is capable of producing gaseous pollutant chemicals (Guthrie, 1967; Rawlinson and Ward, 1973; Francis and Peters, 1980; Hoffman, 1983; Loehr, 1992; Olschewsky and Megna, 1992; Moustafa, 1996; Speight, 1996, 1999). The gaseous emissions are often characterized by chemical species identification, e.g., *inorganic gases* such as sulfur dioxide (SO_2), nitrogen oxides (NO_x), and carbon monoxide (CO) or *organic gases* such as chloroform ($CHCl_3$) and formaldehyde (HCHO). The rate of release or concentrating in the exhaust air stream (in parts per million or comparable units) along with the type of gaseous emission greatly predetermines the applicable control technology.

The three main greenhouse gases that are products of refining are carbon dioxide, nitrous oxide, and methane (Fogg and Sangster, 2003). Carbon dioxide is the main contributor to climate change. Methane is generally not as abundant as carbon dioxide but is produced during refining and, if emitted into the atmosphere is a powerful greenhouse gas and more effective at trapping heat. However, gaseous emissions associated with crude oil refining are more extensive that carbon dioxide and methane and typically include process gases, petrochemical gases, volatile organic compounds (VOCs), carbon monoxide (CO), sulfur oxides (SO_x), nitrogen oxides (NO_x), particulates, ammonia (NH_3), and hydrogen sulfide (H_2S). These effluents may be discharged as air emissions and must be treated. However, gaseous emissions are more difficult to capture than wastewater or solid waste and, thus, are the largest source of untreated wastes released to the environment.

In the refining industry, as in other industries, air emissions include point and non-point sources. Point sources are emissions that exit stacks and flares and, thus, can be monitored and treated. Non-point sources are *fugitive emissions* that are difficult to locate and capture. Fugitive emissions occur throughout refineries and arise from, for example, the thousands of valves, pumps, tanks, pressure relief valves, and flanges. While individual leaks are typically small, the sum of all fugitive leaks at a refinery can be one of its largest emission sources.

The numerous process heaters used in refineries to heat process streams or to generate steam (boilers) for heating or steam stripping, can be potential sources of sulfur oxides (SO_2, and SO_3), nitrogen oxides (NO and NO_2), carbon monoxide (CO), particulates, and emissions of hydrocarbon derivatives. When operating properly and when burning cleaner fuels such as refinery fuel gas, fuel oil or natural gas, these emissions are relatively low. If, however, combustion is not complete, or heaters are fired with refinery fuel pitch or residuals, emissions can be significant.

In addition to the corrosion of equipment of acid gases, the escape into the atmosphere of sulfur-containing gases can eventually lead to the formation of the constituents of acid rain, i.e., the oxides of sulfur (SO_2 and SO_3). Similarly, the nitrogen-containing gases can also lead to nitrous and nitric acids (through the formation of the oxides NO_x, where x = 1 or 2) which are the other major contributors to acid rain. The release of carbon dioxide and hydrocarbon derivatives as constituents of refinery effluents can also influence the behavior and integrity of the ozone layer.

The processes that have been developed to accomplish gas purification vary from a simple once-through wash operation to complex multi-step recycling systems. In many cases, the process complexities arise because of the need for recovery of the materials used to remove the contaminants or even recovery of the contaminants in the original, or altered, form (Kohl and Riesenfeld, 1979; Newman, 1985).

The majority of gas streams exiting each refinery process collected and sent to the gas treatment and sulfur recovery units to recover the refinery fuel gas and sulfur. Emissions from the sulfur recovery unit typically contain some hydrogen sulfide (H_2S), sulfur oxides, and nitrogen oxides. Other emissions sources from refinery processes arise from periodic regeneration of catalysts. These processes generate streams that may contain relatively high levels of carbon monoxide, particulates and volatile organic compounds (VOCs). Before being discharged to the atmosphere, such off-gas streams may be treated first through a carbon monoxide boiler to burn carbon monoxide and any volatile organic compounds, and then through an electrostatic precipitator or cyclone separator to remove particulates.

Analysts need consistent, reliable, and credible methodologies to produce analytical data about gaseous emissions (Patnaik, 2004). To fulfill this need in this text, descriptions are given of this chapter is devoted to descriptions and the various analytical methods that can be applied to identify gaseous emissions from a refinery. Each gas is, in turn, and referenced by its name rather than the generic term *petroleum gas* (ASTM D4150). However, the composition of each gas varies and recognition of this is essential before testing protocols are applied.

Because of the relative complexity of the analytical methods for total petroleum hydrocarbon derivatives, there is a need for devising methods for the determination of total petroleum hydrocarbons. But, the major problem lies in the range of compounds covered by the term *hydrocarbons*. Again, the most notable variation is in the relative volatility and other properties of the hydrocarbon derivatives under investigation. Although instrumental detection methods are available (Sadler and Connell, 2003), another approach involves collection of the contaminated soil and sealing it in a container, where the soil gas can accumulate. This gas is then analyzed by one of several reliable instrumental procedures.

Some methods for determining hydrocarbon derivatives in air matrices usually depend upon adsorption of the constituents of the total petroleum hydrocarbon fraction on to a solid sorbent, subsequent desorption and determination by gas chromatographic methods. Hydrocarbon derivatives within a specific boiling range (n-pentane to n-octane) in occupational air are collected on a sorbent tube, desorbed with solvent, and determined using gas chromatography-flame ionization. Although method precision and accuracy are usually high, performance may be reduced at high humidity.

On the other hand, the complex mixture of crude oil hydrocarbon derivatives potentially present in an air sample can be minimized by is separation of the sample into aliphatic and aromatic fractions, and then these two major fractions are separated into smaller fractions based on carbon number. Individual compounds (e.g., benzene, toluene, ethylbenzene, xylenes, MTBE, naphthalene) are also identified using this method. The range of compounds that can be identified includes C4 (1,3-butadiene) through C 12 (*n*-dodecane).

As a partial compromise between the use of on-site instrumental analysis and laboratory analysis, a passive sampler can be immersed into the soil (at a specified depth or at several depths) to collect the evolved gases that are adsorbed onto a solid phase support. The sampler is then removed to the laboratory, where the gases are transferred by Curie point desorption, directly into the ion source of an interfaced quadrupole mass spectrometer. This procedure has its origin in the crude oil exploration industry and the samplers can be used at a considerable range of depths.

A number of procedures, based on microanalysis of samples for known physical properties (Speight, 2014, 2015) have also been employed. For example, field screening, which uses infrared spectroscopy, employing a portable version of the laboratory procedure, has been used). Field turbidometric methods favor the determination of high boiling hydrocarbon derivatives and are of some use in delineating such pollution within soil. The fluorescence spectra exhibited by the aromatic components provide the basis for laser-induced fluorescence spectroscopy. They allow detection of polycyclic aromatic compounds and thus are able to take account of a fraction not measured by other field screening techniques.

5.1.1.1 Liquefied Petroleum Gas

Liquefied petroleum gas (LPG) is a mixture of the gaseous hydrocarbon derivatives propane ($CH_3CH_2CH_3$, boiling point: –42°C, –44°F) and butane ($CH_3CH_2CH_2CH_3$, boiling point: 0°C, 32°F) that are produced during natural gas refining, crude oil stabilization, and crude oil refining. The propane and butane can be derived from natural gas or from refinery operations but, in this latter case, substantial proportions of the corresponding olefin derivatives will be present and need to be separated. The hydrocarbon derivatives are normally liquefied under pressure for transportation and storage.

The presence of propylene and butylenes in liquefied petroleum gas used as fuel gas is not critical. The vapor pressures of these olefin derivatives are slightly higher than those of propane and butane and the flame speed is substantially higher but this may be an advantage since the flame speeds of propane and butane are slow. However, one issue that often limits the amount of olefin derivatives in liquefied petroleum gas is the propensity of the olefin derivatives to form soot as well as the presence mechanically entrained water (that may be further limited by specifications) (ASTM D1835). The presence of water in liquefied petroleum gas (or in natural gas) is undesirable since it can produce hydrates that will cause, for example, line blockage due to the formation of hydrates under conditions where the water *dew point* is attained (ASTM D1142). If the amount of water is above acceptable levels, the addition of a small methanol will counteract any such effect. Another component of liquefied petroleum gas is propylene ($CH_3CH = CH_2$), which has a significantly lower octane number (ASTM D2623) than propane, so there is a limit to the amount of this component that can be tolerated in the mixture. Analysis by gas chromatography is possible (ASTM D5504, ASTM D6228, IP 405).

Liquefied petroleum gas and liquefied natural gas can share the facility of being stored and transported as a liquid and then vaporized and used as a gas. In order to achieve this, liquefied petroleum gas must be maintained at a moderate pressure but at ambient temperature. The liquefied natural gas can be at ambient pressure but must be maintained at a temperature of roughly –1 to 60°C (30 to 140°F). In fact, in some applications it is actually economical and convenient to use liquefied petroleum gas in the liquid phase. In such cases, certain aspects of gas composition (or quality such as the ratio of propane to butane and the presence of traces of higher molecular weight (higher boiling) hydrocarbon derivatives, water and other extraneous materials) may be of lesser importance compared to the use of the gas in the vapor phase.

For normal (gaseous) use, the contaminants of liquefied petroleum gas are controlled at a level at which they do not corrode fittings and appliances or impede the flow of the gas. For example, hydrogen sulfide (H_2S) and carbonyl sulfide (COS) should be absent although organic sulfur compounds, to the level required for adequate Odorization, are permissible (ASTM D5305). In fact, *stenching* is a normal requirement in liquefied petroleum gas, dimethyl sulfide (CH_3SCH_3) and ethyl mercaptan (C_2H_5SH) are commonly used at a concentration of up to 50 ppm. Natural gas is similarly treated possibly with a wider range of volatile sulfur compounds.

5.1.1.2 Natural Gas

Natural gas is found in crude oil reservoirs as free gas (*associated gas*) or in solution with crude oil in the reservoir (*dissolved gas*) or in reservoirs that contain only gaseous constituents and no (or little) crude oil (*unassociated gas*). The hydrocarbon content varies from mixtures of methane and ethane with very few other constituents (*dry gas*) to mixtures containing all of the hydrocarbon derivatives from methane to pentane and even hexane (C_6H_{14}) and heptane (C_7H_{16}) (*wet gas*) (Speight, 2014). In both cases some carbon dioxide (CO_2) and inert gases, including helium (He), are present together with hydrogen sulfide (H_2S) and a small quantity of organic sulfur.

While the major constituent of natural gas is methane, there are components such as carbon dioxide (CO), hydrogen sulfide (H_2S), and mercaptans (thiols; R-SH), as well as trace amounts of sundry other emissions. The fact that methane has a foreseen and valuable end-use makes it a desirable product, but in several other situations it is considered a pollutant, having been identified as one of several greenhouse gases.

Carbon dioxide (ASTM D1137, ASTM D1945, ASTM D4984) in excess of 3 per cent is normally removed for reasons of corrosion prevention (ASTM D1838). Hydrogen sulfide (ASTM D2420, ASTM D2385, ASTM D2725, ASTM D4084, ASTM D4810) is also removed and the odor of the gas must not be objectionable (ASTM D6273) so mercaptan content (ASTM D1988, ASTM D2385) is important. A simple lead acetate test (ASTM D2420, ASTM D4084) is available for detecting the presence of hydrogen sulfide and is an additional safeguard that hydrogen sulfide not be present (ASTM D1835). The odor of the gases must not be objectionable. Methyl mercaptan, if present, produces a transitory yellow stain on the lead acetate paper that fades completely in less than 5 min. Other sulfur compounds (ASTM D5504, ASTM D6228) present in liquefied petroleum gas do not interfere.

In the lead acetate test (ASTM D2420), the vaporized gas is passed over moist lead acetate paper under controlled conditions. Hydrogen sulfide reacts with lead acetate to form lead sulfide resulting in a stain on the paper varying in color from yellow to black, depending on the amount of hydrogen sulfide present. Other pollutants can be determined by gas chromatography (ASTM D5504, ASTM D6228).

The total sulfur content (ASTM D1072, ASTM D2784, ASTM D3031) is normally acceptably low, and frequently so low that it needs augmenting by means of alkyl sulfide derivatives, mercaptan derivatives, or thiophene derivatives in order to maintain an acceptable safe level of odor.

The hydrocarbon dew point is reduced to such a level that retrograde condensation, i.e., condensation resulting from pressure drop, cannot occur under the worst conditions likely to be experienced in the gas transmission system. Similarly the water dew point is reduced to a level sufficient to preclude formation of C_1 to C_4 hydrates in the system.

The natural gas after appropriate treatment for acid gas reduction, odorization, and hydrocarbon and moisture dew point adjustment (ASTM D1142), would then be sold within prescribed limits of pressure, calorific value and possibly Wobbe Index (cv/(sp. gr.)).

5.1.1.3 Refinery Gas

Refinery gas (process gas) is the non-condensable gas that is obtained during distillation of crude oil or treatment (cracking, thermal decomposition) of crude oil (Speight 2014, 2017). There are also components of the gaseous products that must be removed prior to release of the gases to the atmosphere or prior to use of the gas in another part of the refinery, i.e., as a fuel gas or as a process feedstock.

Refinery gas consists mainly of hydrogen (H_2), methane (CH_4), ethane (C_2H_6), propane (C_3H_8), butane (C_4H_{10}), and olefin derivatives ($RCH = CHR^1$, where R and R^1 can be hydrogen or a methyl group) and may also include off-gases from petrochemical processes. Olefin derivatives such as ethylene (ethene, $CH_2 = CH_2$, boiling point: -104°C, -155°F), propene (propylene, $CH_3CH = CH_2$, boiling point: -47°C, -53°F), butene (butene-1, $CH_3CH_2CH = CH_2$, boiling point: -5°C, 23°F) *iso*-butylene (($CH_3)_2C = CH_2$, -6°C, 21°F), *cis*- and *trans*-butene-2 ($CH_3CH = CHCH_3$, boiling point: ca. 1°C, 30°F) and butadiene ($CH_2 = CHCH = CH_2$, boiling point: -4°C, 24°F) as well as higher boiling olefin derivatives are produced by various refining processes.

Refinery gas varies in composition and volume, depending on crude origin and on any additions to the crude made at the loading point. It is not uncommon to reinject light hydrocarbon derivatives such as propane and butane into the crude before dispatch by tanker or pipeline. This results in a higher vapor pressure of the crude, but it allows one to increase the quantity of light products obtained at the refinery. Since light ends in most crude oil markets command a premium, while in the oil field itself propane and butane may have to be reinjected or flared, the practice of *spiking* crude oil with liquefied petroleum gas is becoming fairly common.

In addition to the gases obtained by distillation of crude oil, more highly volatile products result from the subsequent processing of naphtha and middle distillate to produce gasoline, from desulfurization processes involving hydrogen treatment of naphtha, distillate, and residual fuel; and from the coking or similar thermal treatments of vacuum gas oils and residual fuels. The most common processing step in the production of gasoline is the catalytic reforming of hydrocarbon fractions in the heptane (C_7) to decane (C_{10}) range.

In a series of processes commercialized under Platforming, Powerforming, Catforming, and Ultraforming, paraffinic and naphthenic (cyclic non-aromatic) hydrocarbon derivatives, in the presence of hydrogen and a catalyst are converted into aromatic derivatives, or isomerized to more highly branched hydrocarbon derivatives. Catalytic reforming processes thus not only result in the formation of a liquid product of higher octane number, but also produce substantial quantities of gases. The latter are rich in hydrogen, but also contain hydrocarbon derivatives from methane to butanes, with a preponderance of propane ($CH_3CH_2CH_3$), n-butane ($CH_3CH_2CH_2CH_3$) and isobutane [($CH_3)_3CH$]. Their composition will vary in accordance with reforming severity and reformer feedstock. Since all catalytic reforming processes require substantial recycling of a hydrogen stream, it is normal to separate reformer gas into a propane ($CH_3CH_2CH_3$) and/or a butane [$CH_3CH_2CH_2CH_3/(CH_3)_3CH$] stream, which becomes part of the refinery liquefied petroleum gas production, and a lighter gas fraction, part of which is recycled. In view of the excess of hydrogen in the gas, all products of catalytic reforming are saturated, and there are usually no olefinic gases present in either gas stream.

A second group of refining operations that contributes to gas production is that of the catalytic cracking processes. These consists of fluid-bed catalytic cracking, Thermofor catalytic cracking, and other variants in which heavy gas oils are converted into cracked gas, liquefied petroleum gas, catalytic naphtha, fuel oil, and coke by contacting the heavy hydrocarbon with the hot catalyst. Both catalytic and thermal cracking processes, the latter being now largely used for the production of chemical raw materials, result in the formation of unsaturated hydrocarbon derivatives, particularly ethylene ($CH_2 = CH_2$), but also propylene (propene, $CH_3CH = CH_2$), isobutylene [isobutene, $(CH_3)_2C = CH_2$] and the n-butenes ($CH_3CH_2CH = CH_2$, and $CH_3CH = CHCH_3$) in addition to hydrogen (H_2), methane (CH_4) and smaller quantities of ethane (CH_3CH_3), propane ($CH_3CH_2CH_3$), and butanes [$CH_3CH_2CH_2CH_3$, $(CH_3)_3CH$]. Diolefin derivatives such as butadiene ($CH_2 = CHCH = CH_2$) and are also present.

Additional gases are produced in refineries with coking or visbreaking facilities for the processing of their heaviest crude fractions. In the visbreaking process, fuel oil is passed through externally fired tubes and undergoes liquid phase cracking reactions, which result in the formation of lighter fuel oil components. Oil viscosity is thereby reduced, and some gases, mainly hydrogen, methane, and ethane, are formed. Substantial quantities of both gas and carbon are also formed in coking (both fluid coking and delayed coking) in addition to the middle distillate and naphtha. When coking a residual fuel oil or heavy gas oil, the feedstock is preheated and contacted with hot carbon (coke) which causes extensive cracking of the feedstock constituents of higher molecular weight to produce lower molecular weight products ranging from methane, via liquefied petroleum gas(es) and naphtha, to gas oil and heating oil. Products from coking processes tend to be unsaturated and olefinic components predominate in the tail gases from coking processes.

A further source of refinery gas is hydrocracking, a catalytic high-pressure pyrolysis process in the presence of fresh and recycled hydrogen. The feedstock is again heavy gas oil or residual fuel oil, and the process is mainly directed at the production of additional middle distillates and gasoline. Since hydrogen is to be recycled, the gases produced in this process again have to be separated into lighter and heavier streams; any surplus recycle gas and the liquefied petroleum gas from the hydrocracking process are both saturated.

Both hydrocracker and catalytic reformer tail gases are commonly used in catalytic desulfurization processes. In the latter, feedstocks ranging from light to vacuum gas oils are passed at pressures of 500–1000 psi ($3.5–7.0 \times 10^3$ kPa) with hydrogen over a hydrofining catalyst. This results mainly in the conversion of organic sulfur compounds to hydrogen sulfide (Speight, 2014, 2017):

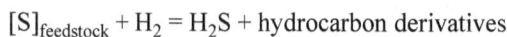

$$[S]_{feedstock} + H_2 = H_2S + \text{hydrocarbon derivatives}$$

The process also produces some lower boiling (lower molecular weight) hydrocarbon derivatives by hydrocracking.

Crude oil refining also produces substantial amounts of carbon dioxide that, with hydrogen sulfide, corrode refining equipment, harm catalysts, pollute the atmosphere, and prevent the use of hydrocarbon components in petrochemical manufacture. When the amount of hydrogen sulfide is high, it may be removed from a gas stream and converted to sulfur or sulfuric acid. Some natural gases contain sufficient carbon dioxide to warrant recovery as dry ice.

Thus refinery streams, while ostensibly being hydrocarbon in nature, may contain large amounts of acid gases such as hydrogen sulfide and carbon dioxide. Most commercial plants employ hydrogenation to convert organic sulfur compounds into hydrogen sulfide. Hydrogenation is effected by means of recycled hydrogen-containing gases or external hydrogen over a nickel molybdate or cobalt molybdate catalyst.

In summary, refinery process gas, in addition to hydrocarbon derivatives, may contain other contaminants, such as carbon oxides (CO_x, where x = 1 and/or 2), sulfur oxides (So_x, where x = 2 and/or 3), as well as ammonia (NH_3), mercaptans (R-SH), and carbonyl sulfide (COS).

Residual refinery gases, usually in more than one stream, which allows a degree of quality control, are treated for hydrogen sulfide removal and gas sales are usually on a thermal content (calorific value, heating value) basis with some adjustment for variation in the calorific value and hydrocarbon type. For fuel uses, gas as specified above presents little difficulty used as supplied. Alternatively a gas of constant Wobbe Index, say for gas turbine use, could readily be produced by the user. Part of the combustion air would be diverted into the gas stream by a Wobbe Index controller. This would be set to supply gas at the lowest Wobbe Index of the undiluted gas.

5.1.1.4 Sulfur Oxides, Nitrogen Oxides, Hydrogen Sulfide, Carbon Dioxide

Sulfur oxides, nitrogen oxides, hydrogen sulfide, and carbon dioxide are commonly produced during refining operations or during use of the refined products. For example, the most common toxic gases present in diesel exhaust include carbon monoxide, sulfur dioxide, nitric oxide, and nitrogen dioxide.

These gases are also classed as primary pollutants because they are emitted directly from the source and then react to produce secondary pollutant, such as acid rain (Speight, 1996). The emissions may include a number of biologically active substances that can pose a major health concern. These gases are classed as pollutants because (1) they may not be indigenous to the location or (2) they are ejected into the atmosphere in a greater-than natural concentration and are, in the current context, the product of human activity. Thus, they can have a detrimental effect on the environment in part or *in toto*.

For these pollutants, the atmosphere has the ability to cleanse itself within hours especially when the effects of the pollutant is minimized by the natural constituents of the atmosphere. For example, the atmosphere might be considered to be cleaning as a result of rain. However, removal of some pollutants from the atmosphere (e.g., sulfates and nitrates) by rainfall results in acid rain that can/will cause serious environmental damage to ecosystems within the water and land systems.

Several methods have been developed to estimate the exposure to such emissions. Most methods are based on either ambient air quality surveys or emission modeling. Exposure to other components of diesel emissions, such as polynuclear aromatic hydrocarbon derivatives, is also higher in occupational settings than it is in ambient environments. The principles of the techniques most often used in exhaust gas analysis include infrared (NDIR and FTIR), chemiluminescence, flame ionization detector (FID and fast FID), and paramagnetic methods.

5.2 Environmental Effects

5.2.1 Gaseous Effluents

Air pollutants are responsible for a number of adverse environmental effects, such as photochemical smog, acid rain, death of forests, or reduced atmospheric visibility. Emissions of greenhouse gases are associated with the global warming. Certain air pollutants, including black carbon, not only contribute to global warming, but are also suspected of having immediate effect on regional climates.

Sulfur is removed from a number of refinery process off-gas streams (sour gas) in order to meet the sulfur oxide emissions limits of the Clean Air Act and to recover saleable elemental sulfur. Process off-gas streams, or sour gas, from the coker, catalytic cracking unit, hydrotreating units and hydroprocessing units can contain high concentrations of hydrogen sulfide mixed with light refinery fuel gases. Before elemental sulfur can be recovered, the fuel gases (primarily methane and ethane) need to be separated from the hydrogen sulfide. This is typically accomplished by dissolving the hydrogen sulfide in a chemical solvent. Solvents most commonly used are amines, such as diethanolamine (DEA). Dry adsorbents such as molecular sieves, activated carbon, iron sponge and zinc oxide are also used. In the amine solvent processes, DEA solution or another amine solvent is pumped to an absorption tower where the gases are contacted and hydrogen sulfide is dissolved in the solution. The fuel gases are removed for use as fuel in process furnaces in other refinery operations. The amine-hydrogen sulfide solution is then heated and steam stripped to remove the hydrogen sulfide gas.

Since the Claus process by itself removes only about 90 percent of the hydrogen sulfide in the gas stream, the Beavon process, the SCOT (Shell Claus Off-gas Treating) process, or the Wellman-Lord process are often used to further recover sulfur. The Claus process consists of partial combustion of the hydrogen sulfide-rich gas stream (with one-third the stoichiometric quantity of air) and then reacting the resulting sulfur dioxide and unburned hydrogen sulfide in the presence of a bauxite catalyst to produce elemental sulfur.

In the Beavon process, the hydrogen sulfide in the relatively low concentration gas stream from the Claus process can be almost completely removed by absorption in a quinone solution. The dissolved hydrogen sulfide is oxidized to form a mixture of elemental sulfur and hydroquinone. The solution is injected with air or oxygen to oxidize the hydroquinone back to quinone. The solution is then filtered or centrifuged to remove the sulfur and the quinone is then reused. The Beavon process is also effective in removing small amounts of sulfur dioxide, carbonyl sulfide, and carbon disulfide that are not affected by the Claus process. These compounds are first converted to hydrogen sulfide at elevated temperatures in a cobalt molybdate catalyst prior to being fed to the Beavon unit. Air emissions from sulfur recovery units will consist of hydrogen sulfide, sulfur oxides, and nitrogen oxides in the process tail gas as well as fugitive emissions and releases from vents.

In the SCOT process, the sulfur compounds in the Claus tail gas are converted to hydrogen sulfide by heating and passing it through a cobalt-molybdenum catalyst with the addition of a reducing gas. The gas is then cooled and contacted with a solution of di-isopropanolamine (DIPA) that removes all but trace amounts of hydrogen sulfide. The sulfide-rich di-isopropanolamine is sent to a stripper where hydrogen sulfide gas is removed and sent to the Claus plant. The di-isopropanolamine is returned to the absorption column.

In the Wellman-Lord process, sodium sulfite is used to capture the sulfur dioxide. The sodium bisulfite thus formed is later heated to evolve sulfur dioxide and regenerate the sulfite scrubbing material. The sulfur dioxide-rich product stream can be compressed or liquefied and oxidized to sulfuric acid, or reduced to sulfur.

Most refinery process units and equipment are manifolded into a collection unit, called the blowdown system. Blowdown systems provide for the safe handling and disposal of liquid and gases that are either automatically vented from the process units through pressure relief valves, or that are manually drawn from units. Recirculated process streams and cooling water streams are often manually purged to prevent the continued buildup of contaminants in the stream. Part or all of the contents of equipment can also be purged to the blowdown system prior to shut down before normal or emergency shutdowns. Blowdown systems utilize a series of flash drums and condensers to separate the blowdown into its vapor and liquid components. The liquid is typically of mixtures of water and hydrocarbon derivatives containing sulfide derivatives, ammonia, and other contaminants, which are sent to the wastewater treatment plant. The gaseous component typically contains hydrocarbon derivatives, hydrogen sulfide, ammonia, mercaptans, solvents, and other constituents, and is either discharged directly to the atmosphere or is combusted in a flare. The major air emissions from blowdown systems are hydrocarbon derivatives in the case of direct discharge to the atmosphere and sulfur oxides when flared.

5.2.2 Liquid Effluents

In terms of waste definition, there are three basic approaches (as it pertains to crude oil, crude oil products, and non-crude oil chemicals) to defining crude oil or a crude oil product as hazardous: (1) a qualitative description of the waste by origin, type, and constituents; (2) classification by characteristics based upon testing procedures; and (3) classification as a result of the concentration of specific chemical substances (Chapter 3).

A wide variety of liquid products are produced from crude oil that varies from the high-volatile naphtha to the low-volatile lubricating oil (Speight, 2014). The liquid products are often characterized by a variety of techniques that include measurement of physical properties and fractionation into group types.

The impact of the release of liquid products on the environment can, in part be predicted from knowledge of the properties of the released liquid. Each part ocular liquid product from crude oil has its own set of unique analytical characteristics (Speight, 1999, 2002). Since these are well documented, there is no need for repetition here. The decision is to include the properties of the lowest boiling liquid product (naphtha) and a high boiling liquid product (fuel oil). From the properties of each product (as determined by analysis) a reasonable estimate can be made of other liquid products but the relationship may not be linear and is subject to the type of crude oil and the distillation range of the product.

Nevertheless, reference is made to the various test methods dedicated to these products and which can be applied to the products boiling in the intermediate range. In the light of the various tests available for composition, such tests will be deemed necessary depending on the environmental situation and the requirements of the legislation as well as at the discretion of the analyst.

5.2.2.1 Naphtha

Naphtha is a liquid crude oil product that boils from about 30°C (86°F) to approximately 200°C (392°F) although there are different grades of naphtha within this extensive boiling range that have different boiling ranges. The term *petroleum solvent* is often used synonymously with *naphtha*.

On a chemical basis, naphtha is difficult to define precisely because it can contain varying amounts of the constituents (paraffin derivatives, naphthene derivatives, aromatic derivatives, and olefin derivatives) in different proportions, in addition to the potential isomers of the paraffin derivatives that exist in the naphtha boiling range (Speight, 2014, 2015). Naphtha is also represented as having a similar boiling range and carbon number to gasoline, being a precursor to gasoline.

The so-called *petroleum ether* solvents are specific boiling range naphtha as is *ligroin*. Thus, the term *petroleum solvent* describes a special liquid hydrocarbon fractions obtained from naphtha and used in industrial processes and formulations (Weissermel and Arpe, 1978). These fractions are also referred to as *industrial naphtha*. Other solvents include *white spirit* that is sub-divided into *industrial spirit* (distilling between 30°C and 200°C, 86°F to 392°F) and *white spirit* (light oil with a distillation range of 135°C to 200°C (275°F to 392°F). The special value of naphtha as a solvent lies in its stability and purity.

Naphtha is produced by any one of several methods that include (1) fractionation of straight-run, cracked, and reforming distillates, or even fractionation of crude oil; (2) solvent extraction; (3) hydrogenation of cracked distillates; (4) polymerization of unsaturated compounds (olefin derivatives); and (5) alkylation processes. In fact, the naphtha may be a combination of product streams from more than one of these processes.

The more common method of naphtha preparation is distillation. Depending on the design of the distillation unit, either one or two naphtha steams may be produced: (1) a single naphtha with an end point of about 205°C (400°F) and similar to straight-run gasoline, or (2) this same fraction divided into a light naphtha and a heavy naphtha. The end point of the light naphtha is varied to suit the subsequent subdivision of the naphtha into narrower boiling fractions and may be of the order of 120°C (250°F).

Sulfur compounds are most commonly removed or converted to a harmless form by chemical treatment with lye, doctor solution, copper chloride, or similar treating agents (Speight, 2014). Hydrorefining processes are also often used in place of chemical treatment. When used as a solvent, naphtha is selected for low sulfur content, and the usual treatment processes, remove only sulfur compounds. Naphtha with a small aromatic content has a slight odor, but the aromatic derivatives increase the solvent power of the naphtha and there is no need to remove aromatic derivatives unless odor-free naphtha is specified.

The variety of applications emphasizes the versatility of naphtha. For example, naphtha is used in paint, printing ink and polish manufacturing and the rubber and adhesive industries as well as in the preparation of edible oils, perfumes, glues and fats. Further uses are found in the dry-cleaning, leather and fur industries and also in the pesticide field. The characteristics that determine the suitability of naphtha for a particular use are volatility, solvent properties (dissolving power), purity, and odor (generally lack thereof).

In order to meet the demands of a variety of uses, certain basic naphtha grades are produced which are identified by boiling range. The complete range of naphtha solvents may be divided, for convenience, into four general categories:

1. Special boiling point spirits having overall distillation range within the limits 30 to 165°C (86 to 329°F),

2. Pure aromatic compounds such as benzene, toluene, ethylbenzene, xylenes, or mixtures (BTEX) thereof;

3. White spirit, also known as mineral spirit and naphtha, usually boiling within 150 to 210°C (302 to 410°F);

4. High boiling crude oil fractions boiling within limits 160 to 325°C (320 to 617°F).

Since the end use dictates the required composition of naphtha, most grades are available in both high and low solvency categories and the various text methods can have major significance in some applications and lesser significance in others. Hence the application and significance of tests must be considered in the light of the proposed end use.

Odor is particularly important since unlike most other crude oil liquids, many of the manufactured products containing naphtha are used in confined spaces, in factory workshops, and in the home.

On the other hand, and at the other end of the spectrum of crude oil liquids, *fuel oil* is applied not only to distillate products, *distillate fuel oil*, but also to residual material that is distinguished from distillate type fuel oil by boiling range and, hence, is referred to as *residual fuel oil* (ASTM D396).

Thus, *residual fuel oil* is fuel oil that is manufactured from the distillation residuum and the term includes all residual fuel oils, including fuel oil obtained by visbreaking as well as by blending residual products from other operations. The various grades of heavy fuel oils are produced to meet rigid specifications in order to assure suitability for their intended purpose.

Detailed analysis of residual products, such as residual fuel oil, is more complex than the analysis of lower molecular weight liquid products. As with other products, there are a variety of physical property measurements that are required to determine of the residual fuel oil meets specification. But the range of molecular types present in crude oil products increases significantly with an increase in the molecular weight (i.e., and increase in the number of carbon atoms per molecule). Therefore, characterization measurements or studies cannot, and do not, focus on the identification of specific molecular structures. The focus tends to be on molecular classes (paraffin derivatives, naphthene derivatives, aromatic derivatives, polycyclic compounds, and polar compounds).

Several tests that are usually applied to the lower molecular weight colorless (or light-colored) products are not applied to residual fuel oil. For example, test methods such as those designed for the determination of the aniline point (or mixed aniline point) (ASTM D611, IP 2) and the cloud point (ASTM D2500, ASTM D5771, ASTM D5772, ASTM D5773) can suffer from visibility effects due to the color of the fuel oil.

Because of the high standards set for naphtha (McCann, 1998) it is essential to employ the correct techniques when taking samples for testing (ASTM D270, ASTM D4057). Mishandling, or the slightest trace of contaminant can give rise to misleading results. Special care is necessary to ensure that containers are scrupulously clean and free from odor. Samples should be taken with the minimum of disturbance so as to avoid loss of volatile components; in the case of low-boiling naphtha may be necessary to chill the sample. And, while awaiting examination samples should be kept in a cool dark place so as to ensure that they do not lose volatile constituents or discolor and develop odors due to oxidation.

The physical properties of naphtha depend on the hydrocarbon types present, in general the aromatic hydrocarbon derivatives having the highest solvent power and the straight-chain aliphatic compounds the lowest. The solvent properties can be assessed by estimating of the amount of the various hydrocarbon types present be made. This method provides an indication of the solvent power of the naphtha on the basis that aromatic constituents and naphthenic constituents provide dissolving ability that paraffinic constituents do not. Another method for assessing the solvent properties of naphtha measures the performance of the fraction when used as a solvent under specified conditions such as, for example, by the Kauri Butanol test method (ASTM D1133). Another method involves measurement of the surface tension from which the solubility parameter is calculated and then provides an indication of dissolving power and compatibility. Such calculations have been used to determine the yield of the asphaltene fraction from crude oil by use of various solvents (Mitchell and Speight, 1973; Speight, 1999; Speight, 200). A similar principal is applied to determine the amount of insoluble material in lubricating oil using *n*-pentane (ASTM D893, ASTM D4055).

Insoluble constituents in lubricating oil can cause wear that can lead to equipment failure. Pentane insoluble materials s can include oil-insoluble materials and some oil-insoluble resinous matter originating from oil or additive degradation, or both. Toluene insoluble constituents arise from external contamination, fuel carbon, and highly carbonized materials from degradation of fuel, oil, and additives, or engine wear and corrosion materials. A significant change in pentane or toluene insoluble constituents indicates a change in oil properties that could lead to machinery failure. The insoluble constituents measured can also assist in evaluating the performance characteristics of used oil or in determining the cause of equipment failure.

Thus, one test (ASTM D893) covers the determination of pentane and toluene insoluble constituents in used lubricating oils using pentane dilution and centrifugation as the method of separation. The other test (ASTM D4055) uses pentane dilution followed by membrane filtration to remove insoluble constituents that have a size greater than 0.8 micron.

The number of potential hydrocarbon isomers in the naphtha boiling range renders complete speciation of individual hydrocarbon derivatives impossible for the naphtha distillation range and methods are used that identify the hydrocarbon types as chemical groups rather than as individual constituents.

The data from the density (specific gravity) test method (ASTM D1298, IP 160) provides a means of identification of a grade of naphtha but is not a guarantee of composition and can only be used to indicate evaluate product composition or quality when used in conjunction with the data from other test methods. Density data are used primarily to convert naphtha volume to a weight basis, a requirement in many of the industries concerned. For the necessary temperature corrections and also for volume corrections the appropriate sections of the crude oil measurement tables (ASTM D1250, IP 200) are used.

The first level of compositional information is group-type totals as deduced by adsorption chromatography (ASTM D1319, IP 156) to give volume percent saturates, olefin derivatives, and aromatic derivatives in materials that boil below 315°C (600°F). In this test method, a small amount of sample is introduced into a glass adsorption column packed with activated silica gel, of which a small layer contains a mixture of fluorescent dyes. When the sample has been adsorbed on the gel, alcohol is added to desorb the sample down the column and the hydrocarbon constituents are separated according to their affinities into three types (aromatic derivatives, olefin derivatives, and saturates). The fluorescent dyes also react selectively with the hydrocarbon types, and make the boundary zones visible under ultraviolet light. The volume percentage of each hydrocarbon type is calculated from the length of each zone in the column.

There are other test methods available. Benzene content and other aromatic derivatives may be estimated by spectrophotometric analysis (ASTM D1017) and also by gas-liquid chromatography (ASTM D2267, ASTM D2600, IP 262). However, two test methods based on the adsorption concept (ASTM D2007, ASTM D2549) are used for classifying oil samples of initial boiling point of at least 200°C (392°F) into the hydrocarbon types of polar compounds, aromatic derivatives, and saturates, and recovery of representative fractions of these types. Such methods are unsuitable for the majority of naphtha samples because of volatility constraints.

An indication of naphtha composition may also be obtained from the determination of aniline point (ASTM D1012, IP 2), freezing point (ASTM D852, ASTM D1015, ASTM D1493), cloud point (ASTM D2500), and the solidification point (ASTM D1493). And although refinery treatment should ensure no alkalinity and acidity (ASTM D847, ASTM D1093, ASTM D1613, ASTM D2896, IP 1) and no olefin derivatives present, the relevant tests using bromine number (ASTM D875, ASTM D1159, IP 130), bromine index (ASTM D2710), and flame ionization absorption (ASTM D1319, IP 156) are necessary to insure low levels (at the maximum) of hydrogen sulfide (ASTM D853) as well as the sulfur compounds in general (ASTM D130, ASTM D849, ASTM D1266, ASTM D2324, ASTM D3120, ASTM D4045, ASTM D6212, IP 107, IP 154) and especially corrosive sulfur compounds such as are determined by the doctor test method (ASTM D4952).

Since aromatic content is a key property of low-boiling distillates such as naphtha and gasoline because the aromatic constituents influence a variety of properties including boiling range (ASTM D86, IP 123), viscosity (ASTM D88, ASTM D445, ASTM D2161, IP 71), stability (ASTM D525, IP 40), and compatibility (ASTM D1133) with a variety of solutes. Existing methods use physical measurements and need suitable standards. Tests such as aniline point (ASTM D611) and Kauri-butanol number (ASTM D1133) are of a somewhat empirical nature and can serve a useful function as control tests. Naphtha composition, however, is monitored mainly by gas chromatography and although most of the methods may have been developed for gasoline (ASTM D2427, ASTM D6296), the applicability of the methods to naphtha is sound.

A multidimensional gas chromatographic method (ASTM D5443) provides for the determination of paraffin derivatives, naphthene derivatives, and aromatic derivatives by carbon number in low olefinic hydrocarbon streams having final boiling points lower than 200°C (392°F). In the method, the sample is injected into a gas chromatographic system that contains a series of columns and switching values. First a polar column retains polar aromatic compounds, bi-naphthene derivatives, and high boiling paraffin derivatives and naphthene derivatives. The eluent from this column goes through a platinum column that hydrogenates olefin derivatives, and then to a molecular sieve column that performs a carbon number separation based on the molecular structure, that is, naphthene derivatives and paraffin derivatives. The fraction remaining on the polar column is further divided into three separate fractions that are then separated on a non-polar column by boiling point. A flame ionization detector detects eluting compounds.

In another method (ASTM D4420) for the determination of the amount of aromatic constituents, a two column chromatographic system connected to a dual filament thermal conductivity detector (or two single filament detectors) is used. The sample is injected into the column containing a polar liquid phase. The non-aromatic derivatives are directed to the reference side of the detector and vented to the atmosphere as they elute. The column is back-flushed immediately before the elution of benzene, and the aromatic portion is directed into the second column containing a non-polar liquid phase. The aromatic components elute in the order of their boiling points and are detected on the analytical side of the detector. Quantitation is achieved by utilizing peak factors obtained from the analysis of a sample having a known aromatic content.

Other methods for the determination of aromatic derivatives in naphtha include a method (ASTM D5580) using a flame ionization detector is also available and methods in which a combination of gas chromatography and Fourier transform infrared spectroscopy (GC-FTIR) (ASTM D5986) and gas chromatography and mass spectrometry (GC-MS) (ASTM D5769).

Hydrocarbon composition is also determined by mass spectrometry—a technique that has seen wide use for hydrocarbon-type analysis of naphtha and gasoline (ASTM D2789) as well as to the identification of hydrocarbon constituents in higher boiling naphtha fractions (ASTM D2425).

One method (ASTM D6379, IP 436) is used to determine the mono-aromatic and di-aromatic hydrocarbon contents in distillates boiling in the range from 50 to 300°C (122 to 572°F). In the method the sample is diluted with an equal volume of hydrocarbon, such as heptane, and a fixed volume of this solution is injected into a high performance liquid chromatograph fitted with a polar column where separation of the aromatic hydrocarbon derivatives from the non-aromatic hydrocarbon derivatives occurs. The separation of the aromatic constituents appears as distinct bands according to ring structure and a refractive index detector is used to identify the components as they elute from the column. The peak areas of the aromatic constituents are compared with those obtained from previously run calibration standards to calculate the %w/w mono-aromatic hydrocarbon constituents and di-aromatic hydrocarbon constituents in the sample.

Compounds containing sulfur, nitrogen, and oxygen could possibly interfere with the performance of the test. Mono-alkenes do not interfere, but conjugated di- and poly-alkenes, if present may interfere with the test performance.

Another method (ASTM D2425) provides more compositional detail (in terms of molecular species) than chromatographic analysis and the hydrocarbon types are classified in terms of a Z-series in which Z (in the empirical formula C_nH_{2n+z}) is a measure of the hydrogen deficiency of the compound. This method requires that the sample be separated into saturate and aromatic fractions before mass spectrometric analysis (ASTM D2549) and the separation is applicable to some fractions not others. For example, the method is applicable to high-boiling naphtha but not to the low-boiling naphtha since it is impossible to evaporate the solvent used in the separation without also losing the lower boiling constituents of the naphtha under investigation.

The percentage of aromatic hydrogen atoms and aromatic carbon atoms can be determined by high-resolution nuclear magnetic resonance spectroscopy (ASTM D5292) that gives the mol percent of aromatic hydrogen or carbon atoms. Proton (hydrogen) magnetic resonance spectra are obtained on sample solutions in either chloroform or carbon tetrachloride using a continuous wave or pulse Fourier transform high-resolution magnetic resonance spectrometer. Carbon magnetic resonance spectra are obtained on the sample solution in chloroform-*d* using a pulse Fourier transform high-resolution magnetic resonance.

The data obtained by this method (ASTM D5292) can be used to evaluate changes in aromatic contents in naphtha as well as kerosene, gas oil, mineral oil, and lubricating oil. However, results from this test are not equivalent to mass- or volume-percent aromatic derivatives determined by the chromatographic methods since the chromatographic methods determine the percent by weight or percent by volume of molecules that have one or more aromatic rings and alkyl substituents on the rings will contribute to the percentage of aromatic derivatives determined by chromatographic techniques.

Low-resolution nuclear magnetic resonance spectroscopy can also be used to determine percent by weight hydrogen in jet fuel (ASTM D3701) and in light distillate, middles distillate, and gas oil (ASTM D4808). As noted above, chromatographic methods are not applicable to naphtha where losses can occur by evaporation.

The nature of the uses found for naphtha demands compatibility with the many other materials employed in formulation, with waxes, pigments, resins, etc.; thus the solvent properties of a given fraction must be carefully measured and controlled. For most purposes volatility is important, and, because of the wide use of naphtha in industrial and recovery plants, information on some other fundamental characteristics is required for plant design.

Although the focus of many tests is analysis of the hydrocarbon constituents of naphtha and other crude oil fractions, heteroatoms compounds that contain sulfur and nitrogen atoms cannot be ignored and methods for their determination are available. The combination of gas chromatography with element selective detection gives information about the distribution of the element. In addition, many individual heteroatomic compounds can be determined.

Nitrogen compounds in middle distillates can be selectively detected by chemiluminescence. Individual nitrogen compounds can be detected down to 100 ppb nitrogen. Gas chromatography with either sulfur chemiluminescence detection or atomic emission detection has been used for sulfur selective detection.

Estimates of the purity of these products were determined in laboratories using a variety of procedures such as freezing point, flame ionization absorbance, ultraviolet absorbance, gas chromatography, and capillary gas chromatography, (ASTM D850, ASTM D852, ASTM D853, ASTM D848, ASTM D849, ASTM D1015, ASTM D1016, ASTM D1078, ASTM D1319, ASTM D2008, ASTM D2236, ASTM D2306, ASTM D2360, ASTM D5917, IP 156).

Gas chromatography (GC) has become a primary technique for determining hydrocarbon impurities in individual aromatic hydrocarbon derivatives and the composition of mixed aromatic hydrocarbon derivatives. Although a measure of purity by gas chromatography is often sufficient, gas chromatography is not capable of measuring absolute purity; not all possible impurities will pass through the gas chromatography column, and not all those that do will be measured by the detector. Despite some shortcomings, gas chromatography is a standard, widely used technique and is the basis of many current test methods for aromatic hydrocarbon derivatives (ASTM D2306, ASTM D2360, ASTM D3054, ASTM D3750, ASTM D3797, ASTM D3798, ASTM D4492, ASTM D4534, ASTM D4735, ASTM D5060, ASTM D5135, ASTM D5713, ASTM D5917, ASTM D6144).

When classes of hydrocarbon derivatives, such as olefin derivatives, need to be measured by techniques such as bromine index are used (ASTM D1492, ASTM D5776).

Impurities other than hydrocarbon derivatives are of concern in the crude oil industry. For example, many catalytic processes are sensitive to sulfur contaminants. Consequently, there is also a series of methods to determine trace concentrations of sulfur-containing compounds (ASTM D1685, ASTM D3961, ASTM D4045, ASTM D4735).

Chloride-containing impurities are determined by various test methods (ASTM D5194, ASTM D5808, ASTM D6069) that have sensitivity to 1 mg/kg, reflecting the needs of industry to determine very low levels of these contaminants.

Water is a contaminant in naphtha and should be measured using the Karl Fischer method (ASTM E-203, ASTM D1364, ASTM D1744, ASTM D4377, ASTM D4928, ASTM D6304), by distillation (ASTM D4006), or by centrifuging (ASTM D96) and excluded by relevant drying methods.

Tests should also be carried out for sediment if the naphtha has been subjected to events (such as oxidation) that could lead to sediment formation and instability of the naphtha and the resulting products. Test methods are available for the determination of sediment by extraction (ASTM D473, IP 285) or by membrane filtration (ASTM D4807, IP 286) and the determination of simultaneously sediment with water by centrifugation (ASTM D96, ASTM D1796, ASTM D2709, ASTM D4007, IP 373, IP 374).

The significance of the measured properties of residual fuel oil is dependent to a large extent on the ultimate uses of the fuel oil. Such uses include steam generation for various processes as well as electrical power generation and propulsion. Corrosion, ash deposition, atmospheric pollution, and product contamination are side effects of the use of residual fuel oil and in particular cases properties such as vanadium, sodium and sulfur contents may be significant.

The character of fuel oil generally renders the usual test methods for *total petroleum hydrocarbons* ineffective since high proportions of the fuel oil (specifically residual fuel oil) are insoluble in the usual solvents employed for the test. In particular, the asphaltene constituents are insoluble in hydrocarbon solvents and are only soluble in aromatic solvents and chlorinated hydrocarbon derivatives (chloroform, methylene dichloride and the like). Residua and asphalt have high proportions of asphaltene constituents that render any test for *total petroleum hydrocarbons* meaningless, unless a suitable solvent is employed in the test method.

Testing residual fuel oil does not suffer from the issues that are associated with sample volatility but the test methods are often sensitive to the presence of gas bubbles in the fuel oil. An air release test is available for application to lubricating oil (ASTM D3427, IP 313) and may be applied, with modification, to residual fuel oil. However, with dark-colored samples, it may be difficult to determine whether all air bubbles have been eliminated. And, as with the analysis and testing of other crude oil products, the importance of correct sampling of fuel oil cannot be over emphasized, because no proper assessment of quality may be made unless the data are obtained on truly representative samples (ASTM D270).

The asphaltene fraction (ASTM D893, ASTM D2006, ASTM D2007, ASTM D3279, ASTM D4124, ASTM D6560) is the highest molecular weight and most complex fraction in crude oil. The asphaltene content gives an indication of the amount of coke that can be expected during exposure to thermal conditions (Speight, 2014, 2015).

In any of the methods for the determination of the asphaltene content, the residual fuel oil is mixed with a large excess (usually > 30 volumes hydrocarbon per volume of sample) low-boiling hydrocarbon such as *n*-pentane or *n*-heptane. For an extremely viscous sample, a solvent such as toluene may be used prior to the addition of the low-boiling hydrocarbon but an additional amount of the hydrocarbon (usually > 30 volumes hydrocarbon per volume of solvent) must be added to compensate for the presence of the solvent. After a specified time, the insoluble material (the asphaltene fraction) is separated (by filtration) and dried. The yield is reported as percentage (% w/w) of the original sample. Furthermore, different hydrocarbon derivatives (such as *n*-pentane or *n*-heptane) give different yields of the asphaltene fraction and if the presence of the solvent is not compensated by use of additional hydrocarbon the yield will be erroneous. In addition, if the hydrocarbon is not present in a large excess, the yields of the asphaltene fraction will vary and will be erroneous (Speight, 2014, 2015).

Another method, not specifically described as an asphaltene separation method, is designed to remove pentane insoluble constituents by membrane filtration (ASTM D4055). In the method, a sample of oil is mixed with pentane in a volumetric flask, and the oil solution is filtered through a 0.8-micron membrane filter. The flask, funnel, and the filter are washed with pentane to completely transfer any particulates onto the filter after which the filter (with particulates) is dried and weighed to give the pentane insoluble constituents as a percent by weight of the sample.

Particulates can also be determined by membrane filtration (ASTM D2276, ASTM D5452, ASTM D6217).

The *precipitation number* is often equated to the asphaltene content but there are several issues that remain obvious in its rejection for this purpose. For example, the method to determine the precipitation number (ASTM D91) advocates the use of naphtha for use with black oil or lubricating oil and the amount of insoluble material (as a % v/v of the sample) is the precipitating number. In the test, 10 ml of sample is mixed with 90 ml of ASTM precipitation naphtha (that may or may not have a constant chemical composition) in a graduated centrifuge cone and centrifuged for 10 min at 600 to 700 rpm. The volume of material on the bottom of the centrifuge cone is noted until repeat centrifugation gives a value within 0.1 ml (the precipitation number). Obviously, this can be substantially different to the asphaltenes content.

If the residual fuel oil is produced by a thermal process such as visbreaking, it may also be necessary to determine if toluene insoluble material is present by the methods, or modifications thereof, used to determine the toluene insoluble of tar and pitch (ASTM D4072, ASTM D4312). In the methods, a sample is digested at 95°C (203°F) for 25 minutes and then extracted with hot toluene in an alundum thimble. The extraction time is eighteen hours (ASTM D4072) or three hours (ASTM D4312). The insoluble matter is dried and weighed.

The composition of residual fuel oils is varied is often reported in the form of four or five major fractions as deduced by adsorption chromatography. In the case of cracked feedstocks, thermal decomposition products (carbenes and carboids) may also be present.

Column chromatography is used for several hydrocarbon type analyses that involve fractionation of viscous oils (ASTM D2007, ASTM D2549), including residual fuel oil. The former method (ASTM D2007) advocates the use of adsorption on clay and clay-silica gel followed by elution of the clay with pentane to separate saturates; elution of clay with acetone-toluene to separate polar compounds; and elution of the silica gel fraction with toluene to separate aromatic compounds. The latter method (ASTM D2549) uses adsorption on a bauxite-silica gel column. Saturates are eluted with pentane; aromatic derivatives are eluted with ether, chloroform, and ethanol.

Correlative methods are derived relationships between fundamental chemical properties of a substance and measured physical or chemical properties. They provide information about oil from readily measured properties (ASTM D2140, ASTM D2501, ASTM D2502, ASTM D3238). One method (ASTM D2501) describes the calculation of the viscosity-gravity coefficient (VGC)—a parameter derived from kinematic viscosity and density that has been found to relate to the saturate/aromatic composition. Correlations between the viscosity-gravity coefficient (or molecular weight and density) and refractive index to calculate carbon type composition in percent of aromatic, naphthenic, and paraffinic carbon atoms are employed to estimate of the number of aromatic and naphthenic rings present (ASTM D2140, ASTM D3238). Another method (ASTM D2502) permits estimation of molecular weight from kinematic viscosity measurements at 38 and 99°C (100 and 210°F) (ASTM D445). It is applicable to samples with molecular weights in the range from 250 to 700 but should not be applied indiscriminately for oils that represent extremes of composition for which different constants are derived (Moschopedis et al., 1976).

A major use for *gas chromatography* for hydrocarbon analysis has been simulated distillation, as discussed previously. Other gas chromatographic methods have been developed for contaminant analysis (ASTM D3524, ASTM D4291).

The aromatic content of fuel oil is a key property that can affect a variety of other properties including viscosity, stability, and compatibility of with other fuel oil or blending stock. Existing methods for this work use physical measurements and need suitable standards. Thus, methods have been standardized using nuclear magnetic resonance (NMR) for hydrocarbon characterization (ASTM D4808, ASTM D5291, ASTM D5292). The nuclear magnetic resonance method is simpler and more precise. Procedures are described that cover light distillates with a 15 to 260°C boiling range, middle distillates and gas oils with boiling ranges of 200 to 370°C and 370 to 510°C, and residuum boiling above 510°C. One of the methods (ASTM D5292) is applicable to a wide range of hydrocarbon oils that are completely soluble in chloroform and carbon tetrachloride at ambient temperature. The data obtained by this method can be used to evaluate changes in aromatic contents of hydrocarbon oils due to process changes.

High ionizing voltage mass spectrometry (ASTM D2786, ASTM D3239) is also employed for compositional analysis of residual fuel oil. These methods require preliminary separation using elution chromatography (ASTM D2549). A third method (ASTM D2425) may be applicable to some residual fuel oil samples in the lower molecular weight range.

The problem of instability in residual fuel oil may manifest itself either as waxy sludge deposited on the soil or as fouling coastlines.

Asphaltene-type deposition may, however, result from the mixing of fuels of different origin and treatment, each of which may be perfectly satisfactory when used alone. For example, straight run fuel oils from the same crude oil are normally stable and mutually compatible whereas fuel oils produced from thermal cracking and visbreaking operations that may be stable but can be unstable or incompatible if blended with straight run fuels and vice versa (ASTM D1661).

Another procedure for predicting the stability of residual fuel oil involves the use of a spot test to show compatibility or cleanliness of the blended fuel oil (ASTM D2781, ASTM D4740). The former method (ASTM D2781) covers two spot test procedures for rating a residual fuel with respect to its compatibility with a specific distillate fuel. Procedure A indicates the degree of asphaltene deposition that may be expected in blending the components and is used when wax deposition is not considered a fuel application problem. Procedure B indicates the degree of wax and asphalt deposition in the mixture at room temperature. The latter method (ASTM D4740) is applicable to fuel oils with viscosities up to 50 cSt at 100°C (212°F) to identify fuels or blends that could result in excessive centrifuge loading, strainer plugging, tank sludge formation, or similar operating problems. In the method, a drop of the preheated sample is put on a test paper and placed in an oven at 100°C. After 1 hour, the test paper is removed from the oven and the resultant spot is examined for evidence of suspended solids and rated for cleanliness using the procedure described in the method. In a parallel procedure for determining *compatibility*, a blend composed of equal volumes of the fuel oil sample and the blend stock is tested and rated in the same way as just described for the *cleanliness* procedure.

For oxidative stability, an important effect after a spill, a test method (ASTM D4636) is available to determine resistance to oxidation and corrosion degradation and their tendency to corrode various metals. The test method consists of one standard and two alternative procedures. In the method, a large glass tube containing an oil sample and metal specimens is placed in a constant temperature bath (usually from 100 to 360°C) and heated for the specific number of hours while air is passed through the oil to provide agitation and a source of oxygen. Corrosiveness of the oil is determined by the loss in metal mass, and microscopic examination of the sample metal surface(s). Oil samples are withdrawn from the test oil and checked for changes in viscosity and acid number as a result of the oxidation reactions. At the end of the test the amount of the sludge present in the oil remaining in the same tube is determined by centrifugation. Also, the quantity of oil lost during the test is determined gravimetrically. Metals used in the basic test and alternative test are aluminum, bronze, cadmium, copper, magnesium, silver, steel, and titanium. Other metals may also be specified as determined by the history and storage of the fuel oil.

5.2.2.2 Wastewater

Wastewaters from crude oil refining consist of cooling water, process water, storm water, and sanitary sewage water. A large portion of water used in crude oil refining is used for cooling. Most cooling water is recycled over and over. Cooling water typically does not come into direct contact with process oil streams and therefore contains less contaminants than process wastewater. However, it may contain some oil contamination due to leaks in the process equipment. Water used in processing operations accounts for a significant portion of the total wastewater. Process wastewater arises from desalting crude oil, steam stripping operations, pump gland cooling, product fractionator reflux drum drains and boiler blowdown. Because process water often comes into direct contact with oil, it is usually highly contaminated. Storm water (i.e., surface water runoff) is intermittent and will contain constituents from spills to the surface, leaks in equipment and any materials that may have collected in drains. Runoff surface water also includes water coming from crude and product storage tank roof drains.

Waste waters are treated in onsite wastewater treatment facilities and then discharged to publicly owned treatment works (POTWs) or discharged to surfaces waters under National Pollution Discharge Elimination System (NPDES) permits. Crude oil refineries typically utilize primary and secondary wastewater treatment. Primary wastewater treatment consists of the separation of oil, water and solids in two stages. During the first stage, an API separator, a corrugated plate interceptor, or other separator design is used. Wastewater moves very slowly through the separator allowing free oil to float to the surface and be skimmed off, and solids to settle to the bottom and be scraped off to a sludge collecting hopper. The second stage utilizes physical or chemical methods to separate emulsified oils from the wastewater. Physical methods may include the use of a series of settling ponds with a long retention time, or the use of dissolved air flotation (DAF). In DAF, air is bubbled through the wastewater, and both oil and suspended solids are skimmed off the top. Chemicals, such as ferric hydroxide or aluminum hydroxide, can be used to coagulate impurities into a froth or sludge that can be more easily skimmed off the top. Some wastes associated with the primary treatment of wastewater at crude oil refineries may be considered hazardous and include: API separator sludge, primary treatment sludge, sludge from other gravitational separation techniques, float from DAF units, and wastes from settling ponds.

After primary treatment, the wastewater can be discharged to a publicly owned treatment works (POTW) or undergo secondary treatment before being discharged directly to surface waters under a National Pollution Discharge Elimination System (NPDES) permit. In secondary treatment, dissolved oil and other organic pollutants may be consumed biologically by microorganisms. Biological treatment may require the addition of oxygen through a number of different techniques, including activated sludge units, trickling filters, and rotating biological contactors. Secondary treatment generates bio-mass waste that is typically treated anaerobically and then dewatered.

Some refineries employ an additional stage of wastewater treatment called polishing to meet discharge limits. The polishing step can involve the use of activated carbon, anthracite coal, or sand to filter out any remaining impurities, such as biomass, silt, trace metals and other inorganic chemicals, as well as any remaining organic chemicals.

Certain refinery wastewater streams are treated separately, prior to the wastewater treatment plant, to remove contaminants that would not easily be treated after mixing with other wastewater. One such waste stream is the sour water drained from distillation reflux drums. Sour water contains dissolved hydrogen sulfide and other organic sulfur compounds and ammonia which are stripped in a tower with gas or steam before being discharged to the wastewater treatment plant.

Wastewater treatment plants are a significant source of refinery air emissions and solid wastes. Air releases arise from fugitive emissions from the numerous tanks, ponds and sewer system drains. Solid wastes are generated in the form of sludge from a number of the treatment units.

Many refineries unintentionally release, or have unintentionally released in the past, liquid hydrocarbon derivatives to ground water and surface waters. At some refineries contaminated ground water has migrated off-site and resulted in continuous "seeps" to surface waters. While the actual volume of hydrocarbon derivatives released in such a manner are relatively small, there is the potential to contaminate large volumes of ground water and surface water possibly posing a substantial risk to human health and the environment.

The overall method for the analysis of wastewater includes sample collection and storage, extraction, and analysis steps. Sampling strategy is an important step in the overall process. Care must be taken to assure that the samples collected are representative of the environmental medium and that they are collected without contamination. There is an extensive list of test methods for water analysis (Tables 5.2, Table 5.3, Table 5.4) that includes numerous modifications of the original methods but most involve alternate extraction methods developed to improve overall method performance for the analysis. Solvent extraction methods with hexane are also in use.

5.2.3 Solid Effluents

In terms of waste definition, there are three basic approaches (as it pertains to crude oil, crude oil products, and non-crude oil chemicals) to defining crude oil or a crude oil product as hazardous: (1) a qualitative description of the waste by origin, type, and constituents; (2) classification by characteristics based upon testing procedures; and (3) classification as a result of the concentration of specific chemical substances (Chapter 3).

Solid effluents are generated from many of the refining processes, crude oil handling operations, as well as wastewater treatment. Both hazardous and non-hazardous wastes are generated, treated and disposed. Refinery wastes are typically in the form of sludge (including sludge from wastewater treatment), spent process catalysts, filter clay, and incinerator ash. Treatment of these wastes includes incineration, land treating off-site, land filling onsite, land filling off-site, chemical fixation, neutralization, and other treatment methods.

A significant portion of the non-crude oil product outputs of refineries is transported off-site and sold as byproducts. These outputs include sulfur, acetic acid, phosphoric acid, and recovered

Table 5.2: Examples of Cost Categories.

Capital Costs	Operating Costs
Site preparation • Site clearing • Site access • Borehole drilling • Permits/licenses • Gas, electricity, and water	*Direct labor* • Direct labor to operate equipment • Direct labor supervision • Payroll expenses • Contract labor • Maintenance direct labor
Structures • Buildings • Platforms • Equipment structures • Equipment shed/warehouse	*Direct materials* • Process materials and chemicals • Utilities • Fuels • Replacement parts
Process equipment and appurtenances • Cost of technology parts and supplies • Materials and supplies to make technology operative	*Overhead* • Plant and equipment maintenance • Equipment rental for operations • Transportation • Licensing
Non-process equipment • Office and administrative equipment • Data processing/computer equipment • Safety equipment • Vehicles	*General and administrative* • Administrative labor • Marketing • Communications • Travel expenses
Utilities • Plumbing • Heating • Security • Vent equipment	*Site Management* • Waste disposal • Health and safety requirements • Analytical services • Regulatory reporting

Table 5.3: Some Elements That Affect the Cost Effectiveness of a Remediation Technology.

Contaminated site properties	Contaminated ground water properties	Contaminant characteristics
Conditions of site access	pH, dissolved oxygen concentration	Concentration profile
Access to power utilities	Total dissolved solids concentration Hardness Iron concentration Concentrations of other contaminants	Character Quantity
Soil classification	Redox potential	
Dimensions Volume	Soil adsorption/desorption properties	

metals. Metals from catalysts and from the crude oil that have deposited on the catalyst during the production often are recovered by third party recovery facilities.

Storage tanks are used throughout the refining process to store crude oil and intermediate process feeds for cooling and further processing. Finished crude oil products are also kept in storage tanks before transport off site. Storage tank bottoms are mixtures of iron rust from corrosion, sand, water, and emulsified oil and wax, which accumulate at the bottom of tanks. Liquid tank bottoms (primarily water and oil emulsions) are periodically drawn off to prevent their continued build up. Tank bottom liquids and sludge are also removed during periodic cleaning of tanks for inspection. Tank bottoms may contain amounts of tetraethyl or tetramethyl lead (although this is increasingly rare due to the phase-out of leaded products), other metals, and phenols. Solids generated from leaded gasoline storage tank bottoms are listed as a RCRA hazardous waste.

Human system	Effects of crude oil hydrocarbons
Nervous system	Single exposure to a moderately high concentration of virtually any hydrocarbon solvent vapor will cause a general depression of Central Nervous System (CNS) which, at high doses, will lead to unconsciousness Short-term exposure by repeated inhalation, to xylene, toluene, white spirit and jet fuel has shown an impairment of concentration Can cause damage to the peripheral nerves
Respiratory system	Cough and/or shortness of breath Inhalation or aspiration may lead to an asthma-like reactive airway syndrome as well as a chemical pneumonitis
Reproductive system	can exert negative effects on various female reproductive sites, including the CNS-pituitary-ovarian axis, their signaling molecules and receptors, ovarian follicles, corpora lutea, oocytes, embryos, oviducts, ovarian cycles, fertility, and the viability of offspring
Renal system	Hydrocarbons (mostly toluene) may cause a metabolic acidosis, leading to renal tubular acidosis, urinary calculi, glomerulonephritis, hyperchloremia, and hypokalemia
Cardiovascular system	Arrhythmias may be induced following exposure
Digestive system	Ingestion may cause irritation of the gastrointestinal tract as well as breakdown of the epithelium, leading to nausea, vomiting, abdominal pain, and hematemesis
Skin	Skin exposure may cause mild irritation

5.2.3.1 Residua and Asphalt

The importance of residua and asphalt to the environmental analyst arises from spillage or leakage in the refinery or on the road. In either case, the properties of these materials are detrimental to the ecosystem in which the release occurred. As with other crude oil products, knowledge of the properties of residua and asphalt can help determine the potential cleanup methods and may even allow regulators trace the product to the refinery where it was produced. In addition, the character of residua and asphalt render the usual test methods for *total petroleum hydrocarbons* ineffective since high proportions of asphalt and residua are insoluble in the usual solvents employed for the test. Application of the test methods for *total petroleum hydrocarbons* to fuel oil is also subject to similar limitations.

Residua are the dark colored near solid or solid products of crude oil refining that are produced by atmospheric and vacuum distillation (Speight, 2014, 2017). Asphalt is usually produced from a residuum and is a dark brown to black cementitious material obtained from crude oil processing and which contains very high molecular weight molecular polar species called asphaltenes that are soluble in carbon disulfide, pyridine, aromatic hydrocarbon derivatives, and chlorinated hydrocarbon derivatives (Chapter 3).

Residua and asphalt derive their characteristics from the nature of their crude oil precursor, the distillation process being a concentration process in which most of the heteroatoms and polynuclear aromatic constituents of the feedstock are concentrated in the residuum (Speight, 2001). Asphalt may be similar to its parent residuum but with some variation possible by choice of manufacturing process. In general terms, residua and asphalt are a hydrocarbonaceous material that consist constituents (containing carbon, hydrogen, nitrogen, oxygen, and sulfur) that are completely soluble in carbon disulfide (ASTM D4). Trichloroethylene or 1,1,1-trichloroethane has been used in recent years as solvents for the determination of asphalt (and residua) solubility (ASTM D2042).

The residua from which asphalt are produced were once considered the garbage of a refinery, have little value and little use, other than as a road oil. In fact, the development of delayed coking (once the so-called the refinery garbage can) was with the purpose of converting residua to liquids (valuable products) and coke (fuel).

Asphalt manufacture involves distilling everything possible from crude oil until a residuum with the desired properties is obtained. This is usually done by stages in which distillation at atmospheric pressure removes the lower boiling fractions and yields an atmospheric residuum (*reduced crude*) that may contain higher boiling (lubricating) oils, wax, and asphalt. Distillation of the reduced crude under vacuum removes the oils (and wax) as overhead products and the residuum remains as a bottom (or residual) product. The majority of the polar functionalities and high molecular weight species in the original crude oil, which tend to be non-volatile, concentrate in the vacuum residuum (Speight, 2000) thereby conferring desirable or undesirable properties on the residuum.

At this stage the residuum is frequently, but incorrectly, referred to as pitch and has a softening point (ASTM D36, ASTM D61, ASTM D2319, ASTM D3104, ASTM D3461) related to the amount of oil removed and increases with increasing overhead removal. In character with the elevation of the softening point, the pour point is also elevated; the more oil distilled from the residue, the higher the softening point.

Propane deasphalting of a residuum also produces asphalt and there are differences in the properties of asphalts prepared by propane deasphalting and those prepared by vacuum distillation from the same feedstock. Propane deasphalting also has the ability to reduce a residuum even further and produce an asphalt product having a lower viscosity, higher ductility, and higher temperature susceptibility than other asphalts. Although, such properties might be anticipated to be very much crude oil dependent. Propane deasphalting is conventionally applied to low-asphalt-content crude oils, which are generally different in type and source from those processed by distillation of higher-yield crude oils. In addition, the properties of asphalt can be modified by air blowing in batch and continuous processes (Speight, 2014, 2017). On the other hand, the preparation of asphalts in liquid form by blending (cutting back) asphalt with a crude oil distillate fraction is customary and is generally accomplished in tanks equipped with coils for air agitation or with a mechanical stirrer or a vortex mixer.

An *asphalt emulsion* is a mixture of asphalt and an anionic agent such as the sodium or potassium salt of a fatty acid. The fatty acid is usually a mixture and may contain palmitic, stearic, linoleic, and abietic acids and/or high molecular weight phenols. Sodium lignate is often added to alkaline emulsions to effect better emulsion stability. Nonionic cellulose derivatives are also used to increase the viscosity of the emulsion if needed. The acid number is an indicator of its asphalt emulsification properties and reflects the presence of high molecular weight asphaltic or naphthenic acids. Diamines, frequently used as cationic agents, are made from the reaction of tallow acid amines with acrylonitrile, followed by hydrogenation. The properties of asphalt emulsions (ASTM D977, ASTM D2397) allow a variety of uses. As with other crude oil products, sampling is an important precursor to asphalt analysis and a standard method (ASTM D140) is available that provides guidance for the sampling of asphalts, liquid and semisolid, at point of manufacture, storage, or delivery.

The properties of residua and asphalt are defined by a variety of standard tests that can be used to define quality and remembering that the properties of residua vary with cut-point (Speight, 2002), i.e., the volume % of the crude oil helps the refiner produce asphalt of a specific type or property (ASTM D496). Roofing and industrial asphalts are also generally specified in various grades of hardness usually with a combination of softening point (ASTM D61, ASTM D2319, ASTM D3104, ASTM D3461) and penetration to distinguish grades (ASTM D312, ASTM D449).

The significance of a particular test is not always apparent by reading the procedure, and sometimes can only be gained through working familiarity with the test. The following tests are commonly used to characterize asphalts but these are not the only tests used for determining the property and behavior of an asphaltic binder. As in the crude oil industry, a variety of tests are employed having evolved through local, or company, use.

Determination of the composition of resids and asphalt has always presented a challenge because of the complexity and high molecular weights of the molecular constituents. The principle behind composition studies is to evaluate resids and asphalt in terms of composition and performance.

The methods employed can be conveniently arranged into a number of categories: (a) fractionation by precipitation; (b) fractionation by distillation; (c) separation by chromatographic techniques; (d) chemical analysis using spectrophotometric techniques (infrared, ultraviolet, nuclear magnetic resource, X-ray fluorescence, emission, neutron activation), titrimetric and gravimetric techniques, elemental analysis; (e) molecular weight analysis by mass spectrometry, vapor pressure osmometry, and size exclusion chromatography.

However, fractional separation has been the basis for most composition analysis and the separation methods are used to produce operationally defined fractions. Three types of separation procedures are now in use: (a) chemical precipitation in which n-pentane separation of an asphaltene fraction is followed by chemical precipitation of other fractions with sulfuric acid of increasing concentration (ASTM D2006); (b) adsorption chromatography using a clay-gel procedure where, after removal of the asphaltene fraction, the remaining constituents are separated by selective adsorption/desorption on an adsorbent (ASTM D2007 and ASTM D4124); (c) size exclusion chromatography in which gel permeation chromatographic (GPC) separation of constituents occurs based on their associated sizes in dilute solutions (ASTM D3593).

The fractions obtained in these schemes are defined operationally or procedurally. The amount and type of the asphaltene constituents are, for instance, defined by the solvent used for precipitating them. Fractional separation of does not provide well-defined chemical components and the separated fractions should only be defined in terms of the particular test procedure (Speight, 1999, 2001). This is analogous to the definition of total crude oil hydrocarbon derivatives in which the composition is defined by the method of extraction (Speight, 2014, 2015). However, these fractions are generated by thermal degradation or by oxidative degradation and are not considered to be naturally occurring constituents. The test method for determining the toluene insoluble constituents of tar and pitch (ASTM D4072, ASTM D4312) can be sued to determine the amount of carbenes and carboids in resids and asphalt.

In the methods, a sample is digested at 95°C (203°F) for 25 minutes and then extracted with hot toluene in an alundum thimble. The extraction time is eighteen hours (ASTM D4072) or three hours (ASTM D4312). The insoluble matter is dried and weighed. Combustion will then show if the material is truly carbonaceous or if it is inorganic ash from the metallic constituents (ASTM D482, ASTM D2415, ASTM D4628, ASTM D4927, ASTM D5185, ASTM D6443).

Another method (ASTM D893) covers the determination of pentane and toluene insoluble constituents in used lubricating oils and can also be applied. Pentane insoluble constituents include oil-insoluble materials and toluene insoluble constituents can come from external contamination and highly carbonized materials from degradation. A significant change in pentane or toluene insoluble constituents indicates a change in properties that could lead to problems in further processing (for resids) or service (for asphalt).

There are two test methods used: Procedure A covers the determination of insoluble constituents without the use of coagulant in the pentane and provides an indication of the materials that can be readily separated from the diluted material by centrifugation. Procedure B covers the determination of insoluble constituents that contains additives and employs a coagulant. In addition to the materials separated by using Procedure A, this coagulation procedure separates some finely divided materials that may be suspended in the resid or asphalt. The results obtained by Procedures A and B should not be compared since they usually give different values. The same procedure should be applied when comparing results obtained periodically when comparing results determined in different laboratories.

In Procedure A, a sample is mixed with pentane and centrifuged after which the resid or asphalt solution is decanted, and the precipitate washed twice with pentane, dried and weighed. For toluene insoluble constituents, a separate sample of the resid or asphalt is mixed with pentane and centrifuged. The precipitate is washed twice with pentane, once with toluene-alcohol solution, and once with toluene. The insoluble material is then dried and weighed. In Procedure B, Procedure A is followed except that instead of pentane, a pentane-coagulant solution is used.

Many investigations of relationships between composition and properties take into account only the concentration of the asphaltene constituents, independently of any quality criterion. However, a distinction should be made between the asphaltene constituents that occur in straight run resids and those which occur in blown asphalts. Remembering that asphaltene constituents are a solubility class rather than a distinct chemical class, means that vast differences occur in the make-up of this fraction when it is produced by different processes.

5.2.3.2 Coke

Coke does not offer the same potential environmental issues as other crude oil products. It is used predominantly as a refinery fuel unless other used for the production of a high-grade coke or carbon are desired. In the former case, the constituents of the coke that will release environmentally harmful gases such as nitrogen oxides, sulfur oxides, and particulate matter should be known. In addition, stockpiling coke on a site where it awaits use or transportation can lead to leachates that the result of rainfall (or acid rainfall) that are highly detrimental. In such a case, application of the toxicity characteristic leaching procedure to the coke (TCLP, EPA SW-846 Method 1311), that is designed to determine the mobility of both organic and inorganic contaminants present in materials such as coke, is warranted before stockpiling the coke in the open is warranted.

Petroleum coke is the residue left by the destructive distillation of crude oil residua in processes such as the delayed coking process. That formed in catalytic cracking operations is usually non-recoverable, as it is often employed as fuel for the process.

Coke is a gray to black solid carbonaceous residue that is produced from crude oil during thermal processing; characterized by having a high carbon content (95% + by weight) and a honeycomb type of appearance and is insoluble in organic solvents (ASTM D121).

Coke occurs in various forms and the terminology reflects the type of coke that can influence behavior in the environment. But no matter what the form, coke usually consists mainly of carbon (greater than 90 percent but usually greater than 95 per cent) and has a low mineral matter content (determined as ash residue). Coke is used as a feedstock in coke ovens for the steel industry, for heating purposes, for electrode manufacture and for production of chemicals. The two most important qualities are *green coke* and *calcined coke*. This latter category also includes *catalyst coke* deposited on the catalyst during refining processes: this coke is not recoverable and is usually burned as refinery fuel.

The test methods for coke are necessary for defining the coke as a fuel (for internal use in a refinery) or for other uses, particularly those test methods where prior sale of the coke is involved. Specifications are often dictated by environmental regulations, if not by the purchaser of the coke.

The test methods outlined below are the methods that are usually applied to petroleum coke but should not be thought of as the only test methods. In fact there are many test methods for coke (ASTM, 2000, Volume 05.06) and these test method should be consulted either when more detail is required or a fuller review is required.

The composition of petroleum coke varies with the source of the crude oil, but in general, large amounts of high-molecular-weight complex hydrocarbon derivatives (rich in carbon but correspondingly poor in hydrogen) make up a high proportion. The solubility of petroleum *coke* in carbon disulfide has been reported to be as high as 50–80%, but this is in fact a misnomer, since the coke is the insoluble, honeycomb material that is the end product of thermal processes.

Carbon and hydrogen in coke can be determined by the standard analytical procedures for coal and coke (ASTM D3178, ASTM D3179). However, in addition to carbon, hydrogen, metallic constituents (*q.v.*), coke also contains considerable amounts of nitrogen and sulfur that must be determined prior to sale or use. These elements will appear as their respectively oxides (NOx, SOx) when the coke is combusted thereby causing serious environmental issues.

A test method (ASTM D5291) is available for simultaneous determination of carbon, hydrogen, and nitrogen in crude oil products and lubricants. There are at least three instrumental techniques available for this analysis, each based on different chemical principles. However, all involve sample combustion, components separation, and final detection.

In one of the variants of the method, a sample is combusted in an oxygen atmosphere, and the product gases are separated from each other by adsorption over chemical agents. The remaining elemental nitrogen gas is measured by a thermal conductivity cell. Carbon and hydrogen are separately measured by selective infrared cells as carbon dioxide and water. In another variant of the method, a sample is combusted in an oxygen atmosphere, and the product gases are separated from each other and the three gases of interest are measured by gas chromatography. In the third variant of the method, a sample is combusted in an oxygen atmosphere, and the product gases are cleaned by passage over chemical agents and the three gases of interest are chromatographically separated and measured with a thermal conductivity detector.

The nitrogen method is not applicable too samples containing < 0.75% by weight nitrogen, or for the analysis of volatile materials such as gasoline, gasoline oxygenate blends, or aviation turbine fuels. The details of the method should be consulted along with those given in an alternate method for the determination of carbon, hydrogen, and nitrogen in coal and coke (ASTM D3179, ASTM D5373).

A test method (ASTM D1552) is available for sulfur analysis and the method covers three procedures applicable to samples boiling above 177°C (350°F) and containing not less than 0.06 mass % sulfur. Thus, the method is applicable to most fuel oils, lubricating oils, residua, and coke, and coke containing up to 8% by weight sulfur can be analyzed. This is particularly important for cokes that originate from heavy oil, and tar sand bitumen where the sulfur content of the coke is usually at least 5% by weight.

In the iodate detection system (ASTM D1552), the sample is burned in a stream of oxygen at a sufficiently high temperature to convert about 97% by weight of the sulfur to sulfur dioxide. The combustion products are passed into an absorber containing an acidic solution of potassium iodide and starch indicator. A faint blue color is developed in the absorber solution by the addition of standard potassium iodate solution. As combustion proceeds, bleaching the blue color, more iodate is added. The sulfur content of the sample is calculated from the amount of standard iodate consumed during the combustion.

In the infrared detection system, the sample is weighed into a special ceramic boat that is then placed into a combustion furnace at 1,370°C (2,500°F) in an oxygen atmosphere. Most of the sulfur present is converted to sulfur dioxide that is then measured with an infrared detector after moisture and dust are removed by traps. The calibration factor is determined using standards approximating the material to be analyzed.

For the iodate method, chlorine in concentrations < 1 mass % does not interfere. The isoprene rubber method can tolerate somewhat higher levels. Nitrogen when present > 0.1 mass % may interfere with the iodate method; the extent being dependent on the types of nitrogen compounds as well as the combustion conditions. It does not interfere in the infrared method. The alkali and alkaline earth metals, zinc, potassium, and lead do not interfere with either method.

Determination of the physical composition can be achieved by any necessary of the of test methods for determining the toluene insoluble constituents of tar and pitch (ASTM D4072, ASTM D4312). Furthermore, a variety of sample can be employed to give a gradation of soluble and insoluble fractions. The coke, of course remains in the extraction thimble (Soxhlet apparatus) and the extracts are freed from the solvent and weight to give percent by weight yield(s).

Finally one aspect that can pay a role in compositional studies is the sieve (screening) analysis. Like all crude oil products, sampling is, or can be, a major issue. If not performed correctly and poor sampling is the result, erroneous and very misleading data can be produced by the analytical method of choice. For this reason, reference is made to standard procedures such as the *Standard Practice for Collection and Preparation of Coke Samples for Laboratory Analysis* (ASTM D346) and the *Standards Test Method for the Sieve Analysis of Coke* (ASTM D293).

5.2.3.3 Particulate Matter

Although typically included in environmental treatises as a gaseous effluents, particulate matter is in fact a solid effluent and because of the way in which it reacts with the environment id considered as a solid in this text.

The general term particulate matter covers solid matter that is a complex effluent that is classified as either *suspended particulate matter*, *total suspended particulate matter*, or simply *particulate matter*. For human health purposes, the fraction of particulate matter that has been shown to contribute to respiratory diseases is termed PM_{10} (i.e., particulate matter with sizes less than 10 microns). From a control standpoint, particulate matter can be characterized as follows: (1) particle size distribution; and (2) particulate matter concentration in the emission (mg/m^3). On occasion, physical property descriptions may also be employed when there are specific control applications.

Traditionally, regulatory and compliance testing requires gravimetric determination of, for example, fuel mass emissions. Instruments utilizing collecting or *in situ* measurement techniques are used for the analysis of various particle parameters for non-regulatory purposes.

In collecting methods, particulate matter emissions are determined through gravimetric analysis of the particulates collected on a sampling filter. Alternatively, the sample can be analyzed using thermal mass analysis (e.g., coulometric analysis). A number of other properties, for instance surface area or biological activity, can be also analyzed. Collecting methods, and especially gravimetric analysis, are well established as the most common method of particulate matter emission determination.

The health effects of particulate matter (a complex mixture of solids and liquids) emissions are not yet well understood but are recognized as major contributors to health problems. Biological activity of particulate matter may be related to particle sizes and/or particle composition. Furthermore, it has generally been concluded that exposure to particulate matter may cause increased morbidity and mortality, such as from cardiovascular disease. Long-term exposure to particulate emissions is also associated with small increase in the relative risk of lung cancer.

5.3 Oil Spill Cleanup at Sea

Oil is one of the most abundant pollutants in the oceans that result from a variety of incidents, including an oil well blowout, a vessel collision or grounding, or a leaking pipeline. However, the cleanup and remediation of an oil spill is a difficult task as no two oil spills are alike. The volume of the spill can be estimated by observing the thickness of the coat of oil and the appearance and color of the oil on the surface of the water.

Chemical and biological methods can be used in conjunction with mechanical means for containing and cleaning up oil spills. The methods developed for oil spill cleanup can be categorized into three main groups which are (1) physical methods such as adsorbents, booms and skimmers, (2) chemical methods such as dispersion, *in situ* burning, and (3) the use of solidifiers (Federici and Mintz, 2014).

5.3.1 Oil Booms

An oil boom, also referred to as a containment boom, is a temporary floating barrier usually made of plastic, metal, or other materials, designed to contain an oil spill. Oil booms are effective in water but are not normally designed for spills on land. This is due to the fact that oil has a lower density than water normally, therefore floats on the surface of the ocean just like the boom. Therefore for clean-up action to be most useful, it should happen very quickly after a spill, before the oil disperses.

Booms vary in type but all have the same general design: a sub-surface skirt that prevents oil from escaping below the floating boom, a flotation device, and tension on the boom. The purpose of booms are to contain surface oil within its boundaries so that collection of the oil can take place. Collection of the oil is often carried out by skimmers. Skimmers recover oil without altering its physical or chemical properties and can do so in a variety of ways, including suction and adhesion. While using booms and skimmers in conjunction is seen as the best option because the oil will be completely removed from the surface, it does face many challenges.

Although there is a great deal of variation in the design and construction of booms, all generally share the following four basic elements: (1) an above-water freeboard to contain the oil and to help prevent waves from splashing oil over the top of the boom, (2) a flotation device, (3) a below-water skirt to contain the oil and help reduce the amount of oil lost under the boom, and (4) a longitudinal support, which is usually in the form of a chain or cable running along the bottom of the skirt, that strengthens the boom against wind and wave action.

There are different varieties of oil booms that are available and each type of boom has been designed for a specific purpose and are (1) fence booms are least effective in rough water where wave and wind action can cause the boom to distort, (2) round or curtain booms have a more circular flotation device and a continuous skirt which perform well in rough water but are more difficult to clean and store than fence booms, and (3) non-rigid or inflatable booms that are easy to clean and store, and they perform well in rough seas.

A disadvantage to the use of a boom is that unless the boom is securely anchored from both sides, strong winds may cause it to distort and allow oil to splash over the boom. Another disadvantage to booms that is they are cumbersome to assemble, deploy or repair and one of the biggest limitations in the use of this method is surface conditions. Wind and waves encourage oil to carry out its natural tendency, to fragment and disperse itself in the water. Rough seas can prevent skimmers from functioning properly and can easily carry oil over the top of booms.

Its effectiveness is based upon a proper evaluation of the location conditions and utilizing the selection criteria for the correct containment boom which includes a review of the float/skirt heights, tensile strength and reserve buoyancy. Therefore, in accordance with local, state and federal guidelines containment booms can be used on most bodies of water based upon using the correct boom selection criteria and deployment techniques to contain a hydrocarbon spill. The boom should never be deployed parallel to the shoreline and islands should never be encircled with boom and personnel should be trained to ensure that there is successful boom deployment and maximum use of available booms.

5.3.2 Skimmers

A skimmer is a mechanical device for recovering spilled material from the water's surface for the purposes of recovery or remediation. In the process, the skimmers physically separate oil from water without the introduction of chemical agents. Unlike dispersants, there is no chemical interaction between skimmers and oil. Skimmers separate oil from water using one of two principles. On one hand, "oleophilic skimmers" employ oil-attracting coatings on the surface of a drum, brush, or other shape to physically attract and separate oil from the underlying water. "Weir skimmers", on the other hand, use gravity to separate oil from water.

Although there are many types of skimmers available for use, the effectiveness and efficiency of a skimmer depends on a number of factors. The most important of which are the viscosity, adhesive properties of the spilt oil, weather conditions, the sea state and level of debris. Different types of skimmers offer advantages and drawbacks depending on the type of oil being recovered, the sea conditions during cleanup efforts, and the presence of ice or debris in the water (Skimmers, US EPA, 2017).

There are three types of skimmers which are (1) weir skimmers, (2) oleophilic skimmers, and (3) suction skimmers. The two major types of oil skimmers used in oil spill cleanup are oleophilic and weir skimmers. Oleophilic skimmers use water repelling materials which limit the intake of water, and greatly increases the concentration of oil collected. This method tends to be beneficial in lowering costs of excess water disposal. Weir skimmers are less selective than the oleophilic skimmers, and are helpful in collecting debris and other undesirable items in the marine environment.

While this means that the oil is less pure than that which is collected by oleophilic skimmers, weir skimmers are easier to transport and can be operated by a single person.

Weir skimmers use gravity to selectively drain oil from the surface of the water and allow liquid to flow over an inclined plane after which the different viscosities of oil and other fluids are used to separate the liquids. Oil floating on top of the water will spill over the dam and be trapped in a well inside, bringing with it as little water as possible and the trapped oil and water mixture can then be pumped out through a pipe or hose to a storage tank for recycling or disposal.

Weir skimmers are hard-wearing, tough, floating, suction-type skimmers extremely reliable and efficient, but the design does substantially disturb water flow and is therefore not suitable for more sensitive oil recovery operations. They are suitable for use on oceans, in ponds, dams, rivers, harbors and marinas, but are prone to becoming jammed and clogged by floating debris.

Oleophilic skimmers have the advantage of flexibility, allowing them to be used effectively on spills of any thickness. On the other hand, suction skimmers operate similarly to a household vacuum cleaner. Oil is sucked up through wide floating heads and pumped into storage tanks. Suction skimmers operate best on smooth water, where oil has collected against a boom or barrier (Skimmers, EPA, 2017). Often the skimmer is only a small floating head connected to an external source of vacuum, such as a vacuum truck. The head of the skimmer is simply an enlargement of the end of a suction hose and a float.

Hundreds of different skimmers by many different manufacturers are available to address the wide variety of potential spill conditions. In a spill, however, the selection of a suitable skimmer depends not only on the spill conditions, but also on the availability of skimmer equipment at the time and location of the spill. In the 2010 Deepwater Horizon oil spill, an estimated 3 percent (~ 147,000 bbl) of spilled oil was recovered by skimmers, when approximately 60 to 80 skimmers operated on a daily basis during cleanup operations. .

5.3.3 Sorbents

Sorbents are materials that soak up liquids and can be used to recover oil through the mechanisms of absorption, adsorption, or both. Sorbents are materials with high attractions for oil and repellent for water and are useful sorbents need to be oleophilic and hydrophobic (water repellent).

Absorbents allow oil to penetrate into pore spaces in the material they are made of, while adsorbents attract oil to their surfaces but do not allow it to penetrate into the material. To be useful in combating oil spills, sorbents need to be both oleophilic and hydrophobic (water-repellent). Although they may be used as the sole cleanup method in small spills, sorbents are most often used to remove final traces of oil, or in areas that cannot be reached by skimmers. Once sorbents have been used to recover oil, they must be removed from the water and properly disposed of on land or cleaned for re-use. Any oil that is removed from sorbent materials must also be properly disposed of or recycled.

Sorbents are either natural or synthetic materials that are used to recover oil through either absorption or adsorption or both and can be divided into three basic categories which are (1) natural organic materials, (2) natural inorganic materials, and (3) synthetic materials. Natural organic sorbents include peat moss, straw, hay, sawdust, ground corncobs, feathers, and other carbon-based products. Organic sorbents can soak up from 3 to 15 times their weight in oil, but they do present some disadvantages. Some organic sorbents tend to soak up water as well as oil, causing them to sink. Many organic sorbents are loose particles, such as sawdust, and are difficult to collect after they are spread on the water. Natural inorganic sorbents include clay, perlite, vermiculite, glass, wool, sand, and volcanic ash. They can absorb from 4 to 20 times their weight in oil. Synthetic sorbents include man-made materials that are similar to plastics, such as polyurethane, polyethylene, and nylon fibers.

Sorbents, or oil-absorbing materials, are commonly used in oil spill remediation, most often for final shoreline cleanup. Sorbents may also be used to clean up the final traces of oil on water, or as a backup to other recovery methods. Sorbent materials may be incorporated into other cleanup technologies; for example, sorbent booms contain oil-absorbing materials to improve the boom's ability to contain oil. A wide variety of materials, both natural and synthetic, have been studied as oil spill sorbents, but most commercially available oil spill sorbents are made from synthetic polymers.

By definition, the process of adsorption involves the adherence of oil to the sorbent material which is dependent upon the viscosity of the oil. On the other hand, the absorption process relies on capillary attraction; oil fills the pores within the material and moves upward (uptake) into the material due to capillary action.

Oil spill cleanup sorbents use the principle of taking up oil but not water and to become quickly saturated by oil, being the most effective in recovering oil. Moreover, the sorbent should be buoyant and remain afloat even when saturated with oil. When choosing sorbents for cleaning up spills, available oil recovery capacity of the sorbent and the oiled sorbent material disposal method should also be given consideration. The characteristics of both sorbents and oil types must be considered when choosing sorbents for cleaning up oil spills (Artemiev and Pinkin, 2008; Sorbents, EPA, 2017) and are (1) the rate of absorption, (2) the rate of adsorption, (3) the amount of oil retention, and (4) the ease of application of the sorbent.

However there are a number of precautions must be considered when using sorbents such as (1) the excessive use of sorbents at a spill scene, especially in a granular or particulate form, can compound cleanup problems and make it impossible to use most mechanical skimmers and, as a result, the sorbent may cause plugging in discharge lines or in the pumps, (2) the sorbent should not sink which may cause harm to the environment, and (3) the recovery and disposal of the oiled sorbent material must be considered.

Sorbent performance is measured according to the quantity of oil it can absorb per unit weight of sorbent material. There are numerous studies in the scientific literature to evaluate novel sorbent materials as several materials are under development to yield higher oil sorption capacity than commercially available polypropylene sorbents.

The following characteristics are worthy of consideration must be considered when choosing sorbents for cleaning up spills: (1) the rate of absorption, which varies with the thickness of the oil, (2) oil retention, which is the weight of recovered oil and can cause a sorbent structure to sag and deform and malfunction, and (3) ease of application insofar as the sorbent may be applied to a spill either manually or mechanically, using blowers or fans.

5.3.4 Burning In situ

In situ burning of surface oil involves a controlled ignition of spilled oil to essentially burn it off the surface of the water. Burning of the spilled material is the oldest technique used in the cleanup of spilled crude oil or crude oil products and involves the controlled burning of oil slicks to eliminate spilled oil offshore (as the example use here) before the spilled material reaches the coastline. The benefits of this methods include extremely efficient and rapid removal of the spilled material and, in addition, if the process is successful, there is need for any further downstream oil separation or removal processes following the burning and, once the burning is complete spilled material is eliminated.

Many factors influence the decision to use *in situ* burning on inland or coastal waters. Elements affecting the use of burning include (1) water temperature, (2) wind direction and speed, (3) slick thickness, (4) oil type, and (5) the amount of oil weathering and emulsification that have occurred.

Briefly, weathering is a measure of the amount of oil that has escaped to the atmosphere through evaporation. Emulsification is the process of oil mixing with water. Oil layer thickness, weathering, and emulsification are usually dependent upon the time period between the actual spill and the start of burn operations.

For spills on open water, *in situ* burning can be accomplished in the following steps (Buist, 1998): (1) two vessels collect a patch of oil in fire-resistant boom that is towed until the oil fills about one-third of the area inside the boom, (2) the boom is towed a safe distance from other patches of oil, (3) the oil inside the boom is ignited, the boom is slowly towed into the wind, to keep the oil toward the back of the boom and so that the smoke will go behind it, (4) the oil burns until the fire goes out, if there is a problem, it is possible to let one end of the boom go, allowing the oil to spread into a thin slick and the fire goes out quickly, (5) any floating oil residue is collected, and the boom is inspected for damage, and (6) the boom is towed to pick up the next batch of oil.

Three factors consistently emerged as the most important in determining the effectiveness of *in situ* burning in a spill which are (1) oil slick thickness, (2) oil properties, such as flash point, volatility, and API gravity, and (3) the tendency of the oil to form an emulsion.

The first steps toward using *in situ* burning at an oil spill are obtaining approval to conduct the burn and developing a burn plan. Information for plan approval should include nature, size, and type of product spilled, weather: current and forecasted, oil trajectories for on-water spills, evaluation of other response options, and feasibility of using *in situ* burning, equipment and personnel requirements and availability. While the burn plan would have data on the amount of oil to be burned, area to be burned, ignition methods, estimated duration of the burn and methods for collecting burn residues. The plan may contain additional information about factors that may affect the burn, safety and health of personnel and the effect on the environment.

To ignite the oil, vaporization (the transition of some oil from the liquid to gas phase) is required to initiate a fire. Many oil properties—such as vapor pressure, volatiles content, flash point, API gravity, and degree of emulsification—are correlated to how well oil will vaporize and ignite. Oils that readily vaporize (oils with high vapor pressure) are easier to ignite than those that do not readily vaporize (oils with lower vapor pressure). For *in situ* burning, however, virtually any type of oil will burn on water if the oil slick is of sufficient thickness.

In situ burning is often considered for Arctic or sub-Arctic environments because of their remoteness and sea ice formations that can constrict deployment of other methods. *In situ* burning has been shown to be less effective when used in ice corridors due to the high rate of melting sea ice. This reduces the thickness of oil, making it hard to ignite (1). *In situ* burning also presents environmental problems. Large clouds of black smoke rise from burning sites that not only pollute the air, but can impact communities both on the coast and far inland. *In situ* burning also leaves a residue that can coat coastlines or even worse, sink to cover benthic organisms.

This technique has been successfully used in many large spill responses, including the 2010 Deepwater Horizon spill, where an estimated 200,000–300,000 barrels of oil were burned at sea (Allen et al., 2011). With the *in situ* burning process, there is no need for handling and disposal of the oil. While a second advantage of *in situ* burning is its relatively high burn efficiency. Burning in the early phase of the spill removes most of the oil before it can cause further damage on the water. Another is that burning reduces the amount of oily wastes for collection and disposal. This factor will have a significant weight in the decision to conduct an *in situ* burning for remote or difficult to access areas.

Better techniques are needed for using *in situ* burning nearshore and along the shoreline on rivers. *In situ* burning could be used for spills on rivers in remote areas or sites with poor access where equipment cannot be deployed. However, techniques needed to be devised and tested for diverting oil from the fast-flowing areas to sites where it can be contained and burned. Also, techniques for controlling and extinguishing an on-water burn need better refinement.

However, while typically effective, the choice to use in situ burning involves many factors, including (1) environmental considerations since the burning process releases toxic fumes), (2) operational constraints, such as the availability of fire-resistant containment booms or lack thereof, and (3) the need for appropriate igniters at the site of the spill.

5.3.5 Dispersants

Dispersants contain molecules with a water-compatible moiety (i.e., they a hydrophilic group) as well as an oil-compatible moiety (i.e., a lipophilic group). These molecules attach to the oil thereby reducing the interfacial tension between oil and water which can result in the break-up of the oil slick. On the other hand, dispersants are chemical agents for treating chemical oil spills and are designed to disperse the oil into water in the form of very small oil droplets.

The effectiveness of a dispersant is determined by the composition of the oil it is being used to treat and the method and rate at which the dispersant is applied. Heavy crude oils do not disperse as well as low-density to medium-density crude oils and crude oil products. Dispersants are most effective when applied immediately following a spill, before the lowest boiling constituents in the spill have evaporated.

The properties of the spilled material (including viscosity and chemical composition) typically determine the effectiveness of chemical dispersion. In fact, the viscosity of the spilled material is often is considered to be the most important factor influencing oil dispersability. Many studies concluded that there exists an upper viscosity limit, above which oil becomes non-dispersable. This upper viscosity limit for dispersability is oil dependent (some oils have a higher viscosity limit than others), but, for all oils, dispersability decreases with increasing viscosity.

Chemical dispersants are employed on oil spills for the purpose of accelerating the natural dispersion of oil in water. The dispersants release the tension between the oil and water and subsequently reduce the droplet size of the oil so it can be distributed in the water column. Once in small enough droplet sizes, the oil can be biodegraded by naturally occurring microorganisms. Chemical dispersants can quickly break up oil slicks and prevent large quantities of oil from covering coastlines and sensitive habitats. This method is considered to be a favorable option for coping with an oil spill. However, as in the other options, there are negative implications when using this method. There is great concern and debate that adding these chemical dispersants to the ocean is poisoning fish, corals, and other marine species.

Dispersants are most effective immediately after a spill and become less effective as oil weathering alters the properties of the oil, decreasing the dispersability of the crude oil or the crude oil product (Nordvik, 1995). Operational parameters (e.g., dispersant droplet size, dispersant concentration, and mixing energy) also factor into the effectiveness of chemical dispersion.

Dispersants are not a direct cleanup method for oil spills because the dispersants do not remove the oil from the spill area or from the genral environment. In some regions dispersants are employed to speed up the natural process of the breakdown of oil. Dispersants can reduce the amount of surface oil, thereby reducing the potential exposure of floral and faunal species to the hazardous constituents in the oil.

5.3.6 Hot Water and High-Pressure Washing

In high pressure hot water flushing, the water is transported to the site by work trucks and is heated and pressurized on site. The water which is heated and pressurized on the work trucks, response boats, or cleanup barges, then it is sent through water hoses to the shore. Used without water flooding, this procedure requires immediate use of vacuum (vacuum trucks or super suckers) to remove the oil/water runoff.

Water flushing at high pressure and/or high temperature is an aggressive methods of shore cleaning and can be effective at removing stranded oil but the process also has the potential to change (i.e., damage) the structure of the shoreline and/or shoreline organisms. Thus, the high pressure and/or high temperature water process can damage plants and animals in the treated zone directly and indirectly in the short-term and in the long-term.

Pressure washing involves rinsing oiled shorelines and rocks using hoses that supply low- or high-pressure water streams and is mainly used in situations where the oil is inaccessible to methods of mechanical removal such as using booms and skimmers.

Washing with high pressure may drive oil from the water surface down into the water column, dispersing or emulsifying the oil, which could have additional environmental effects and require additional recovery methods.

High-pressure, hot-water washing has been evaluated as one of several countermeasures for cleaning oiled shorelines. Water heaters are used to heat up water to around 170°C (340°F), which is then sprayed by hand with high-pressure wands or nozzles from response boats, or cleanup barges.

All attached organisms and plants in the direct spray zone will be removed or killed, even when used properly. Oiled sediment may be transported to shallow nearshore areas, contaminating them and burying benthic organisms. Also, at high operating pressures, the spattering of surfaces adjacent to the work area can be a problem. Areas which may already have been cleaned or which have not been oiled may need to be protected. High levels of 'spatter' can also present the risk of contamination of areas adjacent to the work area.

5.3.7 Chemical Stabilization

Elastomers can be divided in two main categories: thermoset elastomers, as acrylic, butadiene, butyl, chlorinated polyethylene, ethylene propylene, fluorocarbon, isoprene, nitrile, polysulfide, polyurethane, silicone, etc., and thermoplastic elastomers, as thermoplastic urethane elastomers, styrene block copolymers, co-polyether ester elastomers, polyester amide elastomers.

Polyisobutylene (PIB) is a fully saturated, aliphatic polymer which, when dissolved, the macromolecules give rise to a distinct viscoelasticity as well as to a drastic increase in the elongation viscosity of the solution, even at polyisobutylene concentrations of only a few 100 ppm (Bobra et al., 1987).

The key to open water containment and recovery is to change the physical not the chemical character of the oil to be able to better contain and recover it. Since currents and winds quickly spread the oil, and waves emulsify it, therefore making it difficult to contain and recover. Elastol significantly reduce the scale of an offshore oil spill by changing the physical character of the oil and preventing emulsification. It modifies oil behavior without chemical reactions. It improves boom and skimmers effectiveness for faster recovery. Though polyisobutylene is also used in chewing gum and considered nontoxic, it has been recorded to have caused the death of seabirds. When it is caught up in the birds' feathers, it glues them together, preventing them from swimming, flying or catching prey. If environmental friendly chemical stabilization products can be developed to prevent the impact of oil spills without further harming the environment, it would be a step in the direction that Elastol was envisioned.

5.4 Methods for Oil Spill Cleanup on Land

Oil spills have the potential to contaminate the environment and kill plant and animal life. Upon discovery of a crude oil spill, the first step is to stop the spill or leak if this can be done safely.

Oil spill occurrence is often as a result of natural or man-made disasters, human negligence, and acts of terror by certain individuals. As an oil pollution incident may interfere directly with industry or commerce, spoil the enjoyment of amenity pursuits or affect natural processes seemingly unconnected with human affairs. Spilled oil on land prevents water absorption by the soil, and can have adverse effects on plant life.

An oil spills can ruin the infrastructure, plants and animals lives and economy of a particular area with the long-term effects being felt for decades. However, from various tests and studies, a body of general knowledge exists which can assist in minimizing potential impacts through planning of effective oil spill clean-up on land. When dealing with oil spills on land, there are four steps to follow: control the spill, contain the spill, clean up the spill and remediate the soil.

To select the best treatment option for remediation, it is important to comprehend the nature, composition, properties, sources of pollution, type of environment, fate, transport and distribution of the pollutants, mechanism of degradation, interaction and relationships with micro-organisms, the intrinsic and extrinsic factors affecting remediation. This information helps to evaluate and predict the chemical behavior of the pollutants with the short and long-term effects and mitigate the effects of pollution and limit exposure to the pollutants (Ossai et al., 2020).

In reality, the restoration of crude oil-polluted soil is a complex project because of the complex composition of crude oil pollutants, therefore a single repair method often does not work well. Physical and chemistry remediation can repair the oil pollution rapidly and efficiently, but have the disadvantages of high cost, secondary pollution and destroying the soil structure, etc. While biological methods has many benefits, but in actual application it's vulnerable to the natural environmental conditions, leading to unstable and regional repair efficiency. Therefore by analysis of various factors concerning the oil and the environment, the best remediation plan, which is usually a combination of methods, can be created.

5.4.1 Physical Methods

Physical technologies can be used for *in situ* processes or for ex situ processes. The main advantages of *in situ* treatment are that it allows the soil to be remediated without having to excavate or transport the contaminated medium and the process also avoids land disposal restrictions on the redeposition of treated soil.

Physical remediation techniques include excavation, soil washing, vitrification, encapsulation of contaminated areas by impermeable vertical and horizontal layers, electrokinesis, and permeable barrier systems (Barnes et al., 2002). However, an issue that arises with the excavation process that the place from which the layer is removed is made prone to erosion and other environment damaging agents (Araruna et al., 2004).

Spills on land can also be contained for recovery by building berms or dikes in the path of the oil flow, using either soil from the area, sand bags, or other construction materials. Berms are simply mounded hills of soil that are constructed to serve the purpose of preventing entry of oil into a sensitive area or to divert oil to a collection area. Temporary dikes and berms may also be constructed after a discharge is discovered as an active containment measure (or a countermeasure) so long as they can be implemented in time to prevent the spilled oil from reaching surface waters.

Another physical method is the washing of contaminated soil which is often an ex situ treatment process that is applicable to a broad range of organic, inorganic and radioactive contaminants in soil (Anderson, 1993). Different surfactants remove different fractions of crude oil, e.g., artificial surfactant sodium dodecyl sulfate (SDS) removed aliphatic hydrocarbon derivatives while natural surfactants saponin and rhamnolipid removed polycyclic aromatic hydrocarbon derivatives from the contaminated soil (Urum and Pekdemir, 2004).

5.4.2 Chemical Methods

Chemical oxidation involves the injection of oxygen-releasing compounds into soil and groundwater. The oxidants act as an electron acceptor and the contaminants act as an electron donor. Chemicals used to date include: Fenton's reagent, hydrogen peroxide, persulphate, permanganate, percarbonate and ozone. Once introduced into the ground, each has a different lifespan, ranging from one day to three months and involving powerful and rapid reactions effective at treating a number of contaminants. Peroxide, permanganate, persulfate, and ozone are all hazardous chemicals that must be handled properly.

Chemical oxidation is an efficient method to remove crude oil and crude oil products from the soil but the efficiency of this method strongly depends on the soil matrix. Fenton's reagent, a mixture of hydrogen peroxide (H_2O_2) and Ferric ion (Fe^{3+}), is used for chemical oxidation. Hydrogen peroxide is a strong oxidizing agent that generates hydroxyl ions during Fenton's reaction while ferric ion acts as catalyst. Hydroxyl ions are effective agents that destroy the contaminants present in the soil demonstrated that removal of oil from sand at lower pH by using Fenton's reagent is much efficient than at natural pH or peat.

Assessment of the effectiveness of chemical oxidation should include monitoring of groundwater geochemistry (pH, redox, dissolved metals), oxidant concentrations, reaction products such as chloride, and temperature. In addition, post-oxidation monitoring should be conducted to evaluate possible rebound of contaminant concentrations, release of metals, dissipation of oxidants, and rebound of microbial populations.

Chemical oxidation technologies most often fail because of ineffective delivery of oxidants caused by subsurface heterogeneities or by poor delineation of contaminant distribution in the subsurface.

5.5 Bioremediation

The bioremediation process stimulates the growth of specific micro-organisms that use the discharged chemical contaminants as a source of food and energy. In a non-polluted environment, bacteria, fungi, protists (i.e., an organism in which the cells contain a nucleus and that is not an animal, plant, or fungus), and other micro-organisms are constantly at work breaking down organic matter. Bioremediation provides these pollution-eating organisms with fertilizer, oxygen, and other conditions that encourage their rapid growth (Speight and Arjoon, 2012; Speight, 2018).

Bioremediation is considered one of the most sustainable cleanup techniques in comparison with physicochemical methods, which is more cost-effective with less disruptive effect on the environments. Its potential has not been fully exploited in the field because it is too slow to meet the immediate demands of the environment. There are two main approaches to oil spill bioremediation: (1) bioaugmentation involves the addition of oil-degrading bacteria to supplement the existing microbial population, and (2) biostimulation involves the addition of nutrients or growth-enhancing co-substrates and/or improvements in habitat quality to stimulate the growth of indigenous oil degraders.

The choice of nutrient application method employed may be site-specific. For example, in marine environments where tidal and wave action is very low, the water-soluble inorganic nutrients can be applied. If the release rates of water-soluble inorganic nutrients are maintained at an optimum level, the probability of causing eutrophication will be reduced. On the other hand, in marine environments where the tides are high, oleophilic fertilizers can be employed but they are expensive. The slow-release fertilizers may also serve as viable alternatives if there is the possibility of maintaining a prolonged optimum nutrient release. These measures may help to solve the problem of eutrophication (Macaulay and Rees, 2014).

However, like any chemical reaction, for the bioremediation process to be effective, the site must be at the appropriate temperature and nutrients must also be present, otherwise the microbes grow too slowly or die (Atlas and Hazen, 2011).

A single bioremediation approach may be incapable of breaking down all crude oil components (saturates, aromatic derivatives, resins constituents, and asphaltene constituents) within a reasonable time-frame at a minimum cost. Therefore, it may be necessary to employ integrated remediation technologies. However, integrated remediation technologies may attract a high cost of utilization, therefore, the means by which the cost can be reduced to the barest minimum needs to be investigated as this would go a long way to improving oil spill management globally.

Bioremediation exploits natural microbes to degrade organic contaminants to less harmful or innocuous substances. Once contaminants have been degraded the microbes die. The effectiveness of this method is dependent on hydrocarbon concentration, soil characteristics and composition of pollutants (Balba et al., 1998). The degradation process can occur in both aerobic and anaerobic conditions. In aerobic conditions microbes convert contaminants into carbon dioxide and water. In anaerobic conditions, methane as well as some carbon dioxide and hydrogen are produced.

Bioremediation process relies on micro-organisms using a range of microbial groups capable of degrading a wide range of target constituents present in oil contaminated environments as a food source. Hydrocarbon-degrading micro-organisms such as Pseudomonas, Rhodococcus,

Sphingobium and Sphingomonas species are abundant in the ecosystem. These microbes are capable of degrading aromatic hydrocarbon derivatives present in crude oil hydrocarbon derivatives.

The different methods and strategies of bioremediation technologies that can be used are natural attenuation, bioaugmentation, biostimulation and phytoremediation. Natural attenuation utilizes indigenous microbial populations under natural conditions. It is very cost effective but is not always effective and requires extensive long-range observation. Bioaugmentation is the addition of efficient pollutant of hydrocarbon-degrading microbes while Biostimulation is the management of environmental factors (addition of nutrient). Research has shown that bioaugmentation and biostimulation when used together effectively remediate crude oil hydrocarbon derivatives polluted soil.

Phytoremediation is an effective, solar driven and low-cost strategy that uses plants for the removal of contaminants from the soil of large contaminated area. Plants have the ability to grow in polluted soil by metabolizing or accumulating the harmful compounds in their roots or shoots (De Boer and Wagelmans, 2016). Different mechanisms are devised by plants for the removal of contaminants, i.e., phytoaccumulation (absorption of contaminants into the roots or shoots), phytodegradation (degradation of pollutants by utilization of plant enzymes such as laccase, oxygenase and nitro-reductase), phyto-volatilization (release of volatile metabolites into the atmosphere) and phytostabilization (decrease the movement of contaminants).

Soil contamination due to crude oil leakage has adverse effects on human and vegetation growth so its removal is essential. Many methods have been developed to remove crude oil from the soil, i.e., physical, chemical, thermal and biological. These methods have many drawbacks and less acceptable by the society.

5.5.1 Importance of Bioremediation

Industrialization in developed countries has left many types of pollution during the last 150 years and the destination of the pollutants from the various processes is generally soil and groundwater. Crude oil has been a major pollutant over the past few decades due to increase of its demand in society (Chandra et al., 2013), whether as transportation fuels, fuel oils for heating and electricity generation, asphalt and road oil, or feedstock for making the chemicals, plastics, and synthetic materials that are major product.

At an individual level wildlife can be affected by external contamination, in which the oil causes skin burns or changes the condition and physical properties of feathers or fur and/or internal contamination (after inhalation or swallowing), in which the oil may cause internal burns or disturb physiological processes. At the population level wildlife can be affected if the oil spill causes a significant mortality within the breeding stock of a population. On humans, the effects of oil spills may either be direct or indirect, this includes a variety of diseases and negative economic impact.

Combining the traditional approaches with bioremediation can achieve a much more successful cleanup process.

Bioremediation is considered a relatively new commercial process and is often questioned on reliability due to uncontrollable variables in an oil spill, such as the composition of the oil, the indigenous micro-organisms present at the site, the water characteristics such as temperature and energy, and also the available nutrients at the affected site. Among the other applications for which bioremediation is being considered or is currently in use are (1) treatment of non-toxic liquid and solid wastes, (2) treatment of toxic or hazardous wastes, (3) treatment of contaminated groundwater, and (4) grease decomposition.

Crude oil is a very complex mixture of compounds, extending from C6 to C60 that comprises mostly of diverse aliphatic and aromatic hydrocarbon derivatives that regularly escape into the environment. Hydrocarbon derivatives have a general formula of C_nH_{2n+2}, where n is an integer.

Although there is considerable variation between the ratios of organic molecules, the elemental composition of crude oil is well-defined. It is made up of four main elements and usually contains 84% to 87% carbon, 11% to 14% hydrogen, 0.1% to 8% sulfur, 0.1% to 1.8% nitrogen, and 1% to 1.5% oxygen. Also present are metal compounds in the oil (Speight, 2014).

Therefore the use of *in situ* remediation processes, specifically incorporating bioremediation, can (1) significantly reduce the timeline to final closure, (2) reduce or eliminate ongoing costs for long term operation and maintenance, (3) destroy contaminants on-site, (4) reduce long-term monitoring, (5) eliminate off-site disposal, and (6) reduce legacy environmental risks.

Cost measures (Table 5.2) reported by technology providers vary depending on whether the technology treats the contaminants *in situ* or ex situ, whether it is designed for containment or remediation, and whether the contaminated material is soil, other solid material, or ground water. However, no system can account in advance for every detailed cost element of a technology, but a general framework for developing costs can be created of which there are several elements (Table 5.3).

Other than well installation, no major construction is needed and equipment requirements are minimal. Therefore, costs for this process are kept low, compared to other remedial technologies. The capital costs for biological treatment are greatly affected by the method of aeration. In general, capital costs for biological treatment depends on the size, conditions and remediation level required. Total cost may range from but are generally dependent upon the nature and magnitude of the soil impact.

5.6 Issues Related to Use of Bioremedation Technologies

The bioremediation approach has some restrictions insofar as there are various factors, including scientific, non-scientific, and regulatory, that limit the use of bioremediation technologies.

5.6.1 Environmental

The ability of a metal to cause detrimental effects on micro-organisms, can hinder its use in remediating the environment.

There are also concerns that the products if the bioremediation process may be more hazardous than the parent compounds. For example, trichloroethylene (TCE, $CHCl = CCl_2$) is converted to vinyl chloride, a known carcinogen, via a series of biological reactions. This process is known as reductive dehalogenation (Anastasios et al., 2019). Theoretically, under anaerobic conditions, with sufficient quantities of other readily oxidizable substrates and the necessary auxiliary nutrients, methanogenic consortia may be capable of converting trichloroethylene to relatively benign end-products, but in the absence of sufficient oxidizable organic compounds a buildup of vinyl chloride will occur. Additionally, extensive monitoring would be necessary to assure that the final product of this remediation is not vinyl chloride ($CH_2 = CHCl$).

Factors that may affect the success of the Bioremediation process are (1) nutrient limitation, (2) low population or absence of microbes with degradative capabilities, and (3) bioavailability. Also, it might take a longer time to achieve the target level of pollutant concentration, given the remediation process.

In addition, most micro-organisms can only utilize select components of crude oil hydrocarbon derivatives and, as a result, it may require multiple different species of micro-organisms to fully degrade the constituents of crude oil or of a crude oil product.

Also, remediation methods that rely on biological remediation often find it difficult to degrade some of the most toxic oil components such as the polynuclear aromatic hydrocarbon derivatives (PNAs, also known as polycyclic aromatic hydrocarbon derivatives, PAHs). Many polynuclear aromatic hydrocarbon derivatives are (1) toxic, (2) carcinogenic, (3) teratogenic, which are chemicals that can disturb the development of the embryo or fetus, or (4) all of the above.

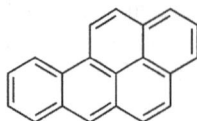

Benzo(a)pyrene

5.6.2 Health

Also micro-organisms that are added to a bioremediation site for bioaugmentation (should) typically die off once contamination and the conditions needed for bioremediation are gone and the chemicals added to stimulate bioremediation are considered safe.

However, if the process is not controlled it is possible the organic contaminants may not be fully degrades thereby resulting in toxic by-products that could be more mobile than the initial contaminant(s) and/or residual levels that can be too high (not meeting regulatory requirements), persistent, and/or toxic. In addition, bioremediation as a process to deal with pollution by heavy metals (i.e., metallic chemical element that has a relatively high density and is toxic or poisonous at low concentrations) is not a very reliable process and often fails to achieve the desired aims of the process. If heavy metals remediation are not monitored and remains in the environment for prolong periods, it can affect different body organs. Gastrointestinal and kidney dysfunction, nervous system disorders, skin lesions, vascular damage, immune system dysfunction, birth defects, and cancer are examples of the complications of heavy metals toxic effects.

Processing activities in the crude oil industry releases hazardous aromatic organic compounds such as polynuclear aromatic derivatives (PNAs), phenol derivatives that are barely degradable by nature, chlorophenol derivatives, and cresol derivatives into the environment (Sayed et al., 2021). If bioremediation is unable to degrade these components into nontoxic constituents, then exposure to oil and oil products either directly or indirectly causes severe health issues in humans.

Crude oil hydrocarbon derivatives have a strong impact on mental health and induce physical/ physiological effects, and they are potentially toxic to genetic, immune, and endocrine systems (Table 5.4). Even though the long-term effects of crude oil constituents in humans are not fully understood yet, certain symptoms may persist for some years of postexposure (Kuppusamy et al., 2020). The polynuclear aromatic hydrocarbon derivatives are known parts of crude oil and crude oil-determined items. This dangerous substances can harm any organ system in the human body such as (1) the sensory system, (2) the respiratory system, (3) the circulatory system, (4) the immune system, (5) the regenerative system, (6) the tactile system, (7) the endocrine system, as well as the liver and kidneys, can cause a wide range of ailments.

Vinyl chloride is readily absorbed by exposure to the chemical and is then rapidly distributed throughout the body. Acute (short-term) exposure to high levels of vinyl chloride in air has resulted in central nervous system effects (CNS), such as dizziness, drowsiness, and headaches in humans while chronic (long-term) exposure to vinyl chloride through inhalation and oral exposure in humans has resulted in liver damage. Cancer is a major concern from exposure to vinyl chloride via inhalation, as vinyl chloride exposure has been shown to increase the risk of a rare form of liver cancer in humans.

5.6.3 Process Evaluation

The economic evaluation of any project involves (1) the identification, (2) measurement, and valuation, and (3) comparison of the costs (inputs) and benefits (outcomes) of two or more alternative treatments or activities. In the evaluation, the costs and consequences of alternative interventions or scenarios are compared to examine the best use of the scarce resources and the issues to be addressed are (1) a comparison of the costs and benefits of a new intervention, (2) a comparison of the costs and benefits between treatment and prevention activities, and (3) a comparison of the costs and benefits between treatment and law enforcement activities.

Considerations that are an integral part when deciding whether an economic evaluation should be undertaken includes the following: (1) whether it will provide the right type of economic evidence to support the investment decision at hand, (2) whether there is already good, relevant economic evidence available, (3) heather evidence of program feasibility and effectiveness is, or will be, available, (4) how important economic evidence is for the investment decision to be made, given other considerations such as equity, (5) the level of upfront investment, and (6) whether there are plans for scaling up the program and whether it will be possible to obtain the data required for the economic evaluation.

The economic evaluation of bioremediation techniques for oil is a tool that will for guide rational investment decisions. As, economic evaluation is primarily about evaluating efficiency. These are two types of efficiency are technical efficiency and allocative efficiency. Technical efficiency refers to the maximum output/outcome obtained for a given program from a given set of resources. Allocative efficiency is about the optimal allocation of resources across a portfolio of programs so as to achieve the maximization of benefits for that portfolio.

5.6.3.1 Feasibility Assessment

A feasibility assessment at the initial design stage of any project helps determine whether the project is viable or not. For an environmental feasibility assessment, both human and environmental health factors are considered, but with businesses, the most important part of a feasibility study is the economics. This type of assessment helps to determine whether the project is both profitable and a logical fit to undertake.

The factors that affect a feasibility study or assessment can be usually divided into five groups: (1) technical feasibility, which focuses on the technical resources needed to meet requirements, (2) economic feasibility, which involves a cost/benefits analysis, (3) the legal feasibility, which involves an investigation of any aspect of the proposed project conflicts with legal requirements, (4) the operational feasibility, which involves a study to analyze and determine whether the organization's needs can be met by completing the project, and the manner in which the project plan satisfies the requirements identified in the requirements analysis phase of system development, and (5) the scheduling feasibility, which estimates the time the project will take for completion and is most important for project success and determines if the project can be completed on time.

Therefore, it is great importance in conducting a bioremediation feasibility (often referred to as bio-feasibility) study prior to committing to a bioremediation cleanup plan.

The plan should also be the means to determine the conditions necessary for achieving set goals. A bio-feasibility study should include, at a minimum, an analysis of the contaminants and their concentrations, the concentration of naturally occurring microbes in the site sample, and a quantitative, direct measure of the microbial capacity to degrade the contaminants. Beyond this minimal study, the bio-feasibility study should be designed to provide such significant information as the biodegradation kinetics, end-point, and range of conditions. The significance of these parameters is discussed, as well as the cost effectiveness of the bio-feasibility study (Chaparian, 1995).

However, the assessment of efficacy and efficiency of bioremediation of a wide range of hydrocarbon-contaminated soils are rather challenging as each site is unique with regard to its chemical and physical characteristics, including temperature, nutrients available to the microbes, bioavailability of the contaminants, pH, moisture content, and other variables. These assessment should include a site analysis, which should describe the hydrological condition and soil qualities, an evaluation of the most appropriate design for the activity, an evaluation of the most appropriate techniques and constructions and a selection of transportation methods.

A feasibility study is the initial design stage to any project or plan and its main purpose of is to assess the viability of an idea or technique. A feasibility study is an in-depth process that determines the practicality of a proposed project plan to determine its success or failure. Projects can be abandoned based on feasibility studies, before any formal authorization of action is initiated. This decision is usually because of it being time-consuming or expensive or that it may have negative consequences and it is possible to choose an alternative method that would have the most beneficial consequences on the environment and on the well-being of living organisms present in the surroundings.

The bioremediation feasibility (or bio-feasibility) study (BFS), typically performed by a qualified laboratory and is geared towards determining the applicability and potential success of bioremediation techniques at the given site. The laboratory must have expertise in applying the principles of microbiology, biochemistry, analytical chemistry, and microbial ecology to the specific conditions at or problems presented by the site and the specific contaminants to be destroyed. It consists of determining, at a minimum: the contaminants and their concentrations; the concentration of naturally occurring microbes in the soil; and a quantitative, direct measurement of the microbial capacity to degrade the contaminants. A key component of this study involves directly measuring the disappearance of contaminants as compared to an abiotic control (Chaparian, 1995).

Key site-specific considerations that will often determine the feasibility of bioremediation include: (1) whether the contaminants are sufficiently biodegradable and there is confidence that the remediation targets will be met within an acceptable timeframe, (2) whether the bioremediated material will be suitable for future use or disposal, taking into account the amendment material added, other contaminants present, and the byproducts and residuals of the treatment, (3) whether the extent and distribution of contamination is sufficiently well known, (4) whether the physiochemical composition and heterogeneity of the soil will allow sufficient uniformity of treatment to meet the remediation targets, and (5) whether biodegrading organisms are naturally present or need to be added.

Appropriate remediation data must be collected to evaluate the applicability of a bioremediation technology. If there is reasonable confidence that the selected bioremediation method will achieve the required treatment outcome, then other issues will need to be considered to determine if it is likely to be an appropriate technology for the site. These include: (1) are there sufficient microorganisms present and is the contaminant bioavailability sufficient to enable degradation?, (2) will the relevant regulatory agencies accept the bioremediation technology as a viable means of remediation?, (3) will it be able to be confirmed that the contaminants have degraded, and have not been simply diluted by the material added or mixing operations, or volatilized and lost in ambient air, if these loss mechanisms are not acceptable to the regulatory agency?, (4) are there planning or regulatory approvals required to use these technologies?, (5) will the treated material be of a form and with contaminant concentrations that will allow the material to be reused as backfill on the site or as clean fill elsewhere, or will subsequent treatment (e.g., stabilization) or landfill disposal be required?, (6) could there be remnant biodegradable material present that would give rise to methane or carbon dioxide concerns, or a geotechnical concern (physical stability), (7) what are the degradation products of the parent compound/s?, (8) are the degradation product more toxic than the parent compound/s and does this risk require additional assessment?, (9) do the breakdown products require a different treatment method (such as the production of vinyl chloride during reductive dechlorination of polychloroethylene, PCE (Figure 5.1), and (10) is there any risk of contamination migrating to other environmental segments through the use of this technology (e.g., incorrect controls during land farming resulting in transfer of contaminants from soil to the atmosphere)?

5.6.3.2 Bioremediation Services

These services make available the ability to improve the environmental state of land and has the benefit of utilizing the top trained and experienced consultants to see the job to completion. Specialists from remediation organizations have specific skill sets far beyond what the company needing the service may have.

There are three critical, interdependent elements in the unambiguous specification the experts remediation services are required to develop for a site: (1) identifying the objective, (2) determining the appropriate metric(s) to measure achievement of the objective, and (3) determining the status of the objective.

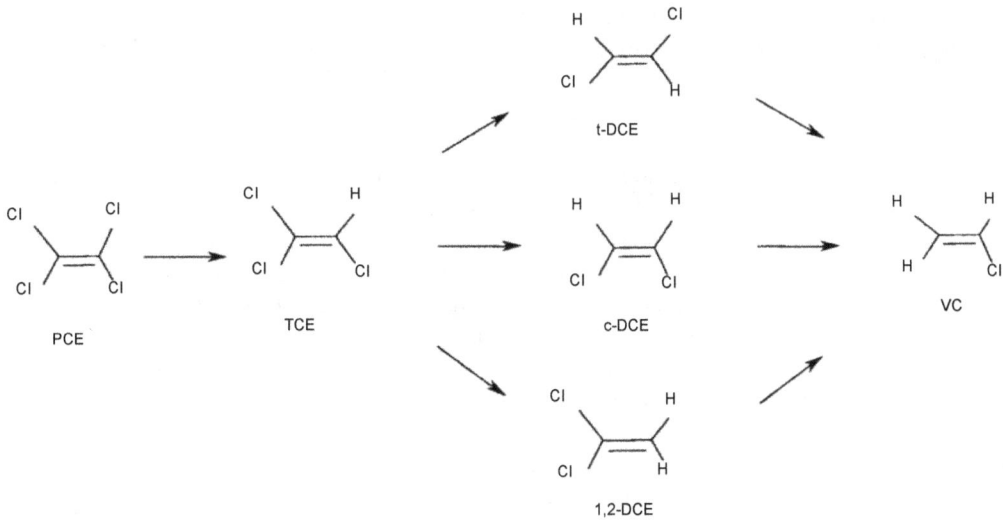

Figure 5.1: Anaerobic Reductive Dechlorination of Chlorinated Ethylene Derivatives (Mallants, 1999).

The four primary types of environmental remediation that usually needs to be taken into consideration are (1) groundwater remediation, (2) soil remediation, (3) surface water remediation, and (4) sediment remediation.

In terms of groundwater remediation, there is always the possibility that the groundwater contains some level of contamination from pollutants.

Another aspects is the sediment is the soil, sand, organic material or other materials that accumulate at the bottom of any body of water. Here, toxic materials can settle and gather in high volumes to pose major environmental and health threats since a pollutants can affect both the water and the soil throughout the property, making sediment remediation extremely important.

Businesses face strict environmental regulations from a complex web of federal, state and local government bodies and as to greatly improve the odds of hiring the right environmental remediation services company, detailed evaluation must be done before a service company is chosen. Some criteria to consider when collaborating with an environmental remediation services provider are:

1. Geographic reach and location—a service provider local to the can provide expedited services when needed.

2. Experience, education and licensure of the staff impact all phases of an environmental remediation project. Also, when it comes to licensing, there is the need to ensure that the environmental remediation services firm is certified in the specific services you currently need and may need in the future.

3. Quality and responsiveness which can mean the difference between success and utter failure and the responsiveness of a company should be explored in ensuring the services and the period in which the service is guaranteed.

4. Extensiveness of services—the four phases of the site remediation process: preliminary assessment, site investigation, remedial investigation, and remedial action should be considered.

5. Insurance, since environmental remediation projects tend to last longer and cost more than organizations plan.

6. Financial stability—to ensure that the appropriate high-quality services will be there when you need them and able to finish what was started, it is best to ensure that your services firm is financially stable.

7. Longevity—start-up companies and fly-by-night business are not acceptable in the environmental services market as it is not worth the risking to your business. Service companies with long, well-established track records of success are usually the best decisions.

5.6.3.3 Barriers to Commercialization

The overall goal of commercialization is to make an impact whether getting ideas to market faster, creating value or generating awareness of the research, inventions and the entrepreneurial culture. A multitude of barriers needs to overcome to lead to successful commercialization of a material. In general, these barriers can be categorized as technical, such as the availability of test procedures and property data, processing and manufacturing technologies, and sensitivity to flaws in materials and processes; regulatory/legal, such as government procurement policies, intellectual property rights, environmental protection, and health and safety; and economic, such as R&D costs, market size, interest rates, cost of capital facilities, and profit goals.

Commercialization of bioremediation services allow for the introduction of an alternative method of environmental clean up to the market. It enables businesses to develop products, techniques with great benefits to become available to society. Vision, business philosophies, and a logical assessment of market opportunities are all-important considerations when developing a commercialization strategy

The successful transition into a green economy hinges on innovative solutions and collaborative action on a global scale (Kant, 2017). However, despite this experience, environmental remediation remains a lengthy and expensive process (Ward, 2004).

Also, financing for environmental technologies is usually quite low. This crisis severely constrains the development of innovative technologies. The decline in venture capital financing for environmental technologies reflects a similar trend in other private and public financing for environmental technology development. This is typically because even if a technology works and is commercially acceptable, it faces additional hurdles in the permitting process, which may create time delays, lack of acceptance or other problems that prevent commercialization. Another reason is concerns market size, and the environmental regulatory system.

If there is no demand for better performance, there is no steady support for firms dedicated to researching and finding better ways of solving environmental problems. Innovative environmental technologies face two approval hurdles, compared to a single approval faced by technologies in other fields. Furthermore, environmental technologies face an approval process by the regulatory authority as well as by potential users. This double approval process makes the use of innovative environmental technologies more difficult and drives financing away from environmental technologies because commercialization is easier in other fields.

Regulatory requirements tend to be a primary influence for remediation projects. However, a variety of factors and considerations also drive and shape any remediation project. Some of these factors are lack of adequate site characterizations, insufficient technology performance or cost data, and cumbersome contracting and procurement requirements. Other factors include a lack of entrepreneurial management or adequate development funding, lack of consistent regulatory enforcement, and limited technology applications.

When the right quality of environmental remediation service is available, it allows for solutions that are efficient, budget-friendly and most important effective.

5.6.3.4 Supporting Research and Development

Bioremediation is a sustainable approach for the environmental pollution management therefore more research in this area is necessary. By assessing the appropriateness of current bioremediation technologies and the professional pool internationally, resources can be targeted towards strengthening any deficiencies and building upon the strengths (Singh et al., 2020).

Therefore, microbiologists and engineers have to understand their respective roles in the development of effective and cost-efficient bioremediation processes and the need for more information in areas that relate to bioremediation. Major new opportunities exist in the development and application of ecological knowledge, biochemical mechanisms and manipulation to greatly enhance existing and novel *in situ* processes.

The efficient utilization of this approach in the field is restricted by a lack of knowledge of the basic principles underlying the phenomenon. Hence, research and investigation into the following areas develops the remediation strategy offered by a company:

- Evaluation of the severity of the problem;
- Determination whether there is a need to remediate;
- Evaluation of the remediation alternatives, including their feasibility;
- Cost and risk reduction; the design of the selected remediation option;
- Implementation of the remediation option and verification and/or monitoring of the remediation performance.

5.6.3.5 Technical Regulations

Technical regulations are legally binding prescriptions that must be applied by all parties, whether they are big or small, regardless of the introduction costs. They aim to achieve the protection of human beings, animals and the environment against dangers and negative influence of all kinds (Inklaar, 2009). Inadequate regulations may cause substantial economic, social and even political damage. Technical regulations are part of the total collection of legal norms of a country or a region. The enforcement of technical regulations, too, is the sole responsibility of the state and its authorities.

While a technical regulation is a Government document that lays down product characteristics or their related processes and production methods, including the applicable administrative provisions, with which compliance is mandatory. In fact, increased environmental concerns internationally, due to rising levels of air, water and soil pollution, have led many governments to adopt regulations aimed at protecting the environment (Meijer et al., 2019).

5.6.3.6 Economic Assessment

Economic assessment is not only a key factor for decisions in the field of site rehabilitation, but also one of the integrating tools for damage assessment. There is a range of different approaches and tools that can be used to undertake economic assessment, calculates the potential costs, and assigns values to the anticipated benefits of a proposed project, program or policy.

The economic valuation of environmental resources should be based on neoclassical economic welfare analysis (Dixon et al., 1997). There are many techniques available to estimate the economic value of environmental goods and services. A summary is shown in the table below (Spurgeon, 2002).

Various implementation techniques can assist in reducing the barrier created in commercialization of bioremediation techniques. These include:

- reduce the tax burden and grant privileges to facilitate their expansion into foreign markets
- create regional and national innovation clusters in priority areas of economic development
- increase to a reasonable amount the R&D allocation in the government budget
- create an exchange of small knowledge-intensive companies to present the existing companies or their projects to external investors.

This is also a reason for the lack of success of new ventures in bringing remediation technologies to the market; the economic factor in participating in the remediation market is difficult. Therefore, the implementation of various techniques can assist remediation companies survive and make a positive difference in the remediation of the environment.

5.6.3.7 Potential for Future Implementation

Ecosystems are constantly faced with the challenges of abundant release of toxic compounds into the environment due to a wide range of anthropogenic activities. Nature provides great immense possibilities for isolation of microbial communities and products that can be utilized in the various application fields such as crude oil industry, detergents, pharmaceutical companies, agriculture, and personal health care products.

With the advances in microbiology, bioremediation has the potential to restore contaminated environments inexpensively and effectively (Table 5.5). Also affecting the process is that the persons responsible for the pollution are generally most interested in controlling costs and maximizing profits, while government agencies may have other priorities of national interest.

The drive for industrial sustainability has initiated the interest of biosurfactants to the top of the agenda of many companies based on its diverse nature. Biosurfactants are aptly and rightfully described as the multifunctional biomolecules/materials of the 21st Century (Olasanmi and Thring, 2018).

While microbiology has been the clear driver of the technology, bioremediation is interdisciplinary, also involving engineering, geology, ecology, and chemistry therefore new knowledge from these areas also affects the development and commercialization of bioremediation technologies.

Therefore, it is necessary to explore the hidden knowledge of these organisms. Advanced practices and sophisticated up-to-date technologies are highly desirable to understand the genetic and molecular biology of microorganisms. Newly developed molecular techniques are offer promising approaches to address the in-depth characterization of microbial communities from molecule to gene. Recent omics technologies such as metagenomics, transcriptomics, and proteomics are helpful in obtaining information about nucleic acids, enzymes, catabolic genes, plasmids, and metabolic machineries and metabolites generated during the biodegradation process. However, the solitary employment of any individual omics technology is not sufficient to explore or illustrate secret information regarding microbial-remediation practices. Therefore, an interdisciplinary application of multiple omics studies highlights the perspectives of system biology for providing an integrative understanding between genes, proteins, and environmental factors responsible for the whole microbial-degradation process, and gives a new array of novel technologies (Mishra et al., 2021). Improvements in the technology will allow an increase in productivity. Further progress to this technology is necessary to make this technology more economically viable.

Understanding microbial processes, supplying materials that stimulate microorganisms that can push the boundaries of bioremediation and the promotion of engineering advances that increases the efficiency and effectiveness of the contact between contaminants and microbes will increase the acceptance in the future of improved cleanup technologies.

Table 5.5: Environmental Valuation Techniques

Category of technique	Name of Technique	Description of approach
Market price based	Market values	Value based on market prices
	Change in productivity	Based on the change in quality and/or quantity of a marketed good net market value
	Damage costs avoided	Equivalent to the value of the economic activity or assets that it protects
	Substitute/surrogate prices	Based on the market value of an alternative product
	Expected values	Based on potential revenues multiplied by probability of occurrence
Cost based	Replacement cost	Based on the cost of replacing the environmental function
Revealed preference or surrogate market	Travel cost method	Inferred from the cost of travel to a site
	Hedonic price	Based on the value of individual components

References

Abbas, S.H., Ismail, I.M., Mostafa, T.M., and Sulaymon, A.H. 2014. Biosorption of heavy metals: a review. J. Chem. Sci. Tech., 3(4): 74–102.

Alzahrani, A.M., and Rajendran, P. 2019. Petroleum Hydrocarbon and Living Organisms, Hydrocarbon Pollution and its Effect on the Environment, Muharrem Ince and Olcay Kaplan Ince, Intech Open, DOI: 10.5772/intechopen.86948. https://www.intechopen.com/chapters/67795.

Anastasios, I.Z., Panagiotis, A., Moussas, S., and Psaltou, G. 2019. Groundwater and Soil Pollution: Bioremediation. Encyclopedia of Environmental Health Second Edition. Jerome Nriagu (Editor) Elsevier BV, Amsterdam, The Netherlands. Page 369–381.

Anderson, W.C. 1993. Thermal Desorption, Innovative site Technology, 6: 8–10.

Araruna, J.T., Portes, V.L.O., Soares, A.P.L., Silva, M.G., Sthel, M.S., Schramm, D.U., Tibana, S., and Vargas, H. 2004. Oil spills debris clean up by thermal desorption. Journal of Hazardous Materials, 110(1-3): 161–171.

Artemiev, A.V., and Pinkin, A.V. 2008. Sorption techniques to clean up water from oil pollution. Water: Chemistry and Ecology, 1: 19–25.

ASTM F2709. 2020. Standard Test Method for Determining Nameplate Recovery Rate of/Stationary Oil Skimmer Systems. Annual Book of Standards. ASTM International, West Conshohocken, Pennsylvania.

ASTM F631. 2020, ASTM, Guide for Collecting Skimmer Performance Data in Controlled Environments. Annual Book of Standards. ASTM International, West Conshohocken, Pennsylvania.

Atlas, R., and Hazen, T. 2011. Oil biodegradation and bioremediation: a tale of the two worst spills in US history. Environmental Science and Technology, 45: 6709–6715. 10.1021.

Balba, M.T., Al Awadhi, N., and Al Daher, R. 1998, Bioremediation of oil-contaminated soil: microbiological methods for feasibility assessment and field evaluation. Journal of Microbiological Methods, 32(2): 155–164.

Barnes, D., Laderach, S., and Showers, C. 2002. Treatment of Petroleum-Contaminated Soil in Cold, Wet, Remote Regions. Forest Service Technology & Development Program. United States Department of Agriculture, Washington, DC. September. https://www.fs.fed.us/t-d/pubs/pdfpubs/pdf02712801/pdf02712801_300dpi.pdf.

Bobra, M.A., Kawamura, P.I., Fingas, M., and Velicogna, D. 1987. Laboratory and Tank Test Evaluation of Elastol. Proceedings. 10th Artic and Marine Oil Spill Program Technical Seminar. June 9–11, 1987, Edmonton, Alberta, Canada. Page 223.

Buist, A. 1998. Window of Opportunity for *In-situ* Burning. Proceedings. Workshop on *In-situ* Burning of Oil Spills. New Orleans, Louisiana. November 2 – 4, 1998. National Institute of Standards and Technology, Gaithersburg, Maryland, NIST Special Publication 935. Page 21–30.

CFR. 2004. Code of Federal Regulations, Unites States Government, Washington, DC. The Code of Federal Regulations (CFR) is the codification of the general and permanent rules published in the Federal Register by the executive departments and agencies of the Federal Government. It is divided into 50 titles that represent broad areas subject to Federal regulation. Each volume of the CFR is updated once each calendar year and is issued on a quarterly basis.

Chandra, S., Sharma, R., Singh, K. et al. 2013. Application of bioremediation technology in the environment contaminated with petroleum hydrocarbon. Ann. Microbiol., 63: 417–431. https://doi.org/10.1007/s13213-012-0543-3.

Chaparian, M. 1995. Successful bioremediation: The importance of the bio-feasibility study. United States.

Cumo, F., Gugliermetti, F., and Guidi, G. 2007. Best available techniques for oil spill containment and cleanup in the mediterranean sea. Water Resources Management, Transactions on Ecology and the Environment, 103: 527–535.

De Boer, J., and Wagelmans, M. 2016, Polycyclic aromatic hydrocarbons in soil–practical options for remediation. Clean-Soil, Air, Water, 44(6): 648–653.

Dixon, J.A., Scura, L.F., Carpenter, R.A., and Sherman, P.B. 1997. Economic analysis of environmental impacts, Earthscan Publications Ltd., London United Kingdom.

Federici, C., and Mintz, J. 2014. Oil Properties and Their Impact on Spill Response Options. Report No. IRM-2014-U-007490. CNA: Resource Analysis Division, Arlington, Virginia. May. https://www.bsee.gov/sites/bsee.gov/files/osrr-oil-spill-response-research/1017aa.pdf.

Igiri, B.F., Okoduwa, S.I.R., Idoko, G.O., Akabuogu, E.P., Adeyi, A.O., and Ejiogu, I.K. 2018. Toxicity and bioremediation of heavy metals contaminated ecosystem from tannery wastewater: a review. Journal of Toxicology, vol. 2018, Article ID 2568038, 16 pages. https://doi.org/10.1155/2018/2568038.

Inklaar, A. 2009. Technical Regulations, Recommendations for their elaboration and enforcement, Guide No. 1/2009.

Kant, M. 2017. Overcoming Barriers to Successfully Commercializing Carbon Dioxide Utilization. Front. Energy Res., 5: 22. doi: 10.3389/fenrg.2017.00022.

Kapahi, M., and Sachdeva, S. 2019. Bioremediation options for heavy metal pollution. Journal of Health & Pollution, 9(24): 191–203. https://www.ncbi.nlm.nih.gov/pmc/articles/PMC6905138/pdf/i2156-9614-9-24-191203.pdf.

Kohli, R. 2019. Application of Microbial Cleaning Technology for Removal of Surface Contamination. In: Developments in Surface Contamination and Cleaning: Applications of Cleaning Techniques. Volume 11. Chapter 15. Page 591–617. https://www.sciencedirect.com/science/article/pii/B9780128155776000153.

Kuppusamy, S., Maddela, N.R., Megharaj, M., and Venkateswarlu, K. 2020. Impact of total petroleum hydrocarbons on human health. pp. 139–165. *In*: Kuppusamy, S., Maddela, N.R., Megharaj, M., and Venkateswarlu, K. (Authors). Total Petroleum Hydrocarbons: Environmental Fate, Toxicity, and Remediation. Springer BV, Amsterdam, The Netherlands. https://www.sciencedirect.com/science/article/pii/B9780128155776000153.

Kuiper, I., Lagendijk, E., Bloemberg, G., and Lugtenberg, B. 2004. Rhizoremediation: a beneficial plant-microbe interaction. Mol. Plant Microbe Interact., 17(1): 6–15.

Macaulay, B., and Rees, D. 2014. Bioremediation of oil spills: a review of challenges for research advancement. Annals of Environmental Sciences, 8: 9–37.

Mallants, D. 1999. Numerical simulation of degradation and transport of chlorinated hydrocarbons using CHAIN_2D. Hydrological Processes, 13: 2847–2859.

Meijer, L.L.J., Huijben, J.C.C.M., Boxstael, A., and Van Romme, A.G.L. 2019. Barriers and drivers for technology commercialization by SMEs in the Dutch sustainable energy sector.

Mishra, S., Lin, Z., Pang, S., Zhang, W., Bhatt P., and Chen, S. 2021. Recent Advanced Technologies for the Characterization of Xenobiotic-Degrading Microorganisms and Microbial Communities, Frontiers in Bioengineering and Biotechnology. https://www.frontiersin.org/article/10.3389/fbioe.2021.632059.

Nordvik, A. 1995. The technology windows-of-opportunity for marine oil spill response as related to oil weathering and operations. Spill Science and Technology Bulletin, 2(1): 17–46.

Olasanmi, I.O., and Thring, R.W. Thring. 2018. The Role of Biosurfactants in the Continued Drive for Environmental Sustainability 10, no. 12: 4817. https://doi.org/10.3390/su10124817.

Ossai, I., Chukwunonso, A., Aziz, H., Auwalu, H., and Fauziah, S. 2020. Remediation of Soil and Water Contaminated with Petroleum Hydrocarbons: A Review, Environmental Technology and Innovation, Volume 17, 100526, ISSN 2352-1864. https://doi.org/10.1016/j.eti.2019.100526.

Sayed, K., Baloo, L., and Sharma, N.K. 2021. Bioremediation of total petroleum hydrocarbons (TPH) by bioaugmentation and biostimulation in water with floating oil spill containment booms as bioreactor basin. Int. J. Environ. Res. Public Health, 18: 2226. https://www.ncbi.nlm.nih.gov/pmc/articles/PMC7956214/pdf/ijerph-18-02226.pdf.

Sharma, I. 2020. Bioremediation Techniques for Polluted Environment: Concept, Advantages, Limitations, and Prospects, Trace Metals in the Environment. IntechOpen. https://www.intechopen.com/chapters/70661.

Singh, P., Singh, V.K., Singh, R., Borthakur, A., Madhav, S., Ahamad, A., K., Ajay, P.D.B., Tiwary, D., and Mishra, P.K. 2020. Bioremediation: a sustainable approach for management of environmental contaminants, Editor(s): Pardeep Singh, Ajay Kumar, Anwesha Borthakur, Abatement of Environmental Pollutants, Elsevier. Chapter 1. Page 1–23. https://doi.org/10.1016/B978-0-12-818095-2.00001-1.

Singh, M., Pant, G., and Hossain, K. 2017, Green remediation: A Tool for Safe and Sustainable Environment: A Review. Appl. Water Sci., 7: 2629–2635. https://doi.org/10.1007/s13201-016-0461-9.

Speight, J.G., and Arjoon, K.K. 2012. Bioremediation of Petroleum and Petroleum Products. Scrivener Publishing, Beverly, Massachusetts, 2012.

Speight, J.G. 2014. The Chemistry and Technology of Petroleum 5th Edition. CRC Press, Taylor & Francis Group, Boca Raton, Florida.

Speight, J.G. 2015. Handbook of Petroleum Product Analysis 2nd Edition. John Wiley & Sons Inc., New York.

Speight, J.G. 2017. Handbook of Petroleum Refining. CRC Press, Taylor & Francis Group, Boca Raton, Florida.

Speight, J.G. 2018. Reaction Mechanisms in Environmental Engineering. Analysis and Prediction Butterworth-Heinemann, Elsevier, Oxford, United Kingdom.

Spurgeon, James. 2002. Rehabilitation, Conservation and Sustainable Utilization of Mangroves in Egypt, Ministry of Agriculture & Land Reclamation, Ministry of State For Environment, Food and Agriculture Organization of the United Nations.

Urum, K., and Pekdemir, T. 2004. Evaluation of Biosurfactants for Crude Oil Contaminated Soil Washing. Chemosphere, 57(9): 1139–1150.

US EPA. 2010b. Green Remediation. United States Environmtnal Protection Agency, Washngton DC. www.clu-in.org/greenremediation.

Vilsack, R. 2009. Green Workforce Needs of Minnesota Businesses. In: Minnesota Employment Review, Minnesota Department of Employment and Economic Development, St. Paul, Minnesota. www.positivelyminnesota.com/Data_Publications/Employment_Review_Magazine/June_2009_Edition/Green_Workforce_Needs_of_Minnesota_Businesses.aspx.

Ward, O.P. 2004. The industrial sustainability of bioremediation processes, Journal of Industrial Microbiology and Biotechnology, Volume 31, Issue 1, 1 January 2004, Pages 1–4, https://doi.org/10.1007/s10295-004-0109-x.

PART II
Bioremediation, Biodegradation, and Site Cleanup

CHAPTER 6

Bioremediation and Biodegradation

One of the major and continuing environmental problems is hydrocarbon contamination resulting from the activities related to crude oil and crude oil products. Soil contamination with hydrocarbon derivatives causes extensive damage of local system since accumulation of pollutants in animals and plant tissue may cause death or mutations.

While conventional methods to remove, reduce, or mitigate the effects of toxic chemical in nature are available include (1) pump and treat systems, (2) soil vapor extraction, (3) incineration, and (4) containment, each of these conventional methods of treatment of contaminated soil and/ or water suffers from recognizable drawbacks and may involve some level of risk. In short, these methods, depending upon the chemical constituents of the spilled material, may limited effectiveness and can be expensive (Speight, 1996; Speight and Lee, 2000; Speight, 2005).

Although the effects of bacteria (microbes) on hydrocarbon derivatives has been known for decades, this technology (now known as *bioremediation*) has shown promise and, in some cases, high degrees of effectiveness for the treatment of these contaminated sites since it is cost-effective and will lead to complete mineralization. Bioremediation functions basically on biodegradation, which may refer to complete *mineralization* of the organic contaminants into carbon dioxide, water, inorganic compounds, and cell protein or transformation of complex organic contaminants to other simpler organic compounds that are not detrimental to the environment. In fact, unless they are overwhelmed by the amount of the spilled material or it is toxic, many indigenous microorganisms in soil and/or water are capable of degrading hydrocarbon contaminants.

The commercial practice of biodegradation focuses primarily on the cleanup of crude oil hydrocarbon derivatives. However, bioremediation is only as successful as the ability of the microorganisms to convert the crude oil hydrocarbon derivatives to benign products and return the affected (contaminated) area to its original condition. This, in turn, is dependent upon the extent of the contamination of the soil and the ability of the microorganisms to convert the spilled oil. Thus, before and during the bioremediation process, analyses are necessary to determine (1) the horizontal extent of the contaminated area, (2) the vertical extent of the contaminated area, (3) the concentration of oil in the contaminated area, and (4) ensuring that the amount of microorganisms to be used are adequate to the clean-up task (Table 6.1, Table 6.2) (Boopathy, 2000; Vidali, 2001; Bennet et al., 2002; US EPA, 2003; Koning et al., 2008; Diez, 2010; Adams et al., 2015; Azubuike et al., 2016; Luka et al., 2018).

Furthermore, the purpose of environmental remediation is to safeguard the environment and protect the floral and faunal inhabitants of the environment by removing contaminants and pollutants from surface water, groundwater, sediment, soil, and other parts of an ecosystem. To accomplish this, the bulk movement of the contaminants is used to achieve remediation, which

Table 6.1: Essential Factors for Microbial Bioremediation.

Factor	Optimal Conditions
Microbial population	Suitable kinds of organisms that can biodegrade all of the contaminants
Oxygen	Sufficient to aerobic biodegradation (ca. 2% v/v oxygen in the gas phase)
Water	Should be from 50 to 70% v/v of the water holding capacity of the soil
Nutrients	Nitrogen, phosphorus, sulfur, and other nutrients support microbial growth
Temperature	Appropriate temperatures for microbial growth (0 to 40°C; 32 to 104°F)
pH	Appropriate range is from 6.5 to 7.5

Table 6.2: Critical Factors that should be Considered when Evaluating the use of Bioremediation for Site Cleanup.

Magnitude, toxicity, and mobility of contaminants:
Site characterization parameters:
  1. horizontal and vertical extent of contamination
  2. the kinds and concentrations of contaminants at the site
  3. the mobility of the contaminants which depends in part on the geological characteristics of the site.

Proximity of human and environmental receptors: Determine if bioremediation is the appropriate cleanup remedy for a site. Dependent on the rate and extent of contaminant degradation.
Degradability of contaminants: Contaminants with a high molecular weight, particularly those with complex ring structures and halogen substituents, degrade more slowly than simpler hydrocarbon derivatives or low molecular weight compounds.
Planned site use: A critical factor is whether the rate and extent of contaminant degradation is sufficient to reduce risks to acceptable levels.
Ability to monitor the process: Biological processes are dynamic and may lack the predictability of more conventional remediation technologies. Risks may include migration of contaminants to previously uncontaminated media.

is typically implemented with stringent monitoring programs. Moreover, the process of pollutant removal depends primarily on the nature of the pollutant, which may include any one or more of the following (1) agrochemicals, such as fertilizers and pesticides, (2) chlorinated compounds, (3) dyes, (4) heavy metal derivatives, (5) hydrocarbon derivatives, (6) plastics, (7) sewage, and (8) nuclear waste.

Global surveys have examined the use of bioremediation technologies for addressing environmental problems, where respondents showed a preference for environmentally friendly remediating methods even though current practice appears to be the opposite.

By screening the microorganisms in order to select cultures with a the ability for sulfate-reducing, denitrogenation, and biodegradation of organic matter to develop a multiple bioremediation system for the biological removal of sulfate derivatives, nitrate derivatives and organic matter present (Lindstrom et al., 1991; Ranalli et al., 2000). Bioremediation can also be applied to recover brownfield sites (former industrial or commercial sites) for development, and for preparing contaminated industrial liquid waste prior to discharge into waterways.

The spillage of crude oil and crude oil products may cause serious problems, especially to the ambient environment and can take years for the spill site to fully recover. Thus, it is not surprising that enhanced bioremediation has emerged as a promising technology for combating spills of crude oil and crude oil products in the oceans of the world.

The constituents of crude oil and crude oil products, like many organic chemical entities, have the potential to be biodegradable into the raw materials of nature and disappear into the environment. However, sustainable biodegradation of crude oil and crude oil products is not as straightforward as it may seem. Crude oil will biodegrade in its natural state, but once it is converted into saleable products, unsustainable pollution problems can arise—instead of returning to the natural cycle, these products pollute and litter the land, water, and air.

The release of many types of contaminants is causing serious harm to all life-forms due to increasing global economic growth. Pollutants such as crude oil constituents, heavy metals (metallic chemical elements that has a relatively high density and is toxic or poisonous at low concentrations; examples of heavy metals include mercury, Hg, cadmium, Cd, arsenic, As, chromium, Cr, thallium, Tl, and lead, Pb) and pesticides are environmentally harmful, causing serious impacts on the ecosystems. To reclaim the purpose of the contaminated environment, the remediation of contaminated sites is essential and the process is directed at the cleanup of polluted environments, including soils, groundwater, and marine environments.

Bioremediation is a sustainable approach for the many types of contaminants released into the environment. Contaminants such as crude oil constituents and the constituents of crude oil products can have serious impacts on the health of (and life forms in) an ecosystem.

However, as a side note, not all crude oil products are harmful to health and the environment. There are records of the use of *crude oil spirit* for medicinal purposes. This was probably a higher boiling fraction of or than *naphtha* or a low boiling fraction of *gas oil* that closely resembled the modern-day *liquid paraffin*, for medicinal purposes. In fact, the so-called liquid paraffin has continued to be prescribed up to modern times as a means for miners to take in prescribed dozes to lubricate the alimentary tract and assist coal dust, taken in during the working hours, to pass though the body. Nevertheless, there are however, those constituents of crude oil that are extremely harmful to health and the environment. Indeed, crude oil constituents either in the pure form or as the components of a fraction have been known to belong to the various families of carcinogens and neurotoxins. Whatever the name given to these compounds they are extremely toxic.

As a result, once a spill has occurred, every effort must be made to rid the environment of the toxins. The chemicals of known toxicity range in degree of toxicity from low to high and represent considerable danger to human health and must be removed (Frenzel et al., 2009). Many of these chemicals substances come in contact with, and are sequestered by, soil or water systems. While conventional methods to remove, reduce, or mitigate the effects of toxic chemical in nature are available include (1) pump and treat systems, (2) soil vapor extraction, (3) incineration, and (4) containment, each of these conventional methods of treatment of contaminated soil and/or water suffers from recognizable drawbacks and may involve some level of risk. In short, these methods, depending upon the chemical constituents of the spilled material, may limited effectiveness and can be expensive (Speight, 1996; Speight and Lee, 2000; Speight, 2005).

Although the effects of bacteria (microbes) on hydrocarbon derivatives has been known for decades, this technology (now known as *bioremediation*) has shown promise and, in some cases, high degrees of effectiveness for the treatment of these contaminated sites since it is cost-effective and will lead to complete mineralization. Bioremediation functions basically on biodegradation, which may refer to complete *mineralization* of the organic contaminants into carbon dioxide, water, inorganic compounds, and cell protein or transformation of complex organic contaminants to other simpler organic compounds that are not detrimental to the environment. In fact, unless they are overwhelmed by the amount of the spilled material or it is toxic, many indigenous microorganisms in soil and/or water are capable of degrading hydrocarbon contaminants.

Bioremediation technologies have been developed in the last five decades and are increasingly being used to mitigate environmental accidents and systematic contamination. In fact, bioremediation continues to be the favored approach for processing spoils of crude oil and crude oil products and may also play an increasing role in concentrating metals and radioactive materials to avoid toxicity or to recover metals for reuse.

The capabilities of micro-organisms and plants to degrade and transform contaminants provides benefits in the cleanup of pollutants from spills and storage sites. These remediation ideas have provided the foundation for many *ex situ* waste treatment processes (including sewage treatment) and a host of *in situ* bioremediation methods that are currently in practice. Thus, bioremediation—the use of living organisms to reduce or eliminate environmental hazards resulting from accumulations of toxic chemicals and other hazardous wastes—is an option that offers the possibility to destroy or render harmless various contaminants using natural biological activity (Gibson and Sayler, 1992). In addition, bioremediation can also be used in conjunction with a wide range of traditional physical and chemical technology to enhance their effectiveness (Vidali, 2001).

A bioremediation process can be used simultaneously with other chemical treatment methods as well as with physical treatment methods for complete management of diverse group of environmental contaminants but unlike the chemical and physical methods of cleanup, bioremediation does not disrupt the natural habitat of the site. Thus, bioremediation of a contaminated site typically works in one of two ways which are (1) methods which enhance the growth of the indigenous contaminant-degrading microbes at the contaminated site, and (2) the addition of specialized microbes which will degrade the contaminants (Speight, 2018).

In the current context, bioremediation of crude oil and crude oil fractions (or products) is the cleanup of crude oil spills or crude oil product spills by the use of microbes to breakdown the crude oil constituents (or other organic contaminants) into less harmful (usually lower molecular weight) and easier-to-remove products (biodegradation). The microbes transform the contaminants through metabolic or enzymatic processes, which vary greatly, but the final product is usually harmless and includes carbon dioxide, water, and cell biomass. Thus, the emerging science and technology of bioremediation offers an alternative method to detoxify crude oil-related soil and water contaminants.

Thus, the bioremediation approach offers an option to completely destroy the pollutants or at least to transform the pollutants to less toxic materials, such as carbon dioxide, methane, water, and inorganic salts—remembering the need to trap the carbon dioxide and methane which are noted greenhouse gases.

Because biological processes are often highly specific and important site factors required for success include (1) the presence of metabolically capable microbial populations, (2) a suitable environment for the growth of micro-organisms, and (3) appropriate levels of nutrients. In fact, as already mentioned (above), bioremediation can often be carried out on site without, in many cases, causing a major disruption of site activities. This also eliminates the need to transport quantities of waste off site and the potential threats to human health and the environment that can arise during transportation (Leahy and Colwell, 1990; Kensa, 2011; Macaulay and Rees, 2014). Most bioremediation systems operate under aerobic conditions but anaerobic conditions are also applicable which can enable the degradation of unresponsive molecular contaminants by using specific microorganisms.

The anaerobic metabolism involves microbial reactions occurring in the absence of oxygen and encompasses many processes including fermentation, methanogenesis, reductive dechlorination, sulfate-reducing activities, and denitrification. Depending on the contaminant of concern, a subset of these activities may be cultivated. In anaerobic metabolism, nitrate, sulfate, carbon dioxide, oxidized metals, or organic compounds may replace oxygen as the electron acceptor. For example, in anaerobic reductive dechlorination, chlorinated solvents may serve as the electron acceptor.

In addition, bioremediation is an option that offers the possibility to destroy or render harmless various contaminants using natural biological activity through the stimulation and enhanced microbial activity (bioaugmentation) or the use of amendments (biostimulation), such as air, organic substrates or other electron donors/acceptors, nutrients, and other compounds that can be added (Da Silva and Alvarez, 2000; Adams et al., 2015; Azubuike et al., 2016; Goswami et al., 2018). Also, bioremediation technologies are typically classified as (1) *in situ* technologies or (2) *ex situ* technologies. The *in situ* bioremediation technologies involve the treatment of the contaminate site at the site while the *ex situ* technologies involve the removal of the contaminated site material (such as soil) to be treated elsewhere (Aggarwal et al., 1990; Madsen, 1991; Lucian and Gavrilescu, 2008; Shah, 2014; Whelan et al., 2016).

Bioremediation

    In situ

        Natural attenuation

            Uses indigenous microbes

        Bioaugmentation

            Uses indigenous microbes
            Non-indigenous microbes added

        Biostimulation

            Uses indigenous microbes
            Water and nutrients added

Enhanced

          Biosparging
          Bioslurping
          Bioventing

Ex situ

        Biopile
        Bioreactor
        Land farming
        Windrow

Three different types of *in situ* bioremediation process are (1) bioattenuation which depends on the natural process of degradation, (2) biostimulation where intentional stimulation of degradation of chemicals is achieved by addition of water, nutrient, electron donors or acceptors, and (3) bioaugmentation chemicals is achieved by addition of water, nutrient, electron donors or acceptors, and bioaugmentation where the microbial members with proven capabilities of degrading or transforming the constituents of crude oil and crude oil products are added (Madsen, 1991).

Bioattenuation is the process in which biodegradation occurs naturally without human intervention and which relies on natural processes to dissipate contaminants through biological transformation, during which the indigenous microbial populations degrade recalcitrant compounds or xenobiotics compounds based on their metabolic processes.

Bioattenuation or natural attenuation processes are often categorized as destructive or non-destructive. Natural attenuation processes may reduce contaminant mass (through destructive processes such as biodegradation and chemical transformations); reduce contaminant concentrations (through simple dilution or dispersion); or bind contaminants to soil particles so the contamination does not spread or migrate very far (adsorption).

Nutrient deficiency severely impairs the catabolic activity of indigenous micro-organisms in hydrocarbon rich environments (HREs) and limits the rate of intrinsic bioremediation. Therefore biostimulation can be used to overcome this effect by the addition of materials to enhance the growth and activity of indigenous micro-organisms in a contaminated site, but is dependent on the indigenous organisms and thus requires that they be present and that the environment be capable of being altered in a way that will have the desired bioremediation effect.

The use of bioaugmentation to accelerate the reductive dechlorination process, achieve remediation targets and the process generally falls in two main strategies which are (1) bioaugmentation by enrichment with indigenous micro-organisms, and (2) bioaugmentation by enrichment with nonindigenous micro-organisms. One potential limitation to bioaugmentation is that effective treatment is contingent upon adequate distribution of the degradative bacteria within the treatment area. Before implementing bioaugmentation, or any *in situ* technology, an evaluation is necessary to consider site-specific characteristics and to determine the most effective treatment technology based on current contaminant and hydrogeochemical conditions and site access. Bioaugmentation can overcome these challenges, as one of its main advantages is that treatment can be tailored to a specific pollutant that is dominant in the environment.

Ex situ bioremediation is a biological process in which excavated soil is placed in a lined above-ground treatment area and aerated following processing to enhance the degradation of organic contaminants by the indigenous microbial population.

After excavation, the polluted soil is then place in layers no more than 0.4 meters thick and oxygen is added and mixing occurs via plowing, harrowing, or milling. Nutrients and moisture may also be added to aid the remediation process. The pH of the soil is also regulated (keeping it near 7.0) using crushed limestone or agricultural lime.

Bio-piling involves placing the contaminated soil into piles that are well aerated and nutrients are added to speed up the bioremediation process.

As one of *ex situ* bioremediation techniques, windrows rely on periodic turning of piled polluted soil to enhance bioremediation by increasing degradation activities of indigenous and/or transient hydrocarbonoclastic bacteria present in polluted soil.

In a bioreactor, the degradation process is a natural process that is accomplished by use of existing and/or added populations of micro-organisms. The design of the bioreactor design depends on contaminant type (soil, sludge, and water), cost, oxygen transfer and mixing and there are two major bioreactors for soil remediation which are (1) dry bioreactors and (2) slurry bioreactors. Dry bioreactors treat soil with microbes and nutrients without the use of any form of amendments. Adequate moisture is maintained for microbial growth by either a sprinkler system or by rainfall. The main advantage of this approach is accurate control of the bioremediation process.

The suitability of a site for the application of a bioremediation technology depends not only on the biodegradability of the contaminant(s) but also on the geological and chemical characteristics of the site. In addition, the types of site conditions that favor bioremediation differ for intrinsic and engineered bioremediation.

Intrinsic bioremediation is the process of converting environmental pollutants into the non-toxic forms through the actions of naturally occurring microbial population. This process is most effective in the soil and water as these biomes always have high chance of being fully contaminated by contaminants and toxins and is usually employed for underground places where tanks have been used to store crude oil or crude oil products. There are several site properties that promote intrinsic bioremediation such as (1) the flow of groundwater throughout the year, (2) the presence of carbonate minerals to buffer the acidity produced during biodegradation, and (3) the ability to dispense of electron acceptors and nutrients for microbial growth. Other environmental factors such as pH, concentration, temperature and nutrient availability decides whether or not biotransformation takes place must also be given consideration.

However, intrinsic bioremediation may not be suitable when site conditions are not a match to the microbial growth requirement because, intrinsic bioremediation is a slow process as growth and availability of micro-organisms are not adequate, there is limited capacity of electron acceptor and nutrients, cold temperature and high concentration of contaminants.

Since bioremediation is an evolving technical area and is evolving into an extremely important process that communities (especially coastal communities) have a plan in place for when a spill of crude oil (or a crude oil product) occurs. Having a response plan that can be implemented immediately after the spill and (1) will not cause much damage to the environment, and (2) will allow the community to recover as quickly as possible.

The major goal of bioremediation is to reduce the exposure of the floral and faunal species at a spill site as well as to reduce the effects of the contaminants on any other aspects of the site. Bioremediation techniques are used with the ultimate goal of being effective in restore polluted environments, while being considered as an eco-friendly approach at a very low cost. Bioremediation is the practice of natural purification of the earth. Put simply, instead of using expensive environmental remediation equipment to remove untreated toxic materials and dispose of them elsewhere, bioremediation techniques use micro-organisms to achieve the cleanup of the site.

Micro-organisms are capable of performing almost any detoxification reaction. Therefore bioremediation can be used for the removal of contaminants, pollutants, and toxins from soil, water, and other environments. In addition, after the microbes have eliminated any potential for continued contamination of the soil and water, these same micro-organisms can then stimulate the growth of other micro-organisms that consume those same impurities and provide long-term protection for the site. In addition, a major reason that bioremediation is preferred over the more conventional methods of site cleanup is that once the contaminants are treated, neutralized, or removed, the products of the process may then be recycled.

In many cases, natural bioremediation can occur at a spill site but often over many years and there are now bioremediation methods (through the use of cultured micro-organisms) that can accelerate the process through knowledge of (1) the types of micro-organisms that will best perform the cleanup, (2) an intimate knowledge of the chemical and physical characteristics of the contaminants, and (3) an investigation of the specific characteristics of the site.

The biological mechanism has the distinct advantage of being a naturally occurring mechanism with a high success rate and when assisted and stimulated through human effort, the prospects are increased. Thus:

Microbes attack contaminant
Microbes convert contaminant to intermediate products
Intermediate products are converted to carbon dioxide and water
Carbon dioxide and water emitted

Bioremediation technology is invaluable for reclaiming polluted soil and water, as it can be as the practice of natural purification of the earth. The process is less harmful to the environment with few (or no) byproducts while conventional physical and chemical treatments maybe somewhat inefficient because of the formation of byproducts that are more toxic than the original contaminants. In addition, when a spill of crude oil (or a crude oil products) occurs, some organisms would die while some would survive while bioremediation provides the organisms with the means to survive (Riccardi et al., 2005).

In the biosphere of the Earth, micro-organisms grow in a wide range of ecosystems and the micro-organisms can grow at sub-zero temperature as well as at extreme heat in the presence of hazardous compounds or any waste stream. Moreover, the two characteristics of microbes are (1) adaptability and (2) their biological system which makes the microbes suitable for use in remediation process. The absolute numbers and the ability of the microbes to degrade (convert) a wide range of chemicals to less toxic products make micro-organisms suitable candidates use in mitigating pollution.

Thus, bioremediation is an important aspect of environmental cleanup because the process does not require the use of chemicals and is capable of removing toxic contaminants from soil and from groundwater. In fact, the bioremediation process is a biological mechanism for recycling and converting contaminants to another form that can used by other organisms and the process provides a key alternative solution to overcome environmental challenges.

In summary, micro-organisms can survive (and thrive) in many places within the biosphere because of the variable metabolic activity that allow them to overcome wide contaminated sites can ultimately lead to undesired effects on the flora and fauna (including serious health effects for humans), appropriate actions must be taken to ensure that this does not happen and the need for environmental remediation is necessary. Each year agricultural effluents, industrial residues, and industrial accidents contaminate surface water systems and soil. Therefore the purpose of the environmental remediation efforts becomes more than merely eliminating the sources of the contaminants but also protecting the floral and faunal species against potential harmful effects (Tilman and Lehman, 2001).

However, it is important to recognized and understand that remediation actions need to be justified and optimized insofar as the adopted actions must do more good than harm. For this reason, a thorough evaluation of the situation followed by the formulation of the desired goal is prerequisite.

Bioremediation is an environmental tool that amplifies natural biological actions to remedy or remediate polluted groundwater and contaminated soil through the use of selected microorganisms which must be suited to the task of contaminant destruction through the use of enzymes that allow the microorganisms to use the environmental contaminants as food and sustenance.

Thus, the use of biotechnology as a toll for remediation is suitable for (1) assessing the wellbeing of ecosystems, (2) the transformation of contaminants into relatively benign products, (3) the generation of biodegradable materials from renewable sources, and (4) the development of environmentally safe manufacturing and disposal processes. Also, environmental biotechnology may employ genetic engineering to improve the efficiency of the cleanup process, which is a key factor in the exploitation of microorganisms to reduce the environmental burden of toxic substances (Sasikumar and Papinazath, 2003; Macaulay and Rees, 2014).

The bioremediation process is a biological process (i.e., an intervention, the purpose of which is to alleviate pollution) that stimulates suitable microbial organisms to use the environmental contaminants as a source of food and energy. For example, there are microorganisms the ingest toxic chemicals and pathogens and eliminate (the toxic chemicals and pathogens) by changing their composition into gases such as ethane (C_2H_6) and carbon dioxide (CO_2). Moreover, there are contaminated soil sites and water systems that have indigenous organisms that can convert toxic contaminants to more benign products. Thus, bioremediation technologies utilize the metabolic potential of indigenous and non-indigenous microorganisms to achieve the decontamination of contaminated sites and, moreover, the organisms that are adequate to the task must be identified and (when necessary) isolated.

However, it is not only the identification of microorganisms that are suitable for the task but also the development of strategies for cleaning up an crude oil spill or the spill of a crude oil product which are affected by a variety of factors, such as (1) the type of crude oil or the crude oil product, (2) the geology and topology of the spill site, and (3) the type of climatic conditions on a seasonal basis. In addition to the successful decontamination of spill sites, which have a long-term impact on the health of ecosystems, this technology has countless other applications which include (1) the environmentally friendly disposal of radioactive waste, (2) the decontamination of disused coal mine sites, and (3) the decontamination metal ore mine sites.

6.1 The Origin of Bioremediation

By definition, bioremediation is the use of living organisms (such as bacteria or other micro-organisms) to degrade wastes or, in the current context, crude oil and crude oil products that have been released into the environment. Thus, bioremediation is a method for dealing with contamination by crude oil, crude oil products, and crude oil waste streams by the biodegradation of the constituents of the spilled material. The process typically occurs through the degradation of crude oil or a crude oil product through the action of microorganisms (biodegradation) (Bartha, 1986a, 1986b; Hoff, 1993; Atlas, 1995; Alexander, 1999; Okoh, 2006). The process is a feasible method of dealing with widespread environmental contaminations because it utilizes indigenous materials when compared to customary (physical and chemical) remediation methods. Also, the microorganisms engaged are capable of performing almost any detoxification reaction. The process offers a method of getting rid of wastes by breaking down organic matter into nutrients that can be used by other organisms.

Microorganism have always been in the environment and fossil records indicate that mounds of bacteria once covered young Earth. Some of the microorganisms evolved to the stage of producing their own food using the carbon dioxide in the atmosphere and energy they derived from the Sun. In over three billion years (3×10^9) of evolution, microorganisms have colonized nearly all ecological niches, including some of the most extreme environments, such as the Arctic (and Antarctic) environment as well as equatorial deserts and environments in between these two extremes of climate, which also include geysers, rocks, and the depths of the oceans—including extremely salty bodies of water, such as the Dead Sea—have allowed for the evolution of organisms known as extremophiles that have been able to adapt to the extreme conditions. It is due to their ability to exits in these different environments that microorganisms play a major role in a variety of biogeochemical cycles (i.e., the cycles that describe the flow of chemical elements between living organisms and the environment) which affect the productivity of the soil or the quality of the water.

From time immemorial (a point of time in the past that was so long ago that humans have no knowledge or memory of it), nature has been rectifying itself to achieve the conditions needed for survival and reduce the greenhouse effect created by humans—for example, plants absorb carbon dioxide by means of photosynthesis and create oxygen. Also, the earth has natural environmental remediation systems which use microorganisms to decompose, recycle, and rectify imbalanced chemical conditions in the soil and the water systems. Moreover, these natural processes influence the regions of the Earth where different plants and animals are not only able to survive but also to thrive. These are the micro-organisms consume contaminants in soils, groundwater, surface waters, and air though digestion and metabolization into products that are much less harmful (i.e., benign) to the environment.

In the modern sense, the history of bioremediation has been generally recognized as beginning in the decade of the 1940s. However, in the sixth century 600 BC, the Romans had the idea of waste management and in wastewater treatment facilities—whether or not it was known that microorganisms were the operative agents is subject to much speculation. The Romans had built aqueducts to help fulfil the water supply needs which led to excessive amount of wastewater and the problem was solved by the construction of a sophisticated sewage system—the Cloaca Maxima—which was constructed in Rome around 510 BC to drain marshes and transport wastes to the Tiber River, where the water was drained and (knowingly or unknowingly) cleaned by natural biodegradation. Also, Roman engineers planned, designed and constructed sewage systems, which collected polluted water in collection vats and lagoons where it was observed that after a period of time the water was cleaned for further use. There is also evidence of kitchen middens (ancient household garbage dumps) and compost piles from that time period.

However, not all crude oil constituents and the constituents of crude oil products are harmful to health and the environment. There are records of the use of *petroleum spirit* for medicinal purposes. This was probably a higher boiling fraction of or than *naphtha* or a low boiling fraction of *gas oil* that closely resembled the modern-day *liquid paraffin*, for medicinal purposes. In fact, the so-called liquid paraffin has continued to be prescribed up to modern times as a means for miners to take in prescribed dozes to lubricate the alimentary tract and assist coal dust, taken in during the working hours, to pass though the body.

There are however, those constituents of crude oil that are extremely harmful to health and the environment. Indeed, crude oil constituents either in the pure form or as the components of a fraction have been known to belong to the various families of carcinogens and neurotoxins. Whatever the name given to these compounds they are extremely toxic.

As a result, once a spill has occurred, every effort must be made to rid the environment of the toxins. The chemicals of known toxicity range in degree of toxicity from low to high and represent considerable danger to human health and must be removed (Frenzel et al., 2009). Many of these chemicals substances come in contact with, and are sequestered by, soil or water systems. While conventional methods to remove, reduce, or mitigate the effects of toxic chemical in nature are available include (1) pump and treat systems, (2) soil vapor extraction, (3) incineration, and (4) containment, each of these conventional methods of treatment of contaminated soil and/or water suffers from recognizable drawbacks and may involve some level of risk. In short, these methods, depending upon the chemical constituents of the spilled material, may limited effectiveness and can be expensive (Speight, 1996; Speight and Lee, 2000; Speight, 2005).

Although the effects of bacteria (microbes) on hydrocarbon derivatives has been known for decades, this technology (now known as *bioremediation*) has shown promise and, in some cases, high degrees of effectiveness for the treatment of these contaminated sites since it is cost-effective and will lead to complete mineralization. Bioremediation functions basically on biodegradation, which may refer to complete *mineralization* of the organic contaminants into carbon dioxide, water, inorganic compounds, and cell protein or transformation of complex organic contaminants to other simpler organic compounds that are not detrimental to the environment. In fact, unless they are overwhelmed by the amount of the spilled material or it is toxic, many indigenous microorganisms in soil and/or water are capable of degrading hydrocarbon contaminants.

6.2 The Mechanism of Bioremediation

As already stated, bioremediation is an environmentally friendly technique used to restore soil and water to its original state by using indigenous microbes to break down and eliminate contaminants. Furthermore, the microorganisms used for bioremediation may be indigenous to a contaminated area or they may be isolated from elsewhere and brought to the contaminated site. Thus, for bioremediation to be effective, microorganisms must convert the pollutants and convert them to harmless products. As bioremediation can be effective only where environmental conditions permit microbial growth and activity, its application often involves the manipulation of environmental parameters to allow microbial growth and degradation to proceed at a faster rate. However, as is the case with other technologies, bioremediation has its limitations and there are several disadvantages that must be recognized (Table 6.3).

Biodegradation involves chemical transformations mediated by microorganisms that: (1) satisfy nutritional requirements, (2) satisfy energy requirements, (3) detoxify the immediate environment, (4) or occur fortuitously such that the organism receives no nutritional or energy benefit (Stoner, 1994). On a structural basis, the hydrocarbon derivatives in crude oil are classified as alkane derivatives (*normal* or *iso*), cycloalkane derivatives, and aromatic derivatives. Alkene derivatives are rare in crude oil but occur in many refined crude oil products as a consequence of the cracking process (Speight, 2014). Increasing carbon numbers of alkane derivatives (homology), variations in carbon chain branching (*iso*-alkane derivatives), ring condensations, and interclass combinations, such as phenyl alkane, account for the high numbers of hydrocarbon derivatives that occur in crude oil.

In addition, smaller amounts of oxygen-containing compounds (phenol derivatives, naphthenic acids), nitrogen-containing compounds (pyridine derivatives, pyrrole derivatives, indole derivatives), sulfur-containing compounds (thiophene derivatives), and the high molecular weight polar asphalt fraction also occur in crude oil but not in refined crude oil products (Speight, 2014).

The inherent biodegradability of these individual components is a reflection of their chemical structure, but is also strongly influenced by the physical state and toxicity of the compounds. As an example, while n-alkane derivatives as a structural group are the most biodegradable crude oil hydrocarbon derivatives, the C_5 to C_{10} homologs have been shown to be inhibitory to the majority of hydrocarbon degraders. As solvents, these homologs tend to disrupt lipid membrane structures of microorganisms. Similarly, alkane derivatives in the C_{20} to C_{40} range are hydrophobic solids at physiological temperatures. Apparently, it is this physical state that strongly influences their biodegradation (Bartha and Atlas, 1977).

Primary attack on intact hydrocarbon derivatives requires the action of oxygenases and, therefore, requires the presence of free oxygen. In the case of alkane derivatives, mono-oxygenase attack results in the production of alcohol. Most microorganisms attack alkane derivatives terminally whereas some perform sub-terminal oxidation. The alcohol product is oxidized finally into an aldehyde. Extensive methyl branching interferes with the beta-oxidation process and necessitates terminal attack or other bypass mechanisms. Therefore, n-alkane derivatives are degraded more readily than iso alkane derivatives.

Cycloalkane derivatives are transformed by an oxidase system to a corresponding cyclic alcohol, which is dehydrated to ketone after which a mono-oxygenase system lactonizes the ring, which is subsequently opened by a lactone hydrolase. These two oxygenase systems usually never occur in the same organisms and hence, the frustrated attempts to isolate pure cultures that grow on

Table 6.3: Examples of the Advantages and Disadvantages of Bioremediation.

Advantages	Disadvantages
A natural process	May be slower than non-natural other technologies
Contaminants are degraded	Potential for plugging injection wells by microbial growth or mineral precipitation
Toxic chemicals are destroyed	Heavy metals may not be removed
In situ technologies can be used	Requires soil with high permeability
Generally less costly than other options.	May require continuous monitoring and maintenance
Can be combined with other technologies	May only occur in more permeable layer or channels within the aquifer

cycloalkane derivatives (Bartha, 1986b). However, synergistic actions of microbial communities are capable of dealing with degradation of various cycloalkane derivatives quite effectively.

As in the case of alkane derivatives, the monocyclic compounds, cyclopentane, cyclohexane, and cycloheptane have a strong solvent effect on lipid membranes and are toxic to the majority of hydrocarbon degrading microorganisms. Highly condensed cycloalkane compounds resist biodegradation due to their relatively complex structure and physical state (Bartha, 1986a).

Condensed polycyclic aromatic derivatives are degraded, one ring at a time, by a similar mechanism, but biodegradability tend to decline with the increasing number of rings and degree of condensation (Atlas and Bartha, 1998). Aromatic derivatives with more than four condensed rings are generally not suitable as substrates for microbial growth, though, they may undergo metabolic transformations. The biodegradation process also declines with the increasing number of alkyl substituents on the aromatic nucleus.

Asphaltic constituents of crude oil tend to increase during biodegradation in relative and sometimes absolute amounts. This would suggest that they not only tend to resist biodegradation but may also be formed *de novo* by condensation reactions of biodegradation and photo-degradation intermediates.

In crude oil as well as in refined crude oil products, the hydrocarbon derivatives occur in complex mixtures and influence each other's biodegradation. The effects may go in negative as well as positive directions. Some iso-alkane derivatives are apparently spared as long as n-alkane derivatives are available as substrates, while some condensed aromatic derivatives are metabolized only in the presence of more easily utilizable crude oil hydrocarbon derivatives, a process referred to as co-metabolism (Wackett, 1996).

Biodegradation of poorly water-soluble liquid hydrocarbon derivatives is often limited by low availability of the substrate to microbes. Adhesion of microorganisms to an oil–water interface can enhance this availability, whereas detaching cells from the interface can reduce the rate of biodegradation. The capability of microbes to adhere to the interface is not limited to hydrocarbon degraders, nor is it the only mechanism to enable rapid uptake of hydrocarbon derivatives, but it represents a common strategy. The general indications are that microbial adhesion can benefit growth on and biodegradation of very poorly water-soluble hydrocarbon derivatives such as n-alkane derivatives and large polycyclic aromatic hydrocarbon derivatives dissolved in a non-aqueous phase. Adhesion is particularly important when the hydrocarbon derivatives are not emulsified thereby giving limited interfacial area between the two liquid phases.

When mixed communities are involved in biodegradation, the ability of cells to adhere to the interface can enable selective growth and enhance bioremediation with time. The critical challenge in understanding the relationship between growth rate and biodegradation rate for adherent bacteria is to accurately measure and observe the population resides at the interface of the hydrocarbon phase.

The control and optimization of bioremediation processes is a complex system of many factors. These factors include: the existence of a microbial population capable of degrading the pollutants;

the availability of contaminants to the microbial population; the environment factors (type of soil, temperature, pH, and the presence of oxygen or other electron acceptors, and nutrients).

The success of the process for the bioremediation of the spills of crude oil and crude oil products depends the establishment and maintenance of conditions that favor enhanced oil biodegradation rates in the contaminated environment (Das and Chandran, 2011). A particularly important requirement is the presence of microorganisms with the appropriate metabolic capabilities which, if present, can assure optimal rates of growth and the biodegradation of the hydrocarbon derivatives which can be sustained by ensuring that adequate concentrations of nutrients and oxygen are present and that the pH is between 6 and 9. The physical and chemical characteristics of the crude oil and the crude oil product are also important determinants of bioremediation success. There are the two main approaches to oil spill bioremediation: (a) bioaugmentation, in which known oil-degrading bacteria are added to supplement the existing microbial population, and (b) biostimulation, in which the growth of indigenous oil degraders is stimulated by the addition of nutrients or other growth-limiting co-substrates.

Thus, an important factor in biological removal of hydrocarbon derivatives from a contaminated environment is their bioavailability to an active microbial population, which is the degree of interaction of chemicals with living organisms or the degree to which a contaminant can be readily taken up and metabolized by a bacterium (Harms et al., 2010). Moreover, the bioavailability of a contaminant is controlled by factors such as the physical state of the hydrocarbon *in situ*, its hydrophobicity, water solubility, sorption to environmental matrices such as soil, and diffusion out of the soil matrix. When contaminants have very low solubility in water, as in the case of n-alkane derivatives and polynuclear aromatic hydrocarbon derivatives, the organic phase components will not partition efficiently into the aqueous phase supporting the microbes.

In the case of soil, the contaminants will also partition to the soil organic matter and become even less bioavailable. Two-phase bioreactors containing an aqueous phase and a non-aqueous phase liquid (NAPL) have been developed and used for bioremediation of hydrocarbon-contaminated soil to address this very problem, but the adherence of microbes to the non-aqueous phase liquid-water interface can still be an important factor in reaction kinetics. Similarly, two-phase bioreactors, sometimes with silicone oil as the non-aqueous phase, have been proposed for biocatalytic conversion of hydrocarbon derivatives like styrene ($C_6H_5CH = CH_2$) (Osswald et al., 1996) to make the substrate more bioavailable to microbes in the aqueous phase. When the carbon source is in limited supply, then its availability will control the rate of metabolism and hence biodegradation, rather than catabolic capacity of the cells or availability of oxygen or other nutrients.

In the case of the bioremediation of waterways, similar principles apply. Under enhanced conditions (1) certain fuel hydrocarbon derivatives can be removed preferentially over others, but the order of preference is dependent upon the geochemical conditions, and (2) augmentation and enhancement via electron acceptors to accelerate the biodegradation process. For example, with regard to the aromatic benzene-toluene-ethylbenzene-xylenes (BTEX): (1) toluene can be preferentially removed under intrinsic bioremediation conditions, (2) biodegradation of benzene is relatively slow, (3) augmentation with sulfate can preferentially stimulated biodegradation of o-xylene, and (4) ethylbenzene may be recalcitrant under sulfate-reducing conditions but readily degradable under denitrifying conditions (Cunningham et al., 2000).

One of the important factors in biological removal of hydrocarbon derivatives from a contaminated environment is the bioavailability of the hydrocarbon derivatives to an active microbial population, which is the degree of interaction of chemicals with living organisms or the degree to which a contaminant can be readily taken up and metabolized by a bacterium (Harms et al., 2010). Two-phase bioreactors, sometimes with silicone oil as the non-aqueous phase, have been proposed   · for biocatalytic conversion of hydrocarbon derivatives like styrene (Oswald et al., 1996) to make the substrate more bioavailable to microbes in the aqueous phase.

In the case of the bioremediation of waterways, similar principles apply. Under enhanced conditions (1) certain fuel hydrocarbon derivatives can be removed preferentially over others, but the order of preference is dependent upon the geochemical conditions, and (2) augmentation and enhancement via electron acceptors to accelerate the biodegradation process. For example, with regard to the aromatic benzene-toluene-ethylbenzene-xylenes (BTEX): (i) toluene can be preferentially removed under intrinsic bioremediation conditions, (ii) biodegradation of benzene is relatively slow, (iii) augmentation with sulfate can preferentially stimulated biodegradation of o-xylene, and (iv) ethylbenzene may be recalcitrant under sulfate-reducing conditions but readily degradable under denitrifying conditions (Cunningham et al., 2000).

While the basic concept of crude oil biodegradation seems straightforward, there are five factors to consider in determining the biodegradability of crude oil and crude oil products.

1. The first factor relates to the inherent biodegradability of the constituents of crude oil and its products. A natural product, such as crude oil, has the potential to return to nature as long as it remains in its relatively natural form. Any plant-based or animal-based natural product has the capability to biodegrade, but crude oil products, having undergone chemical changes may require advanced biodegradation chemistry since they may not biodegrade readily.

2. The second factor relates to the time it takes for crude oil constituents to biodegrade. In nature, different materials biodegrade at different rates. The rate for crude oil and crude oil products to biodegrade depends on the composition of the material.

3. The third factor relaters to the products of biodegradation and whether or not toxic substances are formed during biodegradation or as the end result of biodegradation. To be truly biodegradable, crude oil constituents should undergo complete mineralization (i.e., complete degradation to carbon dioxide and water) but this is not ways the case even though crude oil is a natural product.

4. The fourth factor relates to the characteristics of the environment that the substance or material is in can also affect its ability to biodegrade. Crude oil constituents and the constituents of crude oil-based products might biodegrade in an aerobic environment but not in an anaerobic environment such as swamps, flooded soils, or water-bearing sediments.

5. The fifth factor relates to the capacity of the ecosystem to accommodate the invasive chemical. Once it is determined that crude oil constituents or a crude oil product will biodegrade under specific environmental conditions, there is the issue of the amount of the contaminant that can be sustained by the ecosystem that is receiving it and the sustainable rate of biodegradation, which is dependent upon the amount of the contaminant that an ecosystem can absorb as a nutrient, and if necessary, render harmless. Spills of crude oil and crude oil products are devastating not because some of the constituents do not at an acceptable rate but because the amount of crude oil is much greater than the size of the microbial colony available for biodegradation. Indeed, as much as it is necessary to consider the biodegradability of crude oil and its products, it is necessary to consider the capacity of the system the biodegradable substance or material is being placed into.

Consideration of the above factors leads to many issues which can, however, be alleviated somewhat by knowing the character of crude oil and the history of the processes by which crude oil is refined into saleable products. There are also those chemicals that are not crude oil products but as used in the production of products and as often designated as refinery chemicals or refinery wastes (Chapter 10).

Because a critical aspect of assessing the toxic effects of the release of crude oil and crude oil products is the measurement of the compounds in the environment, the approach is not only to understand the composition and properties of crude oil itself (Chapter 2) but also the processes by which crude oil is refined to products as well as the and properties of the various fractions and products derived from crude oil (Chapter 1, Chapter 2).

Contamination by hydrocarbon derivatives can (and, in many cases, does) cause extensive damage of an ecosystem because the accumulation of hydrocarbon pollutants in animals and plant tissue may cause death or mutations. However, it must be noted that not all crude oil products are harmful to health and the environment. In fact, historical documents record the successful use of a liquid referred to as *crude oil spirit* for medicinal purposes. This liquid was probably a higher boiling fraction of *naphtha* or a low boiling fraction of *gas oil* that closely resembled the modern-day *liquid paraffin*. In fact, the so-called liquid paraffin has continued to be prescribed up to modern times (post World War II) to be taken in prescribed dozes by miners to lubricate the alimentary tract and assist coal dust, taken in during the working hours, to pass through, and out of, the body.

There are however, those constituents of crude oil and crude oil products that are extremely detrimental to an ecosystem and the health (or continued existence) of the local flora and fauna. Indeed, the constituents of crude oil and crude oil products either in the pure form or as the components of a mixture have been known to belong to the various families of carcinogens (which are chemicals with the capacity to cause cancer in humans) and neurotoxins poisons which act on the nervous system. Whatever the name given to these chemicals, they are extremely toxic.

Many of these chemicals substances come in contact with, and are sequestered in the soil or within a water system. Conventional methods to remove or mitigate the effects of such chemicals in nature are available and include (1) pump and treat systems, (2) soil vapor extraction, (3) incineration, and (4) containment, each of these conventional methods of treatment of contaminated soil and/or water may suffer from a disadvantage as well as posing some level of risk. In short, these methods, depending upon the chemical constituents of the spilled material, may limited effectiveness and can be expensive (Speight, 1996; Speight and Lee, 2000; Speight, 2005).

Although the effects of bacteria (microbes) on hydrocarbon derivatives has been known for decades (if not a century, considering the work of Louis Pasteur), this technology involving the use of microbes to change to character and properties of spilled crude oil derivatives (now known as *bioremediation*) has been shown to be an effective method for the treatment of a contaminated site since it is cost-effective and will lead to complete mineralization, i.e., conversion—often referred to as biodegradation—of the hydrocarbon derivatives to carbon dioxide (CO_2) and water (H_2O). In fact, unless there is an overwhelming amount of the spilled hydrocarbon derivative(s) or the spilled material is toxic to the microorganism, many indigenous microorganisms in soil and/or water are capable of degrading hydrocarbon contaminants (Shelton and Tiedje, 1984).

Briefly, and by way of clarification, biodegradation (biotic degradation, biotic decomposition) is the process that occurs (as indicated by the prefix "bio") by the action of microorganisms under typical environmental conditions. In the process, a contaminant undergoes conversion by bacteria or other biological means in which a contaminant is converted into more benign environmentally acceptable products such as water, carbon dioxide, and any form of carbonaceous products that are not harmful to the environment. Typically, most bioremediation systems occur under aerobic conditions, but a system under anaerobic conditions may permit microbial organisms to degrade chemical species that are otherwise non-responsive to aerobic treatment; and vice versa.

Thus, by means of the process organic (carbon-based, carbonaceous material insofar as the material may contain constituents that contain nitrogen oxygen, sulfur, and metals) is converted through chemical processes from a complex contaminant into simpler molecules, eventually returning the molecules into the environment. In many cases, the biodegradation process is a natural process (or a series of processes) by which the constituents of spilled crude oil hydrocarbon derivatives are degraded into products, such as nutrients that can be used by other organisms. On the other hand, bioremediation is a collective term that may include several types of biodegradation processes.

Thus, bioremediation—i.e., the use of living organisms to reduce or eliminate environmental hazards resulting from the accumulation of toxic chemicals and other hazardous waste constituents—is an option that offers the possibility to destroy or render harmless various contaminants using natural biological activity (Gibson and Sayler, 1992; Vidali, 2001: Speight and Arjoon, 2012; Thapa et al., 2012).

In the process, the microorganisms either consume the contaminants (as an energy source) or convert the contaminants to benign products thereby rendering the ecosystem contaminant-free.

By way of clarification, a heavy metals is any metallic chemical element that has a relatively high density and is toxic or poisonous at low concentrations. Examples of heavy metals include mercury (Hg), cadmium (Cd), arsenic (As), chromium (Cr), thallium (Tl), and lead (Pb).

In the current context, the bioremediation of the constituents of crude oil and crude oil or products is the cleanup of a spill of crude oil spills or a spill of a crude oil product by the use of microbes which degrade the constituents of the spill into less harmful (usually more benign lower molecular weight) and easier-to-remove products. The contaminants are transformed by metabolic or enzymatic processes (which vary according to the type of microbial organism) into relatively final products, which include carbon dioxide, water, and cell biomass (Figure 6.1).

The success of the process in terms of the degradation of the constituents of crude oil and crude oil products depends on the ability to determine these conditions and establish them in the contaminated environment. Thus, important site factors required for the success of the process include (1) the presence of metabolically capable and sustainable microbial populations, (2) suitable environmental growth conditions, such as the presence of oxygen, (3) temperature, which is an important variable—keeping a substance frozen or below the optimal operating temperature for microbial species, can prevent biodegradation—most biodegradation occurs at temperatures between 10 and 35°C (50 and 95°F), (4) the presence of water, (5) appropriate levels of nutrients and contaminants, and (6) favorable acidity or alkalinity (Table 6.1).

In regard to the last parameter (i.e., favorable acidity or alkalinity), pH of the soil is extremely important because most microbial species can survive only within a certain pH range—typically, the biodegradation of the crude oil constituents is optimal at a pH equal to 7 (neutral, neither acidic, pH < 7, or alkaline, pH > 7) and the *acceptable* (or optimal) pH range is on the order of 6 to 8. Furthermore, the acidity or alkalinity of the soil or of the groundwater pH can affect availability of nutrients.

In addition to the acidity and alkalinity of the surrounding medium, the temperature also influences rate of biodegradation by controlling rate of enzymatic reactions within the microorganisms. Generally, the rate of an enzymatic reaction approximately doubles for each 10°C (18°F) rise in

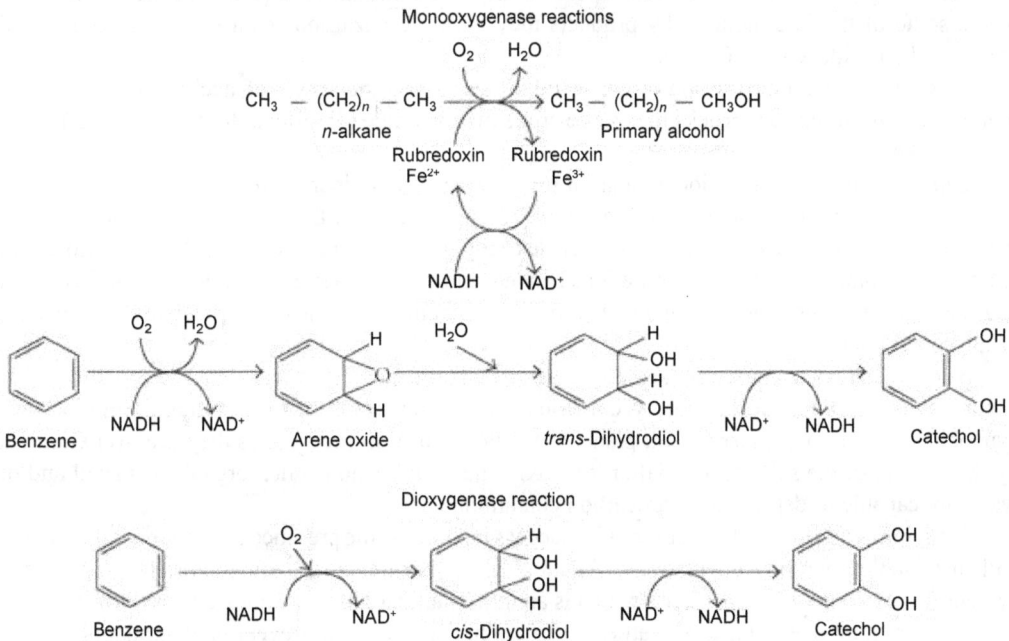

Figure 6.1: Enzymatic Reactions Involved in Hydrocarbon Degradation (Das and Chandran, 2011).

temperature. However, there is an upper limit to the temperature that microorganisms can withstand and most bacteria that occur in soil—including the bacteria that degrade the hydrocarbon derivatives on crude oil and in crude oil products—are mesophile organisms that have an optimum working temperature range on the order of 25 to 45°C (77 to 113°F) (Nester et al., 2001). Thermophilic bacteria (those which survive and thrive at relatively high temperatures and which are typically found in hot springs and compost heaps) may be indigenous to cool soil environments and can be activated to degrade hydrocarbon derivatives with an increase in temperature to 60°C (140°F). This indicates the potential for natural attenuation in cool soils through thermally enhanced bioremediation techniques (Perfumo et al., 2007).

In order the enhance microbial activity there are two other bioremediation technologies that offer useful options for cleanup of spills of crude oil and crude oil products: (1) fertilization and (2) seeding.

Fertilization (*nutrient enrichment*) is the method of adding nutrients such as phosphorus and nitrogen to a contaminated environment to stimulate the growth of the microorganisms capable of biodegradation. Limited supplies of these nutrients in nature usually control the growth of native microorganism populations. When more nutrients are added, the native microorganism population can grow rapidly, potentially increasing the rate of biodegradation. *Seeding* is the addition of microorganisms to the existing native oil-degrading population. Some species of bacteria that do not naturally exist in an area will be added to the native population. As with fertilization, the purpose of seeding is to increase the population of microorganisms that can biodegrade the spilled oil.

Microorganisms degrade or transform contaminants by a variety of mechanisms. Crude oil hydrocarbon derivatives (particularly alkane derivatives) for example are converted to carbon dioxide and water:

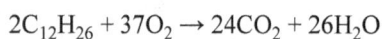

$$2C_{12}H_{26} + 37O_2 \rightarrow 24CO_2 + 26H_2O$$

When the hydrocarbon derivatives are chlorinated (as might occur in several additives to improve the performance of crude oil products) the degradation takes place as a secondary or co-metabolic process rather than a primary metabolic process. Under aerobic conditions, a chlorinated solvent such as trichloroethylene (ClCH = CCl$_2$), which may have been mixed with the crude oil product during processing or during use, can be degraded through a sequence of metabolic steps, where some of the intermediary by-products may be more hazardous than the parent compound (e.g., vinyl chloride, CH$_2$ = CHCl).

Most spills of crude oil and/or crude oil products are the result of accidents at oil wells, pipelines, ships, trains, and trucks that move crude oil and crude oil product from wells to refineries to the market.

Thus, put simply, in the biodegradation processes, a contaminant is transformed or eliminated by the action of living organisms (often referred to as microbes). By way of further clarification, the biodegradation process can be sub-divided divided into three stages: (1) biodeterioration, (2) biofragmentation, and (3) assimilation. Biodeterioration is often referred to as a surface-level degradation or the initial degradation that modifies the chemical physical, and mechanical properties of the contaminant.

Bioremediation because it takes advantage of natural processes and relies on microbes that occur naturally or can be laboratory cultivated; these consist of bacteria, fungi, actinomycetes, cyanobacteria and to a lesser extent, plants (US EPA, 2006). In fact, unless they are overwhelmed by the amount of the spilled material or it is toxic, many indigenous microorganisms in soil and/or water are capable of degrading hydrocarbon contaminants.

Thus, important site factors required for success include (1) the presence of metabolically capable and sustainable microbial populations, (2) suitable environmental growth conditions, such as the presence of oxygen, (3) temperature, which is an important variable—keeping a substance frozen or below the optimal operating temperature for microbial species, can prevent biodegradation—most biodegradation occurs at temperatures between 10 and 35°C (50 and 95°F), (4) the presence of water, (5) and appropriate levels of nutrients and contaminants (Table 6.4).

In regard to the last parameter, soil pH is extremely important because most microbial species can survive only within a certain pH range—generally the biodegradation of crude oil hydrocarbon derivatives is optimal at a pH 7 (neutral) and the *acceptable* (or optimal) pH range is on the order of 6 to 8. Furthermore, soil (or water) pH can affect availability of nutrients.

Furthermore, biodegradation also alters subsurface oil accumulations of crude oil (Winters and Williams, 1969; Speight, 2014). In addition, shallow reservoirs and deposits such as heavy crude oil reservoirs and tar sand deposits (in which the reservoir temperature or the deposit temperature is low-to-moderate, < 80°C, < 176°F) are commonly found to have undergone some degree of biodegradation (Barker et al., 1995; McAllister et al., 1995; Speight and Lee, 2000; Speight, 2014).

These technologies are especially valuable where the contaminated soils are fragile, and prone to erosion. However, when the above parameters are not conducive to bacterial activity, the bacteria (1) grow too slowly, (2) die, or (3) create more harmful chemicals.

Table 6.4: Micro-organisms and Hydrocarbon (Organic Compound) Interaction (Abatenh et al., 2017).

Micro-organisms	Compound	References
Penicillium chrysogenum	Monocyclic aromatic hydrocarbons, benzene, toluene, ethyl benzene and xylene, phenol compounds	Pedro et al., 2014 Abdulsalam et al., 2013
P. alcaligenes P. mendocina and P. putida P. veronii, Achromobacter, Flavobacterium, Acinetobacter	Petrol and diesel polycyclic aromatic hydrocarbons toluene	Safiyanu et al., 2015 Sani et al., 2015
Pseudomonas putida	Monocyclic aromatic hydrocarbons, e.g., benzene and xylene.	Safiyanu et al., 2015 Sarang et al., 2013
Phanerochaete chrysosporium	Biphenyl and triphenylmethane	Erika et al., 2013
A. niger, A. fumigatus, F. solani and *P. funiculosum*	Hydrocarbon	Al-Jawhari, 2014
Coprinellus radians	PAHs, methylnaphthalenes, and dibenzofurans	Aranda et al., 2010
Alcaligenes odorans, Bacillus subtilis, Corynebacterium propinquum, Pseudomonas aeruginosa	phenol	Singh et al., 2013
Tyromyces palustris, Gloeophyllum trabeum, Trametes versicolor	hydrocarbons	Karigar and Rao, 2011
Candida viswanathii	Phenanthrene, benzopyrene	Hesham et al., 2012
cyanobacteria, green algae and diatoms and *Bacillus licheniformis*	naphthalene	Sivakumar, et al., 2012 Lin et al., 2010
Acinetobacter sp., *Pseudomonas* sp., *Ralstonia* sp. and *Microbacterium* sp.,	aromatic hydrocarbons	Simarro et al., 2013
Gleophyllum striatum	striatum Pyrene, anthracene, 9-methyl anthracene, Dibenzothiophene Lignin peroxidase	Yadav et al., 2011
Acinetobacter sp., *Pseudomonas* sp., *Ralstonia* sp. and *Microbacterium* sp.,	aromatic hydrocarbons	Simarro et al., 2013
Gleophyllum striatum	Pyrene, anthracene, 9- methyl anthracene, Dibenzothiophene Lignin peroxidase	Yadav et al., 2011
Acinetobacter sp., *Pseudomonas* sp., *Ralstonia* sp. and *Microbacterium* sp.,	aromatic hydrocarbons	Simarro et al., 2013
Gleophyllum striatum	striatum Pyrene, anthracene, 9-methyl anthracene, Dibenzothiophene Lignin peroxidase	Yadav et al., 2011

Crude oil hydrocarbon derivatives (particularly alkane derivatives) for example are converted to carbon dioxide and water:

$$2C_{12}H_{26} + 37O_2 \rightarrow 24CO_2 + 26H_2O$$

When the hydrocarbon derivatives are chlorinated (as might occur in several additives to improve the performance of crude oil products) the degradation takes place as a secondary or co-metabolic process rather than a primary metabolic process. Under aerobic conditions, a chlorinated solvent such as trichloroethylene ($ClCH = CCl_2$), which may have been mixed with the crude oil product during processing or during use, can be degraded through a sequence of metabolic steps, where some of the intermediary by-products may be more hazardous than the parent compound (e.g., vinyl chloride, $CH_2 = CHCl$).

Indigenous and enhanced organisms have been shown to degrade industrial solvents, polychlorinated biphenyls (PCBs), explosives, and many different agricultural chemicals.

6.3 Types of Bioremediation

Bioremediation is the most effective, economical, eco-friendly management tool that can be used by today's society to manage the polluted environment. Also, all the bioremediation techniques have its own advantage and disadvantage because of its own specific applications.

For bioremediation to be effective for a particular spill, the first phase would be to decide whether the particular spill is a candidate for bioremediation treatment, after which a decision needs to be made whether bioremediation can be implemented effectively, given the logistics of application and monitoring. This has been difficult to establish on dynamic, heterogeneous marine shorelines. Therefore, the evidence that should be noted for the effectiveness of oil bioremediation is that the disappearance of oil in treated areas should be greater and faster than that in untreated areas.

The bioremediation method is an effective method for the cleanup of organic pathogens, arsenic, fluoride, nitrate, volatile organic compounds, metals and many other pollutants like ammonia and phosphates. It is also effective for cleaning insecticides and herbicides, as well as saltwater intrusion into aquifers. Bioremediation also allows for savings that is unlike traditional methods and cleans the contaminated site without the constant need for attention.

Bioremediation is a completely natural process with almost no harmful side effects. It detoxifies hazardous substances instead of merely transferring contaminants from one environmental medium to another (Sharma and Reddy, 2004).

The process is less disruptive to the environment than excavation-based processes and consumes less energy compared to incineration and landfilling. However, bioremediation is limited to the types of contaminants that it can remove effectively. The variables that are necessary for bioremediation to take place may just as well be the cause of its limitations.

The methods (presented in the subsection below) utilizes indigenous bacteria (microbes) compared to the customary (physical and chemical) remediation methods. Also, the microorganisms engaged can perform almost any detoxification reaction. Biodegradation studies provide information on the fate of a chemical or mixture of crude oil-derived chemicals (such as oil spills and process wastes) in the environment thereby opening the scientific doorway to develop further methods of cleanup by (1) analyzing the contaminated sites, (2) determining the best method suited for the environment, (3) optimizing the cleanup techniques which lead to the emergence of new processes. Bioremediation technology exploits various naturally occurring mitigation processes: (1) natural attenuation, (2) biostimulation, and (3) bioaugmentation.

6.3.1 Natural Bioremediation

In the natural bioremediation process, naturally occurring organisms are used to break down (biodegrade) environmental contaminants materials into less toxic or non-toxic materials as the organisms carry out their normal life functions (Dzionek et al., 2016). Natural biodegradation/bioremediation typically involves the use of molecular oxygen (O_2). Thus:

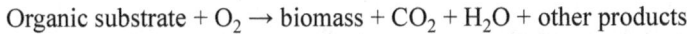

Organic substrate $+ O_2 \rightarrow$ biomass $+ CO_2 + H_2O +$ other products

In the absence of oxygen, some microorganisms obtain energy from fermentation and anaerobic oxidation of organic carbon. Many anaerobic organisms (anaerobes) use nitrate, sulfate, and salts of iron (III) as practical alternates to oxygen acceptor as, for example, in the anaerobic reduction process of nitrates, sulfates, and salts of iron (III):

$$2NO_3^- + 10e^- + 12H^+ \rightarrow N_2 + 6H_2O$$

$$SO_4^{2-} + 8e^- + 10H^+ \rightarrow H_2S + 4H_2O$$

$$Fe(OH)_3 + e^- + 3H^+ \rightarrow Fe^{2+} + 3H_2O$$

In the process, the immobilization of the microorganisms capable of degrading specific contaminants significantly promotes the processes chemistry and also allows for the multiple use of biocatalysts. Among the developed methods of immobilization, adsorption on the surface is the most common method in bioremediation, due to the simplicity of the procedure and its non-toxicity. The choice of carrier is an essential element for successful bioremediation. It is also important to consider (1) the type of process (*in situ* or *ex situ*), (2) the properties of the contaminant, and the properties of the microorganisms.

6.3.2 Traditional Bioremediation

Natural attenuation bioremediation (which occurs without human intervention) relies on natural conditions and behavior of soil microorganisms that are indigenous to soil. Methods for the cleanup of pollutants have usually involved removal of the polluted materials, and their subsequent disposal by land filling or incineration (so called dig, haul, bury, or burn methods) (Speight, 1996; Speight and Lee, 2000; Speight, 2005). Furthermore, available space for landfills and incinerators is declining.

Conventional bioremediation methods that have been, and are still, used are (1) composting, (2) land farming, (3) biopiling and (4) use of a bioslurry reactor (Cassidy and Irvine, 1995; Speight, 1996; Speight and Lee, 2000; Semple et al., 2001).

The use of traditional methods of bioremediation continues but there is also method evolution such as (1) isolating and characterizing naturally-occurring microorganisms with bioremediation potential, (2) laboratory cultivation to develop viable populations, (3) monitoring and measuring the progress of bioremediation through chemical analysis and toxicity testing in chemically-contaminated media, as well as (4) field applications of bioremediation techniques using either *in situ* stimulation of microbial activity by the addition of microorganisms and nutrients and the optimization of environmental factors at the contaminated site itself or *ex situ* restoration of contaminated material in specifically designated areas by land-farming and composting method (Speight, 1996; Speight and Lee, 2000; Maila and Cloete, 2004; Marin et al., 2006).

Bioremediation which occurs without human intervention other than monitoring is often called *natural attenuation*. This natural attenuation relies on natural conditions and behavior of soil microorganisms that are indigenous to soil. Methods for the cleanup of pollutants have usually involved removal of the polluted materials, and their subsequent disposal by land filling or incineration (so called *dig, haul, bury, or burn* methods) (Speight, 1996; Speight and Lee, 2000; Speight, 2005). Furthermore, available space for landfills and incinerators is declining. Perhaps one of the greatest limitations to traditional cleanup methods is the fact that in spite of their high costs, they do not always ensure that contaminants are completely destroyed. Conventional bioremediation methods that have been, and are still, used are (1) composting, (2) land farming, (3) biopiling and (4) use of a bioslurry reactor (Speight, 1996; Speight and Lee, 2000; Semple et al., 2001).

Composting is a technique that involves combining contaminated soil with nonhazardous organic materials such as manure or agricultural wastes; the presence of the organic materials allows the development of a rich microbial population and elevated temperature characteristic of composting. *Land farming* is a simple technique in which contaminated soil is excavated and spread over a prepared bed and periodically tilled until pollutants are degraded.

Biopiling is a hybrid of land farming and composting, it is essentially engineered cells that are constructed as aerated composted piles. A *bioslurry reactor* can provide rapid biodegradation of contaminants due to enhanced mass transfer rates and increased contaminant-to-microorganism contact. These units are capable of aerobically biodegrading aqueous slurries created through the mixing of soils or sludge with water. The most common state of bioslurry treatment is batch; however, continuous-flow operation is also possible.

The technology selected for a particular site will depend on the limiting factors present at the location (Okoh, 2006). For example, where there is insufficient dissolved oxygen, bioventing or sparging is applied, biostimulation or bioaugmentation is suitable for instances where the biological count are low. On the other hand, application of the composting technique, if the operation is unsuccessful, it will result in a greater quantity of contaminated materials. Land farming is only effective if the contamination is near the soil surface or else bed preparation is required. The main drawback with slurry bioreactors is that high-energy input is required to maintain suspension and the potential needed for volatilization.

Other techniques are also being developed to improve the microbe-contaminant interactions at treatment sites so as to use bioremediation technologies at their fullest potential. These bioremediation technologies consist of monitored natural attenuation, bioaugmentation, biostimulation, surfactant addition, anaerobic bioventing, sequential anaerobic/aerobic treatment, soil vapor extraction, air sparging, enhanced anaerobic dechlorination and bioengineering (Speight, 1996; Speight and Lee, 2000).

The use of traditional methods of bioremediation continues but there is also method evolution, which may involve the following steps: (1) isolating and characterizing naturally-occurring microorganisms with bioremediation potential, (2) laboratory cultivation to develop viable populations, (3) studying the catabolic activity of these microorganisms in contaminated material through bench scale experiments, (4) monitoring and measuring the progress of bioremediation through chemical analysis and toxicity testing in chemically-contaminated media, and (5) field applications of bioremediation techniques using either/both of the following steps: *in situ* stimulation of microbial activity by the addition of microorganisms and nutrients and the optimization of environmental factors at the contaminated site itself and/or *ex situ* restoration of contaminated material in specifically designated areas by land-farming and composting method.

6.3.3 Enhanced Bioremediation

Enhanced bioremediation is a process in which indigenous or inoculated microorganisms (e.g., fungi, bacteria, and other microbes) degrade (metabolize) organic contaminants found in soil and/or ground water and convert the contaminants to innocuous end products. Put simply, the process relies on the availability of naturally occurring microbes to consume the contaminants.

The process relies on general availability of naturally occurring microbes to consume contaminants as a food source (crude oil hydrocarbon derivatives in aerobic processes) or as an electron acceptor (chlorinated solvents, which may be waste materials from crude oil processing). In addition to microbes being present, in order to be successful, these processes require nutrients such as carbon, nitrogen, and phosphorus). Enhanced bioremediation involves the addition of microorganisms (e.g., fungi, bacteria, and other microbes) or nutrients (e.g., oxygen, nitrates) to the subsurface environment to accelerate the natural biodegradation process. The process is enhanced by the use of additional technologies such as (1) biostimulation, (2) bioaugmentation, and (3) phytoremediation.

Biostimulation utilizes the indigenous microbial population to remediate a contaminated soil. The process involves the addition of nutrients such as phosphorus-containing and/or nitrogen-containing nutrients to catalyze the natural attenuation processes. In the process, the additives are typically usually added to the subsurface through injection wells and control of the amount id nutrient(s) added controls the growth of populations of the microorganisms. The, addition of nutrients causes rapid growth of the indigenous microorganism population thereby increasing the rate of biodegradation (Zhu et al., 2004).

Bioaugmentation involves the introduction of non-indigenous (exogenic) microorganisms that are capable of detoxifying a particular contaminant, sometimes employing genetically altered microorganisms. The indigenous microorganisms that are typically usually present in very small quantities and may not be able to prevent the spread of a contaminant or are not suitable to degrade a particular contaminant. In such cases, the bioaugmentation process offers a way to provide specific microbes in sufficient numbers to complete the biodegradation (Atlas, 1977, 1991; Compeau et al., 1991; Leavitt and Brown, 1994; Chhatre et al., 1996; Mishra et al., 2001; Vasudevan and Rajaram, 2001). The biodegradation rates of the contaminants are subject to (1) the distribution of the contaminants, (2) the concentration, of the contaminants, (3) the indigenous microbial populations, (4) the reaction kinetics, and (5) the site parameters such as pH, the moisture content, the nutrient supply, and the temperature (US EPA, 2006).

Phytoremediation, which is often referred to as enhanced remediation, involves the use of living plants for the removal of contaminants and metals from soil and is an *in situ* treatment of contaminated soils, sediments, and water. In fact, terrestrial plants, aquatic plants, wetland plants, and algae can be used for the process where there has been contamination by hydrocarbon derivatives (Brown, 1995; Nedunuri et al., 2000; Radwan et al., 2000; Magdalene et al., 2009). The contaminants that are amenable to the various subcategories of phytoremediation include hydrocarbon derivatives that are amenable to the following sub-processes: (1) phytotransformation, which involves the breakdown of organic contaminants sequestered by plants, (2) rhizosphere bioremediation, which involves the use of rhizosphere microorganisms to degrade organic pollutants, (3) phytostabilization, which involves the containment process by using plants—often in combination with soil additives—to assist plant installation, to mechanically stabilizing the site and reducing pollutant transfer to other ecosystem compartments and the food chain, (4) phytoextraction, which involves the ability of some plants to accumulate metals/metalloids in their shoots, (5) rhizofiltration or the combination of phytovolatilization and rhizovolatilization, which involves processes employing metabolic capabilities of plants and associated rhizosphere microorganisms to transform pollutants into volatile compounds that are released to the atmosphere (Korade and Fulekar, 2009). Typically, the phytotransformation technology and the rhizosphere bioremediation technology are applicable to sites that have been contaminated with organic pollutants, including pesticides (Brown, 1995).

Overall, the phyto-technologies are valuable when the contaminated soil is fragile, non-cohesive, and prone to erosion. The establishment of a stabilized system prevents erosion which is especially relevant to the type of soil where removal of large volumes of soil can destabilize the soil system thereby causing extensive erosion. However, in many cases, the above parameters are not conducive to bacterial activity because the bacteria (1) grow too slowly, (2) die, or (3) create more harmful chemicals.

6.3.4 *Monitored Natural Attenuation*

The term *natural attenuation* refers to a variety of physical, chemical, or biological processes that, under favorable conditions, act without human intervention to reduce the mass, toxicity, mobility, volume, or concentration of contaminants in soil or groundwater. These *in situ* processes include biodegradation, dispersion, dilution, sorption, volatilization, and chemical stabilization, or biological stabilization, transformation, or destruction of the contaminant(s). These are processes in which microorganisms contribute to pollutant degradation.

Typically, the method is effective when (1) the subsurface soil is highly permeable, (2) the soil horizon to be treated exists within approximately twenty to thirty five feet a depth below the surface and of and the groundwater depth is also shallow—on the order of less than thirty five feet below the ground surface.

6.4 Factors that Affect Bioremediation

The bioremediation of any crude oil constituents (and the constituents of crude oil products) is defined by the biological, physical, and chemical characteristics of the environment and the process can be carried out by an *in situ* process or by an *ex situ* process, depending on factors such as (but not limited to) (1) the site characteristics, and (2) the type and concentration of contaminant(s). These factors also include other factors such as (1) the existence of a microbial population capable of degrading the pollutants, (2) the availability of contaminants to the microbial population, (3) the environment factors such as the type of soil, the temperature, the pH (i.e., the acidity or alkalinity of (for example) the soil, and (4) the presence of oxygen and nutrients at the site (Okoh, 2006; Naik and Duraphe, 2012).

Biodegradation involves chemical transformations mediated by microorganisms that: (1) satisfy nutritional requirements, (2) satisfy energy requirements, (3) detoxify the immediate environment, (4) or occur fortuitously such that the organism receives no nutritional or energy benefit (Stoner, 1994).

There are various method for determining the chemistry of the biodegradation of pollutants, such as the determination of the amount of carbon dioxide or methane, for anaerobic cases (Zengler et al., 1999) produced during a specified period as well as methods in which the loss of dissolved organic carbon (for substances which are water soluble) is determined as well as methods which are used to determine the uptake of oxygen by the activities of microorganisms. In fact, the biodegradation of crude oil constituents and the constituents of crude oil products can occur under both aerobic (oxic) and anaerobic (anoxic) conditions (Zengler et al., 1999), albeit by the action of different consortia of organisms. Thus, two classes of biodegradation reactions are: (1) aerobic biodegradation and (2) anaerobic biodegradation.

Aerobic bioremediation is most effective in treating non-halogenated organic compounds. Thus:

Organic substrate + O_2 → biomass + CO_2 + H_2O + other inorganic products

Some micro-organisms (chemo-autotrophic aerobes or litho-trophic aerobes) oxidize reduced inorganic compounds (NH_3, Fe^{2+}, or H_2S) to gain energy and fix carbon dioxide to build cell carbon:

NH_3 (or Fe^{2+} or H_2S) + CO_2 + H_2 + O_2 → biomass + NO_3 (or Fe or SO_4) + H_2O

Various chemical species such as nitrate (NO_3^-), iron (Fe^{3+}), manganese (Mn^{4+}), sulfate (SO_4^{2+}), and carbon dioxide (CO_2) can act as electron acceptors if the organisms present have the appropriate enzymes (Sims, 1990).

Anaerobic biodegradation is the microbial degradation of organic substances in the absence of free oxygen. While oxygen serves as the electron acceptor in aerobic biodegradation processes forming water as the final product, degradation processes in anaerobic systems depend on alternative acceptors such as sulfate, nitrate or carbonate yielding, in the end, hydrogen sulfide, molecular nitrogen, and/or ammonia and methane (CH_4), respectively.

In the absence of oxygen, some micro-organisms obtain energy from fermentation and anaerobic oxidation of organic carbon. Many anaerobes use nitrate, sulfate, and salts of iron (III) as practical alternates to oxygen acceptor. The anaerobic reduction process of nitrates, sulfates and salts of iron is an example:

$2NO_3^- + 10e^- + 12H^+ → N_2 + 6H_2O$

$SO_4^{2-} + 8e^- + 10H^+ → H_2S + 4H_2O$

$Fe(OH)_3 + e^- + 3H^+ → Fe^{2+} + 3H_2O$

Anaerobic biodegradation is a multistep process performed by different bacterial groups.

In contrast to the strictly anaerobic sulfate-reducing and methanogenic bacteria, the nitrate-reducing micro-organisms as well as many other decomposing bacteria are mostly facultative anaerobic insofar as these micro-organisms able to grow and to degrade organic substances under aerobic as well as anaerobic conditions. Thus, aerobic and anaerobic environments represent the two extremes of a continuous spectrum of environmental habitats which are populated by a wide variety of micro-organisms with specific biodegradation abilities.

Anoxic conditions may represent an intermediate stage where oxygen supply is limited, still allowing a slow (aerobic) degradation of organic compounds (Speight and Arjoon, 2012).

6.4.1 Chemistry

The biodegradation of the constituents of crude oil and crude oil products is a multi-faceted process insofar as the fluid properties of the spilled material change because different classes of the constituents of the spilled material have different susceptibilities to biodegradation (Goodwin et al., 1983; Okoh, 2006; Speight and Arjoon, 2012). The inherent biodegradability of the constituents of crude oil and crude oil products is not only related to the chemical structure of the constituents but also by the physical state of the constituents. For example, while the n-alkane derivatives are the most biodegradable of the hydrocarbon derivatives, the C5 to C10 homologs may also acts as inhibit the biodegradation process insofar as these particular constituents can also act as solvents which can disrupt the lipid membrane structure of a microorganisms. Also, the alkane derivatives in the C20 to C40 range (more commonly referred to as waxes) are hydrophobic solids at physiological temperatures.

The early stages of the biodegradation of the constituents of crude oil biodegradation are characterized by the loss of n-paraffin derivatives (n-alkane derivatives) followed by the degradation of acyclic hydrocarbon derivatives. Compared with those compound groups, other compound classes (such as highly branched hydrocarbon s and cyclic saturated hydrocarbon derivatives as well as aromatic derivative) are more resistant to biodegradation. However, even those more-resistant compound classes are eventually destroyed as biodegradation proceeds.

The hydrocarbon derivatives in crude oil and in crude oil products are classified as alkane derivatives (normal alkane derivatives or iso-alkane derivatives), cycloalkane derivatives, and aromatic derivatives (Chapter 1). Alkene derivatives (olefin derivatives) are rare in crude oil (because of the inherent tendency to oxidize to, for example, higher molecular weight derivatives over time) but do occur in many crude oil products as a consequence of the cracking process (Speight, 2014, 2017, 2019). The increasing carbon numbers of the alkane derivatives (referred to as homology), variations in carbon chain branching (the iso-alkane derivatives), ring condensations, and interclass combinations, such as phenyl alkane derivatives, also add to the various types of hydrocarbon derivatives that occur in crude oil and crude oil products.

The alkane derivatives are plentiful in crude oil and crude oil products whereas alkene derivatives are rare (if present at all) in crude oil but do occur in many refined crude oil products as a consequence of the cracking process (Speight, 2014, 2017). Increasing carbon numbers of the alkane derivatives (homology), variations in carbon chain branching (*iso*-alkane derivatives), ring condensations, and interclass combinations, such as alkylated aromatic ring systems that account for the high numbers of hydrocarbon derivatives that occur in crude oil and crude oil products.

As a side note, olefin derivatives may be present in crude oil that has been recovered from the reservoir by a thermal (enhanced recovery) process.

In addition, smaller amounts of oxygen-containing compounds (phenol derivatives, naphthenic acids), nitrogen-containing compounds (pyridine derivatives, pyrrole derivatives, indole derivatives), sulfur-containing compounds (thiophene derivatives), and the high molecular weight polar asphalt fraction also occur in crude oil but not in refined crude oil products (Speight, 2014, 2017).

Cycloalkane derivatives are transformed by an oxidase system to a corresponding cyclic alcohol, which is dehydrated to ketone after which a mono-oxygenase system lactonizes the ring, which is subsequently opened by a lactone hydrolase.

As in the case of alkane derivatives, the monocyclic compounds, cyclopentane, cyclohexane, and cycloheptane have a strong solvent effect on lipid membranes and are toxic to the majority of hydrocarbon degrading microorganisms.

The biodegradation of the constituents of crude oil and crude oil products can occur under both aerobic (oxic) and anaerobic (anoxic) conditions (Zengler et al., 1999), albeit by the action of different consortia of organisms. Thus, two classes of biodegradation reactions are: (1) aerobic biodegradation and (2) anaerobic biodegradation.

In the aerobic biodegradation process, the use of molecular oxygen (O_2)—as the terminal electron acceptor—receives electrons transferred from an organic contaminant, which can be represented simply by the equation:

Organic contaminant + O_2 → biomass + CO_2 + H_2O + other inorganic products

At the same time, there are microorganisms (chemo-autotrophic aerobes or litho-trophic aerobes) that oxidize reduced inorganic compounds (such as ammonia, NH_3, ferrous iron, Fe^{2+}, or hydrogen sulfide, H_2S) to gain energy and fix carbon dioxide to build cell carbon (Sims, 1990):

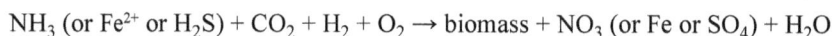

NH_3 (or Fe^{2+} or H_2S) + CO_2 + H_2 + O_2 → biomass + NO_3 (or Fe or SO_4) + H_2O

In the anaerobic biodegradation process, the microbial degradation of organic substances occurs in the absence of free oxygen. This process (in the lack of oxygen as the electron acceptor) depends on alternative acceptors such as sulfate ions ($-SO4^{2-}$), nitrate ions ($-NO_3^-$), or carbonate ions ($-CO_3^{2-}$) yielding hydrogen sulfide (H_2S), molecular nitrogen (N_2), and/or ammonia (NH_3) and methane (CH_4), respectively.

Condensed polycyclic aromatic derivatives are degraded one ring at a time but biodegradability tends to decline with the increasing number of rings and the degree of condensation (Bartha, 1986a, 1986b; Atlas and Bartha, 1998). Aromatic derivatives with more than four condensed rings are generally not suitable as substrates for microbial growth, though, they may undergo metabolic transformations.

The asphaltic constituents of crude oil crude oil—often referred collectively to as the reason constituents and the asphaltene constituents (Speight, 2014)—tend to increase during biodegradation in relative and sometimes absolute amounts. This would suggest that they not only tend to resist biodegradation but may also be formed by condensation reactions of biodegradation and photo-degradation intermediates.

The kinetics aspects of the bioremediation process are not simple due to the fact that the primary function of the microbial metabolism (whether bacterial or fungal) is to grow and sustain more of the microorganism (Cutright, 1995; Rončević et al., 2005; Pala et al., 2006; Maletić et al., 2009).

One example of the empirical approach is represented by the equation:

$dC/dt = kC^n$

In this equation, C is the concentration of the substrate, t is time and k is the degradation rate constant of the compound and n is a fitting parameter (most often taken to be unity) (Hamaker 1972; Rončević et al., 2005; Wethasinghe et al., 2006).

6.4.2 Types of Microbes

In many cases, after an oil spill, the natural microbial systems for degrading the oil are overwhelmed. Therefore molecular engineers are constructing starvation promoters to express heterologous genes needed in the field for survival and adding additional bioremediation genes that code for enzymes able to degrade a broader range of compounds present in the contaminated environments. Various bacterial strains are also being developed, where each strain is specific for a certain organic compound present in oil spills. This will help increase the speed of bioremediation and allow detailed cleanup to take place where no organic contamination remains in the environment. Thus, the decision to bioremediate a site is dependent on cleanup, restoration, and habitat protection objectives and the factors that are present that would have an impact on success. If the circumstances are such that no amount of nutrients will accelerate biodegradation, then the decision should be made on the need to accelerate oil disappearance to protect a vital living resource or simply to speed up restoration of the ecosystem. These decisions are clearly influenced by the circumstances of the spill.

The choice of non-indigenous micro-organisms is influenced by the type of pollutant and temperature also exerts a strong impact on the microbial degradation of crude oil because if the influence of temperature on the physical state of crude oil as well as on the chemical composition of the crude oil, the metabolism of the micro-organisms, and the composition of microbial consortium. However, chlorinated hydrocarbon derivatives (which may be present in some crude oil products) or other aromatic derivatives are (in many such cases) (1) resistant to biodegradation or (2) are toxic to the micro-organisms or (3) produce products that are toxic to the micro-organisms or (4) are, at best, biodegraded very slowly (Sales da Silva, 2020).

Most of the indigenous microbes have the ability to successfully bring up the environmental restoration via oxidizing, immobilizing, or transforming the contaminants (Crawford et al., 2005; Karigar and Rao, 2011). Mycoremediation uses the digestive enzymes of fungi to degrade contaminants such as pesticides, hydrocarbon derivatives, and heavy metals.

Therefore, bioremediation can be tailored to the needs of a specific contaminant and the actions of specific microbes needed to break down the pollutant(s) can be enhanced by selecting the limiting factor needed to promote the growth. Because several different types of contaminants can be present at a polluted site, various types of micro-organisms are required for effective remediation. Some types of micro-organism are able to degrade crude oil constituents (especially the hydrocarbon derivatives) as a source of carbon and energy. However, the choice of the micro-organisms depends on the chemical nature of the contaminant(s) and should be selected carefully (Table 6.4) as the micro-organisms only survive in the presence of a limited range of chemical contaminants (Coelho et al., 2015).

Bioremediation using bacterial species can include using *Pseudomonas* species which are potent bacteria that are capable of degrading hydrocarbon derivatives from petrol and diesel, thereby reducing the impact of oil spills. Other Bacteria that help in bioremediation are Achromobacter, Flavobacterium, Acinetobacter, etc. (Idris et al., 2015).

Both, anaerobic and aerobic bacteria are capable of biotransforming polychlorobiphenyl derivatives (PCBs). Higher chlorinated polychlorobiphenyl derivatives are subjected to reductive dehalogenation by anaerobic micro-organisms.

The desulfurization, dehalogenation, denitrification, ammonification, hydroxylation, biotransformation, and biodegradation of various aromatic and aliphatic compounds are catalyzed by monooxygenase enzymes which occur in soil bacteria and are involved in the transformation of aromatic precursors into aliphatic products (Figure 6.2). This type of enzyme is involved in the degradation of hydrocarbon such as substituted methane derivatives, alkane derivatives, cycloalkane derivatives, alkene derivatives, haloalkene derivatives (i.e., halogenated olefin derivatives), ether derivatives, and aromatic hydrocarbon derivatives, and heterocyclic hydrocarbon derivatives.

Among the biological agents, laccases represent an interesting group of ubiquitous, oxidoreductase enzymes that show promise of offering great potential for biotechnological and bioremediation applications (Gianfreda et al., 1999).

Many micro-organisms produce intra and extracellular laccases capable of catalyzing the oxidation of ortho and p-phenol derivatives, p-amino phenol derivatives, polyphenol derivatives, polyamine derivatives, lignin derivatives, and aryl diamine derivatives as well as some inorganic ions (Figure 6.2). Laccases not only oxidize phenol derivatives but also phenolic acids and methoxy-

$$C_6H_5OH + O_2 \rightarrow \text{ortho-}HOC_6H_4OH$$

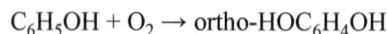

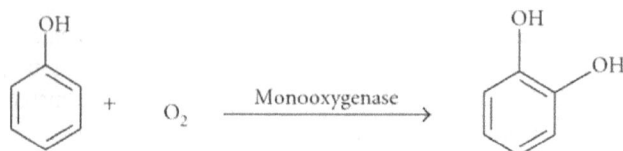

Figure 6.2: Degradation of Phenol by Mono-oxygenase (Arora et al., 2010).

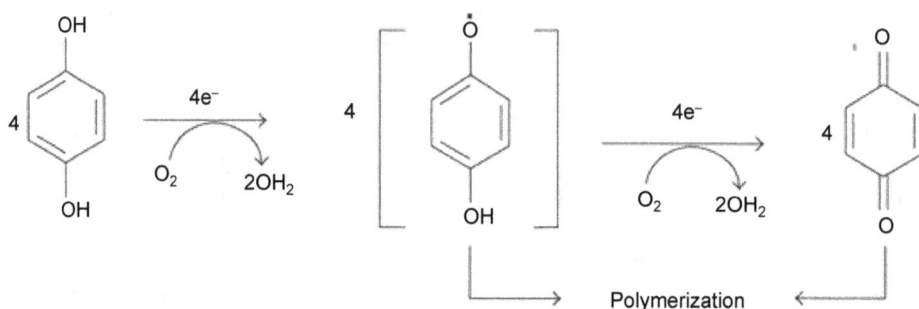

Figure 6.3: General reaction mechanism for phenol oxidation by laccase (Cirino and Arnold, 2002).

phenolic acids. The substrate specificity and affinity of laccase can vary with changes in pH (Karigar and Rao, 2011).

Peroxidase enzymes have been classified into many types based on its source and activity. In fact, some peroxidase enzymes are able to directly oxidize manganese (Mn^{2+}), methoxybenzene derivatives, phenolic aromatic substrates (Karigar and Rao, 2011). In addition, lipase enzymes degrades lipids derived from a large variety of micro-organisms, animals and plants—a lipid is any of a class of organic compounds that are fatty acids or their derivatives and are insoluble in water but soluble in organic solvents and include many natural oils, waxes, and steroids.

6.4.3 Types of Contaminants

Contaminants (as they pertain to the environment) are any physical, chemical, biological, or radiological substances or matter that has an adverse effect on the air, water, soil, or floral and faunal species. The contaminants may include any one or more of the following (1) agrochemicals, such as fertilizers, (2) chlorinated compounds, such as polychlorobiphenyl derivatives as well as chlorobenzene derivatives and chlorinated aliphatic derivatives, (3) dyes, (4) pharmaceuticals, (5) heavy metal derivatives, (6) hydrocarbon derivatives, (7) wood treating chemicals, such as creosote and pentachlorophenol, (8) chlorinated aromatic compounds and chlorinated aliphatic compounds such as trichloroethylene, (9) plastics, (10) sewage, and (11) nuclear waste.

The contaminants can accidentally or deliberately enter the environment often, but not always, as a result of anthropogenic activities and have been changing ecosystems through the effects on wildlife and the food chain. Most enter the environment from activities such as (1) landfill sites, (2) industrial and commercial facilities, (3) spills of crude oil and crude oil products, (4) spills of chemical spills, (5) and wastewater treatment plants, and (6) sewage processing plants. The process can be particularly well adapted to the biodegradation of organic contaminants such as (focusing on the current context) crude oil constituents.

Some pollutants resist breakdown and accumulate in the food chain also harmful chemical pollution and excess nutrient runoff are serious threats to the coastal environment. Therefore being able to identify contaminants and clean up and restore areas that have been impacted by these contaminants is of imperative importance in maintaining the delicate balance in the ecosystem.

Table 6.5: Examples of Micro-organisms Involved in Heavy Metal Detoxification (Tegene and Tenkegna, 2020).

Heavy metals	Sources	Heavy metal degrader
Cr	Tanneries, steel industries, fly ash from the burning of coal	*Pseudomonas mendoca, Cellulosmicrobium cellulans, Oedogoniumrivulare, Saccharomyces cerevisiae, Oscillatoria* sp., *Arthrobacter* sp., *Agrobacter* sp., *Pseudomonas aeruginosa* S128, *Chlamydomonas sp.* (algae), *Chlorella vulgaris* (algae), *Zoogloea ramigera*
Pb	Herbicides, batteries, insecticides, aerial emissions from petrol	*Pseudomonas aeruginosa, Oedogonium rivulare, Saccharomyces cerevisiae*
Hg	Medical waste, coal burning, and Au-Ag mining	*Trichoderma viride* and *Himicola insolens Saccharomyces cerevisiae, Pseudomonas* sp., *Escherichia* sp., *Bacillus* sp., *Clostridium* sp.
Ni	Battery manufacturing, steel alloys, kitchen appliances, surgical instruments, industrial effluents	*Pseudomonas aeruginosa, Oedogonium rivulare*
Cu	Pesticides and fertilizers usage	*Bacillus* species, *Pseudomonas aeruginosa*
Cd	Electroplating, plastic burning, phosphate fertilizer, paints and pigments	*Bacillus* species, *Pseudomonas aeruginosa, Micrococcus roseus*
As	Wood storage and pesticides	*Bacillus* spp.
Zn	Priming paints for metals, varnishes and pigments in aerospace paints	*Escherichia coli, Aspergillus niger*

Contamination caused by the spillage of crude oil and crude oil products is a danger to all floral and faunal species. Crude oil and crude oil products are complex mixtures of paraffin derivatives, alicyclic derivatives (often referred to as naphthenes or naphthene derivatives) and aromatic hydrocarbon derivatives and a smaller proportion of nonhydrocarbon compounds such as naphthenic acids, phenols, thiol, heterocyclic nitrogen, sulfur compounds as well as metallo-porphyrin derivatives (Speight, 2014, 2017; Fowzia and Fakhruddin, 2018). Moreover, many of the hydrocarbon constituents have been known to belong to the family of carcinogens and neurotoxic organic pollutants.

Heavy metals at certain concentrations can have long-term toxic effects within ecosystems and micro-organisms are important in heavy metal remediation (Table 6.5) because they can adopt different mechanisms to interact and survive in the presence of inorganic metals.

In the context of this book, the candidates for bioremediation are the ecological systems that have been subject to the spillage of crude oil and/or crude oil products. Therefore, a physical and chemical analysis of the contaminants is essential if the efficiency and effectiveness of the process is to be improved.

6.4.3.1 Physical Properties

Pollution of the environment with toxic metals—such as arsenic, lead, cadmium, nickel, mercury, chromium, cobalt, zinc and selenium are highly toxic even in minor quantities—has increased suddenly since the onset of the industrial revolution.

Pertinent to the present context, the behavior of a crude oil or a crude oil product released into the environment is determined not only by its chemical composition but also by the physical properties of the material. The toxicity of the contaminants in the environment depends on the solubility and the bioavailability of the crude oil (or crude oil product) constituents (Speight, and Arjoon, 2012; Arjoon and Speight, 2020). In addition, the effects of such a spill depend on a variety of factors which include (1) the quantity and type of the spilled material and (2) the manner by which the constituents of the spilled material interact with the environment. Also, the prevailing climate conditions will influence the physical characteristics and behavior of the crude oil (or the crude oil product). Other key factors include (1) the biological and ecological attributes of the area,

(2) the ecological significance of the floral and faunal species, (3) the sensitivity of the floral and faunal species, (4) the constituents of the crude oil of to the constituents of the crude oil product, (5) the time of year, which determines the climate conditions.

6.4.3.2 Chemical Properties

The chemical properties of the contaminants are an important aspect of determination the character and behavior of the spilled material. Also, during the biodegradation process the micro-organisms transform the contaminants into other chemical forms and microbial cell reproduction by facilitating thermodynamically advantageous reduction/oxidation (redox) reactions which involve the transfer of electrons from electron donors to electron acceptors.

Thus, the some of the constituents of crude oil and the crude oil-derived products (especially the hydrocarbon derivatives) are converted by the naturally occurring (indigenous) soil micro-organisms to carbon dioxide, water, bacterial cells (biomass), and humus materials—which are the organic components of soil.

On a chemical basis, crude oil (and to a lesser extent) crude oil product constituents can be divided into four classes which are (1) the saturates fraction, (2) the aromatic derivatives fraction, (3) the resin fraction, and (4) the asphaltene fraction (Colwell et al., 1977; Speight, 2014, 2017, 2017). The chemical structures in the spilled contaminant determines the partitioning characteristics of the contaminant constituents and each chemical type of the constituents can exhibit very different behavior. For example, the n-normal (straight-chain) alkane derivatives of low molecular weight (C_8H1_8 to $C_{22}H_{46}$) are metabolized most rapidly, followed by the iso-alkane derivatives and higher-molecular-weight n-alkane derivatives, olefin derivatives, monoaromatic compounds (benzene derivatives), and polynuclear aromatic hydrocarbon derivatives (often designated by the acronyms PNAs or the name polyaromatic hydrocarbon derivatives, PAHs) (Speight and Arjoon, 2012; Arjoon and Speight, 2020). The polynuclear aromatic hydrocarbon derivatives are aromatic hydrocarbon derivatives with two or more fused benzene rings with natural as well as anthropogenic sources. They are widely distributed environmental contaminants that have detrimental biological effects, toxicity, mutagenicity, and carcinogenicity (Atlas, 1995; Haritash and Kaushik, 2009). They are formed during the thermal decomposition of organic molecules and their subsequent recombination. Incomplete combustion at high temperature (500–800°C) or subjection of organic material at low temperature (100 to 300°C, 212 to 570°F) for long periods result in the production of these species. They occur as colorless, white/pale yellow solids with low solubility in water, high melting point, high boiling point, and low vapor pressure (Table 6.6) (Atlas, 1995; Haritash and Kaushik, 2009). As the molecular weight increase the solubility of the polynuclear aromatic hydrocarbon derivatives in water decreases and increases in the melting point, the boiling point but there is a concomitant decrease in the vapor pressure (Clar, 1964; Patnaik, 1999).

Some compounds of interest in crude oil due to their harmful effect on living things are N-heptane, cyclohexane, methyl tertiary butyl ester (MTBE), benzene toluene, ethyl benzene, and xylenes (BTEX). Therefore bioremediation of these products will result in the production of products that are more environmentally friendly than the starting material. For example, the chemical equation for the degradation of toluene (C_7H_8) is:

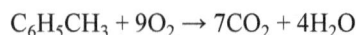

$$C_6H_5CH_3 + 9O_2 \rightarrow 7CO_2 + 4H_2O$$

Another example is the dechlorination of trichloroethane ($C_2H_3Cl_3$, CHCl = CH$_2$, or TCE) to dichloroethane ($C_2H_4Cl_2$, CH$_2$ClCH$_2$Cl, or DCE) by hydrogen-oxidizing anaerobic bacteria:

$$CHCl = CC_{12} + H_2 \rightarrow CH_2ClCH_2Cl + H^+ + Cl^-$$

In this process, the trichloroethane and hydrogen (H_2) decrease as the dichloroethane the hydrogen ion (H^+), and chloride ion (Cl^-) increase.

Toxic heavy metals discharged into the environment as waste either by industries or human activities affect millions of people and animals around the world. Heavy metals are non-biodegradable,

Table 6.6: Examples of Polynuclear Aromatic Hydrocarbon Derivatives.

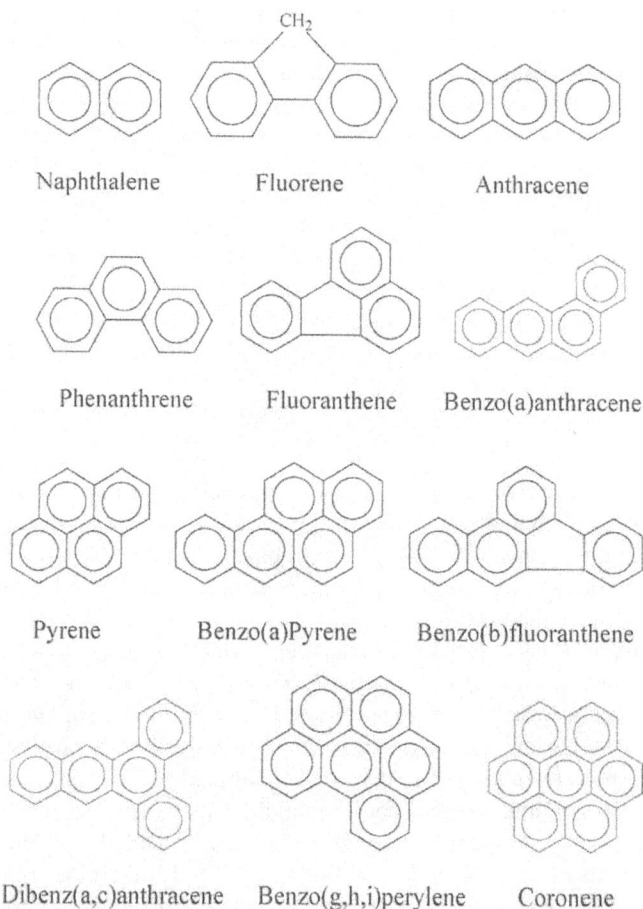

Naphthalene   Fluorene   Anthracene

Phenanthrene   Fluoranthene   Benzo(a)anthracene

Pyrene   Benzo(a)Pyrene   Benzo(b)fluoranthene

Dibenz(a,c)anthracene   Benzo(g,h,i)perylene   Coronene

but they can be transformed through sorption, methylation, complexation and changes in valence state (Adeniji, 2004). Heavy metals such as iron (Fe), chromium (Cr), and mercury (Hg) undergo oxidation and reduction cycles where these transformations affect the mobility and bioavailability of metals.

Micro-organisms may initiate metal mobilization/immobilization by redox reactions; and hence, impact bioremediation processes. More specifically, bioremediation is facilitated by converting an element from the insoluble and stationary form in sediments into a soluble and mobile form. As an example, the divalent mercurous form of (Hg^{2+}) is reduced to the elemental and more volatile form of mercury (Hg^0) and hexavalent chromium (Cr^{6+})—the most toxic form of chromium—is reduced to trivalent chromium (Cr^{3+}) and is immobilized in the hydroxide-oxide-forms. Reduction can also enhance the solubility of ions such as the ferric ion (Fe^{3+}) by reduction to the ferrous ion (Fe^{2+}) which can facilitate leaching of these ions from soil.

6.4.3.3 Site Specific Issues

Remediation of polluted sites using a microbial process (bioremediation) has proven effective and reliable due to its eco-friendly features and the process can either be carried out *ex situ* or *in situ*, depending on several factors, which include but not limited to (1) site characteristics, (2) the type of contaminants, and (3) the concentration of the contaminants. Therefore, choosing appropriate bioremediation technique, which will effectively reduce pollutant concentrations to an innocuous state, is crucial for a successful bioremediation project (Azubuike et al., 2016).

Thus, the process is affected by the characteristics of the contaminated site and parameters worthy of note include (1) soil texture, (2) soil permeability, (3) soil pH, (4) the water holding capacity of the soil, (5) soil temperature, (6) nutrient availability, and (7) oxygen content. Thus, the primary objectives of site characterization are not only those characteristics enumerated above but also the identification of the nature and the extent of contamination. Therefore, choosing appropriate bioremediation technique, which will effectively reduce pollutant concentrations to an innocuous state, is crucial for a successful bioremediation project. Furthermore, the two major approaches to enhance bioremediation are biostimulation and bioaugmentation provided that environmental factors, which determine the success of bioremediation, are maintained at optimal range Azubuike et al., 2016).

6.4.3.3.1 Geology of the Site

The geological and chemical nature of the site (site characterization) is (or should be) conducted when a contaminant has been released and there is potential for the contamination to reach people or adversely affect the natural ecology (Benson and Yuher, 2016).

The knowledge of the geological environment is a source of information that can explain some ecosystem response to various micro-organisms to help predict its nature and rate of activity. Many countries build geological and geographical inventories which can assist in understanding patterns in the behavior of micro-organisms for various environmental parameters. Geologic and hydrologic data usual consist of topography, soil profile up to bed rock, information on aquifer, groundwater depth and flow direction and permeability of soil.

The geologic characteristics also include a number of parameters such as (1) pH, (2) conductivity, (3) acidity, (4) alkalinity, (5) total dissolved solids, (6) hardness, (7) the presence of a wide variety of mineral chemical, and last but by no means least (8) the ability of the minerals to adsorb (or absorb) the chemical contaminants. Other parameters that need to be considered are the chemical characteristics of soil water quality, the type of contamination, concentration of contaminants, spatial extent of contamination, depth of contamination, contaminant retention characteristics and contaminant transport characteristics. For example, the bioavailability of hydrocarbon derivatives (such as the constituents of crude oil or the constituents of crude oil products) is largely a function of concentration and physical state, as well as their hydrophobicity, sorption onto soil particles, volatilization and solubility greatly affect the extent of their biodegradation.

To provide a detailed picture of the nature of the contaminated site, an assessment should include (1) the distribution of the contaminants in the various media, e.g., whether the contamination is limited to the surface soil, is distributed to significant depths, or whether the groundwater is affected, (2) the chemical and physical characteristics of the contaminants and their immediate environment, and (3) determination of the contaminant(s) under any of the relevant environmental conditions, and if so, what the likely source term would be.

6.4.3.3.2 Chemistry of the Site

The chemistry of a site (or mineralogy of the site) is not often defined with any degree of accuracy or certainty until the different minerals have been identified. While a study of the geology of a site, the knowledge of the geological environment is a source of information that can explain some ecosystem response to various micro-organisms to help predict its nature and rate of activity. However, the chemistry of a site is the science that is concerned with the composition, properties, and structure of mineral matter (i.e., the geological constituents) of the site and the manner in which a mineral can not only change from one form to another but also the manner in which the mineral matter can catalyze a change in the nature of the contaminant(s). A focus on the chemical composition of the site can help the environmental scientist and engineer gain an deep understanding of the different chemical processes and reactions that become active when a contaminant is introduced on to the site.

Thus, a study of the chemistry of a site can provide an important perspective that, with the knowledge of the geology of the site, is particularly important in understanding the potential for changes in the properties of the contaminant(s). Generally, as much information as possible should be acquired and, as a consequence, the better informed are the site evaluations and thus the greater chance of success for the planned remediation system.

6.5 Site Remediation

Bioremediation technology exploits various naturally occurring mitigation processes: (1) natural attenuation, (2) biostimulation, and (3) bioaugmentation.

Natural attenuation is a form of bioremediation that occurs without human intervention. This natural attenuation relies on natural conditions and behavior of soil microorganisms that are indigenous to soil. Biostimulation also utilizes indigenous microbial populations to remediate contaminated soils and consists of adding nutrients and other substances to soil to catalyze natural attenuation processes. Bioaugmentation involves the introduction of non-indigenous microorganisms that are capable of detoxifying a particular contaminant, sometimes employing genetically altered microorganisms.

6.5.1 Method Parameters

Several factors that affect the decision of which method is chosen are (1) the nature of the contaminants, (2) the location of contaminated site, cost of cleanup, (3) the time allotted to the cleanup, (4) effects on humans, animals and plants, and last but by no means least (5) the cost of the cleanup.

For a bioremediation method to be successful in soil and water cleanup, the physical, chemical and biological environment must be feasible. Parameters that affect the bioremediation process are (1) low temperatures, (2) preferential growth of microbes obstructive to bioremediation, (3) high concentrations of chlorinated organics, heavy metals and heavy oils poisoning the microorganisms, (4) preferential flow paths severely decreasing contact between injected fluids and contaminants throughout the contaminated zones, and (5) the soil matrix prohibiting contaminant-microorganism contact.

6.5.2 In situ and ex situ Bioremediation

Bioremediation technology falls into two broad categories: (1) *in situ* remediation and (2) *ex situ* remediation which, in turn can be subdivided into sub-categories. Thus:

In situ bioremediation
<blockquote>
Intrinsic bioremediation
Engineered bioremediation
<blockquote>
Biosparging
Bioventing
Bioslurping
Biosurfactant Remediation
Rhizosphere Bioremediation
</blockquote>
</blockquote>

Ex situ bioremediation
<blockquote>
Slurry phase bioremediation
<blockquote>
Bioreactor
</blockquote>
Solid phase bioremediation
<blockquote>
Biopiling
Composting
Land farming
</blockquote>
</blockquote>

The *in situ* bioremediation process involves treatment of the contaminated soil or groundwater in the location in which it exists while *ex situ* bioremediation process requires excavation of contaminated soil or pumping the groundwater to surface tanks before treatment of the soil or water.

By way of explanation, an *in situ* bioremediation process is applied to the treatment of the contaminated soil or the contaminate groundwater in the location in which it exists. On the other hand, an *ex situ* bioremediation processes require excavation of the contaminated soil or pumping of the groundwater into tankage (or a suitable storage pond) before the soil or the water can be treated.

Bioremediation can be used as a cleanup method for both contaminated soil and water. Its applications fall into two broad categories: *in situ* or *ex situ*. *In situ* bioremediation treats the contaminated soil or groundwater in the location in which it was found while *ex situ* bioremediation processes require excavation of contaminated soil or pumping of groundwater before they can be treated.

In recent years, *in situ* bioremediation concepts have been applied in treating contaminated soil and ground water. Removal rates and extent vary based on the contaminant of concern and site-specific characteristics. Removal rates also are affected by variables such as contaminant distribution and concentration; co-contaminant concentrations; indigenous microbial populations and reaction kinetics; and parameters such as pH, moisture content, nutrient supply, and temperature. Many of these factors are a function of the site and the indigenous microbial community and, thus, are difficult to manipulate. Specific technologies may have the capacity to manipulate some variables and may be affected by other variables as well (US EPA, 2006).

When oxygen is limited in supply or absent, as in saturated or anaerobic soils or lake sediment, anaerobic (without oxygen) respiration prevails. Generally, inorganic compounds such as nitrate, sulfate, ferric iron, manganese, or carbon dioxide serve as terminal electron acceptors to facilitate biodegradation.

In situ technologies do not require excavation of the contaminated soils so may be less expensive, create less dust, and cause less release of contaminants than *ex situ* techniques. Also, it is possible to treat a large volume of soil at once. *In situ* techniques, however, may be slower than *ex situ* techniques, may be difficult to manage, and are only most effective at sites with permeable soil.

The typical time frame for an *in situ* bioremediation project can be in the order of 12 to 24 months depending on the levels of contamination and depth of contaminated soil. Due to the poor mixing in this system it becomes necessary to treat for long periods of time to ensure that all the pockets of contamination have been treated.

In situ bioremediation is a very site specific technology that involves establishing a hydrostatic gradient through the contaminated area by flooding it with water carrying nutrients and possibly organisms adapted to the contaminants. Water is continuously circulated through the site until it is determined to be clean.

In situ bioremediation of groundwater speeds the natural biodegradation processes that take place in the water-soaked underground region that lies below the water table. One limitation of this technology is that differences in underground soil layering and density may cause re-injected conditioned groundwater to follow certain preferred flow paths. On the other hand, *ex situ* techniques can be faster, easier to control, and used to treat a wider range of contaminants and soil types than *in situ* techniques. However, they require excavation and treatment of the contaminated soil before and, sometimes, after the actual startup of the bioremediation process.

In situ bioremediation is the preferred method for large sites and is used when physical and chemical methods of remediation may not completely remove the contaminants, leaving residual concentrations that are above regulatory guidelines. This method has the potential to provide advantages such as complete destruction of the contaminant(s), lower risk to site workers, and lower equipment/operating costs. *In situ* bioremediation can be used as a cost-effective secondary treatment scheme to decrease the concentration of contaminants to acceptable levels or as a primary treatment method, which is followed by physical or chemical methods for final site closure.

Generally, a contaminant is more easily and quickly degraded if it is a naturally occurring compound in the environment, or chemically similar to a naturally occurring compound, because microorganisms capable of its biodegradation are more likely to have evolved. Crude oil hydrocarbon derivatives are naturally occurring chemicals; therefore, microorganisms which are capable of attenuating or degrading hydrocarbon derivatives exist in the environment. Development of biodegradation technologies of synthetic chemicals such chlorocarbon derivatives or chlorohydrocarbon derivatives is dependent on outcomes of research that searches for natural or genetically improved strains of microorganisms to degrade such contaminants into less toxic forms.

In summary, bioremediation is increasingly viewed as an appropriate remediation technology for hydrocarbon-contaminated polar soils. As for all soils, the successful application of bioremediation depends on the presence or availability of the appropriate for the biodegradation reactions and the environmental conditions *in situ*. Laboratory studies have confirmed that hydrocarbon-degrading bacteria typically assigned to the genera *Rhodococcus, Sphingomonas* or *Pseudomonas* are present in contaminated polar soils. However, as indicated by the persistence of spilled hydrocarbon derivatives, environmental conditions *in situ* are suboptimal for biodegradation in polar soils.

Therefore, it is likely that *ex situ* bioremediation will be the method of choice for ameliorating and controlling the factors limiting microbial activity, i.e., low and fluctuating soil temperatures, low levels of nutrients, and possible alkalinity and low moisture (Okoh, 2006; Speight and Arjoon, 2012). Care must be taken when adding nutrients to the coarse-textured, low-moisture soils prevalent in continental Antarctica and the high Arctic because excess levels can inhibit hydrocarbon biodegradation by decreasing soil water potentials. Bioremediation experiments conducted on site in the Arctic indicate that land farming and biopiles may be useful approaches for bioremediation of polar soils (Aislabie et al., 1998; Nugroho et al., 2010).

In many cases, after an oil spill, the natural microbial systems for degrading the oil are overwhelmed. Therefore, molecular engineers are constructing starvation promoters to express heterologous genes needed in the field for survival and adding additional bioremediation genes that code for enzymes able to degrade a broader range of compounds present in the contaminated environments. Various bacterial strains are also being developed, where each strain is specific for a certain organic compound present in oil spills. This will help increase the speed of bioremediation and allow detailed cleanup to take place where no organic contamination remains in the environment.

Thus, the decision to bioremediate a site is dependent on cleanup, restoration, and habitat protection objectives and the factors that are present that would have an impact on success. If the circumstances are such that no amount of nutrients will accelerate biodegradation, then the decision should be made on the need to accelerate oil disappearance to protect a vital living resource or simply to speed up restoration of the ecosystem. These decisions are clearly influenced by the circumstances of the spill.

Various methods exist for the testing of biodegradability of substances. Biodegradability is assessed by following certain parameters which are considered to be indicative of the consumption of the test substance by microorganisms, or the production of simple basic compounds which indicate the mineralization of the test substance.

Hence there are various biodegradability testing methods which measure the amount of carbon dioxide (or methane, for anaerobic cases) produced during a specified period; there are those which measure the loss of dissolved organic carbon for substances which are water soluble; those that measure the loss of hydrocarbon infrared bands and there are yet others which measure the uptake of oxygen by the activities of microorganisms (biochemical oxygen demand, BOD).

Finally, evidence for the effectiveness of crude oil bioremediation and crude oil product bioremediation should include: (1) faster disappearance of oil in treated areas than in untreated areas, and (2) a demonstration that biodegradation was the main reason for the increased rate of oil disappearance. To obtain such evidence, the analytical procedures must be chosen carefully, and careful data interpretation is essential but there are disadvantages and errors when the method is not applied correctly (Speight, 2005).

In situ bioremediation is a site specific technology that involves establishing a hydrostatic gradient through the contaminated area by flooding it with water carrying nutrients and possibly organisms adapted to the contaminants. Water is continuously circulated through the site until it is determined to be clean.

In situ bioremediation of groundwater speeds the natural biodegradation processes that take place in the water-soaked underground region that lies below the water table. One limitation of this technology is that differences in underground soil layering and density may cause re-injected conditioned groundwater to follow certain preferred flow paths. On the other hand, *ex situ* techniques can be faster, easier to control, and used to treat a wider range of contaminants and soil types than *in situ* techniques. However, they require excavation and treatment of the contaminated soil before and, sometimes, after the actual bioremediation step.

In situ bioremediation is the preferred method for large sites and is used when physical and chemical methods of remediation may not completely remove the contaminants, leaving residual concentrations that are above regulatory guidelines. This method has the potential to provide advantages such as complete destruction of the contaminant(s), lower risk to site workers, and lower equipment/operating costs. *In situ* bioremediation can be used as a cost-effective secondary treatment scheme to decrease the concentration of contaminants to acceptable levels or as a primary treatment method, which is followed by physical or chemical methods for final site closure.

Finally, evidence for the effectiveness of crude oil bioremediation and crude oil product bioremediation should include: (1) faster disappearance of oil in treated areas than in untreated areas, and (2) a demonstration that biodegradation was the main reason for the increased rate of oil disappearance. To obtain such evidence, the analytical procedures must be chosen carefully and careful data interpretation is essential but there are disadvantages and errors when the method is not applied correctly (Chapter 9) (Speight, 2005).

Solid phase treatment includes organic wastes, domestic wastes, industrial wastes and agricultural wastes. On the other hand, slurry phase bioremediation is a process in which an aqueous slurry is created by combining soil, sediment, or sludge with water and other additives. Upon completion of the process, the slurry is dewatered and the treated soil is sent for disposal. The process is a controlled bioreactor treatment process that involves the excavation of the contaminated soil, mixing it with water and placing it into the bioreactor (Lucian and Gavrilescu, 2008). Factors affecting slurry phase biodegradation are (1) moisture in the soil, (2) the pH, (3) the temperature, (4) the nutrients in the soil, (5) the concentration of toxic pollutants, (6) the microbial population, and (7) aeration of the soil.

The most effective means of implementing *in situ* bioremediation depends on the hydrology of the subsurface area, the extent of the contaminated area and the nature (type) of the contamination. In general, this method is effective only when the subsurface soils are highly permeable, the soil horizon to be treated falls within a depth of 8–10 m and shallow groundwater is present at 30 feet or less below ground surface. The depth of contamination plays an important role in determining whether or not an *in situ* bioremediation project should be employed. If the contamination is near the groundwater but the groundwater is not yet contaminated then it would be unwise to set up a hydrostatic system. It would be safer to excavate the contaminated soil and apply an on-site method of treatment away from the groundwater.

6.5.3 Biostimulation and Bioaugmentation

Other forms of treatment that can be applied to contaminated sites include (1) biostimulation and (2) bioaugmentation.

The biostimulation process involves adding nutrients such as phosphorus-containing compounds and nitrogen-containing compounds to a contaminated site in order to stimulate the growth of the microorganisms that degrade the constituents of the crude oil or the crude oil product. The nutrients stimulate the growth of the indigenous microorganisms after which the population undergoes rapid growth thereby (potentially) increasing the rate of biodegradation.

The bioaugmentation process involves the addition of pre-grown microbial cultures to enhance indigenous (native) microbial populations at a site to improve contaminant cleanup and reduce the cleanup time. In some cases, an indigenous microbe population may not have the ability to degrade a particular contaminant and the bioaugmentation process offers a way to provide specific microbes in sufficient numbers to complete the biodegradation.

Bioaugmentation adds highly concentrated and specialized populations of laboratory-prepared specific microbes (often prepared in a remote laboratory or in an on-site laboratory) to the contaminated area while biostimulation is dependent on appropriate indigenous microbial population and organic material being present at the site.

The microbial inocula to be used for the process are typically prepared from soil or groundwater microbes either from the site where they are to be used or from another site where the biodegradation of the chemicals of interest is known to be occurring. One of the main environmental applications for bioaugmentation is at sites with chlorinated solvents and specific microbes usually perform reductive dechlorination of solvents such as tetrachloroethylene ($CCl_2 = CCl_2$, also called perchloroethylene) and trichloroethylene ($CHCl = CCl_2$).

6.5.4 Miscellaneous Processes

In addition, to the major bioremediation methods, there is a variety of related methods that are also applied to site cleanup. As always, the choice of a suitable method is site dependent and contaminant dependent.

These alternate methods (listed alphabetically in the subsections below) are also capable of use to reduce, detoxify, degrade, mineralize or transform more toxic pollutants to a less toxic. The pollutant removal process depends mainly on the pollutant nature, which includes pesticides, agrochemicals, chlorinated compounds, heavy metals, xenobiotic compounds, organic halogens, greenhouse gases, hydrocarbon derivatives, nuclear waste, dyes, plastics, sludge as well as a wide range of chemical wastes and physical hazardous materials through the all-inclusive and action of microorganisms.

6.5.4.1 Bioslurping

Bioslurping (bio-slurping) is an *in situ* remediation technology that combines the two remedial approaches of bioventing and vacuum-enhanced free-product recovery. It is faster than the conventional remedy of product recovery followed by bioventing. The system is made to minimize groundwater recovery and drawdown in the aquifer. Bio-slurping was designed and is being tested to address contamination by crude oil products with a floating lighter non-aqueous phase liquids (LNAPL) layer.

The process enhances natural *in situ* bioremediation of vadose zone (which extends from the top of the ground surface to the water table) soils and may be the only feasible remediation technology at low-permeability sites.

The *soil vapor extraction* process is used to remove contaminants (in the form of vapors) from the soil above the water table by applying a vacuum to draw out the vapor(s).

6.5.4.2 Biosparging

The *biosparging* process is used to increase the biological activity in the soil by increasing the oxygen supply via sparging air or oxygen into the soil. In some instances air injections are replaced by pure oxygen to increase the degradation rates but it can be a disadvantage if the toxicity is sufficiently to microorganisms even at low concentrations (Schlegel, 1977; Flathman et al., 1991; Lee et al., 1988; Lu, 1994; Lu and Hwang, 1992; Pardieck et al., 1992; Brown and Norris, 1994; Norris, 1994; Scragg, 1999).

In the air sparging process, air pumped into the wells disturbs the groundwater which assists the pollutants to vaporize. The vapors rise into the drier soil above the groundwater and are pulled out of the ground by extraction wells. The harmful vapors are removed in the same way as soil vapor extraction. The air used in soil vapor extraction and air sparging also helps clean up pollution by encouraging the growth of microorganisms.

The *air sparging* process also employs uses a vacuum to extract the vapors. Crude oil products such as the lower-boiling naphtha and gasoline constituents (i.e., benzene, C_6H_6, toluene, $C_6H_5CH_3$, ethylbenzene, $C_6H_5CH_2CH_3$, and the xylene isomers, $H_3CC_6H_4CH_3$, also referred to collectively as BTEX), because they readily transfer from the dissolved to the gaseous phase, but is less applicable to diesel fuel and kerosene.

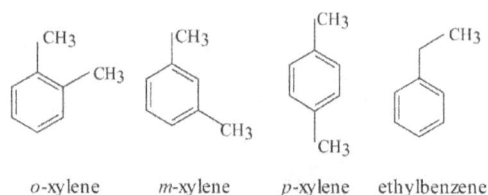

o-xylene   m-xylene   p-xylene   ethylbenzene

In the air sparging process, air is pumped underground to enable more rapid evaporation of these types of chemicals. Methane can be used as an amendment to the sparged air to enhance co-metabolism of chlorinated organic chemicals.

Anaerobic sparging is a process that depends on the delivery of an inert gas (nitrogen or argon) with low ($< 2\%$) levels of hydrogen. Co-metabolic air sparging is the delivery of oxygen containing gas with enzyme-inducing growth substrate (such as methane or propane).

6.5.4.3 Biosurfactant Treatment

Biosurfactants are microbially-produced surface-active compounds which are amphiphilic molecules which are chemical compound possessing both hydrophilic and lipophilic properties. These properties cause the molecules to aggregate at interfaces between fluids with different polarities found in oil spills. Many of the known biosurfactant producers are hydrocarbon-degrading organisms.

Also, biosurfactants are biodegradable and are not a pollution threat and are generally non-toxic to microorganisms and are, therefore, unlikely to inhibit biodegradation of nonpolar organic contaminants. These surfactants are also effective in many diverse geologic formations and are compatible with many existing remedial technologies (such as pump and treat rehabilitation, air sparging, and soil flushing), and significantly accelerate innovative approaches including microbial, natural attenuation enhanced soil flushing and bio-slurping.

6.5.4.4 Bioventing

Bioventing is a process that is used to stimulate the natural *in situ* biodegradation of any aerobically degradable compounds in soil by providing oxygen to existing soil microorganisms. In contrast to soil vapor vacuum extraction, bioventing uses low airflow rates to provide only enough oxygen to sustain microbial activity. Two basic criteria have to be satisfied for successful bioventing (1) the air must be able to pass through the soil in sufficient quantities to maintain aerobic conditions, and (2) natural hydrocarbon-degrading microorganisms must be present in concentrations large enough to obtain reasonable biodegradation rates.

6.5.4.5 Rhizosphere Bioremediation

Rhizosphere bioremediation (sometimes referred to as phytoremediation) is the interaction between plants and microorganisms and is also known as phytostimulation or plant-assisted bioremediation (Wenzel, 2009).

In the process, plants provide oxygen, bacteria, and organic carbon to encourage the degradation of organics in the soil. The microorganisms in the environment created by the plants, together with the roots of the plants, can degrade more contaminants than could occur in a purely microbial system.

Also, plants help with microbial conversions where certain bacteria that metabolize pollutants are able to encourage degradation of chemicals in the soil, so allowing bioremediation to occur with less retardation.

The complexity and heterogeneity polluted soils will require the design of integrated approaches of rhizosphere management such as (1) combining co-cropping of phytoextraction and rhizodegradation crops, (2) inoculation of microorganisms, and (3) soil management.

6.6 Bioremediation of Land Ecosystems

When a site is selected for cleanup it is extremely important to link *biodegradation* with *bioremediation*. The former is a component of oil weathering and is a natural process whereby bacteria or other microorganisms alter and break down organic molecules into other substances, eventually producing fatty acids and carbon dioxide (Hoff, 1993). On the other hand, *bioremediation* is the acceleration of this process through the addition of exogenous microbial populations, through the stimulation of indigenous populations or through manipulation of the contaminated media using techniques such as aeration or temperature control.

Many microorganisms possess the enzymatic capability to degrade crude oil hydrocarbon derivatives. Some microorganisms degrade alkane constituents; others degrade aromatic constituents while others will degrade both alkane and aromatic constituents. Often the normal alkane derivatives in the range C_{10} to C_{26} are viewed as the most readily degraded, but low-molecular-weight aromatic derivatives, such as benzene, toluene and xylene, which are among the toxic compounds found in crude oil, are also biodegraded by many marine microorganisms.

In fact, it has been reported that biodegradation rates of hydrophobic organic compounds decreased in the presence of organic matrices. The biodegradation of n-alkane derivatives in a contaminated soil is inhibited in the presence of unresolved complex mixture of hydrocarbon derivatives originating from fuel oil residue. It was further suggested that inhibition of n-alkane dissolution into water by an unresolved complex mixture of hydrocarbon derivatives resulted in slow biodegradation. Therefore, oil residue in soil or sediment may decrease the biodegradation rate of hydrophobic organic compounds, such as polynuclear aromatic hydrocarbon derivatives.

In addition to the varying rates of biodegradation of organic molecules, a lag time (chemically: an induction period) occurs after spillage of crude oil or a crude oil product before indigenous microbes begin to degrade the crude oil-related constituents. This lag time is related to presence of volatile constituents of the spilled material (and which also may exhibit toxicity to the indigenous microbes), which evaporate in the first few days after the spill. In order to be effective, the indigenous microbial populations must begin to use oil and increase their population density before measurable degradation takes place. This period usually lasts several days and is an important aspect when the appropriateness of application of a bioremediation technology is considered.

There are several different bioremediation techniques and the underlying idea of selecting a bioremediation technique is to accelerate the rates of *natural* hydrocarbon biodegradation by overcoming the rate-limiting factors.

Thus, several techniques can lead to the desired results: (1) indigenous populations of microbial bacteria can be stimulated through the addition of nutrients or other materials, (2) exogenous microbial populations can be introduced in the contaminated environment (bio-augmentation), (3) genetically altered bacteria can be used, and (4) the contaminated site (soil or water) can be manipulated by, for example, aeration or temperature control.

One approach often considered for the bioremediation of crude oil pollutants after an oil spill is the addition of microorganisms (seeding) that are able to degrade hydrocarbon derivatives. Most microorganisms considered for seeding are obtained by enrichment cultures from previously contaminated sites. However, because hydrocarbon-degrading bacteria and fungi are widely distributed in marine, freshwater and soil habitats, adding seed cultures has proven less promising for treating oil spills than adding fertilizers and ensuring adequate aeration. Most tests have indicated that seed cultures are likely to be of little benefit over the naturally occurring microorganisms at a contaminated site for the biodegradation of the bulk of crude oil contaminants.

A thorough site assessment that includes an evaluation of the current condition of the property, the operational history of the facility, and the processes used to discover this information should be conducted early in the revitalization process. Accurate characterization of environmental conditions is required to estimate potential costs associated with site cleanup and regulatory negotiation. Environmental costs can represent a significant financial issue impacting the success or start of a revitalization project. In the event that significant contamination exists, communication with local regulatory agencies is important and should include discussion of impacts to project schedule.

Typically, site assessment (sometimes called environmental assessment) is a method used to identify any potential risk to the environment and the resulting report is usually divided into three separate phases or types of work: (1) qualitative investigation and report, (2) quantitative investigation and report, and (3) remediation and environmental management.

In the first phase—a qualitative investigation and report—visual observations and site history are researched and relevant regulatory records and databases are consulted to determine the reasonable probability of environmental risk. At this stage, sampling and testing are not usually performed but a detailed waste classification and a determination of the origin of the waste assist in planning the removal activities and in utilizing the appropriate funding instrument. A thorough *paper review* and site history must be conducted to establish Oil Pollution Act (OPA) of 1990 or Comprehensive Environmental Response and Liability Act (CERCLA) authority. Typically, an oil refinery waste consists of both OPA wastes, CERCLA wastes, and Resource Conservation and Recovery Act (RCRA) wastes. A careful waste classification is required to use appropriate funding to remediate the site. If a risk assessment report is required, this may include some minor soil sampling and testing as a precautionary measure, but this does not necessarily qualify the report as an environmental assessment.

The second phase—a quantitative investigation and report—sampling of materials (such crude oil and crude oil products) suspected of contamination are analyzed and characterized. It is also opportune to analyze unspoiled sample to determine potential sources of the spilled material (if these sources have not already been identified). The object of this phase is to determine as much as reasonable the extent of any environmental contamination present at the site.

The third and final phase—the remediation and environmental management project—includes study of any reports from the two prior phases as well as an environmental site assessment in order to formulate an appropriate solution to the environmental problem. The remediation or bioremediation project is planned and estimated in terms of time, costs, and objectives.

Other phase may also be introduced and these phases will include a timeline for site examination (say after one year with continuous weekly/monthly monitoring) to determine the success/failure of the remediation effort. After the first year, the timeline may be shortened or lengthened appropriately depending on the first-year data.

Reference was made above to the identification of the source of the crude oil-related pollutants, if the source was not identified at the time of the spill. This can be done through *finger printing* of the spilled crude oil or crude oil product can be used to identify the origin of the spill.

The finger-printing approach requires collection and analysis to search for specific chemical biomarkers in the crude oil or chemical markers specific to the crude oil product. All nearby potential sources of crude oil and crude oil products that might be responsible for the spill must be included in the collection and analysis. When a positive correlation of spilled material and the relevant facility has been identified, the origin of the spill has been determined. If the spilled oil has undergone weathering, water washing, and/or any level of biodegradation, then biomarker analysis and/or other techniques may be required to identify the spill source.

Biomarkers are a group of compounds, primarily hydrocarbon derivatives, found in oils, rock extracts, recent sediment extracts, and soil extracts. Biomarkers are distinguished from other compounds and are generally referred to as *molecular fossils*—they are structurally similar to specific natural products. Typically, even though diagenetic alteration of the original mileages has occurred, biomarkers retain all or most of the original carbon skeleton of the original natural product. Biomarkers reflect the type and age of the source rock that generated the crude oil.

Since different potential sources of a spill may involve oil derived from different basins, biomarker distribution (such as the sterane/hopane ratio or the ratio of steroid, diterpenoid, triterpenoid and hopanoid biological molecules) can be used to either point to the source or rule out the potential sources of a spill and can be used to determine if oil in a contaminated area actually represents more than one spill. Specific chemical compounds (not necessarily biomarkers since these may have been formed in the refining process) can also be used to assess the origin of some refined hydrocarbon products.

Polynuclear aromatic hydrocarbon derivatives are another group of compounds present in crude oil and crude oil products that are especially useful in identifying the source(s) of a spill. A subset of the polynuclear aromatic hydrocarbon derivatives in crude oil is products of the diagenesis of steroid, diterpenoid, triterpenoid and hopanoid biological molecules originally deposited in sediments (biomarkers, see above). Several of these biomarkers are present as fully or partially aromatized compounds with multiple aromatic rings; therefore, they are polycyclic aromatic hydrocarbon derivatives. These polynuclear aromatic hydrocarbon derivatives are resistant to biodegradation conditions typically encountered in spill situations and have proved useful for defining a unique *fingerprint* characteristic of a specific crude oil (or crude oil product), which can be used to correlate a biodegraded oil to a sample of its non-degraded equivalent, and hence can be used to identify the source(s) of a crude oil-related release.

Many land-based environments are characterized by low or elevated temperatures, acidic or alkaline pH, high salt concentrations, or high pressure. Extremophile microorganisms (organisms that thrives in physically or geochemically extreme conditions that are generally detrimental to most terrestrial life forms) are adapted to grow and thrive under these adverse conditions. Hydrocarbon degrading extremophiles are thus ideal candidates for the biological treatment of polluted extreme habitats. In this review we summarize the recent developments, obtained both in laboratory and field studies, in biodegradation and bioremediation of hydrocarbon contaminants that are of environmental concern in extreme habitats.

A wide range of hydrocarbon derivatives that contaminate the environment, mainly due to accidental release or industrial processes, has been shown to be biodegraded (mineralized or transformed) in various extreme environments characterized by low or elevated temperatures, acidic or alkaline pH, high salinity or high pressure, emphasizes the metabolic capacities of extremophiles. When soil is the medium in which treatment will take place, it is of utmost importance to evaluate its properties as well as the general properties of the contaminated site.

Soil is heterogeneous and varies widely in physical, chemical, and biological properties. The soil may be acidic or basic, may have high or low nutrients, and may exhibit a different exchange capacity at different locations in the same area. The characteristics important in the design and operation of a land treatment site include the slope, the soil classification (texture and permeability), soil moisture content, pH, the cation exchange capacity (CEC), and salinity. If the initial soil properties are not ideal for the biodegradation of hydrocarbon derivatives, they can be optimized.

The pH of the soil not only affects the growth of microorganisms, but also has a noticeable effect on the availability of nutrients, mobility of metals, rate of abiotic transformation of organic waste constituents, and soil structure. Usually, a pH range of 6 to 8 units is considered optimum for biodegradation activities—the pH of the soil can be adjusted by addition of chemical reagents. For example, if the soil is too acidic soils (pH < 6), agricultural lime may be used to raise the pH while aluminum sulfate or ferrous sulfate may be used to lower the pH of alkaline soils. However, caution should be used to avoid *over correction* of the pH, and further consultation may be used to help calculate optimum quantities.

Biological activity is regulated by soil temperature, and an ideal temperature range is between 24 to 35°C (75 and 95°F). Since the soil temperature is difficult to control under field conditions, the waste loading rates should be adjusted according to temperatures affect temperature. This adjustment should also be performed during the change in season since the biodegradation rates are lower in the spring and the autumn compared to summer.

In addition to the pH of the soil, *cation exchange capacity* is an important variable that can be used to monitor in order to optimize the degradation process. The cation exchange capacity is an indication of the capacity of the soil to retain metallic ions (the cation exchange capacity is usually obtained through laboratory testing) and is measured in milli-equivalents per one hundred grams of dry soil (meq/100 g). A cation exchange capacity value greater than 25 is an indication that the soil contains more nutrients and has high clay content, whereas values less than 5 indicate a sandy soil with little ion retention. Most metals found in oily wastes are not readily soluble in water; however, variations of pH may change that property and when treating land where the soil has a low cation exchange capacity, care must be taken to manage subsurface of metal ions. With proper pH management, metals remain immobilized in the treatment zone even with low cation exchange capacity values.

Soil salinity results from accumulation of neutral soluble salts (mainly due to neutral salts of sodium, calcium, magnesium, and potassium) in the upper soil horizon following capillary movement of the water, which evaporates and leaves the crystalline form of the salt, which is often indicated by a white crust. Elevated concentrations of the salts can be lethal to many microorganisms. Assessing the feasibility of biodegradation in relation to salinity is achieved by measuring electrical conductivity deciSiemens/meter, (dS/m), which is a general measure of soil salinity. At conductivity values above 1 dS/m, biological growth is hindered, and values above 6 dS/m generally indicate that the soil is sterile.

Finally, the soil should be analyzed for heavy metal content since a high metal concentration could be toxic to microbial survival and growth. Metals in the soil are difficult (if not impossible) to remediate by indigent soil bacteria. Therefore, if the heavy metal concentrations in soil exceed the acceptable residual levels as determined by federal and state regulations (http://www.cleanuplevels.com/), bioremediation is not a viable option.

The action of indigenous bacteria in the soil can account for up to 80% of waste degradation in soil, the remainder being due to evaporation, photo-oxidation, and solubilization in water. This is true as long as environmental conditions such as the presence of oxygen, adequate moisture, moderate temperatures, neutral pH, low to moderate salinity, and excess nutrients, are present to allow bacteria can to grow exponentially. In addition, the spilled material should not be excessive insofar as it overwhelms the indigenous microbes.

While bacteria which degrade hydrocarbon derivatives are ubiquitous, the populations of hydrocarbon-degrading bacteria exposed to hydrocarbon derivatives increase rapidly when given adequate aeration, moisture, favorable pH, and excess nutrients.

Typically, hydrocarbon-degrading bacteria are found in the range of 10^5 to 10^6 bacteria per gram of soil under no oil spill conditions, and when exposed to crude oil, that number increases to 10^6 to 10^8 bacteria per gram of soil. For reference, one gram of rich agricultural soil generally contains 2.5×10^9 bacteria (heterotrophic count), 5×10^5 fungi, 5×10^4 algae, and 3×10^4 protozoa. Soil samples should be analyzed for enumeration of both heterotrophic and hydrocarbon-utilizing bacteria population to verify population densities. The population of microorganisms could be assessed in soil by plate count, most probable number technique, phospholipid fatty acid analysis, or denaturing gradient gel electrophoresis (Zhu et al., 2004). There are several methods that can be applied to the bioremediation of crude oil-contaminated sites, but it is necessary to note that there is no single species of bacteria that can metabolize all the components of crude oil.

However, the nature of the crude oil-related spill and the character of the site may require addition of non-indigenous microorganisms to enhance the native populations. Thus, whenever a spill has occurred and bioremediation is the best means to approach the cleanup after other options (Speight, 1996; Speight and Lee, 2000; Speight, 2005) have been dismissed, investigators should, after suitable test methods have been applied to determine the physical and chemical character of the soil, consider one of two options which include (1) bioaugmentation and (2) biostimulation.

In the *bioaugmentation approach*, crude oil-degrading bacteria are added to the existing bacterial population in the soil to increase the rate of oil consumption. In the *biostimulation approach*, nutrients are added and optimization of environmental conditions to improve the biodegradation efficiency of indigenous bacteria is practiced.

The nutrients nitrogen, phosphorus, and potassium (N, P, and K) are normally added during land treatment in order to enhance microbial activities, which decompose carbon (C) compounds in the soil. Nitrogen, when added through the ammonium salts, can be toxic to microorganisms due to the possibility of generation of ammonia in the soil; the ammonium ion can also promote the increase of oxygen demand. A commonly used strategy is to add nutrients that provide a stoichiometric ratio of carbon: nitrogen: phosphorus: potassium of 100:5:1:1. However, when nitrogen and phosphorus are introduced in excessive quantities biodegradation can be inhibited (Trindade et al., 2002).

For optimum biodegradation, nutrients can be added to the soil using organic or inorganic fertilizers, and their concentration should be closely monitored and supplemented as they are depleted during the biodegradation nitrate and process. Agriculture fertilizers such as ammonium nitrate, urea, ammonium phosphate, and potassium phosphate may be added to increase nutrient concentrations in the soil.

Fertilizers should be added gradually to the soil to minimize pH changes. The amount and frequency of fertilizer addition depend upon field conditions. However, evidence from documented land farming has shown that an appropriate fertilizer dosage that could be repeated, depending upon field conditions, are 500 pounds of nitrogen per acre or 1,100 pounds of urea or 1,500 pounds of ammonium nitrate per acre and 250 pounds of phosphorus per acre.

Organic amendments such as wood chips, sawdust, straw, hay, and animal manure are used to improve soil structure and oxygen infiltration, and to increase the moisture-holding capacity of sandy soils. In general, animal manure should be applied at the rate of about 3 to 4% by weight of soil and should be analyzed for nitrogen and phosphorus before its application. Bulking agents like hay, palm husks, rice hulls, and straw are added to clayey soils to increase pore space and hence, air exchange. The bulking agent should be blended into the soil until a porous structure is obtained and visual evidence of oil is eliminated. A general rule to follow is to add hay in contaminated media is 5 standard hay bales per 1,000 square feet of impacted soils. However, the source of bulking agent should be checked for residual substances (such as pesticides or heavy metals) for toxicity.

Temperature plays a significant role in controlling the nature and the extent of microbial hydrocarbon metabolism, which is of special significance for *in situ* bioremediation (Margesin and Schinner, 1997a). Bioavailability and solubility of less soluble hydrophobic substances, such as aliphatic and polyaromatic hydrocarbon derivatives, are temperature dependent. Increase in temperature affects a decrease in viscosity while a decrease in temperature affects an increase in viscosity, thereby affecting the degree of distribution, and an increase (decrease) in diffusion rates of organic compounds. Therefore, higher reaction rates due to smaller boundary layers are expected at elevated temperatures. The increased (decreased) volatilization and solubility of some hydrocarbon derivatives at elevated (low) temperature affects toxicity and allows bio-transformations with high (low) substrate concentrations (Müller et al., 1998; Whyte et al., 1998; Niehaus et al., 1999).

Cold habitats possess sufficient indigenous microorganisms, psychrotrophic bacteria being predominant, which adapt rapidly to the contamination, as demonstrated by significantly increased numbers of oil degraders shortly after a pollution event. In fact, cold-adapted, certain types of (psychrophilic and psychrotrophic) microorganisms that have been adapted to cold temperature are able to grow at temperatures around 0°C (32°F). They are widely distributed in nature since a large part of the Earth's biosphere is at temperatures below 5°C (41°F) (Margesin and Schinner, 1996a, 1996b). Psychrophilic micro-organisms (organisms that are capable of growth and reproduction in cold temperatures) have an optimum growth temperature of approximately 15°C (59°F) and do not grow above 20°C (68°F), whereas psychrotrophs (cold-tolerant) have optimum and maximum growth temperatures above 15°C (59°F) and 20°C (68°F), respectively (Morita, 1975). Cold-adapted indigenous microorganisms play a significant role in the *in situ* biodegradation of hydrocarbon derivatives in cold environments, where ambient temperatures often coincide with their growth temperature range.

Most studies showed that the indigenous microbial populations degrade hydrocarbon derivatives more efficiently than the introduced strains introduced in an attempt at bioaugmentation (Margesin and Schinner, 2001a). However, bioaugmentation may result in a shorter hydrocarbon acclimation period. Inoculation of contaminated Arctic soils with consortia (Whyte et al., 1999) or with alkane-degrading *Rhodococcus* sp. (Whyte et al., 1998) decreased the induction period and increased the rate of C_{16} mineralization at 5°C (41°F). However, nitrogen-phosphorus-potassium fertilization alone had a comparable effect on hydrocarbon loss like fertilization plus bioaugmentation. This has been shown both in chronically oil-polluted Arctic soil (Whyte et al., 1999) and in artificially diesel oil-contaminated alpine soils (Margesin and Schinner, 1997a, 1997b).

The decreased bioavailability of the long-chain alkane derivatives at low temperature (many form crystals at 0°C, 32°F) may be responsible for their increased recalcitrance, which affects *in situ* bioremediation in cold climates. The application of cold-active solubilizing agents could be useful for enhancing hydrocarbon bioavailability. For example, Antarctic marine bacteria produced bio-emulsifiers when grown with n-alkane derivatives as the sole carbon source (Chugunov et al., 2000).

Benzene, toluene, ethyl benzene, xylene, and polynuclear aromatic hydrocarbon derivatives are frequent soil contaminants but little is known about biodegradation of polynuclear aromatic hydrocarbon derivatives in cold climates. Cold-tolerant isolates (*Sphingomonas*, *Pseudomonas* spp.) from oil-polluted Antarctic soils utilized as benzene, toluene, ethyl benzene, xylene, naphthalene, phenanthrene, and fluorene as the sole carbon and energy source (Aislabie et al., 1998, 2000). In a field study on the biodegradation of dispersed crude oil in cold and icy seawater (–1.8 to 5.5°C, 19 to 42°F), half-life times of polynuclear aromatic hydrocarbon derivatives ranged from two days (naphthalene) to eight days (phenanthrene) at a temperature above 0°C (32°F) and with effective chemical dispersion (Siron et al., 1953).

On the other hand, cold climates are not the only extreme environments and a higher-temperature environment is also classed as an extreme environment. Microorganisms that grow optimally above 40°C are designated as thermophiles—most thermophiles known are moderate and show an upper temperature border of growth between 50 and 70°C (122 to 158°F). Another class of thermophiles, the hyperthermophiles experience optimal growth hyperthermophiles occurs at temperatures in excess of 80°C (176°F) (Stetter, 1998). Thermophiles, predominantly bacilli, possess a substantial potential for the conversion of environmental pollutants, including all major classes (Müller et al., 1998).

Hydrocarbon mineralization is favored by near neutral pH values (pH = 6 to 8) and it is an established practice to add lime to bioremediate acid soils containing hydrocarbon pollutants. However, biodegradation has been reported to proceed in aquifers where the acidity is obvious (pH = <6) (Norris, 1994; Amadi et al., 1996), Acidophiles (organisms that have their optimum growth rate at least 2 pH units below neutrality) are metabolically active in highly acidic environments, and often have a high heavy metal resistance (Norris and Johnson, 1998). Heterotrophic acidophilic microorganisms are an interesting potential for the bioremediation of acidic environments that contain both heavy metals and organic compounds, such as crude oil-polluted acidic drainage waters (Stapleton et al., 1998).

Alkaliphiles (organisms that have their optimum growth rate at least 2 pH units above neutrality) are able to grow or survive at pH values above 9, but their optimum growth rate is around neutrality or less (Kroll, 1990).

Microorganisms requiring salt for growth are referred to as halophiles, whereas microorganisms that are able to grow in the absence as well as in the presence of salt are designated halo-tolerant (Kushner, 1978; Grant et al., 1998). A range of organic pollutants has been shown to be mineralized or transformed by microorganisms able to grow in the presence of salt (Oren et al., 1992; Margesin and Schinner, 2001b).

Halophilic archaea maintain an osmotic balance with the hypersaline environment by accumulating high salt concentrations, which requires salt adaptation of the intracellular enzymes. Eubacteria are more promising degraders than archaea as they have a much greater metabolic diversity. Their intracellular salt concentration is low, and their enzymes involved in biodegradation may be conventional (i.e., not salt-requiring) enzymes similar to those of non-halophiles (Oren et al., 1992).

An inverse relationship between biodegradation of crude oil hydrocarbon derivatives and salinity has been assumed because enrichment cultures from the Great Salt Lake were not able to grow on mineral oil and to mineralize hexadecane in the presence of salt concentrations above 20% w/v (Ward and Brock, 1978; Whitehouse, 1984; Means, 1995). The inhibitory effect of salinity was found to be greater for the biodegradation of aromatic and polar fractions than of the saturated fraction of crude oil hydrocarbon derivatives (Mille et al., 1991) and there are reports of about microorganisms able to oxidize crude oil hydrocarbon derivatives even in the presence of 30% w/v sodium chloride. Among such microorganisms are crude oil-degrading *Streptomyces albaxialis* (Kuznetsov et al., 1992), and an alkane-degrading member of the *Halobacterium* group (Kulichevskaya et al., 1992). In terms of the bioremediation of crude oil-contaminated soil, an inhibitory effect of artificial salinity on mineralization of used lubricating oil has been reported (Rhykerd et al., 1995). Thus, the removal of salt from oil-contaminated soils may reduce the time required for bioremediation. However, different results may be obtained when investigating naturally salt-containing soils, since indigenous microorganisms in such environments are expected to be salt-adapted. In fact, seasonal variations of the mineralization potential of crude oil (520 g kg^{-1} soil) in soil samples from salt marshes has been reported to be high (Jackson and Pardue, 1997).

6.7 Bioremediation of Water Ecosystems

In a water ecosystem a spill of crude oil or a crude oil product represents the introduction of a non-aqueous phase liquid composed of a large number of organic matrices in addition to other toxic constituent, such as polynuclear aromatic hydrocarbon derivatives as occur in resin constituents and asphaltene constituents.

Hydrocarbon biodegradation in marine environments is often limited by abiotic environmental factors such as molecular oxygen, phosphate and nitrogen (ammonium, nitrate and organic nitrogen) concentrations. Rates of degradation of crude oil constituents are negligible in anaerobic sediments because molecular oxygen is required by most microorganisms for the initial step in hydrocarbon metabolism. Oxygen, however, is not limiting in well aerated (high energy) marine environments. Furthermore, marine ecosystems typically have low (sometime zero) concentrations of nitrogen, phosphorus and various mineral nutrients that are needed for the incorporation into cellular biomass, and the availability of these within the area of hydrocarbon degradation is critical.

Obviously, there are many different processes acting on a marine oil spill that can result in modifying the physical and chemical characteristics of the spilled material, such as (1) biodegradation, (2) scattering, (3) evaporation, (4) dissolution, (5) dispersion, (6) sedimentation, and (7) photo-oxidation. These processes occur simultaneously but at different speeds, depending on the physical and chemical properties of the oil and the ambient conditions, like the environmental temperature and energy. On several occasions, however, these processes can be so slow that it is necessary to interfere in order to hasten degradation of the pollutants.

Oil type, weather, wind and wave conditions, as well as air and sea temperature, all play important roles in ultimate fate of spilled oil in marine environment. After oil is discharged in the environment, a wide variety of physical, chemical and biological processes begin to transform the discharged oil. These chemical and physical processes are collectively called weathering and act to change the composition, behavior, route of exposure and toxicity of discharged oil. The weathering processes are described as follows:

- *Spreading and Advection*: Spreading, which dominates the initial stages of the spill and involves the whole oil and is the movement of the entire oil slick horizontally on the surface of water due to effects of gravity, inertia, friction, viscosity and surface tension—on calm water spreading occurs in a circular pattern outward from the center of the release point. Advection is the movement of oil due to overlying winds or underlying currents, which increase the surface area of the oil, thereby increasing its exposure to air, sunlight and underlying water. The advection effects are not uniform and do not affect the chemical composition of the oil.

- *Evaporation*: Evaporation, which is subject to atmospheric conditions such as temperature, is the preferential transfer of lower-boiling constituents of the crude oil-related spill from the liquid phase to the vapor phase. The chemical composition of the spilled material is physically altered and although the volume decreases, the remaining components have higher viscosity and specific gravity leading to thickening of the slick and formation of entities such as tar balls.

- *Dissolution*: Dissolution is the transfer of constituents from the spilled material into the aqueous phase. Once dissolved these constituents are bioavailable and if exposed to marine life can cause environmental impacts and injuries. The largest concentration is found near the surface or the release point and hence the effect on marine life may be localized.

- *Natural Dispersion*: Natural dispersion is the process of forming small oil droplets that become incorporated in the water in the form of a dilute oil-in-water suspension and occurs when breaking waves mix the oil in the water column. This phenomenon reduces the volume at the spill surface but does not change the physicochemical properties of the spilled material. It is estimated to range from 10 to 60% per day for first three days of the spill, depending on the condition of the ecosystem and may be independent of the type of crude oil or crude oil product.

- *Emulsification*: Emulsification is the mixing of sea water droplets into oil spilled on water surface forming a water-in-oil emulsion.

- *Photo-oxidation*: Photo-oxidation occurs when sunlight in the presence of oxygen transforms hydrocarbon derivatives (by increasing the oxygen content of a hydrocarbon) into new by-products. This results in changing in the interfacial properties of the oil, affecting spreading and emulsion formation, and may result in transfer of toxic by-products into the water column due to the by-product's enhanced water solubility.

- *Sedimentation and Shoreline Stranding*: Sedimentation is the incorporation of oil within both bottom and suspended sediments. Shoreline stranding is the visible accumulation of crude oil along the water's edge (shoreline) following a spill. It is affected by proximity of the spill to the affected shoreline, intensity of current and wave action on the affected shoreline and persistence of the spilled product. Sedimentation can begin immediately after the spill but increases and peaks after several weeks, whereas shoreline stranding is a function of the distance of the shoreline from the spill and chemical nature of spilled oil.

- *Biodegradation*: Biodegradation occurs on the water surface, in the water, in the sediments, and at the shoreline. It can begin (depending on the lag time or induction period) after a spill and can continue as long as degradable hydrocarbon derivatives are present.

Weathering processes occur simultaneously—one process does not stop before the other begins. The order in which these processes presented is instantaneous, the relative significance of these processes may change if the spill occurred below the water surface or in tropical or ice conditions. The spill chronology may also vary if the spill is near the shoreline in which case it can contaminate the soil and groundwater even before the weathering or cleanup processes start. Also, there are many onshore and offshore operations in a crude oil industry that can cause soil pollution and aquifer contamination.

Aquifer contamination can also take place because of the migration of the oil through the porous media and its subsequent adsorption on the rock surface. Many technologies have been proposed for the treatment of oil contaminated sites; these can be performed by two basic processes *in situ* and *ex situ* treatment using different cleaning technologies such as thermal treatment, biological treatment, chemical extraction and soil washing, and aeration accumulation techniques (Speight, 1996; Speight and Lee, 2000).

Hydrocarbon degradation rates in the water ecosystems generally follow the order: n-alkane derivatives > branched alkane derivatives > low molecular weight aromatic derivatives > cyclic alkane derivatives. The process is usually aerobic, requiring terminal or sub-terminal oxidation of the alkane derivatives, while aromatic hydrocarbon ring structures are broken through hydroxylation and carboxylation processes.

The conditions of the contaminated area play a major role on whether bioremediation is the appropriate method of cleanup for the given oil spill. The success of bioremediation is dependent upon physical conditions and chemical conditions. Physical parameters include temperature, surface area of the oil, and the energy of the water. Chemical parameters include oxygen and nutrient content, pH, and the composition of the oil.

Temperature affects bioremediation by changing the properties of the oil and also by influencing the crude oil-degrading microbes. When the temperature is lowered, the viscosity of the crude oil is increased which changes the toxicity and solubility of the oil, depending upon its composition (Zhu et al., 2004). Temperature also has an effect on the growth rate of the microorganisms, as well as the degradation rate of the hydrocarbon derivatives, depending upon their characteristics.

Water temperature affects the physical and chemical properties of oil and the rate of biodegradation. Colder temperatures slow the rate; warmer temperatures increase the rate. Temperature can affect the physical and chemical characteristics of the discharged oils, hydrocarbon derivatives with high pour points could be expected to show more temperature-related biodegradation than contaminants with low pour points. However, globally low temperatures have been reported to play a significant role in controlling the nature and extent of microbial hydrocarbon metabolism. Mineralization of isotopically marked (^{14}C) hexadecane has been observed in Arctic soil at 5°C (41°F), but mineralization occurred faster and to a greater extent at 23°C (74°F).

In open water, oil hydrocarbon derivatives undergo aerobic biodegradation by bacteria that use oxygen dissolved in the water. Scientists have monitored dissolved oxygen levels around the spill since it occurred. Early measurements of the deep plume showed a rate of 30 percent oxygen depletion, which demonstrated the presence of biodegrading microbes. Sediment on the ocean floor and along the coast is, for the most part, *anoxic*: it does not contain oxygen. Hydrocarbon derivatives that settle into sediments on the ocean floor and along coasts undergo anaerobic biodegradation, a much slower process. Therefore, onshore oil lingers longer than oil at sea and can become a chronic pollutant.

In addition to carbon and oxygen, bacteria need nitrogen and phosphorus to survive. These nutrients are found naturally in the ocean environment. Nitrogen and phosphorus-based fertilizers from farms and gardens on land also enter marine waters through storm water runoff.

The type of crude oil or the crude oil product, its concentration and the types of hydrocarbon derivatives it contains influence the rate of biodegradation. The Gulf spill released light, sweet crude oil, which is more readily broken down than heavy, sour oil. *Mousse*, tar balls, or oil slicks that wash onshore are concentrated compared to dispersed oil, more protected from wind and wave action than oil in open water and have less surface area for microbes to access. Smaller droplets of oil are more biodegradable.

Microbes adapt to gradual exposure to oil. The more oil a microbial community has been exposed to in the past, the greater its capacity and availability to biodegrade oil in the future. In one study, microbes from sediments previously contaminated with oil were able to metabolize oil 10 to 400 times faster than those from sediments that had never been contaminated. Once a species of bacteria is exposed to oil and metabolizes it, the next generations inherit that ability, a concept known as *genetic adaptation*. This has been studied in a particular species of *Vibrio* in the northwestern Gulf of Mexico.

Biodegradation occurs more readily when oil droplets are dispersed. Chemical dispersants break oil into smaller droplets, which increases the surface area available for bacteria to access. The exact effects of dispersants such as Corexit on the rate of biodegradation are unknown, especially in deep water. The effectiveness of dispersants also depends on the type and consistency of oil and the oil-dispersant ratio.

Oil discharges generate a dynamic situation on the oil–water interphases, caused by a variety of physical and chemical processes, including wind, currents, oil weathering, film generation, and oil dispersion. Thus, immiscible oil surfaces are constantly washed with seawater, and new microbes are continuously contacting the oil film or droplet surfaces. Simulation of these dynamic conditions may be achieved rather in flow-through systems than by closed static experiments. Various controlled flow-through systems have been described which simulate biodegradation under marine conditions, including microcosms, and large tank-based mesocosms. However, since dispersed oil was used in most of these systems it was impossible to differentiate between HC processes in the oil and water phases.

In more recent studies with immobilized oil in flow-through systems may give valuable contributions to the determination and understanding of various processes after oil discharges to aquatic environments. For example, hydrocarbon dissolution and biodegradation from oil films, and the potential impacts in the water column, may be predicted after an oil spill in combination with remote sensing of oil films thickness. Further, if depletion from films and droplets are comparable, the system described here may be used for gaining new insights into dissolution, biodegradation and microbiological processes of dispersed oil. This may also have implications for studies of oil spill treatment with chemical dispersants.

The deep sea (as well as other habitats, such as deep groundwater, deep sediments or oilfields) is influenced by high pressure. Barophiles (piezophiles) are microorganisms that require high pressure for growth, or grow better at pressures higher than atmospheric pressure.

Pollutants with densities greater than that of marine waters (such as oxidized crude oil-water emulsions) will be expected to sink to a level commensurate with the density and even to the deep benthic zone, where the hydrostatic pressure is notably high. A combination of high pressure and low temperatures in the deep ocean results in low microbial activity.

References

Abatenh, E., Gizaw, B., Tsegaye, Z., and Wassie, M. 2017. The role of micro-organisms in bioremediation—a review. Open J. Environ. Biol., 2(1): 38–46.

Abdulsalam, S., Adefila, S.S., Bugaje, I.M., and Ibrahim, S. 2013. Bioremediation of soil contaminated with used motor oil in a closed system. Bioremediation and Biodegradation, 3: 100–172.

Adams, G., Fufeyin, P., Okoro, S., and Ehinomen, I. 2015. Bioremediation, biostimulation, and bioaugmentation: a review. International Journal of Environmental Bioremediation & Biodegradation, 3(1): 28–39.

Adeniji, A. 2004. Bioremediation of Arsenic, Chromium, Lead and Mercury. Report No. P1-43. United States Environmental Protection Agency, Office of Solid Waste and Emergency Response. Technology Innovation Office, Washington, DC, USA. https://clu-in.org/download/studentpapers/bio_of_metals_paper.pdf.

Alexander, M. 1999. Biodegradation and Bioremediation 2nd Edition. Academic Press, London, United Kingdom.

Aranda, E., Ullrich, R., and Hofrichter, M. 2010. Conversion of polycyclic aromatic, methyl naphthalenes and dibenzofuran by two fungal peroxygenases. Biodegradation, 21: 267–281.

Arjoon, K.K., and Speight, J.G. 2020. Chemical and Physical Analysis of a Petroleum Hydrocarbon Contamination on a Soil Sample to Determine Its Natural Degradation Feasibility. Inventions, 5. 43. 10.3390/inventions5030043. https://www.mdpi.com/2411-5134/5/3/43. https://www.researchgate.net/publication/343779125_Chemical_and_Physical_Analysis_of_a_Petroleum_ Hydrocarbon_Contamination_on_a_Soil_Sample_to_Determine_Its_Natural_Degradation_Feasibility.

Arora, P.K., Srivastava, A., and Singh, V.P. 2010. Application of monooxygenases in dehalogenation, desulphurization, denitrification, and hydroxylation of aromatic compounds. Journal of Bioremediation & Biodegradation, 1: 1–8.

Atlas, R.M. 1977. Stimulated petroleum biodegradation. Critical Reviews in Microbiology, 5: 371–386.

Atlas, R.M. 1991. Bioremediation: Using Nature's Helpers-Microbes and Enzymes to Remedy Mankind's Pollutants. Proceedings. Biotechnology in the Feed Industry. Alltech's Thirteenth Annual Symposium. T.P. Lyons and K.A. Jacques (Eds.). Alltech Technical Publications, Nicholasville, Kentucky. pp. 255–264.

Atlas, R.M. 1995. Petroleum biodegradation and oil spill bioremediation. Marine Pollution Bulletin, 31(4–12): 178–182.

Atlas, R.M., and Bartha, R. 1998. Fundamentals and applications. pp. 523–530. *In*: Microbial Ecology 4th Edition. Benjamin/Cummings Publishing Company Inc., California.

Azubuike, C.C., Chikere, C.B., and Okpokwasili, G.C. 2016. Bioremediation techniques-classification based on site of application: principles, advantages, limitations and prospects. World Journal of Microbiology and Biotechnology, 32(11): 180. https://doi.org/10.1007/s11274-016-2137-x. https://link.springer.com/article/10.1007/s11274-016-2137-x.

Barker, G.W., Raterman, K.T., Fisher, J.B., Corgan, J.M., Trent, G.L., Brown, D.R., and Sublette, G.L. 1995. Assessment of natural hydrocarbon bioremediation at two gas condensate production sites. pp. 181–188. *In*: Hinchee, R.E., Wilson, J.T., and Downey, D.C. (Eds.). Intrinsic Bioremediation. Battelle Press, Columbus, Ohio.

Bartha, R. 1986a. Microbial Ecology: Fundamentals and Applications. Addisson-Wesley Publishers, Reading, Massachusetts.

Bartha, R. 1986b. Biotechnology of petroleum pollutant biodegradation. Microb. Ecol., 12: 155–172.

Bennet, J.W., Wunch, K.G., and Faison, B.D. 2002. Use of fungi biodegradation. pp. 960–971. *In*: Hurst, C.J. (Ed.). Manual of Environmental Microbiology 2nd Edition. ASM Press Washington, DC. Chapter 87.

Benson, R.C., and Yuhr, L.B. 2016. Site Characterization in Karst and Pseudokarst Terraines. Springer, Dordrecht. The Netherlands.

Boopathy, R. 2000. Factors limiting bioremediation technologies. Bioresource Technology, 74(1): 63–67.

Brown, K.S. 1995. The green clean: the emerging field of phytoremediation takes root. BioScience, 45: 579–582.

Brown, R.A., and Norris, R.D. 1994. The evolution of a technology: hydrogen peroxide in *In situ* bioremediation. pp. 148–162. *In*: Hinchee, R.E., Alleman, B.C., Hoeppel, R.E., and Miller, R.N. (Eds.). Hydrocarbon Bioremediation. CRC Press, Boca Raton, Florida.

Cassidy, D., and Irvine, R. 1995. Biological treatment of Soils Contaminated with Hydrophobic Organics Using Slurry- and Solid-Phase Techniques. Proceedings of SPIE - The International Society for Optical Engineering. Page SPIE, Volume 2504, page 195–208. https://www.researchgate.net/publication/252928814_Biological_treatment_of_soils_contaminated_with_hydrophobic_organics_using_slurry-_and_solid-phase_techniques.

Chhatre, S., Purohit, H., Shanker, R., and Khanna, P. 1996. Bacterial consortia for petroleum spill remediation. Water. Sci. Technol., 34: 187–193.

Cirino, P.C., and Arnold, F.H. 2002. Protein engineering of oxygenases for biocatalysis. Current Opinion in Chemical Biology, 6(2): 130–135.

Clar, E. 1964. Polycyclic Hydrocarbons, Academic Press Inc. New York.

Colwell, R.R., Walker, J.D., and Cooney, J.J. 1977. Ecological aspects of microbial degradation of petroleum in the marine environment. Critical Reviews in Microbiology, 5(4): 423–445.

Compeau, G.C., Mahaffey, W.D., and Patras, L. 1991. Full-scale bioremediation of a contaminated soil and water site. Page 91–110. *In*: Sayler, G.S., Fox, R., and Blackburn, J.W. (Eds.). Environmental Biotechnology for Waste Treatment. Plenum Press, New York.

Conan, G. 1982. The long-term effects of the amoco cadiz oil spill. Phil. Trans. R. Soc. London, B297: 323–333.

Crawford, R.L., Lynch, J., and Crawford, D.L. 2005. Bioremediation: principles and applications. Cambridge University Press, Cambridge, United Kingdom.

Cunningham, J.A., Hopkins, G.D., Lebron, C.A., and Reinhard, M. 2000. Enhanced anaerobic bioremediation of groundwater contaminated by fuel hydrocarbons at seal beach, California. Biodegradation, 11: 159–170.

Cutright, T.J. 1995. Polycyclic aromatic hydrocarbon biodegradation and kinetics using *Cunninghamella echinulatu var. elegans*. International Biodeterioration & Biodegradation, 35(4): 397–408.

Das, N., and Chandran, P. 2011. Microbial degradation of petroleum hydrocarbon contaminants: an overview. Biotechnology Research International, Volume 2011, Article ID No. 941810. https://www.hindawi.com/journals/btri/2011/941810/.

Da Silva, M., and Alvarez, P. 2010. Bioaugmentation. *In*: McGenity, T.J., van der Meer, J.R., and De Lorenzo, V. (Eds.). Handbook of Hydrocarbon and Lipid Microbiology. Springer-Verlag, Heidelberg, Germany.

DeCola, E., and Fletcher, S. 2006, An assessment of the role of human factors in oil spills from vessels. Report to Prince William Sound RCAC. Nuka Research & Planning Group, LLC. P.O. Box 175 Seldovia, Alaska.

Diez, M.C. 2010. Biological aspects involved in the degradation of organic pollutants. Journal of Soil Science and Plant Nutrition, 10(3): 244–267.

Dzionek, A., Wojcieszyńska, D., and Guzik, U. 2016. Natural carriers in bioremediation: a review. Electronic Journal of Biotechnology, 23: 28–36.

EPA. 1999. Preparing For Oil Spills: Contingency Planning. Office of Emergency and Remedial Response, United States Environmental Protection Agency, Washington DC.

Flathman, P.E., Carson, J.H. Jr., Whitenhead, S.J., Khan, K.A., Barnes, D.M., and Evans, J.S. 1991. Laboratory evaluation of the utilization of hydrogen peroxide for enhanced biological treatment of petroleum hydrocarbon contaminants in soil. pp. 125–142. *In*: Hinchee, R.E., and Olfenbuttel, R.F. (Eds.). *In situ* Bioreclamation: Applications and Investigations for Hydrocarbon and Contaminated Site Remediation. Butterworth-Heinemann, Stoneham, Massachusetts.

Fowzia, A., and Fakhruddin, M. 2018. A review on environmental contamination of petroleum hydrocarbons and its biodegradation. Int. J. Environ. Sci. Nat Res., 11(3): 63–69.

Frenzel, M., James, P., Burton, S.K., Rowland, S.J., and Lappin-Scott, H.M. 2009. Towards bioremediation of toxic unresolved complex mixtures of hydrocarbons: identification of bacteria capable of rapid degradation of alkyltetralins. J. Soils Sediments, 9: 129–136.

Gianfreda, L., Xu, F., and Bollag, J.M. 1999. Laccases: a useful group of oxidoreductive enzymes. Bioremediation Journal, 3(1): 1–25.

Gibson, D.T., and Sayler, G.S. 1992. Scientific Foundation for Bioremediation: Current Status and Future Needs. American Academy of Microbiology, Washington, DC.

Goodwin, N.S., Park, P.D.J., and Rawlinson, A.P. 1983. Crude oil biodegradation under simulated and natural condition. pp. 650–658. *In*: Bjorøy, M. (Ed.). Advances in Organic Geochemistry 1981: John Wiley & Sons Inc. New York.

Gordon, R.P.E. 1998. The contribution of human factors to accidents in the offshore oil industry. Reliability Engineering and Systems Safety, 61: 95–108.

Goswami, M., Chakraborty, P., and Mukherjee, K. 2018. Bioaugmentation and biostimulation: a potential strategy for environmental remediation. J. Microbiol. Exp., 6(5): 223–231.

Hamaker, W. 1972. Decomposition: Quantitative aspects. *In*: Hamaker, J.W., and Thomson, J. (Eds.). Organic Chemicals in the Soil Environment. C.A.I. Goring. Marcel Dekker Inc., New York.

Haritash, A.K., and Kaushik, C.P. 2009. Biodegradation aspects of polycyclic aromatic hydrocarbons (PAHs): a review. Journal of Hazardous Materials, 169: 1–15.

Harms, H., Smith, K.E.C., and Wick, L.Y. 2010. Problems of hydrophobicity/bioavailability. pp. 1439–1450. *In*: Timmis, K.N. (Ed.). Handbook of Hydrocarbon and Lipid Microbiology. Springer, Berlin. Chapter 42.

Hesham, A., Khan, S., Tao, Y., Li, D., and Zhang, Y. 2012. Biodegradation of high molecular weight PAHs using isolated yeast mixtures: application of metagenomic methods for community structure analyses. Environ. Sci. Pollut. Res. Int., 19: 3568–3578.

Hoff, R.Z. 1993. Bioremediation: an overview of its development and use for oil spill cleanup. Marine Pollution Bulletin, 29: 476–481.

Idris, S., Isah, A., Abubakar, U., and Majumdar, R.S. 2015. Review on comparative study on bioremediation for oil spills using microbes. Research Jounral of Pharmaceutical, Biological, and Chemical Sciences, 6: 783–790. https://www.researchgate.net/publication/298170128_Review_on_comparative_study_on_bioremediation_for_oil_spills_using_microbes.

Jernelöv, A., and Lindén, O. 1981. Ixtoc I: a case study of the world's largest oil spill. Ambio, 10(6): 299–306.

Karigar, C.S., and Rao, S. 2011. Role of microbial enzymes in the bioremediation of pollutants: a review. Enzyme Research, vol. 2011, Article ID No. 805187. https://www.hindawi.com/journals/er/2011/805187/.

Kensa, V. 2011. Bioremediation—an overview. J. of Industrial Pollution Control, 27(2): 161–168.

Kessler, J.D., Valentine, D.L., Redmond, M.C., Mengran, D., Chan, E.W., Mendes, S.D., Quirozm, E.W., Villaneuva, C.J., Shusta, S.S., Werra, L.M., Yvon-Lewis, S.A., and Weber, T.C. 2011. A persistent oxygen anomaly reveals the fate of the spilled methane in the deep gulf of Mexico. Science, 331: 312–315.

King, M.W.G., Barker, J.F., and Hamilton, L.K. 1995. Natural attenuation of coal tar organics in groundwater. pp. 171–180. *In*: Hinchee, R.E., Wilson, J.T., and Downey, D.C. (Eds.). Intrinsic Bioremediation, Battelle Press, Columbus, Ohio.

Koning, M., Hupe, K., and Stegmann, R. 2008. Thermal processes, scrubbing/extraction, bioremediation and disposal. pp. 304–317. *In*: Rehm, H-J., and Reed, G. (Eds.). Biotechnology: Environmental Processes II, 2nd Edition. Wiley VCH Verlag GmbH, Weinheim, Germany.

Korade, D.L., and Fulekar, M.H. 2009. Development and evaluation of mycorrhiza for rhizosphere bioremediation. Journal of Applied Biosciences, 17: 922–929.

Leahy, J.J., and Colwell, R.R. 1990. Microbial degradation of hydrocarbons in the environment. Microbiol. Rev., 54(3): 305–315.

Leavitt, M.E., and Brown, K.L. 1994. Bioremediation versus bioaugmentation—three case studies. pp. 72–79. *In*: Hinchee, R.E., Alleman, B.C., Hoeppel, R.E., and Miller, R.N. (Eds.). Hydrocarbon Bioremediation. CRC Press, Inc., Boca Raton, Florida.

Lee, M.D., Thomas, J.M., Borden, R.C., Bedient, P.B., and Ward, C.H. 1988. Biorestoration of aquifers contaminated with organic compounds. CRC Crit. Rev. Environ. Control, 18: 29–89.

Lin, C., Gan, L., and Chen, Z.L. 2010. Biodegradation of naphthalene by strain *Bacillus fusiformis* (BFN). J. Hazard. Mater., 182: 771–777.

Lindstrom, J.E., Prince R.C., Clark, J.C., Grossman, M.J., Yeager, T.R., Braddock, J.F., and Brown, E.J. 1991. Microbial populations and hydrocarbon biodegradation potentials in fertilized shoreline sediments affected by the T/V exxon valdez oil spill. Applied and Environmental Microbiology, 57: 2514–2522.

Lu, C.J., and Hwang, M.C. 1992. Effects of hydrogen peroxide on the *in situ* biodegradation of chlorinated phenols in groundwater. Proceedings. Water Environ. Federation 65th Annual Conference, New Orleans, Louisiana. September 20–24.

Lu, C.J. 1994. Effects of hydrogen peroxide on the *in situ* biodegradation of organic chemicals in a simulated groundwater system. pp. 140–147. *In*: Hinchee, R.E., Alleman, B.C., Hoeppel, R.E., and Miller, R.N. (Eds.). Hydrocarbon Bioremediation. CRC Press, Inc., Boca Raton, Florida.

Lucian, P., and Gavrilescu, M. 2008. Overview of *ex situ* decontamination techniques for soil cleanup. Environmental Engineering and Management Journal, 7: 815–834.

Luka, Y., Highina, B.K., and Zubairu, A. 2018. 2018. Bioremediation—a solution to environmental pollution—a review. American Journal of Engineering Research, 7(2): 101–109.

Macaulay, B.M., and Rees, D. 2014. Bioremediation of oil spills: a review of challenges for research advancement. Annals of Environmental Science, 8: 9–37.

Madsen, E.L. 1991. Determining *in situ* biodegradation: Facts and Challenges. Environ. Sci. Technol., 25: 1663–73.

Magdalene, O.E., Ufuoma, A., and Gloria, O. 2009. Screening of four common nigerian weeds for use in phytoremediation of soil contaminated with spent lubricating oil. African Journal of Plant Science, 3(5): 102–106.

Maila, M.P., and Cloete, T.E. 2004. Bioremediation of petroleum hydrocarbons through landfarming: are simplicity and cost-effectiveness the only advantages? Reviews in Environmental Science & Bio/Technology, 3: 349–360.

Maletić, S., Dalmacija, B., Rončević, S., Agbaba, J., and Petrović, O. 2009. Degradation kinetics of an aged hydrocarbon-contaminated soil. Water Air Soil Pollut., 202: 149–159.

Mambra, S. 2020. The Complete Story of the Exxon Valdez Oil Spill, Maritime History. https://www.marineinsight.com/maritime-history/the-complete-story-of-the-exxon-valdez-oil-spill/.

Marın, J.A., M, J.L., Hernandez, T., and Garcıa, C. 2006. Bioremediation by composting of heavy oil refinery sludge in semiarid conditions. Biodegradation, 17: 251–261.

Martinelli, M., Luise, A., Tromellini, E., Sauer, T.C., Neff, J.M., and Douglas, G.S. 1995. The M/C Haven oil spill: Environmental assessment of exposure pathways and resource injury. IOSC Papers. Proceedings. International Oil Sp[ill Confernece, 1: 679–685. https://www.researchgate.net/publication/273991890_The_MC_Haven_oil_spill_Environmental_assessment_of_exposure_pathways_and_resource_injury.

McAllister, P.M., Chiang, C.Y., Salanitro, J.P., Dortch, I.J., and Williams, P. 1995. Enhanced aerobic bioremediation of residual hydrocarbon sources. pp. 67-76. *In*: Hinchee, R.E., Wilson, J.T., and Downey, D.C. (Eds.). Intrinsic Bioremediation. Battelle Press, Columbus, Ohio.

Mishra, S., Jyot, J., Kuhad, R.C., and Lal, B. 2001. *In situ* bioremediation potential of an oily sludge-degrading bacterial consortium. Current Microbiol., 43: 328–335.

Moldan, A.G.S., Jackson, L.F., McGibbon, S., and Van Der Westhuizen, J. 1985. Some aspects of the Castillo de bellver oil spill. Marine Pollution Bulletin, 16(3): 97–102.

Naik, M.G., and Duraphe, M.D. 2012. Review paper on parameters affecting bioremediation. International Journal of Life Science and Pharma Research, 2(3): 77–80.

Nedunuri, K.V., Govundaraju, R.S., Banks, M.K., Schwab, A.P., and Z. Chen. 2000. Evaluation of phytoremediation for field scale degradation of total petroleum hydrocarbons. J. Environ. Eng., 126: 483–490.

Nester, E.W., Anderson, D.G., Roberts, C.E. Jr., Pearsall, N.N., and Nester, M.T. 2001. Microbiology: A Human Perspective 3rd Edition. McGraw-Hill, New York.

Norris, R.D. 1994. Handbook of Bioremediation. CRC Press, Taylor & Francis Group, Boca Raton, Florida.

Okoh, A.I. 2006. Biodegradation alternative in the cleanup of petroleum hydrocarbon pollutants. Biotechnology and Molecular Biology Review, 1(2): 38–50.

Overton, E.B., T.L. Wade, J.R. Radović, B.M. Meyer, M.S. Miles, and S.R. Larter. 2016. Chemical composition of macondo and other crude oils and compositional alterations during oil spills. Oceanography, 29(3): 50–63.

Pala, D.M., de Carvalho, D.D., Pinto, J.C., and Sant'Anna, G.L. Jr. 2006. A suitable model to describe bioremediation of a petroleum-contaminated soil. International Biodeterioration & Biodegradation, 58(3-4): 254–260.

Pardieck, D.L., Bouwer, E.J., and Stone, A.T. 1992. Hydrogen peroxide use to increase oxidant capacity for *in situ* bioremediation of contaminated soils and aquifers: a review. J. Contaminant Hydrol., 9: 221–242.

Patnaik, P. 1999. A Comprehensive Guide to the Properties of Hazardous Chemical Substances 2nd Edition. John Wiley & Sons Inc., Hoboken, New Jersey.

Perfumo, A., Banat, I.M., Marchant, R., and Vezzulli, L. 2007. Thermally enhanced approaches for bioremediation of hydrocarbon-contaminated soils. Chemosphere, 66: 179–184.

Radwan, S.S., Al-Mailem, D., El-Nemr, I., and Salamah, S. 2000. Enhanced remediation of hydrocarbon contaminated desert soil fertilized with organic carbons. Int. Biodet. Biodeg., 46: 129–132.

Ranalli, G., Matteini M., Tosini I., Zanardini E., and Sorlini C. 2000. Bioremediation of cultural heritage: removal of sulfates, nitrates, and organic substances. *In*: Ciferri, O., Tiano, P. and Mastromei, G. (Eds.). Of Microbes and Art. Springer, Boston, Massachusetts.

Riccardi, C., Papacchini, M., Mansi, A., Ciervo, A., Petrucca, A., LaRosa, G., Marianelli, C., Muscillo, M., Marcelloni, A.M., and Spicagilla, S. 2005. Characterization of bacterial population coming from a soil contaminated by Polycyclic Aromatic Hydrocarbons (PAHs) able to degrade pyrene in slurry phase. Ann. Microbiol., 55(2): 85–90.

Rončević, S., Dalmacija, B., Ivančev-Tumbas, I., Petrović, O., Klašnja, M., and Agbaba, J. 2005. Kinetics of degradation of hydrocarbons in the contaminated soil layer. Archives of Environmental Contamination and Toxicology, 49(1): 27–36.

Safiyanu, I., Isah, A.A., Abubakar, U.S., and Singh, R.M. 2015. Review on comparative study on bioremediation for oil spills using microbes. Research Journal of Pharmaceutical, Biological and Chemical Sciences, 6: 783–790.

Sales da Silva, I.G., Gomes de Almeida, F.C., Padilha da Rocha e Silva, N.M., Casazza, A.A., Converti, A., and Asfora Sarubbo, L. 2020. Soil bioremediation: overview of technologies and trends. Energies, 13(18): 4664–4688. https://www.mdpi.com/1996-1073/13/18/4664.

Sarang, B., Richa, K., and Ram, C. 2013. Comparative study of bioremediation of hydrocarbon fuel. International Journal of Biotechnology and Bioengineering Research, 4: 677–686.

Sasikumar, C.S., and Papinazath, T. 2003. Environmental management:- bioremediation of polluted environment. Proceedings. pp. 465–469. *In*: Bunch, M.J., Suresh, V. and Vasantha Kumaran, T. (Eds.). Third International Conference on Environment and Health. Chennai, Department of Geography, University of Madras, India. December 15–17.

Schlegel, H.G. 1977. Aeration without air: oxygen supply by hydrogen peroxide. Biotechnol. Bioeng., 19: 413.

Scragg, A. 1999. Environmental Biotechnology. Pearson Education Limited, Harlow, Essex, England.

Semple, K.T., Reid, B.J., and Fermor, T.R. 2001. Impact of composting strategies on the treatment of soils contaminated with organic pollutants. Environmental Pollution, 112: 269–283.

Shah, M.P. 2014. Environmental bioremediation: a low cost nature's natural biotechnology for environmental clean-up. J. Pet. Environ. Biotechnol., 5: 191–213.

Shelton, D.R., and Tiedje, J.M. 1984. Isolation and partial characterization of bacteria in an anaerobic consortium that mineralizes 3-chlorobenzoic acid. Appl. Environ. Microbiol., 48: 840–848.

Simarro, R., Gonzalez, N., Bautista, L.F., and Molina, M.C. 2013. Assessment of the efficiency of *in situ* bioremediation techniques in a creosote polluted soil: change in bacterial community. J. Hazard. Mater., 262: 158–167.

Sims, R.C. 1990. Soil remediation techniques at uncontrolled hazardous waste sites. J. Air Waste Mgmt. Assoc., 40(5): 703–732.

Singh, A., Kumar, A., and Srivastava, J.N. 2013. Assessment of bioremediation of oil and phenol contents in refinery wastewater via bacterial consortium. J. Pet. Environ. Biotechnol. 4: 1–4.

Sivakumar, G., Xu, J., Thompson, R.W., Yang, Y., and Randol-Smith P. 2012. Integrated green algal technology for bioremediation and biofuel. Bioresource Technol., 107: 1–9.

Soto L.A., Botello, A.V., Licea-Durán, S., Lizárraga-Partida, M.L., and Yáñez-Arancibia, A. 2014. The environmental legacy of the Ixtoc-I oil spill in Campeche sound, southwestern gulf of Mexico. Frontiers in Marine Science, 1(57): 1–9. https://www.researchgate.net/publication/269334260_The_environmental_legacy_of_the_Ixtoc-I_oil_spill_in_Campeche_Sound_southwestern_Gulf_of_Mexico.

Speight, J.G. 1996. Environmental Technology Handbook. Taylor & Francis, Washington, DC.

Speight, J.G., and Lee, S. 2000. Environmental Technology Handbook. 2nd Edition. Taylor & Francis, New York.

Speight, J.G. 2005. Environmental Analysis and Technology for the Refining Industry. John Wiley & Sons Inc., Hoboken, New Jersey.

Speight, J.G., and Arjoon, K. 2012. Bioremediation of Petroleum and Petroleum Products. Scrivener Publishing, Beverly, Massachusetts, 2012.

Speight, J.G. 2014. The Chemistry and Technology of Petroleum 5th Edition. CRC Press, Taylor and Francis Group, Boca Raton, Florida.

Speight, J.G. 2015. Handbook of Petroleum Product Analysis 2nd Edition. John Wiley & Sons Inc., Hoboken, New Jersey.

Speight, J.G. 2017. Handbook of Petroleum Refining. CRC Press, Taylor and Francis Group, Boca Raton, Florida.

Speight, J.G. 2018. Reaction Mechanisms in Environmental Engineering: Analysis and Prediction. Butterworth-Heinemann, Elsevier, Oxford, United Kingdom, 2018.

Speight, J.G. 2019. Handbook of Petrochemical Processes. CRC Press, Taylor & Francis Group, Boca Raton, Florida.

Stoner, D.L. 1994. Biotechnology for the Treatment of Hazardous Waste. CRC Press, Boca Raton, Florida.

Tegene, B., and Tenkegna, T. 2020. Mode of action, mechanism and role of microbes in bioremediation service for environmental pollution management. Journal of Biotechnology & Bioinformatics Research, 2(3): 1–18.

Thapa, B., Kumar, A., and Ghimire, A. 2012. A review on bioremediation of petroleum hydrocarbon contaminants in soil. Kathmandu University Journal of Science, Engineering and Technology, 8(1): 164–170.

Tilman, D., and Lehman, C. 2001. Human-caused environmental change: impacts on plant diversity and evolution. Proceedings. National Academy of Sciences of the United States of America, 98(10): 5433–5440.

US EPA. 2003. Aerobic Biodegradation of Oily Wastes: A Field Guidance Book For Federal On-scene Coordinators. United States Environmental Protection Agency, Region 6 South Central Response and Prevention Branch, Dallas, Texas.

US EPA. 2006. *In situ* and *Ex situ* Biodegradation Technologies for Remediation of Contaminated Sites. Report No. EPA/625/R-06/015. Office of Research and Development National Risk Management Research Laboratory, United States Environmental Protection Agency, Cincinnati, Ohio.

USCG. 2011. United States Coast Guard. Report of Investigation into the Circumstances Surrounding the Explosion, Fire, Sinking and Loss of Eleven Crew Members Aboard the MOBILE Offshore Drilling Unit Deepwater Horizon In the Gulf of Mexico April 20-22, 2010. Bureau of Safety and Environmental Enforcement, Washington, DC. https://www.bsee.gov/sites/bsee.gov/files/reports/safety/2- deepwaterhorizon-roi-uscg-volume-i-20110707-redacted-final.pdf.

Vasudevan, N., and Rajaram, P. 2001. Bioremediation of oil sludge-contaminated soil. Environ. Int., 26: 409–411.

Vidali, M. 2001. Bioremediation: An Overview. Pure and Applied Chemistry, 73(7): 1163–1172.

Wang, Z., Hollebone, B., Fingas, M., Fieldhouse, B, Sigouin, L., Landriault, M., Smith, P., Noonan, J., Thouin, G., and Weaver, J. 2004. Characteristics of Spilled Oils, Fuels, and Petroleum Products: 1. Composition and Properties of Selected Oils. Report No. EPA EPA/600/R-03/072. United States Environmental Protection Agency, Research Tringle Park North Carolina and Emergencies Science and Technology Division Environmental Technology Centre Environment Canada, Ottawa, Ontario, Canada.

Wenzel, W. 2009. Rhizosphere processes and management in plant-assisted bioremediation (Phytoremediation) of soils. Plant and Soil, 321(1-2): 385–408.

Wethasinghe, C., Yuen, S.T.S., Kaluarachchi, J.J., and Hughes, R. 2006. Uncertainty in biokinetic parameters on bioremediation: health risks and economic implications. Environment International, 32(3): 312–323.

Whelan., M.J., Coulon, F., Hince, G., Rayner, J., McWaters, R., Spedding, T., and Snape, I. 2016. Fate and transport of petroleum hydrocarbons in engineered biopiles in polar regions. Chemosphere, 131: 232–240.

Winters, J.C., and Williams, J.A. 1969. Microbiological alteration of petroleum in the reservoir. Preprints. Division of Petroleum Chemistry, American Chemical Society. Volume, 14(4): E22–E31.

Yadav, M., Singh, S.K., Sharma, J.K., and Yadav, K.D.S. 2011. Oxidation of polyaromatic hydrocarbons in systems containing water miscible organic solvents by the lignin peroxidase of Gleophyllum striatum MTCC-1117. Environ. Technol., 32: 1287–1294.

Zengler, K., Richnow, H.H., Rossello-Mora, R., Michaelis, W., and Widdel, F. 1999. Methane formation from long-chain alkanes by anaerobic microorganisms. Natr. (401): 266–269.

Zhu, X., Venosa, A.D., and Suidan, M.T. 2004. Literature Review on the Use of Commercial Bioremediation Agents for Clean-up of Oil Contaminated Estuarine Environments. Report No. EPA/600/R-04/075. National Risk Management Research Laboratory, Environmental Protection Agency, Cincinnati, Ohio.

CHAPTER 7

Site Evaluation and the Impact of an Oil Spill

In recent decades, bioremediation in recent years has often been preferred as a clean sustainable approach regarding pollution that has been caused by crude oil hydrocarbon derivatives to the environment. It is a process in which living organisms are employed to reduce or eliminate the environmental hazards that result from the accumulation of toxic chemicals and other hazardous waste. The bioremediation process is an option that offers the possibility to destroy or render harmless various contaminants using natural biological activity (Gibson and Sayler, 1992). In addition, bioremediation can also be used in conjunction with a wide range of traditional physical and chemical technology to enhance their effectiveness (Vidali, 2001; Speight and Arjoon, 2012).

The bioremediation of crude oil and crude oil hydrocarbon derivatives is the cleanup of the spills of crude oil or spills of crude oil product spills by the use of microbes to breakdown the crude oil constituents (or other organic contaminants) into less harmful (usually lower molecular weight) and easier-to-remove products (biodegradation). The microbes transform the contaminants through metabolic or enzymatic processes, which vary, but the final product is usually harmless and includes carbon dioxide, water, and cell biomass.

Biodegradation is known to be a natural process (or a series of processes) by which the constituents of spilled crude oil hydrocarbon derivatives are degraded (broken down) into nutrients that can be used by other organisms. Bioremediation functions basically on biodegradation. It is applied to a contaminated site to clean up organic matter and other substances by using microbes with the biodegradation process. Although the effects of the bacteria (microbes) on hydrocarbon derivatives have been known for decades, this technology (now known as *bioremediation*) has shown promise and, in some cases, high degrees of effectiveness for the treatment of these contaminated sites since it is cost-effective and will lead to complete mineralization.

One of the major and continuing environmental problems is hydrocarbon contamination resulting from the activities related to crude oil and crude oil products. Soil contaminated with hydrocarbon derivatives causes extensive damage of local system since accumulation of pollutants in animals and plant tissue may cause death or mutations.

However, not all crude oil products are harmful to health and the environment. There are records of the use of *petroleum spirit* for medicinal purposes. However, those constituents of crude oil those are extremely harmful to health and the environment. Indeed, crude oil constituents either in the pure form or as the components of a fraction have been known to belong to the various families of carcinogens and neurotoxins. Whatever the name given to these compounds they are extremely toxic.

As a result, once a spill has occurred, every effort must be made to rid the environment of the toxins. The chemicals of known toxicity range in degree of toxicity from low to high and represent considerable danger to human health and must be removed (Frenzel et al., 2009). Many of these chemicals substances come in contact with, and are sequestered by, soil or water systems. While conventional methods to remove, reduce, or mitigate the effects of toxic chemical in nature are available include (i) pump and treat systems, (ii) soil vapor extraction, (iii) incineration, and (iv) containment, each of these conventional methods of treatment of contaminated soil and/or water suffers from recognizable drawbacks and may involve some level of risk. In short, these methods, depending upon the chemical constituents of the spilled material, may limited effectiveness and can be expensive (Speight, 1996; Speight and Lee, 2000; Speight, 2005).

The United States Environmental Protection Agency (US EPA) uses bioremediation because it takes advantage of natural processes and relies on microbes that occur naturally or can be laboratory cultivated; these consist of bacteria, fungi, actinomycetes, cyanobacteria and to a lesser extent, plants (US EPA, 2006). These micro-organisms either consume and convert the contaminants or assimilate within them all harmful compounds from the surrounding area, thereby, rendering the region virtually contaminant-free. In fact, unless they are overwhelmed by the amount of the spilled material or it is toxic; many indigenous microorganisms in soil and/or water are capable of degrading hydrocarbon contaminants.

The contaminants of concern in crude oil can degrade under appropriate conditions but the success of the process depends on the ability to determine these conditions and establish them in the contaminated environment. Thus, important site factors required for success include (1) the presence of metabolically capable and sustainable microbial populations, (2) suitable environmental growth conditions, such as the presence of oxygen, (3) temperature, which is an important variable—keeping a substance frozen or below the optimal operating temperature for microbial species, can prevent biodegradation—most biodegradation occurs at temperatures between 10 and 35°C (50 and 95°F), (4) the presence of water, (5) appropriate levels of nutrients and contaminants, and (6) favorable acidity or alkalinity (Table 11.1).

In regard to the last parameter, soil pH is extremely important because most microbial species can survive only within a certain pH range – generally the biodegradation of crude oil hydrocarbon derivatives is optimal at a pH 7 (neutral) and the acceptable (or optimal) pH range is on the order of 6 to 8. Furthermore, soil (or water) pH can affect availability of nutrients.

Thus, through biodegradation processes, living microorganisms (primarily bacteria, but also yeasts, molds, and filamentous fungi) can alter and/or metabolize various classes of compounds present in crude oil and crude oil products. Furthermore, biodegradation also alters subsurface oil accumulations of crude oil (Speight, 2014). Shallow oil accumulations, such as heavy oil reservoirs and tar sand deposits, where the reservoir temperature is low-to-moderate, < 80°C, < 176°F) are commonly found to have undergone some degree of biodegradation (Speight and Lee, 2000).

Intrinsic bioremediation is the combined effect of natural destructive and non-destructive processes to reduce the mobility, mass, and associated risk of a contaminant. Non-destructive mechanisms include sorption, dilution and volatilization. Destructive processes are aerobic and anaerobic biodegradation.

Intrinsic aerobic biodegradation is well documented as a means of remediating soil and groundwater contaminated with fuel hydrocarbon derivatives. In fact, intrinsic aerobic degradation should be considered an integral part of the remediation process (Barker et al., 1995; McAllister et al., 1995). There is growing evidence that natural processes influence the immobilization and biodegradation of chemicals such as aromatic hydrocarbon derivatives, mixed hydrocarbon derivatives, and chlorinated organic compounds (Ginn et al., 1995; King et al., 1995).

These technologies are especially valuable where the contaminated soils are fragile, and prone to erosion. The establishment of a stable vegetation community stabilizes the soil system and prevents erosion. This aspect is especially relevant to certain types of soil where removal of large volumes of soil destabilizes the soil system, which leads to extensive erosion. However, when the above parameters are not conducive to bacterial activity, the bacteria (1) grow too slowly, (2) die, or (3) create more harmful chemicals.

Micro-organisms degrade or transform contaminants by a variety of mechanisms. Crude oil hydrocarbon derivatives (particularly alkane derivatives) for example are converted to carbon dioxide and water:

$$2C_{12}H_{26} + 37O_2 \rightarrow 24CO_2 + 26H_2O$$

Also, the hydrocarbon derivatives in the spilled material may be used as a primary food source by the bacteria, which use the energy to generate new cells.

Some contaminants, such as chlorinated organic or high aromatic hydrocarbon derivatives, are generally resistant to microbial attack. They are degraded either slowly or not at all, hence it is not easy to predict the rates of clean-up for a bioremediation exercise; there are no rules to predict if a contaminant can be degraded.

When the hydrocarbon derivatives are chlorinated (as might occur in several additives to improve the performance of crude oil products) the degradation takes place as a secondary or co-metabolic process rather than a primary metabolic process. In such a case enzymes which are produced during aerobic utilization of carbon sources such as methane, degrade the chlorinated compounds. Under aerobic conditions, a chlorinated solvent such as trichloroethylene ($ClCH = CCl_2$), which may have been mixed with the crude oil product during processing or during use, can be degraded through a sequence of metabolic steps, where some of the intermediary by-products may be more hazardous than the parent compound (e.g., vinyl chloride, $CH_2 = CHCl$).

Since many of the contaminants of concern in crude oil and crude oil-related products oil are readily biodegradable under the appropriate conditions, the success of oil-spill bioremediation depends mainly on the ability to establish these conditions in the contaminated environment using the technology to optimize the total efficiency of the microorganisms.

Over the past two decades, opportunities for applying bioremediation to a much broader set of contaminants have been identified. Indigenous and enhanced organisms have been shown to degrade industrial solvents, polychlorinated biphenyls (PCBs), explosives, and many different agricultural chemicals. Pilot, demonstration, and full-scale applications of bioremediation have been carried out on a limited basis. However, the full benefits of bioremediation have not been realized because processes and organisms that are effective in controlled laboratory tests are not always equally effective in full-scale applications. The failure to perform optimally in the field setting stems from a lack of predictability due, in part, to inadequacies in the fundamental scientific understanding of how and why these processes work.

Human beings form an integral part of the environment and activities such as drilling for and the recovery of natural gas and crude oil can have a serious impact on floral and faunal. Drilling projects operate around the clock and generate pollutants which, if not mitigated, can disrupt wildlife, damage to public lands that were set aside to benefit all people, and global climate change (Speight, 2020). However, natural gas and crude oil have long been considered the lifeblood of the industrialized nations and have been major energy sources since the post-World War II era.

In the context of this book, crude oil and the many refined crude oil products are a complex mixture of hundreds of chemicals, each one with a distinct set of behavior with a potentially serious effect when released into the environment. In addition, because of the varying nature of crude oil (Speight, 2014, 2017), each spill is unique and unrepeatable because of the unlimited possible combination of natural and anthropogenic factors at a spill location. This gives rise to a series of challenges when crude oil or a crude oil product enters an environment thereby giving rise, in many cases, to a series of post-spill events that are not always predicable. Thus, a spill of crude oil or a crude oil product can impact an ecosystem (including the flora and fauna therein) by one or more of the following mechanisms (1) physical smothering, which can have an impact on physiological functions of the flora and fauna, (2) chemical toxicity, which can give rise to lethal or sub lethal effects or causing impairment of cellular functions, (3) ecological changes, which can be the loss of key organisms from a community and the takeover of habitats by opportunistic species, and (4) indirect effects, such as the loss of habitat or shelter and the consequent elimination of ecologically important species as well as disruption to the natural floral and faunal lifecycles. Refining crude oil and the production of crude oil products are parts of a large industry on a worldwide basis and the potential environmental hazards associated with refineries have caused increased concern for communities in close proximity to them. Briefly, crude oil and the production of products refining involves a series of steps that includes separation and blending of crude oil products (Table 7.1).

Table 7.1: Categorization of Crude Oil Refining Processes.

Separation processes: These processes involve separating the different constituents into fractions based on their boiling point differences. Additional processing of these fractions is usually needed to produce final products to be sold within the market.
Conversion processes Coking and cracking are conversion processes used to break down higher molecular weight constituents into lower molecular weight product by heating and by use of catalysts.
Treating Treating processes are used to remove the undesirable components and impurities such as sulfur, nitrogen and heavy metals from the products. This involves processes such as hydrotreating, deasphalting, acid gas removal, desalting, hydrodesulfurization, and sweetening.
Blending/combination processes Refineries use blending/combination processes to create mixtures with the various petroleum fractions to produce a desired final product, such as gasoline with different octane ratings.
Auxiliary processes Refineries also have other processes and units that are vital to operations by providing power, waste treatment and other utility services, such as boilers, wastewater treatment, and cooling towers. Products from these facilities are usually recycled and used in other processes within the refinery and are also important in regard to minimizing water and air pollution.

7.1 Site Evaluation

Site evaluation is the practice of investigating, evaluating and reporting basic soil and site conditions which apply to the onsite treatment and disposal of contaminants. The preliminary site evaluation analyzes a site in regard to public transportation access, vehicular access, community access, facility conditions, zoning constraints, and site development issues. The site evaluation is key to the design process and provides sufficient information to select a suitable, cost-effective treatment system. To this end, the site evaluation should be a systematic process that provides information with enough detail to be useful for the design.

When a site is selected for cleanup it is extremely important to link *biodegradation* with *bioremediation*. The former is a component of oil weathering and is a natural process whereby bacteria or other microorganisms alter and break down organic molecules into other substances, eventually producing fatty acids and carbon dioxide (Hoff, 1993). On the other hand, *bioremediation* is the acceleration of this process through the addition of exogenous microbial populations, through the stimulation of indigenous populations or through manipulation of the contaminated media using techniques such as aeration or temperature control (Atlas, 1995; Hoff, 1993; Swannell et al., 1996).

Many microorganisms possess the enzymatic capability to degrade crude oil hydrocarbon derivatives. Some microorganisms degrade alkane constituents; others degrade aromatic constituents while others will degrade both alkane and aromatic constituents. Often the n-alkane derivatives in the range C_{10} to C_{26} are viewed as the most readily degraded, but low-molecular-weight aromatics, such as benzene, toluene and xylene, which are among the toxic compounds found in crude oil, are also biodegraded by many marine microorganisms. More complex structures are more resistant to biodegradation and the rates of biodegradation of these more complex structural entities are lower than biodegradation rates of the simpler hydrocarbon structures found in crude oil (Atlas, 1995).

In fact, it has been reported that biodegradation rates of hydrophobic organic compounds decreased in the presence of organic matrices. The biodegradation of n-alkane derivatives in a contaminated soil is inhibited in the presence of unresolved complex mixture of hydrocarbon derivatives originating from fuel oil residue (Jonge et al., 1997). It was further suggested that inhibition of n-alkane dissolution into water by an unresolved complex mixture of hydrocarbon derivatives resulted in slow biodegradation. Therefore, oil residue in soil or sediment may decrease the biodegradation rate of hydrophobic organic compounds, such as polynuclear aromatic hydrocarbon derivatives.

In addition to the varying rates of biodegradation of organic molecules, a lag time (chemically: an induction period) occurs after spillage of crude oil or a crude oil product before indigenous microbes begin to degrade the crude oil-related constituents (Hoff, 1993). This lag time is related to presence of volatile constituents of the spilled material (and which also may exhibit toxicity to the indigenous microbes), which evaporate in the first few days after the spill. In order to be effective, the indigenous microbial populations must begin to use oil and increase their population density before measurable degradation takes place. This period usually lasts several days and is an important aspect when the appropriateness of application of a bioremediation technology is considered (Hoff, 1993).

There are several different bioremediation techniques and the underlying idea of selecting a bioremediation technique is to accelerate the rates of *natural* hydrocarbon biodegradation by overcoming the rate-limiting factors.

Thus, several techniques can lead to the desired results: (1) indigenous populations of microbial bacteria can be stimulated through the addition of nutrients or other materials, (2) exogenous microbial populations can be introduced in the contaminated environment (bio-augmentation), (3) genetically altered bacteria can be used, and (4) the contaminated site (soil or water) can be manipulated by, for example, aeration or temperature control.

One approach often considered for the bioremediation of crude oil pollutants after an oil spill is the addition of microorganisms (seeding) that are able to degrade hydrocarbon derivatives. Most microorganisms considered for seeding are obtained by enrichment cultures from previously contaminated sites. However, because hydrocarbon-degrading bacteria and fungi are widely distributed in marine, freshwater and soil habitats, adding seed cultures has proven less promising for treating oil spills than adding fertilizers and ensuring adequate aeration. Most tests have indicated that seed cultures are likely to be of little benefit over the naturally-occurring microorganisms at a contaminated site for the biodegradation of the bulk of crude oil contaminants (Atlas, 1995).

Hydrocarbon biodegradation in marine environments is often limited by abiotic environmental factors such as molecular oxygen, phosphate and nitrogen (ammonium, nitrate and organic nitrogen) concentrations. Rates of degradation of crude oil constituents are negligible in anaerobic sediments because molecular oxygen is required by most microorganisms for the initial step in hydrocarbon metabolism. Oxygen, however, is not limiting in well aerated (high energy) marine environments (Atlas, 1995; Stelmaszewski, 2009).

Furthermore, marine ecosystems typically have low (sometime zero) concentrations of nitrogen, phosphorus and various mineral nutrients that are needed for the incorporation into cellular biomass, and the availability of these within the area of hydrocarbon degradation is critical.

7.1.1 Land Ecosystems

The bioremediation process typically begins with an environmental site assessment, but may include many steps (Sweed et al., 1996).

A thorough site assessment that includes an evaluation of the current condition of the property, the operational history of the facility, and the processes used to discover this information should be conducted early in the revitalization process. Accurate characterization of environmental conditions is required to estimate potential costs associated with site cleanup and regulatory negotiation. Environmental costs can represent a significant financial issue impacting the success or start of a revitalization project. In the event that significant contamination exists, communication with local regulatory agencies is important and should include discussion of impacts to project schedule.

Typically, site assessment (sometimes called environmental assessment) is a method used to identify any potential risk to the environment and the resulting report is usually divided into three separate phases or types of work: (1) qualitative investigation and report, (2) quantitative investigation and report, and (3) remediation and environmental management.

In the first phase—a qualitative investigation and report—visual observations and site history are researched and relevant regulatory records and databases are consulted to determine the reasonable probability of environmental risk. At this stage, sampling and testing are not usually performed but a detailed waste classification and a determination of the origin of the waste assist in planning the removal activities and in utilizing the appropriate funding instrument. A thorough *paper review* and site history must be conducted to establish Oil Pollution Act (OPA) of 1990 or Comprehensive Environmental Response and Liability Act (CERCLA) authority. Typically, an oil refinery waste consists of both OPA wastes, CERCLA wastes, and Resource Conservation and Recovery Act (RCRA) wastes). A careful waste classification is required to use appropriate funding to remediate the site. If a risk assessment report is required, this may include some minor soil sampling and testing as a precautionary measure but this does not necessarily qualify the report as an environmental assessment.

The second phase—a quantitative investigation and report—sampling of materials (such crude oil and crude oil products) suspected of contamination are analyzed and characterized. It is also opportune to analyze an unspoiled sample to determine potential sources of the spilled material (if these sources have not already been identified). The object of this phase is to determine as much as reasonable the extent of any environmental contamination present at the site.

The third and final phase—the remediation and environmental management project—includes study of any reports from the two prior phases as well as an environmental site assessment in order to formulate an appropriate solution to the environmental problem. The remediation or bioremediation project is planned and estimated in terms of time, costs, and objectives.

Other phase may also be introduced and these phase will include a timeline for site examination (say after one year with continuous weekly/monthly monitoring) to determine the success/failure of the remediation effort. After the first year, the timeline may be shortened or lengthened appropriately depending on the first-year data.

Reference was made above to the identification of the source of the crude oil-related pollutants, if the source was not identified at the time of the spill. This can be done through *finger printing* of the spilled crude oil or crude oil product can be used to identify the origin of the spill (Michelsen and Petito Boyce, 1993).

The finger-printing approach requires collection and analysis to search for specific chemical biomarkers in the crude oil or chemical markers specific to the crude oil product. All nearby potential sources of crude oil and crude oil products that might be responsible for the spill must be included in the collection and analysis. When a positive correlation of spilled material and the relevant facility has been identified, the origin of the spill has been determined. If the spilled oil has undergone weathering, water washing, and/or any level of biodegradation, then biomarker analysis and/or other techniques may be required to identify the spill source.

Biomarkers are a group of compounds, primarily hydrocarbon derivatives, found in oils, rock extracts, recent sediment extracts, and soil extracts. Biomarkers are distinguished from other compounds and are generally referred to as *molecular fossils*—they are structurally similar to specific natural products. Typically, even though diagenetic alteration of the original mileages has occurred, biomarkers retain all or most of the original carbon skeleton of the original natural product. Biomarkers reflect the type and age of the source rock that generated the crude oil.

Since different potential sources of a spill may involve oil derived from different basins, biomarker distribution (such as the sterane/hopane ratio or the ratio of steroid, diterpenoid, triterpenoid and hopanoid biological molecules) can be used to either point to the source or rule out the potential sources of a spill and can be used to determine if oil in a contaminated area actually represents more than one spill (Stout et al., 2001, 2005; Speight, 2014). Specific chemical compounds (not necessarily biomarkers since these may have been formed in the refining process) can also be used to assess the origin of some refined hydrocarbon products (Peters et al., 1992; Stout et al., 2005; Speight, 2014).

Polynuclear aromatic hydrocarbon derivatives are another group of compounds present in oil that are especially useful in identifying the source(s) of a spill. A subset of the polynuclear aromatic hydrocarbon derivatives in crude oil is products of the diagenesis of steroid, diterpenoid, triterpenoid and hopanoid biological molecules originally deposited in sediments (biomarkers, see above). Several of these biomarkers are present as fully or partially aromatized compounds with multiple aromatic rings; therefore they are polycyclic aromatic hydrocarbon derivatives. These polynuclear aromatic hydrocarbon derivatives are resistant to biodegradation conditions typically encountered in spill situations and have proved useful for defining a unique *fingerprint* characteristic of a specific crude oil (or crude oil product), which can be used to correlate a biodegraded oil to a sample of its non-degraded equivalent, and hence can be used to identify the source(s) of a crude oil-related release (Burns et al., 1997; Stout et al., 2001, 2005).

Many land-based environments are characterized by low or elevated temperatures, acidic or alkaline pH, high salt concentrations, or high pressure. Extremophile microorganisms (organisms that thrives in physically or geochemically extreme conditions that are generally detrimental to most terrestrial life forms) are adapted to grow and thrive under these adverse conditions. Hydrocarbon degrading extremophiles are thus ideal candidates for the biological treatment of polluted extreme habitats. In this review we summarize the recent developments, obtained both in laboratory and field studies, in biodegradation and bioremediation of hydrocarbon contaminants that are of environmental concern in extreme habitats.

A wide range of hydrocarbon derivatives that contaminate the environment, mainly due to accidental release or industrial processes, has been shown to be biodegraded (mineralized or transformed) in various extreme environments characterized by low or elevated temperatures, acidic or alkaline pH, high salinity or high pressure, emphasizes the metabolic capacities of extremophiles (Margesin and Schinner, 2001b). Those adapted to more than one extreme offer a special potential for the biological decontamination of habitats where various different extreme conditions prevail simultaneously and several examples of such behavior by micro-organisms are presented below.

7.1.1.1 Soil Evaluation

When soil is the medium in which treatment will take place, it is of utmost importance to evaluate its properties as well as the general properties of the contaminated site. Soil is an important aspect of site evaluation since different types of soils present widely different properties, and therefore the response to each use bioremediation process will differ. Soil evaluation enables predictions on the biophysical and economic behavior of land for a bioremediation process.

Soil is heterogeneous and varies widely in physical, chemical, and biological properties. The soil may be acidic or basic, may have high or low nutrients, and may exhibit a different exchange capacity at different locations in the same area. The characteristics important in the design and operation of a land treatment site include the slope, the soil classification (texture and permeability), soil moisture content, pH, the cation exchange capacity (CEC), and salinity. If the initial soil properties are not ideal for the biodegradation of hydrocarbon derivatives, they can be optimized (US EPA, 2003).

7.1.1.2 Physical Properties

A gently sloped terrain can help minimize earthwork, but slopes in excess of 5% are not recommended for land treatment facilities due to erosion problems and less than ideal surface drainage and run-off control capabilities. However, physical manipulation of the land may produce the appropriate slope incline.

A survey should be performed to classify the indigenous soil present on-site. A soil engineer or scientist may be consulted to perform soil classification. Soil particle analysis allows the identification of soil type and is inexpensive to conduct. A general soil classification scheme based on standard sieve analysis is available and provides the unified soil classification system (USCS) (Table 7.2).

Thus, if more than 50% of the soil is retained on No. 200 sieve, it is considered coarse-grained soil; otherwise, it will be fine-grained soil. Coarse-grained soils permit rapid infiltration of liquids and allow good aeration; they are considered to be very permeable. However, they may not control containment of waste and nutrients added to the soil as well as fine-grained soils, which would be considered impermeable. The oxygen (air) transfer rate and substrate availability are greater in coarse-grained soils than in fine-grained soils due to more air pore space and thus favor aerobic conditions desirable for biodegradation. Coarse-grained soils are also more desirable since they can be more favorably loaded with hydrocarbon derivatives. Fine-grained soils should be loaded more lightly in a shallower depth and will generally require more tilling for equivalent performance.

Another important variable that should be assessed during soil characterization is the moisture content, or the amount of water the soil can retain. Saturation, field capacity, wilting point, and oven dry are the four conditions that will help evaluate the irrigation needs of the treated soil. Saturation is undesirable, as it decreases oxygen availability and limits site access for nutrient application and tilling. Approximately 50 to 70% of soil field capacity is ideal for microbial activities, and adequate drainage can help manage that range. Soil field capacity can be determined by saturating the soil, draining it for 24 hours under gravity, then by weighing followed by oven drying at 105°C (221°F) to attain a constant weight. Thus:

$$W_{ds} - W_{ods} = W_{fc}$$

$$W_{\%fc} = W_w/W_s \times 100$$

W_{ds} is the weight of drained soil, W_{ods} is the weight of oven dry soil, W_{fc} is the weight of water in the soil at field capacity, $W_{\%fc}$ is the % of water in soil at field capacity, W_w is the weight of water, and W_s is the dry weight of soil.

The *infiltration rate* should also be assessed because application of a liquid at a rate greater than that rate will result in flooding and erosion. At water levels greater than the field capacity, water may accumulate and result in flooding and erosion. Below the wilting point, the soil becomes too dry, slowing down microbial activities.

Table 7.2: Soil Classification by Particle Size.

Soil type	US Sieve Size	Particle Size
Coarse grained		
Gravel soil	Retained on No. 4	> 4.75 mm
Sand soil	No. 4 through 200	4.75–0.075 mm
Fine grained		
Clay soil	Passing No. 200	< 0.075 mm
Silt soil	Passing No. 200	< 0.075 mm

Table 7.3: Soil Characteristics for Effective Bioremediation.

Soil type	Moisture holding capacity	Permeability
Sand-type	high	high
Clay-type	low	high

Soil moisture, pH, nutrients, oxygen transfer, presence of metals and toxics, and salinity are the utmost controlling factors, which must be monitored and can be optimized to achieve time-efficient biodegradation rates at a given site (Chapter 1). Another important factor is the climate, but it is beyond the control of the responder. Desirable soil parameters ranges that should be maintained to conduct a time-efficient bioremediation in the land treatment unit are: (1) moisture content, which is calculated as a percentage of the field capacity and typically falls in the 50 to 70% range, (2) acidity or alkalinity, which is usually on the order of pH 6 to 8 but may be site specific, (3) temperature, typically 24 to 35°C (75 to 95°F), and (4) nutrient ratio, i.e., the ratio of carbon/nitrogen/phosphorus/potassium, which is generally on the order of 100:5:1:1.

The capacity of the soil to retain water at specific levels has been the most neglected area in bioremediation land-farming operations. Generally, a soil is at field capacity when soil micropores are filled with water and macropores are filled with air but the water holding capacity depends upon the nature of the soil (Table 7.3) Too much water or too little water can be detrimental to an aerobic bioremediation operation. For example saturation (too much water) will inhibit oxygen infiltration and dry conditions (too little water) will decrease the rates of microbial activity or cause cessation of the biodegradation process if a wilting point is reached. A desirable range is between 70 to 80% of field capacity thereby allowing the degrading bacteria access to both air and water, which are very much needed for microbial activity.

The pH of the soil not only affects the growth of microorganisms, but also has a noticeable effect on the availability of nutrients, mobility of metals, rate of abiotic transformation of organic waste constituents, and soil structure. Usually, a pH range of 6 to 8 units is considered optimum for biodegradation activities—the pH of the soil can be adjusted by addition of chemical reagents. For example, if the soil is too acidic soils (pH < 6), agricultural lime may be used to raise the pH while aluminum sulfate or ferrous sulfate may be used to lower the pH of alkaline soils. However, caution should be used to avoid *over correction* of the pH, and further consultation may be used to help calculate optimum quantities.

Biological activity is regulated by soil temperature, and an ideal temperature range is between 24 to 35°C (75 and 95°F). Since the soil temperature is difficult to control under field conditions, the waste loading rates should be adjusted according to temperatures affect temperature. This adjustment should also be performed during the change in season since the biodegradation rates are lower in the spring and the autumn compared to summer.

7.1.1.3 Chemical Properties

In addition to the pH of the soil, *cation exchange capacity* is an important variable that can be used to monitor in order to optimize the degradation process. The cation exchange capacity is an indication of the capacity of the soil to retain metallic ions (the cation exchange capacity is usually obtained through laboratory testing) and is measured in milli-equivalents per one hundred grams of dry soil (meq/100 g). A cation exchange capacity value greater than 25 is an indication that the soil contains more nutrients and has high clay content, whereas values less than 5 indicate a sandy soil with little ion retention. Most metals found in oily wastes are not readily soluble in water: however, variations of pH may change that property and when treating land where the soil has a low cation exchange capacity, care must be taken to manage subsurface of metal ions. With proper pH management, metals remain immobilized in the treatment zone even with low cation exchange capacity values.

Soil salinity results from accumulation of neutral soluble salts (mainly due to neutral salts of sodium, calcium, magnesium, and potassium) in the upper soil horizon following capillary movement of the water, which evaporates and leaves the crystalline form of the salt, which is often indicated by a white crust. Elevated concentrations of the salts can be lethal to many microorganisms. Assessing the feasibility of biodegradation in relation to salinity is achieved by measuring electrical conductivity (deciSiemens/meter, (dS/m)), which is a general measure of soil salinity. At conductivity values above 1 dS/m, biological growth is hindered, and values above 6 dS/m generally indicate that the soil is sterile.

Finally, the soil should be analyzed for heavy metal content since a high metal concentration could be toxic to microbial survival and growth. Metals in the soil are difficult (if not impossible) to remediate by indigent soil bacteria. Therefore, if the heavy metal concentrations in soil exceed the acceptable residual levels as determined by federal and state regulations (http://www.cleanuplevels.com/), bioremediation is not a viable option.

7.1.1.4 Biological Properties

The action of indigenous bacteria in the soil can account for up to 80% of waste degradation in soil, the remainder being due to evaporation, photo-oxidation, and solubilization in water. This is true as long as environmental conditions such as the presence of oxygen, adequate moisture, moderate temperatures, neutral pH, low to moderate salinity, and excess nutrients, are present to allow bacteria can grow exponentially. In addition, the spilled material should not be excessive insofar as it overwhelms the indigenous microbes.

While bacteria which degrade hydrocarbon derivatives are ubiquitous, the populations of hydrocarbon-degrading bacteria exposed to hydrocarbon derivatives increase rapidly when given adequate aeration, moisture, favorable pH, and excess nutrients.

Typically, hydrocarbon-degrading bacteria are found in the range of 10^5 to 10^6 bacteria per gram of soil under no oil spill conditions, and when exposed to crude oil, that number increases to 10^6 to 10^8 bacteria per gram of soil. For reference, one gram of rich agricultural soil generally contains 2.5×10^9 bacteria (heterotrophic count), 5×10^5 fungi, 5×10^4 algae, and 3×10^4 protozoa (US EPA, 2003). Soil samples should be analyzed for enumeration of both heterotrophic and hydrocarbon-utilizing bacteria population to verify population densities. The population of microorganisms could be assessed in soil by plate count, most probable number technique, phospholipid fatty acid analysis, or denaturing gradient gel electrophoresis (Zhu et al., 2001; Zhu et al., 2004). There are several methods that can be applied to the bioremediation of crude oil-contaminated sites (Chapter 5) but it is necessary to note that there is no single species of bacteria that can metabolize all the components of crude oil.

However, the nature of the crude oil-related spill and the character of the site may require addition of non-indigenous microorganisms to enhance the native populations. Thus, whenever a spill has occurred and bioremediation is the best means to approach the cleanup after other options (Speight, 1996; Speight and Lee, 2000; Speight, 2005) have been dismissed, investigators should, after suitable test methods have been applied to determine the physical and chemical character of the soil, consider one of two options which include (1) bioaugmentation and (2) biostimulation.

In the *bioaugmentation approach* (Chapter 1), crude oil-degrading bacteria are added to the existing bacterial population in the soil to increase the rate of oil consumption. In the *biostimulation approach* (Chapter 1), nutrients are added and optimization of environmental conditions to improve the biodegradation efficiency of indigenous bacteria is practiced.

The nutrients nitrogen, phosphorus, and potassium (N, P, and K) are normally added during land treatment in order to enhance microbial activities, which decompose carbon (C) compounds in the soil. Nitrogen, when added through the ammonium salts, can be toxic to microorganisms due to the possibility of generation of ammonia in the soil; the ammonium ion can also promote the increase of oxygen demand. A commonly used strategy is to add nutrients that provide a stoichiometric ratio of carbon/nitrogen/phosphorus/potassium of 100:5:1:1. However, when nitrogen and phosphorus are introduced in excessive quantities biodegradation can be inhibited (Trindade et al., 2002).

For optimum biodegradation, nutrients can be added to the soil using organic or inorganic fertilizers, and their concentration should be closely monitored and supplemented as they are depleted during the biodegradation nitrate and process. Agriculture fertilizers such as ammonium nitrate, urea, ammonium phosphate, and potassium phosphate may be added to increase nutrient concentrations in the soil.

Fertilizers should be added gradually to the soil to minimize pH changes. The amount and frequency of fertilizer addition depend upon field conditions. However, evidence from documented land farming has shown that an appropriate fertilizer dosage that could be repeated, depending upon field conditions, are 500 pounds of nitrogen per acre or 1,100 pounds of urea or 1,500 pounds of ammonium nitrate per acre and 250 pounds of phosphorus per acre (McMillen et al., May 2002).

Organic amendments such as wood chips, sawdust, straw, hay, and animal manure are used to improve soil structure and oxygen infiltration, and to increase the moisture-holding capacity of sandy soils. In general, animal manure should be applied at the rate of about 3 to 4% by weight of soil and should be analyzed for nitrogen and phosphorus before its application. Bulking agents like hay, palm husks, rice hulls, and straw are added to clayey soils to increase pore space and hence, air exchange. The bulking agent should be blended into the soil until a porous structure is obtained and visual evidence of oil is eliminated. A general rule to follow is to add hay in contaminated media is 5 standard hay bales per 1,000 square feet of impacted soils. However, the source of bulking agent should be checked for residual substances (such as pesticides or heavy metals) for toxicity.

7.1.1.5 Temperature

Temperature plays a significant role in controlling the nature and the extent of microbial hydrocarbon metabolism, which is of special significance for *in situ* bioremediation (Margesin and Schinner, 2001a). Bioavailability and solubility of less soluble hydrophobic substances, such as aliphatic and polyaromatic hydrocarbon derivatives, are temperature dependent. Increase in temperature affects a decrease in viscosity while a decrease in temperature affects an increase in viscosity, thereby affecting the degree of distribution, and an increase (decrease) in diffusion rates of organic compounds. Therefore, higher reaction rates due to smaller boundary layers are expected at elevated temperatures. The increased (decreased) volatilization and solubility of some hydrocarbon derivatives at elevated (low) temperature affects toxicity and allows bio-transformations with high (low) substrate concentrations (Müller et al., 1998; Whyte et al., 1998; Niehaus et al., 1999).

Cold habitats possess sufficient indigenous microorganisms, psychrotrophic bacteria being predominant, which adapt rapidly to the contamination, as demonstrated by significantly increased numbers of oil degraders shortly after a pollution event. In fact, cold-adapted, certain types of (psychrophilic and psychrotrophic) microorganisms that have been adapted to cold temperature are able to grow at temperatures around 0°C (32°F). They are widely distributed in nature since a large part of the Earth's biosphere is at temperatures below 5°C (Margesin and Schinner, 1999a, 1999b). Psychrophilic micro-organisms (organisms that are capable of growth and reproduction in cold temperatures) have an optimum growth temperature of approximately 15°C (59°F) and do not grow above 20°C (68°F), whereas psychrotrophs (cold-tolerant) have optimum and maximum growth temperatures above 15°C (59°F) and 20°C (68°F), respectively (Morita, 1975). Cold-adapted indigenous microorganisms play a significant role in the *in situ* biodegradation of hydrocarbon derivatives in cold environments, where ambient temperatures often coincide with their growth temperature range.

Most studies showed that the indigenous microbial populations degrade hydrocarbon derivatives more efficiently than the introduced strains introduced in an attempt at bioaugmentation (Margesin and Schinner, 2001a). However, bioaugmentation may result in a shorter hydrocarbon acclimation period. Inoculation of contaminated Arctic soils with consortia (Whyte et al., 1999) or with alkane-degrading *Rhodococcus* sp. (Whyte et al., 1998) decreased the induction period and increased the rate of C_{16} mineralization at 5°C (41°F). However, nitrogen-phosphorus-potassium fertilization alone had a comparable effect on hydrocarbon loss like fertilization plus bioaugmentation. This has been shown both in chronically oil-polluted Arctic soil (Whyte et al., 1999) and in artificially diesel oil-contaminated alpine soils (Margesin and Schinner, 1997a, 1997b).

The decreased bioavailability of the long-chain alkane derivatives at low temperature (many form crystals at 0°C, 32°F) may be responsible for their increased recalcitrance, which affects *in situ* bioremediation in cold climates. The application of cold-active solubilizing agents could be useful for enhancing hydrocarbon bioavailability. For example, Antarctic marine bacteria produced bio-emulsifiers when grown with n-alkane derivatives as the sole carbon source (Yakimov et al., 1999; Chugunov et al., 2000).

Benzene, toluene, ethyl benzene, xylene, and polynuclear aromatic hydrocarbon derivatives are frequent soil contaminants but little is known about biodegradation of polynuclear aromatic hydrocarbon derivatives in cold climates. Cold-tolerant isolates (*Sphingomonas, Pseudomonas* spp.) from oil-polluted Antarctic soils utilized as benzene, toluene, ethyl benzene, xylene, naphthalene, phenanthrene, and fluorene as the sole carbon and energy source (Aislabie et al., 1998, 2000). In a field study on the biodegradation of dispersed crude oil in cold and icy seawater (−1.8 to 5.5°C, 19 to 42°F), half-life times of polynuclear aromatic hydrocarbon derivatives ranged from two days (naphthalene) to eight days (phenanthrene) at a temperature above 0°C (32°F) and with effective chemical dispersion (Siron et al., 1995).

On the other hand, cold climates are not the only extreme environments and a higher-temperature environment is also classed as an extreme environment. Microorganisms that grow optimally above 40°C are designated as thermophiles—most thermophiles known are moderate and show an upper temperature border of growth between 50 and 70°C (122 to 158°F). Another class of thermophiles, the hyperthermophiles experience optimal growth hyperthermophiles occurs at temperatures in excess of 80°C (176°F) (Stetter, 1998). Thermophiles, predominantly bacilli, possess a substantial potential for the conversion of environmental pollutants, including all major classes (Müller et al. 1998).

7.1.1.6 Acidity and Alkalinity

The acidity and alkalinity (determined as the pH of a site) because of the effects on physical and chemical properties of the site. In terms of the pH of the soil, the results of soil pH are reported on a logarithmic scale; a soil with a pH of 6 is 10 times more acidic than a soil with a pH of 7, and a pH of 5 is 100 times more acidic than a pH of 7. As an example, crops typically grow best when pH is between 6 (slightly acidic) and 7.5 (slightly alkaline).

In the current context of bioremediation, the mineralization of hydrocarbon derivatives is favored by near neutral pH values (pH = 6 to 8) and it is an established practice to add lime to bioremediate acid soils containing hydrocarbon pollutants (Alexander, 1999). However, biodegradation has been reported to proceed in aquifers where the acidity is obvious (pH = < 6) (Norris, 1994; Amadi et al., 1996).

Acidophiles (organisms that have their optimum growth rate at least 2 pH units below neutrality) are metabolically active in highly acidic environments, and often have a high heavy metal resistance (Norris and Johnson, 1998). Heterotrophic acidophilic microorganisms are an interesting potential for the bioremediation of acidic environments that contain both heavy metals and organic compounds, such as crude oil-polluted acidic drainage waters (Stapleton et al., 1998).

Alkaliphiles (organisms that have their optimum growth rate at least 2 pH units above neutrality) are able to grow or survive at pH values above 9, but their optimum growth rate is around neutrality or less (Kroll, 1990).

7.1.1.7 Salinity

The salinity if soil or water is the quality or condition of being salty (the degree of saltiness of the soil or water). Ocean currents are driven to move by differences in the temperature and salinity of the water (i.e., a measurement of the amount of salt in a given solution) and adding fresh water to a marsh lowers salinity (Mille et al., 1991; Ulrich et al., 2009).

Microorganisms requiring salt for growth are referred to as halophiles, whereas microorganisms that are able to grow in the absence as well as in the presence of salt are designated halo-tolerant (Kushner, 1978; Grant et al., 1998). A range of organic pollutants has been shown to be mineralized or transformed by microorganisms able to grow in the presence of salt (Oren et al., 1992; Margesin and Schinner, 2001b).

Halophilic archaea maintain an osmotic balance with the hypersaline environment by accumulating high salt concentrations, which requires salt adaptation of the intracellular enzymes. Eubacteria are more promising degraders than archaea as they have a much greater metabolic diversity. Their intracellular salt concentration is low, and their enzymes involved in biodegradation may be conventional (i.e., not salt-requiring) enzymes similar to those of non-halophiles (Oren et al., 1992).

An inverse relationship between biodegradation of crude oil hydrocarbon derivatives and salinity has been assumed because enrichment cultures from the Great Salt Lake were not able to grow on mineral oil and to mineralize hexadecane in the presence of salt concentrations above 20% w/v (Ward and Brock, 1978; Whitehouse, 1984; Means, 1995). The inhibitory effect of salinity was found to be greater for the biodegradation of aromatic and polar fractions than of the saturated fraction of crude oil hydrocarbon derivatives (Mille et al., 1991) and there are reports of about microorganisms able to oxidize crude oil hydrocarbon derivatives even in the presence of 30% w/v sodium chloride. Among such microorganisms are crude oil-degrading *Streptomyces albaxialis* (Kuznetsov et al., 1992), and an alkane-degrading member of the *Halobacterium* group (Kulichevskaya et al., 1992).

In terms of the bioremediation of crude oil-contaminated soil, an inhibitory effect of artificial salinity on mineralization of used lubricating oil has been reported (Rhykerd et al., 1995). Thus, the removal of salt from oil-contaminated soils may reduce the time required for bioremediation. However, different results may be obtained when investigating naturally salt-containing soils, since indigenous microorganisms in such environments are expected to be salt-adapted. In fact, seasonal variations of the mineralization potential of crude oil (520 g kg^{-1} soil) in soil samples from salt marshes has been reported to be high (Jackson and Pardue, 1997).

7.1.2 Water Ecosystems

In a water ecosystem a spill of crude oil or a crude oil product represents the introduction of a non-aqueous phase liquid composed of a large amount of organic matrices in addition to other toxic constituent, such as polynuclear aromatic hydrocarbon derivatives as occur in resin constituents and asphaltene constituents.

Obviously, there are many different processes acting on a marine oil spill, modifying its physical and chemical characteristics, such as biodegradation, scattering, evaporation, dissolution, dispersion, sedimentation and photo-oxidation (Figure 7.1) (Prince, 1993). These processes occur simultaneously but at different speeds, depending on the physical and chemical properties of the oil and the ambient conditions, like the environmental temperature and energy. On several occasions, however, these processes can be so slow that it is necessary to interfere in order to hasten degradation of the pollutants (Rosa and Triguis, 2007).

Oil type, weather, wind and wave conditions, as well as air and sea temperature, all play important roles in ultimate fate of spilled oil in marine environment. After oil is discharged in the environment, a wide variety of physical, chemical and biological processes begin to transform the discharged oil. These chemical and physical processes are collectively called weathering and act to change the composition, behavior, route of exposure and toxicity of discharged oil. The weathering processes are described as follows:

- *Spreading and Advection*: Spreading, which dominates the initial stages of the spill and involves the whole oil and is the movement of the entire oil slick horizontally on the surface of water due

Biodegradation

General site characteristics
Indigenous microbes
Non-indigenous microbes
Temperature
Weather
Soil characteristics
Moisture
Irrigation/tilling (air exchange)
Nutrients
Fertilizers
Toxicity
Heavy metals
Salinity
Acidity-alkalinity (pH

Figure 7.1: Factors Requiring Assessment Prior to and During the Bioremediation of Petroleum-Contaminated Sites (US EPA, 2003).

to effects of gravity, inertia, friction, viscosity and surface tension—on calm water spreading occurs in a circular pattern outward from the center of the release point. Advection is the movement of oil due to overlying winds or underlying currents, which increase the surface area of the oil, thereby increasing its exposure to air, sunlight and underlying water. The advection effects are not uniform and do not affect the chemical composition of the oil.

- *Evaporation*: Evaporation, which is subject to atmospheric conditions such as temperature, is the preferential transfer of lower-boiling constituents of the crude oil-related spill from the liquid phase to the vapor phase. The chemical composition of the spilled material is physically altered and although the volume decreases, the remaining components have higher viscosity and specific gravity leading to thickening of the slick and formation of entities such as tar balls.

- *Dissolution*: Dissolution is the transfer of constituents from the spilled material into the aqueous phase. Once dissolved these constituents are bioavailable and if exposed to marine life can cause environmental impacts and injuries. The largest concentration is found near the surface or the release point and hence the effect on marine life may be localized.

- *Natural Dispersion*: Natural dispersion is the process of forming small oil droplets that become incorporated in the water in the form of a dilute oil-in-water suspension and occurs when breaking waves mix the oil in the water column. This phenomenon reduces the volume at the spill surface but does not change the physicochemical properties of the spilled material. It is estimated to range from 10 to 60% per day for first three days of the spill, depending on the condition of the ecosystem and may be independent of the type of crude oil or crude oil product.

- *Emulsification*: Emulsification is the mixing of sea water droplets into oil spilled on water surface forming a water-in-oil emulsion.

- *Photo-oxidation*: Photo-oxidation occurs when sunlight in the presence of oxygen transforms hydrocarbon derivatives (by increasing the oxygen content of a hydrocarbon) into new by-products. This results in changing in the interfacial properties of the oil, affecting spreading and emulsion formation, and may result in transfer of toxic by-products into the water column due to the by-product's enhanced water solubility.

- *Sedimentation and Shoreline Stranding*: Sedimentation is the incorporation of oil within both bottom and suspended sediments. Shoreline stranding is the visible accumulation of crude oil along the water's edge (shoreline) following a spill. It is affected by proximity of the spill to the affected shoreline, intensity of current and wave action on the affected shoreline and persistence of the spilled product. Sedimentation can begin immediately after the spill but increases and peaks after several weeks, whereas shoreline stranding is a function of the distance of the shoreline from the spill and chemical nature of spilled oil.

- *Biodegradation*: Biodegradation occurs on the water surface, in the water, in the sediments, and at the shoreline. It can begin (depending on the lag time or induction period) after a spill and can continue as long as degradable hydrocarbon derivatives are present.

Weathering processes occur simultaneously—one process does not stop before the other begins. The order in which these processes presented is instantaneous, the relative significance of these processes may change if the spill occurred below the water surface or in tropical or ice conditions. The spill chronology may also vary if the spill is near the shoreline in which case it can contaminate the soil and groundwater even before the weathering or cleanup processes start. Also there are many onshore and offshore operations in a crude oil industry that can cause soil pollution and aquifer contamination.

Aquifer contamination can also take place because of the migration of the oil through the porous media and its subsequent adsorption on the rock surface. Many technologies have been proposed for the treatment of oil contaminated sites; these can be performed by two basic processes *in situ* and *ex situ* treatment using different cleaning technologies such as thermal treatment, biological treatment, chemical extraction and soil washing, and aeration accumulation techniques (Speight, 1996; Speight and Lee, 2000).

7.1.2.1 Biodegradation

As already stated (Chapter 5), biodegradation is the degradation (breakdown) of the contaminants into products that are environmentally acceptable products (examples of such products are water, carbon dioxide, and biomass) by the action of naturally available microorganisms under normal environmental conditions. Related to the current context of the biodegradation of crude oil and crude oil products, the degradation of hydrocarbon derivatives generally follow the order (Perry, 1984):

n-Alkanes > branched alkanes > low molecular weight aromatics > cyclic alkanes

The process is usually aerobic, requiring terminal or sub-terminal oxidation of the alkane derivatives (Harayama et al., 1999), while aromatic hydrocarbon ring structures are broken through hydroxylation and carboxylation processes (Cerniglia, 1992). Hydrocarbon biodegradation in water is associated both with water-soluble or -miscible compounds, and with the oil-water interphases, mainly on droplets and thin oil films with high surface/volume ratio (Bartha and Atlas, 1987; Button et al., 1992; Floodgate, 1984). A significant number of studies have shown that it is difficult to predict the extent and rates of hydrocarbon degradation processes in marine environment, due to the many factors involved in the process (Leahy and Colwell, 1990; Atlas and Bartha, 1992; Margesin and Schinner, 1999).

Biodegradation rates of polynuclear aromatic hydrocarbon derivatives (PAHs) in spilled oil stranded on tidal flats were evaluated using model reactors to clarify the effects of non-aqueous phase liquid (NAPL) on the biodegradation of polynuclear aromatic hydrocarbon derivatives in stranded oil on tidal flat with special emphasis on the relationship between dissolution rates of polynuclear aromatic hydrocarbon derivatives into water and viscosity of the non-aqueous phase liquid. Biodegradation of polynuclear aromatic hydrocarbon derivatives in non-aqueous phase liquids was limited by the dissolution rates of polynuclear aromatic hydrocarbon derivatives into water. Biodegradation rate of chrysene was smaller than that of acenaphthene and phenanthrene due to the smaller dissolution rates. Dissolution rates of polynuclear aromatic hydrocarbon derivatives in heavy fuel oil are typically lower than the dissolution rates of polynuclear aromatic hydrocarbon derivatives crude oil because of the higher viscosity of heavy fuel oil—the solubility of the fuel oil constituents (i.e., penetration of the solvent into the heavy fuel oil) is more than likely diffusion controlled. Hence, biodegradation rates of polynuclear aromatic hydrocarbon derivatives in heavy fuel oil are lower than the biodegradation rates of crude oil.

Biodegradation rates of polynuclear aromatic hydrocarbon derivatives in non-aqueous phase liquids with slow rate of decrease like fuel oil C was slower than those in non-aqueous phase liquids with rapid rate of decrease like crude oil. The smaller rate of decrease of fuel oil C than crude oil was due to the higher viscosity of fuel oil C. Therefore, not only the dissolution rate of polynuclear aromatic hydrocarbon derivatives but also the rates of decrease of non-aqueous phase liquids were important factors for the biodegradation of polynuclear aromatic hydrocarbon derivatives (Kose et al., 2003).

Generally, the data from the above study (Kose et al., 2003) show that the biodegradation of polynuclear aromatic hydrocarbon derivatives in non-aqueous phase liquids was limited by the dissolution rates of polynuclear aromatic hydrocarbon derivatives into water.

7.1.2.2 Bioremediation

The conditions of the contaminated area plays a major role on whether bioremediation is the appropriate method of cleanup for the given oil spill. The success of bioremediation is dependent upon physical conditions and chemical conditions. Physical parameters include temperature, surface area of the oil, and the energy of the water. Chemical parameters include oxygen and nutrient content, pH, and the composition of the oil. Temperature affects bioremediation by changing the properties of the oil and also by influencing the crude oil-degrading microbes (Nedwell, 1999). When the temperature is lowered, the viscosity of the crude oil is increased which changes the toxicity and solubility of the oil, depending upon its composition (Zhu et al., 2001). Temperature also has an effect on the growth rate of the microorganisms, as well as the degradation rate of the hydrocarbon derivatives, depending upon their characteristics.

Biostimulation, the addition of nutrients, is practiced for cleanup of crude oil spills in seawater when there is an existing population of oil degrading microbes present. When an oil spill occurs, the result is a large increase in carbon and this also stimulates the growth of the indigenous crude oil-degrading microorganisms. However, these microorganisms are limited in the amount of growth and remediation that can occur by the amount of available nitrogen and phosphorus.

By adding these supplemental nutrients in the proper concentrations, the hydrocarbon degrading microbes are capable of achieving their maximum growth rate and hence the maximum rate of pollutant uptake. It has been found that when using nitrogen for the supplemental nutrient, a maximum growth rate is achieved by the oil degrading microorganisms (Boufadel et al., 2006). Biostimulation has been proven to be an effective way of achieving increased hydrocarbon degradation by the indigenous microbial population (Coulon et al., 2006).

Supplemental nutrients tended to move downward during rising tides and seaward during falling tides (Boufadel et al., 2006). This is very useful information in determining the proper timing to add nutrients in order to allow for the maximum residence time of the nutrients in the contaminated areas. The results of this experiment concluded that the nutrients should be applied during low tide at the high tide line, which results in maximum contact time of the nutrients with the oil and hydrocarbon degrading microorganisms.

Waves also have an effect on the distribution and movement of the water and dissolved nutrients, which determine the residence time of the nutrients in the oil affected area. The role of waves on solute movement varies whether there is a tide or not. When Research performed by, showed that when a wave is present there is a sharp seaward hydraulic gradient in the backwash zone, and a gentle gradient landward of this area (Boufadel et al., 2007). Furthermore, the contact time of the nutrients is increased when the waves break seaward of their location and the waves increased the dispersion and washout of the nutrients in the tidal zone, and residence time was approximately 75% when a wave was present with a tide, as compared to a tide with no waves.

The marine environment encompassing the vast majority of earth's surface is a repertoire of a large number of microorganisms. The environmental roles of the biosurfactants produced by many such marine microorganisms have been reported earlier (Poremba et al., 1991; Schulz et al., 1991; Abraham et al., 1998; Das et al., 2008).

The rate and extent of microbial biodegradation of crude oil-related spilled material are dependent on several factors.

7.1.2.3 Temperature

As for any chemical reaction, increasing the temperature increases the rate of degradation of crude oil constituents and the constituents of crude oil products by bacteria up to the point where the activity of the microbes decreases because of an excessively high temperature (Nedwell, 1999). However, increasing the concentration of the crude oil or the crude oil product generally decreases the rate of bacterial oil degradation. Also, salinity plays a major role on the acceleration of biodegradation process of crude oil and crude oil products (Siddique et al., 2002). Also, in an aqueous ecosystem, the water temperature affects the physical and chemical properties of the crude oil or the crude oil product and the rate of biodegradation—colder temperatures slow the rate of biodegradation and warmer temperatures increase the rate of biodegradation.

The ubiquitous distribution of crude oil-degrading bacteria in the various ecosystems has already been reported (Leahy and Colwell, 1990). Before contamination, a count indicated that hydrocarbon-degrading microorganisms comprised less than 0.01% of the total number of bacteria but represented 10% of the cultivable heterotrophic microorganisms. These results are consistent with those that indicate that hydrocarbon-degrading microorganisms comprised 1–10% of the total number of saprophytic bacteria in marine bacterial communities (Wright et al., 1997).

In contrast to the results reported by others (Kennicutt, 1990), the introduction of oil into oil-free Antarctic seawater resulted in enrichment in hydrocarbon-degrading microorganisms by several orders of magnitude within a few days. Such a large enhancement of specific micro-organisms has been previously reported (Delille and Vaillant, 1990; Wagner-Döbler et al., 1998; Delille and Delille, 2000).

The addition of a fertilizer had an immediate stimulation effect on both heterotrophic and hydrocarbon-degrading microorganisms. Several studies have reported favorable effects of fertilizers on oil biodegradation at low temperatures in Arctic (Braddock et al., 1997; Whyte et al., 1998, 1999; Mohn et al., 2001), alpine (Margesin and Schinner, 1997a, 1997b; Margesin and Schinner, 2001a; Margesin et al., 2007a) and Antarctic soils (Kerry, 1993; Wardell, 1995; Aislabie et al., 1998, 2006; Delille, 2000; Ferguson et al., 2003; Ruberto et al., 2003; Powell et al., 2006).

Temperature can affect the physical and chemical characteristics of the discharged oils, hydrocarbon derivatives with high pour points could be expected to show more temperature-related biodegradation than contaminants with low pour points. However, globally low temperatures have been reported to play a significant role in controlling the nature and extent of microbial hydrocarbon metabolism (Nedwell, 1999; Gerdes et al., 2005). Mineralization of isotopically-marked (^{14}C) hexadecane has been observed in Arctic soil at 5°C (41°F), but mineralization occurred faster and to a greater extent at 23°C (74°F).

Furthermore, and not surprisingly, during a 2-month-long microcosms experiment in seawater, biodegradation efficiency was higher at 20°C (68°F) than at 4°C (39°F) for two psychrotrophic Antarctic marine isolated from Terra Nova Bay (Ross Sea) (Michaud et al., 2004). However, Mohn and Stewart (2000) reported that a change in incubation temperature from 7 to 22°C (13 to 72°F) did not affect the extent of mineralization of isotopically-marked (^{14}C) dodecane. The behavior of isolated strains do not necessarily parallel the one of a much more complex natural consortium as environmental factors characterizing natural polar ecosystems are impossible to replicate in laboratory.

Other studies indicated that heterotrophic and hydrocarbon-degrading bacteria counts are often similar in cold and warm climates (Aislabie, 1997; Eckford et al., 2002). These results seem in conflict with the typical temperature-related assumption predicting an increase of microbial metabolism with a temperature increase (Leahy and Colwell, 1990; Bossert and Bartha, 1984). According to the Arrhenius equation, any decrease in temperature should cause an exponential decrease of the reaction rates, the magnitude of which depending on the value of the activation energy (Margesin et al., 2007b).

Low temperatures may affect the utilization of substrates comprising a mixture of hydrocarbon derivatives (Atlas, 1986; Baraniecki et al., 2002). Altered growth responses might result from beneficial changes to membrane fluidity. The production and/or activity of cold-active enzymes are usually significantly below the "optimal" growth temperature (as determined from growth rate) of the enzyme producer, which reflects the thermal characteristics of the secretion process (Margesin et al., 2007b).

In another investigation (Coulon, 2006), it was found that when crude oil was added to the test the presence of crude oil-degrading microbes increased by two orders of magnitude at 4°C (39°F), more than three orders of magnitude at 12°C (54°F), and more than four orders of magnitude at 20°C (68°F). The surface area of the oil is also a significant parameter in the success of bioremediation because the growth of oil degrading microorganisms occurs at the interface of the water and oil. The larger the surface area of the oil results in a larger area for growth and hence larger numbers of microbes. The energy of the water is important because rough waters will disperse and dilute essential nutrients for the microorganisms and also spread the oil, contaminating more areas.

A more recent investigation (Delille et al., 2009) provides further evidence of the high potential of indigenous Antarctic bacterial communities for bioremediation action even at low temperatures. Little difference in data obtained under three incubation temperatures and with two different concentrations of oil is clearly indicating that temperature had only a rather limited influence on crude oil degradation in the studied Antarctic seawater. This point is important when considering bioremediation as an efficient means to clean up contaminated soils in remote polar locations.

7.1.2.4 Effect of Oxygen

From a microbial perspective, there are two methods for biodegradation which are (1) aerobic in which organisms use oxygen as part of the respiration for consumption of nutrients and (2) anaerobic in which organisms use other elements such as sulfur, in the process of respiration and consumption of nutrients.

Thus, aerobic biodegradation is the breakdown of crude oil contaminants by microorganisms when oxygen is present. More specifically, the term refers to occurring or living only in the presence of oxygen; therefore, the chemistry of the system, environment, or organism is characterized by oxidative conditions. Thus, while aerobic microorganisms use oxidative reactions, the degradation by anaerobic bacteria takes place by reductive types of reactions (reductive conversions) (Liu et al., 2017).

In open water, crude oil hydrocarbon derivatives and the hydrocarbon derivatives in crude oil products undergo aerobic biodegradation by bacteria that use oxygen dissolved in the water. Sediment on the ocean floor and along the coast is, for the most part, *anoxic*: it does not contain oxygen. Hydrocarbon derivatives that settle into sediments on the ocean floor and along coasts undergo anaerobic biodegradation, a much slower process. Therefore, onshore oil lingers longer than oil at sea and can become a chronic pollutant.

7.1.2.5 Effect of Nutrients

The addition of a nutrient (such as fertilizer) during the bioremediation has been proved very effective and essential to stimulate microbial growth as well as contaminates removal. However, inhibitory effects have also been observed when concentration of nutrients was excessive, mostly due to their toxicity to microbial species.

In addition to carbon and oxygen, bacteria need nitrogen and phosphorus to survive. These nutrients are found naturally in the ocean environment. Nitrogen and phosphorus-based fertilizers from farms and gardens on land also enter marine waters through storm water runoff.

Effective bioremediation requires nutrients to remain in contact with the oiled material, and the concentrations should be sufficient to support the maximal growth rate of the oil-degrading bacteria throughout the cleanup operation. Low nutrient concentrations reduce the rate of biodegradation, whereas high nutrient concentration may be toxic for marine biota and cause eutrophication and red tide. Hydrocarbon-utilizing microorganisms are ubiquitously distributed in marine ecosystems following a spill.

7.1.2.5.1 Marine Environments

With respect to the marine environment (see Section 7.1.2.10), contamination of coastal areas by oil from offshore spills usually occurs in the intertidal zone where the washout of dissolved nutrients can be extremely rapid. Oleophilic and slow-release formulations have been developed to maintain nutrients in contact with the oil, but most of these rely on dissolution of the nutrients into the aqueous phase before they can be used by hydrocarbon degraders. Therefore, design of effective oil bioremediation strategies and nutrient delivery systems requires an understanding of the transport of dissolved nutrients in the intertidal zone.

Transport through the porous matrix of a marine beach is driven by a combination of tides, waves, and flow of freshwater from coastal aquifers. Tidal influences cause the groundwater elevation in the beach and the resulting hydraulic gradients to fluctuate rapidly. Wave activity affects groundwater flow through two main mechanisms. First, when waves run up the beach face ahead of the tide, some of the water percolates vertically through the sand above the water line and flows horizontally when it reaches the water table. Waves can also affect groundwater movement in the submerged areas of beaches by a pumping mechanism that is driven by differences in head between wave crests and troughs.

Nutrients that are released from slow-release or oleophilic formulations will probably behave similarly to the dissolved lithium tracer that was used in the study. Thus, they will not be effective on high-energy beaches unless the release rate is high enough to achieve adequate nutrient concentrations while the tide is out. Subsurface application of nutrients might be more effective on high-energy beaches. Since crude oil does not penetrate deeply into most beach matrices, however, nutrients must be present near the beach surface to effectively stimulate bioremediation. Since nutrients move downward and seaward during transport through the intertidal zone of sandy beaches, nutrient application strategies that rely on subsurface introduction must provide some mechanism for insuring that the nutrients reach the oil-contaminated area near the surface.

7.1.2.5.2 Freshwater Environments

With respect to freshwater shorelines, an oil spill is most likely to have the greatest impact on wetlands or marshes rather than a wide shoreline zone like a marine intertidal zone.

However, the same principles apply to this type of environment as a marine environment, namely, that nutrients must be maintained in contact with the degrading populations for a sufficient period of time to effect the enhanced treatment. There is an added complication in a wetland, however. Oil penetration is expected to be much lower than on a porous sandy marine beach.

Below only a few centimeters of depth, the environment becomes anaerobic, and crude oil biodegradation is likely to be much slower even in the presence of an adequate supply of nitrogen and phosphorus. Technology for increasing the oxygen concentration in such an environment is still undeveloped, other than reliance on the wetland plants themselves to pump oxygen down to the rhizosphere through the root system.

7.1.2.5.3 Soil Environments

Land-farming techniques for treating oil spills on soil have been used extensively for years by crude oil companies and researchers. Again, the same principles apply: maintenance of an adequate supply of limiting nutrients and electron acceptors (nitrogen, phosphorus, and oxygen) in contact with the degrading populations throughout the entire treatment period. For surface contamination, maintenance of an adequate supply of oxygen is accomplished by tilling. The maximum tilling depth is limited to about 15 to 20 inches, however. If the contamination zone is deeper, other types of technologies would have to be used, such as bioventing, composting, or use of biopiles, all of which require addition of an external supply of forced air aeration.

In conclusion, bioremediation is a proven alternative treatment tool that can be used to treat certain aerobic oil-contaminated environments. Typically, it is used as a polishing step after conventional mechanical cleanup options have been applied. It is a relatively slow process, requiring weeks to months to effect cleanup. If done properly, it can be very cost-effective, although an in-depth economic analysis has not been conducted to date. It has the advantage that the toxic hydrocarbon compounds are destroyed rather than simply moved to another environment.

In order to determine whether bioavailability limits the biodegradability of crude oil hydrocarbon derivatives in aged soils, both the biodegradation and abiotic desorption rates of polynuclear aromatics hydrocarbon derivatives have been measured at various time points in six different aged soils undergoing slurry bioremediation treatment.

Alkane biodegradation are inevitably greater than the respective desorption rates, indicating that these saturated hydrocarbon derivatives apparently do not need to be dissolved into the aqueous phase prior to metabolism by soil microorganisms. The biodegradation of polynuclear aromatics hydrocarbon derivatives was generally not mass-transfer rate limited during the initial phase, while it often became so at the end of the treatment period when biodegradation rates equaled abiotic desorption rates. However, in all cases where polynuclear aromatics hydrocarbon biodegradation was not observed or polynuclear aromatics hydrocarbon removal temporarily stalled, bioavailability limitations were not deemed responsible, since the polynuclear aromatics hydrocarbon derivatives desorbed rapidly from the soil into the aqueous phase. Consequently, aged polynuclear aromatics hydrocarbon derivatives that are often thought to be recalcitrant due to bioavailability limitations may not be so and therefore may pose a great risk to environmental receptors.

7.1.2.6 Effect of Spill Characteristics

The type of oil, its concentration and the types of hydrocarbon derivatives it contains influence the rate of biodegradation. The Gulf spill released light, sweet crude oil, which is more readily broken down than heavy, sour oil. *Mousse*, tar balls, or oil slicks that wash onshore are concentrated compared to dispersed oil, more protected from wind and wave action than oil in open water and have less surface area for microbes to access. Smaller droplets of oil are more biodegradable.

7.1.2.7 Effect of Prior Exposure

Microbes adapt to gradual exposure to oil. The more oil a microbial community has been exposed to in the past, the greater its capacity and availability to biodegrade oil in the future. In one study, microbes from sediments previously contaminated with oil were able to metabolize oil 10 to 400 times faster than those from sediments that had never been contaminated. Once a species of bacteria is exposed to oil and metabolizes it, the next generations inherit that ability, a concept known as *genetic adaptation*. This has been studied in a particular species of *Vibrio* in the northwestern Gulf of Mexico.

7.1.2.8 Effect of Dispersants

A dispersant or a dispersing agent is a substance (typically a surfactant-type chemical) that is added to a suspension of solid or liquid particles in a liquid to improve the separation of the particles and to prevent their settling or clumping.

In the present context, a dispersants is a chemical that is sprayed on a surface of the spill of crude oil (or a crude oil product) to break down the oil into smaller droplets that more readily mix with the water. Dispersants do not reduce the amount of the crude oil (or the crude oil product) entering the environment but the dispersants do tend push the effects of the spill underwater.

Biodegradation occurs more readily when oil droplets are dispersed. Chemical dispersants break oil into smaller droplets, which increases the surface area available for bacteria to access. The exact effects of dispersants such as Corexit on the rate of biodegradation are unknown, especially in deep water. The effectiveness of dispersants also depends on the type and consistency of oil and the oil-dispersant ratio.

7.1.2.9 Effect of Flowing Water

Oil discharges generate a dynamic situation on the oil–water interphases, caused by a variety of physical and chemical processes, including wind, currents, oil weathering, film generation, and oil dispersion. Thus, immiscible oil surfaces are constantly washed with seawater, and new microbes are continuously contacting the oil film or droplet surfaces. Simulation of these dynamic conditions may be achieved rather in flow-through systems than by closed static experiments. Various controlled flow-through systems have been described which simulate biodegradation under marine conditions, including microcosms (Swannell and Daniel, 1999; MacNaughton et al., 2003; Röling et al., 2002; Xu et al., 2003), and large tank-based mesocosms (Wade and Quinn, 1979; Siron et al., 1993; Santas and Santas, 2000; Yamada et al., 2003). However, since dispersed oil was used in most of these systems it was impossible to differentiate between HC processes in the oil and water phases.

In more recent studies with immobilized oil in flow-through systems may give valuable contributions to the determination and understanding of various processes after oil discharges to aquatic environments (Brakstad et al., 2004). For instance, hydrocarbon dissolution and biodegradation from oil films, and the potential impacts in the water column, may be predicted after an oil spill in combination with remote sensing of oil films thickness (Brown and Fingas, 2003). Further, if depletion from films and droplets are comparable, the system described here may be used for gaining new insights into dissolution, biodegradation and microbiological processes of dispersed oil. This may also have implications for studies of oil spill treatment with chemical dispersants.

7.1.2.10 Effect of a Deep-Sea Environment

The deep sea (as well as other habitats, such as deep groundwater, deep sediments or oilfields) is influenced by high pressure. Barophiles (piezophiles) are microorganisms that require high pressure for growth, or grow better at pressures higher than atmospheric pressure (Prieur and Marteinsson, 1998).

When oil is spilled into an ocean, or for that matter into any water system, the constituents of the spill undergoes a number of physical and chemical changes, some of which lead to the removal of the spilled material from the surface of the water while other effects cause the spilled material to remain on the surface of the water. Furthermore, an understanding of the processes involved and how they interact to alter the nature, composition and behavior of the constituents of crude oil and crude oil products with time is necessary. This nay allow the cleanup crews to predict when the spilled material will or will not reach vulnerable resources due to natural dissipation, so that when an active response is required, the type of crude oil or the crude oil product and the behavior of the constituents of the spilled material can be used to determine which response options are likely to be most effective.

The fate of the spilled material depends upon factors such as (1) the amount of the spill, (2) the chemical and physical characteristics of the oil, or the crude oil product, (3) the climatic conditions, and (4) whether or not the oil remains at sea or is washed onto the shore. Therefor an understanding of the processes involved in the change and the manner in which these processes interact to alter the composition and properties of the oil is fundamental to a design of the cleanup of the spill.

As an example of the changes that can occur to such spilled crude oil or to a spilled crude oil product, the constituents of the spill with a density greater than that of marine waters will (as oxidized crude oil-water emulsions) be expected to sink to a level commensurate with the density and even to the deep benthic zone, where the hydrostatic pressure is notably high. A combination of high pressure and low temperatures in the deep ocean results in low microbial activity (Alexander, 1999).

In terms of cleanup operations when an ocean spill occurs, the tendency of the tendency of the spilled material to spread and fragment rapidly may place constraints on any response technique. For example, ship-based recovery systems, typically with swath widths of several feet may be unable to encounter any significant quantities of oil once it has spread and scattered over several miles which, in the case of light (low density, low viscosity) crude oils and crude oil products, which can happen in the hours immediately after the spill, and is often a hindrance to the success of recovery operation at sea insofar as the recovery operations may only recovery a fraction of the spilled material.

Thus, the movement of the spilled material and the changes to the chemical (and physical) nature of the constituents as caused by weathering is a determinant whether any response, beyond monitoring the dissipation of the spoil material, is necessary.

In the case of an active response, the occurrent of weathering processes may require that the suitability of any selected cleanup techniques to be re-evaluated and modified as the response progresses and conditions change. For example, the viscosity of the spilled material can increase quickly and if, for example if mechanical recovery systems are deployed (Chapter 5), the type of skimmers and pumps used may need to be changed as the constituents of the spill are changed by weathering processes leading to a rise in viscosity and/or the formation of emulsions.

Thus, in summary, an understanding of the likely fate and behavior of different crude oil and crude oil products oils and the constraints that the constituents impose on cleanup operations is a necessary aspect of the preparations for effective cleanup operations. In addition, in an ocean environment, information on the prevailing winds and currents throughout the year will indicate the most likely movement of the spill and the ecosystems that might be affected in a given location and, thus, assists in the selection of the appropriate cleanup technique(s) and equipment.

7.2 Effects on Flora and Fauna

The spills of crude oil and/or crude oil products can lead to pollution of the air, the water systems, and the land systems either because of (1) accidents, (2) natural disasters, or (3) or deliberate acts due to the of the spoiled material tankers, pipelines, railcars, offshore platforms, drilling rigs, and wells as well as spills of refined crude oil and their byproducts, heavy fuels used by large ships such as bunker fuel oil (often referred to as bunker C), or the spill of any oily refuse or waste oil. The causes of pipeline ruptures are diverse, they include faulty pumping equipment, faulty pipe seam welds, earthquakes, sabotage, or deliberate spillage.

Accidental oil spills or intentional operational discharges of crude oil and crude oil products and wastes often warrant chemical fingerprinting investigations to (1) establish the source of the crude, if such information is not available, (2) confirm the source of the crude oil if a source is suspected, and (3) determine the impact of the crude oil on the environment, both spatially and temporally (Stout and Wang, 2016). Oil spills have disastrous social, economic and environmental consequences. The key factors or variables that influence oil-spill impacts can be identified in several domains: the oil spill itself, disaster management, marine physical environment, marine biology, human health and society, economy, and policy.

On occasion, decisions have to be made on the basis of conflicting environmental issues. These include (1) the type of crude or the crude oil product oil type, (2) the volume of the spilled material, (3) the sire geography, (4) the site geology, (5) the climate because of the season and seasonal changes in the climate, and (6) the various species of the floral and faunal communities, including the availability of micro-organisms which all contribute to the section of the cleanup method.

7.2.1 Effect on the Biosphere

The biosphere plays an integral role in the support of micro-organisms and their mutual interactions and has been the home to biodiversity within ecosystems while providing a reliable source of food on Earth. Living organisms have and continue to adapt to the environment of the biosphere. The biosphere consists of the atmosphere of the Earth, the water system, and the crust (i.e., the land systems). The biosphere provides the necessary environmental conditions for survival and plays an important role in the global carbon cycle during the evolution of the Earth and is a vital element in the regulation of the climate.

The biosphere can be sub-divided into (1) natural areas, include many different natural habitats: deserts, forests and water bodies, and (2) developed areas, which include urban and commercial areas, agricultural areas and transportation. The biosphere includes all living floral and faunal communities on Earth and the distribution of life on Earth reflects variations in the various ecosystem, principally in temperature and the availability of water (Singh, 2016).

However, even though there has been mention of various systems (above), when a spill of crude oil or a crude oil products occurs many elements of the environment can be affected because the spilled material tends to affect everything in the contact area and becomes an undesired part of the ecosystem.

Spilled oil on land prevents water absorption by the soil, spills on agricultural locations or grasslands have the effect of choking off plant life. Environmental damage not only includes pollution of groundwater and nearby rivers or water streams, but also air pollution from the smoky plumes with crude oil residues occurs as well. The air pollution plume usual extends downwind from the spill and contains particles made from the low-boiling hydrocarbon derivatives (commonly referred to as known as volatile organic compounds) (VOCs) that evaporate quickly. This release of such compounds into the atmosphere can be a contributor to the worsening effects of greenhouse gases.

The impact of an oil spill can also ruin the infrastructure and economy of a particular area with the long-term effects being felt for decades. Also, the effects are considered catastrophic because of the effects on the environments and habitats. Moreover, the spilled material effects the hydrosphere because of pollution to the water systems. Moreover, because of the lower density of the oil compared to the density of water, the oil floats on the surface of fresh water and salt water (unless it is a very heavy oil).

The impact of an oil spill in the marine environment can be very significant and includes (1) the loss of species in some areas, (2) a downgrade in the quality of the sediment, and (3) a negative impact on the onshore and offshore fish as well as the crustacean species.

7.2.2 Effect on Micro-organisms

Micro-organisms (also known as microbes and which include bacteria, protozoa, algae, and fungi) are environmentally, economically, and socially important and make up the largest number of living organisms on the Earth.

The release of crude oil from a geological sediment occurs slowly, thereby allowing the local microbial communities to degrade the oil as it enters their ecosystem. However, during anthropogenic spills of crude oil or a crude oil product, the amount of oil can overwhelm the natural microbial community, allowing the oil to spread and wash up onshore which leads to substantial environmental and ecological problems. In the mechanistic sense, as contact is made between the spilled material and the microbes is a reduction of activity due to reduced air availability which leads to the selective destruction of aerobic microbes thus leaving the resistant and more adaptive microbial strains to proliferate.

Although pollution by crude oil and crude oil products is difficult to treat, a series of oil-degrading bacteria have evolved as a result of existing in close proximity to naturally occurring crude oil. In fact, Many studies have revealed that there are large number of hydrocarbon-degrading bacteria in oil-rich environments, such as oil spill areas (Hazen et al., 2010; Yang et al., 2015), and that their abundance and quantity are closely related to the types of crude oil hydrocarbon derivatives and the surrounding environmental factors (Fuentes et al., 2015; Varjani and Gnansounou, 2017).

Bacteria displaying such capabilities can be utilized for the bioremediation of crude oil-contaminated environments.

A number of aerobic micro-organisms have been isolated from oil polluted soil (Okpokwasili and Okorie, 1988). Common among heterotrophic bacteria genera that have been isolated form soil polluted by crude oil are Streptomyces, Bacillus, Micrococcus, Arthrobacter, Flavobacterium, Methano-bacterium Achromobater, Cloctridium, Thiobacillus, pseudomonas, Alcaligenes and filamentous fungi as Penicillium, Aspergillus, Fusarium Cladosporium and Candida (Antai and Mabomo, 1989).

However, it must always be remembered that when there is a spill of crude oil or a crude oil product, the fate of the spilled material depends on the composition of the material. As numerous factors may determine the types of environmental effects that may result from oil spills, not all oil spills, e.g., crude oil, refined products, weathered crude oil are created equal, and it is likely that the characteristics of different environments (such as temperature, salinity, and solar radiation) will affect ecosystem response to oil.

Spilled oil and oil products lead to rapid changes in the composition of the natural bacterial flora and the oil-degrading bacteria that are nearly undetectable before release of the spilled material can become dominant in the system. Also, crude oil can alter the conditions of the water (such as the chemical composition and the food web interactions) to enhance phytoplankton growth and increase their biomass. The phytoplankton assemblage can change as a result of exposure to crude oil (Ozhan et al., 2014).

7.2.3 Effect on Plants

Plants are (collectively) an important part of the environment and form the basis for the sustainability and long-term health of environmental systems and they use energy from sunlight to convert carbon dioxide into glucose (or other sugars) which can be represented simply by the equation:

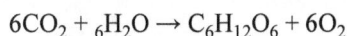

$$6CO_2 + 6H_2O \rightarrow C_6H_{12}O_6 + 6O_2$$

Aquatic plants (1) provide food and cover for fish and aquatic invertebrates, (2) help oxygenate the water, (3) moderate nutrient enrichment, and (4) limit erosion.

In a world with increasing crude oil production, spillage of the oil spill cannot be neglected when addressing plants and the injection of crude oil constituents into the environment has posed serious threats to the marine community. Oil hampers the growth of plants to a great extent by cutting off the air supply and sunlight, thus making it impossible for plants to carry out the photosynthetic reaction (above).

In a plant, the oil constituent(s) may travel in the intercellular spaces and possibly also in the vascular system thereby damaging the cell membranes by penetration of the crude oil (or crude oil product) leading to leakage of cell contents after which the oil constituent(s) may enter the cells. As this occurs, the transpiration rate is reduced by blocking the stomata and intercellular spaces with an accompanying reduction in photosynthetic activities.

A pH value between 7.5 and 7.5 is considered optimum for the growth of many plants. Although many plants respond to an optimum pH, this value usually covers a range from 0.5 units below to 0.5 units above the optimum level. The pH of the soil changes through oil spills thereby increasing the toxicity of the soil. The increase in soil toxicity makes the availability of materials like nutrients to be lowered which leads to stunted growth, death of plant tissue, or yellowing of the leaves caused by a reduced production of chlorophyll. Also, the high coagulation of soil particles due to crude oil pollution is a major cause of poor growth of plants in soil that has been polluted by crude oil or crude oil products causes the plant roots to die due to a lack of oxygen and/or from the toxic hydrogen sulfide generated in the oxygen-depleted soils (Plice, 1948; De Jong, 1980).

Marine algae and seaweed responds variably to oil, and oil spills may result in die-offs for some species while some became more abundant in response to oil spills because the marine environment is highly complex and has natural fluctuations in species composition, abundance, and distribution over time. Within this environment, marine plants have varying degrees of natural resilience to changes within their habitats.

7.2.4 Effect on Animals

Spilled oil causes harm to wildlife through ingestion or inhalation and from external exposure through skin and eye irritation. The severity of the oil-based injury to wildlife depends on the type and quantity of oil spilled, the season and weather, the type of shoreline, and the type of waves and tidal energy in the area of the spill. In small doses—as yet these doses are generally undefined because they depend on the type of organism or plant species—can cause temporary physical harm to animals while large quantities of oil can cause chronic effects and the destruction of life (including human life) can be widespread.

When the oil is released in the water, it is not just the surface wildlife that is affects there is contact with sub-surface wildlife. On the surface, the oil constituents are oxidized which increases the density of the oil to an extent that may be (or often is) greater than the density of the water. At that point the oil begins to sink—contrary to the opinion of some observers, the oil is not removed from the water—and when the oil sinks into the aqueous environment it can have damaging effects on fragile underwater ecosystems, killing or contaminating fish and smaller organisms that are essential links in the global food chain. In some cases, as the chemical change to the oil constituents evolves and degrade to other chemical species, some of the oil constituents may (i.e., can) re-appear on the surface of the water.

Marine ecosystems rely on the various relationships (often complicated and delicate relationships) between organisms and their environment, which can be significantly disrupted by the spillage of crude oil and crude oil products. Food sources contaminated by oil can destabilize entire food chains, and oil spilling onto nesting grounds and breeding sites can cause long-term damage to food chains and life cycles.

Seals, sea lions, walruses, polar bears, sea otters, river otters, beavers, whales, dolphins and porpoises, and manatees, are groups of marine mammals that may be affected by oil spills. In addition, the constituents of crude oil and crude oil products can clog blowholes of whales and dolphins, making it difficult for these animals to breathe and also disrupting the ability of the animals to communicate. Marine mammals living in cold climates (otters, seals, sea lions, and polar bears) are typically more vulnerable than those living in temperate or tropical waters because the likelihood of oil adhering to the fur of the animal will result in it losing its insulating ability which is necessary for keeping the animal warm. Also, the scent that seal and sea lion pups and mothers rely on to identify each other is disguised can lead to rejection, abandonment, and starvation of the pups as well as making the young animals easy prey when oil sticks restricts the use of the flippers making it hard for the young animals to escape predators. Skin lesions are a problem for dolphins and whales when they (unknowingly) swim through areas where oil is present. Absorption of oil or dispersants through the skin can damage the liver and kidneys, cause anemia, suppress the immune system, induce reproductive failure, and in extreme cases kill an animal.

Chronic or delayed responses of marine bird and mammal populations to the constituents of crude oil and crude oil products in the sea can occur because of continued ingestion of oil via contaminated prey, or because of failure of prey populations to recover subsequent to injury. In the years after a marine oil spill, various species of birds and marine mammals may exhibit indirect or delayed responses to the spill, especially in the species such as sea ducks and shorebirds that forage primarily on intertidal and shallow subtidal invertebrates, as well as in several species that forage on small fish caught in inshore waters (Peterson et al., 2001).

As with birds and marine mammals, the number sea turtles reported as being harmed by oil spill may be a serious underestimation of the real true number of animals affected. Sea turtle adults are probably most susceptible to oil spills through inhalation when they surface to breathe, or through ingestion of oil-fouled food and floating tar balls they mistake for food. The loggerhead turtles and leatherback turtles can be impacted as they swim to shore for nesting activities. In the breeding season females may become oiled when they arrive in the contaminated area or when they come ashore to lay eggs and, furthermore, the juvenile turtles may become trapped in oil after hatching when they move towards the sea.

Fish may ingest large amounts of oil through their gills and the fish that have been exposed to oil may suffer from changes in heart and respiratory rate, enlarged liver, reduced growth, and fin erosion as well as a variety of effects at the biochemical and cellular levels that may affect the reproductive capacity negatively and/or result in deformed offspring.

Every spill situation is unique and depending on the particular conditions and circumstances in that area, and on the characteristics of the spill therefore preventing spills is the best way to protect wildlife from oil spills. Catastrophic oil spills can be deadly can forever change sensitive ocean ecosystems.

7.2.5 Effect on Humans

Oil spills also have a huge impact on humans as well since the crude oil products include a lot of common fuels and spills may occur in many locations, particularly in residential areas. The effects on humans may be direct and indirect, depending on the type of contact with the oil spill which depend on the type of oil that was spilled and the location where the spill occurred, such as on land, in a river, or in the ocean. Other factors include the degree of exposure to the spill such as the type of exposure and the length of time that the person was exposed to the oil. The three main ways that the constituents (i.e., the organic constituents) of the crude oil (or the crude oil product) may affect humans are (1) the direct contact of the oil with the skin, (2) the inhalation of volatile constituents of the oil, and (3) the ingestion of water or particles that have been contaminated by the oil constituents.

Crude oil spills are tragic environmental disasters that can cause severe health problems, disturb the ecosystem, and pollute the environment and affect human health through both the chemical exposure and the psychological and socioeconomic impact on the affected individuals and their communities. It is to be expected that the closer the person was to the spill or the more time they spent near the spill, the greater the symptoms and (again, not surprisingly, the people who clean up the spill were at a higher risk and most likely to suffer from adverse health effects.

Some workers are more sensitive to chemicals, and contact with the skin may result in disorders such as skin reddening, swelling, and burning, follicular rashes, defatting or drying of the skin, infections, dermatitis, folliculitis, eye irritation, other allergic reactions, headaches, ataxia, dizziness, nausea, vomiting, cough, respiratory distress, and chest pain (CDC, 2010; McCauley, 2010; Pattillo, 2010; Solomon and Janssen, 2010; Walker, 2010). These effects can be grouped into respiratory damage, liver damage, decreased immunity, increased cancer risk, reproductive damage and higher levels of some toxics (D'Andrea and Reddy, 2014).

Even when odors are not detected, a health risk may exist for some individual compounds if residents are exposed (breath the air) for a long time (Laffon et al., 2006; Perez-Cadahia et al., 2007; Perez-Cadahia et al., 2008a, 2008b; Rodriguez-Trigo et al., 2010; Zock et al., 2011, 2012). The crude oil spills affect human health through their exposure to the inherent hazardous chemicals such as p-phenol derivatives and volatile benzene derivatives. The predicted exposure to chemicals from the oil spill are (1) inhalation, (2) skin contact, (3) food and water ingestion, and (4) contact with the beach sand.

The potential of any given chemical in crude oil to produce chronic or long-latency toxic effects on health varies by chemical structure, which can change over time, and by dose and duration of the exposure. The individuals potentially more susceptible to the effect of contamination and not normally occupationally exposed, such as infants, children, pregnant women, elderly and people with previous health conditions, are also exposed. Direct chemical exposures can occur to a fetus or a child across the placenta or by ingestion, inhalation, or dermal absorption. Children and fetuses are more vulnerable to adverse effects from chemical exposures because developing organs are more sensitive and less efficient at detoxifying and metabolizing chemicals.

The long-term effects seen from the BP oil spill show that time does not eradicate the problem or improve health issues. The exposed cleanup workers from the BP Deepwater Horizon disaster had an increased prevalence of illness symptoms such as shortness of breath, headaches, skin rash, chronic cough, weakness, dizzy spells, painful joints, and chest pain even seven years after their exposure to the oil spill. Deteriorated pulmonary functions had also increased in the oil spill cleanup workers and they had also developed chronic rhinosinusitis and reactive airway dysfunction syndrome as new symptoms years after exposure (D'Andrea and Reddy, 2014, 2018). Fishermen who were re-interviewed in a follow up study 1.5 to 2 years after cleanup work were found to have respiratory symptoms more as compared with unexposed individuals (Rodrigues et al., 2010; Rodríguez Trigo et al., 2010).

Mental health impacts can be observed in the short-term among those affected by oil spills. For example, within four weeks of the Sea Empress spill, there was a worsening in mental health and an increase in anxiety among the urban residents of towns directly impacted by the oil slick (Lyons et al., 1999). Mental health impacts were most often related to income loss or financial uncertainty and, at the community level, cultural losses and deterioration in kin and non-kin relationships and social order. Commonly identified issues included increased anxiety, depression, and posttraumatic stress disorder both in people exposed to the oil, and those helping the clean-up.

The impacts of these disaster must be considered in the timeframe of, not weeks and months, but years.

7.2.6 Effect on the Economy

Typically spills closer to shore and human populations have greater economic impacts and are more expensive to clean while offshore spills have fewer direct economic impacts than near-shore spills in proximity of human population.

With increasing soil infertility due to the destruction of soil micro-organisms, and dwindling agricultural productivity, farmers would be forced to abandon their land and to seek non-existent alternative means of livelihood. Aquatic lives destroyed with the pollution of traditional fishing grounds would lead to the intensifying of hunger and poverty in fishing communities. This can also means that water sources that are intended to provide drinking water can become contaminated. Fishermen and local ship workers can lose their sources of income due to the health problems associated with exposure to oil such as respiratory damage, decreased immunity and increased risk of cancer

Commercial fisheries and aquaculture businesses are usually affected by loss of product, caused by direct mortality or habitat loss, or by loss of access caused by harvesting bans and closures. Oil spills dramatically change the fishing, shrimping and oyster industries in affected areas.

The disturbance caused by oil spills on land and sea traffic can affect import-export activities. Ports can suffer similar problems to marinas and harbors, although it is on a considerably larger scales. Many port authorities demand that the hulls of commercial vessels are cleaned before ships are permitted to sail. There is also considerable disruptions to normal port operations as well as shipyard works and costal construction works due to various influences caused by oil spills. Also, many tourist attractions are shut down, and there is also a psychological factor that can turn people off to going and visiting an area.

Property value may reduce based on the magnitude of the oil spill. When houses decrease in value, the local businesses start to decline. Also, the natural beauty of the coasts are destroyed and thee will be a decrease in tourism after a spill of crude oil or a crude oil product.

An important effect to note is that an oil spill can increase the price of oil because supplies are lowered and the outlook on oil exploration changes. Less oil and increasing demand means that prices would rise that can affect various sectors of the economy.

7.3 Risk Analysis

Chemical analysis provides an accurate breakdown of the chemical composition (chemical structures and concentrations of the components) of a sample. Chemical analysis is important for the identification and verification of materials that may contribute to failure analyses and remediation to ensure that the environment remains sustaining and safe. The method uses of a combination of analytical methods to provide invaluable information, to assist in chemical problem solving and regulatory compliance.

There are two types of chemical analysis which are (1) the determination of those elements and compounds that are present in a sample of unknown material, i.e., qualitative analysis, and (2) the determination of the amount by weight of each element or compound present, i.e., quantitative analysis

The steps taken in the chemical analysis of a material are as follows (1) to understand the problem, (2) design or select an analytical procedure, (3) sampling in which a representative bulk sample of the material to be analyzed is obtained, (4) extracting the smaller lab sample from the bulk sample, (5) sample preparation by which the sample is converted to a state that is suitable for analysis), and (6) chemical analysis and data analysis, reporting and interpretation. Furthermore, each of the above steps must be executed correctly in order for the analytical result to be accurate.

Risk analysis involves the identification and quantification of events that could trigger losses incurred by a process. Once these events have been identified, the risk analysis process is used to estimate the probability of occurrence and the amount of loss associated with each event.

Risk analysis is composed of two parts which are (1) the probability of an error in a process and (2) the negative consequences of the error. Also, there is a number of decision criteria available for stating the accepted risk which are (1) the prevention of adverse effects, (2) mitigating the risk, and (3) balancing the costs and the benefits (Suter, 2007).

Risk assessment is a means for ensuring the health and safety of workers and regulatory requirements create a need to map the environmental risk of potential spills of crude oil and to identify preparedness and response options. However, additional risk reduction can be achieved by ensuring effective response to spill incidents, through the establishment of source control such as well capping and containment solutions and oil spill preparedness to combat a spill of crude oil. Risk assessment is the first component of the four-stage contingency plan for oil spill preparedness and response, which can be used to develop and implement risk-based planning for oil spill incidents.

7.3.1 Oil Spill Risk Assessment

Risk analysis commences with the pre-assessment planning is usually part of the process to ensure that the risk assessment process is feasible and that it will satisfy the objectives of risk management. However, the first step of oil spill risk assessment is to quantify the likelihood of a spill incident based on the information on historic spill records, vessel traffic, and sources of oil pollution.

The ability to identify adverse events that might possibly occur in the future, and to make good and relevant choices from among alternative actions that could be employed to manage such events, is a key factor in modern society. Understanding risk thus allows us to make rational decisions (Bernstein, 1998). Every organization is continuously exposed to an endless number of new or changing threats and vulnerabilities that may affect its operation or the fulfillment of the objectives.

Once the scope has been determined, the process moves on to conducting the risk assessment itself. A clear description should be given of the methodology, models and/or tools used in the risk assessment, including a justification of their use. Results, premises and assumptions should be documented in a manner which enables easy use as input to the response planning (Table 7.4, Table 7.5). Identification of possible risk reducing measures should be performed as a part of the risk assessment process, as well as throughout the preceding steps. The resources/areas of higher risk may differ between seasons due to changes in wind strength and direction and currents or changes in the presence or sensitivity of the resources (Table 7.5) (IPIECA, 2013).

Table 7.4: Elements Typically Documented During Response Planning (IPIECA, 2013).

Scenario element	Notes
Response objectives	Several response objectives can be identified for most oil spills but the two that are generally universal include: • protecting the health and safety of responders and the public (the first priority with all response actions); and • minimizing damage to ecological and socio-economic resources, and reducing the time for resource recovery. Oil spill response actions can be effective, however it is generally unrealistic to indicate that avoidance of all impacts is an achievable objective in the event of a major oil spill. The aim of oil spill response actions is therefore to minimize ecological and socio-economic damages in the context of a net environmental benefit analysis (NEBA) context. This is important in relation to (a) stakeholder/public expectations, (b) setting response objectives during an incident and (c) developing realistic measures for an effective response. A specific scenario will allow overarching aims to be refined into more specific objectives.
Response strategies	In principle the full range of response options should be considered; see Appendix 4 for further information. The legislative framework, national response policy and/or stakeholder needs may either dictate or influence which options are allowed or preferred. However, if response options, such as dispersants, are not considered or if their use is postponed in favor of other options, the overall effectiveness of the response may be greatly reduced and lead to higher impacts. Different strategic options may be proposed at the three tiers; this may stem from the determination of response objectives.
Choosing strategies to minimize damage (i.e., NEBA)	The principle of net environmental benefit analysis should be integral to strategic decision making with respect to selecting the most applicable response strategies for each scenario to achieve the identified objectives. The analysis provides a tool to ensure that strategies are chosen to minimize the potential ecological and socioeconomic damages associated with each scenario. The oil spill risk assessment should provide adequate ecological and socio-economic data to enable the analyses to be undertaken.
Tactical plan	The tactical plan details how the chosen strategies will be implemented and usually forms the most substantial part of the action planning process. The proposed location of equipment deployments relative to the spill source and the associated operating environment(s) for the three tiers, such as (1) open waters close to the source, (2) open waters beyond the vicinity of the source, and (3) the nearshore sheltered waters and shorelines. Substantial oil spill response resources are available from the oil industry's global stockpiles and commercial service providers, and through regional and international agreements. Planning for Tier 3 incidents should take these into account and focus on ensuring the procedural and logistical means to access, mobilize and integrate suitable resources into the theatre of operations.
Sustaining a response	For scenarios entailing a prolonged response, possibly extending for periods of months, an outline description of the means by which the response can be sustained should be included. This should incorporate the establishment of a project management approach including data management, such as incident action planning and a Common Operating Picture, and the maintenance of supply lines for ongoing response actions.

A formal risk assessment process offers several advantages for the management of spills of crude oil and crude oil products and involves (1) the risk assessment process which allows managers to identify the likely causes of oil spills for a specified area, (2) the process allows managers to compare the relative risk between different geographical regions and also provides a basis for the identification of appropriate management strategies to reduce the overall risk for a given region, either through preventative or preparedness strategies, and (3) the process also offers a comparison of potential effects of different technologies and allow for comparison studies to be performed.

Table 7.5: The Steps Involved in an Effective Response to a Spill.

Assess the risk	Before there is an attempt to clean up a spill, everything that might be affected by the spill should be given consideration, including the risks to human health, the environment and property.
Personal protective equipment	The appropriate personal protective equipment to safely respond to the spill should be selected.
Evaluate the incident	The selection of the most appropriate strategy will depend on many factors, including the response resources available, the national and local regulations on oil spill response, the spill scenario and the physical and ecological characteristics of the area impacted by the spill.
Decontaminate the site	This may involve removing and disposing the contaminated media, such as earth, which were exposed during the spill incident.
Complete required report	Complete all notifications and reporting required by local, state, and Federal guidelines for reporting spill incidents.

The probability analysis can established using two methods: quantitative (used in situations where appropriate historical accident and spill data are available) and qualitative (when statistical data is not available). Therefore the purpose of the risk assessment is to provide information to decision makers in a form that allows for the comparison of risk reduction alternatives. It is a straightforward and structured method of ensuring the risks to the health, safety and wellbeing of employees (and others) are suitably eliminated, reduced or controlled. To make a good assessment of potential consequences from various oil spill scenarios, relevant and valid relations between the extent of the oil exposure and the environmental impact should be established for the impact indicators.

A risk assessment should also considers vulnerabilities—weaknesses—that would make an asset more susceptible to damage from a hazard. Appropriate hazard identification tools and methods should be applied. Selected tools should be suited to the objectives of the hazard identification and to the type of hazardous events to be identified. Also, the identified resources should be verifiable and their availability guaranteed to the extent practicable. A vulnerability analysis would be an indicator of the potential impacts of pollutants on the environment, ecology (impacts to the soil, water, air and living organisms within ecosystems), and socioeconomic resources, and is evaluated by considering the importance of the sensitive resources, location, and seasonality as well as exposure, sensitivity and recovery from spilled oil. The vulnerability is evaluated to be higher as exposure to spilled oil is longer, sensitivity of resource is higher, and resource's rate of recovery from the impact of spilled oil is slower.

The risk assessment results can provide intuitive and objective information for priority management and in-depth study of hazard area, they are useful for the preparedness and response of marine pollution accidents. In particular, they can be utilized to develop strategies of oil pollution preparedness and response through scenarios based on oil pollution possibility, oil discharge and prevention method. The final step of risk assessment is to determine the probable impacts of spill incidents by combining both spill probability (the outcome of likelihood analysis) and vulnerability (the outcome of vulnerability analysis). Identification of possible risk reducing measures should be performed as a part of the risk assessment process, as well as throughout the preceding steps (IPIECA, 2013). However, an oil spill incident is unlikely to follow a planning scenario precisely. The response planning should be documented to demonstrate that thorough consideration has been given to the oil spill scenarios.

Typically, risk assessments are driven by management questions. Risk management is the process of identifying, assessing and controlling threats to an organization's capital and earnings. These threats, or risks, could stem from a wide variety of sources, including financial uncertainty, legal liabilities, strategic management errors, accidents and natural disasters. It should be an ongoing

Risk management
    Risk identification
        Risk assessment   Risk analysis
                                    Risk evaluation
    Risk treatment
    Risk reduction/control

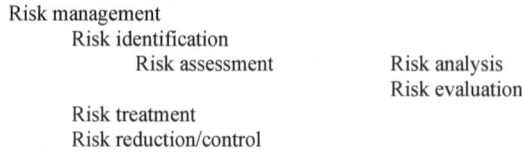

Figure 7.2: A Simplified Representation of the Risk Management Process.

part of a quality management process; a mechanism to review or monitor events on a regular basis once a risk-quality management process has been initiated.

It is the evaluation and implementation of the options for reducing the probability and consequences of spills. Risk assessments and the overall process of risk management are important in order to prioritize and effectively use resources to provide proper decision support.

There are several frameworks that describe the risk management process which are insofar as they are not limited to one specific field of application and they share certain basic steps (Figure 7.2): risk assessment, consisting of risk analysis and risk evaluation, and risk reduction and control collectively complete a full risk management process (BOEM, 2014).

Risk analysis is the process of understanding the nature and determining the level of risk and should provide input for the risk evaluation step, which could act as support when making decisions related to the handling of risks. Risk evaluation involves comparing the results of the risk analysis with risk criteria and evaluating possible actions. This shows whether the risks found are acceptable or not. During the final step, risk treatment, one or more actions are chosen and implemented to mitigate the risk.

The main steps in risk assessment are risk identification, risk analysis and risk evaluation. The quality of the risk assessment will largely depend on how uncertainties are dealt with. Uncertainties must be treated in a comprehensive, repeatable and transparent manner (Burgman, 2005). In order to enter into a dialogue with stakeholders and to ensure an exchange of information with relevant parties, communication and consultation are important during all the risk management phases.

Monitoring and review should also be part of a risk management process, including providing input to improve the process, detecting changes that should be reflected in earlier stages, and identifying emerging risks. Risk management can be a help to decision-makers when making informed actions and when prioritizing between options. More specifically, it is through risk assessments and analyses information to support good decisions can be elicited (Aven, 2012). Besides improving emergency planning, risk assessments will also enable managers to balance economic and environmental factors during oil exploration, production, and decommission activities.

The cost of cleanup after an environmental disaster such as an oil spill is enormously greater. Therefore, environmental risk assessment should be included in the initial project planning in order to create oil development plans that minimize environmental impact. As risk assessment does not set regulations, but rather provides information to decision makers on the components of greatest potential risk, it is up to the individuals and regulators concerned to determine what is an acceptable risk (US EPA, 1998).

7.3.2 Guidelines for Oil Spill Risk Assessment

Using ISO 31000, which is a family of standards relating to risk management codified by the International Organization for Standardization which can assist organizations increase the likelihood of achieving objectives, improve the identification of opportunities and threats and effectively allocate and use resources for risk treatment. The document can be used by to manage risks, make decisions, set and achieve objectives as well as improve performance.

However, the effectiveness of risk management will depend on its integration into the governance of the organization, including decision-making. This requires support from stakeholders, particularly top management. However, because it is impossible to define an actual emergency event in advance, a good response framework will be scalable (capable of dealing with incidents of varying magnitude), flexible (able to respond to changing conditions), and adaptable (able to be modified as needed). The framework for conducting a risk assessment usually follows six steps (1) plan the risk assessment, (2) analyze hazards: identify and describe oil spill scenarios, (3) analyze the probability of spill scenarios, (4) analyze the consequences of spill scenarios, (5) characterize the risks of spill scenarios, and (6) manage the risks.

This would involve the use of existing good practices in the determination of oil spill response resources and would promote consideration, in tactical and logistical detail, of the preferred and viable response strategies to address scenarios covering the range of potential spills of crude oil (and crude oil products). The process would also cover the process of planning and executing the oil spill risk assessment (OSRA), and includes recommendations on how to analyze and evaluate the risk. The level of detail to be implemented and achieved for the oil spill risk assessment should be established prior to conducting the study.

In addition, the oil spill risk assessment process involves gathering the collective knowledge of the team assigned and identification of all of the potential adverse occurrences project at each stage of the project. Thus, the process objectives must identify the specific purpose of the process including what information id needed and the way in which the information will be used.

Since not all risks are created equal and some have a worse outcome than originally foreseen, it is often necessary to create a risk analysis matrix which gives an indication of the likelihood of an event occurring as well as the potential impact the event will have on the project thereby enabling the categorization of the risk. This will provide a reference point for the identification of the risks that which should be prioritized for monitoring. Operating risks are often based around industry best practice which is, in turn, based on a long history of practical experience. Where appropriate information is available, quantitative evaluation is often undertaken to estimate the likelihood of a spill occurring, and the potential size of any spill. As part of the process, an understanding of the migration path of the crude oil (or the crude oil product) enables high risk areas to be identified, and indicates what resources may be impacted by oil under different physical conditions (such as the weather and the season), spill scenarios (such as spill of bunker C oil, vessel grounding, and well blowout) and different response options (such as natural recovery, shoreline cleaning, and the use of chemical dispersants).

The selected tools should be suited to the objectives of the hazard identification and to the type of hazardous events to be identified and it is also important to ensure that personnel involved in the hazard identification process are aware of and understand the basis, to ensure there is sufficient expertise to meet the objectives and scope. Based on the objectives, scope and range of oil spills types identified, spill scenarios can be grouped according to spill volume, equipment source, facility type and spill cause.

Mapping of ecological and socio-economic resources using the selected release scenarios allows for the identification of the resources that can be exposed to the spilled material. Then depending on the outcome of the risk level evaluations necessary measures can be taken to reduce the risk. Identification of possible risk reducing measures should be performed as a part of the risk assessment process. For the purposes of a response to an oil spill, planning and determining oil spill response capability, the results from the oil spill risk assessment will provide important input related to the likelihood of different spill scenarios, the ecological and socio-economic consequences of the scenarios, and the likelihood of exposure and oil volumes in geographical areas (Nissen-Lie et al., 2014). Information and results gathered throughout the oil spill risk assessment process may be useful as an input to oil spill response planning (Table 7.4, Table 7.5).

The primary objective of an oil spill response is to safely undertake actions to minimize the overall environmental and socio-economic damages that are caused, or likely to be caused, by an incident. The advantages and disadvantages of different response strategies need to be compared with each other and with the potential for natural recovery. A variety of strategies may be used throughout the response to an incident. In the case of a major oil spill, it is likely that different strategies may be deployed simultaneously. Various strategies are available for responding to oil pollution on the open sea, near shore or stranded on shorelines. The initiation of a response, or a decision to stop cleaning and leave an area for natural recovery, should be based on an evaluation made both before an oil pollution incident (as part of the response planning process) and through field observations and assessments during ongoing response operations if an incident occurs (IPIECA, 2013). The response planning should be documented to demonstrate that thorough consideration has been given to the oil spill scenarios.

The integrated risk management approach (BOEM, 2014) also necessitates the formulation of overall coordinated strategies involving multi-dimensional elements including technical, locational, social and economic considerations. It is also recommended that there are relevant training and exercising programs for the personnel with allocated roles within the incident management system. It is also important to note that the spill planning approach seeks to allocate resources to those activities that provide the best overall strategy for avoiding, remedying and mitigating oil activities.

Interventions must be evaluated in terms of both feasibility and effectiveness and, in all circumstances, the options selected should achieve a net environmental benefit for the spill. Ideally, available and preferred response options should be identified, with stakeholder input, prior to a spill occurring.

The effectiveness of any spill response is determined early on by the organization's ability to appropriately mobilize to deal with the situation and it is important that the management structure be able to function effectively for a variety of spill sizes, and that a transition process is in place as an incident escalates in size or complexity.

7.3.3 Oil Spill Risk Analysis Model

The oil spill risk assessment model, developed in 1975 by the DOI, is a tool that evaluates offshore oil-spill risks (Smith et al., 1982; LaBelle and Anderson, 1985) which is used to develop probabilistic estimates the probability of oil spills occurring during the production and transportation of a specific volume of oil over the lifetime of the scenario being analyzed. The process uses a spill-rate constant, based on historic accidental spills expressed as a mean number of spills per billion barrels of produced or transported crude oil or crude oil product.

Modeling of the spills of crude oil and crude oil products is fundamental for planning and preparing for, as well as responding to and mitigating, actual spill events and is an approach for extending an understanding of spill risk, community vulnerability, oil behavior, spill outcomes, and impacts (Nelson and Grubesic, 2017). While modeling oil spill risk can be accomplished in several different ways, the overall goal is to determine the likelihood of oil occurrence and the degree of oil-based pollution that an area might experience (Gasparotti, 2010).

At a minimum, spill models must consider surface and/or subsurface transport, which supplies an indication of where oiling is likely to occur. More advanced models incorporate measures of (1) evaporation, (2) dissolution, (3) entrainment, (4) emulsification, (5) biodegradation, and (6) sediment-oil interaction—all of which enables risk and impact assessment researchers to not only understand the final fate of oil, but also to measure the impact severity (Spaulding, 2017). One primary function of the model is to relate oil spill trajectory movements to the locations of wildlife populations, fishing areas, and other potential targets in coastal areas and in continental shelf areas.

Additional consideration for accurately simulating an oil spill is understanding the type of hydrocarbon that is spilled, which is contingent upon the location and underlying reserves associated with the extraction area. For crude oils, many assay databases exist to provide researchers with information on their chemical composition. Aside from this baseline information, ex ante oil spill models require data on spill location, duration, depth, amounts, the flowrate (if modeling a blowout) and coefficients (or solutions) for horizontal and vertical diffusion.

Oil spill trajectory simulations are generated by the oil spill risk assessment model which simulates the likely path (trajectory) of a surface slick, represented as a point started from locations where an accidental spill could occur. The trajectory point is computed based on ocean currents or ice and winds.

These polygons represent locations of various environmental resources. The model compiles the number of contacts to each feature that result from the modeled trajectory simulations from all of the launch points for a specific launch area. The basic products of the oil spill risk assessment model are conditional probabilities, estimated chance of large spill occurrence, and combined probabilities (BOEM, 2014).

Opportunities to continuously improve oil spill risk assessment model are being addressed as possible climate change events make this imperative. With improved modelling data and supporting tools, valuable information for oil spill response plans can be more readily available and effective. By taking spatiotemporal variation in population distributions into account, information on a meaningful scale for decision-making can be provided to support for decision-making environmental risk assessment (Helle et al., 2020).

7.4 The Characteristics of a Spill Response

The factors that affect the consequence of responses are (1) the quantity and properties of the spilled oil, (2) the location of the spill, (3) the environmental and weathering conditions, and (4) the resource status of available response techniques (Ornitz and Champ, 2002).

The response techniques have so-called windows of opportunity which are defined by (1) the type of crude oil product or crude oil product spilled, (2) the initial spill conditions, (3) the rates of oil weathering and emulsion formation, and (4) the different environments and ecosystems that are, or will be, impacted. In fact, for every oil spill, responders presented with a unique set of challenges requiring timely application of appropriate response methods (NOAA, 2013).

Although each incident is unique and may require additional steps in responding to a spill, the following must be accomplished: identify and assess safety hazards, eliminate ignition sources, account for all personnel, secure spill source, initiate required agency and company notifications, activate spill/emergency response teams and activate oil spill response organizations.

An oil spill contingency plan is a detailed oil spill response and removal plan that addresses controlling, containing, and recovering an oil discharge in quantities that may be harmful to navigable waters or adjoining shorelines (Table 7.2) (US EPA, 2019). They are a legal requirement for operations in the offshore oil and gas industry, vessel operations in ports and harbors and the international shipping sector. To ensure a quick response to spills, the EPA emphasizes the importance of preparation and uses several rules and regulations for enforce the importance of preparation.

Spill recovery actions include actions (responsive or proactive) taken to identify and evaluate immediate and/or long-term impacts and steps taken to mitigate and/or remedy those impacts with the overall aim of protecting, recovering and restoring the environment. The proposed recovery actions should address the recovery targets outlined in the recovery targets section of the plan and aim to achieve environmental recovery to the pre-spill condition or better. Actions to remedy impacts to environmental values and associated components range from measures that immediately stabilize the impacted site to measures that bring a site back to full ecosystem structure and function as they existed before the spill incident. Consideration of temporal loss of biodiversity and risks that the desired ecological condition may not be achieved over time are important with respect to any impacted species and ecosystems and to habitats that are critical to species survival. Qualified persons developing recovery actions should be trained and experienced in environmental restoration and remediation and should have support from individuals with local knowledge.

References

Abraham, W.R., Meyer, H., and Yakimov, M. 1998. Novel glycine containing glucose lipids from the alkane using bacterium *Alcanivorax borkumensis*. Biochim. Biophys. Acta, 1393: 57–62.

Aislabie, J. 1997. Hydrocarbon-degrading bacteria in oil-contaminated soils near scott base, Antarctica. pp. 253–258. *In*: Lyons, W.B., Howard-Williams, C., and Hawes, I. (Eds.). Ecosystem processes in Antarctica's ice-free landscape. Balkema Publishers Ltd, Rotterdam, Netherlands.

Aislabie, J., McLeod, M., and Fraser, R. 1998. Potential for biodegradation of hydrocarbons in soil from the ross dependency, Antarctica. Appl. Microbiol. Biotechnol., 49: 210–214.

Aislabie, J., Foght, J., and Saul, D. 2000. Aromatic hydrocarbon-degrading bacteria from soil near scott base, Antarctica. Polar Biol., 23: 183–188.

Aislabie, J., Saul, D.J., and Foght, J.M. 2006. Bioremediation of hydrocarbon-contaminated polar soils. Extremophiles, 10: 171–179.

Alexander, M. 1999. Biodegradation and Bioremediation 2nd Edition. Academic Press, London, United Kingdom.

Amadi, A., Abbey, S.D., and Nma, A. 1996. Chronic effects of oil spill on soil properties and microflora of a rainforest ecosystem in Nigeria. Water Air Soil Pollut., 86: 1–11.

Antai, S.P., and Mgbomo, E. 1989. Distribution of hydrocarbon utilizing bacteria in the Ahoada oil-spill areas. Microbios Letters, 40: 137–143.

Aven, T. 2012. Foundations of Risk Analysis. John Wiley &Sons Inc., Hoboken, New Jersey.

Atlas, R.M. 1995. Petroleum biodegradation and oil spill bioremediation. Marine Pollution Bulletin, 31: 178–182.

Atlas, R.M., and Bartha, R. 1992. Hydrocarbon biodegradation and oil-spill bioremediation. Adv. Microb. Ecol., 12: 287–338.

Baker, J.M. 2019. Oil pollution, encyclopedia of ocean sciences 3rd Edition. Cochran, J.K., Bokuniewicz, H. and Yager, P. (Eds.). Academic Press Inc., London, United Kingdom. pp. 350–358.

Baraniecki, C.A., Aislabie, J., and Foght, J.M. 2002. Characterization of *Sphingomonas* sp. Ant 17, an aromatic hydrocarbon-degrading bacterium isolated from antarctic soil. Microb. Ecol., 43: 44–54.

Barker, G.W., Raterman, K.T., Fisher, J.B., Corgan, J.M., Trent, G.L., Brown, D.R., and Sublette, G.L. 1995. Assessment of natural hydrocarbon bioremediation at two gas condensate production sites. pp. 181–188. *In*: Hinchee, R.E., Wilson, J.T., and Downey, D.C. (Eds.). Intrinsic Bioremediation. Battelle Press, Columbus, Ohio.

Bartha, R., and Atlas, R.M. 1987. Transport and Transformation of petroleum: biological processes. pp. 287–341. Long Term Environmental Effects of Offshore Oil and Gas Development. Applied Science Publishers, London, United Kingdom.

Bernstein, P.L. 1998. Against the Gods. The Remarkable Story of Risk. John Wiley and Sons In., Hoboken, New Jersey.

BOEM. 2014. Ocean Science. The Science and Technology Journal Of The Bureau Of Ocean Energy Management, Volume 11, Issue 3. https://www.boem.gov/sites/default/files/boem-newsroom/Library/Ocean-Science/Ocean-Science-2014-Oct-Nov-Dec.pdf.

Bossert, I., and Bartha, R. 1984. The fate of petroleum in soil ecosystems. pp. 434–476. *In*: Atlas, R.M. (Ed.). Petroleum Microbiology. MacMillan Publishing Co., New York.

Boufadel, M.C., Suidan, M.T., and Venosa, A.D. 2006. Tracer studies in laboratory beach simulating tidal influences. Journal of Environmental Engineering, 132(6): 616.

Boufadel, M.C., Suidan, M.T., and Venosa, A.D. 2007. Tracer studies in a laboratory beach subjected to waves. Journal of Environmental Engineering, 133(7): 722.

Braddock, J.F., Ruth, M.L., Catterall, P.H., Walworth, J.L., and McCarthy, K.A. 1997. Enhancement and inhibition of microbial activity in hydrocarbon-contaminated Artic Soils: implications for nutrient-amended bioremediation. Environ. Sci. Technol., 31: 2078–2084.

Brakstad, O.G., Bonaunet, K., Nordtug, T., and Johansen, Ø. 2004. Biotransformation and dissolution of petroleum hydrocarbons in natural flowing seawater at low temperature. Biodegradation, 15: 337–346.

Brown, C.E., and Fingas, M.F. 2003. Development of airborne oil thickness measurements. Mar. Pollut. Bull., 47: 485–492.

Burgman, M. 2005. Risk and Decisions for Conservation and Environmental Management. Cambridge University Press, Cambridge, United Kingdom.

Burns, W.A., Mankiewicz, P.J., Bence, A.E., Page, D.S., and Parker, K.R. 1997. A principle component and least squares method for allocating polycyclic aromatic hydrocarbons in sediment to multiple sources. Environ. Toxicol. Chem., 16: 1119–1131.

Button, D.K., Robertson, B.B., McIntosh, D., and Jüttner, F. 1992. Interactions between marine bacteria and dissolved-phase and beached hydrocarbons after the Exxon valdez oil spill. Appl. Environ. Microbiol., 58: 243–251.

CDC. 2010. Centers for Disease Control and Prevention. What to Expect from the Oil Spill and How to Protect Your Health. http://emergency.cdc.gov/gulfoilspi1l2010 Iwhauo_expect.asp.

Cerniglia, C.E. 1992. Biodegradation of polycyclic aromatic hydrocarbons. Biodegradation, 3: 351–368.

Chugunov, V.A., Ermolenko, Z.M., Martovetskaya, I.I., Mironava, R.I., Zhirkova, N.A., Kholodenko, V.P., and Urakov, N.N. 2000. Development and application of a liquid preparation with oil-oxidizing bacteria. Appl. Biochem. Microbiol., 36: 577–581.

Coulon, F., McKew, B.A., Osborn, A.M., McGenity, T.J., and Timmis, K.N. 2006. Effects of temperature and biostimulation on oil-degrading microbial communities in temperate estuarine waters. Environmental Microbiology, 9(1): 177–186.

D'Andrea, M.A., and Reddy, G.K. 2014. Crude oil spill exposure and human health risks. Journal of Occupational and Environmental Medicine, 56: 1029–1041.

D'Andrea M.A., and Reddy, G.K. 2018. The development of long-term adverse health effects in oil spill cleanup workers of the deepwater horizon offshore drilling rig disaster. Frontiers in Public Health, Volume 6, Article No. 117. https://www.researchgate.net/publication/324776303_The_Development_of_Long-Term_Adverse_Health_Effects_in_Oil_Spill_Cleanup_Workers_of_the_Deepwater_Horizon_Offshore_Drilling_Rig_Disaster.

Das, P., Mukherjee, S., and Sen, R. 2008. Improved bioavailability and biodegradation of a model polyaromatic hydrocarbon by a biosurfactant producing bacterium of marine origin. Chemosphere, 72: 1229–1234.

De Jong, E. 1980. The effect of a crude oil spill on cereals. Environ. Pollut., 22: 187–196.

Delille, D., and Vaillant, N. 1990. The influence of crude oil on the growth of sub-antarctic marine bacteria. Antarctic Sci., 2: 655–662.

Delille, D. 2000. Response of antarctic soil assemblages to contamination by diesel fuel and crude oil. Microb. Ecol., 40: 159–168.

Delille, D., and Delille, B. 2000. Field observations on the variability of crude oil impact on indigenous hydrocarbon-degrading bacteria from sub-antarctic intertidal sediments. Mar. Environ. Res., 49: 403–417.

Delille, D., Pelletier, E., Rodriguez-Blanco, A., and Ghiglione, J.-F. 2009. Effects of nutrient and temperature on degradation of petroleum hydrocarbons in sub-antarctic coastal seawater. Polar Biol., 32: 1521–1528.

Eckford, R., Cook, F.D., Saul, D., Aislabie, J., and Foght, J. 2002. Free-living nitrogen-fixing bacteria from antarctic soils. Appl. Environ. Microbiol., 68: 5181–5185.

Ferguson, S.H., Franzmann, P.D., Revill, A.T., Snape, I., and Rayner, J.L. 2003. The effects of nitrogen and water on mineralization of hydrocarbons in diesel-contaminated terrestrial antarctic soils. Cold Reg. Sci. Technol., 37: 197–212.

Frenzel, M., James, P., Burton, S.K., Rowland, S.J., and Lappin-Scott, H.M. 2009. Towards bioremediation of toxic unresolved complex mixtures of hydrocarbons: identification of bacteria capable of rapid degradation of alkyltetralins. J. Soils Sediments, 9: 129–136.

Fuentes, S., Barra, B., Caporaso, J.G., and Seeger, M. 2015. From rare to dominant: a fine-tuned soil bacterial bloom during petroleum hydrocarbon bioremediation. Appl. Environ. Microbiol., 82: 888–896.

Gasparotti, C. 2010. Risk assessment of marine oil spills. Environmental Engineering and Management Journal, 9(4): 527–534.

Gerdes, B., Brinkmeyer, R., Dieckmann, G., and Helmke, E. 2005. Influence of crude oil on changes of bacterial communities in Arctic Sea-Ice. FEMS Microbiol Ecol., 53: 129–139.

Gibson, D.T., and Sayler, G.S. 1992. Scientific foundation for bioremediation: current status and future needs. American Academy of Microbiology, Washington, DC.

Ginn, J.S., Sims, R.C., and Murarka, I.P. 1995. *In situ* bioremediation (natural attenuation) at a gas plant waste site. pp. 153–162. *In*: Hinchee, R.E., Wilson, J.T., and Downey, D.C. (Eds.). Intrinsic Bioremediation, Battelle Press, Columbus, Ohio.

Grant, W.D., Gemmell, R.T., and McGenity, T.J. 1998. Halophiles. pp. 93–132. *In*: Horikoshi, K., and Grant, W.D. (Eds.). Extremophiles: Microbial Life in Extreme Environments. Wiley-Liss, New York.

Harayama, S., Kishira, H., Kasai, Y., and Shutsubo, K. 1999. Petroleum biodegradation in marine environments. J. Molec. Microbiol. Biotechnol., 1(1): 63–70.

Hazen, T.C., Dubinsky, E.A., DeSantis, T.Z., Andersen, G.L., Piceno, Y.M., and Singh, N. 2010. Deep-sea oil plume enriches indigenous oil-degrading bacteria. Science, 330: 204–208.

Helle, I., Mäkinen, J., Nevalainen, M., Afenyo, M., and Vanhatalo, J. 2020. Impacts of oil spills on arctic marine ecosystems: a quantitative and probabilistic risk assessment perspective. Environ. Sci. Technol., 54(4): 2112–2121.

Hoff, R.Z. 1993. Bioremediation: an overview of its development and use for oil spill cleanup. Marine Pollution Bulletin, 29: 476–481.

IAEA. 1998. Guidelines for Integrated Risk Assessment and Management in Large Industrial Areas. Report No. IAEA-TECDOC-994. International Atomic Energy Agency, Vienna, Austria. https://www-pub.iaea.org/ MTCD/Publications/PDF/te_994_prn.pdf.

IPIECA. 2013. Oil Spill Risk Assessment and Response Planning for Offshore Installations, Oil Spill Response Joint Industry Project. International Petroleum Industry Environmental Conservation Association IPIECA), London, United Kingdom.

ISO. 2018. ISO 31000 Risk Management—Guidelines. International Organization for Standardization. Geneva, Switzerland. https://www.iso.org/iso-31000-risk-management.html.

Jackson, W.A., and Pardue, J.H. 1997. Seasonal variability of crude oil respiration potential in salt and fresh marshes. J. Environ. Qual., 26: 1140–1146.

Kennicutt, M.C. 1990. Oil spill in antarctica. Environ. Sci. Technol., 24: 620–624.

Kerry, E. 1993. Bioremediation of experimental petroleum spills on mineral soils in the Vesfold Hills, Antarctica. Polar Biol., 13: 163–170.

King, M.W.G., Barker, J.F., and Hamilton, L.K. 1995. Natural attenuation of coal tar organics in groundwater. pp. 171–180. *In*: Hinchee, R.E., Wilson, J.T., and Downey, D.C. (Eds.). Intrinsic Bioremediation. Battelle Press, Columbus, Ohio.

Kose, T., Mukai, T., Takimoto K., and Okada, M. 2003. Effect of non-aqueous phase liquid on biodegradation of PAHs in spilled oil on tidal flat. Water Research, 37: 1729–1736.

Kroll, R.G. 1990. Alkalophiles. pp. 55–92. *In*: Edwards, C. (Ed.). Microbiology of extreme environments. Open University Press, Milton Keynes, United Kingdom.

Kulichevskaya, I.S., Milekhina, E.I., Borzenkov, I.A., Zvyagintseva, I.S., and Belyaev, S.S. 1992. Oxidation of petroleum hydrocarbons by extremely halophilic Archaebacteria. Microbiology, 60: 596–601.

Kushner, D.J. 1978. Life in high salt and solute concentrations. pp. 317–368. *In*: Kushner, D.J. (Ed.). Microbial Life in Extreme Environments. Academic Press, London, United Kingdom.

Kuznetsov, V.D., Zaitseva, T.A., Vakulenko, L.V., and Filippova, S.N. 1992. *Streptomyces albiaxalis* sp. nov.: A new petroleum hydrocarbon-degrading species of thermo- and Halotolerant Streptomyces. Microbiology, 61: 62–67.

LaBelle, R.P., and Anderson, C.M. 1985. The application of oceanography to oil-spill modeling for the outer continental shelf oil and gas leasing program. Marine Technology Society Journal, 19(2): 19–26.

Laffon, B., Fraga-Iriso, R., Perez-Cadahia, B., and Mendez, J. 2006. Genotoxicity associated to exposure to Prestige oil during autopsies and cleaning of oil-contaminated birds. Food Chem. Toxicol., 44(10): 1714–1723.

Leahy, J.G., and Colwell, R.R. 1990. Microbial degradation of hydrocarbons in the environment. Microbiol. Rev., 54: 305–315.

Liu, Q., Tang, J, Gao, K., Gurav, R., and Giesy, J.P. 2017. Aerobic degradation of crude oil by microorganisms in soils from four geographic regions of China Qinglong. Scientific Reports, 7: 14856 | DOI:10.1038/s41598-017-14032-5. https://www.nature.com/articles/s41598-017-14032-5.pdf.

Lyons, R., Temple, J., and Evans, D. 1999. Acute health effects of the sea empress oil spill. J. Epidemiol., Community Health, 53: 306–310.

Margesin, R., and Schinner, F. 1997a. Efficiency of indigenous and inoculated cold-adapted soil microorganisms for biodegradation of diesel oil in alpine soils. Appl. Environ. Microbiol., 63: 2660–2664.

Margesin, R., and Schinner, F. 1997b. Bioremediation of diesel-oil contaminated alpine soils at low temperatures. Appl. Microbiol. Biotechnol., 47: 462–468.

Margesin, R., and Schinner, F. (Ed.). 1999a. Cold-Adapted Organisms. Springer, New York.

Margesin, R., and Schinner, F. 1999b. Biological decontamination of oil spills in cold environments. J Chem. Technol. Biotechnol., 74: 381–389.

Margesin, R., and Schinner, F. 2001a. Biodegradation and bioremediation of hydrocarbons in extreme environments. Appl. Environ. Microbiol., 56: 650–663.

Margesin, R., and Schinner, F. 2001b. Potential of halotolerant and halophilic microorganisms for biotechnology. Extremophiles, 5: 73–83.

Margesin, R., Hämmerle, M., and Tscherko, D. 2007a. Microbial activity and community composition during bioremediation of diesel-oil contaminated soil: effects of hydrocarbon concentration, fertilizers and incubation time. Microb. Ecol., 53: 259–269.

Margesin, R., Neuner, G., and Storey, K.B. 2007b. Cold-loving microbes, plants, and animals—fundamental and applied aspects. Naturwissenschaften, 94: 77–99.

McAllister, P.M., Chiang, C.Y., Salanitro, J.P., Dortch, I.J., and Williams, P. 1995. Enhanced aerobic bioremediation of residual hydrocarbon sources. pp. 67–76. *In*: Hinchee, R.E., Wilson, J.T., and Downey, D.C. (Eds.). Intrinsic Bioremediation. Battelle Press, Columbus, Ohio.

McCauley, L. 2010. Will the BP oil spill affect our health? American Journal of Nursing, 110(9): 54–56.

Means, J.C. 1995. Influence of salinity upon sediment-water partitioning of aromatic hydrocarbons. Mar. Chem., 51: 3–16.

Michaud, L., Giudice, A.L., Saitta, M., De Domenico, M., and Bruni, V. 2004. The biodegradation efficiency on diesel oil by two psychrotrophic antarctic marine bacteria during a two-month-long experiment. Mar. Pollut. Bull., 49: 405–409.

Michelsen, T.C., and Petito Boyce, C. 1993. Cleanup standards for petroleum hydrocarbons. Part 1. Review of methods and recent developments. Journal of Soil Contamination, 2(2): 1–16.

Mille, G., Almallah, M., Bianchi, M., Van Wambeke, F., and Bertrand, J.C. 1991. Effect of salinity on petroleum biodegradation. Fresenius J. Anal. Chem., 339: 788–791.

Mohn, W.W., and Stewart, G.R. 2000. Limiting factors for hydrocarbon degradation at low temperature in arctic soils. Soil Biol. Biochem., 32: 1161–1172.

Mohn, W., Radziminski, C., Fortin, M.-C., and Reimer, K. 2001. On site bioremediation of hydrocarbon-contaminated arctic tundra soils in inoculated biopiles. Appl. Microbiol. Biotechnol., 57: 242–247.

Morita, R.Y. 1975. Psychrophilic bacteria. Bacteriol. Rev., 39: 144–167.

Müller, R., Antranikian, G., Maloney, S., and Sharp, R. 1998. Thermophilic degradation of environmental pollutants. pp. 155–169. *In*: Antranikian, G. (Ed.). Biotechnology of Extremophiles. Advances in Biochemical Engineering/Bio-technology, Volume 61. Springer, New York.

Nedwell, D.B. 1999. Effect of low temperature on microbial growth: lowered affinity for substrates limits growth at low temperature. FEMS Microbiol. Ecol., 30: 101–111.

Nelson, J.R., and Grubesic, T.H. 2017. Oil spill modeling: risk, spatial vulnerability, and impact assessment. Progress in Physical Geography, 42. https://journals.sagepub.com/doi/10.1177/0309133317744737.

Niehaus, F., Bertoldo, C., Kähler, M., and Antranikian, G. 1999. Extremophiles as a source of novel enzymes for industrial application. Appl. Microbiol. Biotechnol., 51: 711–729.

Nissen-Lie, T., Brude, O.W., Aspholm, O., and Taylor, P.M. 2014. Developing a guideline for oil spill risk assessment and response planning for offshore installations. Proceedings. International Oil Spill Conference. Savannah, Georgia. May 5–8. Page 314–327. https://www.researchgate.net/publication/269850901_Developing_A_Guideline_For_Oil_Spill_Risk_Assessment_And_Response_Planning_For_Offshore_Installations.

NOAA. 2013. Characteristics of Response Strategies: A Guide for Spill Response Planning in Marine Environments. United States Department of Commerce, Washington, DC.

Norris, R.D. 1994. Handbook of Bioremediation. CRC Press, Boca Raton, Florida.

Norris, P.R., and Johnson, D.B. 1998. Acidophilic microorganisms. pp. 133–153. *In*: Horikoshi, K., and Grant, W.D. (Eds.). Extremophiles: Microbial Life in Extreme Environments. Wiley-Liss, New York.

Okpokwasili, G.C., and Okorie, B.B. 1988. Biodegradation potentials of micro-organisms isolated from car engine lubricating oil. Tribology Int., 21: 215–220.

Ordinioha, B., and Brisibe, S. 2013. The human health implications of crude oil spills in the Niger delta, Nigeria: An Interpretation of Published Studies. Journal of the Nigeria Medical Association, 54(1): 10–16.

Oren, A., Gurevich, P., Azachi, M., and Hents, Y. 1992. Microbial degradation of pollutants at high salt concentrations. Biodegradation, 3: 387–398.

Ornitz, B., and Champ, M. 2002. Oil Spill First Principles: Prevention and Best Response. Elsevier Science Publishers, Amsterdam, The Netherlands.

Ozhan, K., Parsons, M.L., and Bargu, S. 2014. How were phytoplankton affected by the deepwater horizon oil spill? BioScience, 64(9): 829–836.

Pattillo, R. 2010. Potential health problems related to oil spill in gulf of Mexico. Nurse Educator, 35(5): 185. https://pubmed.ncbi.nlm.nih.gov/20729670/.

Perez-Cadahia, B., Lafuente, A., Cabaleiro, T., Pasaro, E., Mendez, J., and Laffon, B. 2007. Initial study on the effects of prestige oil on human health. Environ. Int., 33(2): 176–185.

Perez-Cadahia, B., Laffon, B., Valdiglesias, V., Pasaro, E., and Mendez, J. 2008a. Cytogenetic effects induced by prestige oil on human populations: the role of polymorphisms in genes involved in metabolism and DNA repair. Mutation Res., 653(1-2): 117–123.

Perez-Cadahia, B., Mendez, J., Pasaro, E., Lafuente, A., Cabaleiro, T., and Laffon, B. 2008b. Biomonitoring of human exposure to prestige oil: effects on DNA and endocrine parameters. Environ. Health Insights, 2: 83–92.

Peters, K.E., Scheuerman, G.L., Lee, C.Y., Moldovan, J.M., and Reynolds, R.N. 1992. Effects of refinery processes on biological markers. Energy and Fuels, 6: 560–577.

Peterson, C.H., McDonald, L.L., Green, R.H., and Erickson, W.P. 2001. Sampling design begets conclusions: the statistical basis for detection of injury to and recovery of shoreline communities after the exxon valdez oil spill. Marine Ecology Progress Series, 210: 255–283. https://www.int-res.com/articles/meps/210/m210p255.pdf.

Plice, M.J. 1948. Some effects of crude petroleum on soil fertility. Soil Sci. Soc. Proc., 413–416.

Poremba, K., Gunkel, W., Lang, S., and Wagner, F. 1991. Marine biosurfactants, III. Toxicity Testing with Marine Microorganisms and Comparison with Synthetic Surfactants. Z. Naturforsch., 46c: 210–216.

Powell, S.M., Harvey, P.McA., Stark, J.S., Snape, I., and Riddle, M.J. 2007. Biodegradation of petroleum products in experimental plots in Antarctic marine sediments is location dependent. Mar. Pollut. Bull., 54: 434–440.

Prieur, D., and Marteinsson, V.T. 1998. Prokaryotes living under elevated hydrostatic pressure. pp. 23–35. *In*: Antranikian, G. (Ed.). Biotechnology of Extremophiles. Advances in Biochemical Engineering/Biotechnology, Volume 61. Springer, New York.

Rhykerd, R.L., Weaver, R.W., and McInnes, K.J. 1995. Influence of salinity on bioremediation of oil in soil. Environ. Pollut., 90: 127–130.

Rodríguez Trigo, G., Zock, J.P., Pozo Rodríguez, F., Gómez, F.P., Monyarch, G., Bouso, L., Coll, M.D., Verea, H., Antó, J.M., and Fuster, C. 2010. Health changes in fishermen 2 years after clean-up of the prestige oil spill. Ann. Intern. Med., 153: 489–498.

Rodrigues, R.V., Miranda-Filho, K.C., Gusmao, E.P., Moreira, C.B., Romano, L.A., and Sampaio, L.A. 2010. Deleterious effects of water-soluble fraction of petroleum, diesel and gasoline on marine Pejerrey Odontesthes argentinensis Larvae. Sci. Total Environ., 408(9): 2054–2059.

Rosa, A.P., and Triguis, J.A. 2007. Bioremediation process on the brazil shoreline. Env. Sci. Pollut. Res., 14(7): 470–476.

Ruberto, L., Vazquez, S.C., and MacCormarck, W.P. 2003. Effectiveness of the natural bacteria *Xora*, biostimulation, and bioaugmentation on the bioremediation of a hydrocarbon-contaminated Antarctic soil. Inter. Biodeter. Biodegrad., 52: 115–125.

Schulz, D., Passeri, A., Schmidt, M., Lang, S., Wagner, F., Wray, V., and Gunkel, W. 1991. Marine Biosurfactants, I. Screening for Biosurfactants among Crude Oil Degrading Marine Microorganisms from the North Sea. Z. Naturforsch., 46c: 197–203.

Siddique, T., Okeke, B.C., Arshad, M., and Frankenberger, W.T. 2002. Temperature and pH effects on biodegradation of hexachlorocyclohexane isomers in water and a soil slurry. J. Agric. Food Chem., 50(18): 5070–5076.

Singh, A.K. 2016. Engineered Nanoparticles. Singh Elsevier BV, Amsterdam, The Netherlands. Page 343–450.

Siron. R., Pelletier, E., Delille, D., and Roy, S. 1993. Fate and effects of dispersed crude oil under icy conditions simulated in Mesocosms. Mar. Environ. Res., 35: 273–302.

Siron, R., Pelletier, E., and Brochu, C. 1995. Environmental factors influencing the biodegradation of petroleum hydrocarbons in cold seawater. Arch. Environ. Contam. Toxicol., 28: 406–416.

Smith, R.A., Slack, J.R., Wyant, T., and Lanfear, K.J. 1982. The oil spill risk analysis model of the U.S. Geological Survey. USGS Professional Paper 1227. United States Geological Survey, Reston, Virginia.

Solomon, G., and Janssen, S. 2010. Health effects of the gulf oil spill. Journal of the American Medical Association, 304(10): 1118–1119. https://jamanetwork.com/journals/jama/article-abstract/186531.

Spaulding, M.L. 2017. State of the art review and future directions in oil spill modeling. Marine Pollution Bulletin, 115(1-2): 7–19.

Speight, J.G. 1996. Environmental Technology Handbook. Taylor & Francis, Washington, DC.

Speight, J.G., and Lee, S. 2000. Environmental Technology Handbook. 2nd Edition. Taylor & Francis, New York.

Speight, J.G. 2005. Environmental Analysis and Technology for the Refining Industry. John Wiley & Sons Inc., Hoboken, New Jersey.

Speight, J.G., and Arjoon, K. 2012. Bioremediation of petroleum and petroleum products. Scrivener Publishing, Beverly, Massachusetts.

Speight, J.G. 2014. The Chemistry and Technology of Petroleum 5th Edition. CRC Press, Taylor & Francis Group, Boca Raton, Florida.

Speight, J.G. 2020. Global Climate Change Demystified. Scrivener Publishing, Beverly, Massachusetts.

Stapleton, R.D., Savage, D.C., Sayler, G.S., and Stacey, G. 1998. Biodegradation of aromatic hydrocarbons in an extremely acidic environment. Appl. Environ. Microbiol., 64: 4180–4184.

Stelmaszewski, A. 2009. Determination of Petroleum Pollutants in Coastal Waters of the Gulf of Gdańsk. Oceanologia, 51(1): 85–92.

Stetter, K.O. 1998. Hyperthermophiles: isolation, classification, and properties. pp. 1–24. *In*: Horikoshi, K., and Grant, W.D. (Eds.). Extremophiles: Microbial Life in Extreme Environments. Wiley-Liss, New York.

Stout, S.A., Uhler, A.D., and McCarthy, K.J. 2001. A Strategy and Methodology for Defensibly Correlating Spilled Oil to Source Candidates: Environmental Forensics, 2: 87–98.

Stout, S.A., Uhler, A.D., and McCarthy, K.J. 2005. Middle Distillate Fuel Fingerprinting using Drimane-Based Bicyclic Sesquiterpanes. Environmental Forensics, 6: 241–251.

Stout, S.A., and Wang, Z. 2016. Chemical fingerprinting methods and factors affecting petroleum fingerprints in the environment. Standard Handbook Oil Spill Environmental Forensics Second Edition. Academic Press Inc., London United Kingdom. Page 61–129.

Suter, G.W. 2007. Ecological Risk Assessment, Florida. CRC Press. Taylor & Francis Group, Boca Raton, Florida.

Swannell, R.P.J., Lee, K., and McDonagh, M. 1996. Field evaluations of marine oil spill bioremediation. Microbiological Reviews, 60: 342–365.

Sweed, H.G., Bedient, P.B., and Hutchins, S.R. 1996. Surface application system for *in situ* groundwater bioremediation site characterization and modeling. Ground Water, 34(2): 211–222.

Trindade, P.V.O., Sobral, L.G., Rizzo, A.C.L., Leite, S.G.F., Lemos, J.L.S., Millioli, V.S., and Soriano, A.U. 2002. Evaluation of the Biostimulation and Bioaugmentation Techniques in the Bioremediation Process of Petroleum Hydrocarbon Contaminated Soil. Proceedings. 9th Annual International Petroleum Environmental Conference, October. Albuquerque, New Mexico.

Ulrich, A.C., Guigard, S.E., Foght, J.M., Semple, K.M., Pooley, K., Armstrong, J.E., and Biggar, K.W.E. 2009. Effect of salt on aerobic biodegradation of hydrocarbons in contaminated groundwater. Biodegradation, 20: 27–38.

US Department of Commerce. 1983. Assessing the Social Costs of Oil Spills: The Amoco Cadiz Case Study. United States Department of Commerce, Washington, DC.

US EPA. 1990. Interim Report. Oil Spill Bioremediation Project. U.S. Environmental Protection Agency, Office of Research and Development, Washington, DC.

US EPA. 1998. Environmental Risk Assessments of Oil and Gas Activities Using National Security and Civilian Data Sources. United States Environmental Protection Agency, Washington, DC. https://cfpub.epa.gov/ncea/risk/hhra/recordisplay.cfm?deid=75587.

US EPA. 2003. Aerobic Degradation of Oily Wastes: A Field Guidance Book for Federal On-Scene Coordinators. Version 1.0. US Environmental Protection Agency Response and Prevention Branch, Region 6 South Central., Dallas Texas.

US EPA. 2006. *In situ* and *Ex Situ* Biodegradation Technologies for Remediation of Contaminated Sites. Report No. EPA/625/R-06/015. Office of Research and Development National Risk Management Research Laboratory, United States Environmental Protection Agency, Cincinnati, Ohio.

US EPA. 2019. What is an Oil Spill Contingency Plan, Part 109 – Criteria for State, Local and Regional Oil Removal Contingency Plans. Report No. 40 CFR. United States Environmental Protection Agency, Washington, DC.

Varjani, S.J., and Gnansounou, E. 2017. Microbial dynamics in petroleum oilfields and their relationship with physiological properties of petroleum oil reservoirs. Bioresour. Technol., 245: 1258–1265.

Velez, M.I., Conde, D., Lozoya, J.P. Rusak, J.A., García-Rodríguez, F., Seitz, C., Harmon, T., Perillo, Gerardo, M.E., Escobar, J., and Vilardy, S.P. 2018. Paleoenvironmental reconstructions improve ecosystem services risk assessment: case studies from two coastal lagoons in South America. Water, 10(10): 1350. https://www.mdpi.com/2073-4441/10/10/1350.

Venosa, A.D., Suidan, M.T., Wrenn, B.A., Strohmeier, K.L., Haines, J.R., Eberhart, B.L., King, D., and Holder, E. 1996. Bioremediation of an experimental oil spill on the shoreline of Delaware Bay. Environmental Science & Technology, 30(5): 1764.

Vidali, M. 2001. Bioremediation: An Overview. Pure and Applied Chemistry, 73(7): 1163–1172.

Wagner-Döbler, I., Bennasar, A., Vancanneyt, M., Strömpl, C., Brümmer, I., Eichner, C., Grammel, I., and Moore, E.R.B. 1998. Microcosm enrichment of biphenyl-degrading microbial communities from soils and sediments. Appl. Environ. Microbiol., 64: 3014–3022.

Walker, B. Jr. 2010. Deepwater horizon oil spill. Environ. Health, 73(4): 49.

Ward, D.M., and Brock, T.D. 1978. Hydrocarbon degradation in hypersaline environments. Appl. Environ. Microbiol., 35: 353–359.

Wardell, L.J. 1995. Potential for bioremediation of fuel-contaminated soil in Antarctica. J. Soil Contam., 4: 111–121.

Whitehouse, B.G. 1984. The effects of temperature and salinity on the aqueous solubility of polynuclear aromatic Hydrocarbons. Mar. Chem., 14: 319–332.

Whyte, L.G., Hawari, J., Zhou, E., Bourbonniere, L., Inniss, W.E., and Greer, C.W. 1998. Biodegradation of variable-chain-length alkanes at low temperature by a Psychrotrophic *Rhodococcus* sp. Appl. Environ. Microbiol., 64: 2578–2584.

Whyte, L.G., Bourbonnière, L., Bellerose, C., and Greer, C.W. 1999. Bioremediation assessment of hydrocarbon-contaminated soils from the high arctic. Biorem. J., 3: 69–79.

Wright, A.L., Weaver, R.W., and Webb, J.W. 1997. Oil Bioremediation in salt marsh mesocosms as influenced by N and P fertilization, flooding and season. Water Air Soil Poll., 95: 179–191.

Yakimov, M.M., Giuliano, L., Bruni, V., Scarfi, S., and Golyshin, P.N. 1999. Characterization of antarctic hydrocarbon-degrading bacteria capable of producing Bioemulsifiers. Microbiologica, 22: 249–256.

Yang, Y., Wang, J., Liao, J., Xie, S., and Huang, Y. 2015. Abundance and diversity of soil petroleum hydrocarbon-degrading microbial communities in oil exploring areas. Appl. Microbiol. Biotechnol., 99: 1935–1946. Society. Volume 14(4): E22–E31.

Zhu, X., Venosa, A.D., Suidan, M.T., and Lee, K. 2001. Guidelines for the Bioremediation of Marine Shorelines and Freshwater Wetlands, US EPA Report, September.

Zhu, X., Venosa, A.D., and Suidan, M.T. 2004. Literature Review on the Use of Commercial Bioremediation Agents for Cleanup of Oil-Contaminated Environments. Report No. EPA/600/R-04/075. National Risk Management Research Laboratory, United States Environmental Protection Agency, Cincinnati, Ohio. July.

Zock, J.-P., Pozo-Rodríguez, F., and Barberà, J.A. 2011. Health effects of oil spills: lessons from the prestige. American Journal of Respiratory and Critical Care Medicine, 184(10): 1094–1096. https://www.atsjournals.org/doi/pdf/10.1164/rccm.201102-0328ED.

Zock, J.P., Rodriguez-Trigo, G., Rodriguez-Rodriguez, E., Espinosa, A., Pozo-Rodriguez, F., Gomez, F., Fuster, C., Castano-Vinyals, G., Anto, J.M., and Barbera, J.A. 2012. Persistent respiratory symptoms in clean-up workers 5 years after the prestige oil spill. Occup. Environ. Med., 69(7): 508–513.

CHAPTER 8

Biodegradation of the Constituents of Crude Oil and Crude Oil Products

The contamination of soils and aquifers by spilled crude oil and crude oil products is a persistent and widespread pollution problem which causes ecological disturbances and the associated health implications (Bundy et al., 2002; Okoh, 2006; Salam et al., 2011). Once crude oil (or a crude oil product) is released and comes into contact with water, air, and the necessary salts, microorganisms present in the environment, the natural process of biodegradation begins (Antai, 1990; Davies et al., 2001; Aluyor and Ori-jesu, 2009). However, some of the crude oil-related pollutants are carcinogenic and mutagenic (Miller and Miller, 1981; Obayori et al., 2009b).

The recognized mechanical and chemical methods for remediation of hydrocarbon-polluted environment are often expensive, technologically complex and lack public accepta.nce ((Speight, 1996; Speight and Lee, 2000; Vidali, 2001; Speight, 2005). Thus, bioremediation is often the method of choice for effective removal of hydrocarbon pollutants from a variety of ecosystems (Okoh and Trejo-Hernandez, 2006). In fact, crude oil and crude oil products are a rich, source of carbon and the reaction of the hydrocarbon derivatives contained therein with aerial oxygen (with the release of carbon-dioxide) is promoted by a variety of microorganisms (Odu, 1977; Atlas, 1981; Atlas and Bartha, 1992; Steffan et al., 1997).

However, the rate of microbial degradation of hydrocarbon derivatives is affected by several physicochemical and biological parameters including (1) the number and species of microorganisms present, (2) the conditions for microbial degradation activity, such as the presence of nutrient, oxygen, (3) the pH and temperature, (4) the quality, quantity and bioavailability or bioaccessibility of the contaminants; and (5) the soil or water characteristics such as particle size distribution (Chapter 5) (Atlas, 1991; Freijer et al., 1996; Margesin and Schinner, 1997a, 1997b; Dandie et al., 2010).

Hydrocarbon degrading bacteria and fungi are mainly responsible for the mineralization of crude oil-related pollutants in which the hydrocarbon derivatives are converted to carbon dioxide and water which are then distributed in the diverse ecosystems (Leahy and Colwell, 1990; Song et al., 1990; Bennet et al., 2002). Furthermore, the population of microorganisms found in a polluted environment will degrade crude oil-related constituents differently and at a different rate than microorganisms in a relatively clean environment (Obire, 1990; Obire, 1993; Obire and Okudo, 1997; Obire and Nwaubeta, 2001).

However, it is uncommon to find organisms that could effectively degrade both aliphatic constituents and aromatic constituents possibly due to differences in metabolic routes and pathways for the degradation of the two classes of hydrocarbon derivatives. There are indications of the existence of bacterial species with propensities for simultaneous degradation of aliphatic hydrocarbon derivatives and aromatic hydrocarbon derivatives (Amund et al., 1987; Obayori et al., 2009b). This rare ability may be as a result of long exposure of the organisms to different hydrocarbon pollutants resulting in genetic alteration and *acquisition* of the appropriate degradative genes.

Moreover, it is essential to recognize that the environmental impact of crude oil spills is dependent on previous hydrocarbon exposures and the adaptive status of the local microbiota (Greenwood et al., 2009). The different structural and functional response of microbial sub-groups to different hydrocarbon derivatives confirms that the overall response of biota is sensitive to crude oil composition. This suggests that the preferred response to anticipated contaminants may be engineered by pre-exposure to representative substrates. The controlled adaption of microbes to a threatening contaminant is the basis of proactive bioremediation technology, including the augmentation of newly contaminated sites with locally remediated soil in which the biota had already been adapted (bioaugmentation; Chapter 1).

Thus, establishing the chemical history of recently contaminated regions is an important aspect of environmental bioremediation. The premise being that microbial species adapted through a history of exposure to more bioavailable crude oil hydrocarbon derivatives is less severely impacted by a spill than microbes with no such pre-exposure or adaptation (Page et al., 1996; Peters et al., 2005). Indeed, the diversity of crude oil hydrocarbon degraders in most natural environments may be significant but, in the absence of a previous pollution history, the numbers of the microbes may be low due to lack of prior stimulus (Swanell et al., 1996).

8.1 Biodegradability

The biodegradability of any crude oil constituent is a measure of the ability of that constituent to be metabolized (or co-metabolized) by bacteria or other microorganisms through a series of biochemical process, which include ingestion by organisms as well as microbial degradation (Payne and McNabb, 1984). The chemical characteristics of the contaminants influence biodegradability; in addition, the location and distribution of crude oil contamination in the subsurface can significantly influence the likelihood of success for bioremediation.

The biodegradability of crude oil and crude oil product is inherently influenced by the chemical and physical composition of the substrate upon which the bacteria are acting (Chapter 1, Chapter 2). For example, crude oil is quantitatively biodegradable and kerosene, which consists almost exclusively of medium chain-length alkane derivatives, is completely biodegradable under suitable conditions but for heavy asphaltic crudes, approximately only 6 to 10% of the material oil may be biodegradable within a reasonable time period, even when the conditions are favorable for biodegradation (Bartha, 1986; Okoh et al., 2002; Okoh et al., 2003; Okoh, 2006). In addition, biodegradation of crude oil constituents and the constituents of crude oil products can be enhanced by use of a consortium of different bacteria compared to the activity of single bacterium species (Ghazali et al., 2004; Milić et al., 2009).

In addition to the composition of the crude oil-related substrate, crude oil and crude oil products introduced to the environment are immediately subject to a variety of changes caused by physical, chemical, and biological effects—collectively (but incorrectly) referred to as weathering. Physical and chemical processes include (1) evaporation, (2) dissolution of crude oil constituents in a water system (or aquifer), (3) dispersion, (4) photochemical oxidation, (5) formation of water-oil emulsions, and (6) adsorption on to suspended particulate material. These processes are not sequential and typically occur simultaneously and cause important changes in the composition and properties of the original pollutant, which in turn may affect the rate or effectiveness of biodegradation.

Specifically, the biodegradation of crude oil may (1) raise the viscosity, which adversely reduces the ability of the degraded product to flow, (2) decreases the API gravity, (3) decreases the hydrocarbon content, thereby increasing the residuum content, (4) increases the concentration of certain metals, (5) increases the sulfur content, (6) increases oil acidity, and (7) adds compounds such as carboxylic acid derivatives and phenol derivatives. All of these changes are seen in the product of the product relative to the unchanged (non-biodegraded) crude oil (Miiller et al., 1987).

The commercial practice of bioremediation focuses primarily on the cleaning up of crude oil hydrocarbon derivatives (Del'Arco and de França, 1999). Thus, successful application of bioremediation technology to a contaminated ecosystem requires knowledge of the characteristics of the site and the parameters that affect the microbial biodegradation of pollutants (Sabate et al., 2004). However, a number of factors have been recognized that have the potential to affect the biodegradation of crude oil constituents (Table 8.1) (Hunkeler et al., 2008).

Despite the difficulty of degrading certain fractions, some hydrocarbon derivatives are among the most easily biodegradable naturally occurring compounds. Biodegradation gradually destroys crude oil-related spills by the sequential metabolism of various classes of compounds present in the oil (Bence et al., 1996). When biodegradation occurs in an oil reservoir, the process dramatically affects the fluid properties (Miiller et al., 1987) and hence the value and producibility of an oil accumulation. Specifically, crude oil biodegradation typically raises viscosity of the residual material (which reduces oil producibility) and reduces the API gravity (which reduces the value of the produced oil). It increases the asphaltene content (relative to the saturated and aromatic hydrocarbon content and the starting material), the concentration of certain metals, the sulfur content and oil acidity. Although the constituents of the asphaltene fraction are traditionally considered to be recalcitrant to microbial alteration and while the amount of asphaltenes may increase because some of the resin constituents are converted to asphaltene-type constituents during the process because of an increase in polarity, it is possible to biodegrade these constituents under the appropriate circumstances (Adarme et al., 1990; Liao et al., 2009; Ali et al., 2012; Tavassoli et al., 2012).

There are indications that crude oil biodegradation involves more biological components than just the microorganisms that directly attack crude oil constituents (the primary degraders) and shows that the primary degraders interact with these components (Head et al., 2006). In addition, primary degraders need to compete with other microorganisms for limiting nutrients and the non-crude oil-degrading microorganisms can be affected by metabolites and other compounds that are released by oil-degrading bacteria and vice versa.

The environment, having been exposed to the input of crude oil and curd oil products for decades, can assimilate the hydrocarbon derivatives under the appropriate conditions. However, areas of particular concern are low energy environments common to estuarine systems. These environments, such as marshes, mud flats, and subtidal areas, are vital to marine fisheries and estuarine productivity and are especially sensitive to contaminant impacts. These systems are particularly vulnerable to impacts of crude oil where research has shown crude oil can persist in these systems for years.

The removal processes for crude oil in wetlands are (1) evaporation, (2) photo-oxidation, (3) dissolution of specific constituents, (4) microbial degradation, and (5) physical flushing. However, once incorporated into the sediment, biodegradation and dissolution are the primary removal mechanisms. Crude oil biodegradation in wetland environments can be limited by anoxia and nutrient availability. Consequently, estuarine wetlands are the most vulnerable of the low-energy intertidal areas to crude oil spills.

Table 8.1: Examples of Common Factors Affecting the Biodegradation of Crude Oil Hydrocarbon Derivatives (Bartha, 1986).

Factor*	Comment
Composition	Amount, structure, toxicity
State	Adsorption, aggregation, dispersion, spreading
Temperature	Degradation rate, evaporation
Weathering	Evaporation, oxidation

* Listed alphabetically and not by influence.

The early stages of crude oil biodegradation (loss of n-paraffin derivatives followed by loss of acyclic derivatives iso-prenoid derivatives) can be readily detected by gas chromatography (GC) analysis of the crude oil. However, in heavily biodegraded crude oils, gas chromatographic analysis alone cannot distinguish differences in biodegradation due to interference of the unresolved complex mixture that dominates the gas chromatographic traces of heavily degraded crude oils. Among such crude oils, differences in the extent of biodegradation can be assessed using gas chromatography-mass spectrometry (GC-MS) to quantify the concentration of biomarkers, especially those biomarkers that have with different resistance to biodegradation (Chapter 9).

During biodegradation, the properties of the crude oil fluid changes because different classes of compounds in crude oil have different susceptibilities to biodegradation (Goodwin et al., 1983). The early stages of biodegradation (in addition to any evaporation effects) are characterized by the loss of n-paraffins (n-alkane derivatives or branched alkane derivatives) followed by loss of acyclic iso-prenoid derivatives (such as norpristane, pristane, and phytane). Compared with those compound groups, other compound classes (such as highly branched and cyclic saturated hydrocarbon derivatives as well as aromatic compounds) are more resistant to biodegradation. However, even the more-resistant compound classes are eventually destroyed as biodegradation proceeds.

8.1.1 Conditions for Biodegradation

The composition and inherent biodegradability of the crude oil hydrocarbon pollutant is, perhaps, the first and most important consideration when the suitability of a cleanup approach is to be evaluated (Atlas, 1975). Heavier crude oil is generally much more difficult to biodegrade than lighter ones, just as heavier crude oils could be suitable for inducing increased selection pressure for the isolation of crude oil hydrocarbon degraders with enhanced efficiency. Also, the amount of heavy crude oil metabolized by some bacterial species increases with increasing concentration of the contaminant while degradation rates may appear to be more pronounced within a specific concentration range (Okoh et al., 2002; Rahman et al., 2002).

An important aspect of the conditions for biodegradation at a spill site is the activity of microorganisms is the ability of the organisms to produce enzymes to catalyze metabolic reactions, which governed by the genetic composition of the organism(s). Enzymes produced by microorganisms in the presence of carbon sources cause initial attack on the hydrocarbon constituents while other enzymes are utilized to complete the breakdown of the hydrocarbon. Thus, lack of an appropriate enzyme either prevents attack or is a barrier to complete hydrocarbon degradation.

Biodegradation of crude oil-related constituents by bacteria can occur under both aerobic (oxic) and anaerobic (anoxic) conditions (Zengler et al., 1999), usually by the action of different consortia of micro-organisms. In the subsurface, biodegradation occurs primarily under anaerobic conditions, mediated by sulfate reducing bacteria in cases where dissolved sulfate is present (Holba et al., 1996), or methanogenic bacteria in cases where dissolved sulfate is low (Bennett et al., 1993). Although subsurface oil biodegradation does not require oxygen, there is a requirement for the presence of essential nutrients (such as nitrogen, phosphorus, potassium), which can be provided by dissolution/ alteration of minerals in the water layer. In the absence of nutrients, the potential for hydrocarbon degradation in anoxic sediments is markedly reduced (Dibble and Bartha, 1976).

In situ groundwater can be an effective medium for the biodegradation of crude oil hydrocarbon derivatives. However, there are exceptions, the most notable of which is such as methyl t-butyl ether, MTBE, which is not a hydrocarbon:

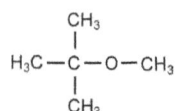

$$H_3C-\underset{\underset{CH_3}{|}}{\overset{\overset{CH_3}{|}}{C}}-O-CH_3$$

Methyl t-butyl ether (MTBE)

The short-chain, low molecular weight, more water-soluble constituents are degraded more rapidly and to lower residual levels than are long-chain, high molecular weight, less soluble constituents. However, as with all bioremediation efforts, crude oil and crude oil products (such as residual fuel oil and asphalt) typically have a high-boiling—often non-volatile—residuum) fraction composed of resin constituents and asphaltene constituents, which is composed of complex, polynuclear aromatic systems (Speight, 1994, 2014).

Microbial utilization of hydrocarbon derivatives (being fully reduced substrates) requires an exogenous electron sink. In the initial attack, this electron sink has to be molecular oxygen. In the subsequent steps too, oxygen is the most common electron sink. In the absence of molecular oxygen, further biodegradation of partially oxygenated intermediates may be supported by nitrate or sulphate reduction.

Uptake and utilization of water insoluble substrates, such as crude oil alkane derivatives, require specific physiological adaptations of the microorganisms. The synthesis of specific amphiphilic molecules (i.e., biosurfactants) is often surmised to be a prerequisite for either specific adhesion mechanisms to large oil drops or emulsification of oil, followed by uptake of submicron oil droplets. In fact, various species of bacteria have been observed to adopt the requisite strategy to deal with water insoluble substrates, such as hydrocarbon derivatives (Rosenberg 1991). Hence, to facilitate hydrocarbon uptake through the hydrophilic outer membrane, many hydrocarbon-utilizing microorganisms produce cell wall-associated or extracellular surface-active agents (Haferburg et al., 1986). This includes such low molecular weight compounds such as fatty acids, triacyl-glycerol derivatives, and phospholipids, as well as the heavier glycolipids, such as *emulsan* (Cirigliano and Carman 1984).

By way of definition, glycolipids are lipids with a carbohydrate attached by a glycosidic (covalent) bond. Their role is to maintain the stability of the cell membrane and to facilitate cellular recognition, which is crucial to the immune response and in the connections that allow cells to connect to one another to form tissues.

Emulsan is the extracellular form of a polyanionic, cell-associated heteropolysaccharide produced by the oil-degrading bacterium *Acinetobacter calcoaceticus* RAG-1 (Rosenberg et al., 1979b; Zuckerberg et al., 1979). The biopolymer stabilizes emulsions of hydrocarbon derivatives in water and has optimal activity when a mixture of aromatic and aliphatic components is present, such as in crude oil (Rosenberg et al., 1979a). The activity of the amphipathic emulsifier is due primarily to its high affinity for the oil-water interface (Zosim et al., 1982) and its ability to orient itself at the interface to form a hydrophilic film around the oil droplets (Zosim et al., 1982; Shabtai et al., 1986; Zosim et al., 1986).

Studies with bacteriophages, antibodies, and emulsan-deficient mutants have demonstrated that: (1) as the cells approach stationary phase, emulsan accumulates on the cell surface before release into the medium (Goldman et al., 1982), (2) cell-bound emulsan serves as a specific receptor and acts as stabilizer for the oil-water interface (Pines and Gutnick, 1981, 1984a), (3) this indicates that the cell-bound form of emulsan is required for growth on crude oil—species without cell-bound emulsan no longer grow well on crude oil-related (Rosenberg et al., 1983a, 1983b; Pines and Gutnick, 1984b, 1986), (4) the affinity of emulsan for the oil-water interface suggests that it might affect microbial degradation of emulsified oils (Gutnick and Minas, 1987).

In another work, crude oil was also treated with purified emulsan, the heteropolysaccharide bioemulsifier produced by *Acinetobacter calcoaceticus* RAG-1. A mixed bacterial population as well as nine different pure cultures isolated from various sources was tested for biodegradation of emulsan-treated and untreated crude oil. Biodegradation was measured both quantitatively and qualitatively. Biodegradation of linear alkane derivatives and other saturated hydrocarbon derivatives, both by pure cultures and by the mixed population, was reduced after emulsan pretreatment. In addition, degradation of aromatic compounds by the mixed population was also reduced in emulsan-treated oil. In sharp contrast, aromatic biodegradation by pure cultures was either unaffected or slightly stimulated by emulsification of the oil (Foght et al., 1989).

8.1.2 Effect of Nutrients

A nutrient a substance that provides nourishment essential for growth and the maintenance of life. Different types of nutrients (primarily nitrogen and phosphorus) have been applied to improve the degradation of hydrocarbon derivatives from crude oil and crude oil products, including classic (water soluble) nutrients and oleophilic and slow-release fertilizers (Delille and Pelletier, 2009).

Bioavailability is one main factor that influences the extent of biodegradation of hydrocarbon derivatives. Generally, hydrocarbon derivatives have low-to-poor solubility in water and, as a result, are adsorbed on to clay or humus fractions, so they pass very slowly to the aqueous phase where they are metabolized by microorganisms. Cyclodextrins are natural compounds that form soluble inclusion complexes with hydrophobic molecules and increase degradation rate of hydrocarbon derivatives *in vitro*.

In the perspective of an *in situ* application, β-cyclodextrin does not increase eluviation (the lateral or downward movement of the suspended material in soil through the percolation of water) of hydrocarbon derivatives through the soil and consequently does not increase the risk of groundwater pollution (Sivaraman et al., 2010). Furthermore, the combination of bioaugmentation and enhanced bioavailability due to β-cyclodextrin was effective for a full degradation (Bardi et al., 2003). Thus, *in situ* bioremediation of polynuclear aromatic hydrocarbon-polluted soil can be improved by the augmentation of degrading microbial populations and by the increase of hydrocarbon bioavailability (Bardi et al., 2007).

Inadequate mineral nutrients, especially nitrogen, and phosphorus, often limit the growth of hydrocarbon utilizers in water and soils. Iron has been reported to be limiting only in clean, offshore seawater (Swannell et al., 1996). Sulfur, in form of sulfate ions, is plentiful in seawater, but could be limiting in some freshwater environments. The slight alkaline pH of seawater seems to be quite favorable for crude oil hydrocarbon degradation, but in acidic soils liming to pH 7.8 to 8.0 had a definite stimulatory effect.

Nutrients are very important ingredients for successful biodegradation of hydrocarbon pollutants, especially nitrogen, phosphorus and in some cases iron. Depending on the nature of the impacted environment, some of these nutrients could become limiting thus affecting the biodegradation processes.

When a major oil spill occurs in freshwater and/or marine ecosystems, the supply of carbon is dramatically increased and the availability of nitrogen and phosphorus generally becomes the limiting factor for oil degradation (Atlas, 1984). This is more pronounced in marine environments, due to the low background levels of nitrogen and phosphorus in seawater (Floodgate, 1984), unlike in freshwater systems that regularly fluctuate in nutrient status as result of perturbations and receipt of industrial and domestic effluents and agricultural runoff. Freshwater wetlands are typically considered to be nutrient limited, due to heavy demand for nutrients by the plants, which can be considered to be *nutrient traps* since a substantial amount of nutrients is often found in the indigenous biomass (Mitsch and Gosselink, 1993).

Generally, the additions of nutrients is necessary to enhance the biodegradation of crude oil-related pollutants (Choi et al., 2002; Kim et al., 2005; Joshi and Pandey, 2011). In fact, even in harsh sub-Arctic climates, it has been observed that the effectiveness of fertilizers for crude oil increases the chemical, microbial and toxicological parameters compared to the used of various fertilizers in a pristine environment (Pelletier et al., 2004).

In another study using poultry manure as organic fertilizer in contaminated soil, biodegradation of crude oil-related hydrocarbon derivatives was reported to be enhanced but the extent of biodegradation was influenced by the incorporation of alternate carbon substrates or surfactants (Okolo et al., 2005). However, excessive nutrient concentrations can inhibit the biodegradation activity (Challain et al., 2006), and there can be a negative effect on the biodegradation of hydrocarbon derivatives in the presence of high nitrogen-phosphorus-potassium levels (Oudot et al., 1998; Chaîneau et al., 2005)—this effect is more pronounced on the bioremediation of aromatic hydrocarbon derivatives (Carmichael and Pfaender, 1997). The biodegradation of various aromatic hydrocarbon derivatives is also sensitive to acidity or alkalinity and also to by-products of the biodegradation of the saturate fraction, which serves to explain the persistence of aromatic crude oil hydrocarbon derivatives in certain ecosystems.

On the other hand, in an investigation of the role of the nitrogen source in biodegradation of crude oil components by a defined bacterial consortium under cold, marine conditions (10°C/50°F), it was observed that nitrate did not affect the pH, whereas ammonium amendment led to progressive acidification, accompanied by an inhibition of the degradation of aromatic (particularly polynuclear aromatic) hydrocarbon derivatives (Foght et al., 1999). However, the aromatic systems were degraded or co-metabolized in the absence of nutrients where the pH remained almost unchanged. The best overall biodegradation was observed in the presence of nitrate without ammonium, plus high phosphate buffering—a disadvantage of nitrate is that significant emulsification of the crude oil occurs. Generally, it is worth bearing in mind that acidity/alkalinity (pH) is an important factor that requires consideration as it affects the solubility of both polynuclear aromatic hydrocarbon derivatives as well as the metabolism of the microorganisms, showing an optimal range for bacterial degradation between 5.5 and 7.8 (Bossert and Bartha, 1984; Wong et al., 2001).

The molar ratio of carbon, nitrogen and phosphorus (C/N/P) is very important for the metabolism of the microorganisms and, therefore, for degradation of polynuclear aromatic hydrocarbon derivatives (Bossert and Bartha, 1984; Alexander, 1994; Kwok and Loh, 2003). A molar ratio 100:10:1 is frequently considered optimal for contaminated soils (Bossert and Bartha, 1984; Alexander, 1994), whilst some authors have reported negative or no effects (Chaîneau et al., 2005). These contradictory results are due to the nutrient ratio required by bacteria that degrade polynuclear aromatic hydrocarbon derivatives, which depends on environmental conditions, type of bacteria and type of hydrocarbon (Leys et al., 2005).

Furthermore, it is not surprising that the chemical form of those nutrients is also important, the soluble forms being (i.e., iron or nitrogen in form of phosphate, nitrate and ammonium) the most frequent and efficient due to their higher availability for microorganisms. Depending on the microbial community and their abundance, another factor that may improve polynuclear aromatic hydrocarbon degradation is the addition of readily assimilated carbon sources, such as glucose (Zaidi and Imam, 1999).

8.1.3 *Effect of Temperature*

Temperature plays an important role in the biodegradation of the hydrocarbon constituents of crude oil and crude oil products because of (1) the direct effect on the chemistry of the pollutants and (2) the effect on the physiology and diversity of the microbial surroundings (Atlas, 1975; Delille and Pelletier, 2009). In short, temperature can play the role of increasing a microbial reaction or inhibiting a microbial reaction in a similar manner to the general rules for the influence of temperature on chemical reactions.

Although the biodegradation rate of hydrocarbon derivatives can occur over a wide range of temperatures, the rate of biodegradation generally decreases with decreasing temperature. Highest degradation rates generally occur in the range of 30 to 40°C (86 to 104°F) in soil environments, 20 to 30°C (68 to 86°F) in some freshwater environments, and 15 to 20°C (69 to 68°F) in marine environments. The effect of temperature is also complicated by other factors such as the composition of the microbial population (Zhu et al., 2001). More typically, biodegradation of crude oil and crude oil products occurs at temperatures less than 80°C (< 176°F) (Conan, 1984; Barnard and Bastow, 1991)—at higher temperatures (unless the microbes are of a specific thermophilic type) many of the microorganisms involved in subsurface oil biodegradation cannot exist.

Although hydrocarbon biodegradation can occur over a wide range of temperatures, the rate of biodegradation generally decreases with decreasing temperature. Highest degradation rates generally occur in the range of 30 to 40°C (86 to 104°F) in soil environments, 20 to 30°C (68 to 86°F) in some freshwater environments, and 15 to 20°C (59 to 68°F) in marine environments (Bossert and Bartha, 1984; Zhu et al., 2001). In fact, the biodegradability of crude oil is highly dependent not only on composition but also on microbial incubation temperature (Atlas, 1975)—at 20°C (68°F) conventional crude oil has higher abiotic losses and is more susceptible to biodegradation than heavy oil. As expected from crude oil chemistry and composition (Speight, 2014), the rate of mineralization for the heavy oil is significantly lower at 20°C (68°F) than for conventional oil.

Thus, the ambient temperature of an environment affects both the properties of spilled crude oil or crude oil product (Speight, 2014, 2015) and the activity or population of microorganisms (Venosa and Zhu, 2003). At low temperatures, the viscosity of the oil increases, while the volatility of toxic low-molecular weight hydrocarbon derivatives is reduced, delaying the onset of biodegradation (Atlas, 1981). Temperature also variously affects the solubility of hydrocarbon derivatives (Foght et al., 1996).

During biodegradation, some preference is shown for removal of the paraffin constituents over the aromatic and asphaltic constituents, especially at low temperatures. Branched paraffins, such as pristane, are degraded at both 10 and 20°C (50 and 68°F). This was confirmed by showing that the residual material (after an incubation period of 42 days) had a lower relative percentage of paraffins and higher percentage of asphaltic constituents (usually resin and asphaltene constituents) than fresh or weathered oil (Atlas, 1975).

Finally, the relative resistance of conventional (light) crude oil and even crude oil distillate products to degradation at low temperatures should be considered in choosing shipping routes for these materials.

8.1.4 Effect of Dispersants

A dispersant or a dispersing agent is a substance, typically a surfactant, which is added to a suspension of solid or liquid particles in a liquid to improve the separation of the particles and to prevent their settling or clumping.

However, the effect of dispersants on the fate of dispersed crude oil and crude oil products oil has often been the subject of conflicting reports insofar as dispersants have had (1) little-to-no effect on the biodegradation of crude oil and crude oil products, (2) there has been a positive effect on the biodegradation of crude oil and crude oil products, and (3) there has been a negative effect on the biodegradation of crude oil and crude oil products (Robichauz and Myrick, 1972; Mulkins-Phillips and Stewart, 1974a; Traxler and Bhattacharya, 1979; Foght and Westlake, 1982; Lee et al., 1985; Litherathy et al., 1989).

On the other hand, and perhaps more specific to the current context, it has been suggested that dispersants tend to increase oil biodegradation by increasing the surface area for microbial attack, and encouraging migration of the droplets through the water column making oxygen and nutrients more readily available (Mulyono et al., 1994). However, dispersants can have a detrimental (toxic) effect on microbial processes thereby retarding the rate of crude oil degradation (Mulyono et al., 1994; Varadaraj et al., 1995). It would appear that the dual capability of dispersants (increasing the surface area of dispersed oil and affecting the growth of hydrocarbon-degraders) is related to the chemistry of the dispersant which influences the effectiveness of dispersants for bioremediation (Varadaraj et al., 1995; Davies et al., 2001).

It is clear that the introduction of external nonionic surfactants (the main components of oil spill dispersants) will influence the alkane degradation rate (Bruheim and Eimhjelle, 1998; Rahman et al., 2003). There are indications that the use of surfactants in situations of crude oil-related contamination may have a stimulatory, inhibitory, or neutral effect on the bacterial degradation of the crude oil constituents (Liu et al., 1995). Thus, there is the need to accurately characterize the roles of chemical and biological surfactants in order that performance in biological systems may be predicted (Rocha and Infante, 1997; Lindstrom and Braddock, 2002).

However, in contrast to chemical dispersants, which caused ecological damage after application for abatement of spilled crude oil-related constituents in marine ecosystems (Smith, 1968), biosurfactants from soil or freshwater microorganisms are less toxic and partially biodegradable (Poremba et al., 1991).

8.1.5 Effect of Weathering

In addition to understanding the physical and chemical properties of crude oil and crude oil products, it is necessary (even critical) critical to understand how the properties change over time during and after a spill. The process by which oil properties change during an oil spill is known as weathering, which is a complex process, and detailed computational models exist to predict how oil properties change during a spill as a result of weathering (Daling and Strøm, 1999). During weathering, the spilled oil typically becomes more viscous through evaporation (the loss of volatile constituents from the spilled material) and by collecting water (a process known as "emulsification"). The degree and rate at which oil properties change as a result of weathering depend on the type of crude oil or the crude oil product and on the conditions surrounding the oil spill (such as weather and location of spill).

Thus, weathering, in the current context, is the deterioration of the constituents of crude oil and crude oil products through contact with water, atmospheric gases, and biological organisms. The weathering process is, in effect, the collective effect of several chemical processes. Thus, weathered crude oil (i.e., crude oil and crude oil related products) that has been exposed to air and oxidized and subjected to other influence such as evaporation offer a challenge to bioremediation efforts.

Furthermore, while many of the physical and chemical characteristics of oil (Chapter 3) influence the choice of a remediation technology, most testing of the properties of the spilled material focus on key properties such as viscosity, density, and API gravity. AS an example, in a study assessing the bioremediation efficiency of a weathered and recently contaminated soil in Brazil (Trindale et al., 2005), the authors reported low biodegradation efficiencies in the weathered soil contaminated with a high crude oil concentration compared to recently contaminated soil. Also, both soils (weathered and recently contaminated) submitted to bioaugmentation and biostimulation techniques presented biodegradation efficiencies approximately twice higher than the ones without natural attenuation.

Several areas at many sites, especially refinery sites with a long history of operation, have weathered oil floating on the surface of the groundwater and oil-saturated soil. However, site conditions can be manipulated to enhance bioremediation and speed up the degradation rates of the contaminants. There are several techniques that can be applied to enhance the biological degradation of contaminants which are (1) supplementation with suitable sources of nitrogen and phosphorus, (2) manipulation of redox potential by the injection of air, oxygen, or nitrate to enhance aerobic biodegradation—the redox potential is a measure of the ease with which a molecule will accept electrons, which means that the more positive the redox potential, the more readily a molecule is reduced, (3) addition of surfactants to make the contaminants bioavailable, (4) site microbial inoculation, and (5) injection of co-substrates such as molasses, or lactate to enhance the biodegradation of chlorinated contaminants;

Oxygen is often the limiting factor in aerobic bioremediation at many sites. The degradation of crude oil hydrocarbon derivatives occurs much faster under aerobic conditions compared to anaerobic conditions. Therefore, the addition of oxygen can significantly increase the remediation rates. Oxygen addition is most frequently used to address dissolved phase contamination, such as total crude oil hydrocarbon derivatives and BTEX, as well as contamination in the capillary fringe zone. Oxygen can only be effective if the hydrocarbon derivatives are bioavailable and there is no nutrient limitation.

The bioremediation of crude oil contaminated soil was investigated using a microscale landfarming (Yudono et al., 2011). The indigenous bacteria, *Pseudomonas pseudoalcaligenes,* *Bacillus megaterium,* and *Xanthobacter autotrophicus* were isolated from the contaminated sites Sungai Lilin Jambi Pertamina Ltd and used further in the bioremediation experiments. The biodegradation rates of crude oil contaminated soil in the presence of the isolated bacteria were studied by using the chemical kinetics approach. The reaction orders were studied by using the differential method and the reaction rate constants were studied by using the integral method. The results showed that the reaction orders were 1.0949, 1.3985, 0.8823, and the reaction rate constants were 0.0189, 0.0204, 0.0324 day-1, respectively. Considering the values of reaction orders and reaction rate constants, the biodegradation rate of contaminated soil by using each bacteria had significantly different value; *Xanthobacter Autotrophicus* bacteria could degrade the crude oil sludge faster than the others.

8.2 Biodegradation of Specific Constituents

Crude oil and crude oil products are complex mixtures of differing molecular species hydrocarbon derivatives and the constituents of these molecular categories are present in varied proportions resulting in high variability in crude oil and crude oil products (Chapter 1, Chapter 2) (Speight, 2014). In terms of bulk fractions (Chapter 2), the resin constituents and the asphaltene constituents are of particular interest (or notoriety) because these constituents typically resist degradation though resistance or toxicity to microorganisms. After a spill, the constituents of crude oil and crude oil products are subjected to physical and chemical processes such as evaporation (the release of volatile constituents of the atmosphere) or photochemical oxidation (as for example, the introduction of oxygen functions into the oxidation-susceptible constituents) which produces changes in the composition and properties of the spilled material (Taghvaei Ganjali et al., 2007; Speight, 2014).

8.2.1 Alkanes

Alkane derivatives are major constituents of conventional crude oil and crude oil products and can be degraded by indigenous or non-indigenous (i.e., added) microorganisms.

Conventional (light) crude oil contains 10 to 40% w/w normal alkane derivatives, but weathered and heavier oils may have only a fraction of a percent. Higher molecular weight alkane derivatives constitute 5 to 20% w/w of light oils and up to 60% w/w of the more viscous oils and tar sand bitumen. Aromatic hydrocarbon derivatives are those characterized by the presence of at least one benzene (or substituted benzene) ring.

The low-molecular-weight aromatic hydrocarbon derivatives are subject to evaporation and, although toxic to much marine life, are also relatively easily degraded. Conventional (light) crude oil typically contain between 2 and 20% w/w low-boiling aromatic compounds, whereas heavy crude oil typically contains less than 2% v/v aromatic compounds. As molecular weight and complexity increase, aromatic derivatives are less readily degraded. Thus, the degradation rate of polynuclear aromatic derivatives (PNAs, also referred to as polyaromatic hydrocarbon derivatives, PAHs) is slower than that the biodegradation of the non-aromatic derivatives.

Of these, the normal alkane series (straight-chain alkane series) is the most abundant and the most quickly degraded. Compounds with chains of up to 44 carbon atoms can be metabolized by microorganisms, but those having 10 to 24 carbon atoms (C_{10} to C_{24}) are usually the easiest to metabolize. Shorter chains (up to approximately C_8) also evaporate relatively easily. Only a few species can use C_1 to C_4 alkane derivatives and C_5 to C_9 alkane derivatives are degradable by some microorganisms but toxic to others.

Branched-chain alkane derivatives are usually more resistant to biodegradation than the linear chain alkane derivatives (n-alkane derivatives) but less resistant than cycloalkane derivatives (naphthene derivatives), which are the alkane derivatives that contain carbon atoms in ring-like central structures. Branched alkane derivatives are increasingly resistant to microbial attack as the number of branches increases. At low concentrations, cycloalkane derivatives may be degraded at moderate rates, but some highly condensed cycloalkane derivatives can persist for long periods after a spill.

Generally, with respect to the molecular composition of the aliphatic constituents of crude oil and crude oil-related products, microbial biodegradation attacks n-alkane derivatives and isoprenoid alkane derivatives. The polycyclic alkane derivatives of sterane and triterpane type ted to be somewhat resistant to biodegradation. Since this is the case even for naphthenic type crude oil (which is originally depleted in *n*-alkane derivatives), it has been concluded that the biodegradation of crude oil type pollutants, under natural conditions, will be restricted to n-alkane derivatives and isoprenoids (Antić et al., 2006).

In aqueous systems, addition of acclimatized naturally occurring microorganisms (bioaugmentation) enhances the biodegradation of hydrocarbon derivatives. Since dissolved hydrocarbon derivatives are more available for microbiological degradation, application of dispersants and surfactants to increase the bioavailability significantly and enhance oil degradation (Mohn and Stewart, 2000; Zhang, 2008; Zahed et al., 2010). Other factors (such salinity and pH) have considerable effects on biodegradation of crude oil hydrocarbon derivatives in the marine environments as well.

For example, the different concentrations of sodium chloride (0 to 5% w/w) exert considerable influence on the biodegradation of crude oil and polynuclear aromatic hydrocarbon derivatives from the heavy crude oil-contaminated soil (Minai-Tehrani, 2009). Not surprisingly, increasing the concentration of sodium chloride in soil has a decreasing effect on crude oil biodegradation and the removal of polynuclear aromatic hydrocarbon derivatives. The biodegradation of total crude oil was higher in the absence of sodium chloride (41%) while the reduction in the biodegradation of polynuclear aromatic hydrocarbon derivatives was observed in the presence of 1% w/w sodium chloride (35%). A lower reduction of crude oil and polynuclear aromatic hydrocarbon derivatives was observed in the presence of 5% w/w sodium chloride (12% and 8%, respectively). The reduction of phenanthrene, anthracene, and pyrene reduction was higher in the presence of 1% w/w sodium chloride, while fluoranthene and chrysene reduction were higher in the absence of sodium chloride.

In a recent study (Lee et al., 2010), *Rhodococcus* sp. EH831, isolated from an oil-contaminated soil, has been shown to degrade a wide range of hydrocarbon derivatives and completely metabolize hexane. EH831 did not lose its activity at medium or low temperatures. Moreover, the biodegradation pathway of hexane by *Rhodococcus* sp. EH831 under aerobic conditions was revealed for the first time, which may be useful for the bioremediation of sites contaminated with various hydrocarbon derivatives or for the treatment of industrial discharge.

8.2.2 Aromatic Hydrocarbons

Spills of aromatic hydrocarbon derivatives such aromatic naphtha and leaks from underground fuel tanks contribute significantly to the contamination of groundwater by aromatic compounds. Non-oxygenated monoaromatic hydrocarbon derivatives, such as benzene, toluene, ethylbenzene, and xylenes (BTEX) are of particular concern—BTEX refers to the chemicals benzene, toluene, the xylene isomers, and ethylbenzene:

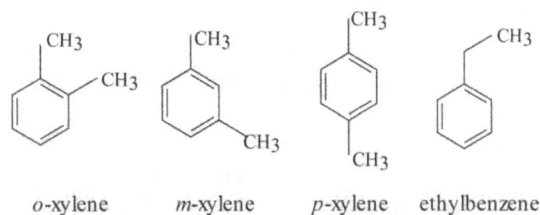

o-xylene   *m*-xylene   *p*-xylene   ethylbenzene

These compounds occur naturally in crude oil and can be found in sea water in the vicinity of natural gas and crude oil reservoirs. The high water solubility of the BTEX species enables them to migrate in the subsurface and contaminate drinking water but are biodegradable under the appropriate conditions (Weiner and Lovley, 1998; Chen and Taylor, 1995, 1997a, 1997b; Taylor et al., 1998; Gieg et al., 1999). All of the BTEX compounds were biodegraded under sulfate-reducing conditions and toluene was also degraded under methanogenic conditions.

The bioremediation process is terminated by lowering the temperature below 40°C (104°F). Such an *in situ* follow-up treatment could also be applied to fuel-contaminated plumes subjected to thermally enhanced vapor stripping as a primary treatment method, or as a stand-alone method, when the initial concentration of volatile organic compounds (VOCs) is low and the subsurface volume to be heated is small. In the latter case, thermophilic hydrocarbon degraders suspended in hot water are pumped into the subsurface.

The biodegradation of alkyl tetralin derivatives has also been studied (Booth et al., 2007a). However, tetralin has been shown to be biodegraded by both mixed cultures of microbes (Strawinski and Stone, 1940; Soli and Bens, 1972) and by some strains able to utilize the compound as sole carbon and energy source (Schreiber and Winkler, 1983; Sikkema and Bont, 1991; Hernáez et al., 1999). It has also been reported that *rhodococci* strains are able to degrade recalcitrant alkyl Tetralin derivatives (Frenzel et al., 2009). The identification of such bacteria capable of the biodegradation of the alkyl tetralin derivatyives may be an important step toward the development of bioremediation strategies for sites contaminated by toxic aromatic hydrocarbon derivatives.

8.2.3 Polynuclear Aromatic Hydrocarbons

Polynuclear aromatic hydrocarbon derivatives, in the current context, are persistent organic compounds with two or more aromatic rings in various structural configurations. Polynuclear aromatic hydrocarbon derivatives constitute a large and diverse class of organic compounds. However, derivatives such as tetralin (1,2,3,4-tetrahydronaphthalene) and decalin (decahydronaphthalene, bicyclo[4.4.0]decane) are not included in this group but are include in the alkane group because of the saturated ring.

Tetralin

Decalin

The chemical properties, and hence the environmental fate, of polynuclear aromatic hydrocarbon derivatives are dependent in part upon both molecular size (i.e., the number of aromatic rings and the pattern of ring linkage). Ring linkage patterns (also known as molecular topology) in polynuclear aromatic hydrocarbon derivatives may occur such that the tertiary carbon atoms are centers of two or three interlinked rings, as in the linear kata-condensed polynuclear aromatic hydrocarbon anthracene or the peri-condensed polynuclear aromatic hydrocarbon pyrene.

However, most polynuclear aromatic hydrocarbon derivatives occur as hybrids encompassing various structural components, such as in the polynuclear aromatic hydrocarbon benzo[*a*]pyrene.

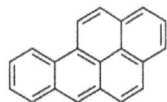

Benzo(a)pyrene

Generally, an increase in the size and angularity of a polynuclear aromatic hydrocarbon molecule results in a concomitant increase in hydrophobicity and electrochemical stability

(Zander, 1983; Harvey, 1997). Polynuclear aromatic hydrocarbon molecule stability and hydrophobicity are two primary factors which contribute to their persistence of in the environment.

Polynuclear aromatic hydrocarbon derivatives are present as natural constituents in fossil fuels, are formed during the incomplete combustion of organic material, and are therefore present in relatively high concentrations in products of fossil fuel refining (Speight, 2012, 2014). Polynuclear aromatic hydrocarbon derivatives released into the environment may originate from crude oil products such as naphtha including gasoline, diesel fuel, and fuel oil (Pavlova and Ivanova, 2003). The concentration of polynuclear aromatic hydrocarbon derivatives in the environment varies widely, depending on the proximity of the contaminated site to the production source, the level of industrial development, and the mode(s) of polynuclear aromatic hydrocarbon transport.

The toxic, mutagenic and carcinogenic properties of polynuclear aromatic hydrocarbon derivatives have resulted in some of these compounds (including naphthalene, phenanthrene and anthracene) to be designated as priority pollutants. In addition, the solubility of polynuclear aromatic hydrocarbon derivatives in aqueous media is very low (Luning Prak and Pritchard, 2002), which affects degradation of these compounds and can lead to and biomagnification within an ecosystem.

Interest in the biodegradation mechanisms and environmental fate of polycyclic aromatic hydrocarbon derivatives (polynuclear aromatic hydrocarbon derivatives) is prompted by their ubiquitous distribution and their potentially deleterious effects on human health (Kanal and Harayama, 2000; Pavlova and Ivanova, 2003; Xia et al., 2006).

Polynuclear aromatic hydrocarbon derivatives (PNAs, also kown as polyaromatic hydrocarbon derivatives (PAHs) are aromatic hydrocarbons with two or more fused benzene rings. They are formed during the thermal decomposition of organic molecules and their subsequent recombination. Incomplete combustion at high temperature (500 to 800°C; 930 to 1,470°F) or subjection of organic material at low temperature (100 to 300°C; 212 to 570°F) for long periods result in the production of theses hydrocarbon derivatives. They occur as colorless, white/pale yellow solids with low solubility in water, high melting point, high boiling point and low vapor pressure. With an increase in molecular weight, the solubility in water decreases; melting and boiling point increase, and the vapor pressure decreases. The common sources of polynuclear aromatic hydrocarbon derivatives in the environment include (1) natural sources, which include forest and rangeland fires, oil seeps, volcanic eruptions and exudates from trees as well as (2) anthropogenic sources, which include burning of fossil fuels, coal tar, wood, garbage, refuse, used lubricating oil, oil filters, municipal solid waste incineration, and spills of crude oil and crude oil products. They are toxic, mutagenic and carcinogenic and do not degrade easily under natural conditions. Persistence in the environment increases with increase in the molecular weight.

The biodegradation of polynuclear aromatic hydrocarbon derivatives by microorganisms and the biodegradation of polynuclear aromatic hydrocarbon derivatives composed of three rings is well documented (Cerniglia, 1984; Gibson and Subramanian, 1984; Cerniglia and Heitkamp, 1989; Cerniglia, 1992; Van der Meer et al., 1992; Shuttleworth and Cerniglia, 1995; Sutherland et al., 1995; Haritash and Kaushik, 2009).

Active bioremediation strategies (such as biostimulation) for application to polynuclear aromatic-contaminated soils can be used to supply nutrients, oxygen, and other amendments to the subsurface to enhance indigenous microbial activity and contaminant biodegradation (Bamforth and Singleton, 2005; Borchert et al., 1995; Mohan et al., 2006). The benefits of adding oxygen and/or nutrients on the biodegradation of polynuclear aromatic hydrocarbon derivatives has been reported for contaminated soils from various sites (Breedveld and Sparrevik, 2000; Eriksson et al., 2000; Li et al., 2005; Liebeg and Cutright, 1999; Lundstedt et al., 2003; Talley et al., 2002). However, only a few studies have focused on the direct effects of biostimulation on the indigenous microbial community and polynuclear aromatic hydrocarbon-degrading bacteria (Ringelberg et al., 2001; Viñas et al., 2005).

Generally, microbial communities present in soils contaminated with polynuclear aromatic hydrocarbon derivatives are enriched by microorganisms able to use them as the only carbon source

(Heitkamp and Cerniglia, 1988; Gallego et al., 2007). However, this process can be affected by a few key environmental factors (Roling-Wilfred et al., 2002; Simarro et al., 2011) that may be optimized to achieve a more efficient process.

Due to their lipophilic nature, polynuclear aromatic hydrocarbon derivatives have a high potential for bio-concentration (Clements et al., 1994; Twiss et al., 1999). In addition to increases in environmental persistence with increasing polynuclear aromatic hydrocarbon molecular size, evidence suggests that in some cases, polynuclear aromatic hydrocarbon toxicity also increases with size, up to at least four or five fused benzene rings (Cerniglia, 1992). The relationship between polynuclear aromatic hydrocarbon environmental persistence and increasing numbers of benzene rings is consistent with the results of various studies correlating environmental biodegradation rates and polynuclear aromatic hydrocarbon molecule size (Banerjee et al., 1995; Shuttleworth and Cerniglia, 1995).

The biodegradation of naphthalene (the simplest polynuclear aromatic hydrocarbon) process was optimized with preliminary experiments in slurry aerobic microcosms (Bestetti et al., 2003). From soil samples collected on a contaminated site, a *Pseudomonas putida* strain (designated as M8), capable to degrade naphthalene was selected. Microcosms were prepared with M8 strain by mixing non-contaminated soil and a mineral medium. Different experimental conditions were tested varying naphthalene concentration, soil/water ratio and inoculum density. The disappearance of hydrocarbon, the production of carbon dioxide, and the ratio of total heterotrophic and naphthalene-degrading bacteria were monitored at different incubation times. The kinetic equation that best fitted the disappearance of contaminant with time was determined. The results showed that the isolated strain enhanced the biodegradation rate with respect to the natural biodegradation.

Of the four-ring polynuclear aromatic hydrocarbon derivatives, fluoranthene, pyrene, chrysene, and benz[*a*]anthracene have been investigated to various degrees.

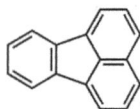

Fluoranthene

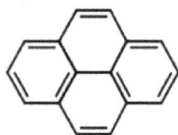

Pyrene

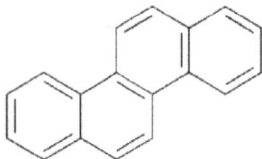

Chrysene

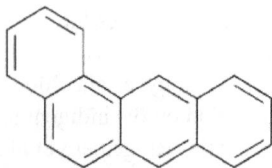

Benz(a)anthracene

Fluoranthene, a polynuclear aromatic hydrocarbon containing a five-membered ring, has been shown to be metabolized by a variety of bacteria, and pathways describing its biodegradation have been proposed (Mueller et al., 1990; Weissenfels, 1990; Weissenfels, 1991; Ye et al., 1996).

Fluoranthene has been used as a model compound in studies which have investigated the effects of surface-active compounds on polynuclear aromatic hydrocarbon biodegradation. Comparisons of the mineralization of fluoranthene by four fluoranthene-degrading strains in the presence of the nonionic surfactants triton x-100 and tween 80 showed that responses differed between strains (Willumsen et al., 1998).

The bacterial degradation of pyrene, a peri-condensed polynuclear aromatic hydrocarbon, has been reported by a number of groups, and some have identified metabolites and proposed pathways (Cerniglia and Heitkamp, 1990). Sediment microcosms inoculated with the mycobacterium showed enhanced mineralization of various polynuclear aromatic hydrocarbon derivatives, including pyrene and benzo(a)pyrene (Headlamp and Cerniglia, 1989).

Measuring the success of bioremediation of crude oil-related spills is based on several parameters, among them the degradation of polynuclear aromatic hydrocarbon derivatives in the crude oil. Though the lower n-alkane derivatives are generally considered the most biodegradable compound class within crude oils (Leahy and Colwell, 1990; Atlas and Bartha, 1992; Prince, 1993), other studies point to exceptional conditions in which polynuclear aromatic hydrocarbon derivatives degrade preferentially to n-alkane derivatives. For example, the biodegradation of alkyl aromatic hydrocarbon derivatives was preferential to that of n-alkane derivatives in crude oil when oil-contaminated sediments were aerobically incubated (Jones et al., 1983). Preferential biodegradation of the aromatic hydrocarbon derivatives was also observed in bitumen of the South Aquitaine Basin in which the n-alkane had not been completely biodegraded (Conan et al., 1980; Conan, 1981).

Generally, aromatic constituents with five or more rings are not easily attacked and may persist in the environment for long periods. High-molecular-weight aromatic derivatives comprise 2 to 10% w/w conventional (light) crude oil and up to 35% w/w of the more viscous crude oil.

Currently, there is only limited information regarding the bacterial biodegradation of polynuclear aromatic hydrocarbon derivatives with five or more rings in both environmental samples and pure or mixed cultures. Most studies have focused on the five-ring benzo(a)pyrene due potential hazards to human health shown by this compound. Many studies have documented the environmental recalcitrance of benzo(a)pyrene to biodegradation (Cerniglia, 1992; Park et al., 1990; Wild and Jones, 1993; Goodin and Webber, 1995; Van Brummelen et al., 1996). Turnover times of in excess of three years in oil-contaminated freshwater sediments and possibly in excess of sixty years in uncontaminated sediments have been reported for the biotransformation of benzo(a)pyrene.

The efficiency of several chemical treatments as potential enhancers of the biodegradation of polynuclear aromatic hydrocarbon derivatives in contaminated soil has been evaluated by analyzing the mineralization of ^{14}C-labeled phenanthrene, pyrene, and benzo(a)pyrene (Piskonen and Itävaara, 2004). The effect of nonionic surfactants with Fenton oxidation and combinations of surfactants with the Fenton oxidation was evaluated in a micro-titer plate assay. The surfactants selected for the study were Tween 80, Brij 35, Tergitol NP-10, and Triton X-100. Phenanthrene mineralization was also positively induced by the Fenton treatments.

However, none of the treatments had a significant effect on benzo(a)pyrene mineralization. Surfactant additions at concentrations of 20% and 80% of the aqueous critical micelle concentration did not significantly affect the mineralization rates. When surfactant addition was combined with the Fenton oxidation, reduced mineralization rates were obtained when compared with mineralization after Fenton's treatment alone. The results indicated that the addition of Fenton's reagent may enhance the mineralization of PAHs in contaminated soil, whereas the addition of surfactants has no significant beneficial effect.

Increases in the understanding of the microbial ecology of polynuclear aromatic hydrocarbon-degrading communities and the mechanisms by which polynuclear aromatic hydrocarbon biodegradation occur will prove helpful for predicting the environmental fate of these compounds and for developing practical polynuclear aromatic hydrocarbon bioremediation strategies in the future (Okerentugba and Ezeronye, 2003).

8.2.4 Phenolic Compounds

Phenolic compounds are common constituents of wastewaters from the oil industry and also have the potential to be formed from benzene derivatives during the weathering process:

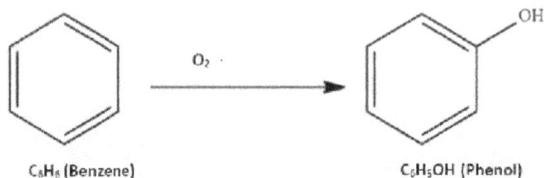

C_6H_6 (Benzene) C_6H_5OH (Phenol)

The microorganisms involved in biodegradation m ay strongly influence the product distribution and may also include catechol isomers which are known to be toxic and carcinogenic for humans, and their contamination of soils and aquifers is of great environmental concern:

Ortho isomer	*Meta* isomer	*Para* isomer
Catechol	Resorcinol	Hydroquinone
Pyrocatechol	1,3-benzenediol	1,4-benzenediol
1,2-benzenediol	*m*-benzenediol	*p*-benzenediol
o-benzenediol	1,3-dihydroxybenzene	1,4-dihydroxybenzene
1,2-dihydroxybenzene	*m*-dihydroxybenzene	*p*-dihydroxybenzene
o-dihydroxybenzene	resorcin	

Soil microorganisms, like *Pseudomonas* spp. and *Mycobacterium*, were found to be capable of transforming and degrading toxic catechol isomers to easily absorbable metabolites. These abilities may be useful in removal of toxic organic compounds from the environment. The successful application of microorganisms to the bioremediation of contaminated sites requires a deeper understanding of how microbial degradation proceeds (Zeyaullah et al., 2009).

Another psychrotrophic *Pseudomonas putida* was reported to remove a wide variety of phenolic compounds from wastewater under aerobic and pH-neutral conditions at temperatures ranging from 1 to 35°C (Pillis and Davis, 1985). This strain can be used in trickling filter systems, activated sludge treatments, and outdoor lagoons, either alone or in combination with other microorganisms conventionally used in waste treatment. It can be cultured in wastewater using either a batch process, a semi-continuous or a continuous process, for a sufficient amount of time (24 h to 4 weeks, depending on the temperature, the volume to be treated, and the concentration of the contamination) to achieve a significant reduction.

The resulting water is suitable for discharge into rivers and streams after conventional processing. Lagoon efficiency is often low in winter when microorganisms are less active. The use of this strain allowed a reduction in the amount of steam required in winter for lagoon heating in order to achieve normal lagoon operation, with a considerable saving in energy costs.

8.2.5 Chlorinated Compounds

Chlorination modifies the physical properties of hydrocarbons in several ways. These compounds are typically denser than water due to the higher atomic weight of chlorine versus hydrogen.

Chlorinated compounds organic compounds generally constructed of a simple hydrocarbon chain (typically one to three carbon atoms in length) that are often used as solvents. They can be divided into three categories based on their structural characteristics: (1) chlorinated methane derivatives, (2) chlorinated ethane derivatives, and chlorinated ethylene derivative (Table 8.2).

Some types of organochlorides have significant toxicity to plants or animals, including humans. Dioxins, produced when organic matter is burned in the presence of chlorine, are persistent organic pollutants (POPs) which pose dangers when they are released into the environment, as are some insecticides (such as DDT). For example, DDT, which was widely used to control insects in the mid-20th Century, also accumulates in food chains and causes reproductive problems (e.g., eggshell thinning) in certain bird species. DDT also posed further issues to the environment as it is extremely mobile, even found in locales where the chemical never been used. Some organochlorine compounds have been used as chemical weapons due to their toxicity.

Chlorinated solvents and many of their transformation products are colorless liquids at room temperature. They are heavier than water with densities greater than 1 gram per cubic centimeter (g/cm^3) which means they can penetrate deeply into an aquifer.

For the contaminated sites containing volatile chlorinated hydrocarbon derivatives, the application of an anaerobic metabolism mode followed by an aerobic phase is more beneficial in bioremediation than using only either anaerobic or aerobic degradation. Polychlorinated biphenyls have widespread environmental occurrence and have been suspected to be carcinogenic.

Table 8.2: Common Organochlorine Compounds.

IUPAC Name	Common Name	Acronym	Molecular Formula
tetrachloromethane	carbon tetrachloride		CCl_4
trichloromethane	chloroform		$CHCl_3$
dichloromethane	methylene chloride		CH_2Cl_2
chloromethane	methyl chloride		CH_3Cl
1,1,1,2-tetrachloroethane			$C_2H_2Cl_4$
1,1,2,2-tetrachloroethane			$C_2H_2Cl_4$
1,1,2-trichloroethane			$C_2H_3Cl_3$
1,1,1-trichloroethane			$C_2H_3Cl_3$
1,2-dichloroethane			$C_2H_4Cl_2$
1,1-dichloroethane			$C_2H_4Cl_2$
chloroethane			C_2H_5Cl
tetrachloroethylene	perchloroethylene	PCE	C_2Cl_4
trichloroethylene		TCE	C_2HCl_3
cis-1,2-dichloroethylene	cis-dichloroethylene	cis-DCE	$C_2H_2Cl_2$
trans-1,2-dichloroethylene	trans-dichloroethylene	trans-DCE	$C_2H_2Cl_2$
1,1-dichloroethylene	vinylidene chloride	1,1-DCE	$C_2H_2Cl_2$
chloroethylene	vinyl chloride	VC	C_2H_3Cl

For this reason there are present research into developing and testing new treatments to break down polychloro-biphenyls (PCBs) into harmless compounds.

Polychloro-biphenyl derivatives can be biodegraded using sequential anaerobic-aerobic reactions. The anaerobic attack is a reductive dechlorination that results in the replacement of a chlorine atom with a hydrogen atom. The aerobic attack occurs on the ring and the outcome is a breakage of the ring and thus destruction of the polychloro-biphenyl derivative. The anaerobic-aerobic attack is used in a sequence because the aerobic attack only works on polychloro-biphenyl derivatives with a relatively low content of chlorine molecules. Thus, the anaerobic attack constitutes a pretreatment to remove chlorine atoms and converts highly chlorinated polychloro-biphenyls to lower chlorinated polychloro-biphenyls, which are destroyed by aerobic treatment.

8.3 Biodegradation of Extra Heavy Crude Oil

One aspect of the bioremediation program that is not sufficiently emphasized is the potential for spillage on to land and/or on to water and, hence, the ensuing biodegradation of the constituents of extra heavy crude oil (including tar sand bitumen). Although extra heavy crude oil and tar sand bitumen are not true members of the crude oil family when the categorization is based on the methods of recovery from a reservoir or a deposit rather than based on a single physical property which can lead to confusion and discrepancies in the definition (Chapter 1) (Speight, 2014, 2015). However, extra heavy crude oil and tar sand bitumen are worthy of mention here because they provide insights into the necessary actions to be taken when the more viscous products of crude oil refining (such as Bunker C oil and asphalt) are released into the environment (Speight, 2013a, 2013b, 2014; Ancheyta, 2016; El-Gendy and Speight, 2016; Speight, 2017).

The largest heavy crude oil and extra-heavy crude oil reserves in the world are found in the Orinoco oil belt of Venezuela as well as tar sand bitumen (API < 10°) in the Athabasca tar sands (also called Athabasca oil sands) in Alberta, Canada, and the Olenik oil sands in Siberia, Russia. This type of oil cannot be produced, transported or refined by conventional methods (Speight, 2011, 2014, 2015; Speight and Islam, 2016). Extra heavy crude oils, and tar sand bitumen are viscous (some are solid to near-solid) black, materials fluids with a high sulfur high nitrogen content and ahigh, metals (predominantly nickel and vanadium) content. These viscous refinery feedstocks also have a high content of aromatic compounds as well as substantial amounts of the polar high molecular weight resin and asphaltene constituents which have a high propensity for coke formation and catalyst poisoning which together with the potential for flocculation and deposition of asphaltene constituents (and thermally changed resin and asphaltene constituents make production and refining processes more difficult leading to an increase in the emissions of the environmentally-problematic sulfur dioxide and nitrogen oxides (Leòn and Kumar, 2005; Speight, 2005; Ancheyta and Speight, 2007; Ayala et al., 2007; Speight, 2014a, 2013b, 2014; Al Bahry et al., 2016; Ancheyta, 2016; Speight, 2017). These viscous feedstocks represent a substantial threat to the environment when not handled in the most appropriate manner. The properties of extra heavy crude oil depend on the major SARA fractions (saturates, aromatic derivatives, resins and asphaltenes) (Figure 8.1) (Speight, 2014, 2015, 2017).

The sulfur, nitrogen, nickel, and vanadium concentrations are reported to be inversely proportional to API gravity (Ancheyta and Speight, 2007; Speight, 2015). High viscosity significantly hampers the pumping, transportation, refining and handling of heavy crude oil and common methods used to overcome problems associated with high viscosity include heating, dilution, and chemical additives. Typically, upgrading these feedstocks has been accomplished with either thermal cracking or by catalytic hydroconversion (Speight, 2014; Hsu and Robinson, 2017; Speight, 2017). Thermal processing ranges from mild cracking (to reduce viscosity) to severe cracking (with the formation of coke). In addition, the chemistry of processes is less selective and requires the supporting infrastructure for the supply of hydrogen and treatment of hydrogen sulfide in cracked off-gases.

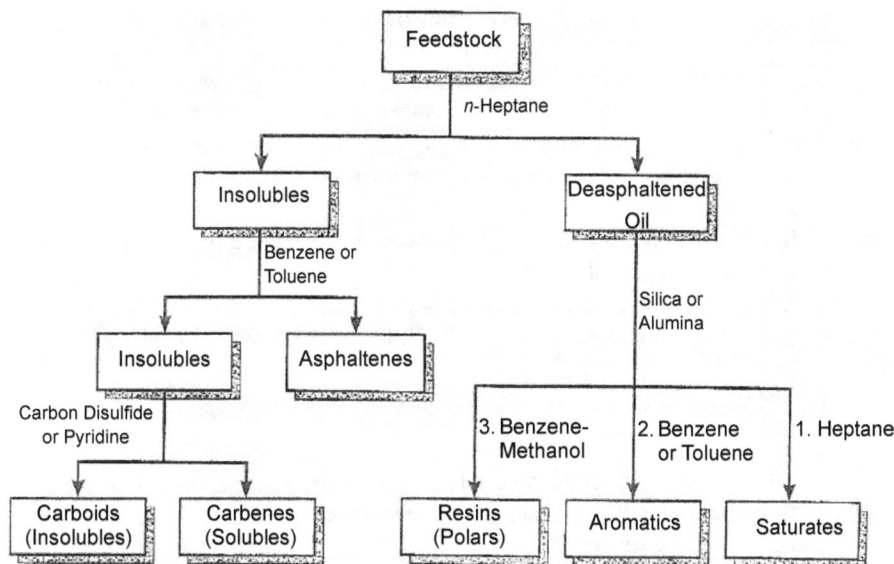

Figure 8.1: Separation Scheme for Various Feedstocks

8.3.1 Microbial Enhanced Oil Recovery

Microbial enhanced oil recovery (MEOR) is a biological based technology consisting in manipulating function or structure, or both, of microbial environments existing in oil reservoirs. The process utilizes microorganisms and their bio-products to increase the production of the difficult-to-produce viscous oil and bitumen oil from reservboirs and deposits (Lazar et al., 2007; Al-Sulaimani et al., 2011a, Al-Sulaimani et al., 2011b; Sun et al., 2011; Speight, 2014, 2016; Speight and El-Gendy, 2018). While microbial enhanced oil recovery is not specifially an on-site refinery process, it can play a role in refining insofar a change made to the viscous feedstock prior to, and during, recovery can have a beneficial effect on the refinability of the viscous material and, therefore, is worthy of mention at this point.

The processes used for microbial enhanced oil recovery involve the use of reservoir microorganisms or specially selected natural bacterial to produce specific metabolic events that lead to enhanced oil recovery (Speight, 2014, 2016; Speight and El-Gendy, 2018). In fact, microbial enhanced oil recovery processes are somewhat akin to *in situ* bioremediation processes (Speight and Arjoon, 2012). Injected nutrients, together with indigenous or added microbes, promote *in situ* microbial growth and/or generation of products which mobilize additional oil and move it to producing wells through reservoir repressurization, interfacial tension/oil viscosity reduction, and selective plugging of the most permeable zones (Bryant et al., 1989, 1996). Alternatively, the oil-mobilizing microbial products may be produced by fermentation and injected into the reservoir.

Typically, nutrients such as sugars, nitrates or phosphates are regularly injected to stimulate the growth of the microbes, which are indigenous to some reservoirs, and aid their performance. The microbes then generate surfactants and carbon dioxide that help to displace the oil in a similar way to other displacement methods. Since growth occurs at exponential rates, the process quickly generates considerable surfactant in a cost-effective manner.

The major mechanisms of microbial enhanced oil recovery include degradation of high-molecular-weight hydrocarbon derivatives in crude oil and production of biosurfactants. Also, microbial enhanced oil recovery is considered as a non-hazardous and economically viable strategy. Other advantages have been attracted attention all over the world, such as high efficiency, easy operation, strong adaptability and pollution-free (Jack, 1991; Al-Sulaimani et al., 2011a, 2011b; Zhang et al., 2014; Gassara et al., 2015). In addition, the microorganisms can be maintained on low cost renewable raw materials and the biosurfactants have excellent emulsification properties (Gao and Zekri, 2011; Karimi et al., 2012; Zhang et al., 2014).

The application of biosurfactants in MEOR, is very promising because of the low toxicity, biodegradability and effectiveness at different conditions of temperature, pressure, salinity and pH (Simpson et al., 2011). The biosurfactants could alter the surface wettability of solids, reduce the interfacial tension between formation water and crude oil (Vaz et al., 2012), and increase the emulsification of crude oil (Zou et al., 2014). Other products including the acid (fatty acids of low molecular weight), gas (carbon dioxide, methane, and hydrogen) and organics (polymer, alcohol, aldehydes) also play significant role of decreasing the oil viscosity. In addition, the high-boiling component of the crude oil can be degraded by the microorganisms and reduce the viscosity of crude oil, thus improve the oil quality (Yang and Lou, 1997; Lazar et al., 2007; Liu et al., 2012).

8.3.2 Biotransformation

Biotransformation is the biochemical modification of one or a mixture chemical compounds. The biotransformation process can be conducted with whole cells, their lysates, or purified enzymes. Increasingly, biotransformation is are effected with purified enzymes.

The efficient recovery and processing of extra heavy crude oil and tar sand bitumen are hampered by the presence of high concentration of asphaltene constituents. The presence of the asphaltene constituents is the one of the main reasons of the increase of the crude oil viscosity (El-Gendy et al., 2006; Bachman et al., 2014; Speight and El-Gendy, 2018). Moreover, it enhances the propensity to form emulsions, polymers and coke (Vazquez-Duhalt et al., 2002; Hernández-López et al., 2015). Thus, the biotransformation or upgrading of asphaltenes and other aromatic compounds contained in petrochemical and other high-boiling hydrocarbon streams is one of the main interests in crude oil industry to maximize the usage of crude oil and minimize the waste (Gupta et al., 2015).

Recently, biological processes have emerged as a cost-effective and environmentally favorable alternative to break asphaltenic structures to obtain high-value low-boiling oils from less-value high-boiling oils (i.e., bio-cracking). The asphaltene constituents can be described as condensed aromatic cores containing alkyl and alicyclic moieties as well as nitrogen, sulfur and metal containing non- and heterocyclic groups (Vazquez-Duhalt et al., 2002; Hernández-López et al., 2015).

The asphaltene constituents have extremely complex and variable molecular structures containing sulfur (0.3–10.3%), oxygen (0.3–4.8%), nitrogen (0.6–3.3%), and metal elements, such as Fe, Ni, and V in a small amount, with an average molecular weight ranged between 600 to 2,000,000 (Tavassoli et al., 2012). Asphaltene is also considered to be the product of complex heteroatomic aromatic macrocyclic structures polymerized through sulfide linkages (El-Gendy et al., 2006; Ali et al., 2012). Breaking the asphaltene constituents into smaller molecules and cutting an internal aliphatic linkage (sulfides, esters and ethers) of an asphaltene molecule can lead to a reduction in viscosity.

The asphaltene constituents are relatively high molecular weight and large and highly hydrophobic, thus mass transfer limitations are expected in aqueous reactions and the biotransformation rates are limited by the mass transfer of target molecules to the biocatalyst and, in the case of whole cells, across the cell membrane (Leòn and Kumar, 2005). Despite these difficulties, there is evidence in the literature for bacterial transformation of these complex, high molecular weight substrates. This is possible because these compounds contain carbon, hydrogen, sulfur, nitrogen and oxygen, which are necessary elements for the survival of microorganisms. The complex structure of the asphaltene constituents offers several possible routes to biotransformation.

Different extremophile bacterial genera, such as *Achromobacter*, *Leptospirillum*, *Pseudomonas*, *Sulfolobus*, *Thiobacillus* have been reported for their capabilities to transform high-boiling oils into lower-boiling products (Premuzic and Lin, 1991a, 1991b; Premuzic et al., 1997, 1999). These genera are adapted to resist high temperatures, pressures, salt and hydrocarbon concentrations. Where, they interact with heteroatoms and organometallic sites in the heavy crude oils, that serve as attachment and initiation points for the microbial activity. The involved reactions probably including oxidation, redistribution, and fragmentation the high-boiling polar fractions (asphaltene constituents) into lower fractions (maltenes). Where asphaltene constituents would probably decompose by rupture at active sites containing heteroatoms, allowing the liberation of the trapped lower molecular weight constituents. The heterocyclic compounds would be possibly oxidized into more soluble compounds that migrated to the aqueous phase. This occurs with a complimentary increase in the concentration of saturated C-chains (C8–C26) and decrease in the higher molecular weight hydrocarbon derivatives, heteroatoms and metal contents. Unfortunately, up till now, the specific microorganisms capable of performing this biotransformation are not known with certainty and there is no available data concerning the involved biochemical reactions and metabolic pathways.

A complex set of multiple biochemical reactions between selected microorganisms and heavy crude oils under controlled conditions have been reported (Premuzic et al., 1999), that led to a significant lowering (24–40%) of the N, S, O, and trace metal contents, with a concurrent redistribution of hydrocarbon derivatives. The reactions are both biocatalyst and crude oil dependent and, in terms of chemical mechanisms, appeared to involve the asphaltene and the associated polar fractions. Asphaltene constituents from a crude oil rich in heavy metals (Castilla crude oil) have been fractionated and the biocatalytic modifications of these fractionated asphaltene constituents by three different hemoproteins: chloro-peroxidase (CPO), cytochrome C peroxidase (Cyt-C), and lignin peroxidase (LPO) have been evaluated in both aqueous buffer and organic solvents. However, only the CPO-mediated reactions were effective in eliminating the Soret peak in both aqueous and organic solvent systems and the chloroperoxidase has been reported to be able to alter components in the high-boiling fractions of crude oil and remove 53 and 27% of total heavy metals (Ni and V, respectively) from petroporphyrin-rich fractions and asphaltene constituents (Mogollón et al., 1998).

Premuzic and Lin (1999a) adapted and modified extremophilic microorganisms (thermophilic, thermo-adapted, barophilic, extreme pH, high salinity and toxic metal adapted microorganisms) such as those belonging to the *Thiobacillus thiooxidans*, *Thiobacillus ferrooxidans*, *Leptospirillum ferrooxidans*, *Acinetobacter calcoaceticus*, *Sulfolobus solfataricus*, *Achromobacter* sp., *Arthrobacter* sp., and *Pseudomonas* sp., for biochemical conversion of a feedstock of heavy crude oil. The upgraded oil feedstock produced by the process are characterized by increased lower-boiling fractions of oils, increased content of saturated hydrocarbon derivatives, decreased content of organic sulfur containing components which have been decreased by at least from approximately 20% to approximately 50%, decreased content of organic nitrogen containing components which have been reduced by approximately 15% to approximately 45% and, a significantly decreased concentration of trace metals by approximately 16% to approximately 60% by weight. Where, the increase in the relative content of the lower-boiling fractions of oil and in the content of saturated hydrocarbon derivatives depends on the chemistry of the starting material. The upgraded oil obtained by the process of the present invention contains an increased content of hydrocarbon surfactants such as emulsifying agents and hydrocarbon-based detergents. Additionally, the upgraded oil also has an increased content of oxygenates which are additives used by gasoline manufacturers to enhance fuel combustion.

The asphaltene constituents have drawn considerable attention due to problems caused by their detrimental effects in the extraction, transportation and processing of residua because of their viscous and flocculating nature and their relative resistance to biodegradation following spills (Speight and Arjoon, 2012; Speight, 2014).

The asphaltene constituents are the highest molecular weight and most polar fraction of crude oil. Despite that the structure of asphaltene constituents' structure has not been fully elucidated; it is widely accepted that they are constituted by interacting systems of polyaromatic sheets bearing alkyl side-chains (Speight, 1994, 2014). Asphaltene molecules have a high content of O, N and S heteroatoms as well as metals (V, Ni and Fe) (Speight, 2014). The problems associated with asphaltene constituents have increased due to the need to extract heavy crude oils, as well as the trend to extract larger amounts of low-boiling fractions out of crude oil by cracking and visbreaking. The degradation of asphaltenes can markedly reduce the viscosity of crude oil and thus enhances oil recovery. Therefore, screening of bacteria with a high ability to degrade asphaltenes and evaluation of their ability to degrade the asphaltene fraction of crude oil are a major and basic component of research regarding microbial enhanced oil recovery in reservoirs containing high-asphaltene heavy crude oil. Moreover, the high asphaltene content can prevent the transfer of biodegradable components in oil droplets toward the oil-bacteria interface and reduce the rate and degree of degradation of biodegradable components (Uraizee et al., 1998).

The focus of many studies has generally been bioremediation of sites contaminated by total crude oil hydrocarbon derivatives (Iturbe et al., 2007; Machackova et al., 2008) and there is a general lack of detailed work has been done on the biodegradation of asphaltene constituents.

In contrast to low-molecular-weight hydrocarbon derivatives, polycyclic aromatic and hydrocarbon derivatives included in the asphaltene fraction are usually considered as being only slightly biodegradable because of their insufficient availability to microbial attack (Gibson and Subramanian, 1984; Cerniglia, 1992; Kanaly and Harayama, 2000). Among the pentacyclic triterpane derivatives, the hopane constituents are so stable that they are commonly used as ubiquitous biomarkers for the assessment of biodegradation levels of crude oil (Ourisson et al., 1999). They were shown to be only slightly biodegraded by specialized microflorae under laboratory conditions (Frontera-Suau et al., 2002).

Microbes reduce the viscosity by degrading high molecular weight constituents into lower molecular weight constituents such as biological surface-active substances, acids, and gases. In addition, anaerobic fermentation leads to the production of acids, carbon dioxide, hydrogen, and alcohols. Anaerobic bacteria produce acetate and butyrate during the initial growth phase (acidogenic phase) of the fermentation process.

8.3.3 Biodegradation and Bioconversion

In the past crude oil biorefining as a branch of crude oil biotechnology, has been related to the production of single cell protein (SCP) from waxy n-alkane derivatives (Hamer and Al-Awadhi, 2000). Nowadays, there is an aim to be applied in upgrading of heavy crude oils. Crude oil biotechnology involves the use of wide range of conditions, milder temperature and pressure, cleaner and selective processes, lower emissions and no-generations of undesirable by-products. Moreover, microbial and enzymatic catalysts can be manipulated and used for more specific applications. Heavy crude oils can be subjected to biorefining to get rid of most of the sulfur, nitrogen, toxic metals and asphaltenes (Le Borgne and Quintero, 2003).

Due to the revolution in protein and genetic engineering, the study of extremophilic microorganisms, biocatalysts in non-aqueous media and nano-biocatalysts, biotechnology found a way in crude oil refining (i.e., biorefining). For example, biodesulfurization (BDS), biodenitrogenation (BDN) and biodemetallization (BDM).

8.3.3.1 Biodesulfurization

Sulfur is the major concern for producers and refiners and has long been a key determinant of the value of crude oils. It is the third most abundant element in crude oil after carbon and hydrogen. Heavy crude oil and tar sand bitumen contain approximately 3 to 6% w/w sulfur that must be removed before the usage as a refinery feedstock. The combustion of S-containing fuels would lead to the increased emissions of sulfur oxides (SOx), the main cause of acid rains and particulate matters (PM) the main cause of black smoke associated with diesel and gasoline vehicles. It has been reported that the total PM emissions from diesel engines are proportional to the diesel sulfur content (Mohebali and Ball, 2016). If desulfurization occurs on the crude feedstocks before they ever enter the refinery system, this will minimize the downstream desulfurization costs. Crude oils with higher viscosities and higher densities usually contain higher amounts of more complex sulfur compounds. The aliphatic acyclic sulfides (thioethers) and cyclic sulfides (thiolanes) are easy to remove during a hydrodesulfurization process or by thermal treatment. On the other hand, sulfur contained in aromatic rings, such as thiophene and the benzologs (e.g., benzothiophene, dibenzothiophene, benzonaphthothiophene) are more resistant to sulfur removal by hydrodesulfurization and thermal conversion (Gray et al., 1996).

Biodesulfurization is a biological method in which microbes or enzymes are used as a catalyst to remove organosulfur compounds, especially the recalcitrant ones, e.g., dibenzothiophene and derivatives. It can be performed aerobically or anaerobically. The main disadvantage is the conversion rate is much slower than hydrodesulfurization, since all the biological reactions are generally slower than the chemical reactions. There are three main routes for aerobic biodesulfurization (i) complete mineralization where the end products are carbon dioxide and water, (ii) the Kodama pathway, where C-C bonds are cleaved and water-soluble by-products are produced, which would significantly inhibit microbial growth dibenzothiophene oxidation and (iii) the 4S-pathway, in which the carbon skeleton is not destroyed and only sulfur is removed (Kilbane and Jackoswki, 1992). The first two pathways are not recommendable for desulfurization of fuels, as the efficiency of biodesulfurization depends on the biocatalyst capabilities to remove sulfur without altering the carbon skeleton or reducing the value of the fuel. But they are recommendable in bioremediation of oil spills and soil or sediments polluted with crude oil hydrocarbon derivatives (Gupta et al., 2005).

The key techno-economic challenge to the ability of biodesulfurization processes is to establish a cost-effective means of implementing the two-phase bioreactor system and de-emulsification steps as well as the product recovery step (Kaufman et al., 1998; McFarland, 1999; Pacheco et al., 1999). Use of multiple-stage air-lift reactors can reduce mixing costs, and centrifugation approaches facilitated de-emulsification, desulfurized oil recovery and recycling of the cells (Ohshiro et al., 1996). Most of the applications of crude oil biotechnology are at the level of laboratory research except biodesulfurization, for which, pilot plants have been established.

The desulfurization of extra heavy crude oil under anaerobic conditions would be attractive because it avoids costs associated with aeration, it has the advantage of liberating sulfur as a gas— hydrogen sulfide be treated with existing refinery desulfurization plants (e.g., Claus process) and does not liberate sulfate as a by-product that must be disposed by some appropriate treatment (Setti et al., 1997; Speight, 2014). Under anaerobic conditions, oxidation of hydrocarbon derivatives to undesired compounds such as colored and gum forming products is minimal (McFarland, 1999). Moreover, anaerobic microorganisms use approximately 10% of the total energy produced while aerobic microorganisms use approximately 50% of the total energy produced. These advantages can be counted as incentives to continue research on reductive biodesulfurization.

The presence of an n-alkane favors the removal of sulfur aromatic compounds which act as a co-substrate ensuring the growth of the culture and also permits the solubilization and the emulsifying of the sulfur aromatic compounds. The microorganism adsorption to the oil-phase is the most likely mechanism for explaining n-alkane biodegradation (Setti et al., 1995). Substrate uptake presumably occurs through diffusion or active transport at the point of contact. It is well known, that, most of the aerobic microorganisms adhere to the n-alkane derivatives (below $n-C_{16}$) which are in a liquid form at room temperature. Where, the n-alkane form a film around the aromatic sulfur compound and as this film is easily attacked by aerobic microorganisms, the bioavailability of sulfur compounds increases. Studies with a strain of *Candida* in two-phase systems (oil-water mixture), suggested that the rate of biodegradation might be also related to the interfacial area because a large part of the biomass, which characteristically is hydrophobic, adheres to the non-aqueous phase layer (NAPL)-water interface as a biofilm (Sarret et al., 1999), the smaller the interfacial tension, the larger the uptake of the dissolved compounds in NAPL and the higher the biodegradation of aliphatic hydrocarbon derivatives present. According to Setti et al. (1997); although aerobic microorganisms can remove high amounts of organic sulfur, the sulfur percentage in the residual high-boiling oil may increase because of simultaneous aliphatic compounds biodegradation.

Oil refineries usually separate crude oil into several fractions and then desulfurize them separately. Moreover, a refinery can make substantial cost savings if most of the sulfur is removed from the crude oil before it is fractionated. Also, it has been suggested that due to the high content of water in crude oil, biodesulfurization of crude oil is more practicable compared to that of diesel oil and gasoline (Zhou and Zhang, 2004). Nowadays, crude oil industry is increasingly dependent on heavy crude oil to meet the domestic demand for gasoline and distillate fuels. Heavy crude oils with high viscosity are expensive to recover transport and process and have a low market value than less viscous oils. Since asphaltene fraction is a major component of heavy crude oils. To use these as a fuel, they must be upgraded by reducing the average molecular weight of the constituents and remove heteroatoms. For bio-modification of asphaltenes, the reactions with the organosulfur moieties could be significant, because sulfur is the third most abundant element in asphaltenes. Sulfur has an important role in the molecular structure of asphaltenes (Sarret et al., 1999). Because of the diversity and complexity of the asphaltene molecular structures that can be attacked, heme proteins were the biocatalysts chosen for investigations on the enzymatic modification of asphaltenes. In a survey of several heme proteins, including horseradish, lignin, manganese chloroperoxidase, and cytochrome c, were found to be able to modify the greatest number of organosulfur compounds including sulfur heterocycles and sulfides, and to have superior specific activity (Vazquez-Duhalt et al., 1993).

The high viscosity of many crude oils is a factor contributing to the underutilization of these valuable natural resources. Viscosity greatly complicates, and may even defeat, the extraction of many types of crude oil from the earth. It remains a concern following extraction, as high viscosity significantly hampers the pumping, transportation, refining and handling of crude oil. Because of this, the crude oil industry has long recognized the need for a safe, economical and effective method for reducing the viscosity of valuable fossil fuel resources.

Under certain circumstances, standard-refining processes such as hydrotreating or hydrodesulfurization can favorably affect the viscosity of crude oil during refining. Some reduction in viscosity is also achieved through the breakdown of complex hydrocarbon derivatives (e.g., aromatic hydrocarbon derivatives), into simpler hydrocarbon derivatives of low molecular weight.

8.3.3.2 Biodenitrogenation

Nitrogen is like sulfur; it is considered as a crude oil contaminant. In general, the nitrogen content of crude oil is low and generally falls within the range 0.1 to 0.9%, although some crude oil may contain up to 2% nitrogen (Speight, 2012, 2014). However, crude oils with no detectable nitrogen or even trace amounts are not uncommon, but in general the more asphaltic the oil, the higher the nitrogen content. Insofar as an approximate correlation exists between the sulfur content and API gravity of crude oils (Ancheyta and Speight, 2007), there also exists a correlation between nitrogen content and the API gravity of crude oil. It also follows that there is an approximate correlation between the nitrogen content and the carbon residue: the higher the nitrogen content, the higher the carbon residue. The presence of nitrogen in crude oil is of much greater significance in refinery operations than might be expected from the small amounts present. Nitrogen compounds can be responsible for the poisoning of cracking catalysts, and they also contribute to gum formation in such products as domestic fuel oil. The trend in recent years toward cutting deeper into the crude to obtain stocks for catalytic cracking has accentuated the harmful effects of the nitrogen compounds, which are concentrated largely in the higher boiling portions (Speight, 2014). Moreover, nitrogen containing compounds (NCCs) coexist with sulfur containing compounds (SCCs) in fossil fuels (Yi et al., 2014). Carbazole (CAR) as an example for the non-basic nitrogenous polyaromatic hydrocarbon derivatives (NPAHs) can directly impact the refining processes, especially during the cracking process, CAR can be converted into basic derivatives, which can be adsorbed to the active sites of the cracking catalysts. It directly inhibits the hydrodesulfurization catalysts. Thus, removal of CAR and other nitrogen-compounds, would significantly increase the extent of catalytic cracking and consequently the gasoline yield. Basic nitrogen compounds are more inhibitory for catalysts than the non-basic ones. However, theses nitrogen compounds can potentially be converted to basic compounds during the refining/catalytic cracking process while will inhibit catalyst activity. Moreover, metals like nickel and vanadium are potent inhibitors for catalysts and in crude oil; metals are typically associated with N-compounds (Hegedus and McCabe, 1981; Mogollón et al., 1998).

Although, specific biodesulfurization of crude oil and the distillates has been reasonably investigated, there is a little information related to biodenitrogenation of oil feed without affecting the calorific value. It has been estimated that biodenitrogenation of crude oil would be beneficial for deep denitrogenation, where the classical hydroprocessing methods are costly and non-selective (Vazquez-Duhalt et al., 2002). It will also eliminate the contribution of fuel nitrogen to NOx emissions. However, the economics of nitrogen-removal processes are affected by the amount of associated hydrocarbon lost from the fuel, during the denitrogenation process.

From the practical point of view, biodenitrogenation and biodesulfurization should be integrated, where, sulfur and nitrogen, would be removed through specific enzymatic attack of the C–N and C–S bonds, respectively, but without C–C bond attack, to preserve the fuel value of the biotreated products. Moreover, a dual microbial process for both selective biodesulfurization and biodenitrogenation, with the overcome of the significant technical hurdles, such as tolerance against solvents, high concentration of nitrogenous compounds, high oil to water ratio, would make microbial refining processes and bio-upgrading of crude oil and the fractions feasible on a large scale (Kayser and Kilbane, 2004). Workers in PETROBRAS, the Brazilian oil company, have isolated *Gordonia* sp. strain F.5.25.8 that can utilize dibenzothiophene through the 4S-pathway and CAR as a sole source of S and N, respectively. F.5.25.8 is the first reported strain that can simultaneously metabolize dibenzothiophene and carbazole (Yu et al., 2006). The F.5.25.8 strain can tolerate up to 42°C, which would add to the advantageous in industrial application of biodesulfurization/ biodenitrogenation as complementary to hydrotreatment process. Moreover, it is reported to have a different genetic organization of the biodesulfurization (dsz) and biodenitrogenation (car) gene clusters relative to *Rhodococcus* erythropolis IGTS8 and *Pseudomonas* sp. IGTN9m, respectively.

Extra heavy crude oil contains metals in the form of salts (zinc, titanium, calcium, and magnesium), petroporphyrins (vanadium, copper, nickel, and iron) and other complexes that exits predominantly within the asphaltene fraction (Ali and Abbas, 2006; Speight, 2014; Hsu and Robinson, 2017). Thus, the higher the asphaltene content of crude oil, the higher the heavy metal content (Speight, 2014). In fact, the accumulation of metallic constituents in the higher molecular weight and polar fractions of crude oils plays a significant role in establishing the refining procedure (Panariti et al., 2000; Speight, 2014, 2017).

Depending on the origin of crude oil, the concentration of the vanadium varies from as low as 0.1 ppm to as high as 1200 ppm, while that of nickel commonly varies from trace amounts to 150 ppm (Ali and Abbas, 2006; Speight, 2014; Hsu and Robinson, 2017). The V/Ni ratio is constant in crude oils of common source rocks and dependent on the geological age of the rocks, where oils from Triassic or older age showing a value higher than unity) (Ball et al., 1960), and this ratio is also used for tracing source effects (El-Gayar et al., 2002). However, the biodegradation of the asphaltene constituents and the resin constituents is reported to influence the Ni/V ratio in these fractions.

Crude oil with a particularly high content of organometallics includes Boscán crude oil, Cerro Negro crude oil, Maya crude oil, Wilmington crude oil, and Prudhoe Bay crude oil (Fish et al., 1984). The metallic constituents often occur as inorganic water-soluble forms and are easily removed during the crude desalting process, in which they are concentrated in the aqueous phase. However, the metallo-porphyrins are embedded in the extremely complex structure of the asphaltene constituents and thus, metal removal from petroporphyrins and complexes that is necessary to be addressed.

Currently, demetallization occurs during the hydrocracking process or during hydrotreating process (Ancheyta and Speight, 2007; Speight, 2014, 2017). However, feedstocks that have relatively high metals contents (> 300 ppm) substantially increase catalyst consumption because the metals poison the catalyst, thereby requiring frequent catalyst replacement. The usual desulfurization catalysts are relatively expensive for these consumption rates but there are catalysts that are relatively inexpensive and can be used in the first reactor to remove a large percentage of the metals. Subsequent reactors downstream of the first reactor would use normal hydrodesulfurization catalysts. Since the catalyst materials are proprietary, it is not possible to identify them here. However, it is understood that such catalysts contain little or no metal promoters, i.e., nickel, cobalt, molybdenum. Metals removal on the order of 90% has been observed with these materials.

Thus, one method of controlling demetallization is to employ separate smaller *guard reactors* just ahead of the fixed-bed hydrodesulfurization reactor section (Speight, 2014, 2017). The preheated feed and hydrogen pass through the guard reactors that are filled with an appropriate catalyst for demetallization that is often the same as the catalyst used in the hydrodesulfurization section. The advantage of this system is that it enables replacement of the most contaminated catalyst (*guard bed*), where pressure drop is highest, without having to replace the entire inventory or shut down the unit. The feedstock is alternated between guard reactors while catalyst in the idle guard reactor is being replaced.

The presence of metal contaminants in the fluid catalytic cracking (FCC) feeds presents another and potentially more serious problem because although sulfur can be converted to gaseous forms which can be readily handled in an FCC unit, the nonvolatile metal contaminants tend to accumulate in the unit and during the cracking process they are deposited on the catalyst together with the coke.

Moreover, since, both nickel and vanadium exhibit dehydrogenation activity, thus, their presence on the catalyst particles tends to promote dehydrogenation reactions during the cracking sequence and this result in increased amounts of coke and gases at the expense of gasoline production (Ali and Abbas, 2006; Speight, 2014; Hsu and Robinson, 2017; Speight, 2017). Heavy metals can be liberated into the environment, during fuel combustion in the form of ash with high concentrations of toxic metal oxides. That leads to undesirable by-products and the need for disposal (Xu, 1997).

Moreover, heavy metals (mostly Ni and V) are furthermore corrosive (Montiel et al., 2009). When crude oil is vacuum-distilled, the metallo-porphyrins tend to be entrained in the vapors and are carry over the vacuum gas oil fraction. The typical demetallization process in crude oil industry is the solvent deasphalting process, where the lower-boiling oils are physically separated from the asphaltene constituents by mixing the heavy crude oil/residue with a very low boiling solvent such as propane and/or butane (Farag et al., 1989; Speight, 2000; Ancheyta and Speight, 2007; Speight, 2014, 2017). Other options for demetallization include a low-temperature coking process in which the metals collect in the initially-formed coke (Speight, 2014, 2017).

The distillation process separates crude oils into fractions according to boiling point, so that each of the processing units following will have the feedstock that meets the required specifications. The metallic constituents concentrate in the residues (Reynolds et al., 1987). The thermal processes, such as the visbreaking and coking (Speight, 2000, 2014, 2017) typically involve redistribution of the hydrogen in the residue to produce lower-boiling products containing more hydrogen while the asphaltene constituents and metals are removed in the form of coke or visbreaking residue. Filtration using a porous membrane is reported to be effective for removal of nitrogen, sulfur, nickel, and vanadium as well as any other metallic constituents that have accumulated during production and transportation (Kutowy et al., 1989). There is also a method for upgrading heavy crude oils by solvent dissolution and ultra-filtration at high pressure (Osterhuber, 1989). The process is especially suitable for removing trace metals (mainly Ni and V) thereby reducing the Conradson carbon residue (CCR) of the resulting oil.

There are also processes for the selective removal of metals from the organic moieties with a minimal conversion of the remaining crude oil. For example, metal removal using a solvent (Savastano, 1981), oxidative demetallization of crude oil asphaltene constituents and residua (Gould, 1980) and the hydro-demetallization process (Adarme et al., 1990; Bartholdy and Hannerup, 1990; Piskorz et al., 1996) are also available for demetallization of crude oils. However, these methods increase the cost of refining and usually produce secondary pollution in the environment (Hernandez et al., 1998).

The destruction of porphyrin ring systems using a chlorination reaction using *Caldariomyces fumago* chloroperoxidase enzyme leading to the biocatalytic demetallization of petroporphyrins and asphaltene constituents with a reduction in nickel octaethyl porphyrin and vanadyl octaethyl porphyrin of approximately 93 and 53%, respectively, is possible (Fedorak et al., 1993). But, chlorination produces the chlorinated compounds which have negative impact to the environment. Biocatalytic removal of Ni and V from petroporphyrins by chloro-peroxidase (CPO) has been also reported (Mogollón et al., 1998).

The biodemetallization of petroporphyrins and crude oil, through the oxidation of prophyrinic rings using hemoproteins (cytochrome C reductases) from Bacillus megaterium and Catharanthus roseuse occurs in the presence of the cofactor; NADPH (the reduced form of nicotinamide adenine dinucleotide phosphate NADP) (Xu et al., 1998). In contrast to peroxidases, such as chloroperoxidase, oxygenases from *Escherichia coli*, animal cells (such as liver or kidney cells), plant cells (such as from mung beans or *Arabidopsis thaliana*) or yeast cells (such as, Candida tropiculis) can degrade porphyrin molecules without subjecting the hydrocarbon to chlorine or peroxide, and the metals which can be removed by the present method include nickel, vanadium, cobalt, copper, iron, magnesium, and zinc. The produced metals can be readily removed by extraction (such as, a de-salt wash), distillation, ion exchange and/or column chromatography. The main advantageous of this process, is it can be operated in a batch, semi-continuous or continuous mode alone or in combination with one or more additional biorefining processes (such as biodesulfurization), in a sealed or open vessel.

The removal of vanadyl porphyrins from the crude oils is reported to be of a great importance. There is also the possibility of microbial demetallization of crude oil using *Aspergillus* sp. MS-100 isolated from a polluted soil of Isfahan refinery for the ability to consume vanadium oxide octaethyl porphyrin (VOOEP) as a sole carbon source (Salehizadeh et al., 2007). Horse myoglobin and plant peroxidase were chosen as the proteins used to synthesize the agents for removal of porphyrins from un-cracked fuels (Paul and Smith, 2009). Also, since most of the metals in crude oil are associated with organonitrogen compounds, it would be possible to simultaneously perform the biodenitrogenation and biometallization processes to reduce the nitrogen and metals content in one process instead to do it in two separate processes.

8.4 Biodegradation of Other Products

In addtion to the conventional distillation products from crude oil and the various petrochemical products (Parkash, 2003; Gary et al., 2007; Speight, 2014; Hsu and Robinson, 2017; Speight, 2017), there is a variety of other products (commonly referred to as waste) that must also receive attention when spilled into the environment. Examples of such products and the processes by which they are created (Chapter 2) are (1) acid sludge, (2) spent acid, (3) spent catalyst, (4) spent caustic, (5) sulfonic acids, and (6) wastewater.

8.4.1 Acid Sludge

Typically, acid sludge is the material that separates from a crude oil on treatment with sulfuric acid and is a composite of toxic liquid residues that contains unsaturated hydrocarbon derivatives, sulfuric acid, water, and various other components. This material is generated during various refining process and is mostly stored in open lagoons that caused a serious pollution problem to the soil and groundwater.

The sludge has a very variable composition between refineries and the concentration of sulfuric acid may be 50% v/v or higher. Treatment of the sludge at the refinery includes neutralizing at an early stage, so the corrosion of the equipment is completely eliminated and the treatment process may be designed to extract hydrocarbon constituents from the sludge in the form of a commercial-grade fuel oil. Storage of the acid sludge on site leads to huge, often irreparable damage to the environment and the environmental fines.

8.4.2 Spent Acid

Spent acid, which is the avid that has been weakened by use: such as (1) a mixed acid that has been used in nitration or (2) an acid that has been used in pickling metal articles. It is typically, in the current context, a mixture of sulfuric acid, nitric acid, water, and the related organic compound is generated mainly through acid-assisted processes. After the acid has performed its function, it becomes partly neutralized (i.e., spent) and also picks up multiple impurities, such as organic compounds and metals derivatives, making it unsuitable for immediate reuse. In some cases, this spent acid may be safely disposed or used in other processes, but usually it must be regenerated back to relatively pure and concentrated acid for reuse in the main process, over and over.

The major processes include alkylation to produce high-octane gasoline, nitration to produce explosives and pesticides, and the acrylonitrile (AN) and methyl methacrylate (MMA) processes. The sulphuric acid regeneration (SAR) plant processes for all of these feeds are essentially the same, with consideration given to the organic and inorganic contaminants, overall water balance, and the desired product specifications. These plants may be readily designed to produce fresh sulphuric acid in concentrations from 93% acid to 40% oleum, with 99.2% sulfuric acid typical for alkylation use.

8.4.3 Spent Catalyst

Catalysts gradually lose their catalytic activity, usually through structural changes, poisoning, overheating, or the deposition of extraneous material such as coke. The inhibitions in catalytic performance are accounted by different factors such as physical losses, steam, high temperature, time, coke formation and poisoning from metal contaminants in feedstock. This type of deactivated catalyst is referred to as "used" catakyst or "spent" catalyst or equilibrium catalyst or simply "ECAT". Thus, a catalyst is "spent" when it no longer exhibits the necessary activity or specificity required by the user.

Solid catalytic materials play major role in crude oil refining and once a catalyst has completed its use cycle it will be withdrawn from the process. At this stage, the catalyst is considered to be "spent" and the heavy metals, coke, and other poisonous elements that have been deposited on the catalyst during the various process cysles render the catalyst a hazardous waste.

Historically, spent catalysts have been disposed of as landfill in approved dumpsites but currently a spent catalyst can only be disposed of into a landfill only if it can be proven *with certainty* that the landfill met the various criteria as a non-hazardous waste that which also would not cause run-off of any potentially hazardous constituents. Under the current legislative guidelines, this is considered unlikely.

However, there are methods that reduce catalyst waste by using post treatment methods or spent catalyst management such as (1) regeneration, (2) rejuvenation, (3) reuse either in fresh catalyst preparation, or (4) reuse in less severe hydrotreating units by cascading before final disposal.

Another option involves the use of the spent catalyst as raw material for concrete and mortar production as partial replacement of sand and cement powder. Other minor applications include their use as catalysts for plastic and biomass pyrolysis and gasification or for production of synthetic fuels.

8.4.4 Spent Caustic

Spent caustic is a waste industrial caustic solution that has become exhausted and is no longer useful. Spent caustics are made of sodium hydroxide or potassium hydroxide, water, and contaminants. In the crude oil crude oil refining industry, caustic solutions (i.e., NaOH in water) are regularly used to remove H2S and organic sulfur compounds from hydrocarbon streams. Once H2S is reacted with NaOH, the solution becomes known as *spent caustic* or *spent sulfidic caustic*. Spent caustics typically have a pH value (> 12) and high sulfide concentrations (2 to 3% w/w). Depending on the source, spent caustic may also contain phenols, mercaptans, amines, and other organic compounds that are soluble or emulsified in the caustic (Bechok, 1967).

Most spent caustics are sent off-site for commercial recovery or reuse, e.g., in pulp and paper mills, for treatment by wet air oxidation, or for disposal by deep-well injection (Kolhatkar and Sublette, 1996). The mainly method that dealing with caustic is chemical method such as neutralization and oxidation (Chernatskaya, 1974), but these need large investments and have a high operating cost. And the most important is that it will lead to a serious environmental pollution and equipment corrosion (Shen, 1995).

Physico-chemical treatment of spent caustics, in general operated at high temperature and pressure, is characterized by high operating costs and large investment. Biological treatment could be an inexpensive alternative. It is rarely seen the report in use of biological treatment of spent caustic (Sipma et al., 2004). Biotechnology for wastewater treatment has a long history (Matveev, 1996), this method has been widely used due to a soft degradation conditions (Eigenson et al., 1974), low-cost and no secondary pollution. If the spent caustic is directly into the biological treatment system without neutralization, it must affect the normal operation of subsequent biochemical systems because of its toxicity, high total dissolved solid (TDS) and complex composition (Wang and Yang, 2001) will resulting in discharged above standard. Incomplete removal of thiols in spent caustic will lead to severe odor nuisance.

Thus, it is very important to find salt-tolerant microbial with excellent degradation for biotreatment of spent caustic.

8.4.5 Wastewater

Crude oil refineries are among the major consumers of water due to cooling towers and process usage. During the treatment and refining of crude oil, large quantities of wastewater are generated, which require treatment. The quality of this wastewater depends on the grade of the crude oil and the process for treating the oil.

In recent years, alkali/surfactant/polymer (ASP) flooding has been well developed for enhancement of oil recovery in some Chinese oilfields (Deng et al., 2005). This method has been proven to be a promising recovery approach to extract large volume of oil remaining in the ground after conventional oil recovery methods have been implemented.

There are various technologies that have been used to treat the conventional oily wastewater, including membrane filtration (Deriszadeh et al., 2010), adsorption (Fakhru'l-Razi et al., 2009), chemical precipitation and oxidation (Fakhru'l-Razi et al., 2009), and electrochemical treatment (Ramalho et al., 2010). Biological approach has been widely used to degrade oil and remediate oil-contaminated water (Lu et al., 2009; Ilori et al., 2011; Lu et al., 2011; Zekri et al., 2009; Zhou and Shen, 2010). However, it would be difficult to meet the requirement for flooding-produced water treatment in oilfield by using an approach alone, because of the highly complex characteristics of pollutants and large quantity of flooding-produced water. Hence, a combination of physicochemical and biological treatment is needed in order to enhance the treatment efficiency of flooding-produced water.

The application of the combination of hydrolysis acidification-dynamic membrane bioreactor-coagulation process to treating wastewater produced from polymer flooding has been reported to be effective (Zhang et al., 2010). In addition, use of a combination of the zero valent iron/EDTA/air system and activated sludge process as well as coagulation and flocculation to remediate oilfield wastewater produced from polymer flooding has also been claimed to be effective (Asia et al., 2006; Lu and Wei, 2011).

8.5 Rates of Biodegradation

Different results for microbial activity measurements may be obtained in laboratory studies depending on pretreatment and size of the sample, even when the environmental conditions are mimicked (Björklöf et al., 2008). These differences may be related to, amongst other factors, differences in the bioavailability of the contaminant in different analyses.

Thus, modeling chemical reactions to determine the rate of the reaction is a common practice in chemistry. However, blending of chemical reactants, which is a part of modeling studies, can have beneficial or adverse effects on chemical reactions. Blend time predictions are usually based on empirical correlations and when a competitive side reaction is present, the final product distribution may be suspect. In biodegradation, the effect of the microbe in relation to the hydrocarbon stream on the reaction outcome is crucial. Also, the scale up of such reactions from the laboratory to the field may not be straightforward. Thus, there is a need for comprehensive, for optimistic caution when models of biodegradation chemistry are used to predict information such as incubation period, product, byproducts, and whether or not the microbes will survive the duration of the treatment. It is also equally important to recognize the potential for interference of one chemical species with another—a situation that is not often determined in the laboratory when simple model substrates (such as single chemical entities or mixtures of two-to-twelve compounds that are not truly representative of crude oil and crude oil products) are used for the experiments.

The most important critical stage of crude oil degradation during the first 48 hours of a spill is usually evaporation, the process by which the lower molecular weight (lower-boiling constituents) constituents of crude oil volatilize into the atmosphere. Evaporation can be responsible for the loss of one- to two-thirds of the mass of spilled material (assuming that the spilled material is conventional crude oil or a distillate product) during this period, with the loss rate decreasing rapidly over time. The constituents of heavy oil, tar sand bitumen, and asphalt do not evaporate to the same extent—the lower boiling constituents being generally absent from these materials.

Evaporative loss is controlled by the composition and physical properties of the crude oil or crude oil derivative, the surface of the spill, wind velocity, and temperature (Payne and McNabb, 1984). Derivation of a universal model for such a process may insert inaccuracies because of the complex and changing nature of crude oil and crude oil products. In addition, the material left behind is richer in metals (mainly nickel and vanadium), waxes, resin constituents, and asphaltene constituents than the original oil. With evaporation, the specific gravity and viscosity of the original oil also increase—after several days, spilled conventional crude oil may begin to resemble Bunker C oil (heavy fuel oil) in composition and properties (Mielke, 1990).

Although, there has been a reported case of lack of correlation between degradation rates, specific growth rates and concentration of the starter oil (Thouand et al., 1999), in such a case, it would appear that biomass was required only to a particular threshold enough to produce the appropriate enzyme system that carry through the degradation process even when biomass production had ceased (Pitter and Chudoba, 1990), where production of variance with the theory of microbial growth in batch cells is totally dependent on the consumed carbon source.

Many reports on the effect of sunlight irradiation so far published have focused on the physicochemical changes on intact crude oil other than to biodegraded crude oil (Jacquot et al., 1996; Nicodem et al., 1998). Recent studies have reported that photo-oxidation increases the biodegradability of crude oil hydrocarbon by increasing its bioavailability and thus enhancing microbial activities (Maki et al., 2005).

In fact, bioremediation is a multi-variable process and optimization through classical methods is subject to question. To overcome the disadvantages of the process, response surface methodology (RSM) has been advocated for analyzing the effects of several independent variables on the bioremediation process in order to assess the optimum conditions for the process (Nasrollahzadeh et al., 2007; Huang et al., 2008; Pathak et al., 2009; Vieira et al., 2009; Mohajeri et al., 2010; Zahed et al., 2010).

The outcome is the suggestion that the rates of biodegradation can be increased by modifying selected physical and chemical conditions that control biodegradation in multiphase systems, namely, (1) bioavailability and (2) terminal electron acceptor availability (Sandrin et al., 2006). With respect to bioavailability, model simulations suggest that: (1) increasing the interfacial area between the aqueous and solid phases, (2) increasing the rate of contaminant solubilization, and (3) minimizing the accumulation of the contaminant in non-aqueous phase liquids will result in significantly higher rates of biodegradation.

Possible approaches to increasing contaminant bioavailability will depend on the system in question. However, any approach that either increases the physical mixing of the non-aqueous phase liquids and aqueous phases or that increases the solubility of the contaminant in the aqueous phase will result in increased biodegradation. Such approaches include the addition of surfactants or co-solvents (to increase solubility), aeration, or hydraulic pulsing, to increase mixing of the two phases.

The complex array of factors that influence biodegradation of crude oil-related is not realistic to expect a simple rate model or kinetic model to provide precise and accurate descriptions of concentrations during different seasons and in different environments. Therefore, it is nearly impossible to predict the rates of biodegradation of such a process. To give a final answer on how much time remediation processes require and what the final mineral oil concentrations will be, experiments should be continued until the biodegradation processes have stopped completely. In future, it will be necessary to use complex models to yield a more exact assessment of soil remediation to the desired level (Maletić et al., 2009).

Finally, when predicting by modeling the behavior of an aged contaminant it is relevant to adapt the models in use to correspond to conditions relevant at the contaminated sites. As with all efforts at modeling the outcome of complex processes, the variable parameters used in the models must be based on (1) the properties of the material (in this case the crude oil-based contaminant), (2) data retrieved about the conditions of the actual site, and on (3) experiments performed using the original aged contaminant without any additions (model compounds or analytical *spikes*).

8.6 Application to Spills

The success of oil spill bioremediation depends on optimization of various physical, chemical, and biological conditions in the contaminated environment. There are two main approaches to bioremediation of crude oil-related spills: (1) bioaugmentation, in which oil-degrading microorganisms are added to supplement the existing microbial population, and (2) biostimulation, in which the growth of indigenous oil degraders is stimulated by the addition of nutrients or other growth-limiting co-substrates and/or habitat alteration (Chapter 1).

The first steps in assessing the potential of a crude oil-contaminated site (or crude oil product-contaminated site) as a potential bioremediation site is a detailed waste classification and a determination of the origin of the waste assist to assist in planning the removal activities.

Typically, an oil refinery waste consists of both Oil Pollution Act (OPA) and Comprehensive Environmental Response and Liability Act (CERCLA) wastes (oily pits from crude oil, refined products, tank bottoms, asbestos, corrosives, small laboratory containers), using Oil Pollution Act or wastewater treatment wastes, as well as waste listed under the Resource Conservation and Recovery Act, and a careful waste classification is required to decide on the appropriate cleanup technology to remediate the site. Once this has been completed, representative samples of the land treatment unit (LTU) soil/waste should be collected and analyzed for, but not necessarily limited to, volatile organic compounds (VOCs), polynuclear aromatic hydrocarbon derivatives (PAH), total crude oil hydrocarbon derivatives (TPH), metals, and naturally occurring radioactive material (NORM) in order to determine the most appropriate method for cleanup.

Bioremediation products have been applied to clean up crude oil hydrocarbon contamination in various ecosystems and under a wide range of environmental conditions. Their applications include *in situ* remediation of hydrocarbon contaminated marine shorelines, soil environments, surface water, groundwater, and water, and *ex situ* treatment of hydrocarbon contaminated soil (e.g., use of land treatment units or other types of reactor systems such as compost piles, biopiles, slurry reactors, etc.) and water (e.g., in a bioreactor).

Bioremediation technology is typically used as a secondary polishing step after conventional mechanical cleanup options have been applied to remove free oil product. However, many case studies have demonstrated that bioremediation can also be used as a primary response strategy, especially for the cleanup of environmentally sensitive areas that are not amenable to conventional cleanup techniques and/or low-level crude oil hydrocarbon contamination.

Bioaugmentation appears to have little benefit for the treatment of spilled oil in an open environment. Microbial addition has not been shown to work better than nutrient addition alone in many field trials. However, case studies provided by vendors seem to suggest that application of bioaugmentation products could still have some potential in the treatment of specific oil components, isolated spills in confined areas, or certain environments where oil-degrading microorganisms are deficient. Unfortunately, the evidence for such a conclusion is not strong and in most cases scientifically deficient.

Biostimulation has been proven to be a promising tool to treat certain aerobic oil-contaminated marine shorelines. One of the key factors for the success of oil biostimulation is to maintain an optimal nutrient level in the interstitial pore water. In other words, background nutrient concentrations at the contaminated site should be one of the primary determining factors in the decision to apply nutrients, and biostimulation might not always be necessary if sufficient nutrients are naturally present at a spill site to supply non-limiting concentrations to the degrading populations. However, effects of nutrients are also highly site-specific. For example, the availability of oxygen rather than nutrients is often the limiting factor in wetland environments, where addition of nutrient products has not been successful in enhancing oil biodegradation (although it has been successful in accelerating the restoration of the affected plant biomass to an abundant and rich recovery).

Different nutrient products have shown variable effectiveness, depending on oil properties, the nature of the nutrient products, and the characteristics of the contaminated environments. Based on limited field trials, it appears that slow-release fertilizers may be an excellent choice if the nutrient release rates are balanced against physical loss rates; water-soluble fertilizers may be more cost-effective in low-energy shorelines and fine-grained sediments where water transport is limited; and oleophilic fertilizers may be more suitable for use on hard, rocky shorelines, although further research is still required to confirm this suggestion. In general, commercial oleophilic nutrient products have not shown clear advantages over common agricultural fertilizers in stimulating oil biodegradation.

Bioremedial approaches may have a role in treating hydrocarbon contamination for non-point sources. Limited available information appears to suggest that application of bioremediation agents could show promise for the treatment of hydrocarbon contamination in stormwater, especially used in conjunction with other stormwater countermeasures, such as wet detention ponds. Bioremediation agents may also be effective for the treatment of bilge water, although, due to the lack of any systematic investigation into its effectiveness, it is still uncertain whether this approach could compete with other existing technologies. Further field tests are needed to provide stronger evidence on the potential of this strategy.

The extreme uncertainty associated with the efficacy of bioremediation agents is due in large part to the poorly designed field tests that have been conducted to demonstrate efficacy. Much of the reported literature lacked proper controls and treatment randomization and replication, or the data were incorrectly analyzed. If there is any hope for advancement of commercial bioremediation for the environments described in this report, especially estuaries, experiments based on sound scientific principles are needed. Unfortunately, resources for field-testing commercial bioremediation agents are scarce, and field studies are extremely expensive to carry out. That's why it is best to rely on laboratory microcosm or mesocosm studies to provide needed data to support this technology. When spills occur and the on-scene coordinator in conjunction with the Regional Response Team decides to implement commercial bioremediation for cleanup, they should try to set aside control areas if at all possible to allow a more effective evaluation of treatment success (Zhu et al., 2004).

Generally, indigenous hydrocarbon-degrading marine micro-organisms promote oil dispersion and biodegradation of crude oil provided that sufficient nutrients are available for microbial growth. However, this may be a relatively slow process, depending on the nature of the oil tested. The application of dispersants to an oil slick at sea can have a dual environmental benefit, stimulating both oil dispersion and oil biodegradation (Mulyono et al., 1994; Varadaraj et al., 1994).

The success of a dispersant at stimulating oil biodegradation is dependent on the ability of the dispersant to promote the growth of indigenous hydrocarbon-degrading micro-organisms. The presence of suspended clay particles substantially increases the rate of oil dispersion and subsequent biodegradation, presumably by providing surfaces on which the micro-organisms can grow and maintain higher populations than that observed in seawater in the absence of clay (Lunel et al., 1995; Lee et al., 1997a; Bragg and Yang, 1995). These mixtures are neutrally-buoyant and may promote the decomposition of oil residues.

Since the contaminants of concern in crude oil are readily biodegradable under appropriate conditions, the success of oil-spill bioremediation depends on our ability to establish those conditions in the contaminated environment. The most important requirement is that bacteria with appropriate metabolic capabilities must be present. If they are, their rates of growth and hydrocarbon biodegradation can be maximized by ensuring that adequate concentrations of nutrients and oxygen are present and that the pH is between about 6 and 9. The physical and chemical characteristics of the oil are also important determinants of bioremediation success. Heavy crude oils that contain large amounts of resin and asphaltene compounds are less amenable to bioremediation than are light-or medium-weight crude oils that are rich in aliphatic components. Finally, the oil surface area is extremely important because growth of oil degraders occurs almost exclusively at the oil-water interface (Atlas and Bartha, 1972).

Obviously, some of these factors can be manipulated more easily than others. For example, nothing can be done about the chemical composition of the oil, and no adequate engineering approaches are currently available for providing oxygen to oil-contaminated surficial sediments in the intertidal zone. Therefore, the two main approaches to oil-spill bioremediation are: (1) *bioaugmentation*, in which oil-degrading bacteria are added to supplement the existing microbial population, and (2) *biostimulation*, in which nutrients or other growth-limiting co-substrates are added to stimulate the growth of indigenous oil degraders. Since oil-degrading bacteria usually grow at the expense of one or more components of crude oil, and these organisms are ubiquitous (Mulkins-Phillips and Stewart, 1974b; Roubal and Atlas, 1978; Atlas, 1981), there is usually no reason to add hydrocarbon degraders unless the indigenous bacteria are incapable of degrading one or more important contaminants. The size of the hydrocarbon-degrading bacterial population usually increases rapidly in response to oil contamination, and it is very difficult, if not impossible, to increase the microbial population over that which can be achieved by biostimulation alone (Westlake et al., 1978; Lee et al., 1997b).

The carrying capacity of most environments is probably determined by factors such as predation by protozoans, the oil surface area, or scouring of attached biomass by wave activity that are not affected by bioaugmentation, and added bacteria seem to compete poorly with the indigenous population (Tagger et al., 1983; Lee and Levy, 1989). Therefore, it is unlikely that they will persist in a contaminated beach even when they are added in high numbers. As a result, bioaugmentation has never been shown to have any long-term beneficial effects in shoreline cleanup operations.

Biostimulation involves the addition of rate-limiting nutrients to accelerate biodegradation by indigenous microorganisms. When an oil spill occurs, it results in a huge influx of carbon into the impacted environment. Carbon is the basic structural component of living matter, and in order for the indigenous microorganisms to be able to convert this carbon into more biomass, they need significantly more nitrogen and phosphorus than is normally present in the environment. Both of these elements are essential ingredients of protein and nucleic acids of living organisms. The main challenge associated with biostimulation in oil-contaminated coastal areas or tidally influenced freshwater rivers and streams is maintaining optimal nutrient concentrations in contact with the oil.

References

Adarme, R., Sughrue, E.L., Johnson, M.M., Kidd, D.R., Phillips, M.D., and Shaw, J.E. 1990. Demetallization of asphaltenes: thermal and catalytic effects with small pore catalysts. J. Am. Chem. Soc., 35: 614–618.

Aislabie, J.M., McLeod, M., and Fraser, R. 1998. Potential for biodegradation of in soil from the ross dependency, Antarctica. Appl. Microbiol. Biotechnol., 49: 210–214.

Aislabie, J.M., Foght, J., and Saul, D. 2000. Aromatic hydrocarbon-degrading bacteria from soil near scott base, Antarctica. Polar Biol., 23: 183–188.

Aislabie, J.M., Balks, M.R., Foght, J.M., and Waterhouse, E.J. 2004. Hydrocarbon spills on antarctic soils: effects and management. Environ. Sci. Technol., 38(5): 1265–74.

Al-Darbi, M.M., Saeed, N.O., Islam, M.R., and Lee, K. 2005. Biodegradation of natural oils in seawater. Energy Sources, 27: 19–34.

Alexander, M. 1994. Biodegradation and Bioremediation. Academic Press Inc., New York.

Ali, H.R., El-Gendy, N.Sh., Moustafa, Y.M., Mohamed, I., Roushdy, M.I., and Hashem, A.I. 2012. Degradation of asphaltenic fraction by locally isolated halotolerant bacterial strains. ISRN Soil Sci. ID 435485.

Al-Sulaimani, H. Al-Wahaibi, Y., Al-Bahry, S., Elshafie, A., Al-Bemani, A., Joshi, S., and Zargari, S. 2011a. Optimization and partial characterization of biosurfactants produced by *Bacillus* species and their potential for *ex situ* enhanced oil recovery. Society of Petroleum Engineers SPE J., 16(3): 672–682.

Al-Sulaimani, H., Joshi, S., Al-Wahaibi, Y., Al-Bahry, S., Elshafie, A., and Bemani, A.A. 2011b. Microbial biotechnology for enhancing oil recovery: current developments and future prospects. Biotechnol. Bioinf. Bioeng., 1(2): 147–158.

Aluyor, E.O., and Ori-jesu, M. 2009. Biodegradation of mineral oils—a review. African Journal of Biotechnology, 8(6): 915–920.

Amund, O.O., Adewale, A.A., and Ugoji, E.O. 1987. Occurrence and characterization of hydrocarbon utilizing bacteria in nigerian soils contaminated with spent motor oil. Indian J. Microbiol., 27: 63–87.

Ancheyta, J., and Speight, J.G. 2007. Hydroprocessing of Heavy Oils and Residua. CRC Press, Taylor & Francis Group, Boca Raton, Florida.

Ancheyta, J. 2016. Deactivation of Heavy Oil Hydroprocessing Catalysts: Fundamentals and Modeling. Wiley, Hoboken, New Jersey.

Antai, S.P. 1990. Biodegradation of Bonny light crude oil by *Bacillus* sp. and *Pseudomonas* sp. Waste Management, 10: 61–64.

Antić, M.P., Jovančićević, B.S., Ilić, M., Vrvić, M.M., and Schwarzbauer, J. 2006. Petroleum pollutant degradation by surface water microorganisms. Environ. Sci. Pollut. Res., 13(5): 320–327.

Asia, I.O., Enweani, I.B., and Eguavoen, I.O. 2006. Characterization and treatment of sludge from the petroleum industry. African Journal of Biotechnology, 5(5): 461–466.

Atlas, R.M., and Bartha, R. 1972. Degradation and mineralization of petroleum in sea water: limitation by nitrogen and phosphorus. Biotech. Bioeng., 14: 309–318.

Atlas, R.W. 1975. Effects of Temperature and crude oil composition on petroleum biodegradation. Applied Microbiology, 30(3): 396–403.

Atlas, R.M. 1978. An assessment of the biodegradation of petroleum in the Arctic. pp. 86–90. *In*: Loutit, M.W., and Miles, J.A.R. (Eds.). Microbial Ecology. Springer-Verlag, Berlin.

Atlas, R.M. 1981. Microbial Degradation of Petroleum Hydrocarbons: An Environmental Perspective. Microbiological Rev., 45: 180–209.

Atlas, R.M. 1991. Microbial hydrocarbon degradation—bioremediation of oil spills. J. Chem. Technol.-Biotechenol., 52: 149–156.

Atlas, R.M., and Bartha, R. 1992. Hydrocarbon Biodegradation and oil spill bioremediation. Adv. Microbial. Ecol., 12: 287–338.

Atlas, R.M. 1995. Petroleum biodegradation and oil spill bioremediation. Mar. Pollut. Bull., 31: 178–182.

Ayala, M., Verdin, J., and Vazquez-Duhalt, R. 2007. The prospects for peroxidase-based biorefining of petroleum fuels. Biocat. Biotransform., 25: 114–129.

Bachman, R.T., Johnson, A.C., and Edyvean, R.G.J. 2014. Biotechnology in the petroleum industry: An overview. International Biodeterioration and Biodegradation., 86: 225–237.

Bagherzadeh-Namazi, A., Shojaosadati, S.A., and Hashemi-Najafabadi, S. 2008. Biodegradation of used engine oil using mixed and isolated cultures. Int. J. Environ. Res., 2: 431–440.

Ball, J.S., Wenger, W.J., Hyden, H.J., Horr, C.A., and Myers, A.T. 1960. Metal content of twenty-four petroleums. J. Chem. Eng. Data., 5: 553–557.

Bamforth, S.M., and Singleton, I. 2005. Bioremediation of polycyclic aromatic hydrocarbons: current knowledge and future directions. J. Chem. Technol. Biotechnol., 80: 723–736.

Bardi, L., Ricci, R., and Mario Marzona, M. 2003. *In situ* bioremediation of a hydrocarbon polluted site with cyclodextrin as a coadjuvant to increase bioavailability. Water, Air, and Soil Pollution: Focus, 3: 15–23.

Bardi, L., Martini, C., Opsi, F., Bertolone, E., Belviso, S., Masoero, G., Marzona, M., and Ajmone Marsan, F. 2007. Cyclodextrin-enhanced *in situ* bioremediation of polyaromatic hydrocarbons-contaminated soils and plant uptake. J. Incl. Phenom. Macrocycl. Chem., 57: 439–444.

Barnard, P.C., and Bastow, M.A. 1991. Hydrocarbon generation, migration, alteration, entrapment and mixing in the Central and Northern North Sea. pp. 167–190. *In*: England, W.A., and Fleet, A.J. (Eds.). Petroleum Migration. Special Publication. Geological Society, London, united Kingdom.

Bartha, R. 1986. Microbial Ecology: Fundamentals and Applications. Addisson-Wesley Publishers, Reading, Massachusetts.

Bartholdy, J., and Hannerup, P.N. 1990. Hydrodemetallation in resid hydroprocessing. J. Am. Chem. Soc., 35: 619–625.

Bechok, M.R. 1967. Aqueous Wastes from Petroleum and Petrochemical Plants. John Wiley & Sons Inc., New York.

Bej, A.K., Saul, D., and Aislabie, J. 2000. Cold-tolerant Alkane-degrading *Rhodococcus* Species from Antarctica. Polar Biol., 23: 100–105.

Bence, A.E., Kvenvolden, K.A., and Kennicutt, M.C. 199. Organic geochemistry applied to environmental assessments of prince william sound, alaska, after the exxon valdez oil spill—a review. Organic Geochemistry, 24: 7–42.

Bennett, P.C., Siegel, D.E., Baedecker, M.J., and Hult, M.F. 1993. Crude oil in a shallow sand and gravel aquifer 1. hydrogeology and inorganic geochemistry. Applied Geochemistry, 8: 529–549.

Bennet, J.W., Wunch, K.G., and Faison, B.D. 2002. Use of fungi biodegradation. pp. 960–971. *In*: Hurst, C.J. (Ed.). Manual of Environmental Microbiology Second Edition. ASM Press Washington, DC. Chapter 87.

Bertrand, J.C., Almallah, M., Acquaviva, M., and Mille, G. 1990. Biodegradation of hydrocarbons by an extremely halophilic archaebacterium. Lett. Appl. Microbiol., 11: 260–263.

Bestetti, G., Collina, E., Di Gennaro, P., Lasagni, M., and Pitea, D. 2003. Kinetic study of naphthalene biodegradation in aerobic slurry phase microcosms for the optimization of the process. Water, Air, and Soil Pollution: Focus, 3: 223–231.

Björklöf, K., Salminen, J., Sainio, P., and Jørgensen, K. 2008. Degradation rates of aged petroleum hydrocarbons are likely to be mass transfer dependent in the Field. Environ. Geochem Health, 30: 101–107.

Borchert, S., Mueller, J., Alesi E., Leins, C., and Haninger, V. 1995. *In situ* bioremediation application strategies for soil and groundwater impacted by PAHs. Land Contam. Reclam., 3: 6-1–6-4.

Bossert, I., and Bartha, R. 1984. The fate of petroleum in soil ecosystems. pp. 453–473. *In*: Atlas, R.M. (Ed.). Petroleum Microbiology. Macmillan, New York.

Braddock, J.F., and McCarthy, K.A. 1996. Hydrologic and microbiological factors affecting persistence and migration of petroleum hydrocarbons spilled in a continuous-permafrost region. Environ Sci. Technol., 30: 2626–2633.

Braddock, J.F., Lindstrom, J.E., Yeager, T.R., Rasley, B.T., and Brown, E.J. 1996. Patterns of microbial activity in oiled and unoiled sediments in prince williams sound. Proceedings. Am. Fish Soc. Symp., 18: 94–108.

Braddock, J.F., Ruth, M.L., Walworth, J.L., and McCarthy, K.A. 1997. Enhancement and inhibition of microbial activity in hydrocarbon-contaminated arctic soils: implications for utrientamended bioremediation. Environ Sci. Technol., 31: 2078–2084.

Bradley, P.M., and Chapelle, F.H. 1995. Rapid toluene mineralization by aquifer microorganisms at adak, alaska: implications for intrinsic bioremediation in cold environments. Environ. Sci. Technol., 29: 2778–2781.

Bragg, J.R. and Yang, S.H. 1995. Clay-oil flocculation and its role in natural cleansing in prince william sound following the exxon valdez oil spill. pp. 178–214. *In*: P.G., Butler, J.N., and Hughes, J.S. (Eds.). Exxon Valdez Oil Spill: Fate and Effects in Alaskan Waters. ASTM STP No. 1219. Wells. American Society for Testing and Materials, West Conshohocken, Pennsylvania.

Breedveld, G.D., and Sparrevik, M. 2000. Nutrient-limited biodegradation of PAH in various soil strata at a creosote contaminated site. Biodegradation, 11: 391–399.

Bruheim, P., and Eimhjelle, K. 1998. Chemically emulsified crude oil as substrate for bacterial oxidation: differences in species response. Can. J. Microb., 44(2): 195–199.

Bryant, R.S., Donaldson, E.C., Yen, T.F., and Chilingarian, G.V. 1989. Microbial enhanced oil recovery. pp. 423–450. *In*: Donaldson, E.C., Chilingarian, G.V., and Yen, T.F. (Eds.). Enhanced Oil Recovery II: Processes and Operations. Elsevier, Amsterdam, Netherlands.

Bryant, R.S., and Lindsey, R.P. 1996. World-wide applications of microbial technology for improving oil recovery. pp. 27–134. *In*: Proceedings. SPE Symposium on Improved Oil Recovery. Society of Petroleum Engineers, Richardson, Texas.

Bryant, R.S., Bailey, S.A., Step, A.K., Evans, D.B., and Parli, J.A. 1998. Biotechnology for heavy oil recovery. SPE Paper No. 36767 Proceedings. SPE/DOE improved oil recovery symposium. Page 1–7.

Carmichael, L.M., and Pfaender, F.K. 1997. The effect of inorganic and organic supplements on the microbial degradation of phenanthrene and pyrene in soils. Biodegradation, 8: 1–13.

Cerniglia, C.E. 1984. Microbial metabolism of polycyclic aromatic hydrocarbons. Adv. Appl. Microbiol., 30: 31–71.

Cerniglia, C.E., and Heitkamp, M.A. 1989. Microbial degradation of polycyclic aromatic hydrocarbons in the aquatic environment. pp. 41–68. *In*: Varanasi, U. (Ed.). Metabolism of Polycyclic Aromatic Hydrocarbons in the Aquatic Environment. CRC Press, Inc., Boca Raton, Florida.

Cerniglia, C.E., and Heitkamp, M.A. 1990. Polycyclic aromatic hydrocarbon degradation by *Mycobacterium*. Methods Enzymol., 188: 148–153.

Cerniglia, C.E. 1992. Biodegradation of polycyclic aromatic hydrocarbons. Biodegradation, 3: 351–368.

Chaîneau, C.H., Rougeux, G., Yepremian, C., and Oudot, J. 2005. Effects of nutrient concentration on the biodegradation of crude oil and associated microbial populations in the soil. Soil. Biol. Biochem., 37: 1490–1497.

Chen, C.I., and Taylor, R.T. 1995. Thermophilic biodegradation of BTEX by Two *Thermus* Species. Biotechnol. Bioeng., 48: 614–624.

Chen, C.I., and Taylor, R.T. 1997a. Batch and fed-batch bioreactor cultivations of a *Thermus* Species with thermophilic BTEX-degrading activity. Appl. Microbiol. Biotechnol., 47: 726–733.

Chen, C.I., and Taylor, R.T. 1997b. Thermophilic biodegradation of BTEX by two consortia of anaerobic bacteria. Appl. Microbiol. Biotechnol., 48: 121–128.

Chernatskaya, N. 1974. Modern methods for determination of contaminants in refinery waste waters. Chemistry and Technology of Fuels and Oils. 10(9): 692–695.

Choi, S.-C., Kwon, K.K., Sohn, J.H., and Kim, S.-J. 2002. Evaluation of fertilizer additions to stimulate oil biodegradation in sand seashore mescocosms. J. Microbiol. Biotechnol., 12: 431–436.

Chugunov, V.A., Ermolenko, Z.M., Martovetskaya, I.I., Mironava, R.I., Zhirkova, N.A., Kholodenko, V.P., and Urakov, N.N. 2000. Development and application of a liquid preparation with oil-oxidizing bacteria. Appl. Biochem. Microbiol., 36: 577–581.

Cirigliano, M.C., and Carman, G.M. 1984. Isolation of a bioemulsifier from *Candida lipolytica*. Appl. Environ. Microb., 48: 747–750.

Clements, W.H., Oris, J.T., and Wissing, T.E. 1994. Accumulation and food chain transfer of fluoranthene and Benzo[*a*]pyrene in *Chironomus riparius* and *Lepomis macrochirus*. Arch. Environ. Contam. Toxicol., 26: 261–266.

Connan, J., Restle, A., and Albrecht, P. 1980. Biodegradation of crude oils in the Aquitaine basin. pp. 1–17. *In*: Douglas, A.G., and Maxwell, J.R. (Eds.). Advances in Organic Geochemistry, 1979. Pergamon Press, Oxford, United kingdom.

Connan, J. 1981. Un exemple de biodegradation preferentielle des hydrocarbures aromatique dans des asphaltes du bassin sud-aquitain (France). Bull. Cent. Rech. Explor-Prod Elf Aquitaine, 5: 151–171.

Connan, J. 1984. Biodegradation of crude oils in reservoirs. pp. 299–335. *In*: Brooks, J., and Welte, D.H. (Eds.). Advances in Petroleum Geochemistry. Academic Press, London, United Kingdom. Volume 1.

Daling, P., and Strøm, T. 1999. Weathering of oils at sea: model/field data comparisons. Spill Science & Technology Bulletin, 5(1): 63–74.

Dandie, C.E., Weber, J., Aleer, S., Adetutu, E.M., Ball, A.S., and Juhasz, A.L. 2010. Assessment of five bioaccessibility assays for predicting the efficacy of petroleum hydrocarbon biodegradation in aged contaminated soils. Chemosphere, 81: 1061–1068.

Das, K., and Mukherjee, A.K. 2007. Crude petroleum-oil biodegradation efficiency of *Bacillus Subtilis* and *Pseudomonas eruginosa* strains isolated from a petroleum-oil contaminated soil from North-East India. Bioresource Technol., 98: 1339–1345.

Davies, L., Daniel, F., Swannell, R., and Braddock, J. 2001. Biodegradability of Chemically-Dispersed Oil. Report No. AEAT/ENV/R0421. AEA Technology Environment, Abingdon, Oxfordshire, United Kingdom.

Del'Arco, J.P., and de França, F.P. 1999. Biodegradation of crude oil in a sandy sediment. International Biodeterioration and Biodegradation, 44: 87–92.

Delille, D., Bassères, A., and Dessommess, A. 1998. Effectiveness of bioremediation for oil-polluted antarctic seawater. Polar Biol., 19: 237–241.

Delille, D., and Delille, B. 2000. Field observations on the variability of crude oil impact in indigenous hydrocarbon-degrading bacteria from sub-antarctic intertidal sediments. Mar. Environ. Res., 49: 403–417.

Delille, D., Coulon, F., and Pelletier, E. 2004. Effects of temperature warming during a bioremediation study of natural and nutrient-amended hydrocarbon-contaminated sub-antarctic soils. Cold Reg. Sci. Tech., 40: 61–70.

Delille, D., and Pelletier, E. 2009. Effects of nutrient and temperature on degradation of petroleum hydrocarbons in sub-antarctic coastal seawater. Polar Biol., 32: 1521–1528.

Delille, D., Pelletier, E., Rodriguez-Blanco, A., and Ghiglione, J. 2009. Effects of nutrient and temperature on degradation of petroleum hydrocarbons in sub-antarctic coastal seawater. Polar Biol., 32: 1521–1528.

Deng, S., Yu, G., Jiang, Z., Zhang, R., and Ting, Y. 2005. Destabilization of oil droplets in produced water from ASP Flooding. Colloids Surf. A., 252: 113–119.

Deriszadeh, A., Husein, M.M., and Harding, T.G. 2010. Produced water treatment by micellar-enhanced ultrafiltration. Environ. Sci. Technol., 44: 1767–1772.

Dibble, J.T., and Bartha, R. 1976. The effect of iron on the biodegradation of petroleum in seawater. Appl. Environ. Microb., 31: 544–550.

Eigenson, S., Loakimis, E.G., and Lukinskayal, N.T. 1974. Prospects for development in water usage and water discharge in petroleum refineries and petrochemical plants. Chemistry and Technology of Fuels and Oils, 10(9): 665–667.

Eja, M.E., Arikpo, G.E., and Udo, S.M. 2006. The bioremediation potentials of fungal species isolated from soils polluted by petroleum products in cross river university of technology, Calabar, Nigeria. International Journal of Natural and Applied Sciences (IJNAS), 1(1): 15–20.

El-Gayar, M., Mostafa, M.S., Abdelfattah, A.E., and Barakat, A.O. 2002. Application of geochemical parameters for classification of crude oils from Egypt into source-related types. Fuel Process. Technol., 79: 13–28.

El-Gendy, N. Sh., Farahat, L.A., Moustafa, Y.M., Shaker, N., and El-Temtamy, S.A. 2006. Biodesulfurization of crude and diesel oil by *Candida parapsilosis* NSh45 isolated from egyptian hydrocarbon polluted sea water. Biosci. Biotechnol. Res. Asia., 3(1a): 5–16.

El-Gendy, N.Sh. 2015. Biodesulfurization of petroleum and its fraction. pp. 655–680. *In*: Pant, K.K., Shishir Sinha, Bajpai, S., and Govil, J.N. (Eds.). Advances in Petroleum Engineering. *In*: Chemical Technology Series. Vol. 4 Advances in Petroleum Engineering II: *Petrochemical*. Studium Press LLC, Houston. Chapter 24.

El-Gendy, N.Sh., and Speight, J.G. 2016. Handbook of Refinery Desulfurization. CRC Press, Taylor & Francis, Boca Raton, Florida.

Eriksson, M., Dalhammar, G., and Borg-Karlson, A.K. 2000. Biological degradation of selected hydrocarbons in an old pah/creosote contaminated soil from a gas work site. Appl. Microbiol. Biotechnol., 53: 619–626.

Fakhru'l-Razi, A., Pendashteh, A., Abdullah, L.C., Biak, D.R.A., Madaeni, S.S., and Abidin, Z.Z. 2009. Review of technologies for oil and gas produced water treatment. J. Hazard. Mater., 170: 530–551.

Farag, A.S., Sif El-Din, O.I., Youssef, M.H. Hassan, S.I., and Farmawy, S. 1989. Solvent demetallization of heavy oil residue. Hung. J. Ind. Chem., 17(3): 289–294.

Fedorak, P.M., Semple, K.M., Vazquez-Duhalt, R., and Westlake, D.W.S. 1993. Chloroperoxidase-mediated modifications of petroporphyrins and asphaltenes. Enzyme Microb. Technol., 15: 429–437.

Fish, R.H., Komlenic, J.J., and Wines, B.K. 1984. Characterization and comparison of vanadyl and nickel compounds in heavy crude petroleums and asphaltenes by reverse-phase and size-exclusion liquid chromatography/ graphite furnace atomic absorption spectrometry. Analytical Chemistry, 56: 2452–2460.

Floodgate, G.D. 1984. The fate of petroleum in marine ecosystems. pp. 355–397. *In*: Atlas, R.M. (Ed.). Petroleum Microbiology. MacMillan, New York.

Foght, J.M., and Westlake, D.W.S. 1982. Effect of the dispersant corexit 9527 on the microbial degradation of prudhoe bay oil. Canadian Journal of Microbiology, 28: 117–122.

Foght, J.M., Gutnick, D.L., and Westlake, D.W.S. 1989. Effect of emulsan on biodegradation of crude oil by pure and mixed bacterial cultures. Applied and Environmental Microbiology, January: 36–42.

Foght, J.M., Westlake, D., Johnson, W.M., and Ridgway, H.F. 1996. Environmental gasoline-utilizing isolates and clinical isolates of *Pseudomonas aeruginosa* are taxonomically indistinguishable by chemotaxonomic and molecular techniques. Microbiol., 142: 1333–1338.

Foght, J., Semple, K., Gauthier, C., Westlake, D.S., Blenkinsopp, S., Sergy, G., Wang, Z., and Fingas, M. 1999. Effect of nitrogen source on biodegradation of crude oil by a defined bacterial consortium incubated under cold, marine conditions. Environ. Technol., 20: 839–849.

Freijer, J.I., de Jonge, H., Bouten, W., and Verstraten, J.M. 1996. Assessing mineralization rates of petroleum hydrocarbons in soils in relation to environmental factors and experimental scale. Biodegradation, 7: 487–500.

Frenzel, M., James, P., Burton, S.K., Rowland, S.J., Lappin-Scott, H.M. 2009. Towards bioremediation of toxic unresolved complex mixtures of hydrocarbons: identification of bacteria capable of rapid degradation of alkyltetralins. J. Soils Sediments, 9: 129–136.

Frontera-Suau, R., Bost, F., McDonald, T., and Morris, P.J. 2002. Aerobic biodegradation of hopanes and other biomarkers by crude oil degrading enrichment cultures. Environ. Sci. Technol., 36: 4585–4592.

Gallego, J.L., García, M.J., Llamas, J.F., Belloch, C., Pelaez, A.I., and Sanchez, J. 2007. Biodegradation of oil tank bottom sludge using microbial consortia. Biodegradation, 18: 269–281.

Gao, C.H., and Zekri, A. 2011. Applications of microbial-enhanced oil recovery technology in the past decade. Energy source. A, Recovery until. Environ. Effects, 33(10): 972–989.

Gassara, F., Suri, N., Stanislav, P., and Voordouw, G. 2015. Microbially enhanced oil recovery by sequential injection of light hydrocarbon and nitrate in low- and high-pressure bioreactors. Environ. Sci. Technol., 49(20): 12594–601.

Ghazali, F.M., Rahman, R.N.Z.A., Salleh, A.B., and Basri, M. 2004. Biodegradation of hydrocarbons in soil by microbial consortium. Intl. Biodeter. Biodegrad., 54: 61–67.

Gibson, D.T., and Subramanian, V. 1984. Microbial degradation of aromatic hydrocarbons. pp. 181–252. *In*: Gibson, D.T. (Ed.). Microbial Degradation of Organic Compounds. Marcel Dekker, Inc., New York, N.Y.

Gieg, L.M., Kolhatkar, R.V., McInerney, M.J., Tanner, R.S., Harris, S.H. Jr,, Sublette, K.L., and Suflita, J.M. 1999. Intrinsic bioremediation of petroleum hydrocarbons in a gas condensate-contaminated aquifer. Environ Sci. Technol., 33: 2550–2560.

Goldman, S., Shabtai, Y., Rubinovitz, C., Rosenberg, E., and Gutnick, D.L. 1982. Emulsan in *Acinetobacter calc oaceticus* RAG-1: Distribution of cell-free and cell-associated cross-reacting material. Appl. Environ. Microbiol., 44: 165–170.

Goodin, J.D., and Webber, M.D. 1995. Persistence and fate of anthracene and Benzo[*a*]pyrene in municipal sludge treated soil. J. Environ. Qual., 24: 271–278.

Goodwin, N.S., Park, P.J.D., and Rawlinson, A.P. 1983. Crude oil biodegradation under simulated and natural conditions. pp. 650–658. *In*: Bjorøy, M. (Ed.). Advances in Organic Geochemistry 1981. John Wiley & Sons Inc. New York.

Gould, K.A. 1980. Oxidative demetallization of petroleum asphaltenes and residua. Fuel, 59(10): 733–736.

Gray, K.A., Pogrebinsky, O.S., Mrachko, G.T., Xi, L., Monticello, D.J., and Squires, C.H. 1996. Molecular mechanisms of biocatalytic desulfurization of fossil fuels. Nat. Biotechnol., 14(13): 1705.

Greenwood, P.F., Wibrow, S., George, S.J., and Tibbett, M. 2009. Hydrocarbon biodegradation and soil microbial community response to repeated oil exposure. Organic Geochemistry, 40: 293–300.

Gupta, N., Roychoudhury, P.K., and Dep, J.K. 2005. Biotechnology of desulfurization of diesel: prospects and challenges. Appl. Microbiol. Biotechnol., 66: 356–366.

Gupta, R.K., and Gera, P. 2015. Process for the upgradation of petroleum residue: Review. International Journal of Advanced Technology in Engineering and Science, 3(2): 643–656.

Gutnick, D.L., and Minas, W. 1987. Perspectives on microbial surfactants. Biochem. Soc. Trans., 15: 22S–35S.

Haferburg, D., Hommel, R., Claus, R., and Kleber, H.P. 1986. Extracellular microbial lipids as biosurfactants. Adv. Biochem. Eng. Biotech., 33: 53–93.

Hamer, G., and Al-Awadhi, N. 2000. Biotechnological applications in the oil industry. Acta Biotechnol., 20: 335–350.

Haritash, A.K., and Kaushik, C.P. 2009. Biodegradation aspects of Polycyclic Aromatic Hydrocarbons (PAHs): a review. Journal of Hazardous Materials, 169: 1–15.

Harvey, R.G. 1997. Polycyclic Aromatic Hydrocarbons. John Wiley & Sons Inc.-VCH, New York.

Head, I.M., Jones, D.M., and Röling, W.F.M. 2006. Marine microorganisms make a meal of oil. Nature Reviews Microbiology, 4: 173–182.

Hegedus, L.L., and McCabe, R.W. 1981. Catalyst poisoning. Catal. Rev.-Sci. Eng., 23: 377–476.

Heitkamp, M.A., and Cerniglia, C.E. 1988. Mineralization of polycyclic aromatic hydrocarbons by a bacterium isolated from sediment below an oil field. Applied and Environmental Microbiology, 54: 1612–1614.

Heitkamp, M.A., and Cerniglia, C.E. 1989. Polycyclic aromatic hydrocarbon degradation by a *Mycobacterium* sp. In Microcosms Containing Sediment and Water from a Pristine Ecosystem. Appl. Environ. Microbiol., 55: 1968–1973.

Hernáez, M.J., Reineke, W., and Santero, E. 1999. Genetic analysis of biodegradation of tetralin by a *Sphingomonas* Strain. Appl. Environ. Microbiol., 65(4): 1806–1810.

Hernandez, A., Mellado, R., and Martinez, J. 1998. Metal accumulation and vanadium induced multidrug resistance by environmental isolates of *Escherichia hermani* and *Enterobacter cloacae*. Appl. Environ. Microbiol., 64: 4377–4320.

Hernández-López, E.L., Ayala, M., and Vazquez-Duhalt, R. 2015. Microbial and enzymatic biotransformations of asphaltenes. Petrol. Sci. Technol., 33(9): 1017–1029.

Hernández-López, E.L., Perezgasga, L., Huerta-Saquero, A., Mouriño-Pérez, R., and Vazquez-Duhalt, R. 2016. Biotransformation of petroleum asphaltenes and high molecular weight polycyclic aromatic hydrocarbons by *Neosartorya fischeri*. Environ. Sci. Pollut. Res., 23: 10773–10784.

Holba, A.G., Dzou, I.L., Hickey, J.J., Franks, S.G., May, S.J., and Lenney, T. 1996. Reservoir geochemistry of south pass 61 Field, Gulf of Mexico: compositional heterogeneities reflecting filling history and biodegradation. Organic Geochemistry, 24: 1179–1198.

Hsu, C.S., and Robinson, P.R. (Eds.). 2017. Handbook of Petroleum Technology. Springer International Publishing AG, Cham, Switzerland.

Huang, L., Ma, T., Li, D., Liang, F., Liu, R., and Li, G. 2008. Optimization of nutrient component for diesel oil degradation by *Rhodococcus erythropolis*, Mar. Pollut. Bull, 56: 1714–1718.

Huesemann, M.H., Hausmann, T.S., and Fortman, T.J. 2004. Does bioavailability limit biodegradation? A comparison of hydrocarbon biodegradation and desorption rates in aged soils. Biodegradation, 15: 261–274.

Hunkeler, D., Rainer, U. Meckenstock, R.U., Lollar, B.S., Schmidt, T.C., and Wilson, J.T. 2008. A Guide for Assessing Biodegradation and Source Identification of Organic Ground Water Contaminants using Compound Specific Isotope Analysis (CSIA). Report No. EPA 600/R-08/148. United States Environmental Protection Agency, Office of Research and Development, National Risk Management Research Laboratory, Ada, Oklahoma.

Ilori, M.O., Adebusoye, S.A., Obayori, O.S., Oyetibo, G.O., Ajidahun, O., James, C., and Amund, O.O. 2011. Extensive biodegradation of nigerian crude oil (Escravos Light) by newly characterized yeast strains. Pet. Sci. Technol., 29: 2191–2208.

Iturbe, R., Flores, C., Castro, A., and Torres, L.G. 2007. Sub-soil contamination due to oil spills in zones surrounding oil pipeline-pump stations and oil pipeline right-of-ways in Southwest-Mexico. Environ. Monit. Assess., 133: 387–398.

Jack, T.R. 1991. Microbial enhancement of oil recovery. Curr Opin. Biotechnol., 2(3): 444–449.

Jacquot, F., Guiliano, M., Doumenq, P., Munoz, D., and Mille, G. 1996. *In vitro* photo-oxidation of crude oil maltenic fractions: evolution of fossil biomarkers and polycyclic aromatic hydrocarbons. Chemosph., 33: 671–681.

Jean, J.M., Lee, Sh., Wang, Chattopadhyay, P., and Maity, J. 2008. Effects of inorganic nutrient levels on the biodegradation of Benzene, Toluene, and Xylene (BTX) by *Pseudomonas* spp. in a laboratory porous media sand aquifer model. Bioresource Technol., 99: 7807–7815.

Jones, D.M., Douglas, A.G., Parkes, R.J., Taylor, J., Giger, W., and Schaffner, G. 1983. The recognition of biodegraded petroleum-derived aromatic hydrocarbons in recent marine sediments. Mar. Poll. Bull., 14: 103–108.

Joshi, P.A., and Pandey, G.B. 2011. Screening of petroleum degrading bacteria from cow dung. Research Journal of Agricultural Sciences, 2(1): 69–71.

Kanal, R.A., and Harayama, S. 2000. Biodegradation of high-molecular-weight polycyclic aromatic hydrocarbons by bacteria. Journal of Bacteriology, 182(8): 2059–2067.

Kanaly, R.A., and Harayama, S. 2000. Biodegradation of high-molecular weight polycyclic aromatic hydrocarbons by bacteria. J. Bacteriol., 182: 2059–2067.

Karimi, M., Mahmoodi, M., Niazi, A., Al-Wahaibi, Y., and Ayatollahi, S. 2012. Investigating wettability alteration during MEOR process, a micro/macro-scale analysis. Colloids Surf. B: Biointerfaces, 95: 129–136.

Kaufman, E.N., Harkins, J.B., and Borole, A.P. 1998. Comparison of batch stirred and electrospray reactors for biodesulfurization of dibenzothiophene in crude oil and hydrocarbon feedstocks. Appl. Biochem. Biotechnol., 73:127–144.

Kayser, K.J., and Kilbane, J.J. 2004. Method for metabolizing carbazole in petroleum. US Patent 6943006.

Kilbane, J.J., and Jackoswki, F. 1992. Biodesulfurization of water-soluble coal-derived material by *Rhodococcus rhodochrous* IGTS8. Biotechnol. Bioeng., 40(9): 1107–1114.

Kim, M., Bae, S.S., Seol, M., Lee, J., and Oh, Y. 2008. Monitoring nutrient impact on bacterial community composition during bioremediation of Anoxic PAH-Contaminated Sediment. J. Microbiol, 46: 615–623.

Kim, S., Choi, D.H., Sim, D.S., and Oh, Y. 2005. Evaluation of bioremediation effectiveness on crude oil-contaminated sand. Chemosph., 59: 845–852.

Knezevich, V.O., Koren, E., Ron, Z., and Rosenberg, E. 2006. Petroleum bioremediation in seawater using guano as the fertilizer. Bioremediation J., 10: 83–91.

Kolhatkar, A., and Sublette, K.L. 1996. Biotreatment of refinery spent sulfidic caustic by specialized cultures and acclimated activated sludge. Appl. Biochem. Biotechnol., 58(57): 945–957.

Kutowy, O., Tweddle, T.A., and Hazlett, J.D. 1989. Method for the molecular filtration of predominantly aliphatic liquids. US Patent No. 4,814,088.

Kwok, C.-K., and Loh, K.-C. 2003. Effects of singapore soil type on the bioavailability of nutrients in soil bioremediation. Advance of Environmental Research, 7: 889–900.

Lazar, I., Petrisor, I., and T. Yen. 2007. Microbial enhanced oil recovery (MEOR). Petrol. Sci. Technol., 25(11): 1353–1366.

Leahy, J.G., and Colwell, R.R. 1990. Microbial degradation of hydrocarbons in the environment. Microbiol. Rev., 54: 305–315.

Le Borgne, S., and Quintero, R. 2003. Biotechnological processes for refining of petroleum. Fuel Process. Technol., 81: 155–169.

Lee, K., Wong, C.S., Cretney, W.J., Whitney, F.A., Parsons, T.R., Lalli, C.M., and Wu, J. 1985. Microbial response to crude oil and corexit 9527: SEAFLUXES enclosure study. Microbial Ecology, 11: 337–351.

Lee, K., and Levy, E.M. 1989. Enhancement of the natural biodegradation of condensate and crude oil on beaches of Atlantic Canada. Proceedings. 1989 Oil Spill Conference. American Petroleum Institute, Washington, DC. Page 479–486.

Lee, K., Lunel, T., Wood, P., Swannell, R., and Stoffyn-Egli, P. 1997a. Shoreline cleanup by acceleration of clay-oil flocculation processes. Proceedings. 1997 International Oil Spill Conference, American Petroleum Institute, Washington, DC. Page 235–240.

Lee, K., Tremblay, G.H., Gauthier, J., Cobanli, S.E., and Griffin, M. 1997b. Bioaugmentation and biostimulation: a paradox between laboratory and field results. Proceedings. 1997 International Oil Spill Conference. American Petroleum Institute, Washington, DC. Page 697–705.

Lee, E.-H., Kim, J., Cho, K-S., Ahn, Y.G., and Hwang, G.-S. 2010. Degradation of hexane and other recalcitrant hydrocarbons by a novel isolate, *Rhodococcus* sp. EH831. Environ Sci. Pollut. Res., 17: 64–77.

Leòn, V., and Kumar, M. 2005. Biological upgrading of heavy crude oil. Biotechnol. Bioprocess Eng., 10: 471–481.

Lepo, J.E., and Cripe, C.R. 1999. Microbial biosystems: new frontiers. Proceedings. 8th international symposium on microbial ecology. Bell, C.R., Brylinsky, M., and Johnson-Green, P. (Eds.). Atlantic Canada Society for Microbial Ecology, Halifax, Nova Scotia, Canada.

Leys, M.N., Bastiaens, L., Verstraete, W., and Springael, D. 2005. Influence of the carbon/nitrogen/phosphorus ratio on polycyclic aromatic hydrocarbons degradation by *Mycobacterium* and *Sphingomonas* in soil. Applied and Microbiology and Biotechnolgy, 66: 726–736.

Li, J., Pignatello, J.J., Smets, B.F., Grasso, D., and Monserrate, E. 2005. Bench-scale evaluation of *in situ* bioremediation strategies for soil at a former manufactured gas plant site. Environ. Toxicol. Chem., 24: 741–749.

Liao, Y., Geng, A., and Huang, H. 2009. The Influence of biodegradation on resins and asphaltenes in the Liaohe Basin. Organic Geochemistry, 40: 312–332.

Liebeg, E.W., and Cutright, T.J. 1999. The investigation of enhanced bioremediation through the addition of macro and micro nutrients in a PAH contaminated soil. Int. Biodeterior. Biodegrad., 44: 55–64.

Lindstrom, J.E., and Braddock, J.F. 2002. Biodegradation of petroleum hydrocarbons at low temperature in the presence of the dispersant corexit 9500. Mar. Poll. Bull., 44: 739–747.

Liu, T., Cao, G., and Ba, Y. 2012. Activation of the indigenous microorganism and field test in Zhan 3 block [J]. Chem. Eng. Oil Gas, 41(4): 411–478.

Liu, Z., Jacobson, A.M., and Luthy, R.G. 1995. Biodegradation of naphthalene in aqueous nonionic surfactant systems. Appl. Environ. Microb., 61: 45–151.

Litherathy, P., Haider, S., Samhan, O., and Morel, G. 1989. Experimental studies on biological and chemical oxidation of dispersed oil in seawater. Water Science and Technology, 21: 845–856.

Lu, M., Zhang, Z., Yu, W., and Zhu, W. 2009. Biological treatment of oilfield-produced water: a field pilot study. Int. Biodeter. Biodegr., 63: 316–321.

Lu, M., and Wei, X. 2011. Treatment of oilfield wastewater 211 containing polymer by the batch activated sludge reactor combined with a zerovalent iron/edta/air system. Bioresour. Technol., 102: 2555–2562.

Lu, M., Wei, X., and Su, Y. 2011. Aerobia treatment of oilfield wastewater with a bio-contact oxidation reactor. Desalination and Water Treatment, 27: 334–340.

Lundstedt, S., Haglund, P., and Oberg, L. 2003. Degradation and formation of polycyclic aromatic compounds during bioslurry treatment of an aged gasworks soil. Environ. Toxicol. Chem., 22: 1413–1420.

Lunel, T. Understanding the mechanism of dispersion through droplet size measurements at sea. *In*: Lane, P. (Ed.). The Use of Chemicals in Oil Spill Response. ASTM STP No. 1252. American Society for Testing and Materials, West Conshohocken, Pennsylvania.

Luning Prak, D.J., and Pritchard, P.H. 2002. Solubilization of polycyclic aromatic hydrocarbon mixtures in micellar nonionic surfactant solution. Water Research, 36: 3463–3472.

Machackova, J., Wittlingerova, Z., Vlk, K., Zima, J., and Ales, L. 2008. Comparison of two methods for assessment of *in situ* jet fuel remediation efficiency. Water Air Soil Pollut., 187: 181–194.

Maila, M.P., and Cloete, T.E. 2004. Bioremediation of petroleum hydrocarbons through landfarming: are simplicity and cost-effectiveness the only advantages? Reviews in Environmental Science & Bio/Technology, 3: 349–360.

Maki, H., Sasaki, T., and Haramaya, S. 2005. Photo-oxidation of biodegradable crude oil and toxicity of the photo-oxidized products. Chemosph., 44: 1145–1151.

Maletić, S., Dalmacija, B., Rončević, S., Agbaba, J., and Petrović, O. 2009. Degradation kinetics of an aged hydrocarbon-contaminated soil. Water Air Soil Pollut., 202: 149–159.

Margesin, R., and Schinner, F. 1997a. Efficiency of indigenous and inoculated cold-adapted soil microorganisms for biodegradation of diesel oil in alpine soils. Appl. Environ. Microbiol., 63: 2660–2664.

Margesin, R., and Schinner, F. 1997b. Bioremediation of diesel-oil-contaminated alpine soils at low temperatures. Appl. Microbiol. Biotechnol., 47: 462–468.

Margesin, R., and Schinner, F. (Eds.). 1999a. Cold-adapted Organisms. Springer, New York.

Margesin, R., and Schinnerm, F. 1999b. Biological decontamination of oil spills in cold environments. J. Chem. Technol. Biotechnol., 74: 381–389.

Margesin, R. 2000. Potential of cold-adapted microorganisms for bioremediation of oil-polluted alpine soils. Int. Biodeterior. Biodegrad., 46: 3–10.

Margesin, R., and Schinner, F. 2001. Potential of halotolerant and halophilic microorganisms for biotechnology. Extremophiles, 5: 73-83.

Marín, J.A., Moreno, J.L., Hernández, T., and García, C. 2006. Bioremediation by composting of heavy oil refinery sludge in semiarid conditions. Biodegradation, 17: 251–261.

Matveev, M.S. 1996. Biological purification of waste water from oil refineries. Chemistry and Technology of Fuels and Oils, 2(5): 329–332.

McFarland, B.L. 1999. Biodesulfurization. Curr. Opin. Microbiol., 2: 257–264.

Mielke, J.E. 1990. Oil in the Ocean: The Short- and Long-Term Impacts of a Spill. Congressional Research Service, 90-356 SPR, July 24, p. 11.

Mihara, C., Yano, T., Kozaki, S., and Imamura, T. 1999. Microbial strain, method for biodegrading organic compounds and method for environmental remediation. United States Patent 5,962,305. October 5.

Miiller, D.E., Holba, A.G., and Huges, W.B. 1987. Effects of biodegradation on crude oils. pp. 233–241. *In*: Meyer, R.F. (Ed.). Exploration for Heavy Crude Oil and Natural Bitumen. AAPG Studies in Geology #25. American Association of Petroleum Geologists, Tulsa, Oklahoma.

Milić, J.S., Beškoski, V.P., Ilić, M.V. Ali, S.A.M., Gojgić-Cvijović, G.Đ., and Vrvić, M.M. 2009. Bioremediation of soil heavily contaminated with crude oil and its products: composition of the microbial consortium. J. Serb. Chem. Soc., 74(4): 455-460.

Miller, E.C., and Miller, J.A. 1981. Search for the ultimate chemical carcinogens and their reaction with cellular macromolecules. Cancer, 47: 2327–2345.

Mills, M.A., Bonner, J.S., McDonald, T.J., Page, C.A., and Autenrieth, R.L. 2003. Intrinsic bioremediation of a petroleum-impacted wetland. Marine Pollution Bulletin, 46: 887–899.

Minai-Tehrani, D., Minoui, S., and Herfatmanesh, A. 2009. Effect of salinity on biodegradation of polycyclic aromatic hydrocarbons (PAHs) of heavy crude oil in soil. Bull. Environ. Contam. Toxicol., 82: 179–184.

Mitsch, W.J., and Gosselink, J.G. 1993. Wetlands 2nd Edition. John Wiley & Sons Inc., New York.

Mogollón, L., Rodríguez, R., Larrota, W., Ortiz, C., and Torres, R. 1998. Biocatalytic removal of nickel and vanadium from petroporphyrins and asphaltenes. Appl. Biochem. Biotechnol., 70–72: 765–777.

Mohebali, G., and Ball, A.S. 2016. Biodesulfurization of diesel fuels—past, present and future perspectives. Int. Biodeter. Biodegrad., 110: 163–180.

Mohajeri, L., Aziz, H.A., Isa, M.H., and Zahed, M.A. 2010. A statistical experiment design approach for optimizing biodegradation of weathered crude oil in coastal sediments. Bioresource Technol., 101: 893–900.

Mohammed, D., Ramsubhag, A.S., and Beckles, D.M. 2007. An assessment of the biodegradation of petroleum hydrocarbons in contaminated soil using non-indigenous, commercial microbes. Water Air Soil Pollut., 182: 349–356.

Mohan, S.V., Kisa, T., Ohkuma, T., Kanaly, R.A., and Shimizu, Y. 2006. Bioremediation technologies for treatment of PAH-contaminated soil and strategies to enhance process efficiency. Rev. Environ. Sci. Biotechnol., 5: 347–374.

Mohn, W.W., and Stewart, G.R. 2000. Limiting factors for hydrocarbon biodegradation at low temperature in arctic soils. Soil Biol. Biochem., 32: 1161–1172.

Montiel, C., Quintero, R., and Aburto, J. 2009. Petroleum biotechnology: Technology trends for the future. Afr. J. Biotechnol., 8(12): 2653–2666.

Mueller, J.G., Chapman, P.J., Blattmann, B.O., and Pritchard, P.H. 1990. Isolation and characterization of a fluoranthene-utilizing strain Of *Pseudomonas paucimobilis*. Appl. Environ. Microbiol., 56: 1079–1086.

Mulkins-Phillips, G.J., and Stewart, J.E. 1974a. Effect of environmental parameters of bacterial degradation on Bunker C Oil, crude oil and hydrocarbons. Applied and Environmental Microbiology, 28: 547–552.

Mulkins-Phillips, G.J., and Stewart, J.E. 1974b. Distribution of hydrocarbon-utilizing bacteria in northwestern atlantic waters and coastal sediments. Can. J. Microbiol., 20: 955–962.

Mulyono, M., Jasjfi, E., and Maloringan, M. 1994. Oil dispersants: do they do any good? Proceedings. 2nd International Conference on Health, Safety, and Environment in Oil and Gas Exploration and Production, Jakarta, 1: 539–549.

Nasrollahzadeh, H.S., Najafpour, G.D., and Aghamohammadi, N. 2007. Biodegradation of phenanthrene by mixed culture consortia in a batch bioreactor using central composite face-entered design. Int. J. Environ. Res., 1: 80–87.

Nicodem, D.E., Guedes, C.L.B., and Correa, R.J. 1998. Photochemistry of petroleum I. Systematic study of a brazilian intermediate crude oil. Mar. Chem., 63: 93–104.

Nikolopouloua, M., and Kalogerakis, N. 2008. Enhanced bioremediation of crude oil utilizing lipophilic fertilizers combined with biosurfactants and molasses. Mar. Pollut. Bull., 56: 1855–1861.

Obayori, O.S., Ilori, M.O., Adebusoye, S.A., Oyetibo, G.O., Omotayo, A.E., and Amund, O.O. 2009. Degradation of hydrocarbons and biosurfactant production by *Pseudomonas* sp. strain LP1. World J. Microbiol. Biotechnol., 25: 1615–1623.

Obire, O. 1990. Bacterial degradation of three different crude oils in Nigeria. Nig. J. Bot. 1: 81–90.

Obire, O. 1993. The suitability of various nigerian petroleum fractions as substrate for bacterial growth. Discov. Innov., 9: 25–32.

Obire, O., and Okudo, I.V. 1997. Effects of crude oil on a freshwater stream in Nigeria. Discov. Innov., 9: 25–32.

Obire, O., and Nwaubeta, O. 2001. Biodegradation of refined petroleum hydrocarbons in soil. J. Appl. Sci. Environ. Mgt., 5(1): 43–46.

Odu, C.T.I. 1977. Pollution and the environment. Bulletin of the Science Association of Nigeria, 3(2): 284–285.

Ohshiro, T., Suzuki, K., and Izumi, Y. 1996. Regulation of dibenzothiophene degrading enzyme activity of *Rhodococcus erythropolis* D-1. Ferm. Bioeng., 82: 121–124.

Okerentugba, P.O., and Ezeronye, O.U. 2003. Petroleum degrading potentials of single and mixed microbial cultures isolated from rivers and refinery effluent in Nigeria. African Journal of Biotechnology, 2(9): 288–292.

Okoh, A.I., Ajisebutu, S., Babalola, G.O., and Trejo-Hernandez, M.R. 2001. Potentials of *Burkholderia cepacia* strain RQ1 in the biodegradation of heavy crude oil. Intl. Microb., 4: 83–87.

Okoh, A.I., Ajisebutu, S., Babalola, G.O., and Trejo-Hernandez, M.R. 2002. Biodegradation of mexican heavy crude oil (Maya) by *Pseudomonas aeruginosa*. J. Trop. Biosci., 2(1): 12–24.

Okoh, A.I. 2003. Biodegradation of bonny light crude oil in soil microcosm by some bacterial strains isolated from crude oil flow stations saver pits in Nigeria. Afr. J. Biotech., 2(5): 104–108.

Okoh, A.I. 2006. Biodegradation alternative in the cleanup of petroleum hydrocarbon pollutants. Biotechnology and Molecular Biology Review, 1(2): 38–50.

Okoh, A.I., and Trejo-Hernandez, M.R. 2006. Remediation of petroleum hydrocarbon polluted systems: exploiting the bioremediation strategies. African J. Biotechnol., 5: 2520–2525.

Okolo, J.C., Amadi, E.N., and Odu, C.T.I. 2005. Effects of soil treatments containing poultry manure on crude oil degradation in a sandy loam soil. Appl. Ecol. Environ. Res., 3(1): 47–53.

Onwurah, I.N.E., Ogugua, V.N., Onyike, N.B., Ochonogor, A.E., and Otitoju, O.F. 2007. Crude oils spills in the environment, effects and some innovative clean-up biotechnologies. Int. J. Environ. Res., 1: 307–320.

Onysko, K.A., Budman, H.M., and Robinson, C.W. 2000. Effect of temperature on the inhibition kinetics of phenol biodegradation by *Pseudomonas putida* Q5. Biotechnol. Bioeng., 70: 291–299.

Osterhuber, E. 1989. Upgrading heavy oils by solvent dissolution and ultrafiltration. US Patent No. 4,7 97,200.

Oudot, J., Merlin, F.X., and Pinvidic, P. 1998. Weathering rates of oil components in a bioremediation experiment in estuarine sediments. Mar. Environ. Res., 45: 113–125.

Ourisson, G., Albrecht, P., and Rohmer, M. 1979. The Hopanoids. Paleochemistry and biochemistry of a group of natural products. Pure Appl. Chem., 51: 709–729.

Pacheco, M.A., Lange, E.A., Pienkos, P.T., Yu, L.Q., Rouse, M.P., Lin, Q., and Linquist, L.K. 1999. Recent advances in desulfurisation of diesel fuel. Paper No. NPRA AM-99-27. Proceedings. NPRA annual Meeting, San Antonio. Texas. Page 1–26.

Page, D.S., Boehm, P.D., Douglas, G.S., Bence, A.E., Burns, W.A., and Mankiewic, P.J. 1996. The natural petroleum hydrocarbon background in subtidal sediments of prince william sound Alaska, USA. Environment Toxicology and Chemistry, 15: 1266–1281.

Panariti, N., Del Bianco, N.A., Del Piero, G., and Marchionna, M. 2000. Petroleum residue upgrading with dispersed catalysts. Part 1. Catalysts activity and selectivity. Appl. Catal. A: Gen., 204(2): 203–213.

Park, K.S., Sims, R.C., DuPont, R.R., Doucette, W.J., and Matthews, J.E. 1990. Fate of polynuclear aromatic hydrocarbon compounds in two soil types: influence of volatilization, abiotic loss, and biological activity. Environ. Toxicol. Chem., 9: 187–195.

Paul, J.A.K., and Smith, M.L. 2009. Method for purification of uncatalyzed natural fuels from metal ions by means of at least one hemeprotein and use of the at least on hemeprotein. United States Patent 8,475,652.

Pavlova, A., and Ivanova, R. 2003. Determination of petroleum hydrocarbons and polycyclic aromatic hydrocarbons in sludge from wastewater treatment basins. J. Environ. Monit., 5: 319–323.

Payne, J.R., and McNabb, GD. Jr. 1984. Weathering of petroleum in the marine environment. Marine Technology Society Journal, 18(3): 24.

Pathak, H., Kantharia, D., Malpani, A., and Madamwar, D. 2009. Naphthalene biodegradation using *Pseudomonas* sp. HOB1: *In vitro* studies and assessment of naphthalene degradation efficiency in simulated microcosms. J. Hazard. Mater., 166: 1466–1473.

Pelletier, E., Delille, D., and Delille, B. 2004. Crude oil bioremediation in sub-antarctic intertidal sediments: chemistry and toxicity of oil residues. Mar. Environ. Res., 57: 311–327.

Peters, K.E., Walters, C.C., and Moldowan, J.M. 2005. The Biomarker Guide – Edition II. Cambridge University Press, Cambridge, United Kingdom.

Pillis, L.J., and Davis, L.T. 1985. Microorganism Capable of Degrading Phenolics. United States Patent 4,556,638. December 3.

Pines, O., and Gutnick, D.L. 1981. Relationship between Phage Resistance and Emulsan Production. Interaction of phages with the cell-surface of *Acinctobacter alcoaceticus* RAG-1. Arch. Microbiol., 130: 129–133.

Pines, O., and Gutnick, D.L. 1984a. Specific binding of a bacteriophage at a hydrocarbon-water interface. Appl. Environ. Microbiol., 157: 179–183.

Pines, O., and Gutnick, D.L. 1984b. Alternate hydrophobic sites on the cell surface of *Acinetobacter calcoaceticuls* RAG-1. FEMS Microbiol. Lett., 22: 307–311.

Pines, O., and Gutnick, D.L. 1986. Role for emulsan in growth of *Acinetobacter calcoaceticus* RAG-1 on crude oil. Appi. Environ. Microbiol., 51: 661–663.

Piskonen, R., and Itävaara, M. 2004. Evaluation of chemical pretreatment of contaminated soil for improved PAH Bioremediation. Appl. Microbiol. Biotechnol., 65: 627634.

Piskorz, J., Radlein, D., Majerski, P., and Scott, D. 1996. Hydrotreating of heavy hydrocarbon oils in supercritical fluids. US Patent No. 5,496,464.

Pitter, P., and Chudoba, J. 1990. Biodegradability of organic substances in the aquatic environment. CRC Press, Boca Raton, Florida. Page 7–83.

Poremba, K., Gunkel, W., Lang, S., and Wagner, F. 1991. Marine biosurfactants. III. Toxicity testing with marine microorganisms and comparison with synthetic detergents. Zeitschrift. Natureforsch., 46c: 210–216.

Prince, R. 1993. Petroleum spill bioremediation in marine environments. Crit. Rev. Microbiol., 19: 217–242.

Premuzic, E.T., and Lin, M.S. 1991a. Biochemical Upgrading of Oils. US patent No. 5,858,766.

Premuzic, E.T., and Lin, M.S. 1991b. Induced biochemical conversions of heavy crude oils. J. Petrol. Sci. Eng., 22(1–3): 171–180.

Premuzic, E.T., Lin, M.S., Lian, H., Zhou, W.M., and Yablon, J. 1997. The use of chemical markers in the evaluation of crude bioconversion products, technology, and economic analysis. Fuel Process. Technol., 52: 207–223.

Premuzic, E.T., Lin, M.S., Bohenek, M., and Zhou, W.M. 1999. Bioconversion reactions in asphaltenes and heavy crude oils. Energy Fuels, 13(2): 297–304.

Rahman, K.S.M., Thahira-Rahman, J., Lakshmanaperumalsamy, P., and Banat, I.M. 2002. Towards efficient crude oil degradation by a mixed bacterial consortium. Biores. Tech., 85: 257–261.

Rahman, K.S.M., Thahira-Rahman, T., Kourkoutas, Y., Petsas, I., Marchant, R., and Banat, I.M. 2003. Enhanced Bioremediation of n-alkanes in petroleum sludge using bacterial consortium amended with rhamnolipid and micronutrients. Biores Tech., 90(2): 159–168.

Ramalho, A.M.Z., Martinez-Huitle, C.A., and Da Silva, D.R. 2010. Application of electrochemical technology for removing petroleum hydrocarbons from produced water using a DSA-Type Anode at Different Flow Rates. Fuel, 89: 531–534.

Ramirez, L.C., Iturbe, R. and Torres, L.G. 2008. Effect of nutrient and surfactant addition on polyaromatic hydrocarbon (PAH) biodegradation in contaminated soils. Land Contamination and Reclamation. 16: 1–11.

Reynolds, G.J., Biggs, W.R., and Bezman, S.A. 1987. Removal of heavy metals from residual oils. ACS Symp. Ser., 344: 205–219.

Reynolds, C.M., Bhunia, P., and Koenen, B.A. 1997. Soil Remediation Demonstration Project: Biodegradation of Heavy Fuel Oils. Special Report 97-20. Cold Regions Research & Engineering Laboratory. US Army Corps of Engineers, Fort Belvoir, Virginia.

Ringelberg, D.B., Talley, J.W., Perkins, E.J., Tucker, S.G., Luthy, R.G., Bouwer, E.J., and Fredrickson, H.L. 2001. Succession of phenotypic, genotypic, and metabolic community characteristics during *in vitro* bioslurry treatment of polycyclic aromatic hydrocarbon-contaminated sediments. Appl. Environ. Microbiol., 67: 1542–1550.

Robichauz, T.J., and Myrick, H.N. 1972. Chemical enhancement of the biodegradation of crude oil pollutants. Journal of Petroleum Technology, 24: 16–20.

Rocha, C., and Infante, C. 1997. Enhanced oily sludge biodegradation by a tensio-active agent isolated from *Pseudomonas aeruginosa* USBCS1. Appl. Microb. Biotech., 47: 615–619.

Roling-Wilfred, F.M., Milner, M., Jones, D.M., Lee, K., Daniel, F., and Swanell-Richard, J.P. 2002. Robust hydrocarbon degradation and dynamic of bacterial communities during nutrients – enhanced oil spill bioremediation. Applied and Environmental Microbiology, 68: 5537–5548.

Rosenberg, E., Perry, A., Gibson, D.T., and Gutnick, D.L. 1979a. Emulsifier of Arthrobacter RAG-1: Specificity of hydrocarbon substrate. Appl. Environ. Microbiol., 37: 409–413.

Rosenberg, E., Zuckerberg, A., Rubinovitz, C., and Gutnick, D.L. 1979b. Emulsifier of Arthrobacter RAG-1: Isolation and emulsifying properties. Appl. Environ. Microbiol., 37: 402–408.

Rosenberg, E., Kaplan, N., Pines, O., Rosenberg, M., and Gutnick, D.L. 1983a. Capsular Polysaccharides Interfere with Adherence of *Acinetobacter calcoaceticus* to Hydrocarbon. FEMS Microbiol. Lett., 17: 157-160.

Rosenberg, E., Gottlieb, A., and Rosenberg, M. 1983b. Inhibition of bacterial adherence to hydrocarbons and epithelial cells by emulsan. Infect. Immun., 39: 1024–1028.

Rosenberg, E. 1991. The hydrocarbon-oxidizing bacteria. pp. 446–458. *In*: Bolows, A., Truger, H.G., Dworkin, M., Schleifer, K.H. and Harder, W. (Eds.). The Prokaryotes. 2nd Edition. Springer-Verlag, New York. Volume 1.

Roubal, G., and Atlas, R.M. 1978. Distribution of hydrocarbon-utilizing microorganisms and hydrocarbon biodegradation potentials in alaskan continental shelf areas. Appl. Environ. Microbiol. 35: 897–905.

Ruberto, L., Dias, R., Lo Balbo, A., Vazquez, S.C., Hernandez, E.A., and Mac Cormack, W.P. 2009. Influence of nutrients addition and bioaugmentation on the hydrocarbon biodegradation of A chronically contaminated Antarctic Soil. J. Appl. Microbiol., 106: 1101–1110.

Sabate, J., Vinas, M., and Solanas, A.M. 2004. Laboratory-scale bioremediation experiments on hydrocarbon-contaminated soils. Internl. Biodeter. Biodegrad., 54: 19–25.

Salam, L.B., Obayori, O.S., Akashoro. O.S., and Okogie, G.O. 2011. Biodegradation of bonny light crude oil by bacteria isolated from contaminated soil. Int. J. Agric. Biol., 13(2): 245–250.

Salas, N., Ortiz, L., Gilcoto, M., Varela, M., Bayona, J.M., Groom, S., Álvarez-Salgado, X.A., and Albaigés, J. 2006. Fingerprinting petroleum hydrocarbons in plankton and surface sediments during the spring and early summer blooms in the Galician Coast (NW Spain) after the Prestige Oil Spill. Mar. Environ. Res., 62: 388–413.

Salehizadeh, H., Mousavi, M., Hatamipour, S., and Kermanshahi, K. 2007. Microbial demetallization of crude oil using *Aspergillus* sp.: vanadium oxide octaethyl porphyrin (VOOEP) as a model of metallic petroporphyrins. Iran. J. Biotechnol., 5(4): 226–231.

Sandrin, T.R., Kight, W.B., Maier, W.J., and Maier, R.M. 2006. Influence of a non-aqueous phase liquid (NAPL) on biodegradation of phenanthrene. Biodegradation, 17: 423–435.

Sarret, G., Connan, J., Kasrai, M., Bancroft, M., Charrié-Duhaut, Lemoine, S., Adam, P., Albrecht, P., and Eybert-Bérard, L. 1999. Chemical forms of sulfur in geological and archaeological asphaltenes from Middle East, France, and Spain determined by sulfur K- and L-edge X-ray absorption near-edge structure spectroscopy. Geochim. Cosmochim. Acta, 63(22): 3767–3779.

Savastano, C.A. 1991. Solvent extraction approach to petroleum demetallation. Fuel Sci. Technol. Int., 9(7): 833–871.

Schreiber, A.F., and Winkler, U.K. 1983. Transformation of Tetralin by Whole Cells of *Pseudomonas stutzeri* As39. Appl. Microbiol. Biotechnol., 18(1): 6–10

Setti, L., Lanzarini, G., and Pifferi, P.G. 1995. Dibenzothiophene biodegradation by a *Pseudomonas* sp. in model solutions. Process Biochem., 30(8): 721–728.

Setti, L., Lanzarini, G., and Pifferi, P.G. 1997. Whole cell biocatalysis for an oil desulfurization process. Fuel Process. Technol., 52: 145–153.

Shabtai, Y., Pines, O., and D.L. Gutnick. 1986. Emulsan: A case study of microbial capsules as industrial products. Dev. Ind. Microbiol., 26: 291–307.

Shen, Z. 1995. Alkaline residue of gasoline from FCC used to purify straight run gas oil. Environmental Protection in Petrochemical Industry, (2): 38–41.

Shuttleworth, K.L., and Cerniglia, C.E. 1995. Environmental aspects of polynuclear aromatic hydrocarbon biodegradation. Appl. Biochem. Biotechnol., 54: 291–302.

Sikkema, J., and Bont, J.A.M. 1991. Isolation and initial characterization of bacteria growing on Tetralin. Biodegradation, 2(1): 15–23.

Simarro, R., González, N., Bautista, L.F., Sanz, R., and Molina, M.C. 2011. Optimization of key abiotic factors of PAH (Naphthalene, Phenanthrene and Anthracene) Biodegradation process by a bacterial consortium. Water Air Soil Pollut., 217: 365–374.

Simpson, D.R., Natraj, N.R., McInerney, M.J., and Duncan, K.E. 2011. Biosurfactant producing Bacillus are present in produced brines from Oklahoma oil reservoirs with a wide range of salinities. Appl. Biochem. Biotechnol., 91: 1083–1093.

Sipma, J., Svitelskaya, A., van, der., Mark., B., Pol, L.W.H., Lettinga, G., Buisman, C.J.N., and Janssen, A.J.H. 2004. Potentials of biological oxidation processes for the treatment of spent sulfidic caustics containing thiols. Water Res., 38: 4331–4340.

Siron, R., Pelletier, E., and Brochu, C. 1995 Environmental factors influencing the biodegradation of petroleum hydrocarbons in cold seawater. Arch. Environ. Contam. Toxicol., 28: 406–416.

Sivaraman, C., Ganguly, A., and Mutnuri, S. 2010. Biodegradation of hydrocarbons in the presence of cyclodextrins. World J. Microbiol. Biotechnol., 26: 227–232.

Smith, J. 1968. Torrey Canyon—Pollution and Marine Life. Report by the Plymouth Laboratory of the Marine Biological Association of the United Kingdom, London. Cambridge University Press, London, United Kingdom. Page 1906.

Soli, G., and Bens, E.M. 1972. Bacteria which attack petroleum hydrocarbons in a saline medium. Biotechnol. Bioeng., 14: 319–330.

Song, H., Wang, X., and Bartha, R. 1990. Bioremediation potential of terrestrial fuel spills. Applied and Environmental Microbiology, 56(3): 652–656.

Sotsky, J.B., and Atlas, R.M. 1994. Frequency of genes in aromatic and aliphatic hydrocarbon biodegradation pathways within bacterial populations from alaskan sediments. Can. J. Microbiol., 40: 981–985.

Speight, J.G. 1996. Environmental Technology Handbook. Taylor & Francis, Washington, DC.

Speight, J.G. 2000. The Desulfurization of Heavy Oils and Residua 2nd Edition. Marcel Dekker, Inc., New York, USA. 1981.

Speight, J.G., and Lee, S. 2000. Environmental Technology Handbook. 2nd Edition. Taylor & Francis, New York.

Speight, J.G. 2005 Environmental Analysis and Technology for the Refining Industry. John Wiley & Sons Inc., Hoboken, New Jersey.

Speight, J.G. 2011. The Refinery of the Future. Gulf Professional Publishing, Elsevier, Oxford, United Kingdom.

Speight, J.G. 2012. Crude Oil Assay Database. Knovel, Elsevier, New York.

Speight, J.G., and Arjoon, K.K. 2012. Bioremediation of Petroleum and Petroleum Products. Scrivener Publishing, Beverly, Massachusetts.

Speight, J.G. 2012. The Chemistry and Technology of Coal 3rd Edition. CRC Press, Taylor & Francis Group, Boca Raton, Florida.

Speight, J.G. 2013a. Heavy Oil Production Processes. Gulf Professional Publishing, Elsevier, Oxford, United Kingdom. 2013a.

Speight, J.G. 2013b. Heavy and Extra Heavy Oil Upgrading Technologies. Gulf Professional Publishing, Elsevier, Oxford, United Kingdom. 2013b.

Speight, J.G. 2014. The Chemistry and Technology of Petroleum 5th Edition. CRC Press, Taylor & Francis Group, Boca Raton, Florida.

Speight, J.G. 2015. Handbook of Petroleum Product Analysis 2nd Edition. John Wiley & Sons Inc., Hoboken, New Jersey.

Speight, J.G. 2016. Introduction to Enhanced Recovery Methods for Heavy Oil and Tar Sands 2nd Edition. Gulf Professional Publishing Company, Elsevier, Oxford, United Kingdom.

Speight, J.G., and Islam, M.R. Peak 2016. Energy – Myth or Reality. Scrivener Publishing, Beverly, Massachusetts.

Speight, J.G. 2017. Handbook of Petroleum Refining. CRC Press, Taylor & Francis Group, Boca Raton, Florida.

Speight, J.G., and El-Gendy, N.Sh. 2018. Introduction to Petroleum Biotechnology. Gulf Professional Publishing Company, Elsevier, Cambridge, Massachusetts.

Steffan, R.J., Mccloy, K., Vainberg, S., Condee, C.W., and Zhang, D. 1997. Biodegradation of the gasoline oxygenates Methyl tert-Butyl Ether, Ethyl tert-Butyl Ether and tert – Amyl Methyl ether by propane-oxidizing bacteria. Appl. Environ. Microbiol., 63(11): 4216–4222.

Strawinski, R.J., and Stone, R.W. 1940. The utilization of hydrocarbons by bacteria. J. Bacteriol., 40(3): 461.

Sun, S., Zhang, Z., Luo, Y., Zhong, W., Xiao, M., Yi, W., Yub, L., and Fu, P. 2011. Exopolysaccharide production by a genetically engineered *Enterobacter* cloacae strain for microbial enhanced oil recovery. Bioresour. Technol., 102: 6153–6158.

Sutherland, J.B., Rafii, F., Khan, A.A., and Cerniglia, C.E. 1995. Mechanisms of polycyclic aromatic hydrocarbon degradation. pp. 269–306. *In*: Young, L.Y., and Cerniglia, C.E. (Eds.). Microbial transformation and degradation of toxic organic chemicals. John Wiley & Sons Inc.-Liss, New York.

Swanell, R.P.J., Lee, K., and McDonagh, M. 1996. Field evaluation of marine oil spill bioremediation. Microb. Rev., June: 342–365.

Swannell, R.P.J., Lepo, J.E., Lee, K., Pritchard, P.H., and Jones, D.M. 1995 Bioremediation of oil-contaminated fine-grained sediments in laboratory microcosms. Proceedings. 2nd International Oil Spill Research and Development Forum, 23–26 May 1995. International Maritime Organisation, London, United Kingdom. Page 45–55.

Tagger, S., Bianchi, A., Juillard, M., LePetit, J., and Roux, B. 1983. Effect of microbial seeding of crude oil in seawater in a model system. Mar. Biol., 78: 13–20.

Taghvaei Ganjali, S., Nahri Niknafs, B., Khosravi, M. et al. 2007. Photo-oxidation of crude petroleum maltenic fraction I natural simulated conditions and structural elucidation of photoproducts. Iran J. Environ. Health Sci. Eng., 4(1): 37–42.

Talley, J.W., Ghosh, U., Tucker, S.G., Furey, J.S., and Luthy, R.G. 2002. Particle-scale understanding of the bioavailability of PAHs in sediment. Environ. Sci. Technol., 36: 477–483.

Tavassoli, T., Mousavi, S.M., Shojaosadati, S.A., and Salehizadeh, H. 2012. Asphaltene biodegradation using microorganisms isolated from oil samples. Fuel, 93: 142–148.

Taylor, T.R., Jackson, K.J., Duba, A.G., and Chen, C.I. 1998. *In situ* Thermally Enhanced Biodegradation of Petroleum Fuel Hydrocarbons and Halogenated Organic Solvents. United States Patent 5,753,122. May 19.

Thouand, G., Bauda, P., Oudot, J., Kirsch, G., Sutton, C., and Vidalie, J.F. 1999. Laboratory evaluation of crude oil biodegradation with commercial or natural microbial inocula. Can. J. Microb., 45(2): 106–115.

Traxler, R.W., and Bhattacharya, L.S. 1979. Effect of a chemical dispersant on microbial utilization of petroleum hydrocarbons. *In*: McCarthy, L.T., Lindblom, G.P., and Walter, H.F. (Eds.). Chemical Dispersants for the Control of Oil Spills. American Society for Testing and Materials, West Conshohocken, Pennsylvania.

Trindade, P.V.O., Sobral, L.G., Rizzo, A.C.L., Leite, S.G.F., and Soriano, A.U. 2005. Bioremediation of a weathered and a recently oil-contaminated soils from brazil: a comparison study. Chemosph., 58: 515–522.

US EPA. 2003. Aerobic Degradation of Oily Wastes: A Field Guidance Book for Federal On-Scene Coordinators. Version 1.0. US Environmental Protection Agency Response and Prevention Branch, Region 6 South Central., Dallas Texas.

Uraizee, F.A., Venosa, A.D., and Suidan, M.T. 1998. A model for diffusion-controlled bioavailability of crude oil components. Biodegradation, 8: 287–296.

Van Brummelen, T.C., Verweij, R.A., Wedzinga, S.A., and Van Gestel, C.A.M. 1996. Enrichment of polycyclic aromatic hydrocarbons in forest soils near a blast furnace plant. Chemosphere, 32: 293–314.

Van der Meer, J.R., de Vos, W.M., Harayama, S., and Zehnder, A.J.B. 1992. Molecular mechanisms of genetic adaptation to xenobiotic compounds. Microbiol. Rev., 56: 677–694.

Varadaraj, R., Robbins, M.L., Bock, J., Pace, S., and MacDonald, D. 1995. Dispersion and biodegradation of oil spills on water. Proceedings. 1995 Oil Spill Conference, American Petroleum Institute, Washington, DC. Page 101–106.

Vaz, D.A., Gudina, E.J., Alameda, E.J., Teixeira J.A., and Rodrigues, L.R. 2012. Performance of a biosurfactant produced by a *Bacillus subtilis* strain isolated from crude oil samples as compared to commercial chemical surfactants. Colloids Surf B Biointerfaces., 89: 167–174.

Vazquez-Duhalt, R., Westlake, D.W.S., and Fedorak, P.M. 1993. Cytochrome c as a biocatalyst for the oxidation of thiophenes and organosulfides. Enzyme Microb. Technol., 15: 494–499.

Vazquez-Duhalt, R., Torres, E., Valderrama, B., and Le Borgne S. 2002. Will biochemical catalysis impact the petroleum refining industry? Energy Fuels, 16: 1239–1250.

Venosa, A.D., Suidan, M.T., Wrenn, B.A., Strohmeier, K.L., Haines, J.R., Eberhart, B.L., King, D., and Holder, E. 1996. Bioremediation of an experimental oil spill on the shoreline of delaware bay. Environmental Sci. and Technol., 30(5): 1764–1775.

Venosa, A.D., and Zhu, X. 2003. Biodegradation of crude oil contaminating marine shorelines and freshwater wetlands. Spill Sci. Tech. Bull., 8(2): 163–178.

Vidali, M. 2001. Bioremediation: an overview. Pure Appl. Chem., 73: 1163–1172.

Vieira, P.A., Faria, S.R., Vieira B., De Franca, F.P., and Cardoso, V.L. 2009. Statistical analysis and optimization of nitrogen, phosphorus, and inoculum concentrations for the biodegradation of petroleum hydrocarbons by response surface methodology. J. Microbiol. Biotechnol., 25: 427–438.

Viñas, M., Sabate, J., Espuny, M.J., and Solanas, A.M. 2005. Bacterial community dynamics and polycyclic aromatic hydrocarbon degradation during bioremediation of heavily creosote-contaminated soil. Appl. Environ. Microbiol., 71: 7008–7018.

Wang, H., and Yang, M. 2001. Screening, identification and domestication of a bacterium able to degrade organic dross. Shandong Science, 14(2): 42–45.

Weiner, J.M., and Lovley, D.R. 1998. Anaerobic benzene degradation in petroleum-contaminated aquifer sediments after inoculation with a benzene-degrading enrichment. Appl. Environ. Microbiol., 64: 775–778.

Weissenfels, W.D., Beyer, M., and Klein, J. 1990. Degradation of phenanthrene, fluorene and fluoranthene by pure bacterial cultures. Appl. Microbiol. Biotechnol., 32: 479–484.

Weissenfels, W.D., Beyer, M., Klein, J., and Rehm, H.J. 1991. Microbial metabolism of fluoranthene: isolation and identification of ring fission products. Appl. Microbiol. Biotechnol., 34: 528–535.

Westlake, D.W.S., Jobson, A.M., and Cook, F.D. 1978. *In situ* Degradation of oil in a soil of the boreal region of the northwest territories. Can. J. Microbiol., 24: 245–260.

Whyte, L.G., Bourbonnière, L., and Greer, C.W. 1997. Biodegradation of petroleum hydrocarbons by psychrotrophic *Pseudomonas* strains possessing both alkane (*alk*) and Naphthalene (*nah*) catabolic pathways. Appl. Environ. Microbiol., 63: 3719–3723.

Whyte, L.G., Hawari, J., Zhou, E., Bourbonnière, L., Inniss, W.E., and Greer, C.W. 1998. Biodegradation of variable-chain-length alkanes at low temperatures by a psychrotrophic *Rhodococcus* sp. Appl. Environ. Microbiol., 64: 2578–2584.

Whyte, L.G., Bourbonnière, L., Bellerose, C., and Greer, C.W. 1999. Bioremediation assessment of hydrocarbon-contaminated soils from the high arctic. Bioremediation J., 3: 69–79.

Wild, S.R., and Jones, K.C. 1993. Biological and abiotic losses of polynuclear aromatic hydrocarbons from soils freshly amended with sewage sludge. Environ. Toxicol. Chem., 12: 5–12.

Williams, W.A., and May, R.J. 1997. Low-temperature microbial aerobic degradation of polychlorinated biphenyls in sediments. Environ. Sci. Technol., 31: 3491–3496.

Willumsen, P.A., U. Karlson, and P.H. Pritchard. 1998. Response of fluoranthene-degrading bacteria to surfactants. Appl. Microbiol. Biotechnol., 50: 475–483.

Wolfe, D.A., Hameedi, M.H., Galt, J.A., Watabayashi, G., Shrot, J., O'Claire, C., Rice, S., Michel, J., Payne, J.R., Braddock, J., Hanna, S., and Sale, D. 1994. The fate of the oil spilled from the exxon valdez. Environ. Sci. Technol., 28: 561A–568A.

Wong, J.W.C., Lai, K.M., Wan, C.K., Ma, K.K., and Fang, M. 2001. Isolation and optimization of PAHs-degradative bacteria from contaminated soil for PAHs bioremediation. Water, Air, and Soil Pollution, 13: 1–13.

Wrenn, B.A., Suidan, M.T., Strohmeir, K.L., Eberhart, B.L., Wilson, G.J., and Venosa, A.D. 1997a. Nutrient transport during bioremediation of contaminated beaches: evaluation with lithium as a conservative tracer. Water Res., 31(3): 515–524.

Wrenn, B.A., Boufadel, M.C., Suidan, M.T., and Venosa, A.D. 1997b. Nutrient transport during bioremediation of crude oil contaminated beaches. *In*: *In situ* and On-Site Bioremediation: Volume 4, pp. 267–272. Battelle Memorial Institute, Columbus, Ohio.

Xia, X.H., Yu, H., Yang, Z.F., and Huang, G.H. 2006. Biodegradation of polycyclic aromatic hydrocarbons in the natural waters of the yellow river: effects of high sediment content on biodegradation. Chemosphere, 65: 457–466.

Xu, G.-W., Mitchell, K.W., and Moticello, D.J. 1997. Process for demetallizing a fossil fuel. United States Patent 5,624,844.

Xu, G.-W., Mitchell, K.W., and Moticello, D.J. 1998. Fuel product produced by demetallizing a fossil fuel with an enzyme. United States Patent 5,726,056.

Yakimov, M.M., Giuliano, L., Bruni, V., Scarfi, S., and Golyshin, P.N. 1999. Characterization of antarctic hydrocarbon-degrading bacteria capable of producing bioemulsifiers. Microbiologica., 22: 249–256.

Yang, C., and Lou, Z. 1997. Screening standard and geological bases for microbial enhanced oil recovery. Petrol. Explor. Develop., 24(2): 69–72.

Ye, D., Siddiqi, M.A., Maccubbin, A.E., Kumar, S., and Sikka, H.C. 1996. Degradation of polynuclear aromatic hydrocarbons by *Sphingomonas paucimobilis*. Environ. Sci. Technol., 30: 136–142.

Yi, N., Xue, G., HogShuai, G., XiangPing, Z., and SuoJiang, Z. 2014. Simultaneous desulfurization and denitrogenation of liquid fuels using two functionalized group ionic liquids. Sci. China, 57(12): 1766–1773.

Yu, L., Meyer, T.A., and Folsom, B.R. 1998. Oil/water/biocatalyst three phase separation process. US Patent No. 5,772,901.

Yu, B., Xu, P., Shi, Q., and Ma, C. 2006. Deep desulfurization of diesel oil and crude oils by a newly isolated *Rhodococcus erythropolis* strain. Appl. Environ. Microbiol., 72: 54–78.

Yudono, B., Said, M., Sabaruddin, N.A., and Fanani, Z. 2011. Kinetics approach of biodegradation of petroleum contaminated soil by using indigenous isolated bacteria. J. Trop. Soils, 16(1): 33–38.

Yumoto, I., Nakamura, A., Iwata, H., Kojima, K., Kusumoto, K., Nodasaka, Y., and Matsuyama, H. 2002. *Dietzia psychralcaliphila* sp. *nov.*, A novel facultatively psychrophilic alkaliphile that grows on hydrocarbons. Int. J. Syst. Evol. Microbiol., 52: 85–90.

Zahed, M.A., Aziz, H.A., Isa, M.H., and Mohajeri, L. 2010. Enhancement of biodegradation of n-alkanes from crude oil contaminated seawater. Int. J. Environ. Res., 4(4): 655–664.

Zaidi, B.R., and Imam, S.H. 1999. Factors affecting microbial degradation of polycyclic aromatic hydrocarbon phenanthrene in caribbean coastal water. Marine Pollution Bulletin, 38: 738–749.

Zander, M. 1983. Physical and chemical properties of polycyclic aromatic hydrocarbons. pp. 1–26. *In*: Bjørseth, A. (Ed.). Handbook of polycyclic aromatic hydrocarbons. Marcel Dekker, Inc., New York.

Zekri, A.Y., Abou-Kassem, J.H., and Shedid, S.A. 2009. A new technique for measurement of oil biodegradation. Pet. Sci. Technol., 27: 666–677.

Zengler, K., Richnow, H.H., Rossello-Mora, R., Michaelis, W., and Widdel, F. 1999. Methane formation from long-chain alkanes by anaerobic microorganisms: Nature, 401: 266–269.

Zeyaullah, Md., Abdelkafe1, A.S., Ben Zabya, W., and Ali, A. 2009. Biodegradation of catechols by micro-organisms—a short review. African Journal of Biotechnology, 8(13): 2916–2922.

Zhang, F., She, Y., Ibrahim M. Banat, Chai, L., Yi, S., Yu, G., and Hou, D. 2014. Potential microorganisms for prevention of paraffin precipitation in a hypersaline oil reservoir. Energy Fuels, 28(2): 1191–1197.

Zhang, G.-L., Wu, Y.-T., Qian, X.-P., and Meng. Q. 2005. Biodegradation of crude oil by *Pseudomonas aeruginosa* in the Presence of rhamnolipids. J Zhejiang Univ. Sci., 6B(8): 725–730.

Zhang, Y., Gao, B., Lu, L., Yue, Q., Wang, Q., and Jia, Y. 2010. Treatment of produced water from polymer flooding in oil production by the combined method of hydrolysis acidification-dynamic membrane bioreactor–coagulation process. J. Petrol., Sci. Eng., 74: 14–19.

Zhao, L., Cao, X.L., Wang, H.Y., Liu, X., and Jiang, S.X. 2008. Determination of petroleum sulfonates in crude oil by column-switching anion-exchange chromatography. Chinese Chem. Lett., 19: 219–222.

Zhou, Q., and Shen, B. 2010. Biodegradation potential and influencing factors of a special microorganism to treat petrochemical wastewater. Pet. Sci. Technol., 28: 135–145.

Zhou, Z.Y., and Zhang, K. 2004. Development situation and prospect of oil fields in China. Petrol. Explor. Develop., 31: 84–87.

Zhu, X., Venosa, A.D., and Suidan, M.T. 2004. Literature Review on the Use of Commercial Bioremediation Agents for Cleanup of Oil-Contaminated Environments. Report No. EPA/600/R-04/075. National Risk Management Research Laboratory, United States Environmental Protection Agency, Cincinnati, Ohio. July.

Zosim, Z., Gutnick, D.L., and Rosenberg, E. 1982. Properties of hydrocarbon-in-water emulsions stabilized by *Acinetobacter* RAG-1 Emulsan. Biotechnol. Bioeng., 24: 281–292.

Zosim, Z., Rosenberg, E., and Gutnick, D.L. 1986. Changes in hydrocarbon emulsification specificity of the polymeric bioemulsifier emulsan: effects of alkanols. Colloid Polym. Sci., 264: 218–223.

Zou, C., Wang, M., Xing, Y., and Lan, G. 2014. Characterization and optimization of biosurfactants produced by *Acinetobacter baylyi* ZJ2 isolated from crude oil contaminated soil sample toward microbial enhanced oil recovery applications. Biochem. Eng. J., 90: 49–58.

Zuckerberg, A., Diver, A., Peeri, Z., Gutnick, D.L., and Rosenberg, E. 1979. Emulsifier of Arthrobacter RAG-1: chemical and physical properties. Appl. Environ. Microbiol., 37: 414–420.

CHAPTER 9

Methods Used to Determine the Progress of Bioremediation

During any incident involving the spill of crude oil or a crude oil product, the properties of the spilled material oil, including the composition of the oil phase composition and oil compositional changes due to weathering, ideally should be known immediately, so that models could be used to predict the environmental impact of the spilled material and the potential alternatives for remediation of the site (Rullkötter and Farrington, 2021). Unfortunately, the properties routinely measured by oil producers and refiners (for example the boiling range distribution, the density, and the sulfur content while being a guide for the section of refinery processes and not the properties that the on-scene responders need to know most urgently to assess the potential for the spilled material will evaporate over time or the changes that affect the behavior and fate of the spilled material in the environment. For example, there is the need to know if the spilled oil likely to sink or submerge in water or adhere to the soil or mineral matter on land or if the use of chemical dispersants can enhance the dispersion; if of the oil or if emulsions will form or the toxicity of the material to aquatic or land-borne organisms (Wang et al., 2003).

As already noted (Chapter 1, Chapter 2), the chemical composition of crude oil and crude oil products is complex and to make it even more complicated, the composition of the crude oil-related spilled material will change over time following release into the environment. These factors make it essential that the most appropriate analytical methods are selected for environmental samples (Dean, 1998; Miller, 2000; Budde, 2001; Sunahara et al., 2002; Dean, 2003; Smith and Cresser, 2003; Aichberger et al., 2005; Speight, 2005). In order to track the progress of a bioremediation project, the physical properties data of crude oils and crude oil products are needed as input to environmental fate simulations. Measured physical properties data include adhesion, boiling point distribution, density, emulsion formation, evaporation, flash point, pour point, sulfur content, surface and interfacial tension, viscosity, and water content (Chapter 3). The chemical composition of the unspilled and spilled material should also be quantified for biomarker concentration, hydrocarbon types including n-alkane distribution and alkylated polynuclear aromatic hydrocarbon derivatives, commonly known as PNAs, and volatile organic compounds.

Thus, the challenge of analyzing crude oil of any origin, native or altered, arises from the complexity of this substrate on the molecular level. Attempts made many decades ago to achieve a complete inventory of all constituents of crude oil has proved to be difficult, if not impossible and subject to much speculation (Speight, 2014, 215). The task is to simplify the mixture by fractionation and then search these fractions for characteristic constituents of high significance, such as biological markers as indicators of origin of the crude oil or selected polynuclear hydrocarbon derivatives as indicators of source of spilled material and/or toxicity.

Analytical methods are specific, precise, accurate and reliable procedures that are used for qualitative and quantitative determination of concentration of a compound by using equipment in laboratories. Analytical procedures play a critical role in equivalence and risk assessment, management and help in the establishment of product-specific acceptance criteria and stability of results.

In addition, during any spill incident involving crude oil or crude oil products, the properties of the spilled oil must be known immediately. Unfortunately, the properties routinely measured by the producers and refiners of crude oil are not the ones that on-scene bioremediation scientists and engineers need to know most urgently. The data from analytical test methods carried out for the refining industry typically focus on the properties and characteristics of the crude oil as they pertain to refining and product slate. Similarly, the data from analytical test methods applied to crude oil products are focused on whether or not the product meets specifications and is suitable for sale.

There is a potential for significant soil and groundwater contamination to have arisen at crude oil refineries. Such contamination consists of (1) crude oil hydrocarbon derivatives including lower boiling, very mobile fractions (paraffins, cycloparaffins and volatile aromatic derivatives such as benzene, toluene, ethylbenzene and xylenes) typically associated with gasoline and similar boiling range distillates, (2) middle distillate fractions (paraffins, cycloparaffins and some polynuclear aromatic derivatives) associated with diesel, kerosene, and lower boiling fuel oil, which are also of significant mobility, (3) higher boiling distillates (long-chain paraffins, cycloparaffins and polynuclear aromatic derivatives that are associated with lubricating oil and heavy fuel oil, (4) various organic compounds associated with crude oil hydrocarbon derivatives or produced during the refining process, e.g., phenols, amines, amides, alcohols, organic acids, nitrogen and sulfur containing compounds, (5) other organic additives, e.g., anti-freeze (glycols), alcohols, detergents and various proprietary compounds, (6) organic lead, associated with leaded gasoline and other heavy metals.

Many of the specific chemicals in crude oil are hazardous because of their chemical reactivity, fire hazard, toxicity, and other properties. In fact, a simple definition of a hazardous chemical (or hazardous waste) is that it is a chemical substance (or chemical waste) that has been inadvertently released, discarded, abandoned, neglected, or designated as a waste material and has the potential to be detrimental to the environment. Alternatively, a hazardous chemical may be a chemical that may interact with other (chemical) substances to give a product that is hazardous to the environment. Thus, it is essential to obtain analytical data relating to the properties and behavior of the spilled material.

Furthermore, the biodegradation processes, which influence the presence and the analysis of crude oil hydrocarbon at a particular site, can be very complex. The extent of biodegradation is dependent on factors such as (1) the type of microorganisms present, (2) the environmental conditions, such as temperature, oxygen levels, and moisture, and (3) the predominant types of hydrocarbon derivatives. In fact, the primary factor controlling the extent of biodegradation is the molecular composition of the crude oil contaminant. Multiple ring cycloalkane derivatives are hard to degrade, while polynuclear aromatic hydrocarbon derivatives display varying degrees of degradation. Straight-chain alkane derivatives biodegrade rapidly with branched-chain alkane derivatives and single saturated ring compounds degrading more slowly.

Analytical data related to properties such as (1) evaporation rate, (2) the viscosity of the residual oil as lower boiling constituents evaporate, (3) to what extent dispensability will occur naturally, or be enhanced with dispersants, (4) oxidation by aerial oxygen, (5) emulsion formation, (6) the density of the oil and any emulsions relative to the ecosystem—this is especially relevant for aqueous or marine spills—a indicate whether the oil is likely to sink or submerge, (7) the viscosity of the oil, at ecosystem temperatures, (8) health hazards to on-site personnel from volatile organic compounds, and the toxicity to the indigenous flora and fauna.

While all of the above properties are important and relevant to bioremediation efforts, the fate of crude oil and crude oil products spilled into rivers, lacustrine water, and marine water is often difficult to predict—because of (amongst other issues) oxidation (Chapter 2). Crude oil and crude oil products oxidize on the surface after which the form oil-in-water emulsions. The inclusion of polar functions such as hydroxyl groups (-OH) or carbonyl groups ($>C = O$) (a result of the oxidation process) causes an increase in the density of the emulsion (relative to the original unoxidized crude oil) and with an increased propensity to form emulsions. As a result, the emulsion and sinks to various depths or even to the seabed, depending on the extent of the oxidation and the resulting density. This may give the appearance that the crude oil spill (as evidenced from the crude oil remaining on the surface of the water) is less than it actually was. The so-called *missing* oil will undergo further chemical changes and eventually reappear on the water surface or on a distant beach.

Analytical methods are specific, precise, accurate and reliable procedures that are used for qualitative and quantitative determination of concentration of a compound by using equipment in laboratories. There is a significant number of crude oil-hydrocarbon impacted sites and evaluation and remediation of these sites may be difficult arise from the complexity of the issues (analytical, scientific, and regulatory not to mention economic) regarding impacted water and soil.

Preliminary tests to assess feasibility of biotreatment of hydrocarbon contaminations are typically performed in the laboratory (for example degradation experiments) or in the field (chiefly *in situ* respiration tests). In general, the quantity of soil used in laboratory experiments is relatively small compared to the quantity of soil that has to be treated on-site and thus spatial heterogeneity is rarely considered in laboratory experiments (Aichberger et al., 2005). This might in particular be true for small scale experiments and/or for inhomogeneous sites.

Furthermore, caution must be taken to ensure that the samples may be disturbed during sampling, storage and pre-treatment through, for example, oxidation or any process that affects the homogeneity of the samples (Frijer et al., 1996). Any such changes can change the rate of biodegradation as determined in the laboratory experiments than the rate of biodegradation determined from data from the field (Davis et al., 2003; Höhener et al., 2003). Laboratory tests to determine biodegradability potential indirectly via oxygen consumption or carbon dioxide may require corrections to the rates then have to be corrected by background respiration result from biodegradation of organic substances and other oxygen consuming processes (Balba et al., 1998; Baker et al., 2000). However, the manner in which the corrections are applied must be beyond reproach or claims of falsification of the data will be the most likely result (Speight and Foote, 2011).

Field tests must take into consideration the site conditions comprising local heterogeneity and changing environmental conditions, which can affect the permeability of the sub-soil (Davis et al., 2003). By the application of preliminary field tests, relevant information can be deducted on the definite system design including actual flow rates, radii of influence, blower and well layout (Leeson and Hinchee, 1997; Baker et al., 2000; Höhener et al., 2003).

Preliminary site assessments account for three critical prerequisites for bioremediation namely for (1) hydrocarbon biodegradability of the predominating compounds, (2) contaminant bioavailability to indigenous microbial populations, and (3) environmental conditions present at a site.

In the case of hydrocarbon biodegradation under specific environmental conditions, the degree of is mainly affected by the type of hydrocarbon derivatives in the contaminant matrix (Huesemann, 1995). Of the various crude oil fractions, n-alkane hydrocarbon derivatives and branched-chain alkane hydrocarbon derivatives of intermediate length (C_{10} to C_{20}) are the preferred substrates to microorganisms and tend to be most readily degradable. Higher molecular weight alkane derivatives ($> C_{17}$) are hydrophobic solids that are difficult to bio-degrade due to the inherent poor response to biodegradation as well as the solubility in water (Chaîneau et al., 1995). Furthermore, cycloalkane derivatives are degraded more slowly than the corresponding n-alkane (unbranched-chain) derivatives and branched-chain alkane derivatives.

Second, contaminant bioavailability may be determined by a series of preliminary tests. Strong interactions between soil matrix and hydrophobic pollutants can evolve causing pollutant retention or even irreversible binding to sorbents. This phenomenon known as ageing increases with time and significantly reduces bioavailability of hydrophobic contaminants in soil (Hatzinger and Alexander, 1995). Pollutant retention over time is governed by physical-chemical characteristics of the pollutant and by soil characteristics. Strong or even irreversible sorption onto soil is in general attributed to the soil organic matter (Luthy et al., 1997; Huang et al., 2003). However, the degree of hydrocarbon degradation is affected mainly by the type of hydrocarbon contaminants matrix and only to a lesser extent by soil characteristics (Huesemann, 1995; Nocentini et al., 2000; Breedveld and Sparrevik, 2001). This might be true in particular for soils derived from further depths in the subsoil, where relatively low amounts of soil organic matter are present (Nierop and Verstraten, 2003).

Third, environmental conditions have to be regarded and include factors such as: temperature, pH, moisture content, availability of mineral nutrients, and contaminant concentration. Most microorganisms can degrade specific types of hydrocarbon derivatives under extreme environmental conditions even if rates might be lower and/or degradation might be incomplete (Margesin et al., 1997; Mohn and Stewart, 2000). Most crude oil-related hydrocarbon derivatives are readily degraded by means of aerobic microorganisms although hydrocarbon derivatives can be degraded in the absence of oxygen (anaerobic conditions) if alternate electron acceptors such as nitrate, manganese (IV), iron (III), sulfate and carbon dioxide hydrocarbon derivatives are present—however, the rate anaerobic degradation may be substantially lower than the rate of aerobic degradation (Holliger and Zehnder, 1996; Heider et al., 1999; Grishchenkov et al., 2000; Boopathy, 2002; Massias et al., 2003).

The addition of nutrients has been reported to have a beneficiary effect on hydrocarbon degradation in soil (Dott et al., 1995; Breedveld and Sparrevik, 2001; Chaîneau et al., 2003)— typically a carbon/nitrogen/phosphorous (C/N/P) ratio of 100/10/1 is commonly proposed (Oudot and Dutrieux, 1989; Atagana et al., 2003) although water is often needed to promote optimal microbial activity (Margesin et al., 2000).

Above all, the long-term fate of crude oil hydrocarbon derivatives in areas where spills have occurred needs to be determined. This is only possible through knowledge of the constituents of crude oil and crude oil products as well as application of the relevant text methods at the time of the spill. The data can then be used to determine whether or not hydrocarbon contamination will persist indefinitely.

The methods that are employed measure the concentration of total crude oil hydrocarbon derivatives generate a single number that represents the combined concentration of all crude oil hydrocarbon derivatives in a sample that are measurable by the particular method (Speight, 2005). Therefore, the determination of the total crude oil hydrocarbon derivatives in a sample is method dependent. On the other hand, methods that measure a crude oil group type concentration separate are used to quantify different categories of hydrocarbon derivatives (e.g., saturates, aromatic derivatives, and polars/resins) (Speight, 2014). The results of crude oil group type analyses can be useful for product identification because products such as, for example, gasoline, diesel fuel, and fuel oil have characteristic levels of various hydrocarbon structural moieties groups. Thus, the methods that measure identifiable crude oil fractions can be used to indicate and/or quantify the changes that have occurred through weathering of the sample.

Although these methods measure different crude oil hydrocarbon categories, there are several basic steps that are common to the analytical processes for all methods, no matter the method type or the environmental matrix. In general, these steps are: (1) sample collection and preservation— requirements specific to environmental matrix and analytes of interest, (2) extraction—separation of the analytes of interest from the sample matrix, (3) concentration—enhances the ability to detect analytes of interest, (4) cleanup, dependent upon the need to remove interfering compounds, and (5) measurement, or quantification, of the analytes (Dean, 1998). Each step affects the final result, and a basic understanding of the steps is vital to data interpretation.

Various methods exist for the testing of biodegradability of substances. Biodegradability is assessed by following certain parameters which are considered to be indicative of the consumption of the test substance by microorganisms, or the production of simple basic compounds which indicate the mineralization of the test substance.

Hence there are various biodegradability testing methods which measure the amount of carbon dioxide (or methane, for anaerobic cases) produced during a specified period; there are those which measure the loss of dissolved organic carbon for substances which are water soluble; those that measure the loss of hydrocarbon infrared bands and there are yet others which measure the uptake of oxygen by the activities of microorganisms (biochemical oxygen demand, BOD).

The purpose of this chapter is to describe well-established analytical methods that are available for detecting, and/or measuring, and/or monitoring total crude oil hydrocarbon derivatives and its metabolites, as well as other biomarkers of exposure and effect of total crude oil hydrocarbon derivatives. The intent is not to provide an exhaustive list of analytical methods. Rather, the intention is to identify well-established methods that are used as the standard methods approved by federal agencies and organizations such as the Environmental Protection Agency and the National Institute for Occupational Safety and Health (NIOSH) or methods prescribed by state governments for water and soil analysis. Other methods presented are those that are approved by groups such as the ASTM International (ASTM, 2004).

9.1 The Spilled Material

Briefly, for environmental purposes, crude oil constituents are sub-divided into two classes: (1) organic constituents and (2) inorganic constituents. Furthermore sub-categorization can be taken to the next level insofar as organic chemicals are classified as (1) volatile organic compounds, (2) semi-volatile *organic compounds*, and (3) non-volatile compounds. The same sub-categorization is also applied to the inorganic constituents of crude oil.

The first class of organic compounds, the *volatile organic compounds* (VOCs), is sub-divided into *regulated compounds* and *unregulated compounds*. Regulated compounds have maximum contaminant levels, but unregulated compounds do not. Regulated generally (but not always) have low boiling points, or low boiling ranges, and some are gases.

Principal sources of releases to air from refineries include: (1) combustion plants, emitting sulfur dioxide, oxides of nitrogen and particulate matter, (2) refining operations, emitting sulfur dioxide, oxides of nitrogen, carbon monoxide, particulate matter, volatile organic compounds, hydrogen sulfide, mercaptans and other sulfurous compounds, (3) bulk storage operations and handling of volatile organic compounds (various hydrocarbon derivatives).

Many of these chemicals can be detected at extremely low levels by a variety of instrumentation, including the human nose! In the case of the crude oil industry, sources for volatile organic compounds typically are crude oil refineries, fuel stations, naphtha (i.e., dry cleaning solvents, paint thinners, cleaning solvents for auto parts) and, in some cases, refrigerants that are manufactured from petrochemicals.

The second class of organic compounds, the *semi-volatile compounds*, typically have high boiling points, or high boiling ranges, and are not always easily detected by the instrumentation that may be used to detect the volatile organic compounds (including the human nose). Some of the common sources of contamination are high boiling crude oil products (e.g., lubricating oils), pesticides, herbicides, fungicides, wood preservatives, and a variety of other chemicals that can be linked to the refining industry.

The third group of constituents, the *non-volatile compounds* offer a group of constituents that are difficult to define chemically. These are the extremely high-boiling constituents that continue to defy identification by a wide variety of techniques leaving their chemical structure and behavior open to much speculation (Speight, 2007a).

Regulations are in place that set the maximum contamination concentration levels that are designed to ensure public safety. There are primary and secondary standards for inorganic chemicals. Primary standards are those chemicals that cause neurological damage, cancer, or blood disorders.

Secondary standards are developed for other environmental reasons. In some instances, the primary standards are referred to as the *Inorganic Chemical Group*. The secondary standards are referred to as the *General Mineral Group* and *General Physical Testing Group*.

In the context of crude oil, the inorganic chemical group includes nickel, vanadium, iron, and copper, as well as sodium and potassium (from reservoir brines) and a variety of trace metals that vary depending upon the source of the crude oil (Speight, 2007a). The mineral group includes calcium, magnesium, sodium, potassium, bicarbonate, carbonate, chloride, sulfate, pH, alkalinity, hardness, electrical conductivity, total dissolved solids, surfactants, copper, iron, manganese, and zinc. The physical group includes turbidity, color, and odor. Many of these chemicals arise from desalting residues and from other processes where catalysts are used. A high level of any of these three chemicals in the soil or in the water is an indication that one or more specific processes (identified from the chemicals that have been released) or pollution prevention processes are not performing according to operational specifications.

Another source of toxic compounds is combustion (Speight, 2007a). In fact, some of the greater dangers of fires are from toxic products and by-products of combustion. The most obvious of these is carbon monoxide (CO), which can cause serious illness or death because it forms carboxyhemoglobin with hemoglobin in the blood so that the blood no longer carries oxygen to body tissues. Toxic sulfur dioxide and hydrogen chloride are formed by the combustion of sulfur compounds and organic chlorine compounds, respectively. A large number of noxious organic compounds such as aldehydes are generated as by-products of combustion and, in addition to forming carbon monoxide, combustion under oxygen-deficient conditions produces polynuclear aromatic hydrocarbon derivatives consisting of fused-ring structures. Some of these compounds, such as benzo(a)pyrene are pre-carcinogenic compounds, insofar as they are acted upon by enzymes in the body to yield cancer-producing metabolites.

While most investigations involving crude oil hydrocarbon derivatives are regulated by various agencies that may require methodologies, action levels, and cleanup criteria that are differ in some respects, the complex chemical composition of crude oil and crude oil products can make it extremely difficult to select the most appropriate analytical test methods for evaluating environmental samples and to accurately interpret and use the data.

Accordingly, general methods of environmental analysis (Smith, 1999), i.e., analysis for the determination of crude oil or crude oil products that have been released, are available. The data determine whether or not a release of such chemicals will be detrimental to the environment and may lead to regulations governing the use and handling of such chemicals. But first, sample collection, preservation, preparation, and handling protocols must be followed to the letter. This, of course, includes *chain of custody* or *sampling handling* protocols that will be defensible if and when legal issues arise. Thus, an accurate sample handling and storage log should be maintained and should include the basic necessary information (Table 7.1). Attention to factors such as these enables standardized comparisons to be made when subsequent samples are taken (Dean, 1998; Speight, 2005, 2015).

Whatever the case, methods of analysis must be available to determine the nature of the released chemical (waste) and from the data predict the potential hazard to the environment.

Thus, during any oil spill incident, the properties of the spilled oil must be known immediately which will allow event (spill) investigators. In addition to the typical physical properties of the spilled material (Speight, 2015) and additional data that are needed include (1) the rate of evaporation of the spilled material, (2) the ability of the spilled material to disperse naturally, (3) the ability of the spilled material to disperse with the aid of dispersants, (4) the tendency of the spilled material to form emulsions, (5) the tendency of the spilled material or the emulsions to sink or submerge, (6) the viscosity of the spilled material at ambient temperatures, (7) the changing viscosity of the spilled material as the more volatile constituents evaporate, and (8) health hazard of the spilled material to on-site personnel as well as the toxicity of the spilled material to land-based or aquatic flora and non-human fauna.

The importance of each of these properties is dependent upon the purity of the crude oil or crude oil product spilled as well as of the properties of the site (for example, soil composition, which amongst other properties is site specific) and the prevailing climatic conditions (arctic, sub-arctic, temperate, and tropical—which are also site specific). Because of this site specificity and the complex nature of crude oil and crude oil products (the latter is often refinery specific), generalizations of the interactions of crude oil and crude oil products on various sites are to be avoided.

Crude oil and crude oil products released into the environment undergo weathering processes with time. These processes include evaporation, leaching (transfer to the aqueous phase) through solution and entrainment (physical transport along with the aqueous phase), chemical oxidation, and microbial degradation. The rate of weathering is highly dependent on environmental conditions. For example, gasoline, a volatile product, will evaporate readily in a surface spill but gasoline released below 10 feet of clay topped with asphalt will tend to evaporate slowly (weathering processes may not be detectable for years).

An understanding of weathering processes is valuable to environmental test laboratories. Weathering changes product composition and may affect testing results, the ability to bio-remediate, and the toxicity of the spilled product. Unfortunately, the database available on the composition of weathered products is limited.

Thus, in order to estimate the impact of a crude oil or crude oil product spill several non-conventional (non-typical) properties must be assessed and these properties have been presented elsewhere in this text (Chapter 1) and must be chosen depending upon several factors which are (1) the property of the spilled material, (2) the potential of the constituents of the spilled material to undergo chemical changes—such as oxidation—after the spill, and (3) geology of the site into which the material is spilled in terms of the mineralogy and the potential of the site minerals to interact with the constituents of the spilled material. In terms of Item No. 2, often when oil is spilled on to a water system, the oil tends to remain for a while then "disappear" into the water and, on occasion settle towards the bottom of the water. In this case, the crude oil or crude oil product—having a density less than the density of the water—will float on the surface. With time, the crude oil constituents may oxidize (by aerial oxidation) which can cause in increase in density and the presence of oxygen functional groups (such as -OH) in the constituents which renders the constituent more hydrophilic than constituents and the constituents also undergo a change in the interfacial tension properties. Depending on the degree of oxidation, the oxidized constituents will sink to different levels in the water system and eventually end up on the bottom. At some stage, any movement of the water can move the oxidized product to the beach. Hence, the apparent disappearance of spilled oil (for a period of time) and its reappearance on a beach.

Thus, the choice of a specific method should be based on compatibility with the particular type of hydrocarbon contamination to be measured and, furthermore, the choice may depend upon local or regional regulatory requirements for the type of hydrocarbon contamination that is known, or suspected, to be present. Furthermore, the risk at a specific site will change with time as contaminants evaporate, dissolve, biodegrade, and/or become sequestered.

9.2 Sample Collection and Preparation

Despite the nature of the environmental regulations and the precautions taken by the refining industry, the accidental release of non-hazardous chemicals and hazardous chemicals into the environment has occurred and, without being unduly pessimistic, will continue to occur (by all industries—not wishing to select the refining industry as the only industry that suffers accidental release of chemicals into the environment). It is a situation that, to paraphrase *chaos theory*, no matter how well one prepares, the unexpected is always inevitable.

It is, at this point that the environmental analyst has to identity the nature of the chemicals and their potential effects on the ecosystem(s) (Smith, 1999). Although crude oil itself and its various products are complex mixtures of many organic chemicals (Chapters 2 and 3), the predominance of one particular chemical or one particular class of chemicals may offer the environmental analyst or scientist an opportunity for predictability of behavior of the chemical(s).

9.2.1 Crude Oil and Crude Oil Products

Briefly, for environmental purposes, chemicals are sub-divided into two classes: (1) *organic chemicals* and (2) *inorganic chemicals*. Furthermore classification occurs insofar as organic chemicals are classified as *volatile organic compounds* or *semi-volatile organic compounds* (on occasion, the word *chemicals* is substituted for the word *compounds* without affecting the definition).

The first class of organic compounds, the *volatile organic compounds* (VOCs), is sub-divided into *regulated compounds* and *unregulated compounds*. Regulated compounds have maximum contaminant levels, but unregulated compounds do not. Regulated generally (but not always) have low boiling points, or low boiling ranges, and some are gases. Many of these chemicals can be detected at extremely low levels by a variety of instrumentation, including the human nose! In the case of the crude oil industry, sources for volatile organic compounds typically are crude oil refineries, fuel stations, naphtha (i.e., dry cleaning solvents, paint thinners, cleaning solvents for auto parts) and, in some cases, refrigerants that are manufactured from petrochemicals.

The second class of organic compounds, the *semi-volatile compounds*, typically have high boiling points, or high boiling ranges, and are not always easily detected by the instrumentation that may be used to detect the volatile organic compounds (including the human nose). Some of the common sources of contamination are high boiling crude oil products (e.g., lubricating oils), pesticides, herbicides, fungicides, wood preservatives, and a variety of other chemicals that can be linked to the refining industry.

Regulations are in place that set the maximum contamination concentration levels that are designed to ensure public safety. There are primary and secondary standards for inorganic chemicals. Primary standards are those chemicals that cause neurological damage, cancer, or blood disorders. Secondary standards are developed for other environmental reasons. In some instances, the primary standards are referred to as the *Inorganic Chemical Group*. The secondary standards are referred to as the *General Mineral Group* and *General Physical Testing Group*.

The inorganic chemical group includes aluminum, antimony, arsenic, barium, beryllium, cadmium, chromium, lead, mercury, nickel, selenium, silver, fluoride, nitrate, nitrite, and thallium. The mineral group includes calcium, magnesium, sodium, potassium, bicarbonate, carbonate, chloride, sulfate, pH, alkalinity, hardness, electrical conductivity, total dissolved solids, surfactants, copper, iron, manganese, and zinc. The physical group includes turbidity, color, and odor. Many of these chemicals arise from desalting residues and from other processes where catalysts are used. A high level of any of these three chemicals in the soil or in the water is an indication that one or more specific processes (identified from the chemicals that have been released) or pollution prevention processes are not performing according to operational specifications.

Another source of toxic compounds is combustion (Chapter 4). In fact, some of the greater dangers of fires are from toxic products and by-products of combustion. The most obvious of these is carbon monoxide (CO), which can cause serious illness or death because it forms carboxyhemoglobin with hemoglobin in the blood so that the blood no longer carries oxygen to body tissues. Toxic sulfur dioxide and hydrogen chloride are formed by the combustion of sulfur compounds and organic chlorine compounds, respectively. In addition, a large number of noxious organic compounds such as aldehydes are generated as by-products of combustion. In addition to forming carbon monoxide, combustion under oxygen-deficient conditions produces polynuclear aromatic hydrocarbon derivatives consisting of fused-ring structures. Some of these compounds, such as benzo(a)pyrene are pre-carcinogenic compounds, insofar as they are acted upon by enzymes in the body to yield cancer-producing metabolites.

Most investigations involving crude oil hydrocarbon derivatives are regulated by various agencies that may require methodologies, action levels, and cleanup criteria that are different.

Table 9.1: Suggested Items for Inclusion in a Sampling Log.

1.	The precise (geographic or other) location (or site or refinery or process) from which the sample was obtained.
2.	The identification of the location (or site or refinery or process) by name.
3.	The character of the bulk material (solid, liquid, or gas) at the time of sampling.
4.	The means by which the sample was obtained.
5.	The means and protocols that were used to obtain the sample.
6.	The date and the amount of sample that was originally placed into storage.
7.	Any chemical analyses (elemental analyses, fractionation by adsorbents or by liquids, functional type analyses) that have been determined to date.
8.	Any physical analyses (API gravity, viscosity, distillation profile) that have been determined to date.
9.	The date of any such analyses included in items 7 and 8.
10.	The methods used for analyses that were employed in 7 and 8.
11.	The analysts who carried out the work in 7 and 8.
12.	A log sheet showing the names of the persons (with the date and the reason for the removal of an aliquot) who removed the samples from storage and the amount of each sample (aliquot) that was removed for testing.

Indeed, the complex chemical composition of crude oil and crude oil products can make it extremely difficult to select the most appropriate analytical test methods for evaluating environmental samples and to accurately interpret and use the data.

Accordingly, general methods of environmental analysis (Smith, 1999), i.e., analysis for the determination of crude oil or crude oil products that have been released, are available. The data determine whether or not a release of such chemicals will be detrimental to the environment and may lead to regulations governing the use and handling of such chemicals. But first, sample collection, preservation, preparation, and handling protocols must be followed to the letter. This, of course, includes *chain of custody* or *sampling handling* protocols that will be defensible if and when legal issues arise. Thus an accurate sample handling and storage log should be maintained and should include the basic necessary information (Table 9.1) (Dean, 2003; Speight, 2015). Attention to factors such as these enables standardized comparisons to be made when subsequent samples are taken.

In summary, many of the specific chemicals in crude oil are hazardous because of their chemical reactivity, fire hazard, toxicity, and other properties. In fact, a simple definition of a hazardous chemical (or hazardous waste) is that it is a chemical substance (or chemical waste) that has been inadvertently released, discarded, abandoned, neglected, or designated as a waste material and has the potential to be detrimental to the environment. Alternatively, a hazardous chemical may be a chemical that may interact with other (chemical) substances to give a product that is hazardous to the environment.

Whatever the case, methods of analysis must be available to determine the nurture of the released chemical (waste) and from the data predict the potential hazard to the environment.

9.2.2 Sample Collection and Preparation

The ability to collect and preserve a sample that is representative of the site is a critically important step (Dean, 2003; Patnaik, 2004). In terms of sampling environmental items such as soils, the sample must be chosen from a population for investigation. A *random sample* is one chosen by a method involving an unpredictable component. Random sampling can also refer to taking a number of independent observations from the same probability distribution, without involving any real population. A probability sample is one in which each item has a known probability of being in the sample (Arjoon and Speight, 2018).

Obtaining representative environmental samples is always a challenge due to the heterogeneity of different sample matrices. Additional difficulties are encountered with crude oil hydrocarbon derivatives due to the wide range in volatility, solubility, biodegradation, and adsorption potential of individual constituents. And the procedures used for sample collection and preparation must be legally defensible.

The sampling methods used for crude oil hydrocarbon derivatives are generally thought of as methods for determination of the total crude oil hydrocarbon derivatives. In part due to the complexity of the components of the total crude oil hydrocarbon derivatives fractions, little is known about their potential for health or environmental impacts. As gross measures of crude oil contamination, the total crude oil hydrocarbon derivatives data simply show that crude oil hydrocarbon derivatives are present in the sampled media. Measured total crude oil hydrocarbon derivatives values suggest the relative potential for human exposure and, therefore, the relative potential for human health effects.

Although most site investigations to determine the assessment of contamination of an ecosystem by crude oil hydrocarbon derivatives are regulated by local or by regional (state) governments, sample collection and preservation recommendations follow strict guidelines (Table 9.1) thereby reducing the potential for sample compromise. Therefore, before a sample is collected, the particular sample collection and preservation requirements must be investigated. And, because of holding (storage) time considerations, the laboratory must be selected and notified prior to the collection of the samples.

9.2.2.1 Sample Collection

The value of any analysis is judged by the characteristics of the sample as determined by laboratory tests. The sample used for the test(s) must be representative of the bulk material or data will be produced that are not representative of the material and will, to be blunt, incorrect no matter how accurate or precise the test method. In addition, the type and cleanliness of sample containers are important if the container is contaminated or is made of material that either reacts with the product or is a catalyst, the test results may be wrong.

Thus, the importance of the correct sampling of any sample destined for analysis should always be over emphasized. Incorrect sampling protocols can lead to erroneous analytical data from which decisions about regulatory issues cannot be accurately made. In addition, adequate records of the circumstances and conditions during sampling have to be made, for example, in sampling from storage tanks, the temperatures and pressures of the separation plant and the atmospheric temperature would be noted.

At the other end of the volatility scale, samples that contain, or are composed of, high molecular weight paraffin hydrocarbon derivatives (wax) that are also in a solid state, may require judicious heating (to dissolve the wax) and agitation (homogenized, to ensure thorough mixing) before sampling. If room temperature sampling is the *modus operandi* and product cooling causes wax to precipitate, homogenization to ensure correct sampling is also necessary.

The first task in any analysis is to separate the analytes from the bulk of the sample. The traditional liquid extraction is the most common means employed but alternate methods are also available (Dean, 1998). A portion of the sample is mixed with an organic solvent into which the analyte is preferentially partitioned. The idea of partitioning is key to the success of this procedure. No organic analyte is completely removed from a sample by a single washing with an organic solvent.

Representative samples are prerequisite for the laboratory evaluation of any type of environmental sample and many precautions are required in obtaining and handling representative samples (ASTM D270, ASTM D1265). The precautions depend upon the sampling procedure, the characteristics (low boiling or high boiling constituents) of the product being sampled, and the storage tank, container, or tank carrier from which the sample is obtained. In addition, the sample container must be clean and the type to be used depends not only on the product but also on the data to be produced.

However, obtaining a representative sample is not just a matter of relying on partition coefficients. Partitioning is thermodynamically controlled and to achieve equilibrium requires a certain amount of intimate contact time between the sample matrix, the analyte molecules, and the organic solvent. Swirling an aqueous sample with the solvent is not effective for achieving a thermodynamic equilibrium. Mixing the two phases together as thoroughly as possible and then allowing the phases to separate achieves the desired equilibrium. The procedure should be repeated with additional portions of solvent. If equilibrium is not established, the efficiency of the extraction procedure is low and the analytical data will be subject to error and indefensible (Dean, 1998).

The basic objective of each procedure is to obtain a truly representative sample or, more often, a composite of several samples that can be considered to be a representative sample. In some cases, because of the size of the storage tank and the lack of suitable methods of agitation, several samples are taken from large storage tanks in such manner that the samples represent the properties of the bulk material from different locations in the tank and, thus, the composite sample will be representative of the entire lot being sampled. This procedure allows for differences in sample that might be due to the stratification of the bilk material due to tank size or temperature at the different levels of the storage tank. Solid samples require a different protocol that might involve melting (liquefying) the bulk material (assuming that thermal decomposition is not induced) followed by homogenization. On the other hand, the protocol used for coal sampling (ASTM D346, ASTM D2013) might also be applied to sampling crude oil products, such as coke, that are solid and for which accurate analysis is required before sales.

Once the sampling procedure is accomplished, the sample container should be labeled immediately, to indicate the product, time of sampling, location of the sampling point, and any other information necessary for the sample identification. And, if the samples were taken from different levels of the storage tank, the levels from which the samples were taken and the amounts taken and mixed into the composite, should be indicated on the sample documentation.

Sampling records for any procedure must be complete and should include, but is not restricted to, items relating to the origin of the sample, methods of storage, analytical tests performed, the test methods used, and the analyst(s) who performed the test methods (Table 9.1).

In summary, there must be a means to identify the sample history as carefully as possible so that each sample is tracked and defined in terms of source and activity. Thus, the accuracy of the data from any subsequent procedures and tests for which the sample is used will be placed beyond a *reasonable doubt*.

9.2.2.1.1 Volatile Compounds

A volatile substance is as one whose boiling point, or sublimation temperature, is such that it exists to a significant extent in the gaseous phase under ambient conditions.
Some Commonly Encountered Volatile Hydrocarbon derivatives

Aliphatic	*Aromatic*
Pentane derivatives	Benzene derivatives
Hexane derivatives	Toluene derivatives
Heptane derivatives	Ethylbenzene derivatives
Octane derivatives	Xylene derivatives
Nonane derivatives	Naphthalene derivatives
Decane derivatives	Phenanthrene derivatives
	Anthracene derivatives
	Acenaphthylene derivatives

There are several sampling procedures that are applicable to volatile compounds but often method application depends upon the compound(s) to be sampled (Dean, 2003). Part of the issue of sampling volatile compounds arises because some volatile substances sublime rather than boil whereas other volatile substances emit significant quantities of vapor well below their boiling point.

For sampling volatile hydrocarbon derivatives in the field, two procedures are generally recommended, viz. zero headspace and solvent extraction. However, these two procedures do not necessarily give equivalent results.

Zero headspace procedures involve the collection of a soil sample with immediate transfer to a container into which the sample fits exactly. The only space for gases is that within the soil pores. The volume of sample collected depends upon the concentration of volatiles in the soil. It is imperative that the container employed can be interfaced directly with the gas chromatograph. Several commercial versions of zero headspace sampling devices are available. The sample is transported to the laboratory at 4°C, where it is analyzed directly by purge and trap gas chromatography (EPA 5035) or other appropriate techniques such as vacuum distillation (EPA 5032) or headspace (EPA 5021).

Solvent extraction procedures involve collection of sample by an appropriate device and subsequent immediate placement into a borosilicate glass vessel, which contains a known quantity of ultra-pure methanol. The bottle is then transported to the laboratory at 4°C, and the methanol fraction analyzed by purge-and-trap gas chromatography (or similar procedure).

In general, the zero headspace procedures are employed when the concentrations of volatiles in the soil are relatively low and the solvent extraction methods are used for more polluted soils. Irrespective of which procedure is used, quantitation of volatiles in soil is subject to serious errors if sufficient care is not taken with the sampling operation. Although direct purge-and-trap methods are frequently advocated for determination of volatiles in samples collected by zero headspace procedures, there are certain problems associated with this technique. Caution is advised since the procedure really only collects that fraction of the volatile that exists in a free form within the soil pore spaces, or is at least in a facile equilibrium with this fraction.

Gas chromatography detectors employed for the determination of volatile organics in soil are generally flame ionization detectors (FID), photo-ionization detectors (PID), or mass spectrometry. Flame ionization detectors will respond to all carbon compounds in the sample, whereas the photo-ionization detectors is capable of some sensitivity by virtue of the energy of the lamp employed. A 10.2 or 10.0 eV lamp yields more specific response to unsaturated (including aromatic) hydrocarbon derivatives and may also be employed to give a complete BTEX (benzene, toluene, ethyl benzene and xylene) characterization at sites where this is likely to be an issue. As regards the columns used for the analysis of volatile hydrocarbon derivatives, a wide variety can be used. Wide bore capillary columns of length typically about 105 meters are generally employed and they must be capable of resolving 3-methyl pentane from methanol as well as ethyl benzene from the xylenes. There may be some variation in choice of column, however, according to the resolution required by the authority. There is some debate concerning appropriate limits for the gasoline range and this is reflected in disparate legislation amongst various countries. For instance, the upper range of the gasoline organics may be defined by naphthalene or dodecene ($C_{12}H_{26}$).

Typical gas chromatography conditions involve an oven temperature ramped between 40 and 240°C (104 and 465°F), with a detector maintained at 250°C and an injector at 200°C. There are two methods of calibration for the gas chromatograph. One method consists of analyzing a mixture of individual hydrocarbon derivatives that bracket the gasoline range and calculating an average response factor from the response for each individual component. The other method involves analyzing a standard that contains one or more gasoline.

9.2.2.1.2 Condensate Releases

Condensate release might be equated to the release of volatile constituents but are often named as such because of the specific constituents of the condensate, often with some reference to the gas condensate that is produced by certain crude oil and natural gas well. However, the condensate is often restricted to the benzene, toluene, ethyl benzene, and xylenes (BTEX) family of compounds.

To determine the concentrations of benzene, toluene, ethyl benzene, and xylenes, approved methods (e.g., EPA SW 846 8021B, EPA SW 846 8260) are not only recommended but are insisted upon for regulatory issues. Polynuclear aromatic hydrocarbon derivatives (PAHs) may be present in condensate and evaluation of condensate contamination should include the use of other test methods (EPA SW 846 8270, EPA SW 846 8310) provided that the detection limits are adequate to the task of soil and groundwater protection. Generally, at least one analysis may be required for the most contaminated sample location from each source area. Condensate releases in non-sensitive areas, require analysis for naphthalene only. The analysts should ensure that the method has detection limits that are appropriate for risk determinations.

9.2.2.1.3 Semi-Volatile and Non-Volatile Compounds

In almost all cases of hydrocarbon contamination, some attention will have to be paid to the presence of semi- and non-volatile hydrocarbon derivatives. However, the collection, handling of samples and their ultimate preparation for analysis is entirely different from that used for volatile hydrocarbon derivatives. In general, it is not necessary to take such rigorous procedures to prevent loss of analyte following collection, although the procedures should still be verified using appropriate quality control measures (Dean, 2003).

Before analysis of semi- or non-volatile components can proceed, it is necessary that the hydrocarbon components be brought into solution. In a sample from a contaminated site, semi- and non-volatile molecules may exist in the soil pores in the free form within the pore spaces, but are far more likely to be adsorbed by organic matter attached to the soil. Indeed the probability of such adsorption increases with increasing hydrophobicity of the molecules.

A number of procedures are available to help this dissolution and include Soxhlet extraction (EPA 3540C), ultrasonic extraction (EPA 3550B), thermal extraction (EPA 8275A) and supercritical fluid extraction (EPA 3560, EPA 3561). Although these procedures are well documented, some of their important details are frequently overlooked, with the result that the extraction is unsatisfactory. In the case of ultrasonic extraction, the method (EPA 3550B) stipulates the use of an ultrasonic disrupter of the horn type, with a minimum power of 300 watts. Many laboratories however wrongly interpret this to mean an ultrasonic bath, used for cleaning glassware. Such baths are of far lower energy and are not capable of separating the hydrocarbon derivatives from their association with humic material. As regards the use of supercritical fluid extraction, a methanol modifier is required to achieve complete extraction of polynuclear aromatic hydrocarbon derivatives, whereas supercritical carbon dioxide is sufficient to elute normal hydrocarbon derivatives (EPA 3560, EPA 3561).

For most analyses, it is necessary to separate the analytes of interest from the matrix (i.e., soil, sediment, and water). Extraction of analytes can be performed using one or more of the following methods: (1) extracting the analytes into a solvent, (2) heating the sample, as may be necessary to remove the solvent and for the analysis of volatile compounds, and (3) purging the sample with an inert gas, as is also used in the analyses of volatile compounds.

Soxhlet, sonication, supercritical fluid, sub-critical or accelerated solvent, and purge and trap extraction have been introduced into a variety of methods for the extraction of contaminated soil. Headspace is recommended as a screening method. Shaking/vortexing is adequate for the extraction of crude oil hydrocarbon derivatives in most environmental samples. For these extraction methods, the ability to extract crude oil hydrocarbon derivatives from soil and water samples depends on the solvent and the sample matrix. Surrogates (compounds of known identity and quantity) are frequently added to monitor extraction efficiency. Environmental laboratories also generally perform matrix spikes (addition of target analytes) to determine if the soil or water matrix retains analytes.

Thus, solvents have different extraction efficiencies and, thus, extracting the same sample in the same manner by two different solvents may result in different concentrations. The choice of solvents is determined by many factors such as cost, spectral qualities, method regulations, extraction efficiency, toxicity, and availability. Methylene chloride has been the solvent of choice for many semi-volatile analyses due to its high extraction efficiency. Chlorofluorocarbon solvents such as trichlorotrifluoroethane (Freon 113) have been used in the past for oil and grease analyses because of their spectral qualities (they do not absorb in the 2930 cm^{-1} infrared measurement wavelength) and low human toxicity. The use of chlorofluorocarbons is to be questioned because not all of the crude oil constituents are soluble in such solvents. Furthermore, the use of chlorofluorocarbons solvents is being reduced, even phased out of analytical methods, because of their detrimental effects on stratospheric ozone. Tetrachloroethylene and carbon tetrachloride are possible replacements but caution is advised since these solvents may be sensitive to light and leave residual chlorine in the sample. Methanol is the most common solvent used to preserve and extract volatiles such as benzene, toluene, ethylbenzene and xylene(s) in soil. But methanol is unsuitable for many constituents of crude oil and crude oil products. In short, there is no one solvents that will satisfy all of the criteria necessary of a complete and full extraction or solubility of crude oil and its products.

If the release has contaminated a water system, there are several methods that can be employed for sample separation (Patnaik, 2004, Section 2). Volatile compounds (gasoline, solvents) in water are generally separated from the aqueous matrix by purging with an inert gas and trapping the compounds on a sorbent (EPA 5030, purge and trap analysis). The sorbent is later heated to release the volatile compounds, and a carrier gas sweeps the compounds into a gas chromatograph. Headspace analysis is recommended as a screening method (EPA 3810, EPA 5021), although it performs well in particular situations, especially field analysis. In this method, the water sample is placed in a closed vessel with a headspace and heated to drive volatiles into the gas phase and instrument contamination is minimized because only volatile compounds are introduced into the instrument. Addition of salts or acids may enhance this process.

Samples containing heavy oil, along with the volatile components; can severely contaminate purge and trap instrumentation and caution is advised when interpreting the data. For such sample, it may be advisable to use a separatory funnel for the water extraction method for semi-volatiles involves extraction using a (EPA 3520). In this method, the sample is poured into a funnel-shaped piece of glassware, solvent is added, and the mixture is shaken vigorously. After layer separation, the extract (i.e., the solvent layer) is removed, filtered, dried with a desiccant, and concentrated. Multiple extractions on the same sample may increase overall recovery.

Another commonly used water extraction method for semi-volatiles involves continuous liquid-liquid extraction (EPA 3520). In this method, the sample (rather than being shaken with the solvent), is treated with a continuously heated solvent that is nebulized (broken into small droplets) and sprayed on top of the water. Liquid-liquid extraction is excellent for samples containing emulsion-forming solids, but it is more time-consuming than separatory funnel extractions. Nevertheless, time consuming or not, the method must produce reliable data that can be used without question for monitoring or regulatory purposes.

Solid phase extraction (EPA 3535) also can be used for extraction and concentration of semi-volatile material. The technique involves passing the water sample through a cartridge or disk containing an adsorbent such as silica or alumina. The adsorbent is often coated with compounds that impart selectivity for particular products or analytes such as polynuclear aromatic hydrocarbon derivatives (PAHs or PNAs). After extraction, the analytes are separated from the solid phase by elution with a small amount of organic solvent. A variant of solid phase extraction involves dipping a sorbent-coated fiber into the water (solid phase micro-extraction). Adsorbed analytes are thermally desorbed directly into a heated chromatographic injection port. Generally, the solid phase extraction method requires much less solvent and glassware than separatory funnel and liquid-liquid extraction.

For the separation of samples from contaminated soil, there are also several possible methods depending upon whether the contaminant is volatile or semi-volatile. Volatile compounds (such as BTEX, benzene, toluene, ethyl benzene, xylene and gasoline) may be extracted from soil using, for example methanol (EPA 5035, purge and trap analysis). In the method, the extraction is usually accomplished by mechanically shaking the soil with methanol. A portion of the methanol extract is added to a purge vessel and diluted in reagent grade water. The extract is then purged similar to a water sample.

Headspace analysis, (EPA 3810, EPA 5021), also works well for analyzing volatile crude oil constituents in soil. In the test method, the soil is placed in a headspace vial and heated to drive out the volatiles from the sample into the headspace of the sample container. Salts can be added for more efficient release of the volatile compounds into the headspace. Similar to water headspace analysis, the soil headspace technique is useful when heavy oils and high analyte concentrations are present which can severely contaminate purge and trap instrumentation. Detection limits are generally higher for headspace analysis than for purge and trap analysis.

The simplest method to separate semi-volatile compounds from soil is to shake or vortex (vigorous mechanical stirring) the soil with a solvent. Adding a desiccant to the soil/solvent mixture can help to break up soil and increase the surface area. The word of caution here is to ensure that the drying agent does not adsorb the solute. Assuming that no such adsorption occurs, the extract can be analyzed directly. Simple shaking is quick and easy, making it an excellent field extraction technique. However, extraction efficiency will vary depending on soil type and whether or not clay minerals (excellent adsorbents for many organic compounds) are present.

Soxhlet extraction (EPA SW-846 3540) is a very efficient extraction process that is commonly used for semi-volatiles crude oil constituents. In the method, the solvent is heated and refluxed (recirculated) through the soil sample continuously for 16 hours, or overnight. This method generates a relatively large volume of extract that needs to be concentrated. Thus, it is more appropriate for semi-volatile constituents than for volatile constituents. Sonication extraction (EPA SW-846 3550) can also be used for semi-volatile compounds and, as the name suggests, involves the use of sound waves to enhance analyte transfer from sample to solvent. Sonication is a faster technique than Soxhlet extraction, and it also can require less solvent.

Supercritical fluid extraction (EPA 3540, for total recoverable crude oil hydrocarbon derivatives; EPA 3561 for polynuclear aromatic hydrocarbon derivatives), is applicable to the extraction of semi-volatile constituents. Supercritical fluid extraction involves heating and pressuring a mobile phase to supercritical conditions (where the solvent has the properties of a gas and a liquid). The supercritical fluid is passed through the soil sample, and the analytes are concentrated on a sorbent or trapped cryogenically. The analytes are eluted with a solvent and analyzed with conventional techniques. Carbon dioxide is the most popular mobile phase.

Another method (EPA 3545, Accelerated Solvent Extraction) has been validated using a variety of soil matrices ranging from sand to clay. In the method, conventional solvents such as methylene chloride (or a hexane-acetone mixture) are heated (100°C, 212°F) and pressurized (2000 psi), then passed through the soil sample (this technique is also suitable for application to for crude oil sludge and crude oil sediment). The method has the advantage of requiring smaller solvent volumes than traditional solvent extraction techniques.

In some cases, when crude oil and/or crude oil products are released to the environment, a free phase is formed and sample(s) of the hydrocarbon material can be collected directly for characterization. The ability to analyze free product greatly aids the determination of product type and potential source. The samples may be diluted prior to analysis (EPA SW-846 3580, waste dilution), gives some guidelines for proper dilution techniques. However, caution is advised since, as part of the initial sample collection procedure, water and sediment may be inadvertently included in the sample. There are several protocols involved in initial isolation and cleanup of the sample that must be recognized. In fact, considerable importance attaches to the presence of *water* or *sediment* in crude oil (ASTM D1796, ASTM D4007) for they lead to difficulties in other analyses.

Sediment usually consists of finely divided solids that may be dispersed in the oil or carried in water droplets. The solids may be drilling mud or sand or scale picked up during the transport of the oil, or may consist of chlorides derived from evaporation of brine droplets in the oil. In any event, the sediment can lead to serious plugging of the equipment, corrosion due to chloride decomposition, and a lowering of residual fuel quality.

Water may be found in the crude either in an emulsified form or in large droplets. The quantity is generally limited by pipeline companies and by refiners, and steps are normally taken at the wellhead to reduce the water content as low as possible. However, after a spill, water can be introduced by climatic conditions and the relevant tests (ASTM D96, ASTM D954, ASTM D1796) are regarded as important in crude oil analyses. Prior to analyses, it is often necessary to separate the water from a crude oil sample and this is usually carried out by one of the procedures described in the preliminary distillation of crude oil (IP 24).

Overall, there are several methods that can be employed for organic semi-volatile sample preparation and clean-up procedures.

9.2.2.1.4 Solids

For homogeneous materials, sampling protocols are relatively simple and straightforward, although caution is always advised lest overconfidence cause errors in the method of sampling as well as introduce extraneous material (EPA, 1998). On the other hand, the heterogeneous nature of soil and contaminated soil complicates the sampling procedures. If the soils and the sample are visibly heterogeneous, there is a very strong emphasis on need to obtain representative samples for testing and analysis.

Thus, the variable composition of contaminated soil as well as solid sample such as crude oil coke, offers many challenges to environmental analysts who must ensure that the sample under investigation is representative of the contaminated site or the coke. Furthermore, sample transportation can initiate (due to movement of the sample) processes that result in size and density segregation, in a manner analogous to variations in coal quality from sample-to-sample (ASTM D346; ASTM D2234; ASTM D4702; ASTM D4915; ASTM D4916; ASTM D6315; ASTM D6518; ASTM D6543; ISO 1988).

Therefore, the challenge in sampling solids for environmental analysis is to collect a relatively small portion of the sample that accurately represents the composition of the whole. This requires that sample increments must be collected in such a manner that no piece, regardless of position (or size) relative to the sampling position and implement, is selectively collected or rejected. Optimization of solids sampling is a function of the many variable constituents of coal and is reflected in the methods by which an unbiased sample can be obtained, as is required by coal sampling (ASTM D197).

Thus, in order to test any particular environmental solids sample, there are two criteria that must be followed and these are: (1) obtain a sample of the solid and (2) ensure that the sample is a true representative of the bulk material, and (last but by no means least) (3) to ensure that the sample does not undergo any chemical or physical changes after completion of the sampling procedure and during storage prior to analysis. In short, the reliability of a sampling method is the degree of perfection with which the identical composition and properties of the entire whole coal are obtained in the sample. The reliability of the storage procedure is the degree to which the coal sample remains unchanged thereby guaranteeing the accuracy and usefulness of the analytical data. At this point, a review of the sampling methods applied to coal, and which, under favorable conditions can be applied to environmental solids, is worth of inclusion.

The sampling procedures (ASTM D346; ASTM D2234; ASTM D4702; ASTM D4915; ASTM D4916; ASTM D6315; ASTM D6518; ASTM D6543) are designed to give a precision such that if gross samples are taken repeatedly from a lot or consignment and prepared according to the standard test methods (ASTM D197; ASTM D2013) and one ash determination is made on the analysis sample from each gross sample, the majority (usually specified as 95 out of 100) of these determinations will fall within ± 10% of the average of all of the determinations. When other precision limits are required or when other constituents are used to specify precision, defined special-purpose sampling procedures may need to be employed.

Thus, when a property of the sample (which exists as a large volume of material) is to be measured, there usually will be differences between the analytical date derived from application of the test methods to a *gross lot* or *gross consignment* and the data from the *sample lot*. This difference (the *sampling error*) has a frequency distribution with a mean value and a variance. *Variance* is a statistical term defined as the mean square of errors; the square root of the variance is more generally known as the *standard deviation* or the *standard error of sampling*.

Recognition of the issues involved in obtaining representative samples of coal and minimization of the *sampling error* has resulted in the designation of methods that dictate the correct manner for coal sampling (ASTM D346; ASTM D2234; ASTM D4702; ASTM D4915; ASTM D4916; ASTM D6315; ASTM D6518; ASTM D6543; ISO 1988; ISO 2309).

Finally, and this applied to all samples where separation of the solute or released material from the matrix is necessary, it must be recognized that the organic compounds, in for example soil or in a liquid matrix (water or an organic solvent), have an affinity for the sample matrix to one degree or another. It is this affinity that allows the sample to be retained by the matrix. The affinity may be due to adsorption on the surface of sample particles or solvation in water or other solvent medium.

Therefore, it is necessary in an extraction process to recognize the potential solute-matrix interactions in order to overcome such interactions. A suitable choice of solvent is necessary, making sure that the solvent itself does not reaction with the solid or with the matrix. If the first extraction solvent (e.g., pyridine or carbon disulfide) of choice is too powerful, switching to a solvent of lesser ability, such as pentane, hexane, or crude oil ether, remembering that if it is a high-boiling crude oil sample that is being extracted, these solvents of lesser ability will leave some of the sample unextracted. In other cases, particularly when the sample is a hydrocarbon crude oil product, a polar solvent (e.g., methanol) can be used to extract polar organic materials like the phenols from the matrix.

On the other hand, organic acids (e.g., carboxylic acids, phenols) and organic bases (pyridines derivative) may be isolatable by adjustment of the pH to control the direction of partitioning. For example, acidification (pH < 5) of the sample converts the organic bases into salts that move into the aqueous (water) phase whereas adjusting the pH of the sample to basic (pH > 9–11, depending on the nature of the solute) with a suitable base neutralizes the basic analytes and reverses the direction of partitioning to the organic phase. At the same time, organic acids are converted to water-soluble (hydrophilic) salts.

Thus, the simple expedient of separate analysis of the acid and basic extractions, rather than combining the extracts into a single sample extract, often serves to reduce matrix interference to a manageable level.

Although widely used, solvent extraction procedures have been demonstrated as sensitive to such variables as content of humic matter and moisture within samples. Supercritical fluid extraction appears to be a more robust procedure. Thermal extraction procedures are sensitive to the size of the soil sample and in some cases since the technique can result in cracking higher molecular weight constituents that do not volatilize out of the thermal zone. In the case of solvent extraction procedures, it is necessary to concentrate and also to clean up the samples. With complex mixtures of semi-volatile hydrocarbon derivatives, it is generally advisable to separate the aliphatic and aromatic fractions.

9.2.2.2 Extract Concentration

Extract concentration is the one area in the isolation procedure that has the greatest potential for loss of analytes. The operative physical concept during concentration is vapor pressure, not boiling point. However, boiling is uncontrolled insofar as all of the molecules in the sample are attempting to convert from the liquid state to the gaseous state. Therefore, equating sample concentration to boiling point leads to erroneous conclusions about choice of an appropriate method. On the other hand, vapor pressure is a continuous function that relates to the rate of evaporation, and evaporation is what is desired during sample concentration. Thus, the proper technique is to control the rate of solvent evaporation while minimizing analyte loss. A completely successful sample extraction can be performed that is completely negated by an inappropriate method for sample extract concentration.

In many cases, sample extracts are filtered, dried with desiccant, and concentrated before analysis. Concentration of the extract may allow for lower sample detection limits. Frequently, sample extracts must be concentrated to obtain detection limits low enough to meet regulatory action limits.

The trapping step in a purge and trap analysis is essentially a concentration step. Analytes are purged from the matrix into a gas stream and captured on a sorbent trap. The analytes are released by heating the trap. Cryogenic trapping is also used in place of sorbent trapping. In cryogenic trapping, a very cold material (such as liquid nitrogen) surrounds a sample loop and as analytes are purged and swept through the sample loop, they freeze in the sample loop. The analytes are released when the trap is heated.

Snyder columns are designed to allow highly volatile solvents to escape while retaining semi-volatile analytes of interest. Snyder columns are generally fitted onto the tops of flasks containing extracts and column design permits solvent to escape as the flask is heated. The analytes of interest condense from a gas to a liquid phase and fall back down into the solvent reservoir. The Kuderna-Danish concentrator is a Snyder column with a removable collection tube attached to the bottom. As solvent is evaporated, the extract is collected in the collection tube.

As an alternative to a Snyder column, the sample extracts may be concentrated with nitrogen evaporation by directing a slow stream of the gas over the extract surface at room temperature, resulting in minor loss of volatiles. Placing the extract container in warm water helps to speed the process, but then some loss of volatiles can occur. Concentration by evaporating excess solvent with a vacuum is not very common in environmental laboratories. Many semi-volatile analytes are lost in the procedure. Additionally, evaporating as a means of concentrating the sample cannot be used if the goal is to detect volatile analytes.

Cleanup steps are an important component of infrared (IR)-based and gravimetric methods because these methods are very sensitive to non-crude oil hydrocarbon interferences. Cleanup steps are not always a part of the crude oil analytical process, but when they are necessary, the goals of extract cleanup steps typically include one or more of the following: (1) removal of non-crude oil compounds, (2) isolation of a particular crude oil fraction, and (3) concentration of analytes of interest.

The techniques employed to extract the analytes of interest can frequently extract interfering compounds. Polar compounds such as animal and plant fats, proteins, and small biological molecules may be improperly identified as crude oil constituents. Extract cleanup techniques can be used to remove them. In an ideal situation, only interfering compounds are removed. In reality, some polar crude oil constituents can also be removed.

Two techniques are used to clean crude oil extracts. In one technique, interfering compounds are removed by passing the extract through a glass column filled with sorbent. A second technique is to swirl the extract with loose sorbent, then remove the sorbent by filtration. Other methods involve trapping the interfering compounds on a sorbent column such as alumina (EPA SW-846 3611) that is designed to remove interfering compounds and to fractionate crude oil wastes into aliphatic, aromatic and polar fractions. The fractions can be analyzed separately or combined for measurement of the total crude oil hydrocarbon derivatives. Alternatively, silica gel (EPA SW-846 3630) is commonly used for polynuclear aromatic hydrocarbon derivatives and phenol derivatives. Variations of this technique are used to clean (EPA 418.1) extracts before infrared analysis. In addition, the gel permeation technique has been used for cleanup (EPA SW-846 3640) and works on the principle of size exclusion. Large macromolecules such as lipids, polymers, and proteins are removed from the sample extract. Extracts obtained from soil that have (or have had) high biological activity may be cleaned by this method.

There are two non-column cleanup methods, one of which uses acid partition (EPA SW-846 3650) to separate the base/neutral and acid components by adjusting pH. This method is often used before alumina column cleanup to remove the acid components. The other method (EPA SW-846 3660) is used for sulfur removal and uses copper, mercury, and tetrabutylammonium sulfite as desulfurization compounds. Sulfur is a common interfering compound for crude oil hydrocarbon analysis, particularly for sediments. Sulfur-containing compounds are very common in crude oil and heavy fuel oil. Elemental sulfur is often present in anaerobically biodegraded fuels. Thus, abnormally high levels of sulfur may be measured as part of the total petroleum hydrocarbon content if the cleanup technique is not used.

Even though cleanup procedures are advocated before sample analysis, there can be several limitations to various cleanup steps. The reasons for decreased effectiveness of cleanup procedures include (1) sample loading may exceed the capacity of cleanup columns, (2) non-crude oil compounds may have chemical structures similar to crude oil constituents and may behave like a crude oil constituent, (3) analytes of interest may be removed during the cleanup, and (4) no single cleanup technique removes all of the chemical interferences.

9.2.2.3 Sample Cleanup

The use of any sample clean up in environmental analysis is always accompanied by the possibility of analyte loss. Procedures that depend upon polarity interactions between the eluting solvent, the solid adsorbent, and the target analytes to achieve selective isolation, are particularly prone to having the desired compounds ending up in the wrong fraction. Sources of these errors include mistakes in the preparation of the eluting solvent, use of the wrong or a deactivated absorbent, and the presence of traces of polar solvents in the sample solution, and the structure of the analyte molecule that will determine the behavior of the analytes during separation (Speight, 1999). Attention to detail and procedure are required for successful use of sample cleanup techniques and the aspects of the procedure that need to be examined include (1) suitability of the materials to achieve the clean-up that will be sample dependent, (2) introduction of laboratory contamination, and (3) success of the procedure on the individual sample.

The introduction of laboratory contamination is a significant, but often overlooked concern in sample cleanup. Examples are the introduction of extractable materials from plastics used for joining tubes and contaminated solvents are only two of the potential points of laboratory contamination.

9.2.2.4 Measurement

The issues that face environment analysts include the need to provide higher-quality results. In addition, environmental regulations may influence the method of choice. Nevertheless, the method of choice still depends to a large extent on the boiling range (or carbon number) of the sample to be analyzed. For example, there is a large variation ion the carbon number range and boiling points (of normal paraffins) for some of the more common crude oil products and thus a variation in the methods that may be applied to these products (Speight, 2005, 2015).

The predominant methods of measuring the properties of crude oil products are covered by approximately seven test methods that used in the determination of bulk quantities of crude oil and crude oil products (ASTM D96, ASTM D287, ASTM D1085, ASTM D1086, ASTM D1087, ASTM D1250, ASTM D1298).

Testing for suspended water and sediment (ASTM D96) is used primarily with fuel oils, where appreciable amounts of water and sediment may cause fouling of facilities for handling the oil and give trouble in burner mechanisms. Three standard methods are available for this determination. The centrifuge method gives the total water and sediment content of the sample by volume, the distillation method gives the water only, volumetrically, and the extraction method gives the solid sediment in per cent by weight.

The determination of density of specific gravity (ASTM D287, ASTM D1298) in the measurement and calculation of volume of crude oil products is important since gravity is an index of the weight of a measured volume of the product. Two scales are in use in the crude oil industry, specific gravity and API gravity, the determination being made in each case by means of a hydrometer of constant weight displacing a variable volume of oil. The reading obtained depends upon both the gravity and the temperature of the oil.

Gauging crude oil products (ASTM D1085, discontinued in 12996 but still in use) involves the use of procedures for determining the liquid contents of tanks, ships and barges, tank cars, and tank trucks. Depth of liquid is determined by gauging through specified hatches, or by reading gauge glasses or other devices. There are two basic types of gauges, inn age and outage. The procedures used depend upon the type of tank to be gauged, its equipment and the gauging apparatus.

An innage gauge is the depth of liquid in a tank measured from the surface of the liquid to the tank bottom or to a datum plate attached to the shell or bottom. The innage gauge is used directly with the tank calibration table and temperature of the product to calculate the volume of product (ASTM D1250). On the other hand, an outage gauge is the distance between the surface of the product in the tank and the reference point above the surface, which is usually located in the gauging hatch. The outage gauge is used either directly or indirectly with the tank calibration table and the temperature of the product to calculate the volume of product. The amount of any free water and sediment in the bottom of the tank is also gauged so that corrections can be made when calculating the net volume of the crude oil or the crude oil product.

The liquid levels of products that have a Reid vapor pressure of 40 lb or more are generally determined by the use of gauge glasses, rotary or slip-tube gauges, tapes and bobs through pressure locks, or other types of gauging equipment. The type of gauging equipment depends upon the size and type of the pressure tank.

There are also procedures for determining the temperatures of crude oil and its products when in a liquid state. Temperatures are determined at specified locations in tanks, ships and barges, tank cars, and tank trucks. For a non-pressure tank, a temperature is obtained by lowering a tank thermometer of proper range through the gauging hatch to the specified liquid level. After the entire thermometer assembly has had time to attain the temperature of the product, the thermometer is withdrawn and read quickly. This procedure is also used for low-pressure tanks equipped with gauging hatches or standpipes, and for any pressure tank that has a pressure lock. For tanks equipped with thermometer wells, temperatures are obtained by reading thermometers placed in the wells with their bulbs at the desired tank levels. If more than one temperature is determined, the average temperature of the product is calculated from the observed temperatures. Electrical-resistance thermometers are sometimes used to determine both average and spot temperatures.

In general, the volume received or delivered is calculated from the observed gauge readings. Corrections are made for any *free* water and sediment as determined by the gauge of the water level in the tank. The resultant volume is then corrected to the equivalent volume at 15.6°C (60°F) by use of the observed average temperature and the appropriate volume correction table (ASTM D1250). When necessary, a further correction is made for any suspended water and sediment that may be present in materials such as crude oil and heavy fuel oils.

For the measurement of other crude oil products, a wide variety of tests is available. In fact, there are approximately three hundred and fifty tests (ASTM, 2000) that are used to determine the different properties of crude oil products. Each test has its own limits of accuracy and precision that must be adhered to if the data are to be accepted.

9.2.2.4.1 Accuracy

The *accuracy* of a test is a measure of how close the test result will be to the true value of the property being measured (ASTM, 2004; Patnaik, 2004). As such the accuracy can be expressed as the *bias* between the test result and the true value. However, the *absolute accuracy* can only be established if the true value is known.

In the simplest sense, a convenient method to determine a relationship between two measured properties is to plot one against the other. Such an exercise will provide either a line fit of the points or a spread that may or may not be within the limits of experimental error. The data can then be used to determine the approximate accuracy of one or more points employed in the plot. For example, a point that lies outside the limits of experimental error (a *flyer*) will indicate an issue of accuracy with that test and the need for a repeat determination.

However, the graphical approach is not appropriate for finding the absolute accuracy between more than two properties. The well-established statistical technique of regression analysis is more pertinent to determining the accuracy of points derived from one property and any number of other properties. There are many instances in which relationships of this sort enable properties to be predicted from other measured properties with as good precision as they can be measured by a single test. It would be possible to examine in this way the relationships between all the specified properties of a product and to establish certain key properties from which the remainder could be predicted, but this would be a tedious task.

However, the application of statistical analysis to experiment data is not always straightforward and may be fraught with inconsistencies due to the assumptions that are involved in the statistical development an interpretation of the data. Statistical analysis of the data is only a part of the picture; the decision process has to be viewed as a coherent whole; a decision can only be passed by taking into account the complex interrelations among the chemical species being consumed and formed and the legal, economic, scientific, and environmental characteristics of the analytical process (Baker, 1966; Dixon and Massey, 1969; Alder and Roessler, 1972; Box et al., 1978; Caulcutt and Boddy, 1983; Jaffee and Spirer, 1987; Meier and Zünd, 2000; Patnaik, 2004, Section 3).

An alternative approach to that of picking out the essential tests in a specification using regression analysis is to take a look at the specification as a whole, and extract the essential features (termed *principal components analysis*).

Principal components analysis involves an examination of set of data as points in *n*-dimensional space (corresponding to *n* original tests) and determines (first) the direction that accounts for the biggest variability in the data (*first principal component*). The process is repeated until *n* principal components are evaluated, but not all of these are of practical importance since some may be attributable purely to experimental error. The number of significant principal components shows the number of independent properties being measured by the tests considered.

Following from this, it is necessary to establish the number of independent properties that are necessary to predict product performance in service with the goals of rendering any specification more meaningful and allowing a high degree of predictability of product behavior. On a long-term approach it might be possible to obtain new tests of a fundamental nature to replace, or certainly to supplement, existing tests. In the short-term, selecting the best of the existing tests to define product quality is the most beneficial route to predictability.

9.2.2.4.2 Precision

The *precision* of a test method is the variability between test results obtained on the same material, using a specific test method (ASTM, 2004; Patnaik, 2004). The precision of a test is usually unrelated to its accuracy. The results may be precise, but not necessarily accurate. In fact, the precision of an analytical method is the amount of scatter in the results obtained from multiple analyses of a homogeneous sample. To be meaningful, the precision study must be performed using the exact sample and standard preparation procedures that will be used in the final method. Precision is expressed as repeatability and reproducibility.

The precision of sampling, for example, solids is a function of the size of increments collected and the number of increments included in a gross sample, improving as both are increased, and subject only to the constraint that increment size not be small enough to cause selective rejection of the largest particles present. The manner in which solids sampling is performed as it relates to the precision of the sample thus depends upon the number of increments collected from all parts of the lot and the size of the increments. In fact, the number and size of the increments are operating variables that can, within certain limits, be regulated by the sampler.

The *intra-laboratory precision* or the *within-laboratory precision* refers to the precision of a test method when the results are obtained by the same operator in the same laboratory using the same apparatus. In some cases, the precision is applied to data gathered by a different operator in the same laboratory using the same apparatus. Thus, intra-laboratory precision has an expanded meaning insofar as it can be applied to laboratory precision.

Repeatability or repeatability interval of a test (r) is the maximum permissible difference due to test error between two results obtained on the same material in the same laboratory.

$r = 2.77 \times$ standard deviation of test

The repeatability interval (r) is, statistically, the 95% probability level or the differences between two test results are unlikely to exceed this repeatability interval more than five times in a hundred.

The *inter-laboratory precision* or the *between-laboratory precision* is defined in terms of the variability between test results obtained on the aliquots of the same homogeneous material in different laboratories using the same test method.

The term *reproducibility* or *reproducibility interval* (R) is analogous to the term repeatability but it is the maximum permissible difference between two results obtained on the same material but now in different laboratories. Therefore, differences between two or more laboratories should not exceed the reproducibility interval more than five times in a hundred.

$R = 2.77 \times$ standard deviation of test

The repeatability value and the reproducibility value have important implications for quality. As the demand for clear product specifications, and hence control over product consistency grows, it is meaningless to establish product specifications that are more restrictive than the reproducibility/repeatability values of the specification test methods.

9.2.2.4.3 Method Validation

Method validation is the process of proving that an analytical method is acceptable for its intended purpose. Many organizations provide a framework for performing such validations (ASTM, 2004). In general, methods for product specifications and regulatory submission must include studies on specificity, linearity, accuracy, precision, range, detection limit, and quantitation limit.

The process of method development and validation covers all aspects of the analytical procedure and the best way to minimize method problems is to perform validation experiments during development.

In order to perform validation studies, the approach should be viewed with the understanding that validation requirements are continually changing and vary widely, depending on the type of product under tested and compliance with any necessary regulatory group.

In the early stages of new product development, it may not be necessary to perform all of the various validation studies. However, the process of validating a method cannot be separated from the actual development of the method conditions, because the developer will not know whether the method conditions are acceptable until validation studies are performed. The development and validation of a new analytical method may therefore be an iterative process. Results of validation studies may indicate that a change in the procedure is necessary, which may then require revalidation. During each validation study, key method parameters are determined and then used for all subsequent validation steps.

The first step in the method development and validation cycle should be to set minimum requirements, which are essentially acceptance specifications for the method. A complete list of criteria should be agreed on during method development and the end users before the method is developed so that expectations are clear. Once the validation studies are complete, the method developers should be confident in the ability of the method to provide good quantitation in their own laboratories. The remaining studies should provide greater assurance that the method will work well in other laboratories, where different operators, instruments, and reagents are involved and where it will be used over much longer periods of time.

The remaining precision studies comprise much of what is often referred to as *ruggedness*. *Intermediate precision* is the precision obtained when an assay is performed by multiple analysts using several instruments on different days in one laboratory. Intermediate precision results are used to identify which of the above factors contribute significant variability to the final result.

The last type of precision study is *reproducibility (q.v.)* that is determined by testing homogeneous samples in multiple laboratories, often as part of inter-laboratory crossover studies. The evaluation of reproducibility results often focuses more on measuring bias in results than on determining differences in precision alone. Statistical equivalence is often used as a measure of acceptable inter-laboratory results. An alternative, more practical approach is the use of *analytical equivalence* in which a range of acceptable results is chosen prior to the study and used to judge the acceptability of the results obtained from the different laboratories.

Performing a thorough method validation can be a tedious process, but the reliability of the data generated with the method is directly linked to the application of quality assurance and quality control protocols, which must be followed assiduously (Quevauviller, 2002).

Briefly, to assure quality assurance and quality control, samples are analyzed using standard analytical procedures. A continuing program of analytical laboratory quality control verifies data quality and involves participation in inter-laboratory crosschecks, and replicate sampling and analysis. When applicable, it is advisable, even insisted by the Environmental Protection Agency, that analytical labs must be certified to complete the analysis requested.

However, in many cases, time constraints often do not allow for sufficient method validation. Many researchers have experienced the consequences of invalid methods and realized that the amount of time and resources required to solve problems discovered later exceeds what would have been expended initially if the validation studies had been performed properly. *Putting in time and effort up front* will help any environmental analysts to find a way through the method validation maze and eliminate many of the problems common to inadequately validated analytical methods.

One method that is often used for method validation is the addition of a surrogate to the sample matrix and following the track of the surrogate through the separation procedure. It is appropriate that at this point, comments should be made on such procedures.

However, it is the nature of the sampling protocols, and the location of the site from which the sample are taken, that it should be assumed that no two samples are identical. Factors such as (1) depth from which the sample is taken, (2) distance of the sampling point from the initial location of the spill, and (3) ambient conditions play a role in determining sample character. In fact, each sample is a unique combination of matrix-analyte interactions. Adding surrogate compounds to each sample and then determining the recovery of the surrogates is used as the benchmark for gauging the success of the extraction procedure. The best surrogates are those that are most like the target analytes and example are the isotopically labeled versions of each of the target molecules. Whether or not use of such molecules is justified relates to the data that are required.

As an alternative, other similar chemical compounds can be used but the choice of the surrogate compounds can limit or it can maximize data interpretation. For most analyses, it is possible to choose surrogates that will always generate excellent recoveries, regardless of the complexity of the sample. Choosing surrogate molecules to produce *acceptable quality control* is not realistic. The use of surrogates is to obtain reliable information about the overall strength and weaknesses of the analytical method. In addition, the surrogate(s) should reflect the chemical behavior and physical properties of the analytes. Important chemical behavior and properties includes (1) acidic and basic properties of the analytes, (2) the range of polarity of the analytes, (3) reactivity in chemical derivatization procedures, and (4) the sensitivity of the analytes to decomposition caused by extremes of chemical or physical environment.

Surrogates chosen to monitor any of these areas should ideally bracket the range of the property. However, it should be pointed out that very few individual methods specify surrogates that provide information on all these areas, let alone the idea of bracketing the property. The analyst must determine the suitability of surrogate to reflect the properties and behavior of the analytes a well as the ability of the surrogate to be used for the collection of reliable and defensible data.

9.2.2.5 Quality Control and Quality Assurance

Quality control (QC) and quality assurance (QA) programs are key components of all analytical protocols in all areas of analysis, including environmental, pharmaceutical, and forensic testing, among others (Patnaik, 2004). These programs mandate that the laboratories follow a set of well-defined guidelines to achieve valid analytical results to a high degree of reliability and accuracy within an acceptable range. Although such programs may vary depending on the regulatory authority, certain key features of these programs are more or less the same (see below).

However, there is often confusion between the terms *quality assurance* and *quality control*, perhaps because there is also considerable overlap between certain aspects of quality assurance and quality control programs.

9.2.2.5.1 Quality Control

Quality controls are single procedures that are performed in conjunction with the analysis to help assess in a quantitative manner the success of the individual analysis. Examples of quality controls are blanks, calibration, calibration verification, surrogate additions, matrix spikes, laboratory control samples, performance evaluation samples, determination of detection limits, etc. The success of the quality control is evaluated against an acceptance limit. The actual generation of the acceptance limit is a function of quality assurance; it would not be termed a quality control.

Unlike quality assurance plans that mostly address regulatory requirements involving comprehensive documentation, quality control programs are science-based, the components of which may be defined statistically. The two most important components of quality control are (1) determination of precision of analysis and (2) determination of accuracy of measurement

Whereas *precision* (Section 9.2.2.4.2) measures the reproducibility of data from replicate analyses, the *accuracy* (Section 9.2.2.4.1) of a test estimates how accurate are the data, that is, how close the data would fall to probable true values or how accurate is the analytical procedure to give results that may be close to true values. Both the precision and accuracy are measured on one or more samples selected at random from a given batch of samples for analysis. The precision of analysis is usually deter- mined by running duplicate or replicate tests on one of the samples in a given batch of samples. It is expressed statistically as standard deviation, relative standard deviation (RSD), coefficient of variance (CV), standard error of the mean (M), and the relative percent difference (RPD).

The standard deviation in measurements, however, can vary with the concentrations of the analytes. On the other hand, RSD, which is expressed as the ratio of standard deviation to the arithmetic mean of replicate analyses and is given as a percent, does not have this problem and is a more rational way of expressing precision:

RSD = – (standard deviation/arithmetic mean of replicate analysis) × 100%

The standard error of the mean, M, is the ratio of the standard deviation and the square root of the number of measurements (n):

M = standard deviation/n

This scale, too, will vary in the same proportion as standard deviation with the size of the analyte in the sample.

In routine testing, many repeat analyses of a sample aliquot may not be possible. Alternatively, therefore, the precision of a test may be determined from duplicate analyses of sample aliquots and expressed as RPD:

$$RPD = (a_1 - a_2)/[(a_1 + a_2)/2] \times 100\%$$

Or

$$RPD = (a_2 - a_1)/[(a_1 + a_2)/2] \times 100\%$$

In this equation, a_1 and a_2 are the results of duplicate analyses of a sample. Since only two tests are performed on a selected sample, the RPD may not be as accurate a measure of precision as the relative standard deviation and since the relative standard deviation does not vary with sample size, it should be used whenever possible to estimate precision of analysis from replicate tests.

The accuracy of an analysis can be determined by several procedures. One common method is to analyze a "known" sample, such as a standard solution or a quality control check standard solution that may be commercially available or a laboratory-prepared standard solution made from a neat compound, and compare the test results with the true values (theoretically expected values). Such samples must be subjected to all analytical steps, including sample extraction, digestion, or concentration, similar to regular samples. Alternatively, accuracy may be estimated from the recovery of a known standard solution *spiked* or added into the sample in which a known amount of the same substance that is to be tested is added to an aliquot of the sample, usually as a solution, prior to the analysis. The concentration of the analyte in the spiked solution of the sample is then measured. The percent spike recovery is then calculated. A correction for the bias in the analytical procedure can then be made, based on the percent spike recovery. However, in most routine analysis such bias correction is not required. Percent spike-recovery then may be calculated as follows:

Recovery, % = (measured concentration)/(theoretical concentration) × 100%

The percent spike recovery to measure the accuracy of analysis may also be determined by the U.S. Environmental Protection Agency (EPA) method often used in environmental analysis:

Recovery, % = $[100(X_s - X_u)]/K$

In this equation, X_s is the measured value for the spiked sample and X_u is the measured value for the unspiked sample adjusted for the dilution of the spike and K is the known value of spike in the sample.

9.2.2.5.2 Quality Assurance

Quality assurance is an umbrella term that is correctly applied to everything that the laboratory does to assure product reliability. As the product of a laboratory is information, anything that is done to improve the reliability of the generated information falls under quality assurance.

Quality assurance includes all the quality controls, the generation of expectations (acceptance limits) from the quality controls, plus a great number of other activities such as (1) analyst training and certification, (2) data review and evaluation, (3) preparation of final reports of analysis, (4) information given to clients about tests that are needed to fulfill regulatory requirements, (5) use of the appropriate tests in the laboratory, (6) obtaining and maintaining laboratory certifications/accreditations, (7) conducting internal and external audits, (8) preparing responses to the audit results, (9) the receipt, storage, and tracking of samples, and (10) tracking the acquisition of standards and reagents.

Thus, the objective of quality assurance (usually in the form of a *quality assurance plan*) is to obtain reliable and accurate analytical results that may be stated with a high level of confidence (statistically), so that such results are legally defensible. The key features of any plan involve essentially documentation and record keeping. In short, quality assurance program involves documentation of sample collection for testing, the receipt of samples in the laboratory, and their transfer to the individuals who perform the analyses. The information is recorded on chain-of-custody forms stating the dates and times along with the names and signatures of individuals who carry out these tasks. Also, other pertinent information is recorded, such as any preservatives added to the sample to prevent degradation of test analytes, the temperature at which the sample is stored, the temperature to which the sample is brought prior to its analysis, the nature of the container (which may affect the stability of the sample), and its holding time prior to testing.

In fact, the performance of quality control is just one small aspect of the quality assurance program. The functions of quality assurance are embodied in the terms *analytically valid* and (in these days of perpetual litigation) *legally defensible.*

Analytically valid means that the target analyte has been (1) correctly identified and (2) quantified using fully calibrated tests. In addition the sensitivity of the test (method detection limit) has been established. And the analysts have demonstrated that they are capable of performing the test. The accuracy and precision of the test on the particular sample must also have been determined and the possibility of false positive and false negative results has been evaluated through performance of blanks and other test-specific interference procedures.

9.2.2.5.3 Method Detection Limit

The method detection limit (MDL) is the smallest quantity or concentration of a substance that the instrument can measure (Patnaik, 2004). It is related to the instrument detection limit (IDL) that depends on the type of instrument and its sensitivity, and on the physical and chemical properties of the test substance.

The method detection limit is, in reality, a statistical concept that is applicable only in trace analysis of certain type of substances, such as organic pollutants by gas chromatographic methods. The method detection limit measures the minimum detection limit of the method and involves all analytical steps including sample extraction, concentration, and determination by an analytical instrument. Unlike the instrument detection limit, the method detection limit it is not confined only to the detection limit of the instrument.

In the environmental analysis of organic pollutants, the method detection limit is the minimum concentration of a substance that can be measured and reported with 99% confidence that the analyte concentration is greater than zero and is determined from the analysis of a sample in a given matrix containing the analyte. For determination of the method detection limit, several replicate analyses are performed at the concentration level of the instrument detection limit or at a level equivalent to two to five times the background noise level. The standard deviation of the replicate tests is found. The method detection limit is determined by multiplying the standard derivation by the *t-factor*.

In environmental analysis, however, periodic determination of the method detection limit (e.g., once per year or with any change in personnel, location, or instrument) is part of the quality control requirement.

9.3 Sampling in the Field

Once the spill has occurred, the environmental analyst and bioremediation specialist have to identify the nature of the spilled material and make deductions about the potential effects on the ecosystem(s) (Smith, 1999). Although crude oil itself and its various products are complex mixtures of many organic chemicals (Chapter 3), the predominance of one particular chemical or one particular class of chemicals may offer the environmental analyst or scientist an opportunity to monitor and study levels of pollutants in the ecosystem.

However, all analytical methods and the resulting data are severely affected by spatial heterogeneity which requires using different analytical approaches and techniques so that topological and geographical properties are taken into account. Complex issues arise in spatial analysis, including the limitations of mathematical knowledge, the assumptions required by existing statistical techniques, and problems in computer based calculations. Hence, it is essential that a well-thought out sampling plan be designed according to valid statistical principles involving randomization and replication of treatments in order to ensure that monitored results reflect reality in such a highly heterogeneous environment.

Thus, sampling is the first step in environmental monitoring. Environmental monitoring system (Rahimova et al., 2020): (1) information on the state of the environment, (2) observation and information on the causes of the spill, and (3) the method should result in the collection and analysis of the information on the implementation of the process objectives.

The ability to collect and preserve a sample that is representative of the site is a critically important step (Dean, 2003). Obtaining environmental samples is always a challenge, due to heterogeneity of different sample matrixes. For soil sampling, four variables are generally considered. These are the spatial distribution of samples across the landscape, the depth of sampling, the time of year when samples are taken and how often an area is sampled. Presently, the sampling methodologies on oil spills vary tremendously. This is because of the challenges in sampling solids for environmental analysis to collect a relatively small portion of the sample that accurately represents the composition of the whole (Speight, 2005). Therefore imprecise, inaccurate, and inconsistent soil sampling techniques are a major source of uncertainty in calculations.

In summary, sampling is a method that allows us to get information for monitoring purposes, as well as for research provided the smiling activity is performed in a manner that is applicable to the site and the spilled material, and can be documented as producing a representative sample using a sampling log. Also, there must be a means to identify the sample history as carefully as possible so that each sample is tracked and defined in terms of source and activity (Table 9.1). The accuracy of the data from any subsequent procedures and tests for which the sample is used will be placed beyond a reasonable doubt. If this is not the case, that data should be re-evaluated without personal bias and (if unsuitability prevails) discarded as being inadequate for use.

Representative samples are prerequisite for the laboratory evaluation of any type of environmental sample and many precautions are required in obtaining and handling representative samples (Speight, 2005; Speight and Arjoon, 2012; Speight, 2015; Arjoon and Speight, 2018). The ability to collect and preserve a sample that is representative of the site is a critically important step (Dean, 1998; Weisman, 1998; Dean, 2003; Patnaik, 2004). Obtaining representative environmental samples is always a challenge due to the heterogeneity of different sample matrices. Additional difficulties are encountered with crude oil hydrocarbon derivatives due to the wide range in volatility, solubility, biodegradation, and adsorption potential of individual constituents. And the procedures used for sample collection and preparation must be legally defensible.

Using soil as an example of the medium to be sampled, the samples can be collected using a variety of methods depending on the depth of the desired sample, the type of sample required (disturbed vs. undisturbed), and the soil type. These sampling techniques range from subsampling or two-stage sampling, double sampling, composite sampling, random sampling, and stratified sampling. But the basis for most sampling plans in environmental sampling is the concept of random sampling. With random sampling each sample point within the site has an equal probability of being selected (US EPA, 1992). Therefore, in the selection of a site, the main consideration in physical soil sampling is the method of randomization of the soil samples because the sample distribution usually depends on the degree of variability in a given area. For those situations where there is inadequate information for developing a conceptual model for a site or for stratifying the site, it may be necessary to use a random sampling design.

In terms of sampling environmental items such as soils, the sample must be chosen from a population for investigation. A *random sample* is one chosen by a method involving an unpredictable component. Random sampling can also refer to taking a number of independent observations from the same probability distribution, without involving any real population. A *probability sample* is one in which each item has a known probability of being in the sample. As part of the sampling protocol, a simple random sample is selected so that all samples of the same size have an equal chance of being selected from the population. Alternatively, a self-weighting sample (also known as an *equal probability of selection method*, EPSEM) is one in which every individual, or object, in the population of interest has an equal opportunity of being selected for the sample. Simple random samples are self-weighting. On the other hand, *stratified sampling* involves selecting independent samples from a number of subpopulations, group or strata within the population. Great gains in efficiency are sometimes possible from judicious stratification. Finally, *cluster sampling* involves selecting the sample units in groups. The analysis of cluster samples must take into account the intra-cluster correlation which reflects the fact that units in the same cluster are likely to be more similar than two units picked at random. Whatever method is employed, the sample usually will not be completely representative of the population from which it was drawn—this random variation in the results is known as sampling error. In the case of random samples, mathematical theory is available to assess the sampling error. Thus, estimates obtained from random samples can be accompanied by measures of the uncertainty associated with the estimate. This can take the form of a standard error, or if the sample is large enough for the central limit theorem to take effect, confidence may be calculated.

The random sampling procedure is the basis for all probability sampling techniques used in soil sampling and serves as a reference point from which modifications to increase the efficiency of sampling are evaluated. Where there is a lack of information, as with oil spills, the simple random sampling design is the only design other than the systematic grid that can be used. The simple random sampling is the basis for all probability sampling techniques used in soil sampling and serves as a reference point from which modifications to increase the efficiency of sampling are evaluated.

Efficient sampling has a number of benefits for researchers. However, it is just as important as to know where to sample. Each step of an analysis contributes random error that affects the over-all standard deviation from the actual results.

9.3.1 Sampling Strategies

The goals of a sampling design program can vary widely as different objectives require different sampling strategies. The cost of the study is also a factor to consider as if more samples are analyzed, more money, time, and resources is needed. Although, it allows for a higher precision and accuracy of the results. The steps of the sampling design process are:

- Review the systematic planning outputs—the sampling objectives need to be stated clearly and constraints regarding schedule, funding, special equipment and facilities, and human resources must be considered.

- Develop general sampling design alternatives—Decide whether the approach will involve episodic sampling events (where a sampling design is established and all data for that phase are collected according to that design) or an adaptive strategy (where a sampling protocol is established and sampling units are selected in the field, in accordance with the protocol, based on results from previous sampling for that phase)

- Formulate mathematical expressions for the performance and cost of each design alternative— For each design, develop the necessary statistical model or mathematical formulae needed to determine the performance of the design, in terms of the desired statistical power or width of the confidence interval.

- Determine the sample size that satisfies the performance criteria and constraints—Calculate the optimal sample size (and sample allocation, for stratified designs or other more complex designs).

- Choose the most resource-effective design—Consider the advantages, disadvantages, and trade-offs between performance and cost among designs that satisfy performance specifications and constraints. Consider practical issues, schedule and budget risks as well as health and safety risks to project Final EPA QA/G-5S 24 December 2002 personnel and the community, and any other relevant issues of concern to those involved with the project.

- Document the design in the QA Project Plan—Provide details on how the design should be implemented, contingency plans if unexpected conditions or events arise in the field, and quality assurance (QA) and quality control (QC) that will be performed to detect and correct problems and ensure defensible results.

9.3.2 Acquiring a Representative Sample

The first consideration for any information involving environmental analytical data is whether the samples adequately represent the site being investigated.

If a study generates data with very large errors, then the uncertainty in the results may prevent one from making a sound scientific conclusion. There are many possible sources of uncertainty to consider when processing or analyzing a sample, and the study designs or an analyst's expertise to identify and avoid as many of those sources of error as possible. Many of those error mechanisms can occur when obtaining any types of samples. In order to identify an appropriate sampling method, one must first understand the different types of errors that can arise from collecting different types of samples.

- Random Sampling: Although, a major advantage is its simplicity and lack of bias, bias can still occur under certain circumstances and there is difficulty gaining access to a true of a larger population or contaminant under investigation.

- Systematic Sampling: The samples can represent an entire population quickly and easily and have a higher level of precision than other randomized methods. It also reduces the potential for bias in the information.

- Judgmental Sampling: Judgmental sampling designs involve the selection of sampling units based on expert knowledge or professional judgment.

- Stratified sampling is the process in which the target population is separated into non-overlapping strata, or subpopulations that are known or thought to be more homogeneous (relative to the environmental medium or the contaminant). The strata may be chosen on the basis of spatial or temporal proximity of the units, or on the basis of preexisting information or professional judgment about the site or process.

- Haphazard sample involves the creation of a random sample by haphazardly by choosing criteria and location in order to try to recreate true randomness. It does not usually work, because of selection bias: where you knowingly or unknowingly create unrepresentative samples.

- Continuous Monitoring: An ideal approach for some environmental measurements. A sample collected at a specific location at a certain point in time (grab sample) would have provided information about the environmental dangers only if a sample happened to be taken at the time the release was taking place, as contaminants that might not have been there when sampling was performed.

By creating an appropriate sample design to the environment it is needed, allows for the collection of representative data of the target population and collection of beneficial data.

Once the sample preparation is complete, there are several approaches to the analysis of crude oil constituents in the water and soil: (1) leachability or toxicity of the sample, (2) the amount of the hydrocarbon derivatives in the sample, (3) crude oil group analysis, and (4) fractional analysis of the sample (Speight, 2005, 2014, 2015). These methods measure different constituents of the crude oil and/or the crude oil product that might be present in crude oil-contaminated environmental media.

9.4 Group Analyses

Crude oil group analyses are conducted to determine amounts of the crude oil compound classes (e.g., saturates, aromatics, and resin constituents, i.e., polar constituents) present in crude oil-contaminated samples. This type of measurement is sometimes used to identify fuel type or to track plumes. It may be particularly useful for higher boiling products, such as asphalt. Group type test methods include multidimensional gas chromatography (not often used for environmental samples), high performance liquid chromatography (HPLC), and thin layer chromatography (TLC) (Miller, 2000; Patnaik, 2004).

Test methods that analyze individual compounds (e.g., benzene-toluene-ethylbenzene-xylene mixtures and polynuclear aromatic hydrocarbon derivatives) are generally applied to detect the presence of an additive or to provide concentration data needed to estimate environmental and health risks that are associated with individual compounds. Common constituent measurement techniques include gas chromatography with second column confirmation, gas chromatography with multiple selective detectors and gas chromatography with mass spectrometry detection (GC/MS) (EPA 8240).

Many common environmental methods measure individual crude oil constituents or *target compound* rather than the whole signal from the total crude oil hydrocarbon derivatives. Each method measures a suite of compounds selected because of their toxicity and common use in industry.

For organic compounds, there are three series of target compound methods that must be used for regulatory purposes:

1. EPA 500 series: Organic Compounds in Drinking Water, as regulated under the Safe Drinking Water Act.

2. EPA 600 series: Methods for Organic Chemical Analysis of Municipal and Industrial Wastewater, as regulated under the Clean Water Act.

3. SW-846 series: Test Methods for Evaluating Solid Waste: Physical/Chemical Methods, as promulgated by the US EPA, Office of Solid Waste and Emergency Response.

The 500 and 600 series methods provide parameters and conditions for the analysis of drinking water and wastewater, respectively. One method (EPA SW-846) is focused on the analysis of nearly all matrices including industrial waste, soil, sludge, sediment, and water miscible and non-water miscible wastes. It also provides for the analysis of groundwater and wastewater but is not used to evaluate compliance of public drinking water systems.

Selection of one method over another is often dictated by the nature of the sample and the particular compliance or cleanup program for which the sample is being analyzed. It is essential to recognize that capabilities and requirements vary between methods when requesting any analytical method or suite of methods. Most compound-specific methods use a gas chromatographic selective detector, high performance liquid chromatography, or gas chromatography/mass spectrometry.

More correctly, group analytical methods are designed to separate hydrocarbon derivatives into categories, such as saturates, aromatics, resin constituents, and asphaltene constituents (SARA) or paraffin derivatives, iso-paraffin derivatives, naphthene derivatives, aromatic derivatives, and olefin derivatives (PIANO). These chromatographic, gas chromatographic, and high performance liquid chromatographic methods (HPLC) were developed for monitoring refinery processes or evaluating organic synthesis products. Column chromatographic methods that separate saturates from aromatic derivatives are often used as preparative steps for further analysis by gas chromatography/mass spectrometry. Thin layer chromatography is sometimes used as a screening technique for crude oil product identification.

9.4.1 Gas Chromatography

Gas chromatography (GC, often referred to as gas-liquid chromatography, GLC) uses the principle of a stationary phase and a mobile phase. Much attention has been paid to the various stationary phases and books have been written on the subject as it pertains to crude oil chemistry.

Briefly, gas-liquid chromatography (GLC) is a method for separating the volatile components of various mixtures (Altgelt and Gouw, 1975; Fowlis, 1995; Grob, 1995). The procedure is an efficient fractionating technique that is suitable for the quantitative analysis of mixtures when the possible constituents of the mixture are known.

The mobile phase is the carrier gas, and the gas selected has a bearing on the resolution. Nitrogen has very poor resolution ability, while helium or hydrogen are better choices with hydrogen being the best carrier gas for resolution. However, hydrogen is reactive and may not be compatible with all sets of target analytes. There is an optimum flow rate for each carrier gas to achieve maximum resolution. As the temperature of the oven increases, the flow rate of the gas changes due to thermal expansion of the gas. Most modern gas chromatographs are equipped with constant flow devices that change the gas valve settings as the temperature in the oven changes, so changing flow rates are no longer a concern. Once the flow is optimized at one temperature it is optimized for all temperatures.

Gas chromatographic methods are currently the preferred laboratory methods for measurement of the total hydrocarbon derivatives because they detect a broad range of hydrocarbon derivatives and provide both sensitivity and selectivity. In addition, identification and quantification of individual constituents of the total hydrocarbon mix is possible.

Methods based on gravimetric analysis are also simple and rapid but they suffer from the same limitations as infrared spectrometric methods. Gravimetric-based methods may be useful for oily sludge and wastewaters, which will present analytical difficulties for other more sensitive methods. Immunoassay methods for the measurement of total crude oil hydrocarbon are also popular for field testing because they offer a simple, quick technique for *in situ* quantification of the total crude oil hydrocarbon derivatives.

For methods based on gas chromatography, the total crude oil hydrocarbon derivatives fraction is defined as any chemicals extractable by a solvent or purge gas and detectable by gas chromatography/flame ionization detection (GC/FID) within a specified carbon range. The primary advantage of such methods is that they provide information about the type of crude oil in the sample in addition to measuring the amount. Identification of product type(s) is not always straightforward, however, and requires an experienced analyst of crude oil products (Sullivan and Johnson, 1993; Speight, 2014, 2015). Detection limits are method-dependent as well as matrix-dependent and can be as low as 0.5 mg/L in water or 10 mg/kg in soil.

The methods have in common a boiling point-type column and a flame ionization detector. For example, if gasoline is suspected to be the sole contaminant, the method will use purge/trap sample introduction.

Gas chromatography-based methods can be broadly used for different kinds of crude oil contamination but are most appropriate for detecting nonpolar hydrocarbon derivatives with carbon numbers between C6 and C25 or C36. Many lubricating oils contain molecules with more than 40 carbon atoms. In fact, crude oil itself contains molecules having more than 100 carbons or more. These high molecular weight hydrocarbon derivatives are outside the detection range of the more common gas chromatographic methods, but specialized gas chromatographs are capable of analyzing such high molecular weight constituents.

Accurate quantification depends on adjusting the chromatograph to reach as high a carbon number as possible, then running a calibration standard with the same carbon range as the sample. There should also be a check for mass discrimination, a tendency for higher molecular weight hydrocarbon derivatives to be retained in the injection port. If a sample is suspected to be heavy oil or to contain a mixture of light oil and heavy oil, the most appropriate method must be used.

Gravimetric or infrared methods are often preferred for high molecular weight samples. These methods can even be used as a check on gas chromatographic data if it is suspected that high molecular weight hydrocarbon derivatives are present but are not being detected.

Calibration standards vary. Most methods specify a gasoline calibration standard for volatile range total crude oil hydrocarbon derivatives and a diesel fuel #2 standard for extractable range total crude oil hydrocarbon derivatives. Some methods use synthetic mixtures for calibration. Because most methods are written for gasoline or diesel fuel, total crude oil hydrocarbon derivatives methods may have to be adjusted to measure contamination by heavier hydrocarbon derivatives (e.g., heavy fuel oil, lubricating oil, or crude oil). Such adjustments may entail use of a more aggressive solvent, a wider gas chromatographic window that allows detection of molecules containing up to C36 or more, and a different calibration standard that more closely resembles the constituents of the sample under investigation.

Gas chromatographic methods can be modified and fine-tuned so that they are suitable for measurement of specific crude oil products or group types. These modified methods can be particularly useful when there is information on the source of contamination, but method results should be interpreted with the clear understanding that a modified method was used for detection of a specific carbon range.

Interpretation of gas chromatographic data is often complicated and the analytical method should always be considered when interpreting concentration data. For example, a volatile range analysis may be very useful for quantifying total crude oil hydrocarbon derivatives at a gasoline release site, but a volatile range analysis will not detect the presence of lube oil constituents. In addition, a modified method that has been specifically selected for detection of gasoline-range organics at a gasoline-contaminated site may also detect hydrocarbon derivatives from other crude oil releases because fuel carbon ranges frequently overlap. Gasoline is found primarily in the volatile range. Diesel fuel falls primarily in an extractable range. Jet fuel overlaps both the volatile and semi-volatile ranges. However, the detection of different kinds of crude oil constituents does not necessarily indicate that there have been multiple releases at a site. Analyses of spilled waste oil will frequently detect the presence of gasoline, and sometimes diesel. This does not necessarily indicate multiple spills since all waste oil contain some fuel. As much as 10% of used motor oil can consist of gasoline.

If the type of contaminant is unknown, a fingerprint analysis can help in the identification procedure. A fingerprint or pattern recognition analysis is a direct injection analysis where the chromatogram is compared to chromatograms of reference materials.

For example, chromatograms of gasoline and diesel differ considerably but many hydrocarbon streams may have similar fingerprints. Diesel No. 2 and No. 2 fuel oil both have the same boiling point range and chromatographic fingerprint. A fingerprint can be used to conclusively identify a mixture when a known sample of that mixture or samples of the source materials of the mixture are available. Furthermore, as a fuel evaporates or biodegrades, its pattern can change so radically that identification becomes difficult (Bartha, 1986). Consequently, a gas chromatographic fingerprint is not a conclusive diagnostic tool. The methods must for total crude oil hydrocarbon analysis must stress calibration and quality control, while pattern recognition methods stress detail and comparability.

The gas chromatographic methods usually cannot quantitatively detect compounds below C6 because these compounds are highly volatile and interference can occur from the solvent peak. As much as 25% of fresh gasoline can be below C6 but the problem is reduced for weathered gasoline and/or diesel range contamination because most of the very volatile hydrocarbon derivatives ($< C6$) may no longer be present in the sample. Gas chromatographic methods may also be inefficient for quantification of polar constituents (nitrogen, oxygen, and sulfur containing molecules). Some of the polar constituents are too reactive to pass through a gas chromatograph and thus will not reach the detector for measurement.

Oxygenated gasoline is sometimes analyzed by GC-based methods but it should be noted that the efficiency of purge methods is lower for oxygenates such as ethers and alcohols because detector response to oxygenates is lower relative to hydrocarbon derivatives. Therefore the data will be biased slightly low for ether-containing fuels compared to equivalent amounts of traditional gasoline. Methanol and ethanol elute before hexane and, consequently, they are not quantified and may not even be detected due to co-elution with the solvent.

On the other hand, gas chromatographic methods may overestimate the concentration of total crude oil hydrocarbon derivatives in the sample due to the detection of non-crude oil compounds. Silica gel cleanup may help to remove this interference but may also remove some polar hydrocarbon derivatives.

Because crude oil is made up of so many isomers, many compounds, especially those with more than eight carbon atoms, co-elute with isomers of nearly the same boiling point. These unresolved compounds are referred to as the unresolved complex mixture. They are legitimately part of the crude oil signal, and unless otherwise specified, should be quantified. Quantifying such a mixture requires a baseline-to-baseline integration mode rather than a peak-to-peak integration mode. The baseline-to-baseline integration quantifies all of the crude oil constituents in the sample but the peak-to-peak integration only the individual resolved hydrocarbon derivatives (not including the unresolved complex mixture) are quantified.

For environmental analysis (Bruner, 1993), particularly the volatile samples such found in total crude oil hydrocarbon derivatives, the gas chromatograph is generally interfaced with a purge and trap system as described in the section on gas chromatographic methods. The photoionization detector works by bombarding compounds with ultraviolet (UV) light, generating a current of ions. Compounds with double carbon bonds, conjugated systems (multiple carbon double bonds arranged in a specific manner), and aromatic rings are easily ionized with the ultraviolet light generated by the Photoionization detector lamp, while most saturated compounds require higher energy radiation.

One method (EPA SW-846 Test Method 8020) that is suitable for volatile aromatic compounds is often referred to as benzene-toluene-ethylbenzene-xylene analysis, though the method includes other volatile aromatic derivatives.

For the semi-volatile constituents of crude oil and crude oil products, the gas chromatograph is generally equipped with either a packed or capillary column. Either neat or diluted organic liquids can be analyzed via direct injection, and compounds are separated during movement down the column. The flame ionization detector uses a hydrogen-fueled flame to ionize compounds that reach the detector. For polynuclear aromatic hydrocarbon derivatives, a method is available (EPA SW-846 Test Method 8100) in which injection of sample extracts directly onto the column is the preferred method for sample introduction for this packed-column method.

In the method, a gas chromatography-flame ionization detector system can be used for the separation and detection of nonpolar organic constituents and, in addition, semi-volatile constituents can also be detected using the system.

9.4.2 Gas Chromatography-Mass Spectrometry

A gas chromatography-mass spectrometry system is used to measure concentrations of target volatile and semi-volatile crude oil constituents. It is not typically used to measure the amount of total crude oil hydrocarbon derivatives (EPA SW-846 Test Method 8260D). This technique gives the more complete information available from the total ion chromatogram and the full-mass-range spectrum of each compound. The technique is sometimes used to quantify compounds present at very low concentrations in a complex hydrocarbon matrix. It can be used if the target compound's spectrum has a prominent fragment ion at a mass that distinguishes it from the rest of the hydrocarbon compounds.

The most common method for GC/MS analysis of semi-volatile compounds (EPA SW-846 Test Method 8270) includes 16 polycyclic aromatic compounds, some of which commonly occur in middle distillate to heavy crude oil products. The method also quantifies non-hydrocarbon derivatives such as phenol derivatives and cresol derivatives that sometime occur in crude oil product or in weathered crude oils and crude oil products.

To reduce the possibility of false positives, the intensities of one to three selected ions are compared to the intensity of a unique target ion of the same spectrum. The sample ratios are compared to the ratios of a standard.

9.4.3 High Performance Liquid Chromatography

A high performance liquid chromatography system can be used to measure concentrations of target semi-volatile and non-volatile crude oil constituents. The system only requires that the sample be dissolved in a solvent compatible with those used in the separation. The detector most often used in crude oil environmental analysis is the fluorescence detector which is particularly sensitive for the detection of aromatic constituents, especially polynuclear aromatic hydrocarbon derivatives.

In the method, polynuclear aromatic hydrocarbon derivatives are extracted from the sample matrix with a suitable solvent, which is then injected into the chromatographic system. Usually the extract must be filtered because fine particulate matter can collect on the inlet frit of the column, resulting in high back-pressures and eventual plugging of the column. For most hydrocarbon analyses, reverse phase high performance liquid chromatography (i.e., using a nonpolar column packing with a more polar mobile phase) is used. The most common bonded phase is the octadecyl (C18) phase. The mobile phase is commonly aqueous mixtures of either acetonitrile or methanol.

After the chromatographic separation, the analytes flow through the cell of the detector. A fluorescence detector shines light of a particular wavelength (the excitation wavelength) into the cell. Fluorescent compounds absorb light and reemit light of other, higher wavelengths (emission wavelengths). The emission wavelengths of a molecule are mainly determined by its structure. For polynuclear aromatic hydrocarbon derivatives, the emission wavelengths are mainly determined by the arrangement of the rings and vary greatly between isomers.

Some of the polynuclear aromatic hydrocarbon derivatives such as phenanthrene, pyrene, and benzo(g,h,i)perylene, are commonly seen in products boiling in the middle to heavy distillate range. In a method for their detection and analysis (EPA SW-846 Test Method 8310) an octadecyl column and an aqueous acetonitrile mobile phase are used. Analytes are excited at 280 nm and detected at emission wavelengths of > 389 nm. Naphthalene, acenaphthene, and fluorene must be detected by a less-sensitive UV detector because they emit light at wavelengths below 389 nm. Acenaphthylene is also detected by UV detector.

The methods using fluorescence detection will measure any compounds that elute in the appropriate retention time range and which fluoresce at the targeted emission wavelength(s) (Falla Sotelo et al., 2008). In the case of one method (EPA SW-846 Test Method 8310), the excitation wavelength excites most aromatic compounds. These include the target compounds and also many aromatic derivatives, such as (1) the alkyl aromatic derivatives, phenol derivatives, (2) the aniline derivatives, (3) compounds containing the pyrrole ring system, such as indole and carbazole derivatives, (4) the pyridine ring system, such as quinoline and acridine, (5) the furan ring system, such as benzofuran and naphthofuran, and (6) the thiophene ring system, such as benzothiophene, and naphthothiophene.

9.4.4 Immunoassay

A number of different testing kits based on immunoassay technology are available for rapid field determination of certain groups of compounds such as benzene-toluene-ethylbenzene-xylene (EPA SW-846 4030, 2015) or polynuclear aromatic hydrocarbon derivatives (EPA SW-846 4035, 2015). The immunoassay screening kits are self-contained portable field kits that include components for sample preparation, instrumentation to read assay results, and immunoassay reagents.

The quality of the analysis of polynuclear aromatic hydrocarbon derivatives is often dependent on the extraction efficiency. The presence of clay minerals in soil lowers the ability to extract polynuclear aromatic hydrocarbon derivatives.

9.4.5 Infrared Spectroscopy

Infrared methods measure the absorbance of the C-H bond and most methods typically measure the absorbance at a single frequency (usually 2930 cm^{-1}) that corresponds to the stretching of aliphatic methylene (CH$_2$) groups. Some methods use multiple frequencies including 2960 cm^{-1} (CH$_3$ groups) and 2900 to 3000 cm^{-1} (aromatic C-H bonds).

Therefore, for infrared spectroscopic methods, the total crude oil hydrocarbon derivatives is any chemicals extracted by a solvent that is not removed by silica gel and can be detected by infrared spectroscopy at a specified wavelength. The primary advantage of the infrared-based methods is that they are simple and rapid. Detection limits (such as EPA SW-846 Test Method 418.1) are approximately 1 mg/L in water and 10 mg/kg in soil. However, the infrared method(s) can often suffers from poor accuracy and precision, especially for heterogeneous soil samples. Also, the infrared methods give no information on the type of fuel present in the sample and there is little, often no, information about the presence or absence of toxic molecules, and no specific information about potential risk associated with the contamination.

Samples are extracted with a suitable solvent (i.e., a solvent with no C-H bonds) and biogenic polar materials are removed with silica gel. Some polar crude oil constituents may be removed as part of the silica gel cleanup. The absorbance of the silica gel eluate is measured at the specified frequency and compared to the absorbance of a standard or standards of known crude oil hydrocarbon concentration. The absorbance is a measurement of the sum of all the compounds contributing to the result. However, infrared methods cannot provide information on the type of hydrocarbon contamination.

9.4.6 Thin Layer Chromatography

The thin-layer chromatography method has the advantage of separating constituents that are too high boiling to pass through a gas chromatograph.

The saturated hydrocarbon derivatives that are present in distillates such as seen in diesel fuel (or kerosene) are readily mobile in a hexane mobile phase. On the other hand, polar compounds such as ketone derivatives or alcohol derivatives travel a smaller distance in hexane than saturated hydrocarbon derivatives. When the aromatic content of a sample is high, as with bunker C fuel oil, the detection limit can be near 100 ppm.

The method is considered to be a qualitative and useful procedure for rapid sample screening.

9.5 Gravimetric Analysis

Gravimetric methods measure all chemicals that are extractable by a solvent, not removed during solvent evaporation, and capable of being weighed. Some gravimetric methods include a cleanup step to remove biogenic material. The advantage of gravimetric methods is that they are simple and rapid. Detection limits are approximately 5–10 mg/L in water and 50 mg/kg in soil.

However, gravimetric methods are not suitable for measurement of low boiling hydrocarbon derivatives that volatilize at temperatures below 70 to 85°C (160 to 185°F). They are recommended for use with (1) oily sludge, (2) for samples containing heavy molecular weight hydrocarbon derivatives, or (3) for aqueous samples when hexane is preferred as the solvent.

Gravimetric methods give no information on the type of fuel present, no information about the presence or absence of toxic compounds, and no specific information about potential risk associated with the contamination.

In the method(s), crude oil constituents are extracted into a suitable solvent. The solvent is evaporated and the residue is weighed.

There are a variety of gravimetric oil and grease methods suitable for testing water and soil samples (e.g., EPA SW-846 9070, EPA 413.1, EPA 9071). Technically the result is an oil and grease result because no cleanup step is used. One method (EPA 9071) is used to recover low levels of oil and grease by chemically drying a wet sludge sample and then extracting it using Soxhlet apparatus. Results are reported on a dry-weight basis. The method is also used when relatively polar high molecular weight crude oil fractions are present, or when the levels of non-volatile grease challenge the solubility limit of the solvent. Specifically, the method (EPA SW-846 9071) is suitable for biological lipids, mineral hydrocarbon derivatives, and some industrial wastewater.

Gravimetric methods for oil and grease (e.g., EPA SW-846 9071) measure anything that dissolves in the solvent and remains after solvent evaporation. These substances include hydrocarbon derivatives, vegetable oils, animal fats, waxes, soaps, greases and related biogenic material.

Gravimetric methods for total crude oil hydrocarbon derivatives (EPA SW-846 Test Method 1664) measure any material that dissolves in the solvent and remains after silica gel treatment and solvent evaporation. This method is a liquid/liquid extraction gravimetric procedure that employs n-hexane as the extraction solvent. However, n-hexane is a poor solvent for the high molecular weight constituents of crude oil and some crude oil products (such as, for example, asphalt) (Speight, 2014, 2015).

All gravimetric methods measure any suspended solids that are not filtered from solution, including bacterial degradation products and clay fines. Method 9071 specifies using cotton or glass wool as a filter.

Because extracts are heated to remove solvent, these methods are not suitable for measurement of low boiling low molecular weight hydrocarbon derivatives (i.e., hydrocarbon derivatives having less than fifteen carbon atoms) that volatilize at temperatures below 70 to 85°C (158 to 185°F). Liquid fuels, from gasoline through No. 2 fuel oil, lose volatile constituents are during solvent removal. In addition, soil results that are reported on a dry-weight basis suffer from potential losses of lower boiling hydrocarbon constituents during moisture determination where the matrix is dried at a temperature on the order of 103 to 105°C (217 to 221°F) for several hours in an oven.

9.6 Microbiological Analysis

Microorganisms can be studied using a range of technologies. Each colony is then referred to as a colony forming unit (CFU) and the result of the analysis expressed in terms of CFU/mL or CFU/g of sample. Bacterial growth can be measured by direct cell counts and viable cell counts.

Selection of the right laboratory and the appropriate testing suite are key considerations for anyone who requires microbiological testing of samples, whether the tests are carried out internally, or by a third-party laboratory. It is very important that samples for microbiological testing are not contaminated during the sampling procedure; hence, the process must be conducted aseptically. This means that the aliquot for microbiological testing must be taken first, if a range of analyses is to be conducted on any sample. Samples then need to be kept in the same state, before testing. In addition, the location, type, time and date, relating to a sample, must all be clearly recorded and logged.

Microorganisms can be studied using a range of technologies. The most widely used methods in the study of microorganisms present in different samples is their growth in a liquid medium, followed by dilution of the sample and plating on a solid agar medium. The theory is that one colony arises on a solid agar medium from one microbial cell. Each colony is then referred to as a colony forming unit (CFU) and the result of the analysis expressed in terms of CFU/mL or CFU/g of sample. Except for providing an estimate of bacterial numbers, this procedure allows obtaining pure culture isolates. Bacterial growth can be measured by direct cell counts and viable cell counts.

Direct microscopic count is one of the most rapid methods of determining the number of cells in a suspension, but generally does not distinguish living and dead cells. Viable cell counts are used to quantify the number of cells capable of multiplying. Plate counts and membrane filtration both measure the concentration of cells by determining the number of colonies that arise from a sample added to an agar plate. The two different plating methods pour plate and spread plate, differ in how the suspension of microorganisms is applied to the agar plate. A simple count of the colonies determines how many cells were in the initial sample (Briški and Vuković Domanovac, 2017).

A merit to this method is that the multiplication of microorganisms can easily be seen with the naked eye, in solid selective media (for the microbiological testing of liquids) or liquid media (for the microbiological examination of solids). However, it requires a lot of work on the part of a skilled technician and the need for a laboratory readily equipped to sterilize the material to use while maintaining sterile conditions. The lack of sterility in fact leads to microbial pollution unrelated to the test sample with a consequent loss of significance in the analysis.

A common technique for the isolation of crude oil-utilizing bacteria from oil-contaminated soil is the vapor phase transfer. A mineral salts agar is inoculated with known aliquot of the soil suspension and a filter paper soaked with the crude oil or other hydrocarbons is placed in the lid of the Petri dish. According to Philp et al. (2005), an obvious limitation of this technique is that only a proportion of the total bacterial population that can utilize the vapor phase hydrocarbons can be isolated. This method is likely to underestimate the total crude oil-degrading populations. However, a further criticism of this method is that it cannot be assumed that colonies that appear on the agar plates are really oil degraders. This is because agar can contain some impurities to allow microbial growth; agar can also absorb volatile nutrients from the air in amounts sufficient to support the growth of many non-oil degrading bacteria (Chikere et al., 2011).

A merit to this method is that the multiplication of microorganisms can easily be seen with the naked eye, in solid selective media (for the microbiological testing of liquids) or liquid media (for the microbiological examination of solids). However, it requires a lot of work on the part of a skilled technician and the need for a laboratory readily equipped to sterilize the material to use while maintaining sterile conditions.

9.6.1 Chemical Analysis of Nutrients

Nutrients are necessary for microbial growth hence bioremediation. Microbes grow more slowly when nutrients are limited. Soil respiration, the emission of carbon dioxide by microbial respiration, is a good indicator of microbial biomass, but may not correlate well with SOC or total carbon. It can provide an indication of the microbial activity in the soil which can be an indicator of soil fertility.

Prior to any analyses for total organic carbon (TOC), the soil or sediment sample must be collected and properly handled. However, the total carbon in a sample is made up of various types of carbon-containing compounds. Thus:

Total Carbon = Inorganic Carbon + Organic Carbon

Sample preparation techniques can range from simply weighing the sample prior to analysis through wet chemistry digestion with strong acids. Costs associated with sample preparation are generally low and consist mainly of replacing consumable chemicals. There are two methods in the literature for the structural characterization of organic carbon forms in soils and sediments. One of these qualitative methods is based on nuclear magnetic resonance (NMR) spectroscopy and the other on diffuse reflectance infrared Fourier transform (DRIFT) spectroscopy.

NMR method is based on spectral lines of different atomic nuclei that are excited when a strong magnetic field and a radiofrequency transmitter are applied (Schumacher, 2002). The sample is placed in a magnetic field and the NMR signal is produced by excitation of the nuclei sample with radio waves into nuclear magnetic resonance, which is detected with sensitive radio receivers. The advantage of NMR techniques is that no extraction of organic matter is needed. However, the NMR methods are expensive and time-consuming.

Diffuse reflectance infrared Fourier transform spectroscopy allows collecting spectra from powdered samples with a minimum of sample preparation required. The IR radiation is directed onto the sample and the light reflected by diffuse scattering is collected with a mirror outside the cell. An advantage is the ability to analyze almost any sample form, solid, liquid or gas, as received. It provides a convenient means for obtaining mid-infrared spectra of soils.

There are several factors that one must consider when selecting a method for the determination of total organic carbon. These factors include the ease of use, health and safety concerns, cost, sample throughput, and comparability to standard reference methods. These factors are a concern for both the sample preparation and sample quantitation phases of the determination of the total organic carbon.

For phosphorus, many chemical solutions have been proposed with water probably being the first extracting liquid that researchers applied to measure phosphorus in soils. When extracting solution is added to soil, there are four basic reactions by which P is removed from the solid phase: (1) dissolving action of acids, (2) anion replacement to enhance desorption, (3) complexing of cations binding P, and (4) hydrolysis of cations binding P. Therefore, the selection of a P soil test depends on the chemical forms of P in the soil. For phosphorus in water, the following EPA methods can be used:

1. Method 365.1: Determination Of Phosphorus By Semi-Automated Colorimetry

2. Method 365.2: Phosphorus, All Forms (Colorimetric, Ascorbic Acid, Single Reagent)

3. Method 365.3: Phosphorous, All Forms (Colorimetric, Ascorbic Acid, Two Reagent)

4. Method 365.4: Phosphorous, Total (Colorimetric, Automated, Block Digester AA II)

5. Method 365.5: Determination of Orthophosphate in Estuarine and Coastal Waters by Automated Colorimetric Analysis

Ammonium molybdate and antimony potassium tartrate react in an acid medium with dilute solutions of phosphorus to form an antimony-phospho-molybdate complex, which is reduced with ascorbic acid to form an intense blue-colored complex and the absorbance of the complex is measured.

The EPA-approved method for measuring total orthophosphate is known as the ascorbic acid method.

9.6.2 Chemical Analysis of Crude Oil and Oil Constituents

The objective of oil chemical analysis is to characterize the environmentally important constituents (such as toxic PAHs and their alkylated homologs) in crude oil and in crude oil products, and to determine their concentrations. In addition, to characterize the major (such as n-alkane derivatives and isoprenoids) and minor (such as biomarker triterpane and sterane compounds) constituents in oil (Figure 9.1). The analytical data and results will provide essential information to document oil exposure pathways, to determine extent and degree of oiling, to evaluate the long-term impact of spilled oil, to estimate recoverability of the injured resources, and to suggest effective clean-up strategies.

There now, exist certified laboratories that use certified crude oil hydrocarbon measurement techniques (Speight, 2015). These modern techniques include gas chromatography (GC), mass spectrometry (MS), infrared spectroscopy (IR), ultraviolet (UV) and fluorescence spectroscopy, supercritical fluid chromatography (SFC), as well as hyphenated techniques such as GC/MS, GC/FTIR, and SFC/GC. For optimization, it is important to match the detector to the application. In some cases, spectroscopic methods are employed to determine hydrocarbon types.

Typically, the objective of oil chemical analysis is to characterize the environmentally important constituents (such as toxic polynuclear aromatic hydrocarbon derivatives and their alkylated homologs) in the crude oil or the crude oil product, and to determine the concentration of these constituents. In addition, to characterize the major (such as n-alkanes and isoprenoids) and minor (such as biomarker triterpane and sterane compounds) constituents in oil. Assessing the damage by crude oil or crude oil products when spilled into environment and natural resources (water, soil, and biological resources) requires the design of appropriate and reliable chemical analytical methods for oil samples collected in the study area. The analytical data and results will provide essential information to document oil exposure pathways, to determine extent and degree of oiling, to evaluate the long-term impact of spilled oil, to estimate recoverability of the injured resources, and to suggest effective clean-up strategies.

Because of the wide range of chemical and physical properties, a wide range of tests have been (and continue to be) developed to provide an indication of the means by which a particular feedstock should be processed. The primary method for the analysis of crude oil and crude oil products, as well as for many chemicals in the environment, is gas chromatography. It has become an important analytical tool in virtually every phase of the crude oil industry, from exploration of crude oil and refining of finished products to research on new petrochemicals. Gas chromatography is used to analyze both finished products and in-process samples. This comprise an injection system to add the sample, a chromatography column to allow the components to separate and a detector to sense when a component is exiting the system. There is a range of detectors available to the crude oil industry that makes gas chromatography flexible.

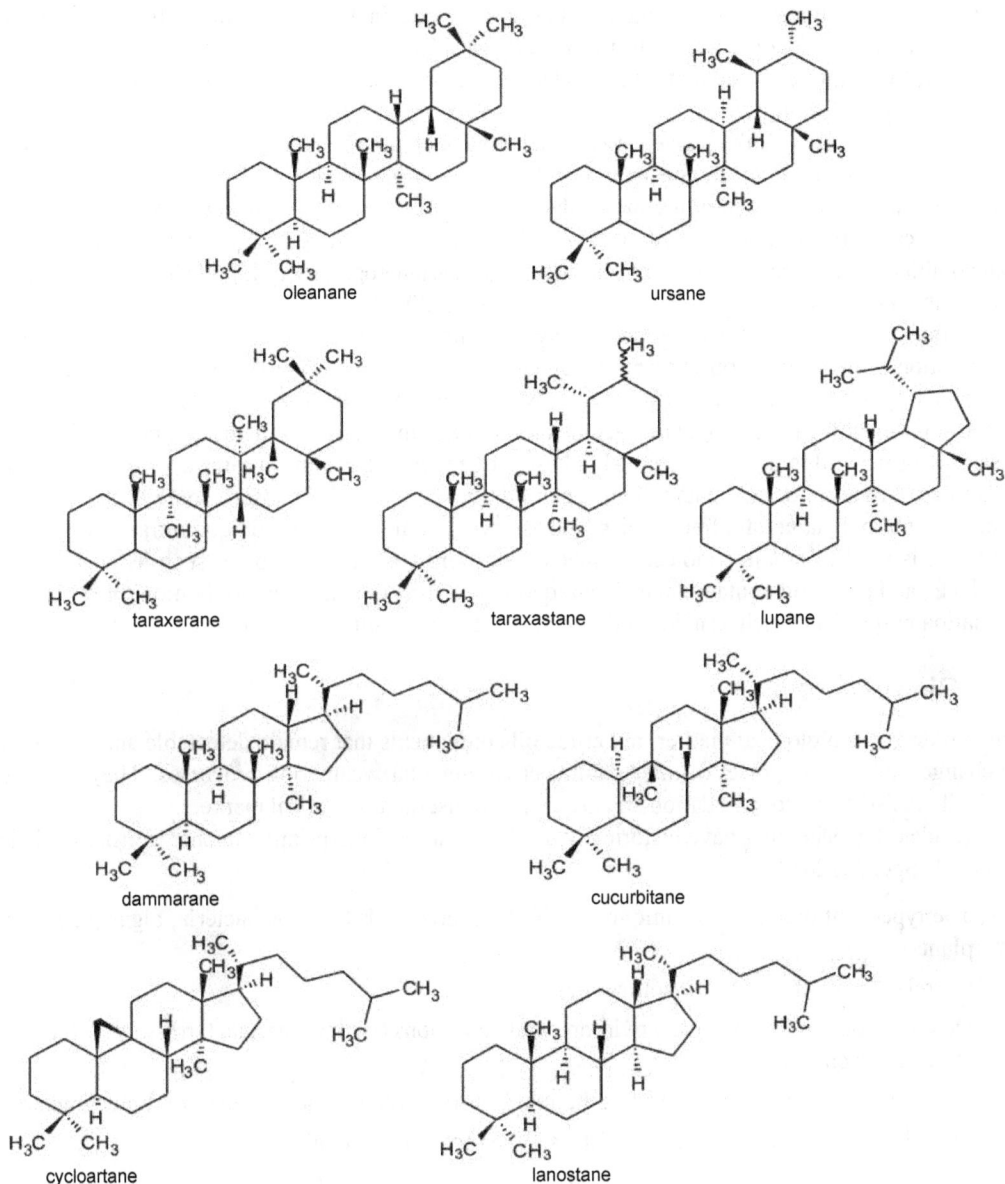

Figure 9.1: Chemical Structures of the Main Subclasses of Triterpenes (Nazaruk and Borzym-Kluczyk, 2015).

For optimization, it is important to match the detector to the application. Gas chromatography uses several types of detectors that are compatible with the crude oil industry: flame photometric detectors for sulphur in residues, photoionization detectors are used for benzene, toluene, and the xylene isomers and other aromatics, thermal conductivity detectors can be used to measure inorganic gases in workspaces, and flame ionization detectors analyze refinery streams.

Molecular spectroscopy has also been applied successfully to determining quantitatively the percentage of each hydrocarbon in mixtures. In some cases, spectroscopic methods are employed to determine hydrocarbon types. For example, ultraviolet spectroscopy may determine aromatics, and mass spectroscopy the content of paraffins, cycloparaffins, olefins, and aromatics in straight-run or cracked fractions in the gasoline range. Even for fractions from lubricating oil distillates (preferably separated as far as practical by other methods) high-temperature mass spectrometry gives useful information concerning the amounts of certain types of hydrocarbons and sulfur compounds. Infrared spectroscopy has also been applied in this molecular-weight range to give values of, for example, the content of methyl groups, as well as of methylene groups in long chains and in cycloparaffin rings.

Mass spectrometry offers a very rapid method for obtaining hydrocarbon type analyses on a wide range of fractions up to and including high boiling gas oil fractions (Speight, 2014, 2015, 2017). It can be used to determine molecular weight of compounds; or using different ionization methods, can provide more structural details through the analysis of fragmentation patterns and can combined with separation techniques such as gas chromatography or liquid chromatography to allow more complex mixtures to be examined (Johnson, 2017).

Infrared (IR) spectroscopy is an extremely versatile technology for oil analysis. IR can provide information on a range of oil characteristics, e.g., contamination, breakdown, additive packages, fluid identity, etc. In all of these cases, the response of the oil to specific regions in the infrared spectrum is examined and weighted, each being unique to the characteristic being analyzed. Firstly, a sample of new oil is tested to establish a baseline reading after which a sample of the used oil is examined. Crude oil-based contaminants and additive molecules will absorb some of the infrared radiation, but only at certain frequencies because every compound has a unique infrared signature. This test is used to monitor molecular species of interest in the oil sample, such as degradation products, additives and contaminants. Degradation products that are commonly monitored include oxidation by-products such as acids and esters, nitration and sulfate by-products

9.7 Biomarkers

Biomarkers (i.e., biological marker) are crude oil components that remain detectable and relatively unchanged in oil residues even after natural environmental weathering processes. They are also typically resistant to biodegradation and are therefore useful as chemical markers.

Crude oil geochemists have historically used biomarker fingerprinting in characterizing oils in terms of (Speight, 2014):

- The type(s) of precursor organic matter in the source rock (such as bacteria, algae, or higher plants)
- Correlation of oils with their source rocks
- Determination of depositional environmental conditions (such as marine, terrestrial, deltaic, or hypersaline environments)
- Assessment of thermal maturity and thermal history of oil and the degree of oil biodegradation
- Providing information on the age of the source rock for crude oil

9.7.1 Types

Biomarkers can be used to assess the nature and the extent of the exposure, to identify alterations occurring within an organism, and to assess underlying susceptibility of an organism (Figure 9.1) (Lionetto et al., 2019).

When coupled with other information (such as toxicity testing results), biomarkers of contaminant exposure can provide a basis for estimating the levels of a chemical stressor that plant and animal communities can and cannot tolerate.

Also, the ability to unambiguously identify spilled crude oil in complex contaminated environmental samples and to link them to the known sources is extremely important in settling questions of environmental impact and environmental liability. The chemical fingerprinting techniques include (1) pattern recognition evaluation of the pattern of the distribution of hydrocarbon derivatives, (2) determination of the constituents of the spilled material, (3) determination of the ratios of source-specific marker compounds such as polynuclear aromatic hydrocarbon derivatives and biomarkers, and (4) analysis of the carbon isotopic ratios. There are also methods for distinguishing naturally-occurring and thermally-derived hydrocarbon derivatives using advanced chemical fingerprinting techniques (Wang and Fingas, 1999).

9.7.2 Commonly Used Biomarkers

Biomarker analysis of aliphatic fractions focused on three groups of compounds, n-alkane derivatives, sterane derivatives and sterene derivatives (i.e., steranes containing double bond in the structure), and pentacyclic triterpane derivatives and their derivatives. Sterane derivatives and sterene derivatives are produced from precursor steroid molecules present in both higher plants and algae, C28-sterane derivatives and C29-sterane derivatives are indicators for the presence of green and C27-sterane derivatives for the presence of red algae, respectively.

Diasterane concentrations are governed mainly by maturity and lithology, and thus provide additional facies (a body of rock with specified characteristics) information. Unusual compounds such as 4-methyl sterane derivatives may occasionally be useful as facies indicators.

Pristane and phytane both occur naturally in crude oil, and are biomarker essential in classifying and identifying crude oil distillates

9.8 Fractionation of the Spilled Material

Rather than quantifying a complex total crude oil hydrocarbon mixture as a single number, crude oil hydrocarbon fraction methods break the mixture into discrete hydrocarbon fractions, thus providing data that can be used in a risk assessment and in characterizing product type and compositional changes such as may occur during weathering (oxidation). The fractionation methods can be used to measure both volatile and extractable hydrocarbon derivatives.

In contrast to traditional methods for total crude oil hydrocarbon derivatives that report a single concentration number for complex mixtures, the fractionation methods report separate concentrations for discrete aliphatic and aromatic fractions. The available crude oil fraction methods are GC-based and are thus sensitive to a broad range of hydrocarbon derivatives. Identification and quantification of aliphatic and aromatic fractions allows one to identify crude oil products and evaluate the extent of product weathering. These fraction data also can be used in risk assessment.

One particular method is designed to characterize C_6 to C_{28+} crude oil hydrocarbon derivatives in soil as a series of aliphatic and aromatic carbon range fractions. The extraction methodology differs from other crude oil hydrocarbon methods because it uses *n*-pentane and not methylene chloride as the extraction solvent. If methylene chloride is used as the extraction solvent, aliphatic and aromatic compounds cannot be separated.

Crude oil hydrocarbon derivatives in this range are extracted efficiently by n-pentane. The whole extract is separated into aliphatic and aromatic crude oil-derived fractions (US EPA SW-846 3611, US EPA SW-846 3630). The aliphatic and aromatic fractions are analyzed separately by gas chromatography, and quantified by summing the signals within a series of specified carbon ranges that represent the fate and transport fractions. The gas chromatograph is equipped with a boiling point column (non-polar capillary column). Gas chromatographic parameters allow the measurement of a hydrocarbon range of n-hexane (C_6) to n-octacosane (C_{28}), a boiling point range of approximately 65 to 450°C.

9.9 Leachability and Toxicity

In terms of composition, hydrocarbon derivatives are by far the most abundant compounds in crude oils and crude oil products, accounting for 50 to 98% v/v. All crude oil blends contain lower-boiling fractions (such as naphtha) as well as wax-like hydrocarbon constituents and higher boiling residua.

When spilled, crude oil and crude oil blends tend to emulsify quickly, forming a stable emulsion (*mousse*). The rate of emulsification is often accelerated by wind—which can accelerate the volatilization of the lower boiling constituents although there are arguments that the emulsification process is related to the wax content of the spilled oil (i.e., the crude oil or the crude oil blend). Typically, 15 to 20% of the oil may evaporate in the first 24 hours of a spill, depending on the wind and sea conditions, and very little oil is dispersed into the water column. The weathered oil then starts to form a stable mousse with up to 75% w/w water, thereby increasing the slick volume four-fold and undergoes dramatic changes in its physical characteristics.

As a result, the viscosity of the oil-in-water mixture increases rapidly and the color usually turns from a dark brown/black to lighter browns and rust colors. As the water content of the emulsion increases, weathering processes (e.g., dissolution and evaporation) slow down. The mousse behaves differently from a fluid and may react to additional weathering forces by forming a surface skin, creating a non-homogenous material with a crust of slightly more weathered mousse surrounding a less weathered core. As the mousse is subject to increased mixing from energetic wave action, the crusts can be torn or ruptured and the less weathered mousse released. The continued exposure of weathered mousse to wave action continues to stretch and tear patches of mousse into smaller bits, resulting in a field of streaks, streamers, small patches and eventually small *tar balls*.

Typically, organisms are not at high risk from crude oil dispersed into the water column, stranded crude tends to smother organisms. In birds, it can cause mortality from ingestion during preening as well as from hypothermia from matted feathers. However, the oil-in-water emulsion is very sticky and makes cleanup and removal more difficult. When stranded on the shoreline, the degree of adhesion varies depending on the substrate type, e.g., this mousse will not penetrate far in finer sediments and may requires stimulation (Abbasnezhad et al., 2011).

Finally, most crude oil blends will emulsify quickly when spilled, creating a stable mousse that presents more persistent cleanup efforts and also represents a significant removal challenge. To complicate matters even further, the means by which the crude oil or crude oil predicts was released into the environment must also be taken into account. Such releases can occur by (1) dispersion, (2) dissolution, (3) emulsification, (4) evaporation, (5) sedimentation or adsorption, and for the purposes of this text, (6) secondary effects such as spreading or dispersion by wind also play a role. These issues have been dealt with in detail elsewhere (Speight, 2005; Speight and Arjoon, 2012) and will not be repeated here.

Once a spill or leakage has occurred, the first step for regulatory and remediation purposes is the application of a standard test method (or standard test methods) to measure the likelihood of toxic substances getting into the environment and causing harm to organisms. The test (required by the United States Environmental Protection Agency) is the *toxicity characteristic leaching procedure* (TCLP, US EPA SW-846 Method 1311), designed to determine the mobility of both organic and inorganic contaminants present in liquid, solid, and multiphase wastes.

The method was developed to estimate the mobility of specific inorganic and organic contaminates that are destined for disposal in municipal landfills. The extraction is performed using acetic as the extraction fluid. The pH of the acetic acid/sodium acetate buffer solution is maintained at 4.93. This sample/acetic acid mixture is subjected to rotary extraction, designed to accelerate years of material exposure in the shortest possible time. After extraction, the resulting liquid is subjected to analysis utilizing a list of contaminants that includes metals, volatile organic compounds, semi-volatile organic compounds, pesticides, and herbicides.

The toxicity characteristic leaching procedure may be subject to misinterpretation if the compounds under investigation are not included in the methods development or the list of contaminants leading to the potential for technically invalid results. However, an alternate procedure, the synthetic precipitation leaching procedure (SPLP, US EPA SW-846 Method 1312) may be appropriate. This procedure is applicable for materials where the leaching potential due to normal rainfall is to be determined. Instead of the leachate simulating acetic acid mixture, nitric and sulfuric acids are utilized in an effort to simulate the acid rains resulting from airborne nitric and sulfuric oxides.

9.10 Monitoring General Site Background Conditions

Environmental monitoring involves the assessment of the quality of the environment to ensure compliance with laws and regulations and to control the risk of pollution and mitigate the harmful effects on the natural environment and protect the health of human beings. It is used as the basis of the production of environmental impact assessments.

Issues related to background are critical when conducting environmental risk assessment (ERA) in sites characterized by contaminated sediments or soils. Background levels have to be considered right from the start. A large portion of the technical discussions and political confrontations about the destiny of the site to be restored and managed will focus on: (1) what is the desired ecological status to be achieved, (2) which levels of contamination should be considered safe, and (3) how such ecological status and concentration levels relate to pre-impact conditions. Background conditions need to be established also for habitat characteristics and for biological parameters, to provide an adequate characterization of exposure and to ensure that remediation effectively translates into a restored ecological status (Pacini, 2008).

Considerations of scale impose the choice of an appropriate specific strategy in establishing a background reference. Background levels defined based on regional, local and unit-specific concentration values have each their own advantages and disadvantages. Factors that need to be taken into account to assess the relevance of regional background values to local situations include sample homogeneity, knowledge of matrix characteristics to assess potential bioavailability and the local and regional variance of the parameter concerned (Pacini, 2008).

Establishing a background concentration is a prerequisite for judging the evolutionary trend of a given substance in the environment and for setting realistic remediation targets. Both the steady natural background, which does not take into account the presence of diffuse anthropogenic sources, and the anthropogenic background, which varies with the environmental history of each contaminant, have a reason to be considered. For many organic contaminants, both background levels may vary due to the progressive refinement of technical instruments which tend to increasingly reduce the limit of detection.

Background reference values assessed from a selection and an elaboration of field data need to be compared and contrasted to risk-based values such as allowable concentration limits, environmental objectives and remediation targets, which are instead derived from laboratory experiments and field-based exposure assessment. Screening against background is safer and it prevents having to elaborate risk-based thresholds through experiments. Background reference levels can explain the natural distribution of species, constitute a relevant part of the characterization of exposure and allow the monitoring of contamination trend-reversal policies (Pacini, 2008).

9.10.1 Oxygen

If dissolved oxygen concentrations drop below a certain level and carbon dioxide levels in water become too high, fish mortality rates will rise.

9.10.2 Acidity-Alkalinity

A higher- or lower-than-normal pH range (typically < 5.5 or > 8.5) in the soil can cause soil infertility and limit the microbial activity. Also, if the pH of water is too high or too low, the aquatic organisms living within it will die.

In a water environment, one of the following methods can be used to measure pH: (1) the colorimetric method or (2) electronic meters.

9.10.3 Temperature

Temperature changes linked to climate change are well documented as it affects many ecosystems.

Growth ranges of psychrophiles, mesophiles, thermophiles and hyperthermophiles are: 0 to 20°C (32 to 68°F), 8 to 48°C (47 to 117°F), 40 to 70°C (104 to 158°F), and 65 to 90°C (149 to 194°F), respectively. As such, temperature fluctuations could influence the structure of the indigenous microbial community, its biodegradation potential, as well as the physicochemical characteristics of the contaminants such as adsorption and solubility.

Temperature, one of the important environmental factor affecting the growth of microorganisms also influences the crude oil degradation not only affecting the activities of crude oil degrading microorganisms, but also changing the properties of crude oil itself. Activities of crude oil-degrading microorganisms could increase with the increase of temperature in some regions, however, toxicity of crude oil components may also increase with temperature while the biodegradation of the fraction of recalcitrant crude oil components may still be poor (Aung et al., 2018).

9.10.4 Salinity

Due to the unfavorable high osmotic potential imposed by salts, the saline condition may limits the activity of some types of microorganisms.

Salinity is a factor that complicates bioremediation of crude oil spills. Additionally, it can lower soil microorganism's biomass and microbial growth rate.

The electrical conductivity or EC of a soil or water sample is influenced by the concentration and composition of dissolved salts. Salts increase the ability of a solution to conduct an electrical current, so a high EC value indicates a high salinity level.

9.11 Assessment of the Methods

Generally, measurement of the total crude oil hydrocarbon derivatives in an ecosystem is performed by the standard method (US EPA 418.1) or by some modification thereof. However, many other methods exist is which the data are also claimed to representative of the total crude oil hydrocarbon derivatives in the ecosystem. In fact, many of the methods for determining the total crude oil hydrocarbon derivatives are prone to (1) producing false negatives (reporting *non-detected* when there was really considerable crude oil hydrocarbon derivatives present), (2) underestimating the extent of crude oil hydrocarbon derivatives present (true of virtually every total crude oil hydrocarbon derivatives methodology), (3) underestimating the overall risk from crude oil hydrocarbon derivatives due to missing significant amounts of some of the compounds of most concern (for example, polynuclear aromatic hydrocarbon derivatives), (4) producing misleading data related to soil hot spots versus areas of less concern due to differing moisture concentrations of otherwise similar samples, (5) producing misleading results because an inappropriate (not close enough to the unknown being sampled) standard (oil) was used in calibration, and (6) producing soil or sediment data which cannot be directly compared with other data for the total petroleum hydrocarbons or guidelines because one is expressed in dry weight and the other in wet weight, and (7) producing relatively accurate dry weight values for heavy crude oil hydrocarbon derivatives but questionable dry weight values for lighter, more volatile compounds (Note: different labs dry the samples different ways and a sample with lots of lighter fraction hydrocarbon derivatives is more prone to hydrocarbon loss; the variable loss of volatile hydrocarbon derivatives in a drying step is therefore an additional area of lab and data variability), (8) producing data which cannot be directly compared with other total crude oil hydrocarbon derivatives data or guidelines because one data set is the result of a Soxhlet extraction method and the other reflects a sonication or other alternative extraction method, (9) producing misleading data related to heavy fraction hydrocarbon derivatives (again such as the higher molecular weight polynuclear aromatic hydrocarbon derivatives) due to loss of the heavier compounds on filter paper, and (10) producing data prone to faulty interpretation of the environmental significance of the results (100 ppm of total crude oil hydrocarbon derivatives from one type of oil may be practically non-toxic while 100 ppm of total crude oil hydrocarbon derivatives from a different type of oil may be very toxic).

Another complication with total crude oil hydrocarbon derivatives values is that crude oil-derived inputs vary considerably in composition; it is essential to bear this in mind when quantifying them in general terms such as *oil* or measurement of the *total crude oil hydrocarbon derivatives*. Crude oil is complex, containing many thousands of compounds ranging from gases to residues boiling about 400°C (750°F).

Furthermore, since different combinations of crude oil hydrocarbons typically contribute to the total petroleum hydrocarbons at different sites, the fate characteristics are also typically different at different sites, even if the concentration of the total petroleum hydrocarbons is the same. Different methods used to generate total crude oil hydrocarbon concentrations, or other similar simple screening measures of crude oil contamination, all produce very different result.

It is not surprising that the data produced as total petroleum hydrocarbons (EPA 418.1) suffers from several shortcomings as an index of potential groundwater contamination or health risk. In fact, the does not actually measure the total petroleum hydrocarbons in the sample but rather measures a specific range of hydrocarbon compounds. This is caused by limitations of the extraction process (solvents used and the concentration steps) and the reference standards used for instrumental analysis. The method specifically states that it does not accurately measure the lighter fractions of gasoline (benzene-toluene-ethylbenzene-xylenes fraction, BTEX) that should include the benzene-toluene-ethylbenzene-xylenes fraction. Further, the method was originally a method for water samples that has been modified for solids, and it is subject to bias.

The total petroleum hydrocarbons represents a summation of the entire hydrocarbon compounds that may be present (and detected) in a soil sample. Because of differences in product composition between, for example, gasoline and diesel, or fresh versus weathered fuels, the types of compounds present at one site may be completely different than those present at another.

Accordingly, the total petroleum hydrocarbons at, for example, a gasoline spill site will be comprised of mostly C_6–C_{12} compounds, while total petroleum hydrocarbons at an older site where the fuel has weathered will likely measure mostly C_8–C_{12} compounds. Because of this inherent variability in the method and the analyte, it is currently not possible to directly relate potential environmental or health risks with concentrations of total petroleum hydrocarbons. The relative mobility or toxicity of contaminants represented by total petroleum hydrocarbons analyses at one site may be completely different from that of another site (for example, C_6 to C_{12} compared to C_{10} to C_{25}). There is no easy way to determine if total petroleum hydrocarbons from the former site will represent the same level of risk as an equal measure of the total petroleum hydrocarbons from the latter. For these reasons, it is clear that total petroleum hydrocarbons offers limited benefits as an indicator measure for cleanup criteria. Its current widespread use as a soil cleanup criterion is a function of a lack of understanding of its proper application and limitations, and its historical use as a simple and inexpensive indicator of general levels of contamination.

When sampling in the environment, it is often impossible to determine which chemical mixtures are causing a total petroleum hydrocarbons reading, which is one of the major weaknesses of the method. At minimum, before using contaminants data from diverse sources, efforts should be made to determine that field collection methods, detection limits, and quality control techniques were acceptable and comparable. This will help the analysts compare the analysis in the concentration range with the benchmark or regulatory criteria concentrations should be very precise and accurate.

Indeed, it must be remembered that quality control field and lab blanks and duplicates will not help in the data quality assurance goal as well as intended if one is using a method prone to false negatives. Methods may be prone to false negatives due to the use of detection limits that are too high, the loss of contaminants through inappropriate handling, or the use of inappropriate methods. The use of inappropriate methods prone to false negatives (or false positives) is particularly common related to total petroleum hydrocarbons and other related crude oil products. This is one reason that more rigorous analyses are often recommended as alternatives to the analysis for the total petroleum hydrocarbons.

In any interpretation of the data for the total petroleum hydrocarbons in a sample, so cannot ignore the amount of moisture cannot be ignored because moisture may block the extraction of petroleum hydrocarbons. Sulfur or phthalate compounds also potentially interfere with the analysis for the total petroleum hydrocarbons. This is similar to the problem of strong interferences from phthalate esters or chlorinated solvents when one is using electron capture methods to look for chlorinated compounds such as polycholorbiphenyl derivatives or pesticides.

Too much reliance on the determination of benzene-toluene-xylenes (BTX) or benzene-toluene-ethylbenzene-xylenes (BTEX) to measure gasoline or diesel contamination may be unaware that more modern gasoline and diesel are better refined and contain fewer of such compounds. It must be remembered that the use of benzene-toluene-xylenes data started as a measure of the more hazardous compounds in gasoline. Modern gasoline and diesel has a higher percentage of straight chain alkane derivatives, non-volatiles, not as many aromatic derivatives, lots of long chain aliphatic compounds, and fewer benzene-toluene-xylenes compounds. In addition, determination of the benzene-toluene-xylenes concentration is not appropriate for aged gasoline characterized by loss of benzene-toluene-xylenes compounds over time. Thus the problem with many analyses for benzene-toluene-xylenes as related to petroleum hydrocarbons is the danger of producing false negatives. For example the test for benzene-toluene-xylenes may indicate no contamination when significant contamination is present.

Total recoverable petroleum hydrocarbons (TRPH), like total petroleum hydrocarbons, is methodologically defined and concentrations given as total petroleum hydrocarbons (or TRPH) alone does not produce much valuable information. To be able to understand the significance of the concentration, the method employed for the determination must be clearly identified (e.g., US EPA 8015 for gasoline, US EPA 8016 for diesel, US EPA 418.1 for total recoverable petroleum hydrocarbons). The data must not be used or interpreted as though various total petroleum hydrocarbons methods were the same as various total recoverable petroleum hydrocarbon methods. When comparing data with soil guideline levels, it is necessary to ascertain which laboratory analysis was done to measure compliance with the current specific guideline.

Additional problems with total petroleum hydrocarbons methods (including method 418.1) include the following:

1. Most methods used to determine the total petroleum hydrocarbons in a sample are inadequate for unknowns because the methods are only as good as the calibration standards. With unknown chemicals present, the precise standards cannot be selected and employing an incorrect calibration standard can lead to erroneous data.

2. Some of the methods that have been used for determination of the total petroleum hydrocarbons also extract vegetable and animal oils that are also present in the sample.

3. The methodology related to volatility can be extremely variable. For example, low boiling oils more susceptible to ambient (and extraction) conditions. The time for evaporation of the oils is a variable and the temperature and heating period is used to calculate dry weight is also a variable) issues. It is preferable to calculate wet weight total petroleum hydrocarbons values first and then very carefully measure percentage moisture in a manner that minimizes losses.

The ASTM method for total petroleum hydrocarbons (ASTM book D7678) is similar to the US EPA standard test method (US EPA 418.1) and calls for extraction with Freon. The estimated the variability of the test method is questionable and may leave room for serious errors in the calculation of the total petroleum hydrocarbons.

Since the determination of the total petroleum hydrocarbons in a sample is subject to many questions, the bias must be defined and alternate reliable and meaningful methods need to be sought. For example, *negative bias* may result when samples are analyzed because of (1) poor extraction efficiency of the solvent (Freon, US EPA 481) or (n-hexane, US EPA 1664) for high molecular weight hydrocarbons, (2) loss of volatile hydrocarbons during extract concentration (Speight, 2005), (3) differences in molar absorptivity between the calibration standard and product type because of the presence of unknown compound types, (4) fractionation of soluble low infrared active aromatic hydrocarbons in groundwater during water washout, (5) removal of five-ring and to six-ring alkylated aromatic derivatives during the silica cleanup procedure—the efficiency of silica gel fractionation varies depending upon the nature of the solute, and (6) preferential biodegradation of n-alkane derivatives.

In addition, *positive bias* is often introduced as a result of (1) product differences in molar absorptivity, (2) partitioning of soluble aromatic derivatives from the bulk product because of oil washout, (3) measurement of naturally occurring saturated hydrocarbon derivatives that exhibit a high molar absorptivity (e.g., plant waxes, $n-C_{25}$, $n-C_{27}$, $n-C_{29}$, and $n-C_{31}$ alkane derivatives), and (4) infrared dispersion of clay particles.

Thus, and to reaffirm earlier statements, there is no one analytical method that is perfect or even adequate, for all cases to determine the amount of total petroleum hydrocarbons in a sample. Different analytical methods have different capabilities and (this is where the environmental analysts plays an important role) it is up to the analysts it is within the purview of the analyst demonstrate that the method applied at specific sites was appropriate.

References

Abbasnezhad, H., Gray, M.R., and Foght, J.R. 2011. Influence of adhesion on aerobic biodegradation and bioremediation of liquid hydrocarbons. Appl. Microbiol. Biotechnol., 92: 653–675.

Aichberger, H., Hasinger, M., Braun, R., and Loibner, A.P. 2005. Potential of preliminary test methods to predict biodegradation performance of crude oil hydrocarbons in soil. Biodegradation, 16: 115–125.

Alder, H.L., and Roessler, E.B. 1972. Introduction to Probability and Statistics. W. H. Freeman, San Francisco, 1972.

Altgelt, K.H., and Gouw, T.H. 1975. In advances in chromatograph. Giddings, J.C., Grushka, E., Keller, R.A., and Cazes, J. (Eds.). Marcel Dekker Inc., New York.

Arjoon, K.K., and Speight, J.G. 2018. The effects of the application of random sampling on the analysis of the biodegradation of oil spills in soil. Petroleum and Chemical Industry International, 2018, 2(1). https://www.opastonline.com/wp-content/uploads/2019/01/the-effects-of-the-application-of-random-sampling-on-the-analysis-of-the-biodegradation-of-oil-spills-in-soil-pcii-19.pdf.

ASTM, 2021. Annual Book of ASTM Standards. ASTM International, West Conshohocken, Pennsylvania.

ASTM D56. 2020. Standard Test Method for Flash Point by Tag Closed Tester. Annual Book of Standards. ASTM International, West Conshohocken, Pennsylvania.

ASTM D86. 2020. Standard Test Method for Distillation of Petroleum Products and Liquid Fuels at Atmospheric Pressure. Annual Book of Standards. ASTM International, West Conshohocken, Pennsylvania.

ASTM D92. 2020. Standard Test Method for Flash and Fire Points by Cleveland Open Cup. Annual Book of Standards. ASTM International, West Conshohocken, Pennsylvania.

ASTM D93. 2020. Standard Test Methods for Flash Point by Pensky-Martens Closed Tester. Annual Book of Standards. ASTM International, West Conshohocken, Pennsylvania.

ASTM D97. 2020. Standard Test Method for Pour Point of Petroleum Products. Annual Book of Standards. ASTM International, West Conshohocken, Pennsylvania.

ASTM D129. 2020. Standard Test Method for Sulfur in Crude oil Products. Annual Book of Standards. ASTM International, West Conshohocken, Pennsylvania.

ASTM D323. 2020. Standard Test Method for Vapor Pressure of Crude oil Products. Annual Book of Standards. ASTM International, West Conshohocken, Pennsylvania.

ASTM D445. 2020. Standard Test Method for Kinematic Viscosity of Transparent and Opaque Liquids. Annual Book of Standards. ASTM International, West Conshohocken, Pennsylvania.

ASTM D971. 2020. Standard Test Method for Interfacial Tension of Oil against Water by the Ring Method. Annual Book of Standards. ASTM International, West Conshohocken, Pennsylvania.

ASTM D1160. 2020. Standard Test Method for Distillation of Petroleum Products at Reduced Pressure. Annual Book of Standards. ASTM International, West Conshohocken, Pennsylvania.

ASTM D2887. 2020. Standard Test Method for Boiling Range Distribution of Petroleum Fractions by Gas Chromatography. Annual Book of Standards. ASTM International, West Conshohocken, Pennsylvania.

ASTM D3710. 2020. Standard Test Method for Boiling Range Distribution of Gasoline and Gasoline Fractions by Gas Chromatography. Annual Book of Standards. ASTM International, West Conshohocken, Pennsylvania.

ASTM D4294. Standard Test Method for Sulfur in Crude oil Products by Energy-Dispersive X-Ray Fluorescence Spectroscopy. Annual Book of Standards. ASTM International, West Conshohocken, Pennsylvania.

ASTM D4486. 2020. Standard Test Method for Kinematic Viscosity of Volatile and Reactive Liquids. Annual Book of Standards. ASTM International, West Conshohocken, Pennsylvania.

ASTM D5185. 2020. Standard Test Method for Determination of Additive Elements, Wear Metals, and Contaminants in Used Lubricating Oils and Determination of Selected Elements in Base Oils by Inductively Coupled Plasma Atomic Emission Spectrometry (ICP-AES). Annual Book of Standards. ASTM International, West Conshohocken, Pennsylvania.

ASTM D6304. 2020. Standard Test Method for Determination of Water in Petroleum Products, Lubricating Oils, and Additives by Coulometric Karl Fischer Titration. Annual Book of Standards. ASTM International, West Conshohocken, Pennsylvania.

ASTM E659. Standard Test Method for Autoignition Temperature of Liquid Chemicals. Annual Book of Standards. ASTM International, West Conshohocken, Pennsylvania.

ASTM F3045. 2020. Standard Test Method for Evaluation of the Type and Viscoelastic Stability of Water-in-oil Mixtures Formed from Crude Oil and Petroleum Products Mixed with Water. Annual Book of Standards. ASTM International, West Conshohocken, Pennsylvania.

Atagana, H.I., Haynes, R.J., and Wallis, F.M. 2003. Optimization of soil physical and chemical conditions for the bioremediation of creosote-contaminated soil. Biodegradation, 14: 297–307.

Aung, M., Li, Q., Takahashi, S., and Motoo, U. 2018. Effect of temperature on hydrocarbon bioremediation in simulated petroleum-polluted seawater collected from Tokyo Bay. Japanese Journal of Water Treatment Biology, 54: 95–104.

Backman, A., Maraha, N., and Jansson, J.K. 2004. Impact of temperature on the physiological status of a potential bioremediation inoculant, Arthrobacter chlorophenolicus A6. Applied and Environmental Microbiology, 70(5): 2952–2958.

Baker, C.C.T. 1966. An Introduction to Mathematics. Arco Publishing Company Inc., New York.

Baker, R.J., Baehr, A.L., and Lahvis, M.A. 2000. Estimation of hydrocarbon biodegradation rates in gasoline-contaminated sediment from measured respiration rates. J. Contam. Hydrol., 41: 175–192.

Balba, M.T., Al-Awadhi, N., and Al-Daher, R. 1998. Bioremediation of oil-contaminated soil: microbiological methods for feasibility assessment and field evaluation. J. Microbiol. Methods, 32: 155–164.

Bartha, R. 1986. Biotechnology of petroleum pollutant biodegradation. Microb. Ecol., 12: 155–172.

Boopathy, R. 2002. Use of anaerobic soil slurry reactors for the removal of petroleum hydrocarbons in soil. Int. Biodeterioration Biodegradation, 52: 161–166.

Box, G., Hunter, W., and Hunter, J. 1978. Statistics for Experimenters. John Wiley & Sons Inc., New York.

Breedveld, G.D., and Sparrevik, M. 2001. Nutrient-limited biodegradation of PAH in various soil strata at a creosote contaminated site. Biodegradation, 11: 391–399.

Briški, F., and Vuković Domanovac, M. 2017. Environmental microbiology. Physical Sciences Reviews, 2(11): 20160118. https://doi.org/10.1515/psr-2016-0118.

Bruner, F. 1993. Gas Chromatographic Environmental Analysis: Principles, Techniques, and Instrumentation. John Wiley & Sons Inc., Hoboken, New Jersey.

Budde, W.L. 2001. The Manual of Manuals. Office of Research and Development, Environmental Protection Agency, Washington, DC.

Cao, J.R. 1992. Microwave Digestion of Crude Oils and Oil Products for the Determination of Trace Metals and Sulphur by Inductively-Coupled Plasma Atomic Emission Spectroscopy, Environment Canada Manuscript Report Number EE-140, Ottawa, Ontario, Canada.

Caulcutt, R., and Boddy, R. 1983. Statistics for Analytical Chemists, 1983. Chapman and Hall, London, England.

Chaîneau, C.H., Morel, J.L., and Oudot, J. 1995. Microbial degradation in soil microcosms of fuel oil hydrocarbons from drilling cuttings. Environ. Sci. Technol., 29: 1615–1621.

Chaîneau, C.H., Yepremian, C., Vidalie, J.F., Ducreux, J., and Ballerini, D. 2003. Bioremediation of a crude oil-polluted soil: biodegradation, leaching and toxicity assessments. Water, Air, Soil Pollut., 144: 419–440.

Chikere, C.B., Okpokwasili, G.C., and Chikere, B.O. 2011. Monitoring of microbial hydrocarbon remediation in the soil. Biotech, 1(3): 117–138.

Davis, C., Cort, T., Dai, D., Illangasekare, T.H., and Munakata-Marr, J. 2003. Effects of heterogeneity and experimental scale on biodegradation of diesel. Biodegradation, 14: 373–384.

Dean, J.R. 1998. Extraction Methods for Environmental Analysis. John Wiley & Sons, Inc., New York.

Dean, J.R. 2003. Methods for Environmental Trace Analysis. John Wiley 7 Sons Inc., Hoboken, New Jersey.

Dixon, W. J., and Massey, F.J. 1969. Introduction to Statistical Analysis, McGraw-Hill, New York.

Dott, W., Feidieker, D., Steiof, M., Becker, P.M., and Kämpfer, P. 1995. Comparison of *ex situ* and *in situ* techniques for bioremediation of hydrocarbon-polluted soils. Int. Biodeterioration & Biodegradation, 301–316.

Dyroff, G.V. (Ed.). 1993. Manual on Significance of Tests for Petroleum Products: 6th Edition, American Society for Testing and Materials, West Conshocken, Pennsylvania.

EPA. 1998. Test Methods for Evaluating Solid Waste - Physical/Chemical Methods. EPA/SW-846, 3rd Edition, 1986, Update 1, 1992, Update II, 1994, Update III, 1996, Update IV, 1998. Environmental Protection Agency, Washington, DC.

EPA. 2004. Environmental Protection Agency, Washington, DC. Web site: http://www.epa.gov.

EPA SW-846 Test Method 418.1. 2015. Identity and Analysis of Total Petroleum Hydrocarbons. United States Environmental; Protection Agency, Washington, DC. https://www.atsdr.cdc.gov/ToxProfiles/tp123-c3.pdf.

EPA SW-846 Test Method 1664. 2015. Analytical Method Guidance for EPA Method 1664A Implementation and Use. United States Environmental; Protection Agency, Washington, DC. https://www.epa.gov/sites/default/files/2015-10/documents/guidance-for-method-1664a_2000.pdf.

EPA SW-846 Test Method 1311. 2015. Toxicity Characteristic Leaching Procedure. United States Environmental; Protection Agency, Washington, DC. https://www.epa.gov/sites/default/files/2015-12/documents/1311.pdf.

EPA SW-846 Test Method 1312. 2015. Synthetic Precipitation Leaching Procedure. United States Environmental; Protection Agency, Washington, DC. https://www.epa.gov/sites/default/files/2015-12/documents/1312.pdf.

EPA SW-846 Test Method 3611B. 2015. Alumina Column Cleanup and Separation of Petroleum Wastes. United States Environmental; Protection Agency, Washington, DC. https://www.epa.gov/sites/default/files/2015-12/documents/3611b.pdf.

EPA SW-846 Test Method 3630. 2015. SW-846 Test Method 3630C: Silica Gel Cleanup. United States Environmental; Protection Agency, Washington, DC. https://www.epa.gov/sites/default/files/2015-12/documents/3630c.pdf.

EPA SW-846 Test Method 4030. 2015. Soil Screening for Petroleum Hydrocarbons by Immunoassay. United States Environmental; Protection Agency, Washington, DC. https://www.epa.gov/sites/default/files/2015-12/documents/4030.pdf.

EPA SW-846 Test Method 4035. 2015. Soil Screening for Polynuclear Aromatic Hydrocarbons by Immunoassay. United States Environmental; Protection Agency, Washington, DC. https://www.epa.gov/sites/default/files/2015-12/documents/4035_0.pdf.

EPA SW-846 Test Method 8020. 2015. Halogenated and Aromatic Volatile Organic Compounds (VOCs) by Gas Chromatography: SW-846 Methods 8010A and 8020A or Method 8021A. United States Environmental; Protection Agency, Washington, DC. https://www.epa.gov/sites/default/files/2015-06/documents/8021.pdf.

EPA SW-846 Test Method 8100. 2015. Polynuclear Aromatic Hydrocarbons. United States Environmental; Protection Agency, Washington, DC. https://www.epa.gov/sites/default/files/2015-12/documents/8100.pdf.

EPA SW-846 Test Method 8260D. 2015. Volatile Organic Compounds by Gas Chromatography/Mass Spectrometry (GC/MS). United States Environmental; Protection Agency, Washington, DC. https://www.epa.gov/sites/default/files/2018-06/documents/method_8260d_update_vi_final_06-11-2018.pdf.

EPA SW-846 Test Method 8270D. 2015. Semivolatile Organic Compounds by Gas Chromatography/Mass Spectrometry (GC/MS). United States Environmental; Protection Agency, Washington, DC. https://19january2017snapshot.epa.gov/sites/production/files/2015-07/documents/epa-8270d.pdf.

EPA SW-846 Test Method 8310. 2015. Polynuclear Aromatic Hydrocarbons. United States Environmental; Protection Agency, Washington, DC. https://www.epa.gov/sites/default/files/2015-12/documents/8310.pdf.

Falla Sotelo, F., Araujo Pantoja, P., López-Gejo, J., Le Roux, J.G.A.C., Quina, F.H., and Nascimento, C.A.O. 2008. Application of fluorescence spectroscopy for spectral discrimination of petroleum samples. Brazilian Journal of Petroleum and Gas, 2(2): 63–71.

Fingas, M.F., Duval, W.S., and Stevenson, G.B. 1979. The Basics of Oil Spill Cleanup, Environment Canada, Ottawa, Ontario, Canada.

Fingas, M.F., Dufort, V.M., Hughes, K.A., Bobra, M.A., and Duggan, L.V. 1989a. Laboratory studies on oil spill dispersants. pp. 207–219. *In*: Flaherty, M. (Ed.). Chemical Dispersants - New Ecological Approaches. Publication No. STP 1084, American Society for Testing and Materials, West Conshohocken, Pennsylvania.

Fingas, M.F., White, B., Stoodley, R.G., and Crerar, I.D. 1989b. Laboratory Testing of Dispersant Effectiveness. Proceedings. 1989 Oil Spill Conference, American Petroleum, Washington, D.C., pp. 365–373.

Fingas, M.F., Kyle, D.A., Bier, I.E., Lukose, A., and Tennyson, E.J. 1991. Physical and Chemical Studies on Oil Spill Dispersants: The Effect of Energy. Proceedings. 14th Arctic and Marine Oil Spill Program Technical Seminar, Environment Canada, Ottawa, Ontario, Canada. Page 87–106.

Fingas, M.F., Kyle, D.A., and Tennyson, E.J. 1992. Physical and Chemical Studies on Oil Spill Dispersants: Effectiveness Variation with Energy. Proceedings. 15th Arctic and Marine Oil Spill Program Technical Seminar, Environment Canada. Ottawa, Ontario, Canada Page 135–142.

Fingas, M.F., Kyle, D.A., and Tennyson, E.J. 1993. Physical and Chemical Studies on Dispersants: The Effect of Dispersant Amount on Energy. Proceedings. 16th Arctic and Marine Oil Spill Program Technical Seminar, Environment Canada, Ottawa, Ontario, Canada. Page 861–876.

Fingas, M.F., Fieldhouse, B., Gamble, L., and Mullin, J. 1995a. Studies of water-in-oil emulsions: stability classes and measurement. Proceedings. 18th Arctic and Marine Oil Spill Program Technical Seminar, Environment Canada, Ottawa, Ontario, Canada. Page 21–42.

Fingas, M.F., Kyle, D.A., Lambert, P., Wang, Z., and Mulling, J. 1995b. Analytical procedures for measuring oil spill dispersant effectiveness in the laboratory. Proceedings. 18th Arctic and Marine Oil Spill Program Technical Seminar, Environment Canada, Ottawa, Ontario, Canada. Page 339–354.

Fingas, M.F., Ackerman, F., Lambert, P., Li, K., Wang, Z., Mullin, J., Hannon, L., Wang, D., Steenkammer, A., Hiltabrand, R. Turpin, R., and Campagna, P. 1995c. The newfoundland offshore burn experiment: further results of emissions measurement. Proceedings. 18th Arctic and Marine Oil Spill Program Technical Seminar, Environment Canada, Ottawa, Ontario, pp. 915–995.

Fingas, M.F. 1995. The evaporation of oil spills. Proceedings. 18th Arctic and Marine Oil Spill Program Technical Seminar, Environment Canada, Ottawa, Ontario, Canada. Page 43–60.

Fingas, M.F. 1998. Studies on the evaporation of crude oil and crude oil products. II. Boundary Layer Regulation. J. Hazardous Materials. 57(1-3): 41–58.

Fowlis, I.A. 1995. Gas Chromatography 2nd Edition. John Wiley & Sons Inc., New York.

Frijer, J.I., De Jonge, H., Bounten, W., and Verstraten, J.M. 1996. Assessing mineralization rates of petroleum hydrocarbons in soils in relation to environmental factors and experimental scale. Biodegradation, 7: 487–500.

Grishchenkov, V.G., Townsend, R.T., McDonald, T.J., Autenrieth, R.L., Bonner, J.S., and Boronin, A.M. 2000. Degradation of petroleum hydrocarbons by facultative anaerobic bacteria under aerobic and anaerobic conditions. Process Biochem., 35: 889–896.

Grob, R.L. 1995. Modern Practice of Gas Chromatography. 3rd Edition. John Wiley & Sons Inc., New York.

Hatzinger, P.B., and Alexander, M. 1995. Effect of ageing chemicals in soil upon their biodegradability and extractability. Environ. Sci. Technol., 29: 537–545.

Heider, J., Spormann, A.M., Beller, H.R., and Widdel, F. 1999. Anaerobic bacterial metabolism of hydrocarbons. FEMS Microbiol. Rev., 22: 459–473.

Höhener, P., Duwig, C., Pasteris, G., Kaufmann, K., Dakhel, N., and Harms, H. 2003. Biodegradation of petroleum hydrocarbon vapors: laboratory studies on rates and kinetics in unsaturated alluvial sand. J. Contam. Hydrol., 1917: 1–23.

Holliger, C., and Zehnder, A.J.B. 1996. Anaerobic biodegradation of hydrocarbons. Curr. Opin. Biotechnol., 7: 326–330.

Huang, W., Pent, P., Yu, Z., and Fu, J. 2003. Effects of organic matter heterogeneity on sorption and desorption of organic contaminants by soils and sediments. Appl. Geochem., 18: 955–972.

Huesemann, M.H. 1995. Predictive model for estimating the extent of petroleum hydrocarbon biodegradation in contaminated soils. Environ. Sci. Technol., 29: 7–18.

Jaffe, A.J., and Spirer, H.F. 1987. Misused Statistics-Straight Talk for Twisted Numbers. Dekker, New York.

Jokuty, P., Fingas, M.F., Whiticar, S., and Fieldhouse, B. 1995. A Study of Viscosity and Interfacial Tension of Oils and Emulsions. Report No. EE-153. Environment Canada, Ottawa, Ontario, Canada.

Jokuty, P., Whiticar, S., McRoberts, K., and Mullin, J. 1996. Oil Adhesion Testing - Recent Results. Proceedings. 19th Arctic and Marine Oil Spill Program Technical Seminar, Environment Canada, Ottawa, Ontario, Canada. Page 9–27.

Leeson, A., and Hinchee, R.E. 1997. Soil Bioventing, Principles and Practice. CRC, Lewis Publishers, Boca Raton, Florida.

Lionetto, M., Caricato, R., and Giordano, M. 2019. Pollution biomarkers in environmental and human biomonitoring. The Open Biomarkers Journal, 9: 1–9.

Luthy, R.G., Aiken, G.R., Brusseau, M.L., Cunningham, S.D., Gschwend, P.M., Pignatello, J.J., Reinhard, M., Traina, S.J., Weber, W.J., and Westall, J.C. 1997. Sequestration of hydrophobic organic contaminants by geosorbents. Environ. Sci. Technol., 31: 3341–3347.

Mackay, D. and Zagorski, W. 1982. Studies of Water-in-Oil Emulsions. Report No. EE-34. Environment Canada, Ottawa, Ontario, Canada.

Margesin, R., Zimmerbauer, A., and Schinner, F. 1997. Efficiency of indigenous and inoculated cold-adapted soil microorganisms for biodegradation of diesel oil in alpine soils. Appl. and Environ. Microbiol., 63: 2660–2664.

Margesin, R., Zimmerbauer, A., and Schinner, F. 2000. Monitoring of bioremediation by soil biological activities. Chemosphere, 40: 339–346.

Massias, D., Grossi, V., and Bertrand, J.C. 2003. *In situ* anaerobic degradation of petroleum alkanes in marine sediments: preliminary results. Geoscience, 335: 435–439.

Meier, P.C., and Zünd, R.E. 2000. Statistical Methods in Analytical Chemistry. John Wiley & Sons Inc., New York.

Miller, M. (Ed.). 2000. Encyclopedia of Analytical Chemistry. John Wiley & Sons Inc., Hoboken, New Jersey.

Mohn, W.M., and Stewart, G.R. 2000. Limiting factors for hydrocarbon biodegradation at low temperature in arctic soils. Soil Biol. Biochem., 32: 1161–1172.

Nazaruk, J., and Borzym-Kluczyk, M. 2015. The role of triterpenes in the management of diabetes mellitus and its complications. Phytochemistry Reviews. Proceedings of the Phytochemical Society of Europe, 14(4): 675–690.

Nierop, K.G.J., and Verstraten, J.M. 2003. Organic matter formation in sandy subsurface horizons of dutch coastal dunes in relation to soil acidification. Org. Geochem., 34: 499–513.

Nocentini, M., Pinelli, D., and Fava, F. 2000. Bioremediation of a soil contaminated by hydrocarbon mixtures: the residual concentration problem. Chemosphere, 41: 1115–1123.

Oudot, J., and Dutrieux, E. 1989. Hydrocarbon weathering and biodegradation in a tropical estuarine ecosystem. Mar. Environ. Res., 27: 195–213.

Pacini, N., 2008, Environmental background assessment: basic principles and practice. Annali dell Instituto Superiore di Sanità. 44. 258–67.

Patnaik, P. (Ed.). 2004. Dean's Analytical Chemistry Handbook. 2nd Edition. McGraw-Hill, New York.

Pavlova, A., and Ivanova, R. 2003. Determination of petroleum hydrocarbons and polycyclic aromatic hydrocarbons in sludge from wastewater treatment basins. J. Environ. Monit., 5: 319–323.

Powell, T., and McKirdy, D. 1973. Relationship between Ratio of Pristane to Phytane, Crude Oil Composition and Geological Environment in Australia. Nature Physical Science, 243: 37–39.

Quevauviller, P. 2002. Quality Assurance for water analysis. John Wiley & Sons Inc., Hoboken, New Jersey, USA.

Rahimova, N.A., Abdullayev, V.H., and Abbasova, V.S. 2020. Development of stages of the implementation of the environmental monitoring program. Glob. J. Ecol., 5(1): 001–004.

Rhodes, I.A., Hinojas, E.M., Barker, D.A., and Poole, R.A. 1994. Pitfalls Using Conventional TPH Methods for Source Identification. Proceedings. Seventh Annual Conference: EPA Analysis of Pollutants in the Environment. Norfolk, VA. Environmental Protection agency, Washington, DC.

Rodríguez, M., García-Gómez, C., Alonso-Blázquez, N., and Tarazona, J. 2014. Soil Pollution Remediation. Encyclopedia of Toxicology. pp. 344–355. 3rd Edition. Elsevier. Inc.

Rullkötter, J., and Farrington, J.W. 2021. What was released? Assessing the physical properties and chemical composition of petroleum and products of burned oil. Oceanography, 34(1): 44–57. https://tos.org/oceanography/article/what-was-released-assessing-the-physical-properties-and-chemical-composition-of-petroleum-and-products-of-burned-oil.

Schramm, L.L. (Ed.). 1992. Emulsions. Fundamentals and Applications in the Petroleum Industry. American Chemical Society, Washington, DC.

Sheng H., and Sun H. 2011. Synthesis, biology and clinical significance of pentacyclic triterpenes: A multi-target approach to prevention and treatment of metabolic and vascular diseases. Nat. Prod. Rep., 28: 543–593.

Smith, R.K. 1999. Handbook of Environmental Analysis, 4th Edition, Genium Publishing, Schenectady, NY.

Smith, K.A., and Cresser, M. 2003. Soil & Environmental Analysis: Modern Instrumental Techniques. Marcel Dekker Inc., New York. 2003.

Speight, J.G. 2005. Environmental Analysis and Technology for the Refining Industry John Wiley & Sons Inc., Hoboken, New Jersey.

Speight, J.G. 2009. Enhanced Recovery Methods for Heavy Oil and Tar Sands. Gulf Publishing Company, Houston, Texas.

Speight, J.G. 2011. An Introduction to Petroleum Technology, Economics, and Politics. Scrivener Publishing, Beverly, Massachusetts.

Speight, J.G., and Foote, R. 2011. Ethics in Science and Engineering. Scrivener Publishing, Beverly, Massachusetts.

Speight, J.G., and Arjoon, K.K. 2012. Bioremediation of Petroleum and Petroleum Products. Scrivener Publishing, Beverly, Massachusetts.

Speight, J.G. 2014. The Chemistry and Technology of Petroleum 5th Edition. CRC Press, Taylor & Francis Group, Boca Raton, Florida.

Speight, J.G. 2015. Handbook of Petroleum Product Analysis 2nd Edition. John Wiley & Sons Inc., Hoboken, New Jersey.

Speight, J.G. 2017. Handbook of Petroleum Refining. CRC Press, Taylor & Francis Group, Boca Raton, Florida.

Sullivan and Johnson. 1993. 'Oil' You Need to know about Crude – Implications of TPH Data for Common Petroleum Products. Soil. May, page 8.

Sunahara, G.I., Renoux, A.Y., Thellen, C., Gaudet, C.L., and Pilon, A. (Eds.). 2002. Environmental Analysis of Contaminated Sites. John Wiley & Sons Inc., New York.

Twardus, E.M. 1980. A Study to Evaluate the Combustibility and Other Physical and Chemical Properties of Aged Oils and Emulsions. Report Number EE-5. Environment Canada, Ottawa, Ontario, Canada.

US EPA. 1992. Preparation of Soil Sampling Protocols: Sampling Techniques and Strategies, EPA/600/R-92/128, Office of Research and Development, Washington D.C., July. Page 3–7.

Wang, Z., Fingas, M.F., Landriault, M., Sigouin, L., and Xu, N. 1995. Identification of Alkyl Benzenes and Direct Determination of BTEX and (BTEX + C3-Benzenes) in Oils by GC/MS. Proceedings. 18th Arctic and Marine Oil Spill Program Technical Seminar, Environment Canada, Ottawa, Ontario, Canada. Page 141–164.

Wang, Z., and Fingas, M. 1999. Identification of the Source(s) of Unknown Spilled Oils. Proceedings. 1999 International Oil Spill Conference. https://www.researchgate.net/publication/256925860_Identification_of_the_Sources_of_Unknown_Spilled_Oils.

Weisman, W. 1998. Analysis of Petroleum Hydrocarbons in Environmental Media. Total Petroleum Hydrocarbons Criteria Working Group Series. Volume 1. Amherst Scientific Publishers, Amherst, MA. (See also: Volume 2: Composition of Petroleum Mixtures, Volume 2, 1998; Volume 3: Selection of Representation Total Petroleum Hydrocarbons Fractions Based on Fate and Transport Considerations, 1997; Volume 4: Development of Fraction-Specific Reference Does and Reference concentrations for Total Petroleum Hydrocarbons, 1997; and Volume 5: human Health Risk-Based Evaluation of Petroleum Contaminated Sites - Implementation of the Working Group Approach, 1999).

Zhendi Wang, Z., Hollebone, B.P., Fingas, M., Fieldhouse, B., Sigouin, L., Landriault, M., Smith, P., Noonan, J., and Thouin, G. 2003. Characteristics of Spilled Oils, Fuels, and Petroleum Products: 1. Composition and Properties of Selected Oils. Report No. EPA/600/R-03/072. National Exposure Research Laboratory Office of Research and Development United States Environmental Protection Agency Research Triangle Park, North Carolina. July. https://nepis.epa.gov/Exe/ZyNET.exe/P1000AE6.TXT?ZyActionD=ZyDocument&Client=EPA&Index=2000+Thru+2005&Docs=&Query=&Time=&EndTime=&SearchMethod=1&TocRestrict=n&Toc=&TocEntry=&QField=&QFieldYear=&QFieldMonth=&QFieldDay=&IntQFieldOp=0&ExtQFieldOp=0&XmlQuery=&File=D%3A%5Czyfiles%5CIndex%20Data%5C00thru05%5CTxt%5C00000013%5CP1000AE6.txt&User=ANONYMOUS&Password=anonymous&SortMethod=h%7C-&MaximumDocuments=1&FuzzyDegree=0&ImageQuality=r75g8/r75g8/x150y150g16/i425&Display=hpfr&DefSeekPage=x&SearchBack=ZyActionL&Back=ZyActionS&BackDesc=Results%20page&MaximumPages=1&ZyEntry=1&SeekPage=x&ZyPURL.

CHAPTER 10

Recommendations for Oil Spill Prevention and Control

Although it may be felt that the environmental analysts must, by definition focus on analysis of refinery products and wastes, knowledge of the various environmental regulations (Table 5.1) is always helpful in determining the analyses that must be performed (Speight, 2005; Speight and Arjoon, 2012).

The definitions of hazardous substances (40 CFR 300.5) and pollutants or contaminants (40 CFR 300.5) specifically exclude petroleum, including crude oil or any fraction thereof unless specifically listed. Although there is no definition of crude oil in Superfund, the Environmental Protection Agency interprets the crude oil exclusion provision to include crude oil and fractions of crude oil, including the hazardous substances, such as benzene, that are indigenous in crude oil and are, therefore, included in the term *crude oil*. The term also includes hazardous substances that are normally mixed with or added to crude oil or crude oil fractions during the refining process, including hazardous substances whose levels are increased during refining. These substances are also part of *crude oil* because their addition is part of the normal oil separation and processing operations at refineries that produce the product commonly understood to be crude oil. However, hazardous substances that are added to crude oil (e.g., mixing of solvents with used oil) or that increase in concentration solely as a result of contamination of the crude oil during use are not part of the crude oil and thus are not excluded from Superfund (Wagner, 1999).

In spite of this exclusion, the refining industry has come under considerable strain because of several important factors and changes in the industry. Over the years, there has been an increased demand for crude oil products and a decrease in domestic production. However, there has been no new major refinery construction in the United States in the last three decades. This lack of infrastructure growth has caused a strain on the industry in meeting existing demand and has resulted in an increase in the amount of crude oil imports to meet the increasing need for liquid fuel.

Furthermore, as a result of the evolving environmental awareness, crude oil refinery operators face more stringent regulation of the treatment, storage, and disposal of hazardous wastes. Under recent regulations, a larger number of compounds have been, and are being, studied. Long-time methods of disposal, such as land farming of refinery waste, are being phased out. New regulations are becoming even more stringent, and they encompass a broader range of chemical constituents and processes.

However, it is not the purpose of this chapter here to enter into any political discussion and the levy of fines for infringement of the environmental laws. The purpose of this chapter is to introduce the reader to an overview of a selection of the many and varied regulations show the types of emissions from refinery processes and the laws that regulate these emissions.

10.1 Refinery Products

Crude oil refineries process crude oil into many different products and the physical and chemical characteristics of crude oil determine how the refineries turn it into the highest value products (Chapter 1, Chapter 2). Thus, a refinery can produce high-value products such as gasoline, diesel fuel, and jet fuel, and the various types of fuel oil.

10.1.1 Bulk Products

The bulk products from crude oil refining are, in contrast to petrochemicals, those bulk fractions that are derived from crude oil and have commercial value as a bulk product. Crude oil is an extremely complex mixture of hydrocarbon compounds, usually with minor amounts of nitrogen-containing, oxygen-containing, and sulfur-containing compounds as well as trace amounts of metal-containing compounds.

The bulk products that are produced by a refinery typically refer to the fractions that are produced by distillation and include (1) gases, from which liquefied petroleum gas and petrochemicals are produced, (2) naphtha, from which gasoline and solvents are produced, (3) kerosene, from which diesel fuel and other liquid fuels are produced, (4) atmospheric and vacuum gas oil, from which the various fuel oils, lubricating oil, white oil, insulating oil, insecticides, grease, and wax are produced, (5) residuum, from which asphalt and coke are produced (Chapter 1, Chapter 2).

10.1.2 Petrochemicals

In the strictest sense, petrochemicals are crude oil products but in contrast to the bulk products are typically individual chemicals that are used as the basic building blocks of the chemical industry (Parkash, 2003; Gary et al., 2007; Speight, 2014; Hsu and Robinson, 2017; Speight, 2017, 2019).

Briefly, for environmental purposes, petrochemicals are sub-divided into two classes: (1) *organic chemicals* and (2) *inorganic chemicals* (Speight, 2019). Furthermore classification occurs insofar as organic chemicals are classified as *volatile organic compounds* or *semi-volatile organic compounds* (on occasion, the word *chemicals* is substituted for the word *compounds* without affecting the definition).

The first class of organic compounds, the *volatile organic compounds* (VOCs), is sub-divided into *regulated compounds* and *unregulated compounds*. Regulated compounds have maximum contaminant levels, but unregulated compounds do not. Regulated generally (but not always) have low boiling points, or low boiling ranges, and some are gases. Many of these chemicals can be detected at extremely low levels by a variety of instrumentation, including the human nose! In the case of the crude oil industry, sources for volatile organic compounds typically are crude oil refineries, fuel stations, naphtha (i.e., dry cleaning solvents, paint thinners, cleaning solvents for auto parts) and, in some cases, refrigerants that are manufactured from petrochemicals).

The second class of organic compounds, the *semi-volatile compounds*, typically have high boiling points, or high boiling ranges, and are not always easily detected by the instrumentation that may be used to detect the volatile organic compounds (including the human nose). Some of the common sources of contamination are high boiling crude oil products (e.g., lubricating oils), pesticides, herbicides, fungicides, wood preservatives, and a variety of other chemicals that can be linked to the refining industry.

10.1.3 Refinery Waste

Another class of chemicals produced by refineries are those chemical that are badly classified as refinery waste.

The com (Parkash, 2003; Gary et al., 2007; Speight, 2014; Hsu and Robinson, 2017; Speight, 2017). During crude oil refining, refineries use and generate an enormous amount of chemicals, some of which are present in air emissions, wastewater, or solid wastes. Emissions are also created through the combustion of fuels, and as byproducts of chemical reactions occurring when crude oil fractions are upgraded.

Waste elimination is common sense and provides several obvious benefits (Table 10.1). Yet waste elimination continues to elude many companies in every sector, including the refinery section, and activity from refinery waste (that is a function of their production system design). It may not matter how a refiner categorizes the waste or how the refiner chooses to pursue waste elimination, one thing remains constant and that is once identified, waste could be eliminated. There are models and structures that allow a refiner to identify and eliminate waste to increase productivity, and hence cost structures, that have a direct impact on refinery operations. Waste elimination though identification (by judicious analysis) and treatment subscribe to the smooth operation of a refinery.

Generally process wastes (emissions) are categorized as gaseous, liquid, and solid. This does not usually include waste from or from accidental spillage of a crude oil feedstock a product.

Briefly, crude oil refining involves a series of steps that includes separation and blending of crude oil products. The five major processes are briefly described below:

- Separation processes: These processes involve separating the different constituents into fractions based on their boiling point differences. Additional processing of these fractions is usually needed to produce final products to be sold within the market.

- Conversion processes: Coking and cracking are conversion processes used to break down higher molecular weight constituents into lower molecular weight product by heating and by use of catalysts.

- Treating: Crude oil-treating processes are used to remove the undesirable components and impurities such as sulfur, nitrogen and heavy metals from the products. This involves processes such as hydrotreating, deasphalting, acid gas removal, desalting, hydrodesulfurization, and sweetening.

- Blending/combination processes: Refineries use blending/combination processes to create mixtures with the various crude oil fractions to produce a desired final product, such as gasoline with different octane ratings.

- Auxiliary processes: Refineries also have other processes and units that are vital to operations by providing power, waste treatment and other utility services, such as boilers, wastewater treatment, and cooling towers. Products from these facilities are usually recycled and used in other processes within the refinery and are also important in regards to minimizing water and air pollution.

Refineries are generally considered a major source of pollutants in areas where they are located and are regulated by a number of environmental laws related to air, land and water (Table 5.1). Thus, refineries are generally considered a major source of pollutants in areas where they are located and are regulated by a number of environmental laws related to air, land and water.

Table 10.1: Benefits of Waste Elimination.

• Solve the waste disposal problems created by land bans
• Reduce waste disposal costs
• Reduce costs for energy, water and raw materials
• Reduce operating costs
• Protect workers, the public and the environment
• Reduce risk of spills, accidents and emergencies
• Reduce vulnerability to lawsuits and improve its public image
• Gen

10.2 Environmental Impact of Crude Oil and Crude Oil Products

Crude oil refining is one of the largest industries in the United States and potential environmental hazards associated with refineries have caused increased concern for communities in close proximity to them. This update provides a general overview of the processes involved and some of the potential environmental hazards associated with crude oil refineries (see also Chapter 3).

Regulations are in place that set the maximum contamination concentration levels that are designed to ensure public safety. There are primary and secondary standards for inorganic chemicals. Primary standards are those chemicals that cause neurological damage, cancer, or blood disorders. Secondary standards are developed for other environmental reasons. In some instances, the primary standards are referred to as the *Inorganic Chemical Group*. The secondary standards are referred to as the *General Mineral Group* and *General Physical Testing Group*.

The inorganic chemical group includes aluminum, antimony, arsenic, barium, beryllium, cadmium, chromium, lead, mercury, nickel, selenium, silver, fluoride, nitrate, nitrite, and thallium. The mineral group includes calcium, magnesium, sodium, potassium, bicarbonate, carbonate, chloride, sulfate, pH, alkalinity, hardness, electrical conductivity, total dissolved solids, surfactants, copper, iron, manganese, and zinc. The physical group includes turbidity, color, and odor. Many of these chemicals arise from desalting residues and from other processes where catalysts are used. A high level of any of these three chemicals in the soil or in the water is an indication that one or more specific processes (identified from the chemicals that have been released) or pollution prevention processes are not performing according to operational specifications.

Another source of toxic compounds is combustion (Chapter 4). In fact, some of the greater dangers of fires are from toxic products and by-products of combustion. The most obvious of these is carbon monoxide (CO), which can cause serious illness or death because it forms carboxyhemoglobin with hemoglobin in the blood so that the blood no longer carries oxygen to body tissues. Toxic sulfur dioxide and hydrogen chloride are formed by the combustion of sulfur compounds and organic chlorine compounds, respectively. In addition, a large number of noxious organic compounds such as aldehydes are generated as by-products of combustion. In addition to forming carbon monoxide, combustion under oxygen-deficient conditions produces polynuclear aromatic hydrocarbon derivatives consisting of fused-ring structures. Some of these compounds, such as benzo(a)pyrene are pre-carcinogenic compounds, insofar as they are acted upon by enzymes in the body to yield cancer-producing metabolites.

Most investigations involving crude oil hydrocarbon derivatives are regulated by various agencies that may require methodologies, action levels, and cleanup criteria that are different. Indeed, the complex chemical composition of crude oil and crude oil products can make it extremely difficult to select the most appropriate analytical test methods for evaluating environmental samples and to accurately interpret and use the data.

Accordingly, general methods of environmental analysis (Smith, 1999), i.e., analysis for the determination of crude oil or crude oil products that have been released, are available. The data determine whether or not a release of such chemicals will be detrimental to the environment and may lead to regulations governing the use and handling of such chemicals. But first, sample collection, preservation, preparation, and handling protocols must be followed to the letter. This, of course, includes *chain of custody* or *sampling handling* protocols that will be defensible if and when legal issues arise. Thus an accurate sample handling and storage log should be maintained and should include the basic necessary information (Table 6.1) (Dean, 2003). Attention to factors such as these enables standardized comparisons to be made when subsequent samples are taken.

In summary, many of the specific chemicals in crude oil are hazardous because of their chemical reactivity, fire hazard, toxicity, and other properties. In fact, a simple definition of a hazardous chemical (or hazardous waste) is that it is a chemical substance (or chemical waste) that has been inadvertently released, discarded, abandoned, neglected, or designated as a waste material and has the potential to be detrimental to the environment. Alternatively, a hazardous chemical may be a chemical that may interact with other (chemical) substances to give a product that is hazardous to the environment.

Whatever the case, methods of analysis must be available to determine the nurture of the released chemical (waste) and from the data predict the potential hazard to the environment.

10.2.1 Air Pollution

Crude oil refineries are a source of hazardous and toxic air pollutants such as BTEX compounds (benzene, toluene, ethylbenzene, and xylene). They are also a major source of criteria air pollutants: particulate matter (PM), nitrogen oxides (NOx), carbon monoxide (CO), hydrogen sulfide (H₂S), and sulfur oxides (SOx). Refineries also release less toxic hydrocarbon derivatives such as natural gas (methane) and other light volatile fuels and oils.

Air emissions can come from a number of sources within a crude oil refinery including: equipment leaks (from valves or other devices); high-temperature combustion processes in the actual burning of fuels for electricity generation; the heating of steam and process fluids; and the transfer of products. These pollutants are typically emitted into the environment over the course of a year through normal emissions, fugitive releases, accidental releases, or plant upsets. The combination of volatile hydrocarbon derivatives and oxides of nitrogen also contribute to ozone formation, one of the most important air pollution problems.

Air emissions include point and non-point sources (Chapter 4). Point sources are emissions that exit stacks and flares and, thus, can be monitored and treated. Non-point sources are *fugitive emissions* that are difficult to locate and capture. Fugitive emissions occur throughout refineries and arise from the thousands of valves, pumps, tanks, pressure relief valves, flanges, etc. While individual leaks are typically small, the sum of all fugitive leaks at a refinery can be one of its largest emission sources.

The numerous process heaters used in refineries to heat process streams or to generate steam (boilers) for heating or steam stripping, can be potential sources of SO_x, NO_x, CO, particulates and hydrocarbon derivatives emissions. When operating properly and when burning cleaner fuels such as refinery fuel gas, fuel oil or natural gas, these emissions are relatively low. If, however, combustion is not complete, or heaters are fired with refinery fuel pitch or residuals, emissions can be significant.

The majority of gas streams exiting each refinery process contain varying amounts of refinery fuel gas, hydrogen sulfide and ammonia. These streams are collected and sent to the gas treatment and sulfur recovery units to recover the refinery fuel gas and sulfur though a variety of add-on technologies (Speight, 1993, 1996). Emissions from the sulfur recovery unit typically contain some hydrogen sulfide, sulfur oxides, and nitrogen oxides. Other emissions sources from refinery processes arise from periodic regeneration of catalysts. These processes generate streams that may contain relatively high levels of carbon monoxide, particulates and volatile organic compounds. Before being discharged to the atmosphere, such off-gas streams may be treated first through a carbon monoxide boiler to burn carbon monoxide and any volatile organic compounds, and then through an electrostatic precipitator or cyclone separator to remove particulates.

Sulfur is removed from a number of refinery process off-gas streams (sour gas) in order to meet the sulfur oxide emissions limits of the Clean Air Act and to recover saleable elemental sulfur. Process off-gas streams, or sour gas, from the coker, catalytic cracking unit, hydrotreating units and hydroprocessing units can contain high concentrations of hydrogen sulfide mixed with light refinery fuel gases.

Before elemental sulfur can be recovered, the fuel gases (primarily methane and ethane) need to be separated from the hydrogen sulfide. This is typically accomplished by dissolving the hydrogen sulfide in a chemical solvent. Solvents most commonly used are amines, such as diethanolamine

(DEA, $HOCH_2CH_2NHCH_2CH_2OH$). Dry adsorbents such as molecular sieves, activated carbon, iron sponge (Fe_2O_3) and zinc oxide (ZnO) are also used (Speight, 1993). In the amine solvent processes, diethanolamine solution or similar ethanolamine solution is pumped to an absorption tower where the gases are contacted and hydrogen sulfide is dissolved in the solution. The fuel gases are removed for use as fuel in process furnaces in other refinery operations. The amine-hydrogen sulfide solution is then heated and steam stripped to remove the hydrogen sulfide gas.

Current methods for removing sulfur from the hydrogen sulfide gas streams are typically a combination of two processes in which the primary process is the Claus Process followed by either the Beavon Process or the SCOT Process or the Wellman-Lord Process.

In the Claus process (Figure 12.1) (Speight, 2014, 2017), the hydrogen sulfide, after separation from the gas stream using *amine extraction*, is fed to the Claus unit, where it is converted in two stages. The first stage is a thermal step in which the hydrogen sulfide is partially oxidized with air in a reaction furnace at high temperatures (1000 to 1400°C, 1830 to 2550°F). Sulfur is formed, but some hydrogen sulfide remains unreacted, and some sulfur dioxide is produced. The second stage is a catalytic stage in which the remaining hydrogen sulfide is reacted with the sulfur dioxide at lower temperatures (200 to 350°C, 390 to 660°F) over a catalyst to produce more sulfur. The overall reaction is the conversion of hydrogen sulfide and sulfur dioxide to sulfur and water:

$2H_2S + SO_2 ==> 3S + 2H_2O$

The catalyst is necessary to ensure that the components react with reasonable speed but, unfortunately, the reaction does not always proceed to completion. For this reason two or three stages are used, with sulfur being removed between the stages. For the analysts, it is valuable to know that carbon disulfide (CS_2) is a by-product from the reaction in the high-temperature furnace. The carbon disulfide can be destroyed catalytically before it enters the catalytic section proper.

Generally, the Claus process may only remove about 90 percent of the hydrogen sulfide in the gas stream and, as already noted, other processes such as the Beavon process, the SCOT process, or Wellman-Lord processes are often used to further recover sulfur.

In the Beavon process, the hydrogen sulfide in the relatively low concentration gas stream from the Claus process can be almost completely removed by absorption in a quinone solution. The dissolved hydrogen sulfide is oxidized to form a mixture of elemental sulfur and hydroquinone. The solution is injected with air or oxygen to oxidize the hydroquinone back to quinone. The solution is then filtered or centrifuged to remove the sulfur and the quinone is then reused. The Beavon process is also effective in removing small amounts of sulfur dioxide, carbonyl sulfide, and carbon disulfide that are not affected by the Claus process. These compounds are first converted to hydrogen sulfide at elevated temperatures in a cobalt molybdate catalyst prior to being fed to the Beavon unit. Air emissions from sulfur recovery units will consist of hydrogen sulfide, sulfur oxides, and nitrogen oxides in the process tail gas as well as fugitive emissions and releases from vents.

The SCOT process is also widely used for removing sulfur from the Claus tail gas. The sulphur compounds in the Claus tail gas are converted to hydrogen sulfide by heating and passing it through a cobalt-molybdenum catalyst with the addition of a reducing gas. The gas is then cooled and contacted with a solution of di-isopropanolamine (DIPA) that removes all but trace amounts of hydrogen sulfide. The sulfide-rich di-isopropanolamine is sent to a stripper where hydrogen sulfide gas is removed and sent to the Claus plant. The di-isopropanolamine is returned to the absorption column.

The Wellman-Lord process is divided into two main stages: (1) absorption and (2) regeneration. In the absorption section, hot flue gases are passed through a pre-scrubber where ash, hydrogen chloride, hydrogen fluoride and sulfur trioxide are removed. The gases are then cooled and fed into the absorption tower. A saturated solution of sodium sulfite is then sprayed into the top of the absorber onto the flue gases; the sodium sulfite reacts with the sulfur dioxide forming sodium bisulfite ($NaHSO_3$). The concentrated bisulfite solution is collected and passed to an evaporation system for regeneration. In the regeneration section, sodium bisulfite is converted, using steam, to sodium sulfite that is recycled back to the flue gas. The remaining product, the released sulfur dioxide, is converted to elemental sulfur, sulfuric acid or liquid sulfur dioxide.

Most refinery process units and equipment are sent into a collection unit, called the blowdown system. Blowdown systems provide for the safe handling and disposal of liquid and gases that are either automatically vented from the process units through pressure relief valves, or that are manually drawn from units. Recirculated process streams and cooling water streams are often manually purged to prevent the continued buildup of contaminants in the stream. Part or all of the contents of equipment can also be purged to the blowdown system prior to shut down before normal or emergency shutdowns. Blowdown systems utilize a series of flash drums and condensers to separate the blowdown into its vapor and liquid components. The liquid is typically composed of mixtures of water and hydrocarbon derivatives containing sulfides, ammonia, and other contaminants, which are sent to the wastewater treatment plant. The gaseous component typically contains hydrocarbon derivatives, hydrogen sulfide, ammonia, mercaptans, solvents, and other constituents, and is either discharged directly to the atmosphere or is combusted in a flare. The major air emissions from blowdown systems are hydrocarbon derivatives in the case of direct discharge to the atmosphere and sulfur oxides when flared.

10.2.2 Water Pollution

Refineries are also potential contributors to ground water and surface water contamination. Some refineries use deep-injection wells to dispose of wastewater generated inside the plants, and some of these wastes end up in aquifers and groundwater. These wastes are then regulated under the Safe Drinking Water Act (SDWA). Wastewater in refineries may be highly contaminated and may arise from various processes (such as wastewaters from desalting, water from cooling towers, storm water, distillation, or cracking). This water is recycled through many stages during the refining process and goes through several treatment processes, including a wastewater treatment plant, before being released into surface waters.

The wastes discharged into surface waters are subject to state discharge regulations and are regulated under the Clean Water Act (CWA). These discharge guidelines limit the amounts of sulfides, ammonia, suspended solids and other compounds that may be present in the wastewater. Although these guidelines are in place, contamination from past discharges may remain in surface water bodies.

Wastewater and Treatment

Wastewaters from crude oil refining consist of process water, cooling water, storm water, and sanitary sewage water (Chapter 4).

Water used in processing operations accounts for a significant portion of the total wastewater. Process wastewater arises from desalting crude oil, steam-stripping operations, pump gland cooling, product fractionator reflux drum drains and boiler blowdown. Because process water often comes into direct contact with oil, it is usually highly contaminated. Most cooling water is recycled over and over. Cooling water typically does not come into direct contact with process oil streams and therefore contains less contaminants than process wastewater. However, it may contain some oil contamination due to leaks in the process equipment. Storm water (i.e., surface water runoff) is intermittent and will contain constituents from spills to the surface, leaks in equipment and any materials that may have collected in drains. Runoff surface water also includes water coming from crude and product storage tank roof drains. Sewage water needs no further explanation of its origins but must be treated as opposed to discharge on to the land or into ponds.

Wastewater is are treated in onsite wastewater treatment facilities and then discharged to publicly owned treatment works (POTWs) or discharged to surfaces waters under National Pollution Discharge Elimination System (NPDES) permits. Crude oil refineries typically utilize primary and secondary wastewater treatment.

Primary wastewater treatment consists of the separation of oil, water and solids in two stages. During the first stage, an API separator, a corrugated plate interceptor, or other separator design is used. Wastewater moves very slowly through the separator allowing free oil to float to the surface and be skimmed off, and solids to settle to the bottom and be scraped off to a sludge collection hopper. The second stage utilizes physical or chemical methods to separate emulsified oils from the wastewater. Physical methods may include the use of a series of settling ponds with a long retention time, or the use of dissolved air flotation (DAF). In DAF, air is bubbled through the wastewater, and both oil and suspended solids are skimmed off the top. Chemicals, such as ferric hydroxide or aluminum hydroxide, can be used to coagulate impurities into a froth or sludge that can be more easily skimmed off the top. Some wastes associated with the primary treatment of wastewater at crude oil refineries may be considered hazardous and include API separator sludge, primary treatment sludge, sludge from other gravitational separation techniques, float from DAF units, and wastes from settling ponds.

After primary treatment, the wastewater can be discharged to a publicly owned treatment works (POTW) or undergo *secondary treatment* before being discharged directly to surface waters under a National Pollution Discharge Elimination System (NPDES) permit. In secondary treatment, microorganisms may consume dissolved oil and other organic pollutants biologically. Biological treatment may require the addition of oxygen through a number of different techniques, including activated sludge units, trickling filters, and rotating biological contactors. Secondary treatment generates biomass waste that is typically treated anaerobically and then dewatered.

Some refineries employ an additional stage of wastewater treatment called *polishing* to meet discharge limits. The polishing step can involve the use of activated carbon, anthracite coal, or sand to filter out any remaining impurities, such as biomass, silt, trace metals and other inorganic chemicals, as well as any remaining organic chemicals.

Certain refinery wastewater streams are treated separately, prior to the wastewater treatment plant, to remove contaminants that would not easily be treated after mixing with other wastewater. One such waste stream is the sour water drained from distillation reflux drums. Sour water contains dissolved hydrogen sulfide and other organic sulfur compounds and ammonia which are stripped in a tower with gas or steam before being discharged to the wastewater treatment plant.

Wastewater treatment plants are a significant source of refinery air emissions and solid wastes. Air releases arise from fugitive emissions from the numerous tanks, ponds and sewer system drains. Solid wastes are generated in the form of sludge from a number of the treatment units.

Many refineries unintentionally release, or have unintentionally released in the past, liquid hydrocarbon derivatives to ground water and surface waters. At some refineries, contaminated ground water has migrated off-site and resulted in continuous *seeps* to surface waters. While the actual volume of hydrocarbon derivatives released in such a manner are relatively small, there is the potential to contaminate large volumes of ground water and surface water possibly posing a substantial risk to human health and the environment.

10.2.3 Soil Pollution

Contamination of soils from the refining processes is generally a less significant problem when compared to contamination of air and water. Past production practices may have led to spills on the refinery property that now need to be cleaned up. Natural bacteria that may use the crude oil products as food are often effective at cleaning up crude oil spills and leaks compared to many other pollutants. Many residuals are produced during the refining processes, and some of them are recycled through other stages in the process. Other residuals are collected and disposed of in landfills, or they may be recovered by other facilities. Soil contamination including some hazardous wastes, spent catalysts or coke dust, tank bottoms, and sludge from the treatment processes can occur from leaks as well as accidents or spills on or off site during the transport process.

Other Waste and Treatment

Solid wastes are generated from many of the refining processes, crude oil handling operations, as well as wastewater treatment (Chapter 4). Both hazardous and non-hazardous wastes are generated, treated and disposed. Solid wastes in a refinery are typically in the form of sludge (including sludge from wastewater treatment), spent process catalysts, filter clay, and incinerator ash. Treatment of these wastes includes incineration, land treating off-site, land filling onsite, land filling off-site, chemical fixation, neutralization, and other treatment methods (Speight, 1996; Woodside, 1999).

A significant portion of the non-crude oil product outputs of refineries is transported off-site and sold as by-products. These outputs include sulfur, acetic acid, phosphoric acid, and recovered metals. Metals from catalysts and from the crude oil that have deposited on the catalyst during the production often are recovered by third party recovery facilities.

Storage tanks are used throughout the refining process to store crude oil and intermediate process feeds for cooling and further processing. Finished crude oil products are also kept in storage tanks before transport off site. Storage tank bottoms are mixtures of iron rust from corrosion, sand, water, and emulsified oil and wax, which accumulate at the bottom of tanks. Liquid tank bottoms (primarily water and oil emulsions) are periodically drawn off to prevent their continued build up. Tank bottom liquids and sludge are also removed during periodic cleaning of tanks for inspection. Tank bottoms may contain amounts of tetraethyl or tetramethyl lead (although this is increasingly rare due to the phase out of leaded products), other metals, and phenols. Solids generated from leaded gasoline storage tank bottoms are listed as a hazardous waste.

10.3 Pollution Prevention

Pollution prevention is everyone's responsibility. Preventing pollution may be a new role for production-oriented managers and workers, but their cooperation is crucial. It will be the workers themselves who must make pollution prevention succeed in the workplace.

Several options have been identified that refineries can undertake to reduce pollution. These include pollution prevention options, recycling options, and waste treatment options. Furthermore, pollution prevention options are often is presented in four different categories, viz.: (1) pollution prevention options, (2) waste recycling, and (3) waste treatment. Either one or the other or any combination of the three options may be in operation in any given refinery.

Despite the nature of the environmental regulations and the precautions taken by the crude oil refining industry (in fact, by all chemical industries that handle chemicals—not wishing to select the refining industry as the only industry that suffers accidental release of chemicals into the environment), the accidental release of non-hazardous chemicals and hazardous chemicals into the environment has occurred and, without being unduly pessimistic, may continue to occur. It is a situation that, to paraphrase *chaos theory*, no matter how well one prepares, the unexpected is always inevitable.

It is, at this point that the refinery operators has to identity the nature of the chemicals and their potential effects on the ecosystem(s) (Smith, 1999). Although crude oil itself and its various products are complex mixtures of many organic chemicals (Chapters 2 and 3), the predominance of one particular chemical or one particular class of chemicals may offer the environmental analyst or scientist an opportunity for predictability of behavior of the chemical(s).

The increasingly long list of chemical contaminants released into the environment on a large scale includes numerous aliphatic and aromatic compounds, such as crude oil hydrocarbon derivatives. The local concentration of such contaminants on the amount present and the rate at which the compound is released, its stability in the environment under both aerobic and anaerobic conditions, the extent of its dilution in the environment, the mobility of the compound in a particular environment and its rate of biological or non-biological degradation (Harayama, 1997; Ellis, 2000; Janssen et al., 2001; Dua et al., 2002).

Remediation technologies that can reduce or remove a contaminant can be classified into four categories based on the process acting on the contaminant. These categories are removal, separation and destruction and containment, which can be either a physical, chemical or biological processes. Physical removal (Speight, 1996; Speight and Lee, 2000; Speight, 2005), isolation, microbial remediation, which includes phytoremediation for the purposes of this text, are the most commonly used remediation techniques.

The physical removal of contaminated soil and groundwater is the most common form of remediation but the process does not eliminate the contamination, rather transfers it to another location. In ideal cases, the other location will be a facility that is specially designed to contain the contamination for a sufficient period of time or treat it as necessary. In this way, proper removal reduces the risk of exposure to the environmental contaminants.

Isolation technology is typically carried out using clay, concrete, manmade liners or a combination of these and is often used a contaminant is difficult or extremely expensive to remove or destroy. The process essentially isolates the contaminant from the affected ecosystem.

In terms of microbial remediation (bioremediation), which involves the breakdown of pollutants by microorganisms, aerobic processes are considered the most efficient and generally applicable— aerobic degradation is dependent on the presence of molecular oxygen and is catalyzed by enzymes that have evolved for the catabolism of natural substrates and exhibit low specificities. Depending upon the type of enzyme catalyzing the reaction, either one (mono-oxygenase) or two (di-oxygenase) oxygen atoms are inserted into the molecule via an electrophilic attack on an unsubstituted carbon atom. Anaerobic degradation proceeds via reductive dehalogenation, wherein an electron transfer to the compound results in hydrogenation (Grant et al., 1998).

Crude oil hydrocarbon derivatives are widespread common environmental pollutants (Megharaj et al., 2000). The search for an effective remediation for crude oil-contaminated soil is a huge challenge to environmental researchers. The requirement of bioremediation is highlighted after the case of Van Daze Oil Spills in 1989 (Bragg et al., 1994). Bioremediation research has recently attracted widespread attention (Atlas, 1995; Grishchenkov et al., 2000; Chen et al., 2008). It has been reported that suitable microbes are available and can be used to effectively remediate crude oil contamination even if at low environmental temperature (Rike et al., 2003; Sanscartier et al., 2009).

Although regulations are strictly enforced in developed countries like the United States and most of the European countries to meet the challenges of crude oil-related contamination, these regulations often remain unenforced in most of the developing countries. Cleaning up such sites is often not only technically challenging but also very expensive. Considerable pressure encourages the adoption of waste management alternatives to burial, the traditional means of disposing of solid and liquid wastes.

Approaches such as air-stripping (to remove volatile compounds) and incineration have been used (Speight, 1996; Speight and Lee, 2000; Speight, 2005). However, where the contaminants infect a large area but are in low (albeit significant concentration), such methods are either very costly or simply not feasible (Blackburn and Hafker, 1993; Singh et al., 2001). In such schemes, microorganisms can provide an effective alternative through the biodegradation of the contaminants.

Since most of the contaminants of concern in crude oil are readily biodegradable under the appropriate conditions, the success of oil-spill bioremediation depends mainly on the ability to establish these conditions in the contaminated environment using the above new developing technologies to optimize the microorganisms' total efficiency. The technologies used at various polluted sites depend on the limiting factor present at the location. For example, where there is insufficient dissolved oxygen, bioventing or sparging is applied, biostimulation or bioaugmentation is suitable for instances where the biological count is low.

The most important issues related to the foregoing text are the means by which pollution prevention can be mitigated. Having defined the process products and emission (Chapters 3 and 4), *pollution prevention* is the operational guideline for refinery operators, process engineers, process chemists, and, for that matter, anyone who handles crude oil and/or crude oil products. It is in this area that environmental analysis plays a major role (EPA, 2004).

Pollution prevention is, simply, reduction or elimination of discharges or emissions to the environment. The limits of pollutants emitted to the atmosphere, the land, and water are defined by various pieces of legislation that have been put into place over the past four decades (Chapter 5) (Speight, 1996; Woodside, 1999). This includes all pollutants such as hazardous and non-hazardous wastes, regulated and unregulated chemicals from all sources. Pollution associated with crude oil refining typically includes volatile organic compounds (volatile organic compounds), carbon monoxide (CO), sulfur oxides (SO_x), nitrogen oxides (NO_x), particulates, ammonia (NH_3), hydrogen sulfide (H_2S), metals, spent acids, and numerous toxic organic compounds (Hydrocarbon Processing, 2003). Sulfur and metals result from the impurities in crude oil. The other wastes represent losses of feedstock and crude oil products. These pollutants may be discharged as air emissions, wastewater, or solid waste. All of these wastes are treated. However, air emissions are more difficult to capture than wastewater or solid waste. Thus, air emissions are the largest source of untreated wastes released to the environment.

Pollution prevention can be accomplished by reducing the generation of wastes at their source (source reduction) or by using, reusing or reclaiming wastes once they are generated (environmentally sound recycling). However, environmental analysis plays a major role in determining if emissions-effluents (air, liquid or solid) fall within the parameters of the relevant legislation. For example, issues to be addressed are the constituents of gaseous emissions, the sulfur content of liquid fuels, and the potential for leaching contaminants (through normal rainfall or through the agency of acid rain) from solid products such as coke.

The purpose of this chapter is to present a description of the methods by which crude oil products-effluents-emissions are treated in an attempt to insure that pollution does not occur and products-effluents-emissions fall within the legislative specifications. Indeed, as already noted, environmental analysis is the major discipline by which the character of the products-effluents-emissions can be determined and, hence monitored.

10.3.1 Options

Pollution prevention options are usually subdivided into four areas: (1) good operating practices, (2) processes modification, (3) feedstock modification, and (4) product reformulation (Lo, 1991). The options described here include only the first three of these categories since product reformulation is not an option that is usually available to the environmental analyst, scientist or engineer.

10.3.1.1 Operating Practices

Good operating practices (Table 10.2) prevent waste by better handling of feedstocks and products without making significant modifications to current production technology. If feedstocks are handled appropriately, they are less likely to become wastes inadvertently through spills or outdating. If products are handled appropriately, they can be managed in the most cost-effective manner.

Table 10.2: A Selection of Improved Operating Practices.

• Specify sludge and water content for feedstock
• Minimize carryover to API separator
• Use recycled water for desalter
• Replace desalting with chemical treatment system
• Collect catalyst fines during delivery
• Recover coke fines

For example, a significant portion of refinery waste arises from oily sludge found in combined process/storm sewers. Segregation of the relatively clean rainwater runoff from the process streams can reduce the quantity of oily sludge generated. Furthermore, there is a much higher potential for recovery of oil from smaller, more concentrated process streams.

Solids released to the refinery wastewater sewer system can account for a large portion of a refinery's oily sludge. Solids entering the sewer system (primarily soil particles) become coated with oil and are deposited as oily sludge in the API oil/water separator. Because a typical sludge has a solids content of five to thirty percent by weight, preventing one pound of solids from entering the sewer system can eliminate several pounds 3 to 20 pounds of oily sludge.

Methods used to control solids include using a street sweeper on paved areas, paving unpaved areas, planting ground cover on unpaved areas, re-lining sewers, cleaning solids from ditches and catch basins, and reducing heat exchanger bundle cleaning solids by using an antifoulant in the cooling water.

Benzene and other solvents in wastewater can often be treated more easily and effectively at the point at which they are generated rather than at the wastewater treatment plant after it is mixed with other wastewater.

10.3.1.2 Process Modifications

The crude oil industry requires very large, capital-intensive process equipment. Expected lifetimes of process equipment are measured in decades. This limits economic incentives to make capital-intensive process modifications to reduce wastes generation. However, some process modifications (Table 10.3) or process improvement (Table 10.4) reduce waste generation.

The crude oil industry has made many improvements in the design and modification of processes and technologies to recover product and unconverted raw materials. In the past, they pursued this strategy to the point that the cost of further recovery could not be justified. Now the costs of end-of-pipe treatment and disposal have made source reduction a good investment. Greater reductions are possible when process engineers trained in pollution prevention plan to reduce waste at the design stage. For example, although barge loading is not a factor for all refineries, it is an important emissions source for many facilities. One of the largest sources of volatile organic carbon emissions is the fugitive emissions from loading of tanker barges. These emissions could be reduced by more

Table 10.3: Options for Process Modifications.

• Add coking operations. 　Certain refinery hazardous wastes can then be used as coker feedstock, reducing the quantity of sludge for disposal.
• Install secondary seals on floating roof tanks. 　Where appropriate, replace with fixed roofs to eliminate the collection of rainwater, contamination of crude oil or finished products, and oxidation of crude oil.
• Where feasible, 　o Replace clay filtration with hydrotreating. 　o Substitute air coolers or electric heaters for water heat exchangers to reduce sludge production. 　o Install tank agitators. This can prevent solids from settling out. 　o Concentrate similar wastewater streams through a common dewatering system.

Table 10.4: Process Improvement.

• Segregate oily wastes to reduce the quantity of oily sludge generated and increase the potential for oil recovery.
• Reuse rinse waters where possible.
• Use optimum pressures, temperatures and mixing ratios.
• Sweep or vacuum streets and paved process areas to reduce solids going to sewers.
• Use water softeners in cooling water systems to extend the useful life of the water.

than ninety percent by installing a vapor loss control system that consists of vapor recovery or the destruction of the volatile organic carbon emissions in a flare.

Fugitive emissions are one of the largest sources of refinery hydrocarbon emissions. A leak detection and repair (LDAR) program consists of using a portable detecting instrument to detect leaks during regularly scheduled inspections of valves, flanges, and pump seals. Older refinery boilers may also be a significant source of emissions of sulfur oxides (SO_x), nitrogen oxides (NO_x), and particulate matter. It is possible to replace a large number of old boilers with a single new cogeneration plant with emissions controls.

Since storage tanks are one of the largest sources of VOC emissions, a reduction in the number of these tanks can have a significant impact. The need for certain tanks can often be eliminated through improved production planning and more continuous operations. By minimizing the number of storage tanks, tank bottom solids and decanted wastewater may also be reduced. Installing secondary seals on the tanks can significantly reduce the losses from storage tanks containing gasoline and other volatile products.

Solids entering the crude distillation unit are likely to eventually attract more oil and produce additional emulsions and sludge. The amount of solids removed from the desalting unit should, therefore, be maximized. A number of techniques can be used such as: using low shear mixing devices to mix desalter wash water and crude oil; using lower pressure water in the desalter to avoid turbulence; and replacing the water jets used in some refineries with mud rakes which add less turbulence when removing settled solids.

Purging or blowing down a portion of the cooling water stream to the wastewater treatment system controls the dissolved solids concentration in the recirculating cooling water. Solids in the blowdown eventually create additional sludge in the wastewater treatment plant. However, minimizing the dissolved solids content of the cooling water can lower the amount of cooling tower blowdown. A significant portion of the total dissolved solids in the cooling water can originate in the cooling water makeup stream in the form of naturally occurring calcium carbonates. Such solids can be controlled either by selecting a source of cooling tower makeup water with less dissolved solids or by removing the dissolved solids from the makeup water stream. Common treatment methods include: cold lime softening, reverse osmosis, or electrodialysis.

In many refineries, using high-pressure water to clean heat exchanger bundles generates and releases water and entrained solids to the refinery wastewater treatment system. Exchanger solids may then attract oil as they move through the sewer system and may also produce finer solids and stabilized emulsions that are more difficult to remove. Solids can be removed at the heat exchanger cleaning pad by installing concrete overflow weirs around the surface drains or by covering drains with a screen. Other ways to reduce solids generation are by using anti-foulants on the heat exchanger bundles to prevent scaling and by cleaning with reusable cleaning chemicals that also allow for the easy removal of oil.

Surfactants entering the refinery wastewater streams will increase the amount of emulsions and sludge generated. Surfactants can enter the system from a number of sources including: washing unit pads with detergents; treating gasoline with an end point over 200°C (> 392°F) thereby producing spent caustics; cleaning tank truck tank interiors; and using soaps and cleaners for miscellaneous tasks. In addition, the overuse and mixing of the organic polymers used to separate oil, water, and solids in the wastewater treatment plant can actually stabilize emulsions. The use of surfactants should be minimized by educating operators, routing surfactant sources to a point downstream of the DAF unit and by using dry cleaning, high pressure water or steam to clean oil surfaces of oil and dirt.

Replacing 55-gallon drums with bulk storage facilities can minimize the chances of leaks and spills. And, just as 55-gallon drums can lead to leaks, underground piping can be a source of undetected releases to the soil and groundwater. Inspecting, repairing or replacing underground piping with surface piping can reduce or eliminate these potential sources.

Finally, open ponds used to cool, settle out solids and store process water can be a significant source of volatile organic carbon emissions. Wastewater from coke cooling and coke volatile organic carbon removal is occasionally cooled in open ponds where volatile organic carbon easily escape to the atmosphere. In many cases, open ponds can be replaced with closed storage tanks.

10.3.1.3 Material Substitution Options

Spent conventional degreaser solvents can be reduced or eliminated through substitution with less toxic and/or biodegradable products. In addition, chromate containing wastes can be reduced or eliminated in cooling tower and heat exchanger sludge by replacing chromates with less toxic alternatives such as phosphates.

Using catalysts of a higher quality will lead in increased process efficiency while the required frequency of catalyst replacement can be reduced. Similarly, the replacement of ceramic catalyst support with activated alumina supports presents the opportunity for recycling the activated alumina supports with the spent alumina catalyst.

Recycling

Recycling is the use, reuse or reclamation of a waste after it is generated. At present the crude oil industry is focusing on recycling and reuse as the best opportunities for pollution prevention (Table 10.5). Although pollution is reduced more if wastes are prevented in the first place, a next best option for reducing pollution is to treat wastes so that they can be transformed into useful products.

Caustic substances used to absorb and remove hydrogen sulfide and phenol contaminants from intermediate and final product streams can often be recycled. Spent caustics may be saleable to chemical recovery companies if concentrations of phenol or hydrogen sulfide are high enough. Process changes in the refinery may be needed to raise the concentration of phenols in the caustic to make recovery of the contaminants economical. Caustics containing phenols can also be recycled on-site by reducing the pH of the caustic until the phenols become insoluble thereby allowing physical separation. The caustic can then be treated in the refinery wastewater system.

Oily sludge can be sent to a coking unit or the crude distillation unit where it becomes part of the refinery products. Sludge sent to the coker can be injected into the coke drum with the quench water, injected directly into the delayed coker, or injected into the coker blowdown contactor used in separating the quenching products. Use of sludge as a feedstock has increased significantly in recent years and is currently carried out by most refineries. The quantity of sludge that can be sent to the coker is restricted by coke quality specifications that may limit the amount of sludge solids in the coke. Coking operations can be upgraded, however, to increase the amount of sludge that they can handle.

Significant quantities of catalyst fines are often present around the catalyst hoppers of fluid catalytic cracking reactors and regenerators. Coke fines are often present around the coker unit and coke storage areas. The fines can be collected and recycled before being washed to the sewers or migrating off-site via the wind. Collection techniques include dry sweeping the catalyst and coke fines and sending the solids to be recycled or disposed of as non-hazardous waste. Coke fines can

Table 10.5: Options for Recycling.

• Use phenols and caustics produced in the refining operations as chemical feeds in other applications.
• Use oily waste sludge as feedstock in coking operations.
• Regenerate catalysts. Extend useful life. Recover valuable metals from spent catalyst. Possibly use catalyst as a concrete admixture or as a fertilizer.
• Maximize slop oil recovery. Agitate sludge with air and steam to recover residual oils.
• Regenerate filtration clay. Wash clay with naphtha, dry by steam heating and feed to a burning kiln for regeneration.
• Recover valuable product from oily sludge with solvent extraction.

also be recycled for fuel use. Another collection technique involves the use of vacuum ducts in dusty areas (and vacuum hoses for manual collection) that run to a small baghouse for collection.

An issue that always arises relates to the disposal of laboratory sample from any process control or even environmental laboratory that is associated with a refinery. Samples from such a laboratory can be recycled to the oil recovery system.

Treatment Options

When pollution prevention and recycling options are not economically viable, pollution can still be reduced by treating wastes so that they are transformed in to less environmentally harmful wastes or can be disposed of in a less environmentally harmful media.

The toxicity and volume of some de-oiled and dewatered sludge can be further reduced through thermal treatment. Thermal sludge treatment units use heat to vaporize the water and volatile components in the feed and leave behind a dry solid residue. The vapors are condensed for separation into the hydrocarbon and water components. Non-condensable vapors are either flared or sent to the refinery amine unit for treatment and use as refinery fuel gas.

Furthermore, because oily sludge makes up a large portion of refinery solid wastes, any improvement in the recovery of oil from the sludge can significantly reduce the volume of waste. There are a number of technologies currently in use to mechanically separate oil, water and solids, including: belt filter presses, recessed chamber pressure filters, rotary vacuum filters, scroll centrifuges, disc centrifuges, shakers, thermal driers and centrifuge-drier combinations.

Waste material such as tank bottoms from crude oil storage tanks constitute a large percentage of refinery solid waste and pose a particularly difficult disposal problem due to the presence of heavy metals. Tank bottoms are comprised of heavy hydrocarbon derivatives, solids, water, rust and scale. Minimization of tank bottoms is carried out most cost effectively through careful separation of the oil and water remaining in the tank bottom. Filters and centrifuges can also be used to recover the oil for recycling.

Spent clay from refinery filters often contains significant amounts of entrained hydrocarbon derivatives and, therefore, must be designated as hazardous waste. Back washing spent clay with water or steam can reduce the hydrocarbon content to levels so that it can be reused or handled as a non-hazardous waste. Another method used to regenerate clay is to wash the clay with naphtha, dry it by steam heating and then feed it to a burning kiln for regeneration. In some cases clay filtration can be replaced entirely with hydrotreating process options.

Decant oil sludge from the fluidized bed catalytic cracking unit can (and often does) contain significant concentrations of catalyst fines. These fines often prevent the use of decant oil as a feedstock or require treatment which generates an oily catalyst sludge. Catalyst fines in the decant oil can be minimized by using a decant oil catalyst removal system. One system incorporates high voltage electric fields to polarize and capture catalyst particles in the oil. The amount of catalyst fines reaching the decant oil can be minimized by installing high efficiency cyclones in the reactor to shift catalyst fines losses from the decant oil to the regenerator where they can be collected in the electrostatic precipitator.

10.4 Adoption of Pollution Reduction Options

Although numerous cases have been documented where crude oil refineries have simultaneously reduced pollution and operating costs, there are often barriers to doing so. The primary barrier to most pollution reduction projects is cost. Many pollution reduction options simply do not pay for themselves. Corporate investments typically must earn an adequate return on invested capital for the shareholders and some pollution prevention options at some facilities may not meet the requirements set by the companies. In addition, the equipment used in the crude oil refining industry are very capital intensive and have very long lifetimes. This reduces the incentive to make process modifications to (expensive) installed equipment that is still useful. It should be noted that pollution prevention techniques are, nevertheless, often more cost-effective than pollution reduction through end-of-pipe treatment.

Of course, facility training programs that emphasize the importance of keeping solids out of the sewer systems will help reduce that portion of wastewater treatment plant sludge arising from the everyday activities of refinery personnel. For example, educating personnel on how to avoid leaks and spills can reduce contaminated soil.

A systematic approach will produce better results than piecemeal efforts. An essential first step is a comprehensive waste audit (Table 10.6). The waste audit should systematically evaluate opportunities for improved operating procedures, process modifications, process redesign and recycling.

Crude oil refinery wastes result from processes designed to remove naturally occurring contaminants in the crude oil, including water, sulfur, nitrogen and heavy metals (Table 10.7). Setting up a pollution prevention program does not require exotic or expensive technologies. Some of the most effective techniques are simple and inexpensive. Others require significant capital expenditures, however many provide a return on that investment.

Biodegradation has shown great promise for cleanup of chemical spills, especially spills of crude oil-related contaminants. It is one of several viable options that are available as a single or piggy-back method for environmental cleanup but considerations such as (1) the characteristics of the site, (2) the properties of the spilled material, (3) the prevailing climatic conditions, and (4) the nature of the micro-organisms that are available must be taken into account.

In the last three decades biodegradation-bioremediation has become one of the most rapidly developing fields of environmental restoration insofar as, in many cases, microorganisms have been used successfully to reduce the concentration and toxicity of chemicals such as crude oil-related hydrocarbon derivatives, polycyclic aromatic hydrocarbon derivatives, and other chemicals such as metals and metallo-organic compounds. However, although simple aromatic compounds are

Table 10.6: Elements of a Waste Audit.

• List all generated waste
• Identify the composition of the waste and the source of each substance
• Identify options to reduce the generation of these substances in the production or manufacturing process
• Focus on wastes that are most hazardous and techniques that are most easily implemented
• Compare the technical and economic feasibility of the options identified
• Evaluate the results and schedule periodic reviews of the program so it can be adapted to reflect changes in regulations, technology, and economic feasibility.

Table 10.7: Options for Waste Reduction.

• Segregate process (oily) waste streams from relatively clean rainwater runoff in order to reduce the quantity of oily sludge
• Generated and increased the potential for oil recovery. Significant portion of the refinery waste comes from oily sludge found in combined process/storm sewers.
• Conduct inspection of petroleum refinery systems for leaks. For example, check hoses, pipes, valves, pumps and seals. Make necessary repairs where appropriate.
• Conserve water. Reuse rinse waters if possible. Reduce equipment-cleaning frequency where beneficial in reducing net waste generation.
• Use correct pressures, temperatures and mixing ratios for optimum recovery of product and reduction in waste produced.
• Employ street sweeping or vacuuming of paved process areas to reduce solids to the sewers.
• Pave runoff areas to reduce transfer of solids to waste systems. Use water softeners in cooling water systems to extend useful cycling time of the water.

biodegradable by a variety of degradative pathways, the halogenated derivatives are more resistant to microbial attack and necessitates the development of suitable degradation pathways.

However, selecting the most appropriate strategy to treat a specific site can be guided by considering three basic principles: (i) the amenability of the pollutant to biological transformation to less toxic products, (ii) the accessibility of the contaminant to microorganisms, and (iii) the opportunity for optimization of microbial activity.

As the processes evolve, bioremediation will be the preferred method for the long-term restoration of polluted systems with the added advantage of environmental compatibility with the system. This will, of necessity, enable process optimization and the more efficient application of biological degradation processes to the removal of chemical contaminants under a variety of climatic conditions. However, many pollutants continue to persist in the environment, a number of issues have been identified as challenges posed to the microorganisms working in contaminated sites, such as: (i) the development of a range of suitable microorganisms for site cleanup, (ii) the bioavailability of the contaminant to microorganisms, (iii) the applicability of microorganisms to a range of chemical pollutants, (iv) an inadequate supply of nutrients for the microorganisms to survive, and (v) sufficient biochemical potential for effective biodegradation.

Thus, selecting the most appropriate strategy to treat a specific site can be guided by considering three basic principles: (i) the amenability of the pollutant to biological transformation to less toxic products, and (ii) the opportunity for bioprocess optimization.

In summary, the potential and success of microorganisms in the remediation of spills of chemicals including the widely-variable crude oil and crude oil-related products has been widely acknowledged. As advances in biotechnology continue, bioremediation will continue to be rapidly growing area of preference for the treatment of hazardous wastes and contaminated sites.

10.5 The Future of Bioremediation

There have been many suggestions about the future of the crude oil industry and the reserves of crude oil that are available. Among these suggestions is one that the bulk of the world's oil and gas has already been discovered and that declining production is inevitable. Another suggestion is that substantial amounts of oil and gas remain to be found. There are also suggestions that fall between these two extremes.

In the last two decades, new fields have indeed been discovered, for example, in Kazakhstan near the Caspian Sea, and the potential for crude oil discoveries have opened up in Eastern Europe, Asia, in Canadian coastal areas, and in Colombia. Potentially the richest discovery has been the finding of vast reserves in deep water in the Gulf of Mexico. These reserves were only beginning to be tapped in the mid-1990s, using floating platforms (Chapter 6) tethered to the sea bottom by steel cables, and such innovative technologies as the use of deep water robotic machines for construction and maintenance.

Liquid fuel sources that still remain to be exploited include tar sand deposits (Chapters 3 and 5), oil shale (Stouten, 1990; Speight, 2008, 2012), and the liquefaction and gasification of coal (Speight, 2013). All attempts to utilize these sources have proved so far to be uneconomic compared to the costs of producing oil and natural gas. Future technologies may, however, find ways of creating viable fuels from these various substances. That being the case, and although oil is now recognized as likely to be abundant into the first fifty years of the twenty first century, environmental concerns will probably impose increasing restrictions on both its production and consumption.

Thus, the general prognosis for emission cleanup is not pessimistic and can be looked upon as being quite optimistic. Indeed, it is considered likely that most of their environmental impact of crude oil refining can be substantially abated. A considerable investment in retrofitting or replacing existing facilities and equipment might be needed. However, it is possible and a conscious goal must be to improve the efficiency with which crude oil is transformed and consumed.

Obviously, much work is needed to accommodate the continued use of crude oil. In the meantime, we use what we have, all the while working to improve efficient usage and working to ensure that there is no damage to the environment. Such is the nature of crude oil refining the expectancy of protecting the environment.

Over the past decade, opportunities for applying bioremediation to a much broader set of contaminants have been identified. Indigenous and enhanced organisms have been shown to degrade industrial solvents, polychlorinated biphenyls (PCBs), explosives, and many different agricultural chemicals. Pilot, demonstration, and full-scale applications of bioremediation have been carried out on a continuing basis. However, the full benefits of bioremediation have not been realized because processes and organisms that are effective in controlled laboratory tests are not always equally effective in full-scale applications. The failure to perform optimally in the field setting stems from a lack of predictability due, in part, to inadequacies in the fundamental scientific understanding of how and why these processes work.

This, if bioremediation is to be effective, the microorganisms must enzymatically attack the pollutants and convert them to non-contaminating products—some microbes may produce products that are not only toxic to themselves but also the ecosystem. Parameters that affect the bioremediation process include temperature, nutrients (fertilizers), and the amount of oxygen present in the soil and/ or the affected water system (Chapter 1, Chapter 9). These conditions allow the microbes to grow and multiply and consume more of the contaminant. When conditions are adverse, microbes grow too slowly or die or they can create more harmful chemicals. In addition, the application of any technology is dependent not only on the availability of the technology but also on the reliability of the technology as well as on the suitability of the technology for the specific site conditions and whether the technology is readily available (i.e., emerging, developing, or proven).

10.5.1 Conventional Bioremediation

Conventional bioremediation methods used are composting, land farming, biopiling and bioslurry reactors. Composting is a technique that involves combining contaminated soil with nonhazardous organic additives such as manure or agricultural wastes; the presence of the organic materials allows the development of a rich microbial population and elevated temperature characteristic of composting. Land-farming is a simple technique in which contaminated soil is excavated and spread over a prepared bed and periodically tilled until pollutants are degraded. While biopiling is a hybrid of land-farming and composting, it is essentially engineered cells that are constructed as aerated composted piles.

Bioslurry reactors can provide rapid biodegradation of contaminants due to enhanced mass transfer rates and increased contaminant-to-microorganism contact. These units are capable of treating high concentrations of organic contaminants in soil and sludge. These reactors can aerobically biodegrade aqueous slurries created through the mixing of soils or sludge with water. The most common state of bioslurry treatment is batch; however, continuous-flow operation can be achieved.

Microorganisms excel at using organic substances, natural or synthetic, as sources of nutrients and energy. Indeed, the diversity of crude oil-related constituents for growth had led to the discovery of enzymes capable of transforming many unrelated natural organic compounds by many different catalytic mechanisms (Butler and Mason, 1997; Ellis, 2000).

However, depending on behavior in the environment, organic compounds are often classified as biodegradable, persistent or recalcitrant. A biodegradable organic compound is one that undergoes a biological transformation (Blackburn and Hafker, 1993; Liu and Suflita, 1993). A persistent organic compound does not undergo biodegradation in certain environments; and a recalcitrant compound resists biodegradation in a wide variety of environments. While partial biodegradation is usually an alteration by a single reaction, primary biodegradation involves a more extensive chemical change.

Mineralization is a parallel term to biodegradation, referring to complete degradation to the end products of carbon dioxide, water, and other inorganic compounds.

Biodegradation and its application in bioremediation of organic pollutants have benefited from the biochemical and molecular studies of microbial processes (Lal et al., 1986; Fewson, 1988; Sangodkar et al., 1989; Chaudhary and Chapalamadugu, 1991; Bollag and Bollag, 1992; Van der Meer et al., 1992; Dickel et al., 1993; Deo et al., 1994; Johri et al., 1996; Kumar et al., 1996; Johri et al., 1999; Faison, 2001; Janssen et al., 2001). Indeed, the biotransformation of organic contaminants in the natural environment has been extensively studied to understand microbial ecology, physiology and evolution for their potential in bioremediation (Bouwer and Zehnder, 1993; Chen et al., 1999; Johan et al., 2001; Mishra et al., 2001; Watanabe, 2001).

As a result, there is a strong demand to increase the adoption of bioremediation as an effective technique for risk reduction on hydrocarbon impacted soils (Diplock et al., 2009). However, the biodegradation effectiveness diminishes with the time extension and the inhibiting effect may become dominant with time. The key solution to bioremediation is to speed up the restoration process and eliminate or delay the inhibitory effect, such as through the selection of specifically targeted strains or microorganisms (Marijke and Vlerken, 1998), or through the alteration of microbial community structure changes (Antizar-Ladislao et al., 2008) during the treatment.

Like bioremediation, phytoremediation has recently been developed as a remedial strategy for organic contaminants. It is believed that phytoremediation is much less disruptive to the environment and may have a high probability of public acceptance as a low cost alternative (Alkorta and Garbisu, 2001). Furthermore, rhizosphere microbes can become contaminant degraders under stress condition (Wenzel, 2009; Gerhart et al., 2009; Korade and Fulekar, 2009).

The recent successful implementation of phytoremediation by a number of researchers (Escalante-Espinosa et al., 2005; Lin et al., 2005; Merkl et al., 2005; Erute Magdalene et al., 2009) has indicated that such an approach is quite promising and can be a viable alternative to the conventional bioremediation. However, the use of phytoremediation is constrained by climate and geological conditions of the sites to be cleaned such as temperature, altitude, soil type, and the accessibility by agricultural equipment (Macek et al., 2000). Plants growing in crude oil-contaminated soil have to cope with the nutrient deficiency and hydrocarbon toxicity the complicated soil characteristics such as the heterogeneous distribution of the crude oil, soil structure, nutrient shortage and transplanting which may cause some unexpected consequences. Furthermore the selection of proper plant species (Salt et al., 1998) is also crucial for the experiment. In general, phytoremediation can be a commercial strategy for low total crude oil hydrocarbon (TPH) contaminated soil, which could improve soil characteristics with minimal risks.

Pollutants sometimes cannot be removed completely by a single remediation process, especially when using biological methods. Therefore multi-process bioremediation provides a promising and environmentally friendly solution, which is cost-effective and pollution-free during the active cleaning process. In addition, multi-process bioremediation may become an effective strategy for rapid biodegradation by altering microbial community structure. It is thus a challenging and rewarding research to search for an innovative solution to speed up TPH reduction for effective environmental cleaning.

10.5.2 Enhanced Bioremediation

The natural processes that drive bioremediation can be enhanced to increase the effectiveness and to reduce time required to meet cleanup objectives. Enhanced bioremediation involves the addition of microorganisms (e.g., fungi, bacteria, and other microbes) or nutrients (e.g., oxygen, nitrates) to the subsurface environment to accelerate the natural biodegradation process—a process in which indigenous degrade (metabolize) organic contaminants found in soil and/or ground water and convert them to innocuous end products. The process relies on general availability of naturally occurring microbes to consume contaminants as a food source (crude oil hydrocarbon derivatives in aerobic processes) or as an electron acceptor (chlorinated solvents). In addition to microbes being present, in order to be successful, these processes require nutrients (carbon: nitrogen: phosphorus).

The potential of microorganisms in the remediation of some of the compounds hitherto known to be non-biodegradable has been widely acknowledged globally. With advances in biotechnology, bioremediation has become a rapidly growing area and has been commercially applied for the treatment of hazardous wastes and contaminated sites. Although a wide range of new microorganisms have been discovered that are able to degrade highly stable, toxic organic xenobiotic, still many pollutants persist in the environment.

Briefly, a xenobiotic is a chemical which is found in an organism, such as a bacterium, but which is not normally produced or expected to be present in the organism. The word is very often used in the context of a pollutant and is a substance that is not indigenous to an ecosystem or a biological system and which did not exist in nature before human intervention.

A number of reasons have been identified as challenges posed to the micro-organisms working in contaminated sites (Dua et al., 2002). Such potential limitations to biological treatments include: poor bioavailability of chemicals, presence of other toxic compounds, inadequate supply of nutrients and insufficient biochemical potential for effective biodegradation. A wide range of bioremediation strategies have been developed for the treatment of contaminated soils using natural and modified microorganisms.

Selecting the most appropriate strategy to treat a specific site can be guided by considering three basic principles: the amenability of the pollutant to biological transformation to less toxic products, the bioavailability of the contaminant to microorganisms and the opportunity for bioprocess optimization. With the help of advances in bioinformatics, biotechnology holds a bright future for developing bioprocesses for environmental applications.

Biotechnological processes for the bioremediation of crude oil-related pollutants offer the possibility of *in situ* treatments and are mostly based on the natural activities of microorganisms. Biotechnological processes to destroy contaminants of the type found in crude oil and crude oil products offer many advantages over physicochemical processes (Table 10.8). When successfully operated, biotechnological processes may achieve complete destruction of crude oil-related pollutants. However, an important factor limiting the bioremediation of sites contaminated with such contaminants is the slow rate of degradation (Iwamoto and Nasu, 2001), which may limit the practicality of using microorganisms in remediating contaminated sites. This is an area where

Table 10.8: Advantages and Disadvantages of Bioremediation.

Advantages
1) A natural process.
2) Toxic chemicals are destroyed or removed from environment.
3) Can affect complete destruction of a wide variety of contaminants.
4) Can transform contaminants to harmless products.
5) Can be carried out on site without causing a major disruption of normal activities.
6) If performed *in situ*, no excavation or transport costs.
7) Less energy is required as compared to other technologies.
8) Microbes do not need manual supervision.

Disadvantages
1) Not all contaminants are biodegradable.
2) The microbial action may be highly specific to one site only.
3) Heavy metals remain on the site.
4) The products may be more toxic than the original contaminant(s).
5) For *in situ* bioremediation, the site must have highly permeable soil.
6) Process may not remove all quantities of contaminants.
7) Extrapolation from laboratory scale to field scale my by subject to inaccuracies.
8) The contaminant may be a complex mixture that is not evenly dispersed.
9) The contaminant may be present as solids, liquids, and gases.

genetic engineering can make a marked improvement. Molecular techniques can be used to increase the level of a particular protein or enzyme or series of enzymes in bacteria with an increase in the reaction rate (Chakrabarty, 1986).

Biosurfactants are surface-active microbial products that have numerous industrial applications (Desai and Banat, 1997; Sullivan, 1998; Sekelsky and Shreve, 1999). Many micro-organisms, especially bacteria, produce biosurfactants when grown on water-immiscible substrates. Most common biosurfactants are glycolipids in which carbohydrates are attached to a long-chain aliphatic acid, while others such as lipopeptide derivatives, lipoprotein derivatives, and hetero-polysaccharide derivatives, are more complex. The most promising applications of biosurfactants are in the cleaning of oil-contaminated tankers, oil-spill management, transportation of heavy crude, enhanced oil recovery, recovery of oil from sludge, and bioremediation of sites contaminated with hydrocarbon derivatives, heavy metals, and other crude oil-related pollutants.

Furthermore, the bioremediation of polynuclear aromatic hydrocarbon derivatives was designed by the addition of surfactants; and mathematical models were constructed to explain the effect of surfactants on biodegradation (Harayama, 1997). With the increasing awareness of the applicability of biosurfactants, the focus is now on the utilization of biosurfactants for the bioremediation of non-aqueous-phase liquids (NAPLs).

Advances in genetic and protein engineering techniques have opened up new avenues to move towards the goal of *genetically engineered microorganisms* to function as biocatalysts, in which certain desirable biodegradation pathways or enzymes from different organisms are brought together in a single host with the aim of performing specific reactions (Masai et al., 1995; Hauschild et al., 1996; Timmis and Piper, 1999; Sayler and Ripp, 2000). A strategy has also been suggested (Timmis and Piper, 1999) for designing organisms with novel pathways and the creation of a bank of genetic modules encoding broad-specificity enzymes or pathway segments that can be combined at will to generate new or improved activities.

Crude oil and crude oil products, as target pollutants are difficult to identify because of the complexity of the crude oil system (Chapter 4, Chapter 5, Chapter 6). The presence of metals and metallo-organic constituents, which are not destroyed biologically but are only transformed from one oxidation state to another, interfere with the bioremediation processes. Genetic engineering allows transferring the heavy metal-resistance genes to identifiable micro-organism hosts that can then serve as excellent the base from which to construct recombinant strains to overcome the challenge of the metal constituents of crude oil.

Methods for the rapid and specific identification of microorganisms within their natural environments continue to be developed. Classic methods are time-consuming and only work for a limited number of microorganisms (Amann et al., 2001). An increasing need to develop new methods for characterization of microorganisms able to degrade the various types of crude oil-related pollutants has led to the use of molecular probes to identify, enumerate, and isolate microorganisms with degradative potential.

Through the genetic engineering of metabolic pathways, it is possible to extend the range of substrates that an organism can utilize. Aromatic hydrocarbon di-oxygenases have broad substrate specificity and catalyze enantio-specific reactions with a wide range of substrates.

10.5.3 Bioremediation in Extreme Environments

The biodegradation of many components of crude oil hydrocarbon derivatives has been reported in a variety of terrestrial and marine cold ecosystems, extreme environments such including as alpine soil (Margesin, 2000), Arctic soil (Braddock et al., 1997), Arctic seawater (Siron et al., 1995), Antarctic soil (Aislabie et al., 1998; Aislabie et al., 2000, 2004), as well as Antarctic seawater and sediments (Delille et al., 1998; Delille and Delille, 2000; Al-Darbi et al., 2005).

In addition, many environments are characterized not only by low or elevated temperatures but also by acidic or alkaline pH, high salt concentrations, or high pressure. Extremophilic microorganisms are adapted to grow and thrive under these adverse conditions. Hydrocarbon degrading extremophiles are thus ideal candidates for the biological treatment of polluted extreme habitats (Margesin and Schinner, 2001). A wide range of hydrocarbon derivatives that contaminate the environment can be biodegraded (mineralized or transformed to carbon dioxide and water) in various extreme environments that are characterized by low or elevated temperatures, acidic or alkaline pH, high salinity or high pressure. This emphasizes the metabolic capacities of extremophilic microorganisms and the microbes that are adapted to more than one extreme offer an increased potential for the biological decontamination of habitats where various different extreme conditions prevail.

Antarctic exploration and research have led to some significant although localized impacts on the environment. Human impacts occur around current or past scientific research stations, typically located on ice-free areas that are predominantly soils. Fuel spills, the most common occurrence, have the potential to cause the greatest environmental impact in the Antarctic through accumulation of aliphatic and aromatic compounds. Effective management of hydrocarbon spills is dependent on understanding how they impact soil properties such as moisture, hydrophobicity, soil temperature, and microbial activity. Numbers of hydrocarbon-degrading bacteria, typically *Rhodococcus*, *Sphingomonas*, and *Pseudomonas* species for example, may become elevated in contaminated soils, but overall microbial diversity declines. Alternative management practices to the current approach of *dig it up and ship it out* are required but must be based on sound information (Aislabie et al., 2004).

Cold-tolerant bacteria, isolated from oil-contaminated soils in Antarctica, were able to degrade *n*-alkanes (C_6 to C_{20}) typical of the hydrocarbon contaminants that persist in Antarctic soil (Bej et al., 2000). Representative isolates were identified as *Rhodococcus* species, they retained metabolic activity at sub-zero temperatures of $-2°C$. A psychrotrophic *Rhodococcus* sp. from Arctic soil (Whyte et al., 1998) utilized a broad range of aliphatic compounds (C_{10} to C_{21} alkanes, branched alkanes, and a substituted cyclohexane) present in diesel oil at $5°C$ ($41°F$). The strain mineralized the short-chain alkanes (C_{10} and C_{16}) to a significantly greater extent than the long-chain alkanes (C_{28} and C_{32}) at 0 and $5°C$ (32 and $41°F$). The decreased bioavailability of the long-chain alkanes at low temperature (many form crystals at $0°C$, $32°F$) may be responsible for their increased recalcitrance, which affects *in situ* bioremediation in cold climates.

In addition, the physical environment is also important for hydrocarbon biodegradation. This has been demonstrated in sub-Antarctic intertidal beaches (Delille and Delille, 2000) and in Arctic soils (Mohn and Stewart, 2000). Soil nitrogen and concentrations of total crude oil hydrocarbon derivatives together accounted for 73% of the variability of the induction period (lag time) for dodecane mineralization at $7°C$ ($45°F$) in Arctic soils. High total carbon concentrations were associated with high mineralization rates; high sand content resulted in longer half-times for mineralization. Dodecane mineralization was limited by both N and P; mineralization kinetics varied greatly among different soils (Mohn and Stewart, 2000).

Cold habitats possess sufficient indigenous microorganisms, psychrotrophic bacteria (bacteria capable of surviving or even thriving in a cold environment) being predominant. They adapt rapidly to the presence of the contaminants, as demonstrated by significantly increased numbers of oil degraders shortly after a pollution event, even in the most northerly areas of the world (Whyte et al., 1999). However, the temperature threshold for significant crude oil biodegradation is approximately $0°C$ ($32°F$) (Siron et al., 1995).

However, the bulk of information on hydrocarbon degradation borders on activities of mesophiles, although significant biodegradations of hydrocarbon derivatives have been reported in psychrophilic environments in temperate regions (Yumoto et al., 2002; Pelletier et al., 2004; Delille et al., 2004). Full-scale *in situ* remediation of crude oil contaminated soils has not yet been used in Antarctica for example, partly because it has long been assumed that air and soil temperatures are too low for an effective biodegradation (Delille et al., 2004). Such omissions in research programs need to be corrected.

The technical feasibility of *in situ* bioremediation of hydrocarbon derivatives in cold groundwater systems has been demonstrated (Bradley and Chapelle, 1995). Rapid aerobic toluene mineralization was demonstrated in sediments from a cold (mean groundwater temperature 5°C/41°F) crude oil-contaminated aquifer in Alaska. The mineralization rate obtained at 5°C in this aquifer was comparable to that measured in sediments from a temperate aquifer (mean temperature 20°C). Rates of overall microbial metabolism in the two sediments were comparable at their respective *in situ* temperatures.

Bioaugmentation of contaminated cold sites with hydrocarbon-degrading bacteria has been tested as a bioremediation strategy. For the most part, indigenous microbial populations degrade hydrocarbon derivatives more efficiently than the introduced microbial strains. However, bioaugmentation (through the use of non-indigenous microbial species) may biodegradation through the onset of a shorter hydrocarbon acclimation period (Mohammed et al., 2007).

Inoculation of contaminated Arctic soils with consortia (Whyte et al., 1999) or with alkane-degrading *Rhodococcus* sp. (Whyte et al., 1998) decreased the lag time and increased the rate of C_{16} mineralization at 5°C (41°F). However, nitrogen-phosphorus-potassium fertilization alone had a comparable effect on hydrocarbon loss like fertilization plus bioaugmentation, this has been shown both in chronically oil-polluted Arctic soil (Whyte et al., 1999) and in artificially diesel oil-contaminated alpine soils (Margesin and Schinner, 1997a, 1997b, 2001; Vieira et al., 2009).

Finally, the application of cold-active solubilizing agents could be useful for enhancing hydrocarbon bioavailability. Two hydrocarbon-degrading Antarctic marine bacteria, identified as *Rhodococcus fascians*, produced bio-emulsifiers when grown with *n*-alkanes as the sole carbon source. The strains utilized hexadecane and biphenyl as sole carbon sources at temperatures ranging from 4 to 35°C (39 to 95°F), the optimum temperature was 15 to 20°C (59 to 68°F) (Yakimov et al., 1999; Chugunov et al., 2000).

In summary, hydrocarbon biodegradation rates in cold groundwater systems are not necessarily lower than in temperate systems but activity measurements should be performed at the prevailing *in situ* temperature in order to obtain a realistic estimate of the naturally occurring biodegradation.

10.6 Advantages and Disadvantages of Bioremediation

Bioremediation has advantages over traditional cleanup methods of crude oil-related spills (Table 15.1). One of the major advantages of bioremediation is the savings in cost and also the savings in the time put forth by workers to clean a contaminated site. The financial savings of bioremediation, when used properly, have tremendous benefits compared to traditional cleanup processes. After the Exxon Valdez spill, the cost to clean 75 miles of shoreline by bioremediation was less than cost to provide physical washing of the shore for one day (Atlas, 1995; Zhu et al., 2004). In addition traditional methods of site cleanup, bioremediation continues to clean the contaminated site without the constant need of workers.

Bioremediation is also advantageous due to its environmentally friendly approach since no foreign or toxic chemicals are added to the site. It is also environmentally friendly because it does not require any disruption to the natural habitat which often occurs from physical and chemical methods of cleanup. Bioremediation allows for natural organisms to degrade the toxic hydrocarbon derivatives into simple compounds which pose no threat to the environment, and this also eliminates the need to remove and transport the toxic compounds to another site. This loss of a need to transport the oil and contaminated soils lowers further risk of additional oil spills, and also saves energy and money which would be put forth in the transportation process. These environmental benefits also make bioremediation a positively viewed method by the general public. With the limited resources in today's world, this is a very much supported technology, which pleases the public and hence is given political support and funding for further research.

One of the greater downsides of bioremediation for crude oil-related spills is that it is a slow process. Such spills can pose a great threat to many different habitats, environments, and industries, and depending upon the urgency of cleanup, bioremediation may not always be the best available option. Also there are many variables that affect whether bioremediation is capable and practical for the cleanup of different spills. Depending on where the spill takes place and the conditions of the soil or water, it may be very difficult to provide proper nutrient concentrations to the oil degrading microorganisms (Delille et al., 2009). If an oil spill occurs offshore, there is typically much more energy and waves, and this can cause for the quick loss and dilution of nutrients provided by biostimulation (Chapter 14). In the case of bioaugmentation, there are other issues—particularly the competition that will develop between the native and foreign microbes, which has the potential to render the bioremediation method unsuccessful.

However, it must be recognized that all of the spilled crude oil or crude oil product removed by microbes in a liner timeframe, since the residual fractions (especially those that contain high molecular eight polar constituents) that are not consumed initially will be more refractory to microbial attack (Speight, 2005, 2007). Conditions in any ecosystem are rarely favorable for maximum biodegradation.

Furthermore, a major issue relates to whether or not degradation that has been demonstrated to occur in the laboratory will occur in the soil or water ecosystems. Biodegradation phenomena reported in laboratory studies of pure culture of microorganisms reflect only the *potential degradation* that *may* occur in any natural ecosystem. Physical-chemical properties of crude oil, crude oil products, and any of the related constituents, the concentration of the crude oil related contaminant(s), as well as the concentration and diversity of the microbial flora of a specific ecosystem are variable factors in the biodegradation process.

Furthermore, there have been various suggestions that when the concentration of the crude oil-related contaminants passes below a threshold level (which is typically not defined) biodegradation ceases because bacterial growth is limited. Surely this relates to the amount of recalcitrant species (such as resin constituents and asphaltene constituents or related polynuclear aromatic systems) remaining in the non-biodegraded material.

The qualitative and quantitative differences in the hydrocarbon and the non-hydrocarbon content of crude oil (Speight, 2005, 2007) influence the susceptibility of crude oil and certain crude oil products to biodegradation. This must be acknowledged as a major consideration in determining eco-toxicological effects of crude oil constituents.

10.7 Conclusion

Biodegradation (with subsequent bioremediation) has shown great promise for cleanup of spills of crude oil-related contaminant caution is advised in extolling its virtues as the panacea for cleanup of all crude oil-related spills. It is, nevertheless, one of several viable options that are available as a single or piggy-back method for environmental cleanup but considerations such as (1) site parameters, (2) the properties of the spilled material, (3) prevailing climatic conditions, and (4) the nature of the micro-organisms that are available must be taken into account. In fact, a number of bioremediation strategies are being developed to treat contaminated wastes and sites. Selecting the most appropriate strategy to treat a specific site can be guided by considering three basic scientific principles which are (1) biochemistry, which is the amenability of the pollutant to biological transformation to less toxic products, (2) bioavailability, which relates to the accessibility of the contaminant to microorganisms, and (3) bioactivity, which relates to the opportunity for optimization of biological activity.

In addition, the continued advances in biotechnology and allowed bioremediation to become one of the most rapidly developing fields of environmental restoration, utilizing microorganisms to reduce the concentration and toxicity of as crude oil-related hydrocarbon derivatives, polycyclic aromatic hydrocarbon derivatives, and other constituents such as metals and metallo-organic compounds. However, although simple aromatic compounds are biodegradable by a variety of degradative pathways, their halogenated counterparts are more resistant to bacterial attacks and often necessitate the evolution of novel pathways (Chakrabarty, 1982; Engasser et al., 1990).

A number of bioremediation strategies have been developed to treat crude oil-contaminated wastes and sites. However, selecting the most appropriate strategy to treat a specific site can be guided by considering three basic principles: the amenability of the pollutant to biological transformation to less toxic products (biochemistry), the accessibility of the contaminant to microorganisms (bioavailability) and the opportunity for optimization of biological activity (bioactivity).

Bioremediation is a preferred method for the long-term restoration of crude oil hydrocarbon polluted systems, with the added advantage of cost efficiency and environmental friendliness (Okoh, 2006). Although exhaustive investigations have been in relation to biodegradation of crude oil and crude oil products, these studies must continue. The identification of active microbial strains is not always ascertained to a sufficient degree, and misidentifications or incomplete identifications are sometimes reported. There is much to be done in terms of the optimization of the process conditions for more efficient application of biological degradation of crude oil and crude oil products under different climatic conditions.

Although a wide range of new microorganisms have been discovered that are able to degrade highly stable, toxic organic chemicals, many pollutants continue to persist in the environment. A number of reasons have been identified as challenges posed to the microorganisms working in contaminated sites, such as: (1) poor bioavailability of chemicals, (2) presence of non-crude oil compounds, (3) inadequate supply of nutrients, and (4) insufficient biochemical potential for effective biodegradation. Thus, selecting the most appropriate strategy to treat a specific site can be guided by considering three basic principles: (1) the amenability of the pollutant to biological transformation to less toxic products, (2) the bioavailability of the contaminant to microorganisms, and (3) the opportunity for optimization of the bioprocess (Macaulay and Rees, 2014).

In summary, the potential and success of microorganisms in the remediation of spills of crude oil and crude oil-related products has been widely acknowledged. With advances in biotechnology, bioremediation has become a rapidly growing area and has been commercially applied for the treatment of hazardous wastes and contaminated sites. Moreover, a positive feature of bioremediation is the diversity of its application to solids, liquids, and liquid–solid mixtures, involving both *in situ* and *ex situ* environments.

In situ and *ex situ* biodegradation technologies are increasingly selected to remediate contaminated sites, either alone or in combination with other source control measures. Bioremediation technologies have proven effective in remediating fuels and volatile organic compounds (VOCs) and are often able to address diverse organic contaminants including semi-volatile organic compounds (SVOCs), polynuclear aromatic hydrocarbon derivatives (PNAs), pesticides, herbicides, and nitro-aromatic compounds (such as, for example, explosives), as well as contamination by heavy metal derivatives.

Site characterization and long-term monitoring are necessary to support system design and sizing as well as to verify continued performance. In addition, there also regulatory requirements to be addressed regarding system design, implementation, operation, and performance, including the disposition of liquid effluents and other wastes resulting from the treatment process.

References

Aislabie, J.M., McLeod, M., and Fraser, R. 1998. Potential for biodegradation of petroleum hydrocarbons in soil from the ross dependency, Antarctica. Appl. Microbiol. Biotechnol., 49: 210–214.

Aislabie, J., Foght, J., and Saul, D. 2000. Aromatic hydrocarbon-degrading bacteria from soil near scott base, Antarctica. Polar Biol., 23: 183–188.

Aislabie, J.M., Balks, M.R., Foght, J.M., and Waterhouse, E.J. 2004. Hydrocarbon spills on antarctic soils: effects and management. Environ. Sci. Technol., 38(5): 1265–74.

Al-Darbi, M.M., Saeed, N.O., Islam, M.R., and Lee, K. 2005. Biodegradation of natural oils in seawater. Energy Sources, 27: 19–34.

Alkorta, I., and Garbisu, C. 2001. Phytoremediation of organic contaminants in soils. Bioresour. Technol., 79: 273–276.

Amann, R., Fuchs, B.M., and Behrens, S. 2001. The identification of microorganisms by fluorescence *in situ* hybridization. Curr. Opin. Biotechnol., 12: 231–236.

Antizar-Ladislao, B., Spanova, K., Beck, A.J., and Russell, N.J. 2008. Microbial community structure changes during bioremediation of PAHs in an aged coal tar-contaminated soil by in-vessel composting. Int. Biodeterior. Biodegrad., 61: 357–364.

Bej, A.K., Saul, D., and Aislabie, J. 2000. Cold-tolerant Alkane-degrading *Rhodococcus* Species from Antarctica. Polar Biol., 23: 100–105.

Blackburn, J.W., and Hafker, W.R. 1993. The impact of biochemistry, bioavailability and bioactivity on the selection of bioremediation techniques. Trends Biotechnol., 11: 328–333.

Bollag, W.B., and Bollag, J.M. 1992. Biodegradation. Encycl. Microbiol., 1: 269–280.

Bouwer, E.J., and Zehnder, A.J.B. 1993. Bioremediation of organic compounds: microbial metabolism to work. Trends Biotechnol., 11: 360–367.

Braddock, J.F., Ruth, M.L., Walworth, J.L., and McCarthy, K.A. 1997. Enhancement and inhibition of microbial activity in hydrocarbon-contaminated arctic soils: implications for utrientamended bioremediation. Environ. Sci. Technol., 31: 2078–2084.

Bradley, P.M., and Chapelle, F.H. 1995. Rapid toluene mineralization by aquifer microorganisms at Adak, Alaska: implications for intrinsic bioremediation in cold environments. Environ Sci. Technol., 29: 2778–2781.

Bragg, J.R., Prince, R.C., Harner, E.J., and Atlas, R.M. 1994. Effectiveness of bioremediation for the exxon valdez oil spill. Nature, 368: 413–418.

Butler, C.S., and Mason, J.R. 1997. Structure, function analysis of the bacterial aromatic ring hydroxylating dioxygenases. Adv. Microb. Physiol., 38: 47–84.

Chakrabarty, A.M. 1982. Genetic mechanisms in the dissimilation of chlorinated compounds. pp. 127–139. *In*: Chakrabarty, A.M. (Ed.). Biodegradation and Detoxification of Environmental Pollutants. CRC Press, Boca Raton, Florida.

Chakrabarty, A.M. 1986. Genetic engineering and problems of environmental pollution. Biotechnology, 8: 515–530.

Chaudhary, G.R., and Chapalamadugu, S. 1991. Biodegradation of halogenated organic compounds. Microbiol. Rev., 55: 59–78.

Chen, W., Bruhlmann, F., Richnis, R.D., and Mulchandani, A. 1999. Engineering of improved microbes and enzymes for bioremediation. Curr. Opin. Biotechnol., 10: 137–141.

Chen, Y.D., Barker, J.F., and Gui, L. 2008. A strategy for aromatic hydrocarbon bioremediation under anaerobic condition and the impacts of ethanol: a microcosm study. J. Contam. Hydrol., 96: 17–31.

Chugunov, V.A., Ermolenko, Z.M., Martovetskaya, I.I., Mironava, R.I., Zhirkova, N.A., Kholodenko, V.P., and Urakov, N.N. 2000. Development and application of a liquid preparation with oil-oxidzing bacteria. Appl. Biochem. Microbiol., 36: 577–581.

Delille, D., Bassères, A., and Dessommess, A. 1998. Effectiveness of bioremediation for oil-polluted antarctic seawater. Polar Biol., 19: 237–241.

Delille, D., and Delille, B. 2000. Field observations on the variability of crude oil impact in indigenous hydrocarbon-degrading bacteria from sub-antarctic intertidal sediments. Mar. Environ. Res., 49: 403–417.

Delille, D., Pelletier, E., Rodriguez-Blanco, A., and Ghiglione, J. 2009. Effects of nutrient and temperature on degradation of petroleum hydrocarbons in sub-antarctic coastal seawater. Polar Biol., 32: 1521–1528.

Deo, P.G., Karanth, N.G., and Karanth, N.G.K. 1994. Biodegradation of hexachlorocyclohexane isomers in soil and food environment. Crit. Rev. Microbiol., 20: 57–78.

Desai, J.D., and Banat, I.M. 1997. Microbial production of surfactants and their commercial potential. Microbiol. Mol. Biol. Rev., 61: 47–64.

Dickel, O., Haug, W., and Knackmus, H,-J. 1993. Biodegradation of nitrobenzene by a sequential anaerobic–aerobic process. Biodegradation, 4: 187–194.

Dua, M., Singh, A., Sethunathan, N., and Johri, A.K. 2002. Biotechnology and bioremediation: successes and limitations. Appl. Microbiol. Biotechnol., 59: 143–152.

Ellis, B.M.L. 2000. Environmental biotechnology informatics. Curr. Opin. Biotechnol., 11: 232–235.

Engasser, K.H., Auling, G., Busse, J., and Knackmus, H.-J. 1990. 3-fluorobenzoate enriched bacterial strain FLB 300 degrades benzoate and all three isomeric monofluoro-benzoates. Arch. Microbiol., 153: 193–199.

Erute Magdalene, O., Ufuoma, A., and Gloria, O. 2009. Screening of four common nigerian weeds for use in phytoremediation of soil contaminated with spent lubricating oil. African Journal of Plant Science, 3(5): 102–106.

Escalante-Espinosa, E., Gallegos-Martinez, M.E., Favela-Torres, E., and Gutierrez-Rojas, M. 2005. Improvement of the hydrocarbon phytoremediation rate by Cyperus Laxus Lam. Inoculated with a microbial consortium in a model system. Chemosphere, 59: 405–413.

Faison, B.D. 2001. Hazardous Waste Treatment. SIM News, 51: 193–208.

Fewson, C.A. 1988. Biodegradation of xenobiotic and other persistent compounds: the causes of recalcitrance. Trends Biotechnol., 6: 148–153.

Gary, J.G., Handwerk, G.E., and Kaiser, M.J. 2007. Petroleum Refining: Technology and Economics, 5th Edition. CRC Press, Taylor & Francis Group, Boca Raton, Florida.

Gerhart, K.E., Huang, X., Glick, B.R., and Greenberg, B.M. 2009. Phytoremediation and rhizoremediation of organic soil contaminants: potential and challenges. Plant Sci., 176: 20–30.

Grant, W.D., Gemmell, R.T., and McGenity, T.J. 1998. Halophiles. pp. 93–132. *In*: Horikoshi, K., and Grant, W.D. (Eds.). Extremophiles—Microbial Life in Extreme Environments. Wiley-Liss, New York.

Grishchenkov, V.G., Townsend, R.T., McDonald, T.J., Autenrieth, R.L., Bonner, J.S., and Boronin, A.M. 2000. Degradation of petroleum hydrocarbons by facultative anaerobic bacteria under aerobic and anaerobic conditions. Process Biochem., 35: 889–896.

Harayama, S. 1997. Polycyclic aromatic hydrocarbon bioremediation design. Curr. Opin. Biotechnol., 8: 268–273.

Hauschild, J.E., Masai, E., Sugiyama, K., Hatta, T., Kimbara, K., Fukuda, M., and Yano, K. 1996. Identification of an Alternative 2,3-Dihydroxybiphenyl 1,2-Dioxygenase In *Rhodococcus* sp. Strain RHA1 and Cloning of the Gene. Appl. Environ. Microbiol., 62: 2940–2946.

Hsu, C.S., and Robinson, P.R. (Eds.). 2017. Handbook of Petroleum Technology. Springer International Publishing AG, Cham, Switzerland.

Iwamoto, T., and Nasu, M. 2001. Current bioremediation practice and perspective. J. Biosci. Bioeng., 92: 1–8.

Janssen, D.B., Oppentocht, J.E., and Poelarends, G. 2001. Microbial dehalogenation. Curr, Opin. Biotechnol., 12: 254–258.

Johan, E.T., Vlieg, V.H., and Janssen, D.B. 2001. Formation and detoxification of reactive intermediates in the metabolites of chlorinated Ethenes. J. Biotechnol., 85: 81–102.

Johri, A.K., Dua, M., Tuteja, D., Saxena, R., Saxena, D.M., and Lal, R. 1996. Genetic manipulations of microorganisms for the degradation of hexachlorocyclohexane. FEMS Microbiol. Rev., 19: 69–84.

Johri, A.K., Dua, M., Singh, A., Sethunathan, N., and Legge, R.L. 1999. Characterization and regulation of catabolic genes. Crit. Rev. Microbiol., 25: 245–273.

Korade, D.L., and Fulekar, M.H. 2009. Development and evaluation of mycorrhiza for rhizosphere bioremediation. Journal of Applied Biosciences, 17: 922–929.

Kumar, S., Mukerjim, K.G., and Lal, R. 1996. Molecular aspects of pesticide degradation by microorganisms. Crit. Rev. Microbiol., 22: 1–26.

Lal, R., Lal, S., and Shivaji, S. 1986. Use of microbes for detoxification of pesticides. Crit. Rev. Microbiol., 3: 1–14.

Lin, Q., and Mendelessohn, I.A. 2009. Potential of restoration and phytoremediation with *Juncus roemerianus* for diesel-contaminated coastal wetlands. Ecol. Eng., 35: 85–91.

Liu, S., and Suflita, J.M. 1993. Ecology and evolution of microbial populations for bioremediation. Trends Biotechnol., 11: 344–352.

Macaulay, B.M., and Rees, D. 2014. Bioremediation of oil spills: a review of challenges for research advancement. Annals of Environmental Science, 8: 9–37.

Macek, T., Mackova, M., and Kas, J. 2000. Exploitation of plants for the removal of organics in environmental remediation. Biotechnol. Adv., 18: 23–34.

Margesin, R., and Schinner, F. 1997a. Efficiency of indigenous and inoculated cold-adapted soil microorganisms for biodegradation of diesel oil in alpine soils. Appl. Environ. Microbiol., 63: 2660–2664.

Margesin, R., and Schinner, F. 1997b. Bioremediation of diesel-oil-contaminated alpine soils at low temperatures. Appl. Microbiol. Biotechnol., 47: 462–468.

Margesin, R. 2000. Potential of cold-adapted microorganisms for bioremediation of oil-polluted alpine soils. Int. Biodeterior. Biodegrad., 46: 3–10.

Margesin, R., and Schinner, F. 2001. Biodegradation and bioremediation of hydrocarbons in extreme environments. Appl. Microbiol. Biotechnol., 56: 650–663.

Marijke, M.A., and van Vlerken, F. 1998. Chances for biological techniques in sediment remediation. Water Sci. Tech., 37: 345–353.

Masai, E., Yamada, A., Healy, J.M., Hatta, T., Kimbara, K., Fukuda, M., and Yano, K. 1995. Characterization of biphenyl catabolic genes of gram positive polychlorinated biphenyl degrader *Rhodococcus* sp. RH1. Appl. Environ. Microbiol., 61: 2079–2085.

Megharaj, M., Singleton, I., McClture, N. C., and Naidu, R. 2000. Influence of petroleum hydrocarbon contamination on microalgae and microbial activities in a long–term contaminated soil. Arch. Environ. Contam. Toxicol., 38: 439–445.

Merkl, N., Schultze-Kraft, R., and Infante, C. 2005. Phytoremediation in the tropics-influence of heavy crude oil on root morphological characteristics of graminoids. Environ. Pollut., 138: 86–91.

Mishra, V., Lal, R., and Srinivasan, S. 2001. Enzymes and operons mediating xenobiotic degradation in bacteria. Crit. Rev. Microbiol., 27: 133–166.

Mohammed, D., Ramsubhag, AS., and Beckles, D.M. 2007. An assessment of the biodegradation of petroleum hydrocarbons in contaminated soil using non-indigenous, commercial microbes. Water Air Soil Pollut., 182: 349–356.

Mohn, W.W., and Stewart, G.R. 2000. Limiting factors for hydrocarbon biodegradation at low temperature in arctic soils. Soil Biol. Biochem., 32: 1161–1172.

Parkash, S. 2003. Refining Processes Handbook. Gulf Professional Publishing, Elsevier, Amsterdam, Netherlands.

Pelletier, E., Delille, D., and Delille, B. 2004. Crude oil bioremediation in sub-antarctic intertidal sediments: chemistry and toxicity of oil residues. Mar. Environ. Res., 57: 311–327.

Rike, A.G., Haugen, K.B., Børresen, M., Engene, B., and Kolstad, P. 2003. *In situ* biodegradation of petroleum hydrocarbons in frozen arctic soils. Cold Reg. Sci. Technol., 37: 97–120.

Salt, D.E., Smith, R.D., and Raskin, I. 1998. Phytoremediation. Annu. Rev. Plant Physiol. Plant Mol. Biol., 49: 643–668.

Sangodkar, U.M.X., Aldrich, T.L., Haugland, R.A., Johnson, J., Rothmel, R.K., Chapman, P.J., and Chakrabarty, A.M. 1989. Molecular basis of biodegradation of chloroaromatic compounds. Acta Biotechnol., 9: 301–316.

Sanscartier, D., Zeeb, B., Koch, I., and Reimer, K. 2009. Bioremediation of diesel-contaminated soil by heated and humidified biopile system in cold climates. Cold Reg. Sci. Technol., 55: 167–173.

Sayler, G.S., and Ripp, S. 2000. Field application of genetically engineered microorganisms for bioremediation processes. Curr. Opin. Biotechnol., 11: 286–289.

Sekelsky, A.M., and Shreve, G.S. 1999. Kinetic model of biosurfactant-enhanced hexadecane biodegradation by *Pseudomonas aeruginosa*. Biotechnol. Bioeng., 63: 401–409.

Singh, A., Mullin, B., and Ward, O.P. 2001. Reactor-based process for the biological treatment of petroleum wastes. Proceedings. Middle East Petrotech 2001 Conference. Bahrain. Page 1–13.

Siron, R., Pelletier, E., and Brochu, C. 1995 Environmental factors influencing the biodegradation of petroleum hydrocarbons in cold seawater. Arch. Environ. Contam. Toxicol., 28: 406–416.

Speight, J.G. 1996. Environmental Technology Handbook. Taylor & Francis, Washington, DC.

Speight, J.G., and Lee, S. 2000. Environmental Technology Handbook. 2nd Edition. Taylor & Francis, New York.

Speight, J.G. 2005. Environmental Analysis and Technology for the Refining Industry. John Wiley & Sons Inc., Hoboken, New Jersey.

Speight, J.G., and Arjoon, K.K. 2012. Bioremediation of Petroleum and Petroleum Products. Scrivener Publishing, Beverly, Massachusetts.

Speight, J.G. 2014. The Chemistry and Technology of Petroleum 5th Edition. CRC Press, Taylor & Francis Group, Boca Raton, Florida.

Speight, J.G. 2017. Handbook of Petroleum Refining. CRC Press, Taylor and Francis Group, Boca Raton, Florida.

Speight, J.G. 2019. Handbook of Petrochemical Processes. CRC Press, Taylor & Francis Group, Boca Raton, Florida.

Sullivan, E.R. 1998. Molecular genetics of biosurfactant production. Curr. Opin. Biotechnol., 9: 263–269.

Timmis, K.N., and Piper, D.H. 1999. Bacteria designed for bioremediation. Trends Biotechnol., 17: 201–204.

Van der Meer, J.R. 1997. Evolution of novel metabolic pathways for the degradation of chloroaromatic compounds. Antoine Van Leeuwenhoek, 71: 159–178.

Van der Meer, J.R., de Vos, W.M., Harayama, S., and Zehnder, A.Z.B. 1992. Molecular mechanisms of genetic adaptation to xenobiotic compounds. Microbiol. Rev., 56: 677–694.

Vieira, P.A., Faria, S.R., Vieira B., De Franca, F.P., and Cardoso, V.L. 2009. Statistical analysis and optimization of nitrogen, phosphorus, and inoculum concentrations for the biodegradation of petroleum hydrocarbons by response surface methodology. J. Microbiol. Biotechnol., 25: 427–438.

Watanabe, K. 2001. Microorganisms relevant to bioremediation. Curr. Opin. Biotechnol., 12: 237–241.

Wenzel, W. 2009. Rhizosphere processes and management in plant-assisted bioremediation (Phytoremediation) of Soils. Plant and Soil, 321(1-2): 385–408.

Whyte, L.G., Hawari, J., Zhou, E., Bourbonnière, L., Inniss, W.E., and Greer, C.W. 1998. Biodegradation of variable-chain-length alkanes at low temperatures by a Psychrotrophic *Rhodococcus* sp. Appl. Environ. Microbiol., 64: 2578–2584.

Whyte, L.G., Bourbonnière, L., Bellerose, C., and Greer, C.W. 1999. Bioremediation assessment of hydrocarbon-contaminated soils from the high arctic. Bioremediation J., 3: 69–79.

Yakimov, M.M., Giuliano, L., Bruni, V., Scarfi, S., and Golyshin, P.N. 1999. Characterization of antarctic hydrocarbon-degrading bacteria capable of producing bioemulsifiers. Microbiologica., 22: 249–256.

Yumoto, I., Nakamura, A., Iwata, H., Kojima, K., Kusumuto, K., Nodasaka, Y., and Matsuyama, H. 2002. *Dietzia psychralcaliphila* sp. *nov.*, A novel facultatively psychrophilic alkaliphile that grows on hydrocarbons. Int. J. Syst. Evol. Microbiol., 52: 85–90.

Zhu, X., Venosa, A.D., and Suidan, M.T. 2004. Literature Review on the Use of Commercial Bioremediation Agents for Cleanup of Oil-Contaminated Environments. Report No. EPA/600/R-04/075. National Risk Management Research Laboratory, United States Environmental Protection Agency, Cincinnati, Ohio. July.

Appendix
The Chemistry of Crude Oil and Crude Oil Products

A1 Abstract

The severity of the spills of crude oil and crude oil products depends on the quantity of material released and its physical and chemical properties. Because crude oil and crude oil products are extremely complex mixtures of many hundreds (even thousands) of gaseous, liquid, and solid constituents, full elucidation of their compositions at the molecular level is impossible with presently available analytical techniques.

A2 Introduction

The significance of any spill of crude oil and/or crude oil products arises from the quantity of crude oil released, uncontrolled and addressing the full impact of the released material on the flora and fauna requires consideration of both the quantity and the chemical composition and properties of the discharged material.

The products that result from the spillage also need consideration and alteration of the constituents of the original crude oil or the crude oil product by (1) physical effects, such as evaporation or dissolution), (2) chemical effects, such as photo-oxidation by sunlight at the Land surface or water surface, and (3) biological processes, such as microbial transformation and degradation which are often collectively referred to as weathering.

A3 Crude Oil and Crude Oil Products

Crude oil is generated from the remnants of (mainly) plant biomass in fine-grained sediments deposited long ago in aquatic environments under low oxygen conditions. Upon progressive burial, the sediments are compacted into crude oil source rocks, typically at depths of several thousand meters. Over millions of years, the organic material in these rocks is transformed into gas and oil under the influence of geothermal heat flow. Increase in pressure due to the conversion of solid material into gases and liquids forces these products out of the source rocks into more porous carrier rocks. The crude oil then migrates upward because its density is lower than that of pore waters. When it reaches a rock formation that is isolated at the top by an impermeable cap rock (such as a stratum of clay minerals), where the crude oil accumulates in a deep reservoir rock.

Crude oil is an extremely complex mixture of many thousands, if not millions, of individual constituents at the molecular level. As used by the oil industry, crude oil is a collective term comprising gaseous (natural gas), liquid (crude oil), and solid (asphalt) components. Due to co-dissolution effects and elevated temperatures, crude oil commonly exists as a single phase or in two phases (gas and liquid) in reservoirs.

A3.1 Chemical Composition

The analysis of spilled oil in terms of its origin and transformation by physical (evaporation, dissolution) or chemical (photo-oxidation, microbial oxidation processes, and the selected incorporation into biomass or other forms of metabolism) processes is in most cases targeted toward its chemical composition rather than its physical properties. The common strategy applied as a first step, after evaporation of the most volatile components (topping), is to separate the complex mixture of oil components into compound classes by polarity using liquid chromatography with various adsorbents on thin-layer plates, in gravity columns, or by medium-pressure or high-performance liquid chromatography. The compound classes usually obtained are saturated hydrocarbons (alkanes), aromatic hydrocarbons (including some heteroaromatic species), resins, and asphaltenes (SARA). Saturates, aromatics, and resins comprise a single solubility fraction, collectively called the maltene fraction (or, income cases, the petrolene fraction) which are soluble in alkane solvents (most commonly n-pentane or n-heptane), while the asphaltene constituents are isolated as a fraction by virtue of the insolubility of these constituents in an alkane solvent (commonly n-pentane or n-heptane). Further subfractions can be obtained by employing a number of more sophisticated techniques.

By definition, the term alkane is a synonym for a saturated hydrocarbon derivative (i.e., a compound that contains carbon and hydrogen only and has no double bonds or aromatic units). Alkane derivatives can be straight chains of methylene (CH_2) groups with methyl (CH_3) groups at the end of the chain. Furthermore, the alkane derivatives can have one or more alkyl side chains (branched alkane derivatives and isoprenoid alkane derivatives), or they can contain one ring or several rings (cyclic and polycyclic alkane derivatives) (Table A.1).

Naphthenes, also known as cycloalkane derivatives, are saturated hydrocarbon derivatives that have at least one ring of carbon atoms. Naphthene derivatives have the general formula C_nH_{2n}—a common example is cyclohexane (C_6H_{12}). As such, naphthene hydrocarbon derivatives in crude oil are relatively stable compounds. Cycloalkane hydrocarbon derivatives can exhibit configurational isomerism insofar as the attached groups can differ in their position relative to the ring. The cycloalkane prefix of *cis* denotes that two groups (other than hydrogen atoms) attached to the ring both lie either above or below the plane of the ring. The prefix *trans* denotes that one of the two groups lies above and the other lies below the plane of the ring. The *cis–trans* isomers (for example, *cis*-,2-dimethylcyclopentane and *trans*-1,2-dimethylcyclopentane) are also called *geometric isomers* but the *cis–trans* isomers are not constitutional isomers because their atoms are bonded in the same sequence.

The boiling point and densities of naphthene hydrocarbon derivatives are higher than those of alkane derivatives having the same number of carbon atoms.

Several of the polycyclic saturated alkanes like (tetracyclic) steranes and pentacyclic triterpanes are classified as biomarkers. Their presence and relative abundance as well as their remarkable stability under weathering allow their use as fingerprints to gain information on the origin of crude oils and to distinguish crude oils from different sources.

Biomarkers (also termed biological markers, molecular fossils, fossil molecules, or geochemical fossils) are organic compounds in natural waters, sediments, soils, fossils, crude oils, or coal that can be unambiguously linked to specific precursor molecules biosynthesized by living organisms. The main reason for this specificity is that the bonds to the four neighboring atoms (carbon or hydrogen) are sterically oriented (tetrahedral). Thus, the rings are not planar as they are in aromatic hydrocarbon derivatives but have a (sometimes slightly skewed) three-dimensional chair or boat configuration.

Table A-1: General Description of the Hydrocarbon Constituents of Crude Oil and Crude Oil Products.

Alkanes (also called *normal paraffins* or *n-paraffins*). These constituents are characterized by unbranched (linear) or branched (non-linear) chains of carbon atoms with attached hydrogen atoms; alkanes contain no carbon-carbon double bonds (hence the designation *saturated*), are generally insoluble in cold water—examples of alkanes are pentane (C_5H_{12}) and heptane (C_7H_{16}).
Cycloalkanes or *cycloparaffins* (also called *naphthenes*). These constituents are characterized by the presence of simple closed rings of carbon atoms (such as the cyclopentane ring or the cyclohexane ring); naphthenes can also contain alky moieties on the ring and are generally stable and relatively insoluble in water—examples are cyclohexane and methyl cyclohexane.
Alkenes (also called *olefins*). These constituents are characterized by the presence of a carbon-carbon double bond ($>C = C<$) and can be unbranched (linear) or branched (non-linear) chains of carbon atoms; typically alkenes are not generally found in crude oil (having reacted over the millennia during which crude oil was formed in the ground) but are common in thermally-produced products, such as naphtha (a precursor to gasoline)—common gaseous alkenes include ethylene ($CH_2 = CH_2$) and propene (also called propylene, $CH_3CH = CH_2$).
Single-ring Aromatics. Aromatic constituents are characterized by the presence of an aromatic ring with six carbon atoms and are considered to be the most acutely toxic component of crude oil constituents because of their association with chronic and carcinogenic effects; the rings can carry alkyl of naphthene substituents further distinguishes the aromatic constituents—low molecular weight aromatic constituents may have a noticeable solubility in water, increasing the potential for exposure to aquatic resources.
Multi-ring Aromatics Aromatic constituents with two or more condensed rings are referred to as polynuclear aromatic hydrocarbons (PNAs, sometime PAHs); the most abundant aromatic hydrocarbon families in crude oil and crude oil products have two and three fused rings with one to four carbon atom alkyl group substitutions—condensed aromatic constituents with more than two condensed rings (three-to-five) are also present in the higher boiling fractions of crude oil but higher condensed systems are unlikely.

Aromatic hydrocarbon derivative contain one or more hexagonal, six-carbon ring structures with the equivalent of three conjugated double bonds; in reality, the electrons are not localized in three separate bonds but are shared among the six carbon atoms. As a consequence of the specific bond type, aromatic hydrocarbons are planar. In heteroaromatic compounds, a carbon atom in the six-membered ring is replaced by a nitrogen atom (as for example in the pyridine molecule). Other heteroatom-containing compounds in crude oil have a sulfur, a nitrogen, or, less commonly, an oxygen atom in a conjugated five-membered ring, either alone or adjacent to one or more aromatic ring(s). These classes of compounds are known as thiophene-type sulfur, pyrrole-type nitrogen, or furan-type oxygen.

Polycyclic aromatic hydrocarbons (PAHs, also referred to as polynuclear Aramaic compounds, PNAs) have two or more fused aromatic rings and can be quite large. These systems occur in crude oil and in crude oil products. They can be toxic, carcinogenic, or mutagenic, and are relatively persistent in the environment. Most aromatic hydrocarbons in crude oil carry one or more alkyl substituents (methyl groups) or longer alkyl chains such as those in the alkylated homologues of naphthalene, phenanthrene, dibenzothiophene, and fluorene.

The constituents of the resin fraction (also called hetero-compounds or N, S, O-compounds) are the most polar maltene fractions obtained by liquid chromatographic separation of crude oil and, in some cases, crude oil products. The precise chemical structures of most of the individual components are not fully defined, but the elevated polarity of the resin fraction is due to the presence of heteroatoms like nitrogen, sulfur, oxygen, and metals (as in the porphyrin derivatives which contain nickel, or vanadium, or iron), as well as the larger molecular size of many of the compounds. Resin fractions are not commonly analyzed in detail, but their relative proportions are determined gravimetrically.

The asphaltene fraction is the most polar fraction and (typically) the highest molecular weight fraction of crude oil (or crude oil products) and is obtained by precipitation of a solution of crude oil (or a crude oil product) in a small amount of a suitable solvent (one volume, equal to the volume of crude oil or the crude oil products when necessary) followed the addition of excess (a minimum of forty volumes) nonpolar solvent (n-pentane or n-heptane) before liquid chromatographic separation. The yields are solvent dependent. Thus, all these sub-fractions of crude oil (or the crude oil product) are defined by the procedures used to separate the material as well as by their chemical nature of the crude oil or the crude oil product. Thus, the asphaltene fraction is defined by the insolubility of the constituents in alkane solvents (i.e., n-pentane or n-heptane) but solubility in aromatic solvents (such as benzene or toluene). Representations of possible asphaltene molecular structures of the asphaltene constituents have been proposed many times but remain highly speculative in spite of claims to the contrary.

A3.2 Physical Properties

The most common physical properties used to describe crude oil and crude oil products are (1) density, (2) viscosity, and (3) boiling point range.

The density of crude oil and crude oil products is usually expressed as API (American Crude oil Institute) gravity, which is inversely related to specific density.

Viscosity, a measure of a fluid's internal resistance to flow at a given temperature and pressure, depends on the chemical composition of the crude oil, including the amount of dissolved gas it contains.

The crude oil refining industry uses the boiling point properties of crude oils to produce distillation fractions (cuts) of defined boiling ranges, each with a mixture of different chemical compound types. These cuts, after further refinement, are the oil fractions known as naphtha (sometimes referred to as gasoline but not yet ready for sale as such), kerosene (sometimes referred to as diesel fuel or jet fuel but not yet ready for sale as such jet fuels), fuel oil, and gas oil.

A4 Analytical Techniques for Oil Spill Sample Analysis

The challenge of analyzing crude oil of any origin, native or altered, arises from the complexity of this substrate on the molecular level. The task is to simplify the mixture by fractionation and then search these fractions for characteristic components of high significance, such as biological markers as indicators of origin or selected PAHs as indicators of source and/or toxicity.

Fractionation uses dead oil (i.e., crude oil from which the most volatile components) have been evaporated so that they do not interfere with gravimetric determination of the proportions of the separated fractions. The process starts with removal of the asphaltene fraction, which may otherwise precipitate during chromatography and cause poor or faulty fractionation.

Column chromatography, a technique for separating mixtures of organic compounds, involves dissolving the mixtures in a mobile phase and passing them through a column filled with a stationary phase. Compounds in the mixture have different affinities for the mobile and stationary phases. They are adsorbed onto the stationary phase and then, as the mobile phase flows through the column, separated and sequentially released and collected. The residence tome of a constituents in the column depends on the stationary and mobile phases, and its boiling point, molecular size, molecular shape, and polarity.

Column chromatography (as a collective term) refers to the use of a vertical glass column filled with silica gel or aluminum oxide, as stationary phase; an organic solvent dripped into the top of the column, as mobile phase; slow percolation through the column under the force of gravity; and separated groups of compounds collected as they drip out the bottom. In a more general sense, thin-layer chromatography (TLC), medium-pressure liquid chromatography (MPLC), high-performance liquid chromatography (HPLC), and preparative gas chromatography (preparative GC) are forms of chromatographic separation with different stationary and mobile phases. This initial step is occasionally followed by further fractionation including separation of straight-chain, branched, and cyclic saturated hydrocarbons by urea (H_2NCONH_2) adduction, thiourea (H_2NCSNH_2) adduction, application of zeolites with different pore sizes, or by separation of aromatic hydrocarbons into classes of constituents based on the number of rings in the molecule.

Even after fractionation, the identification of individual constituents of crude oils and crude oil products is not straightforward. A breakthrough in the 1970s was the development of capillary columns a few tens of meters in length and internally coated with different types of silicone oil as stationary phases. Shortly after, this was followed by the construction of devices for coupling such columns to fast-scanning mass spectrometers with connected computers for data processing. Continuous development of the gas chromatography-mass spectrometry (GC-MS) technique over about two decades was the basis for a rapid increase in understanding of the molecular composition of organic matter in sedimentary rocks and crude oil, and of changes in its composition as a function of geological and environmental conditions.

The straight-chain alkanes (*n*-alkanes) and isoprenoid alkanes pristane and phytane that are abundant in most unaltered crude oils can be identified and quantified by gas chromatography with flame ionization detection (GC-FID; e.g., Peters et al., 2005a). (This rarely applies to other compounds, which may not have the regular retention time pattern on a GC column or are abundant enough for easy detection.) They require the power of mass-specific detection and benefit from the fact that entire groups of saturated hydrocarbon biomarkers have a common mass spectrometric key fragment. Thus, mass chromatograms, which by a factor of about 30 are more sensitive than entire mass spectra, reveal the presence of the full range of these compounds. An even further advanced mass spectrometric detection technique, mass fragmentography (GC-MS-MS), further increases the sensitivity by another factor of up to five and is also more compound-specific than mass chromatography.

Similarly, comprehensive two-dimensional gas chromatography (GC × GC) increases the resolution of complex mixtures. Gaines et al. (1999) published an early application of this technique to a marine oil spill. Using two sequentially coupled gas chromatographic columns with different separation efficiencies related to molecular properties (such as volatility and polarity), GC × GC produces a two-dimensional retention surface, which significantly improves compound identification compared to the one-dimensional retention data of a normal GC column. Compounds with similar chemical structures are grouped together in the chromatogram, allowing rapid preliminary identification with even minor components being separated and detectable. The GC × GC technique can be used to examine changes in the abundance of the constituents of crude oil and crude oil products and the products were ascribed to various categories of physical and chemical weathering.

The two-dimensional retention surfaces from GC × GC analysis of crude oils and their transformation products after an oil spill, despite the additional dimension of separation, are still very complex.

A5 Challenges

The complexity of crude oil and crude oil products represents a challenge even with present advances in analytical chemistry. In fact, an assessment of the full impact of the release of crude oil and crude oil products on the flora and fauna of an ecosystem as well on human health within the ecosystem region, requires determination of the quantities of materials that entered the environment, including the amounts of dispersants used during spill mitigation. In addition, chemical analysis of the composition of the released materials and their relative proportions are of similarly high importance.

Chemical analyses of the released crude oil and the products of physical, chemical, and biological alteration may need to be performed using a wide range of techniques taken from analytical chemistry, organic geochemistry, and environmental chemistry. In addition, more advanced techniques such as the sequential combination of two gas chromatographic columns with different separation efficiencies (GC × GC) and ultrahigh-resolution mass spectrometry (FT-ICR MS) may be the methods of choice for identifying on the high-molecular-weight and polar constituents of crude oil and crude oil products, particularly those newly formed during the photo-oxidation process.

Together, a suitably-chosen suite of analytical methods may provide not only an inventory of the most important constituents of the released crude oil but also a greater understanding of the reaction mechanisms of weathering processes and the identity of the transformation products, as well as the impacts of the changes on physical properties, bioavailability, and toxicity of the discharged material.

Further Reading

Gary, J.G., Handwerk, G.E., and Kaiser, M.J. 2007. Petroleum Refining: Technology and Economics, 5th Edition. CRC Press, Taylor & Francis Group, Boca Raton, Florida.

Hoiberg, A.J. 1964. Bituminous Materials: Asphalts, Tars, and Pitches. John Wiley & Sons Inc., New York.

Hsu, C.S., and Robinson, P.R. (Eds.). 2017. Handbook of Petroleum Technology. Springer International Publishing AG, Cham, Switzerland.

Long, R.B., and Speight, J.G. 1998. The composition of petroleum. *In*: Speight, J.G. (Ed.). Petroleum Chemistry and Refining. Taylor & Francis, Washington, DC. Chapter 2.

Meyer, R.F., and Dietzman, W.D. 1981. World geography of heavy crude oils. *In*: Meyer,R.F., and Steele, C.T. (Eds.). The Future of Heavy Crude and Tar Sands. McGraw-Hill, New York, p. 16.

Meyers, R.A. 1997. Handbook of Crude oil Refining Processes 2nd Edition. McGraw-Hill, New York.

Mokhatab, S., Poe, W.A., and Speight, J.G. 2006. Handbook of Natural Gas Transmission and Processing Elsevier, Amsterdam, Netherlands.

Mushrush, G.W., and Speight, J.G. 1995. Petroleum Products: Instability and Incompatibility. Taylor & Francis, Philadelphia, Pennsylvania.

Parkash, S. 2003. Refining Processes Handbook. Gulf Professional Publishing, Elsevier, Amsterdam, Netherlands.

Peters, K.E., Walters, C.C., and Moldowan, J.M. 2005a. The Biomarker Guide: Volume 1: Biomarkers and Isotopes in the Environment and Human History. Columbia University Press, Columbia University, New York.

Peters, K.E., Walters, C.C., and Moldowan, J.M. 2005b. The Biomarker Guide: Volume 2: Biomarkers and Isotopes in Petroleum Exploration and Earth History. Columbia University Press, Columbia University, New York.

Pfeiffer, J.H. 1950. The Properties of Asphaltic Bitumen. Elsevier, Amsterdam, Netherlands.

Speight, J.G. 1986. Polynuclear aromatic systems in petroleum. Preprints, Am. Chem. Soc., Div. Petrol. Chem., 31(4): 818.

Speight, J.G. 2012a. Crude Oil Assay Database. Knovel, Elsevier, New York. Online version available at: http://www.knovel.com/web/portal/browse/display?_EXT_KNOVEL_DISPLAY_bookid=5485&VerticalID=0.

Speight, J.G. 2014. The Chemistry and Technology of Petroleum 5th Edition. CRC Press, Taylor & Francis Group, Boca Raton, Florida.

Speight, J.G. 2015. Handbook of Petroleum Product Analysis 2nd Edition. John Wiley & Sons Inc., Hoboken, New Jersey.

Speight, J.G. 2017. Handbook of Crude oil Refining. CRC Press, Taylor & Francis Group, Boca Raton, Florida.

Van Nes, K., and van Westen, H.A. 1951. Aspects of the Constitution of Mineral Oils. Elsevier, Amsterdam, Netherlands.

Conversion Factors

1. Concentration Conversions

1 part per million (1 ppm) = 1 microgram per liter (1 μg/L)
1 microgram per liter (1 μg/L) = 1 milligram per kilogram (1 mg/kg)
1 microgram per liter (μg/L) × 6.243×10^8 = 1 lb per cubic foot (1 lb/ft³)
1 microgram per liter (1 μg/L) × 10^{-3} = 1 milligram per liter (1 mg/L)
1 milligram per liter (1 mg/L) × 6.243×10^5 = 1 pound per cubic foot (1 lb/ft³)
I gram mole per cubic meter (1 g mol/m³) × 6.243×10^5 = 1 pound per cubic foot (1 lb/ft³)
10,000 ppm = 1% w/w
1 ppm hydrocarbon in soil × 0.002 = 1 lb of hydrocarbons per ton of contaminated soil

2. Sludge Conversions

1,700 lbs wet sludge = 1 yd³ wet sludge
1 yd³ sludge = wet tons/0.85
Wet tons sludge × 240 = gallons sludge
1 wet ton sludge × % dry solids/100 = 1 dry ton of sludge

3. Weight Conversion

1 ounce (1 oz) = 28.3495 grams (18.2495 g)
1 pound (1 lb) = 0.454 kilogram
1 pound (1 lb) = 454 grams (454 g)
1 kilogram (1 kg) = 2.20462 pounds (2.20462 lb)
1 stone (English, 1 st) = 14 pounds (14 lb)
1 ton (US; 1 short ton) = 2,000 lbs
1 ton (English; 1 long ton) = 2,240 lbs
1 metric ton = 2204.62262 pounds
1 tonne = 2204.62262 pounds

4. Temperature Conversions

°F = (°C × 1.8) + 32
°C = (°F − 32)/1.8
(°F − 32) × 0.555 = °C
Absolute zero = −273.15°C
Absolute zero = −459.67°F

5. Area

1 square centimeter (1 cm²) = 0.1550 square inches
1 square meter 1 (m²) = 1.1960 square yards
1 hectare = 2.4711 acres
1 square kilometer (1 km²) = 0.3861 square miles
1 square inch (1 inch²) = 6.4516 square centimeters
1 square foot (1 ft²) = 0.0929 square meters
1 square yard (1 yd²) = 0.8361 square meters
1 acre = 4046.9 square meters
1 square mile (1 mi²) = 2.59 square kilometers

6. Nutrient Conversion Factor

1 pound phosphorus × 2.3 (1 lb P × 2.3) = 1 pound phosphorous pentoxide (1 lb P_2O_5)
1 pound potassium × 1.2 (1 lb K × 1.2 = 1 pound potassium oxide (1 lb K_2O)

7. Other Approximations

14.7 pounds per square inch (14.7 psi) – 1 atmosphere (1 atmos)
1 kiloPascal (kPa) × 9.8692 × 10^{-3} = 14.7 pounds per square inch (14.7 psi)
1 yd³ = 27 ft³
1 US gallon of water = 8.34 lbs
1 imperial gallon of water – 10 lbs
1 yd³ = 0.765 m³
1 acre-inch of liquid = 27,150 gallons = 3.630 ft³
1 ft depth in 1 acre (in-situ) = 1,613 × (20 to 25 % excavation factor) = ~ 2,000 yd³
1 yd³ (clayey soils-excavated) = 1.1 to 1.2 tons (US)
1 yd³ (sandy soils-excavated) = 1.2 to 1.3 tons (US)

Index

A

Accuracy 12, 123, 124, 129, 130, 132, 187, 260, 381, 386, 390–392, 394–398, 402

Aromatics 5, 8, 12, 13, 16, 18–20, 22–24, 27–30, 34, 37, 44–54, 56, 60–63, 65, 66, 69, 70, 72, 75, 86, 89, 90, 93, 94, 96–100, 102–106, 109, 115, 119, 120, 122–124, 127–130, 134, 143, 146–150, 155, 159–166, 187, 189, 191, 193–198, 200, 204, 216–220, 240–243, 247, 253–259, 267, 269, 272, 273, 275, 283–286, 288, 293, 294, 296, 297, 301, 323–330, 332–338, 340–342, 344–346, 353, 371, 372, 374, 376, 378, 381–385, 387, 388, 399, 400, 403–405, 408, 409, 411, 414, 416, 417, 426, 431, 438, 443, 444, 446, 447

Asphaltene 3, 13, 22, 23, 27, 28, 34, 60, 72, 88, 103, 107, 117, 119, 121, 123, 134, 136, 150, 195, 198–200, 204, 206, 207, 217, 254, 258, 273, 294, 325, 327, 330, 332, 340–344, 346–349, 352, 355, 400, 446

B

Bioattenuation 235

Bioaugmentation 169, 184, 217, 218, 220, 234, 235, 242, 248, 250, 251, 260, 261, 264, 265, 271, 272, 291, 292, 324, 328, 331, 333, 353–355, 432, 445, 446

Bioconversion 344

Biodegradation 42, 98, 102, 105, 108, 109, 120, 121, 136, 145, 153, 154, 160, 168, 175, 176, 183, 184, 221, 226, 229, 231–236, 238–256, 258, 260, 262–276, 282, 283, 285–302, 314, 323–338, 340, 343–345, 347, 349, 351–355, 372, 373, 380, 397, 410, 414, 417, 432, 438–447

Biodenitrogenation 344, 346, 347, 349

Biodesulfurization 344–347, 349

Biomarkers 25, 34, 268, 269, 287, 288, 326, 344, 371, 375, 408, 410, 411

Bioremediation 105, 109, 153, 168, 169, 172, 178, 184, 217–222, 224–226, 229, 231–246, 248–252, 254, 255, 257–273, 275, 282–288, 290–295, 297, 299–301, 305, 323–325, 327, 328, 330–335, 337–341, 344, 345, 352–355, 371–373, 396, 407, 414, 432, 438–447

Bioslurping 168, 235, 261, 265

Biosparging 168, 235, 261, 265

Biostimulation 169, 217, 218, 234, 235, 242, 248, 250, 251, 260, 261, 264, 265, 271, 291, 297, 331, 335, 353–355, 432, 446

Biosurfactant treatment 266

Biotransformation 236, 255, 337, 342, 343, 441

Bioventing 168, 235, 250, 261, 265, 266, 301, 432

Bitumen 3, 6, 8, 11–15, 18, 19, 28, 29, 31, 55, 63, 111, 112, 122, 208, 332, 337, 340–342, 344, 352

C

Chemical stabilization 215, 251

Chlorinated compounds 232, 256, 265, 284, 339, 349, 416

Coke 5, 9, 28, 31, 37, 60, 61, 72–74, 77, 79, 81–83, 88, 98, 100, 116, 135–137, 146, 148, 152, 159, 162, 190, 199, 204, 207, 208, 340, 342, 348, 350, 381, 386, 424, 430, 433, 436

Composition of a spill 154

Containment 97, 99, 159, 160, 169, 170, 174, 181, 209, 210, 213, 215, 216, 219, 231, 233, 239, 244, 251, 283, 289, 309, 432

Crude oil refining 37, 40, 41, 48, 61, 62, 67, 70, 73, 76, 78, 89, 94, 109, 131, 134, 137, 141, 142, 148, 186, 187, 190, 201, 204, 285, 340, 344, 350, 424–426, 429, 431, 433, 437, 439, 440

D

Diesel fuel 3, 5, 17, 39, 42, 48–51, 64, 69, 70, 82, 94, 106, 114, 126, 142, 177, 266, 335, 374, 401, 402, 405, 424

Dispersants 94, 99, 108, 110, 111, 113, 119, 121, 150, 154, 156, 160, 166, 183, 210, 214, 276, 301, 302, 306, 310, 313, 330, 331, 333, 355, 371, 372, 376

Dispersion 93, 98, 103, 110, 112, 113, 119, 121, 122, 144, 145, 154, 160, 166, 209, 214, 235, 251, 272–274, 276, 293–295, 297, 302, 324, 325, 355, 371, 412, 417

Dissolution 25, 27, 30, 34, 98, 99, 109, 121, 122, 144, 145, 154, 160, 267, 273, 274, 276, 286, 294–297, 300, 302, 314, 324–326, 348, 383, 412

Dredging 169–171

E

Economic assessment 225

Effluents 41, 68, 70, 71, 77, 79, 88, 94, 131, 133, 134, 137, 141, 142, 149, 174, 184–186, 191, 192, 202, 209, 237, 257, 328, 433, 447

Emulsification 53, 56, 57, 98, 99, 103, 112, 113, 120, 121, 135, 160, 161, 183, 205, 212, 213, 215, 274, 295, 314, 327, 329, 331, 342, 345, 412

Environment 3, 4, 7–10, 15, 17, 23, 25, 29, 30, 34, 38, 40–43, 61, 69, 86, 89, 90, 92, 93, 95–100, 102–104,

108–112, 114, 116, 118, 120, 122, 131, 133, 135, 136, 141–148, 150, 151, 153–168, 170, 171, 173–175, 181–186, 191, 193, 202, 207, 209, 210, 212–220, 222, 225, 226, 231–240, 242–246, 248, 250, 252–258, 260, 261, 263, 266–269, 272, 273, 275, 276, 282–284, 286–288, 292–294, 296, 297, 299–307, 309–311, 315, 323–325, 328–330, 333, 335, 337–341, 348, 349, 353–355, 371, 372, 376, 377, 379, 385, 389, 393, 396, 399, 408, 410, 412, 413, 415, 425–427, 430–433, 439–445, 447

Evaporation 7, 25, 27, 34, 56, 57, 59, 98–100, 103, 107, 113, 114, 120–122, 130, 144, 145, 154, 160, 161, 183, 197, 212, 266, 270, 273, 274, 291, 294, 295, 314, 324–326, 331, 332, 352, 371, 372, 376, 377, 385, 387, 388, 405, 412, 416, 428

Excavation 168–172, 176, 177, 216, 236, 248, 262, 264, 442

F

Feasibility assessment 221
Fractionation 16, 21, 22, 29, 44, 55, 80, 83, 115, 133, 193, 200, 206, 371, 379, 411, 417
Fuel oil 3, 5, 7, 8, 16, 17, 21, 26, 37–39, 41–43, 49–51, 69, 80, 82–84, 88, 91, 93, 97, 104, 106, 107, 110, 112, 121, 122, 124, 126, 131, 133, 134, 144, 145, 147–149, 154, 155, 159, 165, 186, 190, 193–195, 198–201, 204, 208, 218, 267, 286, 296, 297, 303, 327, 335, 346, 349, 352, 372, 374, 388–390, 401, 402, 405, 406, 424, 427

G

Gas chromatography 21, 24, 115, 118, 120, 123–125, 128, 182, 187, 188, 196–198, 200, 208, 326, 382, 399–401, 403, 408–411
Gravimetry 130
Groundwater remediation 167, 223

H

Heavy crude oil 3, 6, 7, 11, 12, 14, 15, 18, 22, 28, 29, 31, 63, 83, 111, 112, 121, 122, 128, 183, 214, 247, 326, 332, 333, 340, 342–348, 355, 403, 414
High performance liquid chromatography 27, 30, 34, 123, 127, 399, 400, 403, 404

I

Immunoassay 118, 124, 125, 401, 404
Infrared spectroscopy 23, 129, 187, 196, 404, 408, 410
In situ oxidation 169, 171

K

Kerosene 3, 5, 16, 17, 21, 26, 37, 39, 41, 42, 46, 48–51, 56, 62, 64, 69–72, 80, 94, 97, 110, 113, 117, 121, 122, 149, 159, 197, 266, 324, 372, 405, 424

M

Method validation 392, 393
Microbial enhanced oil recovery 341, 344
Microbiological analysis 406

N

Nanoremediation 173
Naphtha 3, 5, 6, 12, 16, 17, 19, 21, 25, 37, 39, 41, 42, 44–47, 50, 51, 54, 56, 59, 62, 65, 66, 69, 76, 80, 86, 92, 107, 110, 113, 114, 117, 121, 122, 133, 145–148, 189, 190, 193–199, 233, 239, 244, 266, 333, 335, 375, 378, 411, 424, 436, 437

O

Oil booms 209, 210

P

Petroleum group analysis 122
Phenolic compounds 338
Phytoremediation 168, 173, 218, 250, 251, 266, 432, 441
Polynuclear aromatic hydrocarbons 19, 20, 27, 51, 52, 65, 66, 89, 90, 93, 94, 98, 104, 123, 124, 127, 128, 143, 146, 155, 160, 164–166, 191, 219, 220, 242, 258, 259, 267, 269, 272, 273, 286, 288, 293, 294, 296, 297, 328, 329, 333–338, 353, 371, 372, 376, 378, 383–385, 388, 400, 403, 404, 408, 411, 414, 426, 443, 447
Precision 123, 124, 129, 187, 386, 390–396, 398, 399, 405

Q

Quality assurance 124, 179, 393–396, 398, 415
Quality control 126, 178, 179, 191, 383, 393–396, 398, 402, 415

R

Refinery processes 21, 37, 38, 41, 42, 68, 77, 81, 85, 91, 109, 110, 123, 131, 133, 142, 152, 186, 371, 400, 423, 427
Refinery products 17, 40, 69, 75, 423, 424, 436
Refinery waste 42, 67, 136, 268, 287, 353, 423–425, 434, 438
Remediation management 178
Representative sample 199, 353, 380, 381, 386, 387, 397, 399
Resins 13, 22, 23, 27, 28, 34, 63, 72, 88, 103, 106, 107, 119, 122, 123, 149, 150, 197, 217, 258, 273, 294, 325, 327, 330, 332, 340, 341, 347, 352, 355, 374, 399, 400, 446
Risk analysis 309, 312–314
Risk assessment 94, 115, 153, 166, 167, 182, 268, 287, 309–315, 372, 411, 413

S

Sample collection 107, 182, 202, 374, 376, 377, 379, 380, 385, 395, 397, 426
Sample preparation 124, 309, 386, 399, 404, 407
Sampling strategies 398
Saturates 13, 22–24, 28, 34, 63, 103, 106, 119, 122, 123, 150, 195–197, 200, 217, 258, 329, 340, 341, 374, 399, 400
Site cleanup 229, 232, 237, 265, 268, 287, 439, 445
Site evaluation 261, 282, 285, 288

Skimmers 107, 113, 150, 209–212, 214, 215, 303
Soil remediation 167–169, 175, 223, 236, 353
Soil vapor extraction 169, 174, 231, 233, 239, 244, 250, 265, 266, 283
Solidification 30, 31, 169, 170, 173, 175, 196
Sorbents 106, 150, 187, 211, 212, 374, 384, 385, 388
Stabilization 57, 169, 170, 173, 175, 187, 215, 222, 251
Surfactant enhanced aquifer remediation 169, 176

T

Technical regulations 225
Test methods 25, 40, 42, 55, 60, 102, 104, 108, 111, 113, 115, 118, 120, 123, 133–135, 154, 193–196, 198, 199, 201, 202, 204, 206–208, 271, 291, 372, 376, 379–381, 383, 384, 386, 389, 391, 392, 399, 400, 403–405, 412, 416, 426
Thermal desorption 169, 176–178
Toxicity 7, 53, 69, 70, 90, 92–94, 96–98, 104, 108, 116, 123, 135, 145, 147, 148, 154, 155, 158–160, 162–167, 169, 174, 176, 184, 185, 207, 232, 233, 239, 240, 249–251, 257, 258, 265, 267, 271, 274, 275, 283, 284, 286, 292, 294, 297, 299, 305, 325, 332, 336, 339, 342, 351, 371, 372, 376, 377, 379, 383, 399, 400, 410–412, 414, 415, 426, 437, 438, 441, 446
Types of spills 154

V

Viscosity 6, 8, 11–16, 25, 27, 32, 33, 39, 48–53, 55, 59, 64, 78, 79, 82, 89, 100, 102, 104, 107–109, 111, 113, 118, 119, 121, 122, 135, 146, 148, 150, 154, 155, 162, 190, 196, 200, 201, 205, 210, 212, 214, 215, 271, 274, 275, 292, 295–297, 303, 324, 325, 330, 331, 340–342, 344, 346, 352, 371, 372, 376, 379, 412
Volatile compounds 89, 90, 93, 120, 128, 164, 165, 183, 251, 375, 378, 381, 383–385, 403, 414, 424, 432
Volatility 6, 8, 9, 15, 21, 25, 32–34, 43, 47, 56, 57, 59, 61, 91, 96, 107, 110, 118, 120, 132, 135, 144, 154, 162, 174, 187, 194, 196, 197, 199, 213, 330, 380, 397, 416

W

Weathering 7, 8, 22, 25, 34, 94, 98–100, 106, 107, 112, 113, 119–122, 153, 160–162, 166, 183, 212, 214, 267, 268, 273, 274, 276, 285, 287, 294, 296, 302, 303, 315, 324, 325, 331, 338, 371, 374, 377, 410–412

About the Authors

Karuna K. Arjoon
Karuna Arjoon possesses a Master of Philosophy degree from the University of Trinidad and Tobago where she served in various academic positions.

She is also the co-author of the book Bioremediation of Petroleum and Petroleum Products and the article The Effects of the Application of Random Sampling on the Analysis of the Biodegradation of Oil Spills in Soil (a Petroleum and Chemical Industry International Journal publication). Karuna has also provided consultant services for several international companies, has certifications in various areas of safety, and is presently part of the education system in the field of chemistry.

Dr. James G. Speight
Dr. James G. Speight has doctorate degrees in Chemistry, Geological Sciences, and Petroleum Engineering. He is the author of more than 80 books in crude oil science, fossil fuel science, petroleum engineering, environmental sciences, and ethics.

He has more than fifty years of experience in areas associated with (i) the properties, recovery, and refining of reservoir fluids, conventional crude oil, heavy crude oil, extra heavy oil, tar sand bitumen, and oil shale, (ii) the properties and refining of natural gas, gaseous fuels, (iii) the production and properties of chemicals from crude oil, coal, and other sources, (iv) the properties and refining of biomass, biofuels, biogas, and the generation of bioenergy, and (v) the environmental and toxicological effects of energy production and fuels use. His work has also focused on safety issues, environmental effects, environmental remediation, and safety issues as well as reactors associated with the production and use of fuels and biofuels.

For Product Safety Concerns and Information please contact our EU
representative GPSR@taylorandfrancis.com
Taylor & Francis Verlag GmbH, Kaufingerstraße 24, 80331 München, Germany

www.ingramcontent.com/pod-product-compliance
Lightning Source LLC
Chambersburg PA
CBHW080120220326
41598CB00032B/4904

9 781032 411156